ERRORLESS

38 YEARS

NTA NEET UG

Chapter-wise & Topic-wise

PHYSICS

Solved Papers (2025 - 1988)

with Value Added Notes

DISHA™
Publication Inc

DISHA Publication Inc.
A-23 FIEE Complex, Okhla Phase II
New Delhi-110020
Tel: 49842349/ 49842350

Edited By
Sanjeev Kumar Jha
Govind Thakur

Typeset By
DISHA DTP Team

Write To Us At
feedback_disha@aiets.co.in
www.dishapublication.com

DISHA™
Publication Inc

Contents

Class XI

* Marked Chapters/Topics are not present in new NCERT books but retained by NTA for NEET.

Class XII

* Marked Chapters/Topics are not present in new NCERT books but retained by NTA for NEET.

NEET 2025 Paper Code 45
Page Reference in Disha's 38 Years Physics

Qn	1	2	3	4	5	6	7	8	9	10	11	12	13	14	15	16	17	18	19	20	21	22	23	24	25
Page	A-112	B-111	B-53	A-39	A-54	B-32	A-56	B-169	B-98	A-37	B-80	B-55	B-108	A-131	A-14	B-88	A-72	B-147	A-117	B-15	A-17	A-39	A-142	A-4	B-223
Qn	26	27	28	29	30	31	32	33	34	35	36	37	38	39	40	41	42	43	44	45					
Page	B-98	B-124	B-1	A-131	B-147	A-93	A-94	B-32	B-150	B-14	B-35	B-135	B-130	A-74	B-167	A-122	A-1	B-124	A-142	A-156					

Disha's **SMART PREP Edge**

CRACK NEET
Deep NEET Analysis
ANALYZING THE EXAM

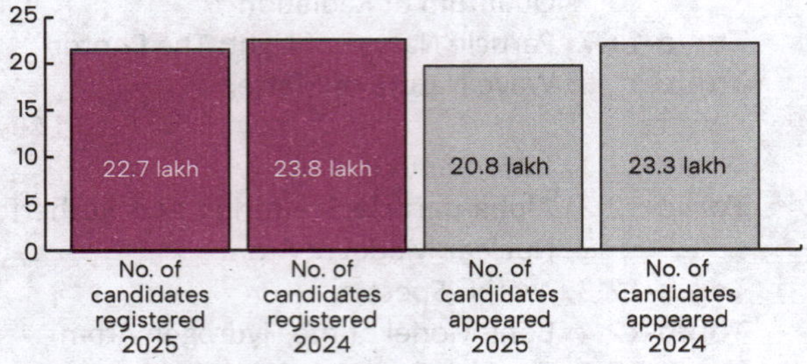

Candidate Appeared 2024 vs 2025

Highest Marks obtained in 2024 ---o **720**

No. of Candidates Securing Highest Marks ---o **67**

NEET Marks vs Rank

Medical aspirants have the chance of predicting their expected NEET 2025 rank by calculating the marks they are likely to score. For convenience in the same NEET marks vs rank data has been provided in the tables that follow.

Final Marks vs Rank in NEET 2025-2022

Marks Range	2025 (Expected Rank)	2024 Rank Range	2023 Rank Range	2022 Rank Range
720-715	1 to 5	1-67	1-19	1-19
714-710	6 to 15	68 - 150	20 - 50	20 - 200
709-700	16 to 50	151 - 250	51 - 200	201 - 512
699-690	51 to 100	251 - 600	201 - 500	513 - 971
689-680	101 to 200	601 - 1000	501 - 1000	972 - 1701
679-670	201 to 300	1001 - 2000	1001 - 1700	1702 - 2751
669-660	301 to 400	2001 - 3500	1701 - 2750	2752 - 4163
659-650	410 to 500	3501 - 5000	2751 - 4160	4164 - 6061
649-640	501 to 600	5001 - 7000	4161 - 6060	6062 - 8522
639-630	601 to 750	7001 - 9500	6061 - 8520	8523 - 11463
629-620	751 to 900	9501 - 11000	8521 - 11460	11464 - 15057
619-610	901 to 1200	11001 - 13500	11461 - 15050	15058 - 19136
609-600	1201 to 1600	13501 - 15000	15051 - 19130	19137 - 23731
599-590	1601 to 2000	15001 - 23730	19131 - 23730	23732 - 28745
589-580	2001 to 3500	23731 - 28740	23731 - 28740	28746 - 34261
579-570	3501 to 6000	28741 - 34260	28741 - 34260	34262 - 40257
569-560	6001 to 10000	34261 - 40250	34261 - 40250	40258 - 46747
559-550	10001 to 15000	40251 - 46740	40251 - 46740	46748 - 53539
549-540	15001 to 25000	46741 - 53530	46741 - 53530	53540 - 60853
539-530	25001 to 35000	53531 - 60850	53531 - 60850	60854 - 68444
529-520	35001 to 45000	60851 - 68440	60851 - 68440	68445 - 76497
519-510	45001 to 60000	68441 - 76490	68441 - 76490	76498 - 85025
509-500	60001 to 75000	76491 - 85020	76491 - 85020	85026 - 93986
499-490	75001 to 100000	85021 - 93980	85021 - 93980	93987 - 103350
489-480	100001 to 115000	93981 - 103340	93981 - 103340	103351 - 113223
479-470	115001 to 123000	103341 - 113220	103341 - 113220	113224 - 123338
469-460	123001 to 134000	113221 - 123330	113221 - 123330	123339 - 133916
459-450	134001 to 146000	123331 - 133910	123331 - 133910	133917 - 144909
449-440	146001 to 148900	133911 - 144900	133911 - 144900	144910 - 156179

Question Type Analysis (NEET 2025) Paper Code 45

Subject	Single Choice	Statement I & II	Multi Statement	Assertion & Reason	Match the Column	Diagram Based	Grand Total
Physics	45	0	0	0	0	0	45
Chemistry	30	4	4	6	0	0	45
Botany	26	5	8	4	6	0	45
Zoology	19	3	10	1	5	3	45
Total	120	12	22	6	17	3	180

SUBJECTWISE DETAILED ANALYSIS - PHYSICS

Subject Wise Difficulty Level (No. of Questions)

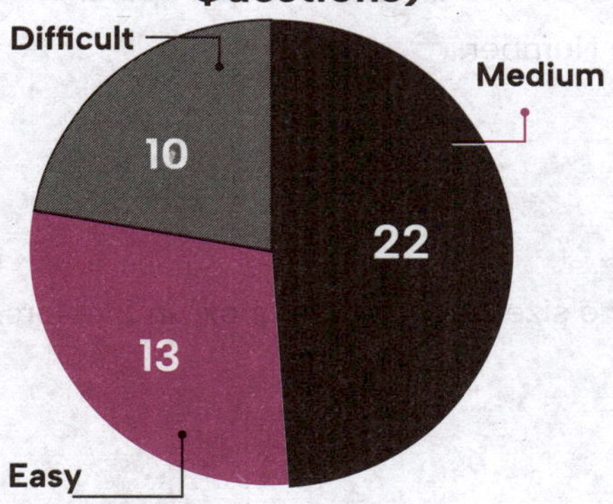

Subjects	11th	12th	11th% age	12% age
Physics	21	24	44.44%	55.56%

Physics Class-wise Question Distribution

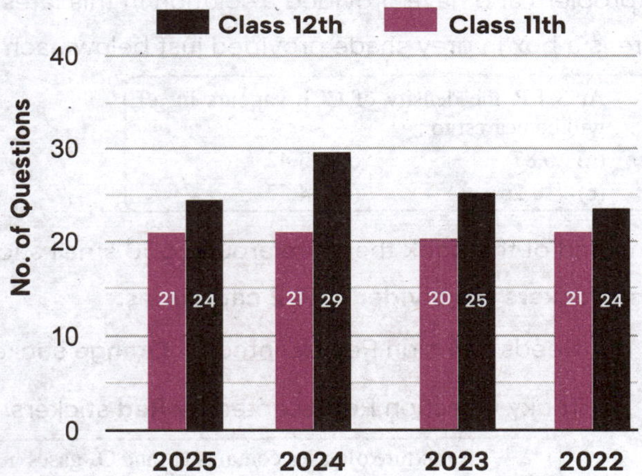

NEET 2025 Physics Chapter-Wise Question Distribution

Class	Chapter Name	No. of Qns
11	Units and Measurements	2
11	Motion in a Straight Line	2
11	Motion in a Plane	0
11	Laws of Motion	3
11	Work, Energy and Power	2
11	System of Particles and Rotational Motion	2
11	Gravitation	2
11	Mechanical Properties of Solids	0
11	Mechanical Properties of Fluids	1
11	Thermal Properties of Matter	1
11	Thermodynamics	1

11	Kinetic Theory	2
11	Oscillations	2
11	Waves	1
12	Electric Charges and Fields	1
12	Electrostatic Potential and Capacitance	2
12	Current Electricity	3
12	Moving Charges and Magnetism	2
12	Magnetism and Matter	0
12	Electromagnetic Induction	1

12	Alternating Current	1
12	Electromagnetic Waves	2
12	Ray Optics and Optical Instruments	2
12	Wave Optics	2
12	Dual Nature of Radiation and Matter	2
12	Atoms	3
12	Nuclei	0
12	Semi-conductor Electronics: Materials, Devices and Simple Circuits	2
11 & 12	Experimental Physics	1
Total		**45**

HOW TO USE THIS BOOK EFFECTIVELY & SCORE 20% MORE MARKS?

Revision is the key to success in any exam. As we prepare for any exam we come across a lot of topics/ questions where we face problem and are not able to solve them on first instance and have to refer solutions. Further there are a lot of tricky questions which require extra skill to solve a question. These questions, if revised before the exam, bolsters your confidence and can definitely help us score more. Disha Publication has identified this problem and have provided a solution in this latest edition of 38 Years NEET Solved Papers.

There is a box in grey shade provided just below each Question Number.

> **1.** At S.T.P. the density of CCl_4 vapours in g/L will be nearest to : 1988
> (a) 6.87 (b) 3.42
> (c) 10.26 (d) 4.57

At the end of the book there are around 650 small stickers (same size as of the grey box) in 2 colors.

These stickers are divided into 2 categories:

1. **NR** Needs Revision Represented by **Orange** stickers.
2. **TQ** Tricky Question Represented by **Red** stickers.

> **2.** NR A mixture of gases contains H_2 and O_2 gases in the ratio of 1 : 4 (w/w). What is the molar ratio of the two gases in the mixture ? 2015
> (a) 4 : 1 (b) 16 : 1
> (c) 2 : 1 (d) 1 : 4

> **6.** TQ The number of water molecules is maximum in :
> NCERT | Page-18 / N-18 | 2015 RS
> (a) 18 molecules of water
> (b) 1.8 gram of water

So whenever you encounter any of these questions just pluck a sticker of the respective category and paste it in the box. Once you are through the whole syllabus you would have a selected 650 questions which you can revise in the last few days before the exam.

Get Exam Ready.

Team Disha

1 Units and Measurements

TREND ANALYSIS

TOPIC	2025		2024		2023		2022		2021		2020	
	Qns.	LOD	Qns.	LOD	Qns.	LOD	Qns.	LOD	Qns.	LOD	Qns.	LOD
The International System of Units												
Significant Figures							1	Easy			1	Easy
Dimensions of Physical Quantities			1	Avg.			3	Avg.	1	Avg.	1	Easy
Dimensional Analysis and its Applications	1	Diff.	1	Avg.			1	Avg.	1	Avg.		
Errors in Measurement	1	Avg.					1 1	Easy Avg.			1	Easy

How to Code Qns for Better & Faster Revision →

Type 1: Able to solve (No code Req)

Type 2: Solved with struggle or by seeing solution **(Use code NR means Needs Revision or orange stickers)**

Type 3: Tricky/ Could not solve on my own. Need help from Teacher/ Friend. **(Use code TQ means Tricky Question or red stickers)**

You can either hand write the code in boxes or else paste coloured stickers given at end of book.

Topic-1.2 The International System of Units

1. The unit of thermal conductivity is :
NCERT | Page-1 | 2019
(a) $J\ m\ K^{-1}$ (b) $J\ m^{-1}\ K^{-1}$
(c) $W\ m\ K^{-1}$ (d) $W\ m^{-1}\ K^{-1}$

2. The unit of permittivity of free space, ε_o is
NCERT | Page-2 | 2004
(a) $Coulomb^2/(Newton\text{-}metre)^2$
(b) Coulomb/Newton-metre
(c) $Newton\text{-}meter^2/Coulomb^2$
(d) $Coulomb^2/Newton\text{-}meter^2$

Topic-1.3 Significant Figures

3. The area of a rectangular field (in m^2) of length 55.3 m and breadth 25 m after rounding off the value of correct significant digits is:
NCERT | Page-5 | 2022
(a) 1382 (b) 1382.5
(c) 14×10^2 (d) 138×10^1

4. Taking into account of the significant figures, what is the value of 9.99 m – 0.0099 m?
NCERT | Page-5 | 2020
(a) 9.98 m (b) 9.980 m
(c) 9.9 m (d) 9.9801 m

Topic-* Errors in Measurement

5. A physical quantity P is related to four observations a, b, c and d as follows : 2025

$$P = \frac{a^3 b^2}{c\sqrt{d}}$$

The percentage errors of measurement in a, b, c and d are 1%, 3%, 2% and 4% respectively. The percentage error in the quantity P is
(a) 10% (b) 2%
(c) 13% (d) 15%

6. In an electrical circuit, the voltage is measured as $V = (200 \pm 4)$ volt and the current is measured as $I = (20 \pm 0.2)$ A. The value of the resistance is: Ph-II 2024
(a) $(10 \pm 4.2)\ \Omega$ (b) $(10 \pm 0.3)\ \Omega$
(c) $(10 \pm 0.1)\ \Omega$ (d) $(10 \pm 0.8)\ \Omega$

7. The diameter of a spherical bob, when measured with vernier callipers yielded the following values: 3.33 cm, 3.32 cm, 3.34 cm, 3.33 cm and 3.32 cm. The mean diameter to appropriate significant figures is:
Manipur 2023
(a) 3.33 cm (b) 3.32 cm
(c) 3.328 cm (d) 3.3 cm

8. The errors in the measurement which arise due to unpredictable fluctuations in temperature and voltage supply are
2023

(a) Instrumental errors
(b) Personal errors
(c) Least count errors
(d) Random errors

9. A metal wire has mass (0.4 ± 0.002) g, radius (0.3 ± 0.001) mm and length (5 ± 0.02) cm. The maximum possible percentage error in the measurement of density will nearly be **2023**

(a) 1.2% (b) 1.3%
(c) 1.6% (d) 1.4%

10. The percentage error in the measurement of g is : [Given that $g = \frac{4\pi^2 L}{T^2}$, L = (10 ± 0.1) cm, T = (100 ± 1)s] **RS 2022**

(a) 7% (b) 2%
(c) 5% (d) 3%

11. Time intervals measured by a clock give the following readings :
1.25 s, 1.24 s, 1.27 s, 1.21 s and 1.28 s
What is the percentage relative error of the observations? **Ph-II 2020**

(a) 1.6% (b) 2%
(c) 4% (d) 16%

12. In an experiment, the percentage of error occurred in the measurement of physical quantities A, B, C an D are 1%, 2%, 3% and 4% respectively. Then the maximum percentage of error in the measurement X, where $X = \frac{A^2 B^{1/2}}{C^{1/3} D^3}$ will be : **2019**

(a) $\left(\frac{3}{13}\right)\%$ (b) 16%
(c) –10% (d) 10%

13. In an experiment four quantities a, b, c and d are measured with percentage error 1%, 2%, 3% and 4% respectively. Quantity P is calculated as follows $P = \frac{a^3 b^2}{cd}$,% error in P is: **2013**

(a) 10% (b) 7%
(c) 4% (d) 14%

14. A student measures the distance traversed in free fall of a body, initially at rest in a given time. He uses this data to estimate g, the acceleration due to gravity. If the maximum percentage errors in measurement of the distance and the time are e_1 and e_2 respectively, the percentage error in the estimation of g is **2010 M**

(a) $e_2 - e_1$ (b) $e_1 + 2e_2$
(c) $e_1 + e_2$ (d) $e_1 - 2e_2$

15. If the error in the measurement of radius of a sphere is 2%, then the error in the determination of volume of the sphere will be: **2008**

(a) 4% (b) 6%
(c) 8% (d) 2%

16. The density of a cube is measured by measuring its mass and length of its sides. If the maximum error in the measurement of mass and length are 4% and 3% respectively, the maximum error in the measurement of density will be **1996**

(a) 7% (b) 9% (c) 12% (d) 13%

17. The percentage errors in the measurement of mass and speed are 2% and 3% respectively. The error in kinetic energy obtained by measuring mass and speed will be **1995**

(a) 12% (b) 10%
(c) 8% (d) 2%

18. A certain body weighs 22.42 gm and has a measured volume of 4.7 cc. The possible error in the measurement of mass and volume are 0.01 gm and 0.1 cc. Then maximum error in the density will be **1991**

(a) 22% (b) 2%
(c) 0.2% (d) 0.02%

Topic-1.4

Dimensions of Physical Quantities

19. The quantities which have the same dimensions as those of solid angle are : **NCERT | Page-7 | 2024**

(a) strain and angle (b) stress and angle
(c) strain and arc (d) angular speed and stress

20. The mechanical quantity, which has dimensions of reciprocal of mass (M^{-1}) is **NCERT | Page-7 | Manipur 2023**

(a) Torque
(b) Gravitational constant
(c) Angular momentum
(d) Coefficient of thermal conductivity

21. Match List-I with List-II **NCERT | Page-7 | 2022**

	List-I		List-II
(A)	Gravitational constant (G)	(i)	$[L^2T^2]$
(B)	Gravitational potential energy	(ii)	$[M^{-1}L^3T^{-2}]$
(C)	Gravitational potential	(iii)	$[LT^{-2}]$
(D)	Gravitational intensity	(iv)	$[ML^2T^{-2}]$

Choose the **correct answer** from the options given below:

(a) A–ii; B–iv; C–i; D–iii
(b) A–ii; B–iv; C–iii; D–i
(c) A–iv; B–ii; C–i; D–iii
(d) A–ii; B–i; C–iv; D–iii

22. The physical quantity that has the same dimensional formula as pressure is: NCERT | Page-7 | 2022
(a) Coefficient of viscosity
(b) Force
(c) Momentum
(d) Young's modulus of elasticity

23. The dimensions $[MLT^{-2}A^{-2}]$ belong to the:
(a) self inductance NCERT | Page-7 | 2022
(b) magnetic permeability
(c) electric permittivity
(d) magnetic flux

24. If E and G respectively denote energy and gravitational constant, then $\frac{E}{G}$ has the dimensions of

NCERT | Page-7 | 2021

(a) $[M^2][L^{-2}][T^{-1}]$ (b) $[M^2][L^{-1}][T^0]$
(c) $[M][L^{-1}][T^{-1}]$ (d) $[M][L^0][T^0]$

25. Dimensions of stress are: NCERT | Page-7 | 2020
(a) $[ML^2T^{-2}]$ (b) $[ML^0T^{-2}]$
(c) $[ML^{-1}T^{-2}]$ (d) $[MLT^{-2}]$

26. The pair of quantities having same dimensions is
NCERT | Page-7 | Kar. 2013
(a) Young's modulus and energy
(b) impulse and surface tension
(c) angular momentum and work
(d) work and torque

27. The dimensions of $(\mu_0 \epsilon_0)^{\frac{-1}{2}}$ are
NCERT | Page-7 | 2012 M, 2011
(a) $[L^{1/2}T^{-1/2}]$ (b) $[L^{-1}T]$
(c) $[LT^{-1}]$ (d) $[L^{-1/2}T^{1/2}]$

28. The dimension of $\frac{1}{2}\varepsilon_0 E^2$, where ε_0 is permittivity of free space and E is electric field, is: NCERT | Page-8 | 2010
(a) $[ML^2T^{-2}]$ (b) $[ML^{-1}T^{-2}]$
(c) $[ML^2T^{-1}]$ (d) $[MLT^{-1}]$

29. Which two of the following five physical parameters have the same dimensions? NCERT | Page-7 | 2008
(A) Energy density (B) Refractive index
(C) Dielectric constant (D) Young's modulus
(E) Magnetic field
(a) (B) and (D) (b) (C) and (E)
(c) (A) and (D) (d) (A) and (E)

30. The ratio of the dimension of Planck's constant and that of the moment of inertia is the dimension of
NCERT | Page-7 | 2005
(a) time
(b) frequency
(c) angular momentum
(d) velocity

31. The dimensions of universal gravitational constant is
NCERT | Page-7 | 2004
(a) $[M^{-2}L^2T^{-1}]$ (b) $[M^{-1}L^3T^{-2}]$
(c) $[ML^2T^{-1}]$ (d) $[M^{-2}L^3T^{-2}]$

32. The dimensions of Planck's constant are same as
NCERT | Page-7 | 2001
(a) energy (b) power
(c) momentum (d) angular momentum

33. Which one of the following groups have quantities that do not have the same dimensions? NCERT | Page-7 | 2000
(a) pressure, stress (b) velocity, speed
(c) force, impulse (d) work, energy

34. The force F on a sphere of radius 'a' moving in a medium with velocity v is given by $F = 6\pi\eta av$. The dimensions of η are NCERT | Page-8 | 1997
(a) $[ML^{-3}]$ (b) $[ML^{-2}]$
(c) $[ML^{-1}]$ (d) $[ML^{-1}T^{-1}]$

35. An equation is given as : $\left(P+\frac{a}{V^2}\right)=b\frac{\theta}{V}$ where P = Pressure, V = Volume & θ = Absolute temperature. If a and b are constants, then dimensions of a will be
NCERT | Page-9 | 1996
(a) $[ML^5T^{-2}]$ (b) $[M^{-1}L^5T^2]$
(c) $[ML^{-5}T^{-1}]$ (d) $[ML^5T^1]$

36. Which of the following will have the dimensions of time
NCERT | Page-7 | 1996
(a) LC (b) $\frac{R}{L}$ (c) $\frac{L}{R}$ (d) $\frac{C}{L}$

37. Turpentine oil is flowing through a tube of length l and radius r. The pressure difference between the two ends of the tube is p. The viscosity of oil is given by

$$\eta = \frac{p\left(r^2 - x^2\right)}{4vl}$$

where v is the velocity of oil at a distance x from the axis of the tube. The dimensions of η are NCERT | Page-9 | 1993
(a) $[M^0L^0T^0]$ (b) $[MLT^{-1}]$
(c) $[ML^2T^{-2}]$ (d) $[ML^{-1}T^{-1}]$

38. Of the following quantities, which one has dimension different from the remaining three? NCERT | Page-7 | 1989
(a) Energy per unit volume
(b) Force per unit area
(c) Product of voltage and charge per unit volume
(d) Angular momentum.

Topic-1.5 Dimensional Formulae and Dimensional Equations

39. The dimensional formula for magnetic flux is
NCERT | Page-7 | 1999

(a) $[ML^2T^{-2}A^{-1}]$ (b) $[ML^3T^{-2}A^{-2}]$
(c) $[M^0L^{-2}T^2A^{-2}]$ (d) $[ML^2T^{-1}A^2]$

40. The dimensional formula for permeability μ is given by
NCERT | Page-7 | 1991

(a) $[MLT^{-2}A^{-2}]$ (b) $[M^0L^1T]$
(c) $[M^0L^2T^{-1}A^2]$ (d) None of the above

41. According to Newton, the viscous force acting between liquid layers of area A and velocity gradient $\Delta V/\Delta Z$ is given by $F = -\eta A\frac{\Delta V}{\Delta Z}$ where η is constant called coefficient of viscosity. The dimensional formula of η is
NCERT | Page-8 | 1990

(a) $ML^{-2}T^{-2}$ (b) $M^0L^0T^0$
(c) ML^2T^{-2} (d) $ML^{-1}T^{-1}$

42. The dimensional formula of pressure is
NCERT | Page-8 | 1990

(a) $[MLT^{-2}]$ (b) $[ML^{-1}T^2]$
(c) $[ML^{-1}T^{-2}]$ (d) $[MLT^2]$

43. The dimensional formula of torque is
NCERT | Page-7 | 1989

(a) $[ML^2T^{-2}]$ (b) $[MLT^{-2}]$
(c) $[ML^{-1}T^{-2}]$ (d) $[ML^{-2}T^{-2}]$

44. Dimensional formula of self inductance is
NCERT | Page-7 | 1989

(a) $[MLT^{-2}A^{-2}]$ (b) $[ML^2T^{-1}A^{-2}]$
(c) $[ML^2T^{-2}A^{-2}]$ (d) $[ML^2T^{-2}A^{-1}]$

45. The dimensional formula for angular momentum is
NCERT | Page-7 | 1988

(a) $[M^0L^2T^{-2}]$ (b) $[ML^2T^{-1}]$
(c) $[MLT^{-1}]$ (d) $[ML^2T^{-2}]$

46. If C and R denote capacitance and resistance, the dimensional formula of CR is
NCERT | Page-7 | 1988

(a) $[M^0L^0T^1]$
(b) $[M^0L^0T^0]$
(c) $[M^0L^0T^{-1}]$
(d) not expressible in terms of M,L,T

Topic-1.6 Dimensional Analysis and its Applications

47. A balloon is made of a material of surface tension S and its inflation outlet (from where gas is filled in it) has small area A. It is filled with a gas of density ρ and takes a spherical shape of radius R. When the gas is allowed to flow freely out of it, its radius changes from R to 0 (zero) in time T. If the speed $v(r)$ of gas coming out of the balloon depends on r as r^a and $T \propto S^\alpha A^\beta \rho^\gamma R^\delta$ then
NCERT | Page-8 | 2025

(a) $a=\frac{1}{2}, \alpha=\frac{1}{2}, \beta=-1, \gamma=+1, \delta=\frac{3}{2}$

(b) $a=-\frac{1}{2}, \alpha=-\frac{1}{2}, \beta=-1, \gamma=\frac{1}{2}, \delta=\frac{5}{2}$

(c) $a=-\frac{1}{2}, \alpha=-\frac{1}{2}, \beta=-1, \gamma=\frac{1}{2}, \delta=\frac{7}{2}$

(d) $a=\frac{1}{2}, \alpha=\frac{1}{2}, \beta=-\frac{1}{2}, \gamma=\frac{1}{2}, \delta=\frac{7}{2}$

48. The potential energy of a particle moving along x-direction varies as $V=\frac{Ax^2}{\sqrt{x}+B}$. The dimensions of $\frac{A^2}{B}$ are:
NCERT | Page-8 | Ph-II 2024

(a) $[M^{3/2}L^{1/2}T^{-3}]$ (b) $[M^{1/2}LT^{-3}]$
(c) $[M^2L^{1/2}T^{-4}]$ (d) $[ML^2T^{-4}]$

49. A force defined by $F = \alpha t^2 + \beta t$ acts on a particle at a given time t. The factor which is dimensionless, if α and β are constants, is:
NCERT | Page-8 | 2024

(a) $\beta t/\alpha$ (b) $\alpha t/\beta$
(c) $\alpha\beta t$ (d) $\alpha\beta/t$

50. Plane angle and solid angle have: NCERT | Page-7 | 2022

(a) Dimensions but no units
(b) No units and no dimensions
(c) Both units and dimensions
(d) Units but no dimensions

51. If force $[F]$, acceleration $[A]$ and time $[T]$ are chosen as the fundamental physical quantities. Find the dimensions of energy.
NCERT | Page-8 | 2021

(a) $[F][A^{-1}][T]$ (b) $[F][A][T]$
(c) $[F][A][T^2]$ (d) $[F][A][T^{-1}]$

52. A physical quantity of the dimensions of length that can be formed out of c, G and $\frac{e^2}{4\pi\varepsilon_0}$ is [c is velocity of light, G is universal constant of gravitation and e is charge]
NCERT | Page-7 | 2017

(a) $c^2\left[G\frac{e^2}{4\pi\varepsilon_0}\right]^{1/2}$ (b) $\frac{1}{c^2}\left[\frac{e^2}{G4\pi\varepsilon_0}\right]^{1/2}$

(c) $\frac{1}{c}G\frac{e^2}{4\pi\varepsilon_0}$ (d) $\frac{1}{c^2}\left[G\frac{e^2}{4\pi\varepsilon_0}\right]^{1/2}$

53. Planck's constant (h), speed of light in vacuum (c) and Newton's gravitational constant (G) are three fundamental constants. Which of the following combinations of these has the dimension of length? **NCERT | Page-8 | Ph-II 2016**

(a) $\sqrt{\frac{hc}{G}}$ (b) $\sqrt{\frac{Gc}{h^{3/2}}}$

(c) $\sqrt{\frac{hG}{c^{3/2}}}$ (d) $\sqrt{\frac{hG}{c^{5/2}}}$

54. If energy (E), velocity (V) and time (T) are chosen as the fundamental quantities, the dimensional formula of surface tension will be: **NCERT | Page-7 | 2015**

(a) $[EV^{-1}T^{-2}]$ (b) $[EV^{-2}T^{-2}]$

(c) $[E^{-2}V^{-1}T^{-3}]$ (d) $[EV^{-2}T^{-1}]$

55. If dimensions of critical velocity υ_c of a liquid flowing through a tube are expressed as $[\eta^x\rho^y r^z]$, where η, ρ and r are the coefficient of viscosity of liquid, density of liquid and radius of the tube respectively, then the values of x, y and z are given by : **NCERT | Page-8 | 2015 RS**

(a) –1, –1, 1 (b) –1, –1, –1

(c) 1, 1, 1 (d) 1, –1, –1

56. If force (F), velocity (V) and time (T) are taken as fundamental units, then the dimensions of mass are : **NCERT | Page-7 | 2014**

(a) $[F\,V\,T^{-1}]$ (b) $[F\,V\,T^{-2}]$

(c) $[F\,V^{-1}\,T^{-1}]$ (d) $[F\,V^{-1}\,T]$

57. The density of material in CGS system of units is 4g/cm^3. In a system of units in which unit of length is 10 cm and unit of mass is 100 g, the value of density of material will be **NCERT | Page-3 | 2011 M**

(a) 0.4 (b) 40 (c) 400 (d) 0.04

58. If the dimensions of a physical quantity are given by $M^aL^bT^c$, then the physical quantity will be: **NCERT | Page-8 | 2009**

(a) Velocity if $a = 1, b = 0, c = -1$

(b) Acceleration if $a = 1, b = 1, c = -2$

(c) Force if $a = 0, b = -1, c = -2$

(d) Pressure if $a = 1, b = -1, c = -2$

59. Dimensions of resistance in an electrical circuit, in terms of dimension of mass [M], of length [L], of time [T] and of current [I], would be **NCERT | Page-8 | 2007**

(a) $[ML^2T^{-2}]$ (b) $[ML^2T^{-1}I^{-1}]$

(c) $[ML^2T^{-3}I^{-2}]$ (d) $[ML^2T^{-3}I^{-1}]$

60. The velocity v of a particle at time t is given by $v = at + \frac{b}{t+c}$, where a, b and c are constant. The dimensions of a, b and c are respectively **NCERT | Page-8 | 2006**

(a) $[L^2$, T and $LT^2]$ (b) $[LT^2$, LT and L$]$

(c) [L, LT and $T^2]$ (d) $[LT^{-2}$, L and T$]$

61. Which of the following is a dimensional constant? **NCERT | Page-7 | 1995**

(a) Refractive index

(b) Poissons ratio

(c) Relative density

(d) Gravitational constant

62. In a particular system, the unit of length, mass and time are chosen to be 10 cm, 10 g and 0.1 s respectively. The unit of force in this system will be equivalent to **NCERT | Page-3 | 1994**

(a) 0.1 N (b) 1 N

(c) 10 N (d) 100 N

63. The time dependence of a physical quantity p is given by $p = p_0 \exp(-\alpha t^2)$, where α is a constant and t is the time. The constant α **NCERT | Page-8 | 1993**

(a) is dimensionless (b) has dimensions T^{-2}

(c) has dimensions T^2 (d) has dimensions of p

64. P represents radiation pressure, c represents speed of light and S represents radiation energy striking unit area per sec. The non zero integers x, y, z such that $P^x S^y c^z$ is dimensionless are **NCERT | Page-8 | 1992**

(a) $x = 1, y = 1, z = 1$ (b) $x = -1, y = 1, z = 1$

(c) $x = 1, y = -1, z = 1$ (d) $x = 1, y = 1, z = -1$

65. The frequency of vibration f of a mass m suspended from a spring of spring constant k is given by a relation of the type $f = c\,m^x k^y$, where c is a dimensionless constant. The values of x and y are **NCERT | Page-9 | 1990**

(a) $x = \frac{1}{2}, y = \frac{1}{2}$ (b) $x = \frac{1}{2}, y = -\frac{1}{2}$

(c) $x = \frac{1}{2}, y = -\frac{1}{2}$ (d) $x = -\frac{1}{2}, y = \frac{1}{2}$

66. If $x = at + bt^2$, where x is the distance travelled by the body in kilometres while t is the time in seconds, then the unit of b is **NCERT | Page-2 | 1989**

(a) km/s (b) kms (c) km/s^2 (d) kms^2

ANSWER KEY

1	(d)	8	(d)	15	(b)	22	(d)	29	(c)	36	(c)	43	(a)	50	(d)	57	(b)	64	(c)
2	(d)	9	(c)	16	(d)	23	(b)	30	(b)	37	(d)	44	(c)	51	(c)	58	(d)	65	(d)
3	(c)	10	(d)	17	(c)	24	(b)	31	(b)	38	(d)	45	(b)	52	(d)	59	(c)	66	(c)
4	(a)	11	(a)	18	(b)	25	(c)	32	(d)	39	(a)	46	(a)	53	(c)	60	(d)		
5	(c)	12	(b)	19	(a)	26	(d)	33	(c)	40	(a)	47	(c)	54	(b)	61	(d)		
6	(b)	13	(d)	20	(b)	27	(c)	34	(d)	41	(d)	48	(c)	55	(d)	62	(a)		
7	(a)	14	(b)	21	(a)	28	(b)	35	(a)	42	(c)	49	(b)	56	(d)	63	(b)		

Hints & Solutions

1. **(d)** In steady state, the amount of heat flowing from one face to the other face in time dt is given by

$$dH = \frac{kA(T_1 - T_2)dt}{\ell} \Rightarrow \frac{dH}{dt} = \frac{kA}{\ell}\Delta T$$

(k = coefficient of thermal conductivity)

$$\therefore\ k = \frac{\ell dH}{A\,dt\,\Delta T}$$

Unit of $k = Wm^{-1}K^{-1}$

2. **(d)** $\varepsilon_o = \frac{q^2}{(r^2)4\pi F}$

$\Rightarrow$ unit of ε_o is (coulomb)^2/ newton-metre^2

3. **(c)** Area = length × breadth

$= 55.3 \times 25 = 1382.5\ m^2$

As, Answer of multiplication is rounded off to the same number of significant figure as present in least precise term.

Here, 25 has least significant figure and it is 2 so, Answer will also have 2 SF.

Therefore, correct option is (c) i.e. 14×10^2

4. **(a)** In subtraction the number of decimal places in the result should be equal to the number of decimal places of that term in the operation which contain lesser number of decimal places.

$$\begin{array}{r} 9.99 \\ -0.0099 \\ \hline 9.9801 \end{array}$$

As the least number of decimal places is 2. So, answer should be 9.98 m.

5. **(c)** Given, $a^3b^2c^{-1}d^{-1/2}$

$$\left(\frac{\Delta P}{P}\times 100\right) = \left(\frac{3\Delta a}{a} + \frac{2\Delta b}{b} + \frac{1\Delta c}{c} + \frac{1}{2}\frac{3\Delta d}{d}\right)\times 100$$

$$= 3(1\%) + 2(3\%) + 1(2\%) + \frac{1}{2}(4\%) = 13\%$$

6. **(b)** Resistance, $R = \frac{V}{I} = \frac{200}{20} = 10\Omega$

$$\frac{\Delta R}{R} = \frac{\Delta V}{V} + \frac{\Delta I}{I} = \frac{4}{200} + \frac{0.2}{20} = \frac{6}{200}$$

$$\Rightarrow \Delta R = \frac{6}{200}\times R = \frac{6}{20} = 0.3\Omega$$

$\Rightarrow$ Resistance $= R \pm \Delta R = (10 \pm 0.3)\ \Omega$

7. **(a)** Mean diameter,

$$d = \frac{[3.33\,cm + 3.32\,cm + 3.34\,cm + 3.33\,cm + 3.32\,cm]}{5}$$

$= 3.328$ cm

After rounding off to 2 decimal places,

$\bar{d} = 3.33$ cm

8. **(d)** Random errors is also called chance error. It occurs due to parameter which are beyound the control of experimeter.

9. **(c)** Density, $\rho = \frac{\text{mass}}{\text{volume}} = \frac{M}{\pi r^2 l}$

$$\frac{\Delta\rho}{\rho} = \frac{\Delta M}{M} + \frac{2\Delta r}{r} + \frac{\Delta l}{l} = \left(\frac{0.002}{0.4} + \frac{2\times 0.001}{0.3} + \frac{0.02}{5}\right)$$

$$\frac{\Delta\rho}{\rho} = 0.0156$$

% error in density, $\frac{\Delta\rho}{\rho}\times 100\% = 1.56\% \approx 1.6\%$

10. **(d)** Given,

$$g = \frac{4\pi^2 L}{T^2}$$

$$\%\text{ error} = \frac{\Delta g}{g}\times 100 = \frac{\Delta L}{L}\times 100 + 2\frac{\Delta T}{T}\times 100$$

$$= \left(\frac{0.1}{10}\times 100\right) + 2\left[\frac{1}{100}\times 100\right] = 1 + 2 = 3\%$$

NOTES In multiplication and division of two measured quantities, the relative error in final result is the sum of relative error in the measured quantities.

11. **(a)** Average value of time interval

$$\bar{T} = \frac{1.25+1.24+1.27+1.21+1.28}{5} = \frac{625}{5} = 1.25 \text{ s}$$

Absolute errors in the measured values of time intervals are

$|\Delta T_1| = 1.25 - 1.25 = 0$

$|\Delta T_2| = |1.25 - 1.24| = 0.01$

$|\Delta T_3| = |1.25 - 1.27| = 0.02$

$|\Delta T_4| = |1.25 - 1.21| = 0.04$

$|\Delta T_5| = |1.25 - 1.28| = 0.03$

Mean absolute error,

$$\Delta\bar{T} = \frac{|\Delta T_1|+|\Delta T_2|+|\Delta T_3|+|\Delta T_4|+|\Delta T_5|}{5}$$

$$= \frac{0+0.01+0.02+0.04+0.03}{5} = \frac{0.1}{5} = 0.02 \text{ s}$$

$\therefore$ Percentage relative error $= \frac{\Delta\bar{T}}{T}\times 100$

$$= \frac{0.02}{1.25}\times 100 = 1.6\%.$$

12. **(b)** Given, $x = \frac{A^2B^{1/2}}{C^{1/3}D^3}$

$$\% \text{ error} = \frac{\Delta x}{x}\times 100 = 2\frac{\Delta A}{A}\times 100$$

$$+\frac{1}{2}\frac{\Delta B}{B}\times 100+\frac{1}{3}\frac{\Delta C}{C}\times 100+3\frac{\Delta D}{D}\times 100$$

$$= 2\times 1\% + \frac{1}{2}\times 2\% + \frac{1}{3}\times 3\% + 3\times 4\%$$

$= 2\% + 1\% + 1\% + 12\% = 16\%$

13. **(d)** Given, $P = \frac{a^3b^2}{cd}$

Therefore, $\frac{\Delta P}{P}\times 100\% = 3\frac{\Delta a}{a}\times 100\%$

$$+2\frac{\Delta b}{b}\times 100\% + \frac{\Delta c}{c}\times 100\% + \frac{\Delta d}{d}\times 100\%$$

$= 3\times 1\% + 2\times 2\% + 3\% + 4\% = 14\%$

NOTES When we multiply or divide two measured quantities, the relative error in the final result is equal to the sum of the relative errors in the measured quantities. And when we add or subtract two measured quantities the absolute error in the final result is equal to the sum of the absolute error in the measured quantities.

14. **(b)** From $h = ut + \frac{1}{2}gt^2 \Rightarrow g = \frac{2h}{t^2}[\because u = 0]$

$\ln g = \ln h - 2\ln t$

$$\left(\frac{\Delta g}{g}\times 100\right)_{max} = \frac{\Delta h}{h}\times 100 + 2\frac{\Delta t}{t}\times 100$$

$= e_1 + 2e_2$

15. **(b)** Given, error in the measurement of radius of a sphere, $\frac{\Delta r}{r}\times 100 = 2\%$

Volume of the sphere, $V = \frac{4}{3}\pi r^3$

$\therefore$ Error in the volume, $\frac{\Delta V}{V}\times 100$

$= 3\cdot\frac{\Delta r}{r}\times 100 = 3\times 2\% = 6\%$

16. **(d)** As we know, density $= \frac{\text{mass}}{\text{volume}}$

$$\rho = \frac{M}{L^3}$$

Maximum error in the measurement of density

$$\therefore \frac{\Delta\rho}{\rho} = \frac{\Delta M}{M} + 3\frac{\Delta L}{L}$$

% error in density = % error in Mass

+3 (% error in length)

= 4 + 3(3) = 13%

17. **(c)** Percentage error in mass $\left(\frac{\Delta m}{m}\times 100\right) = 2\%$ and percentage error in speed $\left(\frac{\Delta v}{v}\times 100\right) = 3\%$

Kinetic energy, $k = \frac{1}{2}mv^2$.

$\therefore$ Error in measurement of kinetic energy $\frac{\Delta K}{K}$

$$= \frac{\Delta m}{m} + 2\left(\frac{\Delta v}{v}\right) = \left(\frac{2}{100}\right) + \left(2\times\frac{3}{100}\right) = \frac{8}{100} = 8\%$$

= 8%.

18. **(b)** Density, $D = \frac{\text{Mass(M)}}{\text{Volume(V)}}$

$$\therefore \frac{\Delta D}{D} = \frac{\Delta M}{M} + \frac{\Delta V}{V} = \left(\frac{0.01}{22.42} + \frac{0.1}{4.7}\right)\times 100$$

= 2%

19. **(a)** Solid angle $d\Omega = \frac{dA}{r^2}$ has dimensions $[M^0L^0T^0]$

Strain $= \frac{\Delta\ell}{\ell}$ has dimensions $[M^0L^0T^0]$

Angle measured in radians $\theta = \frac{\ell}{r}$ is also dimensionless $[M^0L^0T^0]$

Solid angle, strain and angle are dimensionless.

NOTES Angle, trigonometric function ($\cos\theta$, $\sin\theta$, $\tan\theta$), Logarithmic function ($\log x$), exponential function (e^x), Ratio, Pure numbers are dimensionless.

20. **(b)** Gravitational force, $F = \frac{GM_1M_2}{r^2}$

$$\Rightarrow [G] = \frac{[F][r]^2}{[M_1][M_2]} = \frac{[MLT^{-2}][L^2]}{[M][M]}$$

$\Rightarrow [G] = [M^{-1}L^3T^{-2}]$

Torque $T = F \cdot r = [ML^2T^{-2}]$

Angular momentum, $L = Mvr = [ML^2T^{-1}]$

21. **(a)** $[G] = \left[\frac{Fr^2}{m_1m_2}\right] = \frac{[MLT^{-2}L^2]}{[M^2]}$

$= [M^{-1}L^3T^2]$

$[U] = [W] = [ML^2T^{-2}]$

$$[V] = \left[\frac{U}{M}\right] = \frac{[ML^2T^{-2}]}{[M]} = [L^2T^{-2}]$$

$$[I] = \left[\frac{F}{M}\right] = [LT^{-2}]$$

22. **(d)** Pressure $= \frac{F}{A}$ and Young's modulus

$$Y = \frac{\text{Stress}}{\text{Strain}} = \frac{F/A}{\Delta\ell/\ell_0}$$

Hence the dimensions of pressure (P) and Young's modulus (Y) are same.

23. **(b)** For circular loop magnetic field B at centre,

$$B = \frac{\mu_0 i}{2r} \Rightarrow \mu_0 = \frac{2Br}{I}$$

$$[\mu_0] = \frac{[B][r]}{[I]} = \frac{[MLT^{-2}A^{-1}L^{-1}][L]}{[A]}$$

$$= [MLT^{-2}A^{-2}]$$

24. **(b)** Dimensional formula of energy $E = \frac{1}{2}mv^2$

$[E] = [M^1L^2T^{-2}]$ ...(i)

Dimensional formula of gravitational constant

$$G = \frac{Fr^2}{m_1m_2}$$

$[G] = [M^{-1}L^3T^{-2}]$ ...(ii)

From eq. (i) & (ii)

$$\therefore \frac{E}{G} = \frac{[M^1L^2T^{-2}]}{[M^{-1}L^3T^{-2}]} = [M^2L^{-1}T^0]$$

NOTES Numerical constant like, $\frac{1}{2}$, 1, 2 or 2π has no dimension.

25. **(c)** Stress $= \frac{\text{Force}}{\text{Area}}$

$$\therefore \text{Stress} = \frac{[MLT^{-2}]}{[L^2]} = [ML^{-1}T^{-2}]$$

26. **(d)** Work = Force × displacement

$= [MLT^{-2}][L] = [ML^2T^{-2}]$

Torque = Force × Length

= mass × acceleration × length

$= [M] \times [LT^{-2}] \times [L] = [ML^2T^{-2}]$

NOTES If dimensions are given, physical quantity may not be unique as many physical quantities have same dimensions. *e.g.*, If the dimensional formula of a physical quantity is $[ML^2T^{-2}]$ it may be work or energy or torque.

27. **(c)** $(\mu_0\varepsilon_0)^{-1/2} = \frac{1}{\sqrt{\mu_0\varepsilon_0}}$ = speed of light

So, dimensions of $(\mu_0\varepsilon_0)^{-1/2}$ will be $[LT^{-1}]$

28. **(b)** $\frac{1}{2}\varepsilon_0E^2$ represents energy density *i.e.*, energy per unit volume.

$$\Rightarrow \left[\frac{1}{2}\varepsilon_0E^2\right] = \frac{[ML^2T^{-2}]}{[L^3]} = [ML^{-1}T^{-2}]$$

29. **(c)** Energy density $= \frac{\text{Energy}}{\text{Volume}}$

$$= \frac{[ML^2T^{-2}]}{[L^3]} = [ML^{-1}T^{-2}]$$

Refractive index and dielectric constant have no dimensions.

Young's Modulus $= \frac{F}{A} \times \frac{l}{\Delta l}$

$$\Rightarrow \frac{[MLT^{-2}]}{[L^2]} \cdot \frac{[L]}{[L]} = [ML^{-1}T^{-2}].$$

Magnetic field, $B = \frac{F}{i\ell}$

$$= \frac{[MLT^{-2}]}{[A][L]}$$

$= [MT^{-2}A^{-1}]$.

30. **(b)** Dimension formula for the planck's constant, $h = [ML^2 T^{-1}]$

Dimension formula for the moment of inertia, $I = [ML^2]$

So, the ratio between the plank's constant and moment of inertia is

$$\Rightarrow \frac{h}{I} = \frac{[ML^2T^{-1}]}{[ML^2]}$$

$= [T^{-1}] \Rightarrow$ dimension of frequency

31. **(b)** $F = \frac{G\,Mm}{r^2} \Rightarrow G = \frac{F\,r^2}{Mm}$

$\therefore$ Dimension of G is $\frac{[MLT^{-2}][L^2]}{[M][M]}$

$= [M^{-1}L^3T^{-2}]$

32. **(d)** We know that $E = h\nu$

$$h = \frac{E}{\nu} = \frac{[ML^2T^{-2}]}{[T^{-1}]} = [ML^2T^{-1}]$$

Angular momentum = $I\omega$

$= [ML^2][T^{-1}] = [ML^2T^{-1}]$

33. **(c)** Force has dimension $[MLT^{-2}]$ while impulse has dimension $[MLT^{-1}]$, both have different dimensions.

34. **(d)** $F = 6\pi\eta\, av$

$$\eta = \frac{F}{6\pi\, av} = \frac{[MLT^{-2}]}{[L][LT^{-1}]} = [ML^{-1}T^{-1}]$$

35. **(a)** $\left(P + \frac{a}{V^2}\right) = b\frac{\theta}{V}$

According to the principle of homogeneity quantity with same dimension can be added or subtracted.

Hence, Dimension of P = Dimension of $\frac{a}{V^2}$

$\Rightarrow$ Dimension of $\frac{\text{Force}}{\text{Area}}$ = Dimension of $\frac{a}{V^2}$

$$\Rightarrow \left[\frac{MLT^{-2}}{L^2}\right] = \frac{a}{[L^3]^2} \Rightarrow a = [M\,L^5\,T^{-2}]$$

NOTES To get the dimensions of physical constant, we write any formula or equation incorporating the given constant and then by substituting the dimensional formula of all other quantities, we can find the dimensions of the required constant or coefficients.

36. **(c)** $\varepsilon = -L\frac{di}{dt}$ ....(i)

$\varepsilon = iR$ ....(ii)

From (i) & (ii), $iR = -L\frac{di}{dt}$

$\therefore$ Dimension of L.H.S. = Dimension of R.H.S.

$$[A]\,[R] = [L]\,[AT^{-1}] \Rightarrow \left[\frac{L}{R}\right] = [T]$$

NOTES $\frac{L}{R}$ is time constant of R-L circuit so, dimensions of $\frac{L}{R}$ is same as that of time.

37. **(d)** $\eta = \frac{p(r^2 - x^2)}{4vl} = \frac{[ML^{-1}T^{-2}][L^2]}{[LT^{-1}][L]}$

$= [ML^{-1}T^{-1}]$

NOTES According to the principle of homogeneity, the dimensions of each term on the L.H.S. must be equal to the dimensions of the terms on the R.H.S. Only then dimensional equation or formula is dimensionally correct.

38. **(d)** For angular momentum, the dimensional formula is $[ML^2T^{-1}]$. For other three, it is $[ML^{-1}T^{-2}]$.

39. **(a)** Dimension of magnetic flux

= Dimension of magnetic field × Dimension of area

$[ML^0T^{-2}A^{-1}]\,[L^2] = [ML^2T^{-2}A^{-1}]$

40. **(a)** We have

$$B = \frac{\mu_0}{4\pi}\frac{id\,\ell\,\sin\theta}{r^2} \qquad \text{[By Biot-Savart's law]}$$

$$[\mu_0] = \frac{[B][r^2]}{[i][d\,\ell]}$$

$$= \frac{[MLT^{-2}][L^2]}{[AL]\times[A]\times[L]} = [MLT^{-2}A^{-2}]$$

NOTES We can also find the dimensional formula of μ_0 by using the relation.

Speed of light, $c = \frac{1}{\sqrt{\mu_0\varepsilon_0}}$

41. **(d)** $F = -\eta A\frac{\Delta V}{\Delta Z} \Rightarrow \eta = (-1)\frac{F\Delta Z}{A\Delta V}$

So dimensional formula of η

$$= \frac{[MLT^{-2}][L]}{[L^2][LT^{-1}]} = [ML^{-1}T^{-1}]$$

42. **(c)** Pressure = Force / Area

So dimensional formula of pressure

$$= \frac{[MLT^{-2}]}{[L^2]} = [ML^{-1}T^{-2}]$$

43. **(a)** τ = Force × distance

So dimensional formula,

$[\tau] = [MLT^{-2}][L] = [ML^2T^{-2}]$

44. **(c)** $L = \varepsilon\left(\frac{dt}{di}\right) = \frac{W}{q}\left[\frac{dt}{di}\right] = \frac{W}{i \cdot t}\left[\frac{dt}{di}\right]$

or, [Self inductance]

$= \frac{[ML^2T^{-2}][T]}{[AT][A]} = [ML^2T^{-2}A^{-2}]$

45. **(b)** Angular momentum

= Momentum of inertia × Angular velocity

So dimensional formula,

[Angular momentum] = $[ML^2] \times [T^{-1}]$

$= [ML^2T^{-1}]$

46. **(a)** $CR = \left(\frac{q}{V}\right)\left(\frac{V}{i}\right) = \left(\frac{i \cdot t}{i}\right) = t$ = time

$= [T] = [M^0L^0T^1]$

RC is the time constant of the R-C circuit.

47. **(c)** Writing the dimensions of each and every term

T (Time) = $[T_1]$

S (Surface Tension) = $\frac{F}{\ell} = \frac{[M^1L^1T^{-2}]}{[L^1]} = [M^1T^{-2}]$

A (Area) = $[L^2]$

ρ (density) = $[M^1L^{-3}]$

R (Radius) = $[L^1]$

Putting dacu all dimensions on the both side of equation and comparing

$T \propto S^{\alpha}A^{\beta}\rho^{\gamma}R^{\delta}$

$[M^0L^0T^0] = [MT^{-2}][L^2]^{\beta}[ML^{-3}]^{\gamma}[L]^{\delta}$

$[M^0L^0T^0] = [M^{\alpha+\gamma}, L^{2\beta+\delta-3\gamma}, T^{-2\alpha}$

$\Rightarrow -2\alpha = 1$

$\alpha + \delta = 0$

$\Rightarrow \alpha = -\frac{1}{2} \Rightarrow \gamma = -\alpha = \frac{1}{2}$

$2\beta + \gamma - 3\gamma = 0$

$2\beta + \delta - 3\left(\frac{1}{2}\right) = 0 \Rightarrow 2\beta + \delta - \frac{3}{2}$

48. **(c)** Potential energy, $V = \frac{Ax^2}{\sqrt{x} + B}$ From homogeneous rule of dimensions, $B = \sqrt{L}$

$\therefore \quad ML^2T^{-2} = \frac{AL^2}{L^{1/2}}$

$\Rightarrow A = ML^{1/2}T^{-2}$

$\therefore \quad \frac{A^2}{B} = \frac{M^2LT^{-4}}{L^{1/2}} = M^2L^{1/2}T^{-4}$

49. **(b)** From principle of homogeneity

$[F] = [\alpha t^2] = [\beta t] \Rightarrow [\alpha t] = \frac{[F]}{[t]}$ and $[\beta] = \frac{[F]}{[t]}$

$\therefore \quad \frac{[\alpha t]}{[\beta]} = [F^0t^0] \quad \therefore \quad \frac{\alpha t}{\beta}$ = dimensionless

50. **(d)** Plane angle unit is radian, whereas solid angle unit is steradian, but they don't have any dimensions.

> **NOTES** Angle is exceptional physical quantity, which is a ratio of two similar physical quantities angle = $\frac{\text{arc}}{\text{radius}}$ but still requires a unit to specify it.

51. **(c)** Energy, $E \propto F^aA^bT^c$

$[E] = [F^a][A^b][T^c]$

$\Rightarrow [ML^2T^{-2}] = [MLT^{-2}]^a[LT^{-2}]^b[T]^c$

$[ML^2T^{-2}] = [M^aL^{a+b}T^{-2a-2b+c}]$

Comparing dimensions on both sides.

a = 1; a + b = 2 and –2 = – 2a – 2b + c

$\therefore$ b = 1 and –2 = – 2 – 2 + c or, c = 2

$\therefore$ Dimensions of energy = $[FAT^2]$

52. **(d)** Let dimensions of length is related as,

$[L] = [c]^x[G]^y\left[\frac{e^2}{4\pi\varepsilon_0}\right]^z$

$[L] = [LT^{-1}]^x[M^{-1}L^3T^{-2}]^y[ML^3T^{-2}]^z$

$[L] = [L^{x+3y+3z}M^{-y+z}T^{-x-2y-2z}]$

Comparing both sides

$-y + z = 0 \Rightarrow y = z$ ...(i)

$x + 3y + 3z = 1$ ...(ii)

$-x - 4z = 0$ ($\because$ y = z) ...(iii)

From (i), (ii) and (iii)

$z = y = \frac{1}{2}, x = -2$

Hence, $[L] = c^{-2}\left[G \cdot \frac{e^2}{4\pi\varepsilon_0}\right]^{1/2}$

53. **(c)** Let the dimension of length depends on h, G, c as $\ell \alpha h^xG^yc^z$

$M^0L^1T^0 = (ML^2T^{-1})^x(M^{-1}L^3T^{-2})^y(LT^{-1})^Z$
$= M^{x-y}L^{2x+3y+z}T^{-x-2y-Z}$

Equating :

$$\left.\begin{array}{l} x - y = 0 \\ 2x + 3y + z = 1 \\ -x - 2y - z = 0 \end{array}\right\} \begin{array}{l} \Rightarrow x = \frac{1}{2}; y = \frac{1}{2}; z = -\frac{3}{2} \\ \Rightarrow \ell\alpha\frac{\sqrt{hG}}{c^{3/2}} \end{array}$$

54. (b) As we know that, surface tension(S) $= \dfrac{\text{Force}[F]}{\text{Length}[L]}$

So, $[S] = \dfrac{[MLT^{-2}]}{[L]} = [MT^{-2}]$

Energy, (E) = Force × Displacement,
$[E] = [ML^2T^{-2}]$

Velocity (V) $= \dfrac{\text{displacement}}{\text{time}}$

$[V] = [LT^{-1}]$

Let surface tension expressed as,
$S = E^a V^b T^c$ where a, b, c are constant.
Put the value

$$\frac{[MLT^{-2}]}{[L]} = [ML^2T^{-2}]^a \left[\frac{L}{T}\right]^b [T]^c$$

From the principle of homogeneity,
Equating the dimension of LHS and RHS
$[ML^0T^{-2}] = [M^aL^{2a+b}T^{-2a-b+c}]$
$\Rightarrow \quad a = 1, 2a + b = 0, -2a - b + c = -2$
$\Rightarrow \quad a = 1, b = -2, c = -2$
Hence, the dimensions of surface tension are $[E\,V^{-2}\,T^{-2}]$

NOTES Length, mass and time are arbitrarily chosen as fundamental quantities in mechanics. In fact any three quantities in mechanics can be termed as fundamental as all other quantities can be expressed in terms of these. If force (F) and acceleration (a) are taken as fundamental quantities, then mass will be defined as force (F)/acceleration (a) and will be termed as derived quantity.

55. (d) Applying dimensional method :
$v_c = \eta^x \rho^y r^z$

here,

dimension of critical velocity, $v_c = [LT^{-1}]$

co-efficient of viscosity, $\eta = \dfrac{F}{6\pi r v}$

so dimension of $\eta = \dfrac{[MLT^{-2}]}{[L][LT^{-1}]} = [ML^{-1}T^{-1}]$

dimension of density, $\rho = \dfrac{[M]}{[L^3]} = [ML^{-3}]$

dimension of radius, $r = [L]$

Put these values in equation (i),

$[M^0LT^{-1}] = [ML^{-1}T^{-1}]^x\,[ML^{-3}T^0]^y\,[M^0LT^0]^z$

Equating powers both sides

$x + y = 0; -x = -1 \therefore x = 1$

$1 + y = 0 \therefore y = -1$

$-x - 3y + z = 1$

$-1 - 3(-1) + z = 1$

$-1 + 3 + z = 1$

$\therefore z = -1$

56. (d) Force = mass × acceleration

$\Rightarrow \text{Mass} = \dfrac{\text{force}}{\text{acceleration}} = \dfrac{\text{force}}{\text{velocity / time}}$

$= [F\,V^{-1}\,T]$

57. (b) In CGS system, density $d = 4\dfrac{g}{cm^3}$

unit of length = 1 cm
unit of mass = 1 g
And in another system of units
The unit of mass is 100g and unit of length is 10 cm, so substitute these values

$$\text{Density} = \frac{4\left(\dfrac{100g}{100}\right)}{\left(\dfrac{10}{10}cm\right)^3} = \frac{\left(\dfrac{4}{100}\right)(100g)}{\left(\dfrac{1}{10}\right)^3 (10cm)^3}$$

$$= \frac{4}{100}\times(10)^3 \cdot \frac{100g}{(10cm)^3} = 40\,\text{unit}$$

NOTES Apart from fundamental and derived units, we have also used practical units *e.g.*, horse power (h.p.) is a practical units of power. Practical units may or may not belong to a system but can be expressed in any system of units.

58. (d) Pressure $= \dfrac{\text{Force}}{\text{Area}}$

$\Rightarrow \dfrac{[MLT^{-2}]}{[L^2]} = [ML^{-1}T^{-2}]$

$\Rightarrow a = 1, b = -1, c = -2.$

59. (c) As we know that

$R = \dfrac{[V]}{[i]} = \dfrac{[w]}{[q \cdot i]} = \dfrac{[w]}{[i \cdot t \cdot i]}$

Dimension of Resistance

$= \left[\dfrac{ML^2T^{-2}}{I^2T}\right] = [ML^2T^{-3}I^{-2}]$

60. (d) Dimension of $a.t$ = dimension of velocity $\Rightarrow$ $[a.t] = [LT^{-1}] \quad \Rightarrow \quad [a] = [LT^{-2}]$

Dimension of c = dimension of t

(two physical quantity of same dimension can only be added)

So, dimension of $c = [T]$

Dimension of $\frac{b}{t+c}$ = Dimension of velocity

$\frac{[b]}{[T]} = [LT^{-1}] \Rightarrow [b \cdot T^{-1}] = [LT^{-1}]$

$\Rightarrow [b] = [L]$

So, answer is $[LT^{-2}]$, [L] and [T]

NOTES A dimensionally correct equation may or may not be physically correct. In a dimensionally correct equation, the dimensions of each term on both sides of an equation must be the same.

61. **(d)** A quantity which has dimensions and a constant value is called dimensional constant. Therefore, gravitational constant (G) is a dimensional constant and dimension of $G = [M^{-1}L^3T^{-2}]$

NOTES Relative density, refractive index and poisson ratio all the three are ratios, therefore they are dimensionless constants. Angle is an exceptional physical quantity, which is a ratio of two similar physical quantities $\left(\text{angle} = \frac{\text{arc}}{\text{radius}}\right)$, but still requires a unit and no dimensions.

62. **(a)** As we know force = Mass × Acceleration = Mass × length × time^{-2} = (10g)(10 cm)$(0.1\text{s})^{-2}$
$= (10^{-2}\text{ kg})(10^{-1}\text{m})(10^{-1}\text{s})^{-2} = 10^{-1}\text{N}$.

63. **(b)** In $p = p_0 \exp(-\alpha t^2)$, where αt^2 is dimensionless

$\therefore [\alpha] = \frac{1}{[t^2]} = \frac{1}{[T^2]} = [T^{-2}]$

64. **(c)** Let the expression, $\alpha = P^x S^y c^z$ ...(i)

and given that dimension of $\alpha = [M^0L^0T^0]$ ...(ii)

Dimension of radiation pressure $P = \frac{\text{Force}}{\text{Area}}$

$= \frac{[MLT^{-2}]}{[L\ ]} = [ML^{-1}T^{-2}]$

Dimension of radiation energy/unit area per unit time

$S = \frac{\text{Energy}}{\text{Area} \times \text{Time}} = \frac{[ML^2T^{-2}]}{[L^2][T]} = [MT^{-3}]$

Dimension of speed of light, $c = [LT^{-1}]$

By equation (i) we get,

So, the dimension of $\alpha = [ML^{-1}T^{-2}]^x [MT^{-3}]^y [LT^{-1}]^z$

According to equation (ii),

$\Rightarrow [M^0L^0T^0] = [ML^{-1}T^{-2}]^x [MT^{-3}]^y [LT^{-1}]^z$

$\Rightarrow [M^0L^0T^0] = [M^{x+y} L^{-x+z} T^{-2x-3y-z}]$

Applying the principle of homegenity of dimension we get,

$x + y = 0$ ...(iii)

$-x + z = 0$ ...(iv)

$-2x - 3y - z = 0$ ...(v)

After solving above three equation we get,

$x = 1;\ y = -1;\ z = 1$

NOTES Try out the given alternatives. When $x = 1, y = -1, z = 1$

$P^xS^yc^z = P^1S^{-1}c^1 = \frac{Pc}{S}$

$= \frac{[ML^{-1}T^{-2}][LT^{-1}]}{[ML^2T^{-2}/L^2T]} = [M^0L^0T^0]$

65. **(d)** $f = c\, m^x k^y$;

Spring constant k = force/length.

$[M^0L^0T^{-1}] = [M]^x [MT^{-2}]^y$

$= [M^{x+y} T^{-2y}]$

$\Rightarrow x + y = 0, -2y = -1$ or $y = \frac{1}{2}$

Therefore, $x = -\frac{1}{2}$

NOTES The method of dimensions cannot be used to derive relations other than product of power functions.

66. **(c)** Given, $x = at + bt^2$

where, x = distance in km

t = time in sec

By the principle of homogencity, the dimensions of each terms of a dimensional equation on both sides are the same.

So, from equation, $x = at + bt^2$

dimension of left side = $[L]$

dimension of right side should be = [L]

$\therefore [L] = [b][T^2]$

$[b] = \frac{[L]}{[T^2]}$ = dimension of acceleration

so unit of b should be km/sec^2.

2 Motion in a Straight Line

TREND ANALYSIS

TOPIC	2025		2024		2023		2022		2021		2020	
	Qns.	LOD	Qns.	LOD	Qns.	LOD	Qns.	LOD	Qns.	LOD	Qns.	LOD
Average Velocity and Average Speed					1	Easy						
Acceleration	1	Avg.	1	Avg.								
Instantaneous Velocity and Speed												
Circular Motion							1	Easy				
Kinematic Equations for Uniformly Accelerated Motion, Relative Velocity	1	Diff.			2	Avg.	1	Easy	1	Avg.	1	Avg.

How to Code Qns for Better & Faster Revision

Type 1: Able to solve (No code Req)
Type 2: Solved with struggle or by seeing solution **(Use code NR means Needs Revision or orange stickers)**
Type 3: Tricky/ Could not solve on my own. Need help from Teacher/ Friend. **(Use code TQ means Tricky Question or red stickers)**

You can either hand write the code in boxes or else paste coloured stickers given at end of book.

Topic-* Average Velocity and Average Speed

1. A vehicle travels half the distance with speed v and the remaining distance with speed 2v. Its average speed is **2023**

(a) $\frac{v}{3}$ (b) $\frac{2v}{3}$ (c) $\frac{4v}{3}$ (d) $\frac{3v}{4}$

2. A person travelling in a straight line moves with a constant velocity v_1 for certain distance 'x' and with a constant velocity v_2 for next equal distance. The average velocity v is given by the relation **Odisha 2019**

(a) $v=\sqrt{v_1v_2}$ (b) $\frac{1}{v}=\frac{1}{v_1}+\frac{1}{v_2}$

(c) $\frac{2}{v}=\frac{1}{v_1}+\frac{1}{v_2}$ (d) $\frac{v}{2}=\frac{v_1+v_2}{2}$

3. A particle covers half of its total distance with speed v_1 and the rest half distance with speed v_2. Its average speed during the complete journey is **2011M**

(a) $\frac{v_1v_2}{v_1+v_2}$ (b) $\frac{2v_1v_2}{v_1+v_2}$

(c) $\frac{2v_1^2v_2^2}{v_1^2+v_2^2}$ (d) $\frac{v_1+v_2}{2}$

4. A car moves from X to Y with a uniform speed v_u and returns to X with a uniform speed v_d. The average speed for this round trip is **2007**

(a) $\sqrt{v_uv_d}$ (b) $\frac{v_dv_u}{v_d+v_u}$

(c) $\frac{v_u+v_d}{2}$ (d) $\frac{2v_dv_u}{v_d+v_u}$

5. A bus travelling the first one third distance at a speed of 10 km/h, the next one third at 20 km/h and the last one-third at 60 km/h. The average speed of the bus is **1991**

(a) 9 km/h (b) 16 km/h
(c) 18 km/h (d) 48 km/h

6. A car moves a distance of 200 m. It covers the first half of the distance at speed 40 km/h and the second half of distance at speed v. The average speed is 48 km/h. Find the value of v **1991**

(a) 56 km/h (b) 60 km/h
(c) 50 km/h (d) 48 km/h

7. A car covers the first half of the distance between two places at 40 km/h and other half at 60 km/h. The average speed of the car is **1990**

(a) 40 km/h (b) 48 km/h
(c) 50 km/h (d) 60 km/h

Topic-2.2 Instantaneous Velocity and Speed

8. The position-time (x–t) graph for positive acceleration is: **NCERT | Page-15 | RS 2022**

(a)
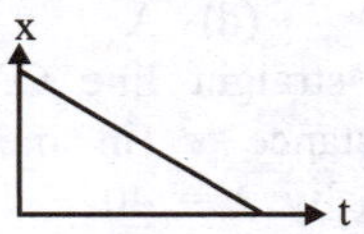

(b)
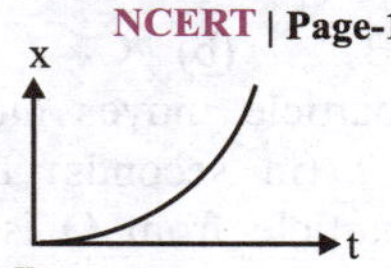

(c)
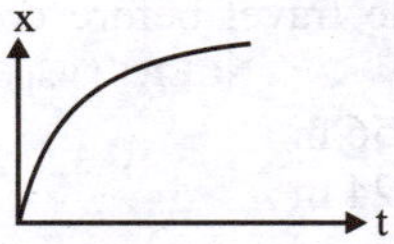

(d)
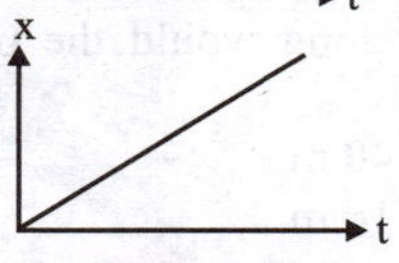

9. The displacement-time graphs of two moving particles make angles of 30° and 45° with the x-axis as shown in the figure. The ratio of their respective velocity is: **NCERT | Page-14 | 2022**

(a) 1 : 1

(b) 1 : 2

(c) $1:\sqrt{3}$

(d) $\sqrt{3}:1$

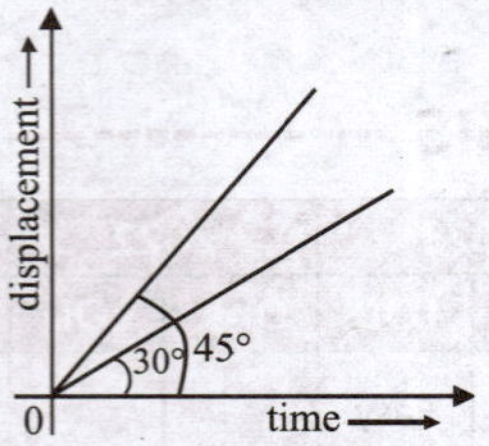

10. Two cars P and Q start from a point at the same time in a straight line and their positions are represented by $x_p(t) = at + bt^2$ and $x_Q(t) = ft - t^2$. At what time do the cars have the same velocity? **NCERT | Page-14 | Ph-II 2016**

(a) $\frac{a+f}{2(1+b)}$ (b) $\frac{f-a}{2(1+b)}$

(c) $\frac{a-f}{1+b}$ (d) $\frac{a+f}{2(b-1)}$

11. If the velocity of a particle is $v = At + Bt^2$, where A and B are constants, then the distance travelled by it between 1s and 2s is : **NCERT | Page-15 | 2016**

(a) $\frac{3}{2}A+4B$ (b) $3A+7B$

(c) $\frac{3}{2}A+\frac{7}{3}B$ (d) $\frac{A}{2}+\frac{B}{3}$

12. The displacement '*x*' (in metre) of a particle of mass '*m*' (in kg) moving in one dimension under the action of a force, is related to time '*t*' (in sec) by $t = \sqrt{x}+3$. The displacement of the particle when its velocity is zero, will be **NCERT | Page-15 | Kar. 2013**

(a) 2 m (b) 4 m (c) zero (d) 6 m

13.

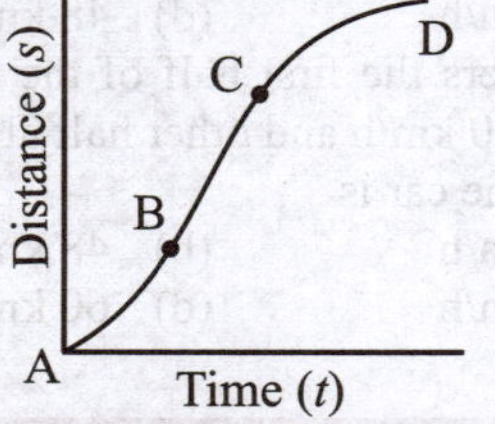

A particle shows distance-time curve as given in this figure. The maximum instantaneous velocity of the particle is around the point: **NCERT | Page-17 | 2008**

(a) B (b) C (c) D (d) A

14. A particle moves along a straight line OX. At a time *t* (in seconds) the distance *x* (in metres) of the particle from O is given by $x = 40 + 12t - t^3$. How long would the particle travel before coming to rest? **NCERT | Page-18 | 2006**

(a) 40 m (b) 56 m

(c) 16 m (d) 24 m

15. The displacement time graph of a moving particle is shown below

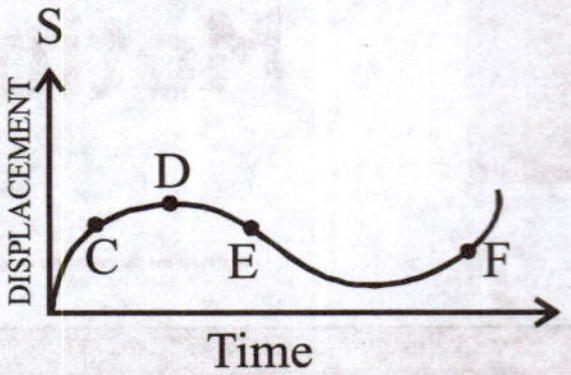

The instantaneous velocity of the particle is negative at the point **NCERT | Page-14 | 1994**

(a) D (b) F (c) C (d) E

16. Which of the following curve does not represent motion in one dimension? **1992**

(a)

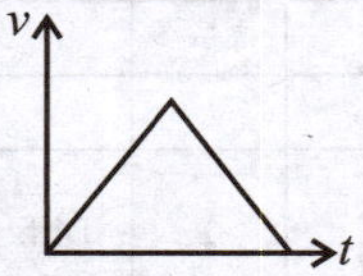

(b)

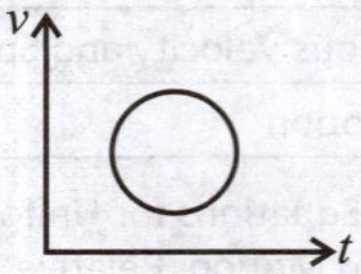

(c)

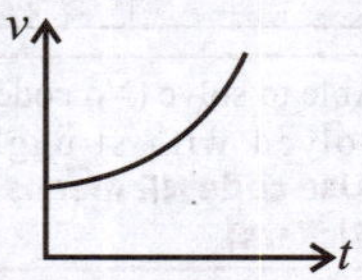

(d)

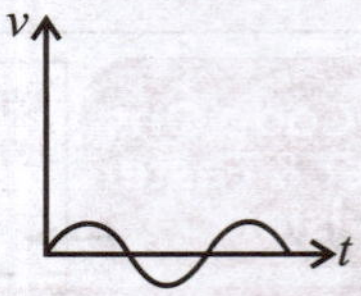

Topic-2.3 Acceleration

17. In some appropriate units, time (*t*) and position (*x*) relation of a moving particle is given by $t = x^2 + x$. The acceleration of the particle is **NCERT | Page-16 | 2025**

(a) $-\frac{2}{(x+2)^3}$ (b) $-\frac{2}{(2x+1)^3}$

(c) $+\frac{2}{(x+1)^3}$ (d) $+\frac{2}{2x+1}$

18. A particle is moving along *x*-axis with its position (*x*) varying with time (*t*) as $x = \alpha t^4 + \beta t^2 + \gamma t + \delta$. The ratio of its initial velocity to its initial acceleration, respectively, is: **NCERT | Page-16 | Ph-II 2024**

(a) $2\alpha : \delta$ (b) $\gamma : 2\delta$

(c) $4\alpha : \beta$ (d) $\gamma : 2\beta$

19. The velocity (v) – time (t) plot of the motion of a body is shown below:

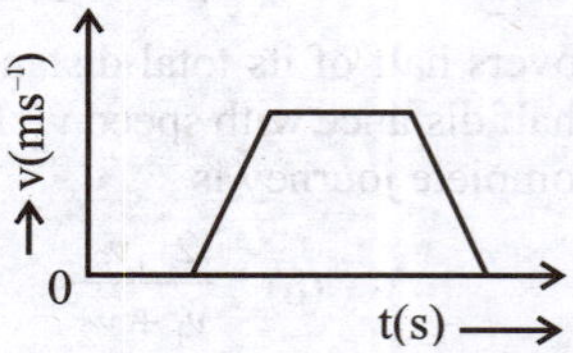

The acceleration (a) – time (t) graph that best suits this motion is : **NCERT | Page-16 | 2024**

(a)

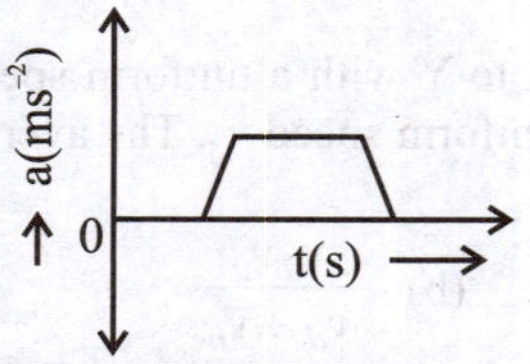

(b)

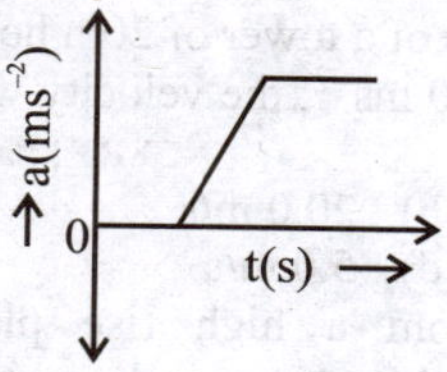

(c)

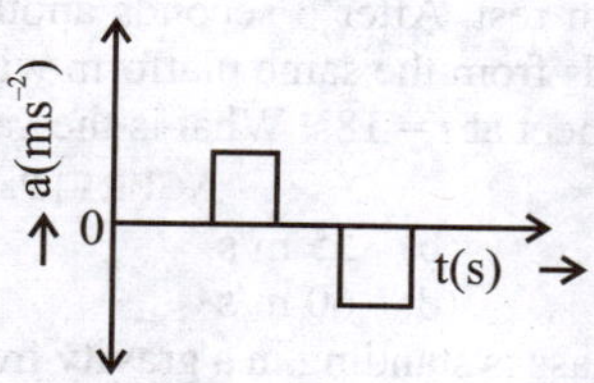

(d)

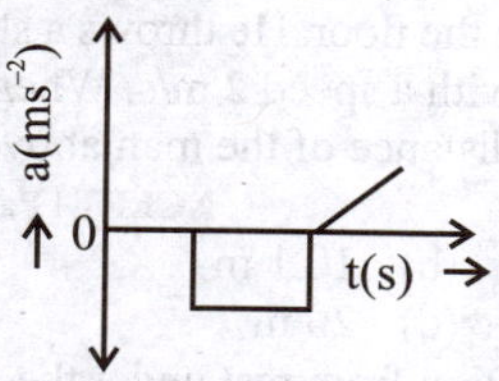

20. A particle of unit mass undergoes one-dimensional motion such that its velocity varies according to $v(x) = bx^{-2n}$
where b and n are constants and x is the position of the particle. The acceleration of the particle as the function of x, is given by: **NCERT | Page-16 | 2015**

(a) $-2nb^2x^{-4n-1}$ (b) $-2b^2x^{-2n+1}$
(c) $-2nb^2e^{-4n+1}$ (d) $-2nb^2x^{-2n-1}$

21. The motion of a particle along a straight line is described by equation :
$x = 8 + 12t - t^3$
where x is in metre and t in second. The retardation of the particle when its velocity becomes zero, is : **NCERT | Page-15 | 2012**

(a) 24 ms^{-2} (b) zero
(c) 6 ms^{-2} (d) 12 ms^{-2}

22. A body is moving with velocity 30 m/s towards east. After 10 seconds its velocity becomes 40 m/s towards north. The average acceleration of the body is **NCERT | Page-16 | 2011**

(a) 1 m/s^2 (b) 7 m/s^2
(c) 4 m/s^2 (d) 5 m/s^2

23. A particle moves a distance x in time t according to equation $x = (t + 5)^{-1}$. The acceleration of particle is proportional to: **NCERT | Page-18 | 2010**

(a) (velocity)$^{3/2}$ (b) (distance)^2
(c) (distance)$^{-2}$ (d) (velocity)$^{2/3}$

24. The position x of a particle with respect to time t along x-axis is given by $x = 9t^2 - t^3$ where x is in metres and t in second. What will be the position of this particle when it achieves maximum speed along the +ve x direction? **NCERT | Page-18 | 2007**

(a) 54 m (b) 81 m
(c) 24 m (d) 32 m

25. A particle moving along x-axis has acceleration f, at time t, given by $f = f_0\left(1 - \frac{t}{T}\right)$, where f_0 and T are constants. The particle at $t=0$ has zero velocity. In the time interval between $t = 0$ and the instant when $f = 0$, the particle's velocity (v_x) is **NCERT | Page-18 | 2007**

(a) $\frac{1}{2}f_0T^2$ (b) f_0T^2
(c) $\frac{1}{2}f_0T$ (d) f_0T

26. The displacement x of a particle varies with time t as $x = ae^{-\alpha t} + be^{\beta t}$, where a, b, α and β are positive constants. The velocity of the particle will **NCERT | Page-18 | 2005**

(a) be independent of α and β
(b) drop to zero when $\alpha = \beta$
(c) go on decreasing with time
(d) go on increasing with time

27. The displacement of a particle is represented by the following equation : $s = 3t^3 + 7t^2 + 5t + 8$ where s is in metre and t in second. The acceleration of the particle at $t = 1$s is **NCERT | Page-18 | 2000**

(a) 14 m/s^2 (b) 18 m/s^2
(c) 32 m/s^2 (d) zero

28. The displacement of a particle varies with time (t) as: $s = at^2 - bt^3$. The acceleration of the particle will be zero at time t equal to **NCERT | Page-18 | 1997**

(a) $\frac{a}{b}$ (b) $\frac{a}{3b}$
(c) $\frac{3b}{a}$ (d) $\frac{2a}{3b}$

29. A car accelerates from rest at a constant rate α for some time, after which it decelerates at a constant rate β and comes to rest. If the total time elapsed is t, then the maximum velocity acquired by the car is **NCERT | Page-18 | 1994**

(a) $\left(\frac{\alpha^2+\beta^2}{\alpha\beta}\right)t$ (b) $\left(\frac{\alpha^2-\beta^2}{\alpha\beta}\right)t$
(c) $\frac{(\alpha+\beta)t}{\alpha\beta}$ (d) $\frac{\alpha\beta t}{\alpha+\beta}$

30. A particle moves along a straight line such that its displacement at any time t is given by
$s = (t^3 - 6t^2 + 3t + 4)$ metres
The velocity when the acceleration is zero is **NCERT | Page-18 | 1994**

(a) 3 ms^{-1} (b) – 12 ms^{-1}
(c) 42 ms^{-2} (d) – 9 ms^{-1}

Topic-2.4 Kinematic Equations for Uniformly Accelerated Motion

31. A bullet from a gun is fired on a rectangular wooden block with velocity u. When bullet travels 24 cm through the block along its length horizontally, velocity of bullet becomes $\frac{u}{3}$. Then it further penetrates into the block in the same direction before coming to rest exactly at the other end of the block. The total length of the block is NCERT | Page-18 | 2023
(a) 27 cm (b) 24 cm
(c) 28 cm (d) 30 cm

32. A horizontal bridge is built across a river. A student standing on the bridge throws a small ball vertically upwards with a velocity 4 m s^{-1}. The ball strikes the water surface after 4 s. The height of bridge above water surface is (Take g = 10 m s^{-2}) NCERT | Page-19 | 2023
(a) 56 m (b) 60 m
(c) 64 m (d) 68 m

33. The ratio of the distances travelled by a freely falling body in the 1st, 2nd, 3rd and 4th second: NCERT | Page-19 | 2022
(a) 1 : 4 : 9 : 16 (b) 1 : 3 : 5 : 7
(c) 1 : 1 : 1 : 1 (d) 1 : 2 : 3 : 4

34. A small block slides down on a smooth inclined plane, starting from rest at time $t = 0$. Let S_n be the distance travelled by the block in the interval $t = n - 1$ to $t = n$. Then, the ratio $\frac{S_n}{S_{n+1}}$ is NCERT | Page-17 | 2021
(a) $\frac{2n}{2n-1}$ (b) $\frac{2n-1}{2n}$
(c) $\frac{2n-1}{2n+1}$ (d) $\frac{2n+1}{2n-1}$

35. A ball is thrown vertically downward with a velocity of 20 m/s from the top of a tower. It hits the ground after some time with a velocity of 80 m/s. The height of the tower is : (g = 10 m/s^2) NCERT | Page-19 | 2020
(a) 340 m (b) 320 m
(c) 300 m (d) 360 m

36. A stone falls freely under gravity. It covers distances h_1, h_2 and h_3 in the first 5 seconds, the next 5 seconds and the next 5 seconds respectively. The relation between h_1, h_2 and h_3 is NCERT | Page-19 | 2013
(a) $h_1 = \frac{h_2}{3} = \frac{h_3}{5}$
(b) $h_2 = 3h_1$ and $h_3 = 3h_2$
(c) $h_1 = h_2 = h_3$
(d) $h_1 = 2h_2 = 3h_3$

37. A boy standing at the top of a tower of 20m height drops a stone. Assuming $g = 10$ ms^{-2}, the velocity with which it hits the ground is NCERT | Page-20 | 2011
(a) 10.0 m/s (b) 20.0 m/s
(c) 40.0 m/s (d) 5.0 m/s

38. A ball is dropped from a high rise platform at $t = 0$ starting from rest. After 6 seconds another ball is thrown downwards from the same platform with a speed v. The two balls meet at $t = 18$s. What is the value of v? (take g = 10 m/s^2) NCERT | Page-19 | 2010
(a) 75 m/s (b) 55 m/s
(c) 40 m/s (d) 60 m/s

39. A man of 50 kg mass is standing in a gravity free space at a height of 10 m above the floor. He throws a stone of 0.5 kg mass downwards with a speed 2 m/s. When the stone reaches the floor, the distance of the man above the floor will be: NCERT | Page-20 | 2010
(a) 9.9 m (b) 10.1 m
(c) 10 m (d) 20 m

40. A particle starts its motion from rest under the action of a constant force. If the distance covered in first 10 seconds is S_1 and that covered in the first 20 seconds is S_2, then: NCERT | Page-17 | 2009
(a) $S_2 = 3S_1$ (b) $S_2 = 4S_1$
(c) $S_2 = S_1$ (d) $S_2 = 2S_1$

41. The distance travelled by a particle starting from rest and moving with an acceleration $\frac{4}{3}ms^{-2}$, in the third second is: NCERT | Page-17 | 2008
(a) 6 m (b) 4 m
(c) $\frac{10}{3}$m (d) $\frac{19}{3}$m

42. A particle moves in a straight line with a constant acceleration. It changes its velocity from 10 ms^{-1} to 20 ms^{-1} while passing through a distance 135 m in t second. The value of t is: NCERT | Page-17 | 2008
(a) 10 (b) 1.8 (c) 12 (d) 9

43. Two bodies, A (of mass 1 kg) and B (of mass 3 kg), are dropped from heights of 16m and 25m, respectively. The ratio of the time taken by them to reach the ground is NCERT | Page-19 | 2006
(a) 12/5 (b) 5/12
(c) 4/5 (d) 5/4

44. A ball is thrown vertically upward. It has a speed of 10 m/sec when it has reached one half of its maximum height. How high does the ball rise? NCERT | Page-19 | 2005, 2001
Take $g = 10$ m/s^2.
(a) 10 m (b) 5 m
(c) 15 m (d) 20 m

45. If a ball is thrown vertically upwards with speed u, the distance covered during the last t seconds of its ascent is NCERT | Page-19 | 2003
(a) $(u + gt)\,t$ (b) ut
(c) $\frac{1}{2}gt^2$ (d) $ut - \frac{1}{2}gt^2$

46. A man throws balls with the same speed vertically upwards one after the other at an interval of 2 seconds. What should be the speed of the throw so that more than two balls are in the sky at any time? [Given g = 9.8 m/s^2] NCERT | Page-19 | 2003

(a) Only with speed 19.6 m/s
(b) More than 19.6 m/s
(c) At least 9.8 m/s
(d) Any speed less than 19.6 m/s

47. A car moving with a speed of 40 km/h can be stopped by applying brakes at least after 2 m. If the same car is moving with a speed of 80 km/h, what is the minimum stopping distance? NCERT | Page-18 | 1998

(a) 8 m (b) 6 m (c) 4 m (d) 2 m

48. If a car at rest accelerates uniformly to a speed of 144 km/h in 20 s, it covers a distance of NCERT | Page-14 | 1997

(a) 2880 m (b) 1440 m
(c) 400 m (d) 20 m

49. If a ball is thrown vertically upwards with a velocity of 40 m/s, then velocity of the ball after two seconds will be (g = 10 m/s^2) NCERT | Page-19 | 1996

(a) 15 m/s (b) 20 m/s
(c) 25 m/s (d) 28 m/s

50. A body is thrown vertically upward from the ground. It reaches a maximum height of 20 m in 5 sec. After what time, it will reach the ground from its maximum height position? NCERT | Page-20 | 1995

(a) 2.5 sec (b) 5 sec
(c) 10 sec (d) 25 sec

51. A stone released with zero velocity from the top of a tower, reaches the ground in 4 sec. The height of the tower is (g = 10m/s^2) NCERT | Page-19 | 1995

(a) 20 m (b) 40 m
(c) 80 m (d) 160 m

52. Three different objects of masses m_1, m_2 and m_3 are allowed to fall from rest and from the same point O along three different frictionless paths. The speeds of the three objects on reaching the ground will be in the ratio of NCERT | Page-19 | 1995

(a) $m_1 : m_2 : m_3$ (b) $m_1 : 2m_2 : 3m_3$
(c) 1 : 1 : 1 (d) $\frac{1}{m_1} : \frac{1}{m_2} : \frac{1}{m_3}$

53. The water drops fall at regular intervals from a tap 5 m above the ground. The third drop is leaving the tap at an instant when the first drop touches the ground. How far above the ground is the second drop at that instant? (Take g = 10 m/s^2) NCERT | Page-19 | 1995

(a) 1.25 m (b) 2.50 m
(c) 3.75 m (d) 5.00 m

54. A body starts from rest, what is the ratio of the distance travelled by the body during the 4th and 3rd seconds? NCERT | Page-17 | 1993

(a) $\frac{7}{5}$ (b) $\frac{5}{7}$ (c) $\frac{7}{3}$ (d) $\frac{3}{7}$

55. A car is moving along a straight road with a uniform acceleration. It passes through two points *P* and *Q* separated by a distance with velocity 30 km/h and 40 km/h respectively. The velocity of the car midway between *P* and *Q* is NCERT | Page-17 | 1988

(a) 33.3 km /h (b) $20\sqrt{2}$ km/ h
(c) $25\sqrt{2}$ km/h (d) 35 km/h

56. A body dropped from top of a tower fall through 40 m during the last two seconds of its fall. The height of tower is (*g* = 10 m/s^2) NCERT | Page-19 | 1991

(a) 60 m (b) 45 m
(c) 80 m (d) 50 m

57. What will be the ratio of the distances moved by a freely falling body from rest in 4th and 5th seconds of journey? NCERT | Page-19 | 1988

(a) 4 : 5 (b) 7 : 9
(c) 16 : 25 (d) 1 : 1

Topic-* Relative Motion

58. Two cities X and Y are connected by a regular bus service with a bus leaving in either direction every T min. A girl is driving scooty with a speed of 60 km/h in the direction X to Y notices that a bus goes past her every 30 minutes in the direction of her motion, and every 10 minutes in the opposite direction. Choose the correct option for the period T of the bus service and the speed (assumed constant) of the buses. 2025

(a) 9 min, 40 km/h (b) 25 min, 100 km/h
(c) 10 min, 90 km/h (d) 15 min, 120 km/h

59. Preeti reached the metro station and found that the escalator was not working. She walked up the stationary escalator in time t_1. On other days, if she remains stationary on the moving escalator, then the escalator takes her up in time t_2. The time taken by her to walk up on the moving escalator will be: 2017

(a) $\frac{t_1 t_2}{t_2 - t_1}$ (b) $\frac{t_1 t_2}{t_2 + t_1}$
(c) $t_1 - t_2$ (d) $\frac{t_1 + t_2}{2}$

60. A bus is moving with a speed of 10 ms^{-1} on a straight road. A scooterist wishes to overtake the bus in 100 s. If the bus is at a distance of 1 km from the scooterist, with what speed should the scooterist chase the bus? 2009

(a) 40 ms^{-1} (b) 25 ms^{-1}
(c) 10 ms^{-1} (d) 20 ms^{-1}

61. A train of 150 metre long is going towards north direction at a speed of 10 m/s. A parrot flies at the speed of 5 m/s towards south direction parallel to the railway track. The time taken by the parrot to cross the train is 1988

(a) 12 sec (b) 8 sec
(c) 15 sec (d) 10 sec

ANSWER KEY

1	(c)	8	(b)	15	(d)	22	(d)	29	(d)	36	(a)	43	(c)	50	(b)	57	(b)
2	(c)	9	(c)	16	(b)	23	(a)	30	(d)	37	(b)	44	(a)	51	(c)	58	(d)
3	(b)	10	(b)	17	(b)	24	(a)	31	(a)	38	(a)	45	(c)	52	(c)	59	(b)
4	(d)	11	(c)	18	(d)	25	(c)	32	(c)	39	(b)	46	(b)	53	(c)	60	(d)
5	(c)	12	(c)	19	(c)	26	(d)	33	(b)	40	(b)	47	(a)	54	(a)	61	(d)
6	(b)	13	(b)	20	(a)	27	(c)	34	(c)	41	(c)	48	(c)	55	(c)		
7	(b)	14	(b)	21	(d)	28	(b)	35	(c)	42	(d)	49	(b)	56	(b)		

Hints & Solutions

1. **(c)** Average speed,

$$v_{avg} = \frac{2v_1v_2}{v_1+v_2} = \frac{2\times v\times 2v}{v+2v} = \frac{4v}{3}$$

2. **(c)** As $t_1 = \frac{x}{v_1}$ and $t_2 = \frac{x}{v_2}$

$$\therefore\ v = \frac{x+x}{t_1+t_2} = \frac{2x}{\frac{x}{v_1}+\frac{x}{v_2}} = \frac{2v_1v_2}{v_1+v_2}$$

$$\therefore\ \frac{2}{v} = \frac{1}{v_1}+\frac{1}{v_2}$$

3. **(b)** Let the total distance covered by the particle be 2s. Then

$$v_{av} = \frac{2s}{\frac{s}{v_1}+\frac{s}{v_2}} = \frac{2v_1v_2}{v_1+v_2}$$

NOTES The average speed of an object is the total distance travelled by the object divides by the elapsed time to cover that distance. It's a scalar quantity which means it is defined only by magnitude. A related concept, average velocity, is a vector quantity. A vector quantity is defined by magnitude and direction both.

4. **(d)** Average speed

$$= \frac{\text{total distance travelled}}{\text{total time taken}}$$

Let s be the distance from X to Y.

$$\therefore \text{ Average speed} = \frac{s+s}{t_1+t_2} = \frac{2s}{\frac{s}{v_u}+\frac{s}{v_d}} = \frac{2v_uv_d}{v_d+v_u}$$

5. **(c)** Average speed $= \dfrac{s}{\frac{s/3}{10}+\frac{s/3}{20}+\frac{s/3}{60}}$

$$= \frac{s}{s/18} = 18 km/h$$

NOTES In case speed is continuously changing with time, then,

$$V_{av} = \frac{\int vdt}{\int dt}$$

6. **(b)** Given, Total distance = 200 m, speed in first half distance = 40 km/hr, speed in second half distance = vkm/hr.

So, $v_{av} = \dfrac{\text{Total distance covered}}{\text{Total time elapsed}}$

$$48 = \frac{200\times10^{-3}}{\left(\frac{100\times10^{-3}}{40}\right)+\left(\frac{100\times10^{-3}}{v}\right)}$$

or $\frac{1}{40}+\frac{1}{v} = \frac{2}{48} = \frac{1}{24}$

or $\frac{1}{v} = \frac{1}{24}-\frac{1}{40} = \frac{2}{120} = \frac{1}{60}$ or $v = 60$ km/h

NOTES $v_{av} = \dfrac{2v_1v_2}{v_1+v_2} \Rightarrow 48 = \dfrac{2\times40\times v}{40+v}$

$\Rightarrow$ v = 60 km/h

7. **(b)** Total distance = s

Total time taken $= \frac{s/2}{40}+\frac{s/2}{60} = \frac{5s}{240} = \frac{s}{48}$

$\therefore$ Average speed $= \dfrac{\text{total distance}}{\text{total time}}$

$= \dfrac{s}{s/48} = 48 km/h$

Alternative method:

$$v_{av} = \frac{2v_1v_2}{v_1+v_2} = \frac{2\times40\times60}{40+60} = 48 \text{ km/h}$$

8. **(b)** For + ve acceleration,

$\frac{dv}{dt} > 0$ so, velocity is increasing therefore, slop of x-t graph is increasing and graph is as shown in figure

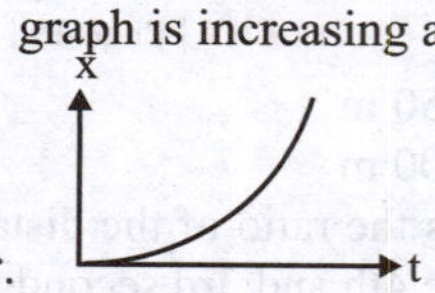

$\therefore$

9. **(c)** $V = \frac{ds}{dt}$ = slope of s – t curve = tan θ

$$\frac{V_{30°}}{V_{45°}} = \frac{\tan 30°}{\tan 45°} = \frac{1}{\sqrt{3}}$$

10. (b) Given that

$x_p(t) = at + bt^2$ $x_Q(t) = ft - t^2$

$v_p = a + 2bt$ $v_Q = f - 2t$

as $v_p = v_Q$

$a + 2bt = f - 2t$

$\Rightarrow t = \dfrac{f-a}{2(1+b)}$

11. (c) Given : Velocity $v = At + Bt^2$

$\Rightarrow \dfrac{dx}{dt} = At + Bt^2$

By integrating we get distance travelled by the particle between 1s and 2s,

$\Rightarrow \int_0^x dx = \int_1^2 (At + Bt^2)\, dt$

$x = \dfrac{A}{2}(2^2 - 1^2) + \dfrac{B}{3}(2^3 - 1^3) = \dfrac{3A}{2} + \dfrac{7B}{3}$

12. (c) $\because t = \sqrt{x} + 3$

$\Rightarrow \sqrt{x} = t - 3 \Rightarrow x = (t-3)^2$ ...(i)

$v = \dfrac{dx}{dt} = 2(t-3) = 0 \Rightarrow t = 3$

From equation (i)

$\therefore x = (3-3)^2 \Rightarrow x = 0$

13. (b) The slope of the graph $\dfrac{ds}{dt}$ is maximum at C and hence the instantaneous velocity is maximum at C.

NOTES The instantaneous velocity of an object at a given instant is first derivative of displacement with respect to time. The slope of displacement-time graph at any instant of time gives the measure of instantanesous velocity of an object at that instant.

14. (b) $x = 40 + 12t - t^3$

$v = \dfrac{dx}{dt} = 12 - 3t^2$

For $v = 0$; $t = \sqrt{\dfrac{12}{3}} = 2$ sec

So, after 2 seconds velocity becomes zero.

Value of x in 2 sec. $= 40 + 12 \times 2 - 2^3$

$= 40 + 24 - 8 = 56$ m

15. (d) At *E*, the slope of the curve or instantaneous velocity of the particle is negative.

NOTES The line bending towards time axis represents decreasing velocity or negative velocity of the particle. It means the particle possesses retardation.

16. (b) In one dimensional motion, the body can have at a time one velocity but not two values of velocities.

17. (b) Given, $t = x^2 + x$

$\Rightarrow \dfrac{dt}{dx} = \dfrac{d}{dx}[x^2 + x]$

$\Rightarrow \dfrac{1}{v} = 2x + 1 \Rightarrow v = (2x+1)^{-1}$

$\therefore$ Acceleration,

$a = \dfrac{vdv}{dx} = (2x+1)^{-1}[-1(2x+1)^{-2} \times 2]$

$= -1\,(2x+1)^{-3} \times 2 = -2\,(2x+1)^{-3}$

$= -\dfrac{2}{(2x+1)^3}$

18. (d) Position, $x = \alpha t^4 + \beta t^2 + \gamma t + \delta$

Velocity, $v = \dfrac{dx}{dt} = 4\alpha t^3 + 2\beta t + \gamma$

Initial velocity $= v_i = \gamma$

Acceleration, $a = \dfrac{dv}{dt} = 12\alpha t^2 + 2\beta$

Initial acceleration $= a_i = 2\beta$

$\therefore \dfrac{v_i}{a_i} = \dfrac{\gamma}{2\beta}$

19. (c) As, acceleration = slope of v-t curve

$\therefore$ Initially, a = 0

After sometime, a = constant (+ve)

Again, a = 0

Again, a = constant (–ve)

$\therefore$ The corresponding a - t graph is

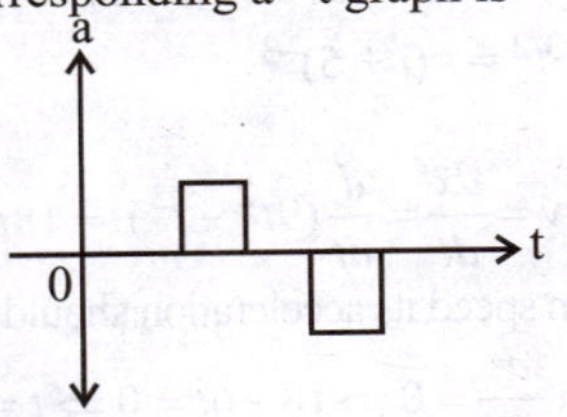

NOTES The area under velocity-time curve gives displacement, not distance.

i.e. Area $= \int_{t_1}^{t_2} \vec{v}\, dt = \vec{x_2} - \vec{x_1}$ = Displacement

20. (a) Given, $v(x) = bx^{-2n}$

$a = \dfrac{dv}{dt} = \dfrac{dv}{dx}\cdot\dfrac{dx}{dt} = v\cdot\dfrac{dv}{dx}$

So, $\dfrac{dv}{dx} = -2\,nb\,x^{-2n-1}$

$\therefore$ Acceleration of the particle as function of x,

$a = v\dfrac{dv}{dx} = bx^{-2n}\,\{b(-2n)x^{-2n-1}\}$

$= -2nb^2x^{-4n-1}$

NOTES For one dimensional motin, the angle between velocity and acceleration is either 0° or 180° and it does not change with time.

21. (d) $x = 8 + 12t - t^3$

The final velocity of the particle will be zero, because it retarded.

$\therefore \quad v = 0 + 12 - 3t^2 = 0$

$\Rightarrow \quad 3t^2 = 12$

$\therefore \quad t = 2$ sec

Now the retardation

$$a = \frac{dv}{dt} = 0 - 6t$$

$a[t = 2] = -12$ m/s^2

$\therefore$ retardation = 12 m/s^2

22. (d) Average acceleration

$$<\vec{a}> = \frac{\text{Change in velocity}}{\text{Total Time}}$$

$$<\vec{a}> = \frac{|40\hat{j} - 30\hat{i}\,|}{10}$$

$$|<\vec{a}>| = \sqrt{4^2 + (-3)^2}$$

$= 5$ m/s^2

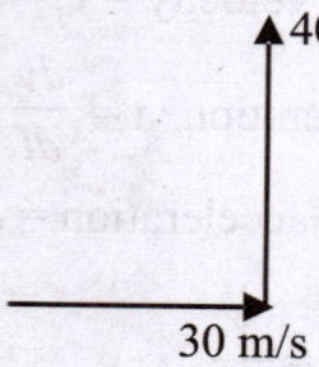

23. (a) Distance $x = \frac{1}{t+5}$

$\therefore$ velocity $v = \frac{dx}{dt} = \frac{-1}{(t+5)^2}$

$\therefore$ acceleration $a = \frac{d^2x}{dt^2} = \frac{2}{(t+5)^3} = 2x^3$

Therefore, $v^{3/2} = -(t+5)^{-3}$

So, $a \propto v^{3/2}$

24. (a) Speed $v = \frac{dx}{dt} = \frac{d}{dt}(9t^2 - t^3) = 18t - 3t^2$

For maximum speed its acceleration should be zero,

Acceleration $\frac{dv}{dt} = 0 \Rightarrow 18 - 6t = 0 \Rightarrow t = 3$

At $t = 3$ its position is

$x_{max} = 81 - 27 = 54$ m

NOTES Position of any point is completely expressed by two factors its distance from the observer and its direction with respect to observer. That is why position is characterised by a vector known as position vector.

25. (c) Here, $f = f_0\left(1 - \frac{t}{T}\right)$

or, $\frac{dv}{dt} = f_0\left(1 - \frac{t}{T}\right)$

If $f = 0$, then

$0 = f_0\left(1 - \frac{t}{T}\right) \Rightarrow t = T$

Hence, particle's velocity in the time interval $t = 0$ and $t = T$ is given by

$$v_x = \int_{v=0}^{v=v_x} dv = \int_{t=0}^{T}\left[f_0\left(1 - \frac{t}{T}\right)\right]dt$$

$$= f_0\left[\left(t - \frac{t^2}{2T}\right)\right]_0^T$$

$$= f_0\left(T - \frac{T^2}{2T}\right) = f_0\left(T - \frac{T}{2}\right) = \frac{1}{2}f_0T.$$

26. (d) $x = ae^{-\alpha t} + be^{\beta t}$

$$V = \frac{dx}{dt} = -a\alpha e^{-\alpha t} + b\beta e^{\beta t}$$

$$a = \frac{d^2x}{dt^2} = a\alpha^2 e^{-\alpha t} + b\beta^2 e^{\beta t}$$

Clearly, $a > 0$ for any value of t.

So, V(t) will always increase with time.

27. (c) Displacement

$s = 3t^3 + 7t^2 + 5t + 8$

Velocity $= \frac{ds}{dt} = 9t^2 + 14t + 5$

Acceleration $= \frac{d^2s}{dt^2} = 18t + 14$

Acceleration at ($t = 1$s)

$= 18 \times 1 + 14 = 18 + 14 = 32$ m/s^2

28. (b) Distance, $s = at^2 - bt^3$

velocity, $v = \frac{ds}{dt} = 2at - 3bt^2$

acceleration $a = \frac{dv}{dt} = 2a - 6bt$

Acceleration is zero at

$2a - 6bt = 0 \quad \Rightarrow \quad t = \frac{a}{3b}$

29. (d)

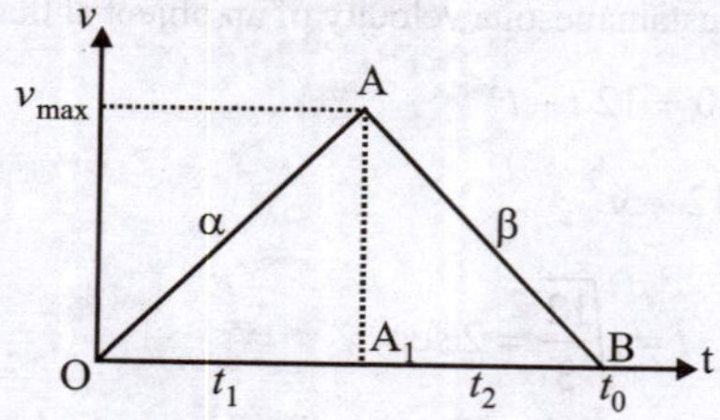

In Fig.

$AA_1 = v_{max.} = \alpha t_1 = \beta t_2$

But $t = t_1 + t_2 = \frac{v_{max}}{\alpha} + \frac{v_{max}}{\beta}$

$$= v_{max}\left(\frac{1}{\alpha} + \frac{1}{\beta}\right) = v_{max}\left(\frac{\alpha + \beta}{\alpha\beta}\right)$$

or, $v_{max} = t\left(\frac{\alpha\beta}{\alpha + \beta}\right)$

NOTES If a body starting from rest accelerates at a constant rate α for certain time and then retards at constant β and comes to rest after t. second from the starting point, then

Distance travelled by the body $= \frac{\alpha\beta t^2}{(2\alpha + 2\beta)}$

30. (d) Velocity, $v = \frac{ds}{dt} = 3t^2 - 12t + 3$

Acceleration, $a = \frac{dv}{dt} = 6t - 12$; For $a = 0$, we have, $0 = 6t - 12$ or $t = 2$s. Hence, at $t = 2$ s the velocity will be $v = 3 \times 2^2 - 12 \times 2 + 3 = -9\ \text{ms}^{-1}$

31. (a)

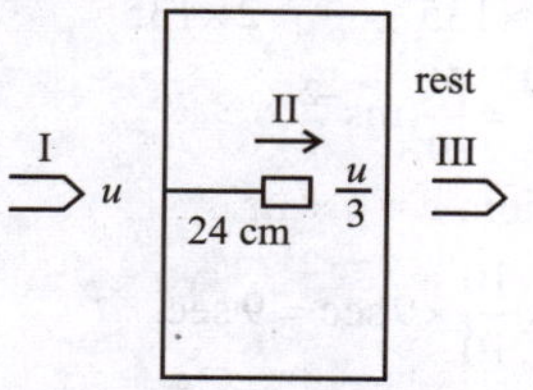

Using third equation of motion between I and II

$v^2 - u^2 = 2as$

$\Rightarrow \left(\frac{u}{3}\right)^2 = u^2 - 2a \times 24$

$\Rightarrow 2a(24) = \frac{8u^2}{9}$ ... (i)

Using third equation of motion again between II and III

$0 = \left(\frac{u}{3}\right)^2 - 2as$

$\Rightarrow \frac{u^2}{9} = 2as$ ... (ii)

From equation (i) and (ii)

$2a24 = 8(2as)$

$\Rightarrow s = 3$ cm

Length of wooden block = 24 + 3 = 27 cm

32. (c) Given,

Initial velocity of ball, $u = 4\ \text{ms}^{-1}$

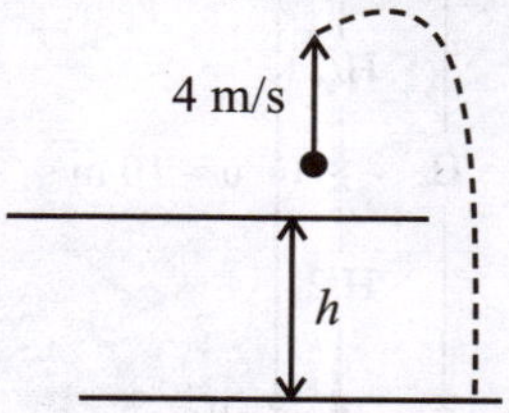

$s = ut - \frac{1}{2}gt^2$

$= 4 \times 4 - \frac{1}{2} \times 10 \times (4)^2 = -64$ m

Height of bridge above water surface = 64 m

33. (b) $S_n = u + \frac{a}{2}(2n-1)$

$S_1 : S_2 : S_3 : S_4 = (2\times 1 - 1) : (2\times 2 - 1) : (2\times 3 - 1) : (2 \times 4 - 1) = 1 : 3 : 5 : 7$

NOTES If a body starts from rest and moves with uniform acceleration then distance covered by the body in n^{th} sec is proportional to (2n – 1) i.e., $S_n \propto (2n-1)$.

34. (c) Acceleration along inclined plane,

$a = g \sin\theta$

S_n = distance travelled by object during n^{th} second.

Initial speed u = 0

$S = u + -(2n-1)$

$S_n = 0 + \frac{g\sin\theta}{2}(2n-1) = \frac{g\sin\theta}{2}(2n-1)$ ..(i)

Distance travelled during $(n + 1)^{th}$ second.

$S_{n+1} = 0 + \frac{g\sin\theta}{2}[2(n+1)-1]$

$= \frac{g\sin\theta}{2}(2n+1)$ ...(ii)

Dividing equations (i) and (ii)

$\frac{S_n}{S_{n+1}} = \frac{\frac{g\sin\theta}{2}(2n-1)}{\frac{g\sin\theta}{2}(2n+1)} = \frac{(2n-1)}{(2n+1)}$

NOTES The distance covered by a body in n^{th} second i.e., $S_n \propto (2n-1)$ if body starts from rest and moves with uniform acceleration.

35. (c)

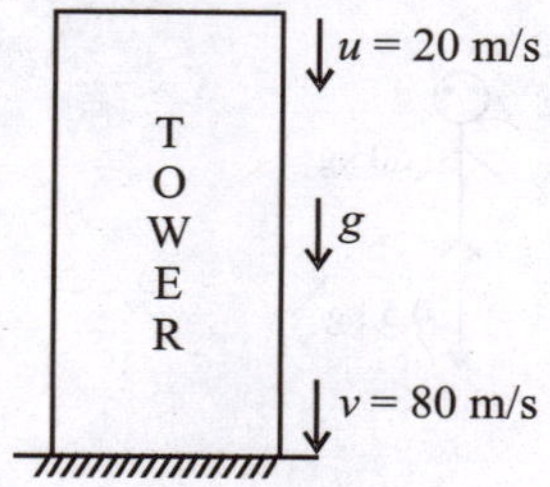

Using $v^2 = u^2 + 2gh$

Height, $h = \frac{v^2 - u^2}{2g} = \frac{(80)^2 - (20)^2}{2 \times 10}$

$= \frac{6400 - 400}{20} = 300$ m

36. (a) $\because\ h = \frac{1}{2}gt^2 \therefore h_1 = \frac{1}{2}g(5)^2 = 125$

$h_1 + h_2 = \frac{1}{2}g(10)^2 = 500 \Rightarrow h_2 = 375$

$h_1 + h_2 + h_3 = \frac{1}{2}g(15)^2 = 1125$

$\Rightarrow h_3 = 625$

$h_2 = 3h_1, h_3 = 5h_1$ or $h_1 = \frac{h_2}{3} = \frac{h_3}{5}$

NOTES The distance covered in time t, 2t, 3t, etc. will be in the ratio of $1^2 : 2^2 : 3^2$ i.e., square of integers i.e., $h \propto t^2$.

37. (b) Here, $u = 0$

We have, $v^2 = u^2 + 2gh$

$\Rightarrow v = \sqrt{2gh} = \sqrt{2 \times 10 \times 20} = 20$ m/s

38. **(a)** Clearly distance moved by 1st ball in 18s = distance moved by 2nd ball in 12s. ....(i)

Now, distance moved in 18 s by 1st ball = $\frac{1}{2} \times 10 \times 18^2$ $= 90 \times 18 = 1620$ m

Distance moved in 12 s by 2nd ball

$= ut + \frac{1}{2}gt^2 \quad \Rightarrow \quad 12v + \frac{1}{2}\times 10 \times (12)^2$

$\Rightarrow \quad 12\,v + 720$

From equation (i)

$\therefore \quad 1620 = 12\,v + 5 \times 144$

$\Rightarrow \quad v = 135 - 60 = 75 \text{ ms}^{-1}$

39. **(b)** No external force is acting, therefore, momentum is conserved.

By momentum conservation,

$50\,u + 0.5 \times 2 = 0$

where u is the velocity of man.

$u = -\frac{1}{50} \text{ms}^{-1}$

Negative sign of u shows that man moves upward.

Time taken by the stone to reach the ground

$= \frac{10}{2} = 5$ sec

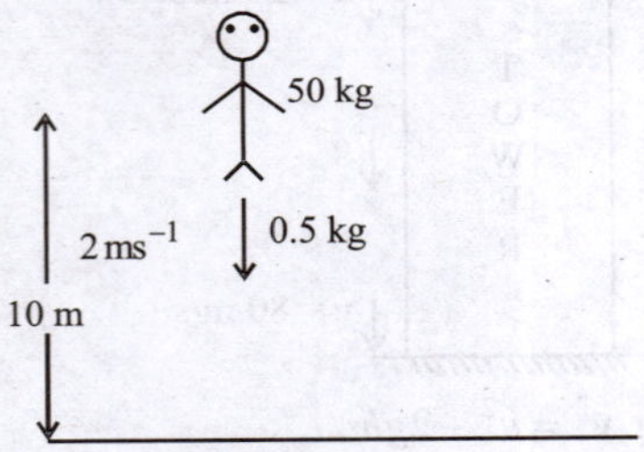

Distance moved by the man $= 5 \times \frac{1}{50} = 0.1 \text{m}$

$\therefore$ when the stone reaches the floor, the distance of the man above floor = 10.1 m

40. **(b)** $u = 0$, $t_1 = 10$s, $t_2 = 20s$

Using the relation, $S = ut + \frac{1}{2}at^2$

Acceleration being the same in two cases,

$S_1 = \frac{1}{2} a \times t_1^2,\ S_2 = \frac{1}{2} a \times t_2^2$

$\therefore \quad \frac{S_1}{S_2} = \left(\frac{t_1}{t_2}\right)^2 = \left(\frac{10}{20}\right)^2 = \frac{1}{4}$

$S_2 = 4S_1$

41. **(c)** Distance travelled in the nth second is given by $S_n = u + \frac{a}{2}(2n-1)$

Given $u = 0$, $a = \frac{4}{3}\text{ms}^{-2}$, $n = 3$

$\therefore S_n = 0 + \frac{4}{3\times 2}(2 \times 3 - 1) = \frac{4}{6}\times 5 = \frac{10}{3}\text{m}$

42. **(d)** Initial velocity, $u = 10 \text{ ms}^{-1}$

Final velocity, $v = 20 \text{ ms}^{-1}$

Distance, $s = 135$ m

Let, acceleration = a

Using the formula, $v^2 = u^2 + 2as$

$a = \frac{v^2 - u^2}{2s} = \frac{(20)^2 - (10)^2}{2\times 135} = \frac{400 - 100}{2\times 135}$

or, $a = \frac{300}{2\times 135} = \frac{150}{135} = \frac{10}{9}\text{ms}^{-2}$

Now, using the relation, $v = u + at$

$t = \frac{v-u}{a} = \frac{20-10}{10/9} = \frac{10}{10}\times 9 \text{ sec} = 9$ sec.

43. **(c)** Let t_1 & t_2 be the time taken by A and B respectively to reach the ground then from the formula,

$h = \frac{1}{2}gt^2$,

For first body, $\quad 16 = \frac{1}{2}gt_1^2$

For second body, $25 = \frac{1}{2}gt_2^2$

$\therefore \quad \frac{16}{25} = \frac{t_1^2}{t_2^2} \Rightarrow \frac{t_1}{t_2} = \frac{4}{5}$.

NOTES If a body is a dropped from some height, the motion is independent of mass of the body.

The time taken to reach the ground $t = \sqrt{2h/g}$ and final velocity $V = \sqrt{2gh}$ and initial velocity, u = 0.

44. **(a)** For part AB

From 3rd equation of motion

$v^2 = u^2 - 2\text{gH}$

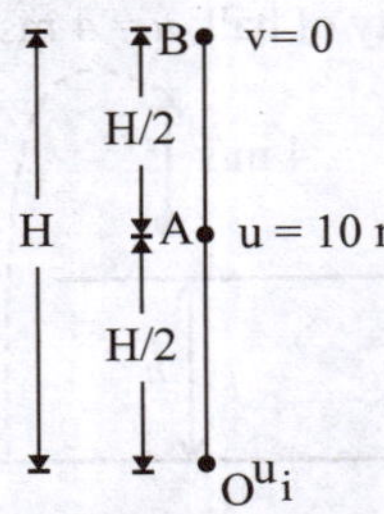

$0 = u^2 - 2\text{g(H/2)} = u^2 - \text{gH}$

$\text{H} = \frac{u^2}{g} = \frac{10^2}{10} = 10$ m

45. **(c)** Let body takes T sec to reach maximum height.

Then $v = u - gT$

$v = 0$, at highest point.

So, $\quad T = \frac{u}{g}$ ...(i)

Velocity attained by body in $(T - t)$ sec $\ v = u - g(T - t)$

$v = u - gT + gt = u - g\frac{u}{g} + gt \quad (\because T = u/g)$

or $v = gt$ ...(ii)

$\therefore$ Distance travelled in last t sec of its ascent

$s = vt - \frac{1}{2}gt^2 \qquad s = (gt)t - \frac{1}{2}gt^2 = \frac{1}{2}gt^2$

NOTES If a body is projected vertically upward, then, final velocity, v = 0; intial speed, u = gt and height attained, $h = \frac{u^2}{2g} = \frac{g^2t^2}{2g} = \frac{1}{2}gt^2$

46. **(b)** Let the required speed of throw be u ms^{-1}. Then time taken to reach maximum height,

$t = \frac{u}{g}$

For two balls to remain in air at any time, t must be greater than 2.

$\therefore \quad \frac{u}{g} > 2 \Rightarrow u > 19.6$ m/s

47. **(a)** $v^2 - u^2 = 2as$

$\Rightarrow a = \frac{v^2 - u^2}{2s} = -\frac{u_1^2}{2s},$

For same retarding force $s \propto u^2$

$\because \quad \frac{s_2}{s_1} = \frac{u_2^2}{u_1^2} \quad \Rightarrow \quad \frac{s_2}{s_1} = \left(\frac{80}{40}\right)^2 = 4$

$\therefore s_2 = 4s_1 = 8$ m

Alternative method :

If F is retarding force and s the stopping distance, then

$\frac{1}{2}mv^2 = Fs$

For same retarding force, $s \propto v^2$

$\therefore \quad \frac{s_2}{s_1} = \left(\frac{v_2}{v_1}\right)^2 = \left(\frac{80\text{km/h}}{40\text{km/h}}\right)^2 = 4$

$\therefore \quad s_2 = 4s_1 = 4 \times 2 = 8$ m

48. **(c)** Initial velocity of car $(u) = 0$

Final velocity of car (v) = 144 km/hr = 40 m/s

Time taken = 20 s

We know that, $v = u + at$

$40 = a \times 20 \Rightarrow a = 2$ m/s^2

Also, $v^2 - u^2 = 2as$

$\Rightarrow s = \frac{v^2 - u^2}{2a}$

$\Rightarrow s = \frac{(40)^2 - (0)^2}{2 \times 2} = \frac{1600}{4} = 400$ m.

49. **(b)** Initial velocity (u) = 40 m/s

Acceleration $a = -g$ m/s^2 = -10 m/s^2

Time = 2 seconds

By Ist equation of motion,

$v = u + at$

$v = 40 - 10(2) = 20$ m/s

50. **(b)** $h_{max} = 20$ m and $t = 5$ sec. Time taken by the body to reach the ground from some height is the same as taken to reach that height. Therefore, time to reach the ground from its maximum height is 5 sec.

51. **(c)** Initial velocity (u) = 0; Time (t) = 4 sec and gravitational acceleration (g) = 10 m/s^2.

Height of tower

$h = ut + \frac{1}{2}gt^2 = (0 \times 4) + \frac{1}{2} \times 10 \times (4)^2 = 80$ m

52. **(c)** The speed of an object, falling freely due to gravity $V = \sqrt{2gh}$, depends only on its height and not on its mass. Since the paths are frictionless and all the objects fall through the same height, therefore, their speeds on reaching the ground will be in the ratio of 1 : 1 : 1.

53. **(c)** Height of tap = 5m and (g) = 10 m/sec^2.

For the first drop, $S = ut + \frac{1}{2}gt^2$

$5 = (0 \times t) + \frac{1}{2} \times 10t^2 = 5t^2$ or $t^2 = 1$ or $t = 1$ sec.

It means that the third drop leaves after one second of the first drop. Or, each drop leaves after every 0.5 sec.

Distance covered by the second drop in 0.5 sec

$= ut + \frac{1}{2}gt^2 = (0 \times 0.5) + \frac{1}{2} \times 10 \times (0.5)^2$

= 1.25 m.

Therefore, distance of the second drop above the ground = 5 – 1.25 = 3.75 m.

54. **(a)** $\frac{S_4}{S_3} = \frac{0 + \frac{a}{2}(2 \times 4 - 1)}{0 + \frac{a}{2}(2 \times 3 - 1)} = \frac{7}{5}$

NOTES Equation of the motion of uniformly accelerated motion, the distance travelled in nth sec is given by,

$S_n = u + \frac{a}{2}(2n - 1)$

55. **(c)** Let PQ = x, then

$u = 30$ km/h

$v = 40$ km/h

$a = \frac{40^2 - 30^2}{2x} = \frac{350}{x} \quad [\because v^2 = u^2 + 2as]$

Also, velocity at mid point is given by v_m,

$\Rightarrow v_m^2 - 30^2 = 2 \times \frac{350}{x} \times \frac{x}{2}$

This gives $v = 25\sqrt{2}$ km/h

56. **(b)** Let the body fall through the height of tower in n seconds. From,

$D_n = u + \frac{a}{2}(2n - 1)$ we have, total distance travelled in last 2 seconds of fall is

$D = D_n + D_{(n-1)}$

$= \left[0 + \frac{g}{2}(2n - 1)\right] + \left[0 + \frac{g}{2}\{2(n-1) - 1\}\right]$

$$=\frac{g}{2}(2n-1)+\frac{g}{2}(2n-3)=\frac{g}{2}(4n-4)$$

$$=\frac{10}{2}\times 4(n-1)$$

or, $40 = 20\ (n-1)$ or $n = 2 + 1 = 3s$

Distance travelled in is $n = 3$ sec

$$s=ut+\frac{1}{2}at^2=0+\frac{1}{2}\times 10\times 3^2=45\text{ m}$$

57. (b) $\dfrac{x(4)}{x(5)}=\dfrac{\frac{g}{2}(2\times 4-1)}{\frac{g}{2}(2\times 5-1)}=\dfrac{7}{9}$

$$\left[\because S_{n\text{th}}=u+\frac{a}{2}(2n-1)\text{ and } u=0, a=g\right]$$

> **NOTES** In the absence of air resistance, all bodies irrespective of the size, weight or composition fall with the same acceleration near the surface of the earth. This motion of a body falling towards the earth from a small altitude (h<<R) is called free fall. Distance travelled in n[th] second $h_n=\frac{g}{2}(2n-1)$

58. (d) Given, $t_1 = 30\text{ min} = \frac{1}{2}$ hr

$t_2 = 10\text{ min} = \frac{1}{6}$ hr

V_B = Speed of Bus

V_S = Speed of scooty (girl)

As, bus and scooty moves same distance d between points X and Y.

$\therefore d = (V_B - V_S)\, t_1 = (V_B + V_S)\, t_2$

$$\Rightarrow (V_B-60)\frac{1}{2}=(V_B+60)\frac{1}{6}$$

$\Rightarrow 3V_B - 180 = V_B + 60 \Rightarrow 2V_B = 240$

$\therefore V_B = 120$ km/h

$\therefore$ Distance, $d = (V_B - V_S)t_1$

$$=(120-60)\frac{1}{2}=30\text{ km}$$

$$\therefore t=\frac{d}{V_B}=\frac{30}{120}=\frac{1}{4}=15\text{ min}$$

59. (b) Let the distance be '*d*' time taken by preeti to travel up the stationary escalator = t_1 Velocity of preeti w.r.t. elevator $v_1=\frac{d}{t_1}$

Since the distance is same let the time taken when preeti stands on the moving escalator = t_2.

Velocity of elevator w.r.t. ground $v_2=\frac{d}{t_2}$

Then net velocity of preeti w.r.t. ground

$v = v_1 + v_2$

$$\frac{d}{t}=\frac{d}{t_1}+\frac{d}{t_2}\ \Rightarrow \frac{1}{t}=\frac{1}{t_1}+\frac{1}{t_2}$$

$\therefore\ t=\dfrac{t_1t_2}{(t_1+t_2)}$ (time taken by preeti to walk up on the moving escalator)

60. (d) 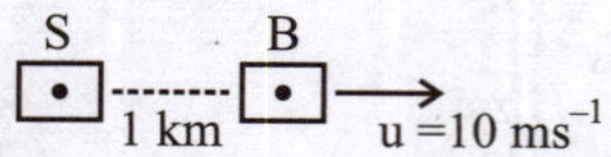

Let v be the relative velocity of scooter w.r.t bus as $v = v_S - v_B$

$$v=\frac{1000}{100}=10\text{ ms}^{-1}, v_B=10\text{ms}^{-1}$$

$\therefore\quad v_S = v + v_B = 10 + 10 = 20\text{ ms}^{-1}$

$\therefore\quad$ velocity of scooter = 20 ms^{-1}

> **NOTES** When two bodies are moving in same directions, relative velocity is difference between individual velocities. $\vec{V}_{AB}=\vec{V}_A-\vec{V}_B$

61. (d) Relative velocity of parrot w.r.t the train $= 5 - (-10) = 15\text{ ms}^{-1}$.

Time taken by parrot to cross the train $=\frac{150}{15}=10\text{ s}$

> **NOTES** When two bodies are moving in opposite directions, then their relative velocity is sum of individual velocities. The distance which bird has to travel to cross the train is equal to the length of train. Now since the train and the bird are travelling in opposite direction, therefore the speed will be added.

3 Motion in a Plane

TREND ANALYSIS

TOPIC	2025		2024		2023		2022		2021		2020	
	Qns.	LOD	Qns.	LOD	Qns.	LOD	Qns.	LOD	Qns.	LOD	Qns.	LOD
Motion in a Plane									1	Avg.		
Relative Velocity in two Dimension												
Projectile Motion					1	Easy	1	Avg.				
Uniform Circular Motion			1	Easy			1	Easy	1	Avg.		

How to Code Qns for Better & Faster Revision

Type 1: Able to solve (No code Req)

Type 2: Solved with struggle or by seeing solution **(Use code NR means Needs Revision or orange stickers)**

Type 3: Tricky/ Could not solve on my own. Need help from Teacher/ Friend. **(Use code TQ means Tricky Question or red stickers)**

You can either hand write the code in boxes or else paste coloured stickers given at end of book.

Topic-3.2 Scalars and Vectors

1. Which of the following statement is **not** true?

NCERT | Page-28 | Manipur 2023

(a) Pressure is a vector quality

(b) Relative density is a scalar quantity

(c) Coefficient of viscosity is a scalar quantity

(d) Surface tension is a scalar quantity

2. Which of the following is not a vector quantity?

NCERT | Page-28 | 1996

(a) displacement

(b) electric field

(c) work

(d) acceleration

3. Which of the following is not a vector quantity?

NCERT | Page-28 | 1995

(a) speed

(b) velocity

(c) torque

(d) displacement

Topic-3.3 Multiplication of Vectors by Real Numbers

4. The resultant of $(\vec{A} \times 0)$ will be equal to

NCERT | Page-33 | 1992

(a) zero (b) A

(c) zero vector (d) unit vector

Topic-3.4 Addition and Subtraction of Vectors-Graphical Method

5. Six vectors, $\vec{a}$ through $\vec{f}$ have the magnitudes and directions indicated in the figure. Which of the following statements is true?

NCERT | Page-33 | 2010

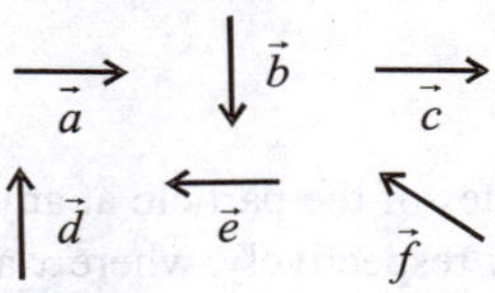

(a) $\vec{b}+\vec{c}=\vec{f}$

(b) $\vec{d}+\vec{c}=\vec{f}$

(c) $\vec{d}+\vec{e}=\vec{f}$

(d) $\vec{b}+\vec{e}=\vec{f}$

Topic-3.5 Resolution of Vectors

If a unit vector is represented by $0.5\hat{i}+0.8\hat{j}+c\hat{k}$, the value of c is

NCERT | Page-32 | 1999

(a) 1

(b) $\sqrt{0.11}$

(c) $\sqrt{0.01}$

(d) 0.39

Topic-3.6
Vector Addition-Analytical Method

7. The angle between two vectors of magnitude 12 and 18 units when their resultant is 24 units, is
NCERT | Page-33 | 1999
(a) 63°51′ (b) 75°52′
(c) 82°31′ (d) 89°16′

Topic-3.7
Motion in a Plane

8. The position of a particle is given by
NCERT | Page-34 | Manipur 2023
$\vec{r}(t) = 4t\hat{i} + 2t^2\hat{j} + 5\hat{k}$
where t is in seconds and r in metre. Find the magnitude and direction of velocity v(t), at t = 1 s, with respect to x-axis.
(a) $3\sqrt{2}ms^{-1}, 30°$ (b) $3\sqrt{2}ms^{-1}, 45°$
(c) $4\sqrt{2}ms^{-1}, 45°$ (d) $4\sqrt{2}ms^{-1}, 60°$

9. A car starts from rest and accelerates at 5 m/s². At t = 4 s, a ball is dropped out of a window by a person sitting in the car. What is the velocity and acceleration of the ball at t = 6 s?
(Take g = 10 m/s²) NCERT | Page-35 | 2021
(a) $20\sqrt{2}$ m/s, 10 m/s²
(b) 20 m/s, 5 m/s²
(c) 20 m/s, 0
(d) $20\sqrt{2}$ m/s, 0

10. The x and y coordinates of the particle at any time are $x = 5t - 2t^2$ and $y = 10t$ respectively, where x and y are in meters and t in seconds. The acceleration of the particle at t = 2s is NCERT | Page-36 | 2017
(a) 5 m/s² (b) –4 m/s²
(c) –8 m/s² (d) 0

11. A particle is moving such that its position coordinate (x, y) are
(2m, 3m) at time t = 0
(6m, 7m) at time t = 2 s and
(13m, 14m) at time t = 5s.
Average velocity vector ($\vec{V}_{av}$) from t = 0 to t = 5s is :
NCERT | Page-35 | 2014
(a) $\frac{1}{5}(13\hat{i} + 14\hat{j})$ (b) $\frac{7}{3}(\hat{i} + \hat{j})$
(c) $2(\hat{i} + \hat{j})$ (d) $\frac{11}{5}(\hat{i} + \hat{j})$

12. A particle has initial velocity $(2\vec{i} + 3\vec{j})$ and acceleration $(0.3\vec{i} + 0.2\vec{j})$. The magnitude of velocity after 10 seconds will be : NCERT | Page-17 | 2012
(a) $9\sqrt{2}$ units (b) $5\sqrt{2}$ units
(c) 5 units (d) 9 units

13. A particle has initial velocity $(3\hat{i} + 4\hat{j})$ and has acceleration $(0.4\hat{i} + 0.3\hat{j})$. It's speed after 10 s is:
NCERT | Page-18 | 2010
(a) 7 units (b) $7\sqrt{2}$ units
(c) 8.5 units (d) 10 units

14. A particle starting from the origin (0, 0) moves in a straight line in the (x, y) plane. Its coordinates at a later time are $(\sqrt{3}, 3)$. The path of the particle makes with the x-axis an angle of NCERT | Page-35 | 2007
(a) 45° (b) 60° (c) 0° (d) 30°

15. A particle moves in a plane with constant acceleration in a direction different from the initial velocity. The path of the particle is NCERT | Page-35 | 2005
(a) an ellipse (b) a parabola
(c) an arc of a circle (d) a straight line

16. A body of 3 kg moves in the XY plane under the action of a force given by $6t^2\hat{i} + 4t\hat{j}$. Assuming that the body is at rest at time t = 0, the velocity of the body at t = 3s is
NCERT | Page-35 | 2002
(a) $6\hat{i} + 6\hat{j}$ (b) $18\hat{i} + 6\hat{j}$
(c) $18\hat{i} + 12\hat{j}$ (d) $12\hat{i} + 18\hat{j}$

17. Two particles A and B are connected by a rigid rod AB. The rod slides along perpendicular rails as shown here. The velocity of A to the left is 10 m/s. What is the velocity of B when angle α = 60°? NCERT | Page-37 | 1998

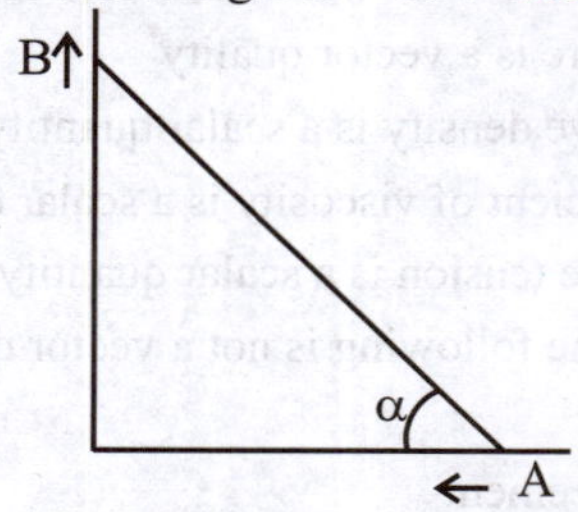

(a) 5.8 m/s (b) 9.8 m/s
(c) 10 m/s (d) 17.3 m/s

18. The position vector of a particle is $\vec{r} = (a\cos\omega t)\hat{i} + (a\sin\omega t)\hat{j}$. The velocity of the particle is NCERT | Page-35 | 1995
(a) directed towards the origin
(b) directed away from the origin
(c) parallel to the position vector
(d) perpendicular to the position vector

19. Two boys are standing at the ends A and B of a ground where $AB = a$. The boy at B starts running in a direction perpendicular to AB with velocity v_1. The boy at A starts running simultaneously with velocity v and catches the other boy in a time t, where t is
NCERT | Page-37 | 2005
(a) $a/\sqrt{v^2 + v_1^2}$ (b) $a/(v + v_1)$
(c) $a/(v - v_1)$ (d) $\sqrt{a^2/(v^2 - v_1^2)}$

20. A bullet is fired from a gun with a speed of 1000 m/s in order to hit a target 100 m away. At what height above the target should the gun be aimed? (The resistance of air is negligible and $g = 10\ m/s^2$) NCERT | Page-37 | 1995

(a) 5 cm (b) 10 cm
(c) 15 cm (d) 20 cm

Topic-*
Relative Velocity in two Dimension

21. The speed of a swimmer in still water is 20 m/s. The speed of river water is 10 m/s and is flowing due east. If he is standing on the south bank and wishes to cross the river along the shortest path, the angle at which he should make his strokes w.r.t. north is given by : 2019

(a) 30°west (b) 0°
(c) 60° west (d) 45° west

22. A ship A is moving westwards with a speed of 10 km h^{-1} and a ship B 100 km South of A, is moving northwards with a speed of 10 km h^{-1}. The time after which the distance between them becomes shortest, is: 2015

(a) 5 h (b) $5\sqrt{2}$ h
(c) $10\sqrt{2}$ h (d) 0 h

23. A person swims in a river aiming to reach exactly opposite point on the bank of a river. His speed of swimming is 0.5 m/s at an angle 120° with the direction of flow of water. The speed of water in stream is 1999

(a) 1.0 m/s (b) 0.5 m/s
(c) 0.25 m/s (d) 0.43 m/s

24. A boat is sent across a river with a velocity of 8 km h^{-1}. If the resultant velocity of boat is 10 km h^{-1}, then the velocity of the river is 1993

(a) 12.8 km h^{-1} (b) 6 km h^{-1}
(c) 8 km h^{-1} (d) 10 km h^{-1}

Topic-3.9
Projectile Motion

25. A ball is projected from point A with velocity 20 ms^{-1} at an angle 60° to the horizontal direction. At the highest point B of the path (as shown in figure), the velocity v m s^{-1} of the ball will be NCERT | Page-39 | Manipur 2023

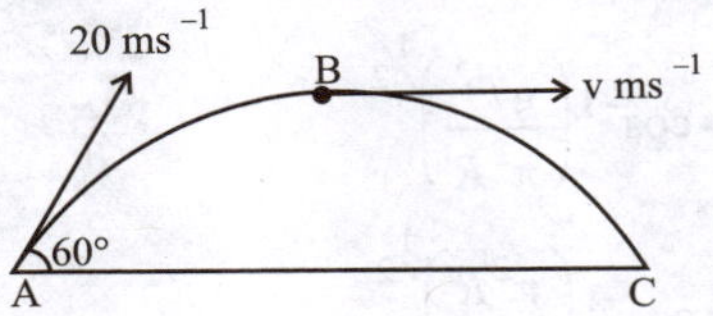

(a) Zero (b) 10
(c) 20 (d) $10\sqrt{3}$

26. A bullet is fired from a gun at the speed of 280 m s^{-1} in the direction 30° above the horizontal. The maximum height attained by the bullet is ($g = 9.8\ m\ s^{-2}$, sin 30° = 0.5) NCERT | Page-39 | 2023

(a) 2800 m (b) 2000 m
(c) 1000 m (d) 3000 m

27. A cricket ball is thrown by a player at a speed of 20 m / s in a direction 30° above the horizontal. The maximum height attained by the ball during its motion is. ($g = 10\ m/s^2$) NCERT | Page-39 | RS 2022

(a) 25 m (b) 5 m (c) 10 m (d) 20 m

28. A ball is projected with a velocity, 10 ms^{-1}, at an angle of 60° with the vertical direction. Its speed at the highest point of its trajectory will be: NCERT | Page-39 | 2022

(a) $5\sqrt{3}ms^{-1}$ (b) 5 ms^{-1}
(c) 10 ms^{-1} (d) Zero

29. A projectile is fired from the surface of the earth with a velocity of 5 ms^{-1} and angle θ with the horizontal. Another projectile fired from another planet with a velocity of 3 ms^{-1} at the same angle follows a trajectory which is identical with the trajectory of the projectile fired from the earth. The value of the acceleration due to gravity on the planet is (in ms^{-2}) given $g = 9.8\ m/s^2$ NCERT | Page-39 | 2014

(a) 3.5 (b) 5.9
(c) 16.3 (d) 110.8

30. The velocity of a projectile at the initial point A is $(2\hat{i}+3\hat{j})$ m/s. It's velocity (in m/s) at point B is NCERT | Page-38 | 2013

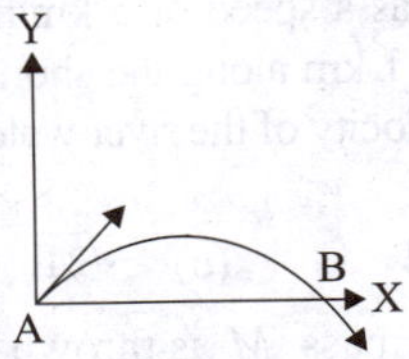

(a) $-2\hat{i}+3\hat{j}$ (b) $2\hat{i}-3\hat{j}$
(c) $2\hat{i}+3\hat{j}$ (d) $-2\hat{i}-3\hat{j}$

31. The horizontal range and the maximum height of a projectile are equal. The angle of projection of the projectiles is : NCERT | Page-40 | 2012

(a) $\theta = \tan^{-1}\left(\frac{1}{4}\right)$ (b) $\theta = \tan^{-1}(4)$
(c) $\theta = \tan^{-1}(2)$ (d) θ = 45°

32. A missile is fired for maximum range with an initial velocity of 20 m/s. If $g = 10\ m/s^2$, the range of the missile is NCERT | Page-40 | 2011

(a) 40 m (b) 50 m
(c) 60 m (d) 20 m

33. A projectile is fired at an angle of 45° with the horizontal. Elevation angle of the projectile at its highest point as seen from the point of projection is **NCERT | Page-39 | 2011 M**

(a) 60° (b) $\tan^{-1}\left(\frac{1}{2}\right)$

(c) $\tan^{-1}\left(\frac{\sqrt{3}}{2}\right)$ (d) 45°

34. The speed of a projectile at its maximum height is half of its initial speed. The angle of projection is **NCERT | Page-39 | 2010 M**

(a) 60° (b) 15°
(c) 30° (d) 45°

35. A particle of mass m is projected with velocity v making an angle of 45° with the horizontal. When the particle lands on the level ground the magnitude of the change in its momentum will be: **NCERT | Page-39 | 2008**

(a) 2mv (b) $mv/\sqrt{2}$
(c) $mv\sqrt{2}$ (d) zero

36. For angles of projection of a projectile (45° – θ) and (45° + θ), the horizontal ranges described by the projectile are in the ratio of **NCERT | Page-39 | 2006**

(a) 1 : 3 (b) 1 : 2 (c) 2 : 1 (d) 1 : 1

37. Two projectiles are fired from the same point with the same speed at angles of projection 60º and 30º respectively. Which one of the following is true? **NCERT | Page-40 | 2000**

(a) Their maximum height will be same
(b) Their range will be same
(c) Their landing velocity will be same
(d) Their time of flight will be same

38. A boat which has a speed of 5 km/hr in still water crosses a river of width 1 km along the shortest possible path in 15 minutes. The velocity of the river water in km/hr is **2000, 1998**

(a) 3 (b) 4 (c) $\sqrt{21}$ (d) 1

39. If a body A of mass M is thrown with velocity v at an angle of 30° to the horizontal and another body B of the same mass is thrown with the same speed at an angle of 60° to the horizontal, the ratio of horizontal range of A to B will be **NCERT | Page-40 | 1992**

(a) 1 : 3 (b) 1 : 1
(c) $1:\sqrt{3}$ (d) $\sqrt{3}:1$

40. The maximum range of a gun of horizontal terrain is 16 km. If $g = 10\ ms^{-2}$, then muzzle velocity of a shell must be **NCERT | Page-40 | 1990**

(a) $160\ ms^{-1}$ (b) $200\sqrt{2}\ ms^{-1}$
(c) $400\ ms^{-1}$ (d) $800\ ms^{-1}$

41. Two bodies of same mass are projected with the same velocity at an angle 30° and 60° respectively. The ratio of their horizontal ranges will be **NCERT | Page-40 | 1990**

(a) 1 : 1 (b) 1 : 2
(c) 1 : 3 (d) $2:\sqrt{2}$

Topic-3.10

Uniform Circular Motion

42. Let ω_1, ω_2 and ω_3 be the angular speed of the second hand, minute hand and hour hand of a smoothly running analog clock, respectively. If x_1, x_2 and x_3 are their respective angular distances in 1 minute then the factor which remains constant (k) is **NCERT | Page-41, 42 | Ph-II 2024**

(a) $\frac{\omega_1}{x_1} = \frac{\omega_2}{x_2} = \frac{\omega_3}{x_3} = k$
(b) $\omega_1 x_1 = \omega_2 x_2 = \omega_3 x_3 = k$
(c) $\omega_1 x_1^2 = \omega_2 x_2^2 = \omega_3 x_3^2 = k$
(d) $\omega_1^2 x_1 = \omega_2^2 x_2 = \omega_3^2 x_3 = k$

43. A particle moving with uniform speed in a circular path maintains: **NCERT | Page-42 | 2024**

(a) Constant velocity
(b) Constant acceleration
(c) Constant velocity but varying acceleration
(d) Varying velocity and varying acceleration

44. A particle is executing uniform circular motion with velocity $\vec{v}$ and acceleration $\vec{a}$. Which of the following is **true**? **NCERT | Page-42 | Manipur 2023**

(a) $\vec{v}$ is a constant; $\vec{a}$ is a constant
(b) is not a constant; $\vec{a}$ is a constant
(c) $\vec{v}$ is a constant; $\vec{a}$ is not a constant
(d) $\vec{v}$ is not a constant; $\vec{a}$ is not a constant

45. The angular speed on a flywheel moving with uniform angular acceleration changes from 1200 rpm to 3120 rpm in 16 seconds. The angular acceleration in rad/s^2 is: **NCERT | Page-42 | 2022**

(a) 4π (b) 12π (c) 104π (d) 2π

46. A particle moving in a circle of radius R with a uniform speed takes a time T to complete one revolution. If this particle were projected with the same speed at an angle 'θ' to the horizontal, the maximum height attained by it equals 4R. The angle of projection, θ, is then given by : **NCERT | Page-39 | 2021**

(a) $\theta = \sin^{-1}\left(\frac{2gT^2}{\pi^2 R}\right)^{1/2}$

(b) $\theta = \cos^{-1}\left(\frac{gT^2}{\pi^2 R}\right)^{1/2}$

(c) $\theta = \cos^{-1}\left(\frac{\pi^2 R}{gT^2}\right)^{1/2}$

(d) $\theta = \sin^{-1}\left(\frac{\pi^2 R}{gT^2}\right)^{1/2}$

47. The angular speed of the wheel of a vehicle is increased from 360 rpm to 1200 rpm in 14 second. Its angular acceleration is,

NCERT | Page-42 | Ph-II 2020

(a) 1 rad/s^2 (b) 2π rad/s^2
(c) 28π rad/s^2 (d) 120π rad/s^2

48. The position vector of a particle $\vec{R}$ as a function of time is given by:

$\vec{R} = 4\sin(2\pi t)\hat{i} + 4\cos(2\pi t)\hat{j}$

Where R is in metre, t in seconds and $\hat{i}$ and $\hat{j}$ denote unit vectors along x-and y-directions, respectively.
Which one of the following statements is wrong for the motion of particle?

NCERT | Page-41 | 2015 RS

(a) Magnitude of acceleration vector is $\frac{v^2}{R}$, where v is the velocity of particle
(b) Magnitude of the velocity of particle is 8 meter/second
(c) path of the particle is a circle of radius 4 metre
(d) Acceleration vector is along $-\vec{R}$

49. A particle moves in a circle of radius 5 cm with constant speed and time period 0.2π s. The acceleration of the particle is

NCERT | Page-41 | 2011

(a) 15 m/s^2 (b) 25 m/s^2
(c) 36 m/s^2 (d) 5 m/s^2

50. A car runs at a constant speed on a circular track of radius 100 m, taking 62.8 seconds in every circular loop. The average velocity and average speed for each circular loop respectively, is

NCERT | Page-41 | 2006

(a) 0, 10 m/s (b) 10 m/s, 10 m/s
(c) 10 m/s, 0 (d) 0, 0

51. A stone tied to the end of a string of 1 m long is whirled in a horizontal circle with a constant speed. If the stone makes 22 revolutions in 44 seconds, what is the magnitude and direction of acceleration of the stone?

NCERT | Page-42 | 2005

(a) π^2 ms^{-2} and direction along the radius towards the centre
(b) π^2 ms^{-2} and direction along the radius away from the centre
(c) π^2 ms^{-2} and direction along the tangent to the circle
(d) $\pi^2/4$ ms^{-2} and direction along the radius towards the centre

52. The circular motion of a particle with constant speed is

NCERT | Page-41 | 2005

(a) periodic but not simple harmonic
(b) simple harmonic but not periodic
(c) periodic and simple harmonic
(d) neither periodic nor simple harmonic

53. A stone is tied to a string of length ℓ and is whirled in a vertical circle with the other end of the string as the centre. At a certain instant of time, the stone is at its lowest position and has a speed u. The magnitude of the change in velocity as it reaches a position where the string is horizontal (g being acceleration due to gravity) is

NCERT | Page-42 | 2004

(a) $\sqrt{2g\ell}$ (b) $\sqrt{2(u^2 - g\ell)}$
(c) $\sqrt{u^2 - g\ell}$ (d) $u - \sqrt{u^2 - 2g\ell}$

54. A particle moves along a circle of radius $\left(\frac{20}{\pi}\right)$ m with constant tangential acceleration. If the velocity of the particle is 80 m/s at the end of the second revolution after motion has begun, the tangential acceleration is

NCERT | Page-41 | 2003

(a) 40π m/s^2 (b) 40 m/s^2
(c) 640π m/s^2 (d) 160π m/s^2

55. Two particles of mass M and m are moving in a circle of radii R and r. If their time-periods are same, what will be the ratio of their linear velocities?

NCERT | Page-41 | 2001

(a) $MR : mr$ (b) M : m
(c) R : r (d) 1 : 1

56. A ball of mass 0.25 kg attached to the end of a string of length 1.96 m is moving in a horizontal circle. The string will break if the tension is more than 25 N. What is the maximum speed with which the ball can be moved?

NCERT | Page-42 | 1998

(a) 14 m/s (b) 3 m/s
(c) 5 m/s (d) 3.92 m/s

57. A body is whirled in a horizontal circle of radius 20 cm. It has an angular velocity of 10 rad/s. What is its linear velocity at any point on circular path

NCERT | Page-42 | 1996

(a) $\sqrt{2}$ m/s (b) 2 m/s
(c) 10 m/s (d) 20 m/s

58. When a body moves with a constant speed along a circle

NCERT | Page-42 | 1994

(a) its velocity remains constant
(b) no force acts on it
(c) no work is done on it
(d) no acceleration is produced in it

59. An electric fan has blades of length 30 cm measured from the axis of rotation. If the fan is rotating at 120 rpm, the acceleration of a point on the tip of the blade is

NCERT | Page-42 | 1990

(a) 1600 ms^{-2} (b) 47.4 ms^{-2}
(c) 23.7 ms^{-2} (d) 50.55 ms^{-2}

ANSWER KEY																			
1	(a)	7	(b)	13	(b)	19	(d)	25	(b)	31	(b)	37	(b)	43	(d)	49	(d)	55	(c)
2	(c)	8	(c)	14	(b)	20	(a)	26	(c)	32	(a)	38	(a)	44	(d)	50	(a)	56	(a)
3	(a)	9	(a)	15	(b)	21	(a)	27	(b)	33	(b)	39	(b)	45	(a)	51	(a)	57	(b)
4	(c)	10	(b)	16	(b)	22	(a)	28	(a)	34	(a)	40	(c)	46	(a)	52	(a)	58	(c)
5	(c)	11	(d)	17	(a)	23	(c)	29	(a)	35	(c)	41	(a)	47	(b)	53	(b)	59	(b)
6	(b)	12	(b)	18	(d)	24	(b)	30	(b)	36	(d)	42	(a)	48	(b)	54	(b)		

Hints & Solutions

1. **(a)** All pressure, relative density, coefficient of viscosity and surface tension are scalar quantity.

2. **(c)** Work is a scalar quantity.

3. **(a)** A vector quantity has both magnitude and direction. In the given options, speed has only magnitude, therefore, it is non-vector or scalar quantity.

NOTES If a physical quantity has magnitude and direction both, then it does not always imply that it is a vector. For it to be a vector, laws of vector addition has to be satisfied.

4. **(c)** When a vector is multiplied with a scalar, the result is a vector.

NOTES The cross product of $\vec{A}$ and $\vec{B}$ is a vector with its direction perpendicular to both vector $\vec{A}$ and $\vec{B}$.
Vector $\vec{A}\times 0$ is a zero vector because 0 is a scalar then also the product is zero but a (scalar × vector) is also vector.

5. **(c)** Using the law of vector addition, $(\vec{d}+\vec{e})$ is as shown in the fig.

$\therefore \quad \vec{d}+\vec{e}=\vec{f}$

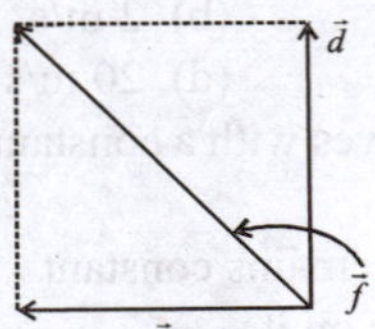

6. **(b)** $\hat{r}=0.5\hat{i}+0.8\hat{j}+c\hat{k}$

$|\hat{r}|=1=\sqrt{(0.5)^2+(0.8)^2+c^2}$

$(0.5)^2+(0.8)^2+c^2=1$

$c^2=0.11 \Rightarrow c=\sqrt{0.11}$

NOTES Unit vector is a vector which has a magnitude of one. It is a vector divided by its magnitude. Unit vector for $\vec{A}$ is $\hat{A}=\frac{\vec{A}}{A}$
Unit vector gives direction.

7. **(b)** We know that, $R^2=A^2+B^2+2AB\cos\theta$

$(24)^2=(12)^2+(18)^2+2(12)(18)\cos\theta$

$\cos\theta=\frac{108}{432} \Rightarrow \theta=75°52'$

8. **(c)** Given, position, $\vec{r}(t)=\left(4t\hat{i}+2t^2\hat{j}+5\hat{k}\right)$m

$\therefore$ Velocity, $\vec{v}=\frac{d\vec{r}(t)}{dt}=4\hat{i}+4t\hat{j}\big|_{t=1}$

$=\left(4\hat{i}+4\hat{j}\right)\text{ms}^{-1}$

$\therefore \ |v|=\sqrt{4^2+4^2}=4\sqrt{2}\text{ ms}^{-1}$

And, $\tan\phi=\frac{v_y}{v_x}=\frac{4}{4}=1$

$\therefore$ Direction with respect to X–axis is 45°

9. **(a)** Velocity of car at t = 4 s = velocity the ball dropped out of a window $v=u+at$

$\Rightarrow v=0+5\times 4=20\text{ ms}^{-1}$

$\therefore$ Velocity of ball along the horizontal and is 20ms^{-1}.

After 2 seconds of motion :

Vertical velocity of ball $(v_y)=u_y+a_yt$

$v_y=0+10\times 2=20\text{ ms}^{-1}$ $(\because a_y=g=10\text{ m/s}^2)$

$\therefore$ Magnitude of velocity of ball

$(v)=\sqrt{v_x^2+v_y^2}=20\sqrt{2}\text{ m/s}$

Since the ball is under free fall.

Acceleration of ball (a) at t = 6s is

$a=g=10\text{ m/s}^2$

NOTES All bodies, in the absence of air resistance, irrespective of the size, weight or composition fall with same acceleration near the surface of the earth is called free fall.

10. **(b)** Given:

$x=5t-2t^2\ y=10t$

$v_x = \frac{dx}{dt} = 5 - 4t \qquad v_y = \frac{dy}{dt} = 10$

$a_x = \frac{dv_x}{dt} = -4 \qquad a_y = \frac{dv_y}{dt} = 0$

$\vec{a} = a_x\hat{i} + a_y\hat{j} \qquad \vec{a} = -4\hat{i}\ \text{m/s}^2$

Hence, acceleration of particle at ($t = 2$ s) = -4m/s^2

11. **(d)** $\vec{v}_{av} = \frac{\Delta\vec{r}\ \text{(displacement)}}{\Delta t\ \text{(time taken)}}$

$= \frac{(13-2)\hat{i} + (14-3)\hat{j}}{5-0} = \frac{11}{5}(\hat{i} + \hat{j})$

> **NOTES** When a point have coordinate (x, y) then its position vector $= x\hat{i} + y\hat{j}$
> When a particle moves from point (x_1, y_1) to (x_2, y_2) then its displacement vector
> $\vec{r} = (x_2 - x_1)\hat{i} + (y_2 - y_1)\hat{j}$

12. **(b)** $\vec{v} = \vec{u} + \vec{a}t$

$= (2\hat{i} + 3\hat{j}) + (0.3\hat{i} + 0.2\hat{j}) \times 10 = 5\hat{i} + 5\hat{j}$

$|\vec{v}| = \sqrt{5^2 + 5^2}$

$|\vec{v}| = 5\sqrt{2}$ units

13. **(b)** Given, $\vec{u} = 3\hat{i} + 4\hat{j}$ and $\vec{a} = 0.4\hat{i} + 0.3\hat{j}$

$\Rightarrow \quad u_x = 3$ units, $u_y = 4$ units

$\Rightarrow \quad a_x = 0.4$ units, $a_y = 0.3$ units

Along x-axis,

$\therefore \quad v_x = u_x + a_x \times 10 = 3 + 4 = 7$ units

Along y-axis,

and $v_y = 4 + 0.3 \times 10 = 4 + 3 = 7$ units

Net final velocity

$\therefore \quad v = \sqrt{v_x^2 + v_y^2} = 7\sqrt{2}$ units

Alternative method:

$\vec{v}_i = 3\hat{i} + 4\hat{j}$ and $\vec{a} = 0.4\hat{i} + 0.3\hat{j}$

time, $t = 10$ sec.

Final velocity $\vec{v}_f$ after time $t = 10$ sec,

$\vec{v}_f = \vec{v}_i + \vec{a}t$

$\therefore\ \vec{v}_f = (3\hat{i} + 4\hat{j}) + (0.4\hat{i} + 0.3\hat{j})(10) = 7\hat{i} + 7\hat{j}$

> **NOTES** The particle speeds up i.e., the speed of the particle increases when the angle between $\vec{a}$ and $\vec{v}$ lies between 0° to 90°. The particle speeds down i.e., the speed of the particle decreases when the angle between $\vec{a}$ and $\vec{v}$ lies between 90° to 180°.

14. **(b)**

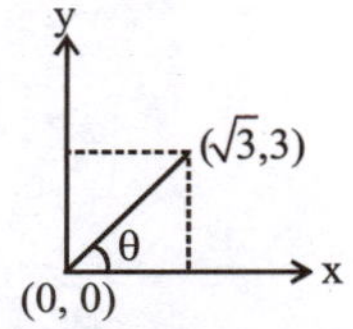

Let θ be the angle which the particle makes with x-axis.

From figure, $\tan\theta = \frac{3}{\sqrt{3}} = \sqrt{3}$

$\Rightarrow \theta = \tan^{-1}(\sqrt{3}) = 60°$

> **NOTES** If a vector $\vec{R}$ in X–Y plane then its orthogonal vector components are $R_X = R\cos\theta$ and $R_Y = R\sin\theta$.
> And $\frac{R\sin\theta}{R\cos\theta} = \tan\theta = \frac{R_Y}{R_X}$.

15. **(b)**

16. **(b)** $\vec{F} = 6t^2\,\hat{i} + 4t\,\hat{j}$

$F_x = 6t^2,\ F_y = 4t$

$a_x = \frac{2t^2}{3} = 2t^2,\ a_y = \frac{4t}{3}$

$\frac{dv_x}{dt} = 2t^2$

$\int_0^{v_x} dV_x = \int_0^t 2t^2 dt \Rightarrow v_x = \frac{2}{3}t^3 = \frac{2}{3}\cdot 3^3 = 18$

$\frac{dV_y}{dt} = \frac{4}{3}t \Rightarrow \int_0^{V_y} dV_y = \frac{4}{3}\int_0^t t\,dt$

$\Rightarrow \quad V_y = \frac{4}{3}\frac{t^2}{2} = \frac{4}{3}\cdot\frac{3^2}{2} = 6$

$\Rightarrow \quad \vec{V} = v_x\hat{i} + v_y\hat{j} = 18\hat{i} + 6\hat{j}$

17. **(a)** Given, $v_A = 10$m/sec

$\alpha = 60°$

let length of the rod = L

From figure,

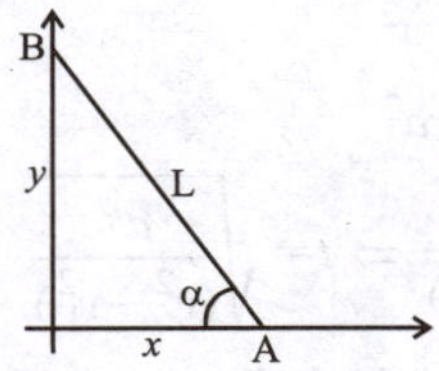

$\Rightarrow \quad x^2 + y^2 = L$

differentiation with respect to time 't'

$\Rightarrow 2x\frac{dx}{dt} + 2y\frac{dy}{dt} = 0$

$\Rightarrow x\frac{dx}{dt} + y\frac{dy}{dt} = 0$

where, $\frac{dx}{dt} = -v_A$ and $\frac{dy}{dt}$

So, $-xv_A + yv_B = 0$

$$v_B = \frac{x}{y} v_A$$

where $\frac{x}{y} = \cot \alpha$

So, $v_B = v_A \cot \alpha$

$v_B = 10 \cot 60°$

$$v_B = \frac{10}{\sqrt{3}} = 5.773 \simeq 5.8 \text{ m/sec}$$

18. (d) Position vector,

$\vec{r} = (a \cos \omega t)\hat{i} + (a \sin \omega t)\hat{j}$

Velocity vector,

$$\vec{v} = \frac{d(\vec{r})}{dt} = \frac{d}{dt}\{(a \cos \omega t)\hat{i} + (a \sin \omega t)\hat{j}\}$$

$= (-a\omega \sin \omega t)\hat{i} + (a\omega \cos \omega t)\hat{j}$

$= \omega[(-a \sin \omega t)\hat{i} + (a \cos \omega t)\hat{j}]$

$$\therefore \ \vec{r}.\vec{v} = \left[(a \cos \omega t)\hat{i} + (a \sin \omega t)\hat{j}\right] \cdot \left[-(a\omega \sin \omega t)\hat{i} + (a\omega \cos \omega t)\hat{j}\right] = 0$$

$\therefore$ velocity is perpendicular to the displacement.

19. (d) According to the given condition,

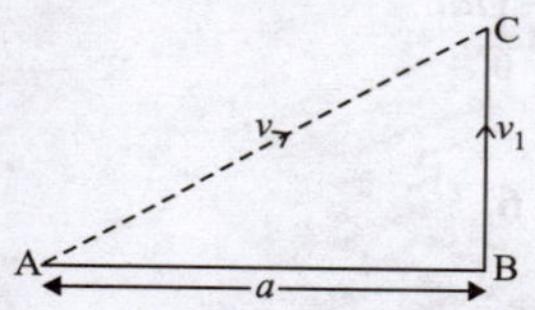

Let both boy meet at C in time t then,

AC = vt and BC = $v_1 t$

From pythagoras theorem,

$AC^2 = AB^2 + BC^2$

$v^2t^2 = a^2 + v_1^2 t^2$

$t^2(v^2 - v_1^2) = a^2$

$$t^2 = \frac{a^2}{(v^2 - v_1^2)} \Rightarrow t = \sqrt{\frac{a^2}{(v^2 - v_1^2)}}$$

20. (a) Speed of the bullet (v) = 1000 m/s and horizontal distance of the target (s) = 100 m.

Time taken to cover the horizontal distance (t)

$$= \frac{100}{1000} = 0.1 \text{ sec}$$

During this time, the bullet will fall down vertically due to gravitational acceleration.

$\therefore$ height $(h) = ut + \frac{1}{2}gt^2$

$$= (0 \times 0.1) + \frac{1}{2} \times 10(0.1)^2 = 0.05\text{m} = 5\text{cm}$$

21. (a) Velocity of swimmer in still water $V_{SR} = 20$ m/s

Velocity of river w.r.t. ground $V_{RG} = 10$ m/s

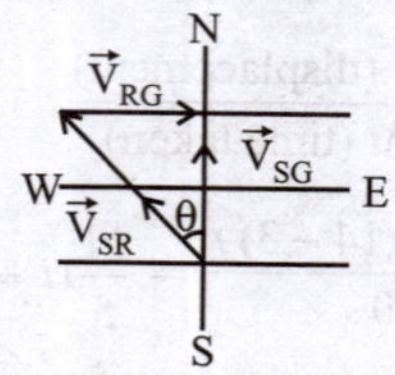

$\vec{V}_{SG} = \vec{V}_{SR} + \vec{V}_{RG}$

$$\sin \theta = \left|\frac{\vec{V}_{RG}}{\vec{V}_{SR}}\right| \Rightarrow \sin \theta = \frac{10}{20}$$

$\Rightarrow \sin \theta = \frac{1}{2} \quad \therefore \theta = 30°$ west

i.e., to cross the river along the shortest path, swimmer should make his strokes 30° west.

22. (a) Let t be the time passed and x be distance covered by the two ships after which the distance between them becomes shortest.

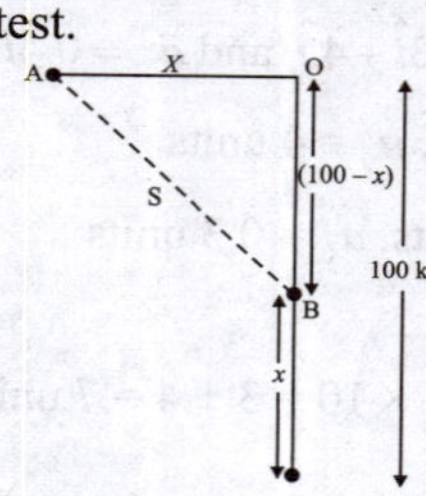

distance(s) between the ships A and B is given by,

$$S = \sqrt{(100 - x)^2 + x^2}$$

For S to be minimum, $\frac{ds}{dt} = 0$

$$\frac{1}{9[(100 - x)^2 + x^2]^{-1/2}}[-2(100 - x) + 2x] = 0$$

$\Rightarrow \ 4x - 200 = 0$

$\Rightarrow \ x = 50$ m

So, after both the ships have covered 50 m, distance between them becomes shortest. Time taken by them will be

$$t = \frac{x}{10} = \frac{50}{10} = 5\text{hr.}$$

23. (c)

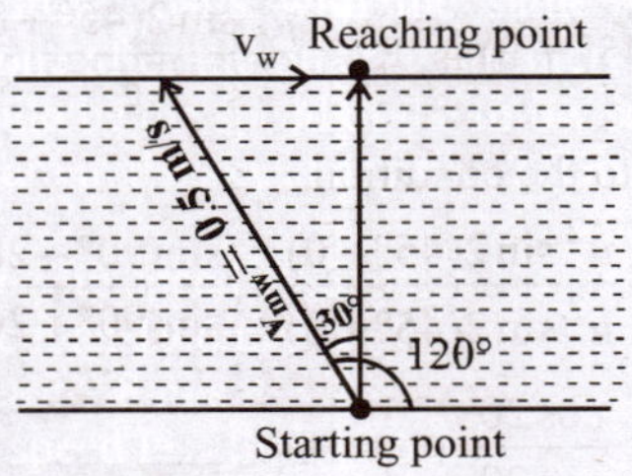

Velocity of person, $v_{mw} = 0.5$ m/s

$\sin 30° = \frac{v_w}{v_m} \Rightarrow v_w = v_{mw} \sin 30°$

$\Rightarrow v_w = \frac{v_{mw}}{2} = \frac{0.5 \text{ m/s}}{2} = 0.25 \text{ m/s}$

24. (b) $v_r = \sqrt{v_R^2 - v_B^2} = \sqrt{10^2 - 8^2} = 6 \text{ km h}^{-1}$

25. (b) At the highest point (B) of a projectile,

$v = v_o \cos 60° = 20 \cos 60° = 20 \times \frac{1}{2} = 10$ m/s

$[\because \text{At B}, V_y = 0]$

26. (c) Maximum height, $H = \frac{u^2 \sin^2 \theta}{2g}$

$= \frac{(280)^2 \times (\sin^2 30°)}{2 \times 9.8} = \frac{280 \times 280 \times 0.5 \times 0.5}{2 \times 9.8}$

$\therefore H = 1000$ m

27. (b)

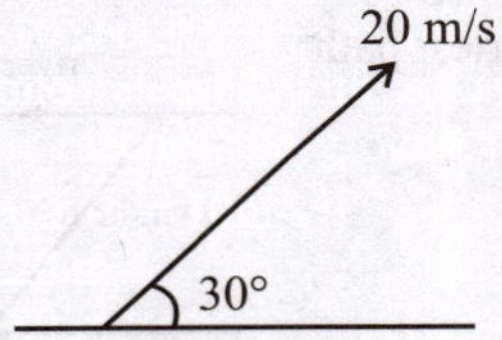

Maximum height,

$H = \frac{u^2 \sin^2 \theta}{2g} = \frac{(20)^2 \times \left(\frac{1}{2}\right)^2}{2 \times 10} = 5\text{m}$

28. (a) $V_{\text{Horizontal}} = 10 \sin 60°$

$= \frac{10 \times \sqrt{3}}{2} = 5\sqrt{3}$

as there is zero acceleration along horizontal direction, at highest point we will have same horizontal velocity also there is no vertical velocity at highest point.

so $|\vec{V}| = |\vec{V}_{\text{Horizontal}}| = 5\sqrt{3}$ m/s

29. (a) Trajectory is identical for both projectiles

$\therefore$ Range, $R = \frac{u^2 \sin 2\theta}{g} \Rightarrow g \propto u^2$

So that, $\frac{g_{\text{planet}}}{g_{\text{earth}}} = \frac{(u_{\text{planet}})^2}{(u_{\text{earth}})^2}$

Therefore, $g_{\text{planet}} = \left(\frac{3}{5}\right)^2 (9.8 \text{ m/s}^2) = 3.5 \text{ m/s}^2$

30. (b) At point B the direction of velocity component of the projectile along *y*-component reverses, while the *x*-component remains unchanged.

Hence, $\vec{V}_B = 2\hat{i} - 3\hat{j}$

31. (b) Horizontal range,

$$R = \frac{u^2 \sin 2\theta}{g} \quad ....(1)$$

Maximum height,

$$H = \frac{u^2 \sin^2 \theta}{2g} \quad ....(2)$$

According to the condition, $R = H$

$\Rightarrow \frac{u^2 \sin 2\theta}{g} = \frac{u^2 \sin^2 \theta}{2g}$

$\Rightarrow 2 \sin \theta \cos \theta = \frac{\sin^2 \theta}{2}$

$2 \cos \theta = \frac{\sin \theta}{2} \Rightarrow \cot \theta = \frac{1}{4}$

$\Rightarrow \tan \theta = 4$

$\Rightarrow \theta = [\tan^{-1}(4)]$

NOTES In case of projectile motion, if range *R* is *n* times the maximum height *H*

i.e., $R = nH \Rightarrow \frac{u^2 \sin 2\theta}{g} = n \frac{u^2 \sin^2 \theta}{2g}$

$\Rightarrow \tan \theta = \frac{4}{n}$ angle of projection $\theta = \tan^{-1}\left(\frac{4}{n}\right)$

32. (a) For maximum range, the angle of projection, $\theta = 45°$

$\therefore R = \frac{u^2 \sin 2\theta}{g}$

$= \frac{(20)^2 \sin(2 \times 45°)}{10}$

$= \frac{400 \times 1}{10} = 40$ m

33. (b)

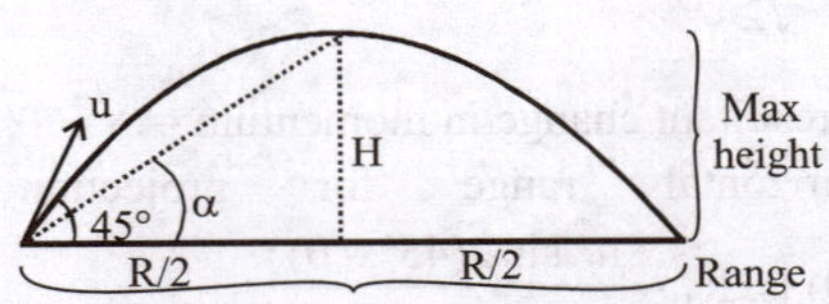

Maximum height,

$$H = \frac{u^2 \sin^2 45°}{2g} = \frac{u^2}{4g} \quad ... (1)$$

$$\text{Range, } R = \frac{u^2 \sin 90°}{g} = \frac{u^2}{g}$$

$$\therefore \frac{R}{2} = \frac{u^2}{2g} \quad ... (2)$$

According to the required condition,

$$\therefore \tan\alpha = \frac{H}{R/2}$$

Put the values from equation (1) and (2)

$$\Rightarrow \tan\alpha = \frac{\left(\frac{u^2}{4g}\right)}{\left(\frac{u^2}{2g}\right)} \qquad \therefore \alpha = \tan^{-1}\left(\frac{1}{2}\right)$$

34. **(a)** Speed of a projectile at maximum height,

$v' = v_0 \cos\theta$

$$\frac{v_0}{2} = v_0 \cos\theta \left[\because v' = \frac{v_0}{2} \text{ given}\right]$$

$$\therefore \cos\theta = \frac{v_0}{2v_0} = \frac{1}{2} \quad \therefore \theta = 60°$$

35. **(c)** The momentum along *x*-axis remains unchanged

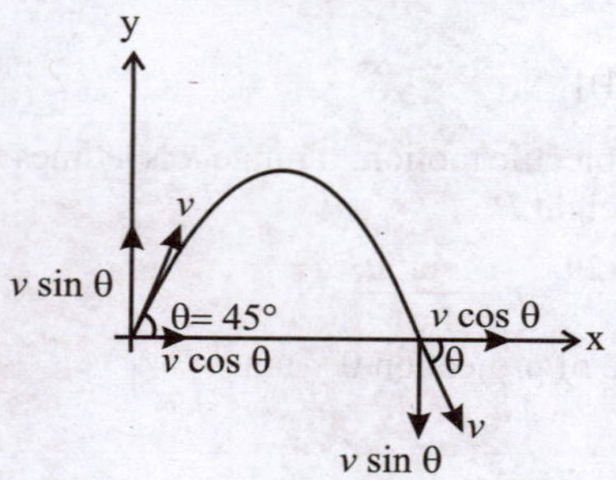

Clearly, change in momentum along *x*-axis $= mv\cos\theta - mv\cos\theta = 0$

Momentum changed only in vertical direction or *y*-axis.

So, $\Delta P = \Delta P_{vertical}$

$\Rightarrow P_{final} - P_{initial}$

$= mv\sin\theta - (-mv\sin\theta)$

$= 2\,mv\sin\theta = 2mv \times \sin 45°$

$= 2mv \times \frac{1}{\sqrt{2}} = \sqrt{2}mv$

Hence, resultant change in momentum $= \sqrt{2}mv$

36. **(d)** Horizontal range for projection angle $(45° - \theta)$ is, $R_1 = \frac{u^2 \sin 2(45° - \theta)}{g}$

Horizontal range projection for angle $(45° + \theta)$ is, $R_2 = \frac{u^2 \sin 2(45° + \theta)}{g}$

According to the condition,

$$\Rightarrow \frac{R_1}{R_2} = \frac{u^2 \sin 2(45° - \theta)}{u^2 \sin 2(45° + \theta)} = \frac{\sin(90° - 2\theta)}{\sin(90° + 2\theta)}$$

$$\Rightarrow \frac{R_1}{R_2} = \frac{\cos 2\theta}{\cos 2\theta} = \frac{1}{1}$$

So, $R_1 : R_2 = 1 : 1$

> **NOTES** The angle of elevation (ϕ) of the highest point of the projectile and the angle of projection θ are related to each other as $\tan\phi = \frac{1}{2}\tan\theta$.

37. **(b)** Given, $u_1 = u_2 = u$, $\theta_1 = 60°$, $\theta_2 = 30°$

In 1st case, we know that range

$$R_1 = \frac{u^2 \sin 2(60°)}{g} = \frac{u^2 \sin 120°}{g}$$

$$= \frac{u^2 \sin(90° + 30°)}{g} = \frac{u^2(\cos 30°)}{g} = \frac{\sqrt{3}u^2}{2g}$$

In 2nd case, when $\theta_2 = 30°$, then

$$R_2 = \frac{u^2 \sin 60°}{g} = \frac{u^2\sqrt{3}}{2g} \Rightarrow R_1 = R_2$$

[we get same value of ranges].

38. **(a)** Speed along the shortest path

$$= \frac{1}{15/60} = 4 \text{ km/hr}$$

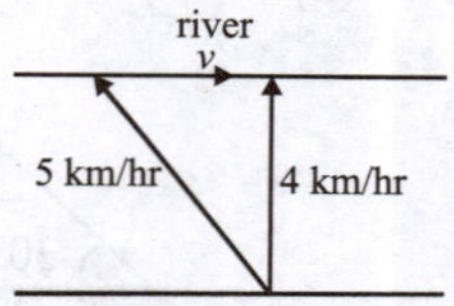

Speed of river $v = \sqrt{5^2 - 4^2} = 3$ km/hr

39. **(b)** Horizontal range is same when angle of projection with the horizonatal is θ and $(90° - \theta)$.

40. **(c)** $R_{max} = \frac{v^2}{g} = 16000$ [16km = 16000m]

or $v = (16000g)^{1/2} = (16000 \times 10)^{1/2} = 400 \text{ ms}^{-1}$

41. **(a)** Horizontal range is same when angle of projection is θ or $(90° - \theta)$.

42. **(a)** $\omega_1 = \frac{2\pi}{60}$;

$$\therefore x_1 = \frac{2\pi}{60} \times 60 = 2\pi$$

$$\omega_2 = \frac{2\pi}{3600};$$

$$\therefore x_2 = \frac{2\pi}{3600} \times 60 = \frac{2\pi}{60}$$

$\omega_3 = \frac{2\pi}{3600 \times 12}$;

$\therefore \quad x_3 = \frac{2\pi}{3600 \times 12} \times 60 = \frac{2\pi}{720}$

$\frac{\omega_1}{x_1} = \frac{\omega_2}{x_2} = \frac{\omega_3}{x_3} = \frac{1}{60} = k$

43. **(d)** When a particle is moving with uniform speed in a circular path then direction changes continuously. Therefore particle maintains varying velocity and varying acceleration.

44. **(d)** In uniform circular motion, due to change in direction, velocity $\vec{V}$ and acceleration $\vec{a}$ are not constant.

45. **(a)** Angular acceleration,

$$\alpha = \frac{\Delta\omega}{\Delta t} = \frac{2\pi(f_2 - f_1)}{\Delta t} = \frac{2\pi(52-20)}{16}$$

$$= \frac{2\pi \times 32}{16} = 4\pi$$

46. **(a)** Speed of particle to complete one revolution,

$(u) = \frac{2\pi R}{T}$ ... (i)

Particle is projected with same speed (u) at an angle θ to the horizontal

$\therefore$ Maximum Height $(H) = \frac{u^2 \sin^2\theta}{2g}$

or, $\frac{u^2 \sin^2\theta}{2g} = 4R \quad [\because H = 4R \text{ given}]$

$\Rightarrow \sin^2\theta = \frac{8gR}{u^2}$

or, $\sin^2\theta = \frac{8gRT^2}{4\pi^2R^2} = \frac{2gT^2}{\pi^2 R}$ (using equation (i))

$\therefore \quad \theta = \sin^{-1}\left(\frac{2gT^2}{\pi^2 R}\right)^{1/2}$

47. **(b)** Initial angular speed of wheel.

$\omega_0 = 2\pi f_0 = 2\pi \times \frac{360}{60}$ rad/s $= 12\pi$ rad/s

Final angular speed of wheel,

$\omega = 2\pi f = 2\pi \times \frac{1200}{60}$ rad/s $= 40\pi$ rad/s

Time taken, $t = 14$ s

From the equation of rotational motion,

$\omega = \omega_0 + \alpha t \Rightarrow \alpha = \frac{\omega - \omega_0}{t}$

$\therefore \alpha = \frac{40\pi - 12\pi}{14} = \frac{28\pi}{14} = 2\pi \text{ rad/s}^2$

48. **(b)** Here, $x = 4\sin(2\pi t)$ ...(i)

$y = 4\cos(2\pi t)$ ...(ii)

Squaring and adding equation (i) and (ii)

$x^2 + y^2 = 4^2 \Rightarrow R = 4$

Motion of the particle is circular motion, acceleration vector is along $-\vec{R}$ and its magnitude $= \frac{V^2}{R}$

Velocity of particle, $V = \omega R = (2\pi)(4) = 8\pi$

49. **(d)** Centripetal acceleration $a_c = \omega^2 r$

$= \left(\frac{2\pi}{T}\right)^2 r = \left(\frac{2\pi}{0.2\pi}\right)^2 \times 5 \times 10^{-2} = 5 \text{ m/s}^2$

50. **(a)** Distance covered in one circular loop $= 2\pi r$

$= 2 \times 3.14 \times 100 = 628$ m

Avg speed $= \frac{628}{62.8} = 10$ m / sec

Displacement in one circular loop = 0

$\because$ Avg velocity $= \frac{0}{\text{time}} = 0$

51. **(a)** $a_r = \omega^2 R \;\&\; a_t = \frac{dv}{dt} = 0$

or, $a_r = (2\pi n)^2 R = 4\pi^2 n^2 R^2 = 4\pi^2 \left(\frac{22}{44}\right)^2 (1)^2$

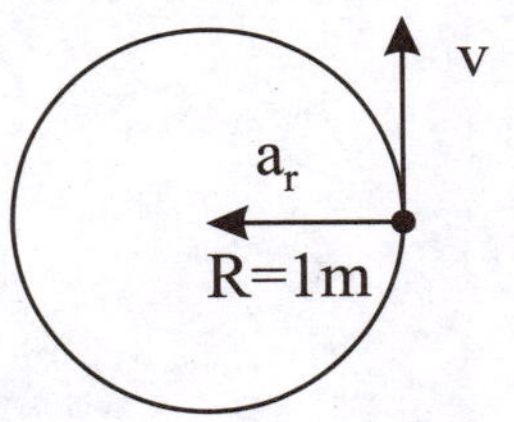

$a_{net} = a_r = \pi^2 \text{ms}^{-2}$ and direction along the radius towards the centre.

> **NOTES** If the force acting on a particle is always perpendicular to the velocity of the particle, then the path of the particle is a circle. The centripetal force (F_c) is always perpendicular to the velocity of the particle, i.e., $F_c \perp v$

52. **(a)** In circular motion of a particle with constant speed, particle repeats its motion after a regular interval of time but does not oscillate about a fixed point. So, motion of particle is periodic but not simple harmonic.

53. **(b)** $W_{mg} = \Delta K$

$\Rightarrow -mg\ \ell = ½ mv^2 - ½ mu^2$

or, $mv^2 = m(u^2 - 2g\ell)$

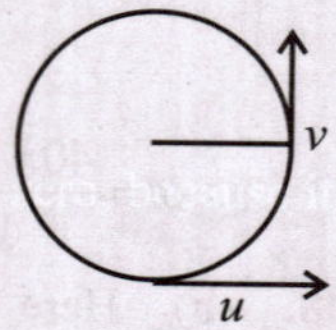

or, $\vec{v} = \sqrt{u^2 - 2g\ell}\ \hat{j}$

$\vec{u} = u\hat{i}$

$\therefore \vec{v} - \vec{u} = \sqrt{u^2 - 2g\ell}\ \hat{j} - u\hat{i}$

$\therefore |\vec{v} - \vec{u}| = [(u^2 - 2g\ell) + u^2]^{1/2} = \sqrt{2(u^2 - g\ell)}$

54. (b) Given, $r = \frac{20}{\pi}$ m

$\omega_f = \frac{v_f}{r} = \frac{80}{20} \times \pi = 4\pi$ rad/s

and, $\theta = 2 \times 2\pi = 4\pi$ rad

From equation,

$\omega_f^2 = \omega_0^2 + 2\alpha\theta$ $\qquad [\because \omega_0 = 0]$

$(4\pi)^2 = 0 + 2.\alpha.4\pi$

$\alpha = 2\pi$ rad/s^2

$\therefore$ Tangential acceleration

$a_t = \alpha.r$

$= 2\pi \cdot \frac{20}{\pi} = 40$ m/sec^2

55. (c) Linear velocity $v = r\omega$

$v_1 = \omega r_1,\ v_2 = \omega r_2$

[ω is same in both cases because time period is same]

$\frac{v_1}{v_2} = \frac{r_1}{r_2} = \frac{R}{r}$

56. (a) $T = \frac{mv^2}{R}$

$v = \sqrt{\frac{TR}{m}} = \sqrt{\frac{25 \times 1.96}{0.25}} = \frac{5 \times 14}{5} = 14\ m/s$

NOTES In the horizontal circular motion of the ball tension in the string is balanced by the centrifugal force $\left(\frac{mv^2}{\ell}\right)$ and hence the maximum tension in the string will be for the maximum speed of the ball.

57. (b) Radius of circular path = 20 cm = $\frac{2}{10}$ m

Angular speed of body = 10 rad/s

Linear velocity = radius × Angular speed

$= \frac{2}{10} \times 10 = 2$ m/s

58. (c) On circular motion, the force acts along the radius and displacement at a location is perpendicular to the radius *i.e.,* $\theta = 90°$

As work done = $\vec{F}.\vec{S} = FS\cos 90° = 0$

59. (b) Centripetal acc. = $\omega^2 r = 4\pi^2 v^2 r$

$= 4 \times (3.14)^2 \times \left(\frac{120}{60}\right)^2 \times \frac{30}{100} = 47.4$ ms^{-2}

$[\because \omega = 2\pi v]$

4 Laws of Motion

TREND ANALYSIS

TOPIC	2025		2024		2023		2022		2021		2020	
	Qns.	LOD	Qns.	LOD	Qns.	LOD	Qns.	LOD	Qns.	LOD	Qns.	LOD
Newton's Second Law of Motion	1	Avg.			1	Avg.			1	Avg.		
Conservation of Momentum							1	Avg.				
Common Forces in Mechanics	2	Avg.			1	Avg.						
Circular Motion			1	Avg.								
Solving Problems in Mechanics			1	Easy							1	Avg.

How to Code Qns for Better & Faster Revision

Type 1: Able to solve (No code Req)
Type 2: Solved with struggle or by seeing solution **(Use code NR means Needs Revision or orange stickers)**
Type 3: Tricky/ Could not solve on my own. Need help from Teacher/ Friend. **(Use code TQ means Tricky Question or red stickers)**

You can either hand write the code in boxes or else paste coloured stickers given at end of book.

Topic-4.4 Newton's First Law of Motion

1. Physical independence of force is a consequence of **NCERT | Page-51, 53, 56 | 1991**
(a) third law of motion
(b) second law of motion
(c) first law of motion
(d) all of these laws

Topic-4.5 Newton's Second Law of Motion

2. A ball of mass 0.5 kg is dropped from a height of 40 m. The ball hits the ground and rises to a height of 10 m. The impulse imparted to the ball during its collision with the ground is (Take $g = 9.8\ m/s^2$) **NCERT | Page-55 | 2025**
(a) 21 NS (b) 7 NS
(c) 0 (d) 84 NS

3. A 1 kg object strikes a wall with velocity 1 ms^{-1} at an angle of 60° with the wall and reflects at the same angle. If it remains in contact with wall for 0.1 s, then the force exerted on the wall is **NCERT | Page-54 | Manipur 2023**
(a) $10\sqrt{3}N$ (b) $20\sqrt{3}N$
(c) $30\sqrt{3}N$ (d) Zero

4. A football player is moving southward and suddenly turns eastward with the same speed to avoid an opponent. The force that acts on the player while turning is **NCERT | Page-57 | 2023**
(a) Along eastward (b) Along northward
(c) Along north-east (d) Along south-west

5. A ball of mass 0.15 kg is dropped from a height 10 m, strikes the ground and rebounds to the same height. The magnitude of impulse imparted to the ball is ($g = 10\ m/s^2$) nearly **NCERT | Page-55 | 2021**
(a) 1.4 kg m/s (b) 0 kg m/s
(c) 4.2 kg m/s (d) 2.1 kg m/s

6. A particle moving with velocity is acted by three forces shown by the vector triangle PQR. The velocity of the particle will: **NCERT | Page-55 | 2019**

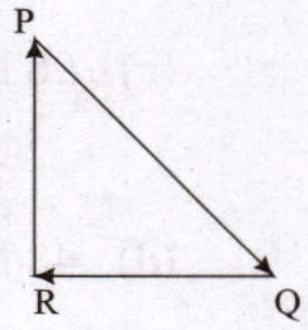

(a) increase
(b) decrease
(c) remain constant
(d) change according to the smallest force

7. The force 'F' acting on a particle of mass 'm' is indicated by the force-time graph shown below. The change in momentum of the particle over the time interval from zero to 8 s is : **NCERT | Page-54 | 2014**

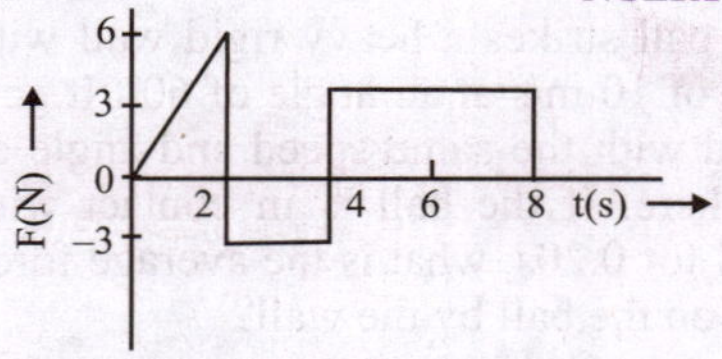

(a) 24 Ns (b) 20 Ns
(c) 12 Ns (d) 6 Ns

8. A stone is dropped from a height *h*. It hits the ground with a certain momentum *P*. If the same stone is dropped from a height 100% more than the previous height, the momentum when it hits the ground will change by : **NCERT | Page-57 | 2012 M**
(a) 68% (b) 41%
(c) 200% (d) 100%

9. A body of mass M hits normally a rigid wall with velocity V and bounces back with the same velocity. The impulse experienced by the body is **NCERT | Page-55 | 2011**
(a) MV (b) 1.5 MV
(c) 2 MV (d) zero

10. A body under the action of a force $\vec{F} = 6\hat{i} - 8\hat{j} + 10\hat{k}$, acquires an acceleration of 1 m/s^2. The mass of this body must be **NCERT | Page-54 | 2009**
(a) 10 kg (b) 20 kg
(c) $10\sqrt{2}$ kg (d) $2\sqrt{10}$ kg

11. Sand is being dropped on a conveyor belt at the rate of M kg/s. The force necessary to keep the belt moving with a constant velocity of v m/s will be: **NCERT | Page-57 | 2008**
(a) Mv newton (b) 2 Mv newton
(c) $\frac{Mv}{2}$ newton (d) zero

12. A 0.5 kg ball moving with speed of 12 m/s strikes a hard wall at an angle of 30° with the wall. It is reflected with the same speed and at the same angle. If the ball is in contact with the wall for 0.25 seconds, the average force acting on the wall is **NCERT | Page-57 | 2006**

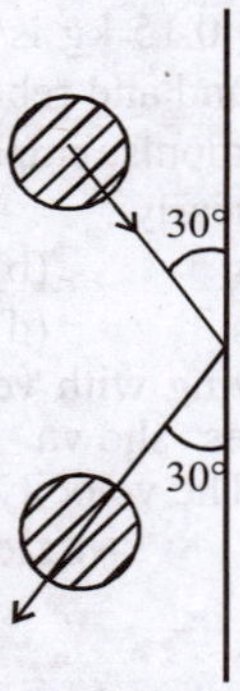

(a) 24 N
(b) 12 N
(c) 96 N
(d) 48 N

13. If *a* cricketer catches a ball of mass 150 gm moving with a velocity of 20 m/s, then he experiences a force of (Time taken to complete the catch is 0.1 sec.) **NCERT | Page-54 | 2001**
(a) 300 N (b) 30 N
(c) 3 N (d) 0.3 N

14. A 3 kg ball strikes a heavy rigid wall with a speed of 10 m/s at an angle of 60°. It gets reflected with the same speed and angle as shown here. If the ball is in contact with the wall for 0.20s, what is the average force exerted on the ball by the wall?

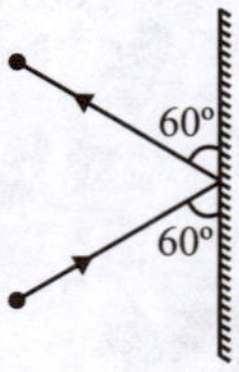

NCERT | Page-57 | 2000
(a) 150 N (b) zero
(c) $150\sqrt{3}N$ (d) 300 N

15. A bullet is fired from a gun. The force on the bullet is given by $F = 600 - 2 \times 10^5\, t$ where, F is in newton and t in second. The force on the bullet becomes zero as soon as it leaves the barrel. What is the average impulse imparted to the bullet? **NCERT | Page-55 | 1998**
(a) 1.8 N-s (b) zero
(c) 9 N-s (d) 0.9 N-s

16. A 5000 kg rocket is set for vertical firing. The exhaust speed is 800 ms^{-1}. To give an initial upward acceleration of 20 ms^{-2}, the amount of gas ejected per second to supply the needed thrust will be ($g = 10$ ms^{-2}) **1998**
(a) 127.5 kg s^{-1} (b) 187.5 kg s^{-1}
(c) 185.5 kg s^{-1} (d) 137.5 kg s^{-1}

17. A 10 N force is applied on a body produces an acceleration of 1 m/s^2. The mass of the body is **NCERT | Page-55 | 1996**
(a) 5 kg (b) 10 kg
(c) 15 kg (d) 20 kg

18. A ball of mass 150 g, moving with an acceleration 20 m/s^2, is hit by a force, which acts on it for 0.1 sec. The impulsive force is **NCERT | Page-56 | 1996**
(a) 0.5 N (b) 0.1 N
(c) 0.3 N (d) 1.2 N

19. If the force on a rocket moving with a velocity of 300 m/sec is 345 N, then the rate of combustion of the fuel, is **1995**
(a) 0.55 kg/sec (b) 0.75 kg/sec
(c) 1.15 kg/sec (d) 2.25 kg/sec

20. A satellite in a force free space sweeps stationary interplanetary dust at a rate (dM/dt) = αv. The acceleration of satellite is **1994**
(a) $\frac{-2\alpha v^2}{M}$ (b) $\frac{-\alpha v^2}{M}$
(c) $\frac{-\alpha v^2}{2M}$ (d) $-\alpha v^2$

21. A 600 kg rocket is set for a vertical firing. If the exhaust speed is 1000 ms^{-1}, the mass of the gas ejected per second to supply the thrust needed to overcome the weight of rocket is **NCERT | Page-55 | 1990**
(a) 117.6 kg s^{-1} (b) 58.6 kg s^{-1}
(c) 6 kg s^{-1} (d) 76.4 kg s^{-1}

22. A particle of mass m is moving with a uniform velocity v_1. It is given an impulse such that its velocity becomes v_2. The impulse is equal to **NCERT | Page-55 | 1990**
(a) $m[|v_2| - |v_1|]$ (b) $\frac{1}{2}m[v_2^2 - v_1^2]$
(c) $m[v_1 + v_2]$ (d) $m[v_2 - v_1]$

Topic-4.7
Conservation of Momentum

23. A shell of mass m is at rest initially. It explodes into three fragments having mass in the ratio 2 : 2 : 1. If the fragments having equal mass fly off along mutually perpendicular directions with speed v, the speed of the third (lighter) fragment is **NCERT | Page-84 | 2022**
(a) $\sqrt{2}$ v (b) $2\sqrt{2}$ v
(c) $3\sqrt{2}$ v (d) v

24. An object flying in air with velocity $(20\hat{i}+25\hat{j}-12\hat{k})$ suddenly breaks into two pieces whose masses are in the ratio 1 : 5. The smaller mass flies off with a velocity $(100\hat{i}+35\hat{j}+8\hat{k})$. The velocity of the larger piece will be NCERT | Page-84 | Odisha 2019

(a) $-20\hat{i}-15\hat{j}-80\hat{k}$ (b) $4\hat{i}+23\hat{j}-16\hat{k}$
(c) $-100\hat{i}-35\hat{j}-8\hat{k}$ (d) $20\hat{i}+15\hat{j}-80\hat{k}$

25. An explosion breaks a rock into three parts in a horizontal plane. Two of them go off at right angles to each other. The first part of mass 1 kg moves with a speed of 12 ms^{-1} and the second part of mass 2 kg moves with speed 8 ms^{-1}. If the third part flies off with speed 4 ms^{-1} then its mass is NCERT | Page-84 | 2013

(a) 5 kg (b) 7 kg (c) 17 kg (d) 3 kg

26. An explosion blows a rock into three parts. Two parts go off at right angles to each other. These two are, 1 kg first part moving with a velocity of 12 ms^{-1} and 2 kg second part moving with a velocity of 8 ms^{-1}. If the third part flies off with a velocity of 4 ms^{-1}, its mass would be: NCERT | Page-84 | 2009

(a) 7 kg (b) 17 kg (c) 3 kg (d) 5 kg

27. A body of mass 5 kg explodes at rest into three fragments with masses in the ratio 1 : 1 : 3. The fragments with equal masses fly in mutually perpendicular directions with speeds of 21 m/s. The velocity of heaviest fragment in m/s will be NCERT | Page-84 | 1989

(a) $7\sqrt{2}$ (b) $5\sqrt{2}$ (c) $3\sqrt{2}$ (d) $\sqrt{2}$

Topic-4.8 Equilibrium of a Particle

28. Three forces acting on a body are shown in the figure. To have the resultant force only along the y-direction, the magnitude of the minimum additional force needed is: NCERT | Page-58 | 2008

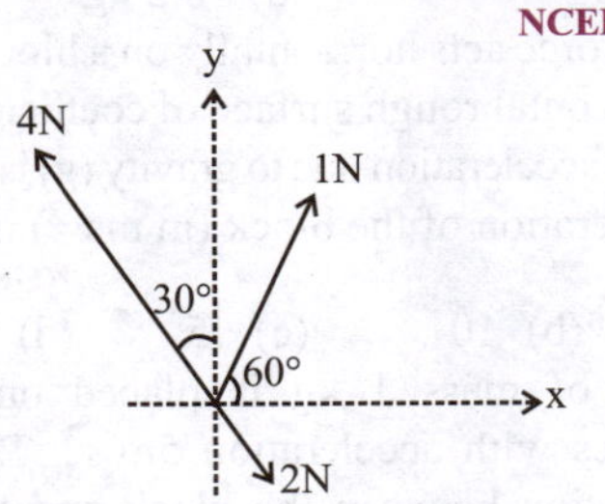

(a) 0.5 N (b) 1.5 N
(c) $\frac{\sqrt{3}}{4}$ N (d) $\sqrt{3}$N

Topic-4.9 Common Forces in Mechanics

29. There are two inclined surface of equal length (L) and same angle of inclination 45° with the horizontal. One of them is rough and the other is perfectly smooth. A given body takes 2 times as much time to slide down on rough surface than on the smooth surface. The coefficient of kinetic friction (μ_k) between the object and the rough surface is close to: NCERT | Page-60, 61 | 2025

(a) 0.25 (b) 0.40
(c) 0.5 (d) 0.75

30. A uniform rod of mass 20 kg and length 5 m leans against a smooth vertical wall making an angle of 60° with it. The other end rests on a rough horizontal floor. The friction force that the floor exerts on the rod is (take g = 10 m/s^2) NCERT | Page-60, 61 | 2025

(a) 100 N (b) $100\sqrt{3}$ N
(c) 200 N (d) $200\sqrt{3}$ N

31. A block of mass 2 kg is placed on an inclined rough surface AC (as shown in figure) of coefficient of friction μ. If g = 10 ms^{-2}, the net force (in N) on the block will be NCERT | Page-59 | Manipur 2023

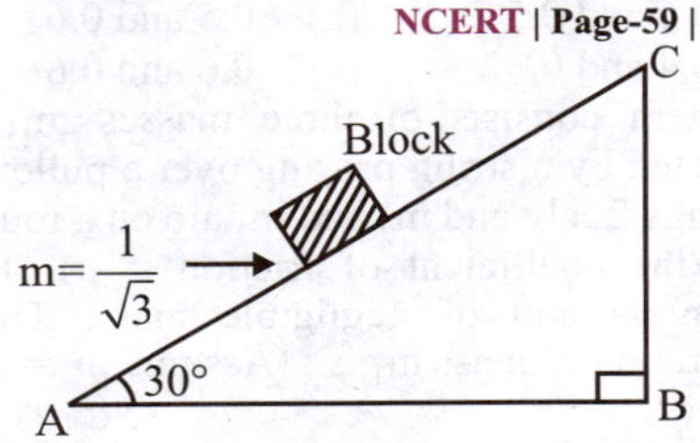

(a) 10 (b) 20
(c) $10\sqrt{3}$ (d) Zero

32. Calculate the maximum acceleration of a moving car so that a body lying on the floor of the car remains stationary. The coefficient of static friction between the body and the floor is 0.15 (g = 10 m s^{-2}). NCERT | Page-60 | 2023

(a) 1.2 m s^{-2} (b) 150 m s^{-2}
(c) 1.5 m s^{-2} (d) 50 m s^{-2}

33. A body of mass m is kept on a rough horizontal surface (coefficient of friction = μ). A horizontal force is applied on the body, but it does not move. The resultant of normal reaction and the frictional force acting on the object is given by F, where F is, NCERT | Page-61 | Odisha 2019

(a) $|\vec{F}| = mg$ (b) $|\vec{F}| = mg + \mu mg$
(c) $|\vec{F}| = \mu mg$ (d) $|\vec{F}| \le mg\sqrt{1+\mu^2}$

34. Which one of the following statements is incorrect? NCERT | Page-60 | 2018

(a) Rolling friction is smaller than sliding friction.
(b) Limiting value of static friction is directly proportional to normal reaction.
(c) Coefficient of sliding friction has dimensions of length.
(d) Frictional force opposes the relative motion.

35. A block A of mass m_1 rests on a horizontal table. A light string connected to it passes over a frictionless pulley at the edge of table and from its other end another block B of mass m_2 is suspended. The coefficient of kinetic friction between the block and the table is μ_k. When the block A is sliding on the table, the tension in the string is NCERT | Page-61 | 2015

(a) $\frac{(m_2 - \mu k m_1)g}{(m_1 + m_2)}$ (b) $\frac{m_1 m_2 (1 + \mu_k)g}{(m_1 + m_2)}$

(c) $\frac{m_1 m_2 (1 - \mu_k)g}{(m_1 + m_2)}$ (d) $\frac{(m_2 + \mu_k m_1)g}{(m_1 + m_2)}$

36. A plank with a box on it at one end is gradually raised about the other end. As the angle of inclination with the horizontal reaches 30º the box starts to slip and slides 4.0 m down the plank in 4.0s. The coefficients of static and kinetic friction between the box and the plank will be, respectively : NCERT | Page-61 | 2015 RS

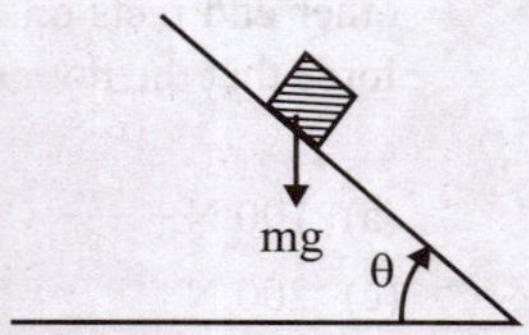

(a) 0.6 and 0.5 (b) 0.5 and 0.6
(c) 0.4 and 0.3 (d) 0.6 and 0.6

37. A system consists of three masses m_1, m_2 and m_3 connected by a string passing over a pulley P. The mass m_1 hangs freely and m_2 and m_3 are on a rough horizontal table (the coefficient of friction = μ). The pulley is frictionless and of negligible mass. The downward acceleration of mass m_1 is : (Assume $m_1 = m_2 = m_3 = m$) NCERT | Page-65 | 2014

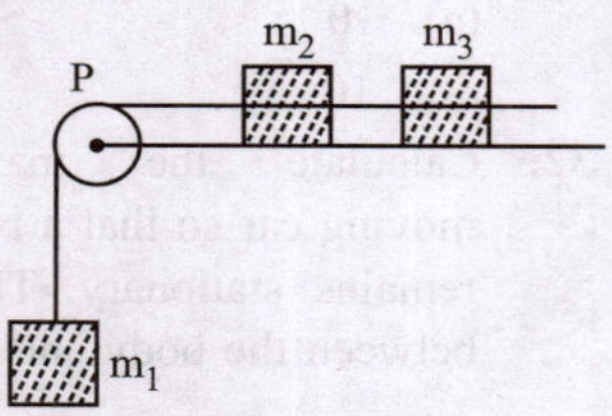

(a) $\frac{g(1 - g\mu)}{g}$

(b) $\frac{2g\mu}{3}$

(c) $\frac{g(1 - 2\mu)}{3}$

(d) $\frac{g(1 - 2\mu)}{2}$

38. The upper half of an inclined plane of inclination θ is perfectly smooth while lower half is rough. A block starting from rest at the top of the plane will again come to rest at the bottom, if the coefficient of friction between the block and lower half of the plane is given by NCERT | Page-61 | 2011

(a) $\mu = \frac{2}{\tan\theta}$ (b) $\mu = 2\tan\theta$

(c) $\mu = \tan\theta$ (d) $\mu = \frac{1}{\tan\theta}$

39. A conveyor belt is moving at a constant speed of 2m/s. A box is gently dropped on it. The coefficient of friction between them is μ = 0.5. The distance that the box will move relative to belt before coming to rest on it taking $g = 10\ ms^{-2}$, is NCERT | Page-60 | 2011 M

(a) 1.2 m (b) 0.6 m
(c) zero (d) 0.4 m

40. A block of mass m is in contact with the cart C as shown in the Figure. NCERT | Page-60 | 2010

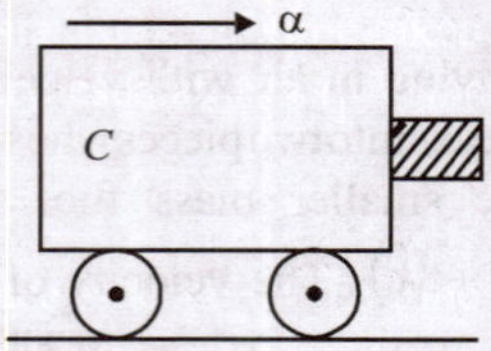

The coefficient of static friction between the block and the cart is μ. The acceleration α of the cart that will prevent the block from falling satisfies:

(a) $\alpha > \frac{mg}{\mu}$ (b) $\alpha > \frac{g}{\mu m}$

(c) $\alpha \geq \frac{g}{\mu}$ (d) $\alpha < \frac{g}{\mu}$

41. A block B is pushed momentarily along a horizontal surface with an initial velocity V. If μ is the coefficient of sliding friction between B and the surface, block B will come to rest after a time NCERT | Page-60 | 2007

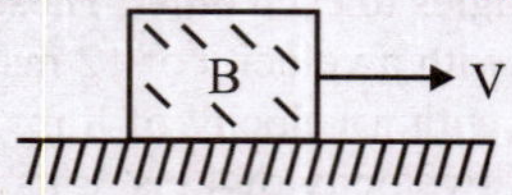

(a) gμ/V (b) g/V
(c) V/g (d) V/(gμ)

42. The coefficient of static friction, μ_s, between block A of mass 2 kg and the table as shown in the figure is 0.2. What would be the maximum mass value of block B so that the two blocks do not move? The string and the pulley are assumed to be smooth and massless. ($g = 10\ m/s^2$) NCERT | Page-65 | 2004

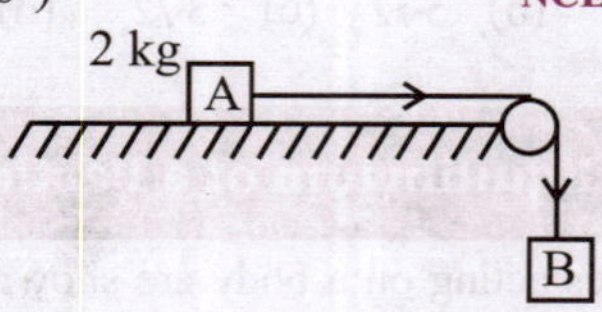

(a) 0.4 kg (b) 2.0 kg
(c) 4.0 kg (d) 0.2 kg

43. A 100 N force acts horizontally on a block of 10 kg placed on a horizontal rough surface of coefficient of friction μ = 0.5. If the acceleration due to gravity (g) is taken as $10\ ms^{-2}$, the acceleration of the block (in ms^{-2}) is NCERT | Page-61 | 2002

(a) 2.5 (b) 10 (c) 5 (d) 7.5

44. A block of mass 1 kg is placed on a truck which accelerates with acceleration $5 m/s^2$. The coefficient of static friction between the block and truck is 0.6. The frictional force acting on the block is NCERT | Page-61 | 2001

(a) 5 N (b) 6 N
(c) 5.88 N (d) 4.6 N

45. A block has been placed on an inclined plane with the slope angle θ, block slides down the plane at constant speed. The coefficient of kinetic friction is equal to NCERT | Page-61 | 1993

(a) sin θ (b) cos θ
(c) g (d) tan θ

46. Consider a car moving along a straight horizontal road with a speed of 72 km/h. If the coefficient of static friction between the tyres and the road is 0.5, the shortest distance in which the car can be stopped is (taking g = 10 m/s^2) **NCERT | Page-61 | 1992**

(a) 30 m (b) 40 m
(c) 72 m (d) 20 m

47. A heavy uniform chain lies on horizontal table top. If the coefficient of friction between the chain and the table surface is 0.25, then the maximum fraction of the length of the chain that can hang over one edge of the table is **NCERT | Page-61 | 1991**

(a) 20% (b) 25% (c) 35% (d) 15%

48. Starting from rest, a body slides down a 45° inclined plane in twice the time it takes to slide down the same distance in the absence of friction. The coefficient of friction between the body and the inclined plane is **NCERT | Page-61 | 1988**

(a) 0.80 (b) 0.75 (c) 0.25 (d) 0.33

Topic-4.10 Circular Motion

49. A bob is whirled in a horizontal circle by means of a string at an initial speed of 10 rpm. If the tension in the string is quadrupled while keeping the radius constant, the new speed is: **NCERT | Page-63 | Ph-II 2024**

(a) 20 rpm (b) 40 rpm
(c) 5 rpm (d) 10 rpm

50. A bob is whirled in a horizontal plane by means of a string with an initial speed of ω rpm. The tension in the string is T. If speed becomes 2ω while keeping the same radius, the tension in the string becomes: **NCERT | Page-63 | 2024**

(a) T (b) 4T (c) $\frac{T}{4}$ (d) $\sqrt{2}T$

51. A block of mass 10 kg is in contact against the inner wall of a hollow cylindrical drum of radius 1 m. The coefficient of friction between the block and the inner wall of the cylinder is 0.1. The minimum angular velocity needed for the cylinder to keep the block stationary when the cylinder is vertical and rotating about its axis, will be : (g = 10 m/s^2) **NCERT | Page-63 | 2019**

(a) $\sqrt{10}$ rad/s (b) $\frac{10}{2\pi}$ rad/s
(c) 10 rad/s (d) 10 π rad/s

52. One end of string of length l is connected to a particle of mass 'm' and the other end is connected to a small peg on a smooth horizontal table. If the particle moves in circle with speed 'v' the net force on the particle (directed towards centre) will be (T represents the tension in the string) : **NCERT | Page-79 | 2017**

(a) $T+\frac{mv^2}{l}$ (b) $T-\frac{mv^2}{l}$
(c) Zero (d) T

53. A car is negotiating a curved road of radius R. The road is banked at an angle θ, the coefficient of friction between the tyres of the car and the road is μ_s. The maximum safe velocity on this road is : **NCERT | Page-63 | 2016**

(a) $\sqrt{gR^2\left(\frac{\mu_s+\tan\theta}{1-\mu_s\tan\theta}\right)}$

(b) $\sqrt{gR\left(\frac{\mu_s+\tan\theta}{1-\mu_s\tan\theta}\right)}$

(c) $\sqrt{\frac{g}{R}\left(\frac{\mu_s+\tan\theta}{1-\mu_s\tan\theta}\right)}$

(d) $\sqrt{\frac{g}{R^2}\left(\frac{\mu_s+\tan\theta}{1-\mu_s\tan\theta}\right)}$

54. Two stones of masses m and 2 m are whirled in horizontal circles, the heavier one in radius $\frac{r}{2}$ and the lighter one in radius r. The tangential speed of lighter stone is n times that of the value of heavier stone when they experience same centripetal forces. The value of n is : **NCERT | Page-63 | 2015 RS**

(a) 3 (b) 4 (c) 1 (d) 2

55. A car is moving in a circular horizontal track of radius 10 m with a constant speed of 10 m/s. A bob is suspended from the roof of the car by a light wire of length 1.0 m. The angle made by the wire with the vertical is **NCERT | Page-63 | Kar. 2013**

(a) 0° (b) $\frac{\pi}{3}$ (c) $\frac{\pi}{6}$ (d) $\frac{\pi}{4}$

56. A car of mass 1000 kg negotiates a banked curve of radius 90 m on a frictionless road. If the banking angle is 45°, the speed of the car is : **NCERT | Page-63 | 2012**

(a) 20 ms^{-1} (b) 30 ms^{-1}
(c) 5 ms^{-1} (d) 10 ms^{-1}

57. A car of mass m is moving on a level circular track of radius R. If μ_s represents the static friction between the road and tyres of the car, the maximum speed of the car in circular motion is given by : **NCERT | Page-63 | 2012 M**

(a) $\sqrt{\mu_s mRg}$ (b) $\sqrt{Rg/\mu_s}$
(c) $\sqrt{mRg/\mu_s}$ (d) $\sqrt{\mu_s Rg}$

58. A 500 kg car takes a round turn of radius 50 m with a velocity of 36 km/h. The centripetal force is **NCERT | Page-63 | 1999**

(a) 250 N (b) 750 N
(c) 1000 N (d) 1200 N

59. What will be the maximum speed of a car on a road turn of radius 30 m if the coefficient of friction between the tyres and the road is 0.4 (Take g = 9.8 m/s^2) **NCERT | Page-63 | 1995**

(a) 10.84 m/s (b) 9.84 m/s
(c) 8.84 m/s (d) 6.84 m/s

60. A particle of mass M is moving in a horizontal circle of radius R with uniform speed V. When it moves from one point to a diametrically opposite point, its
NCERT | Page-63 | 1992
(a) kinetic energy changes by $MV^2/4$
(b) momentum does not change
(c) momentum changes by 2 MV
(d) kinetic energy changes by MV^2

61. When milk is churned, cream gets separated due to
NCERT | Page-63 | 1991
(a) centripetal force (b) centrifugal force
(c) frictional force (d) gravitational force

Topic-4.11
Solving Problems in Mechanics

62. A box of mass 5 kg is pulled by a cord, up along a frictionless plane inclined at 30° with the horizontal. The tension in the cord is 30 N. The acceleration of the box is (Take $g = 10 \text{ m s}^{-2}$)
NCERT | Page-65 | Ph-II 2024
(a) 2 m s^{-2} (b) Zero
(c) 0.1 m s^{-2} (d) 1 m s^{-2}

63. A horizontal force 10 N is applied to a block A as shown in figure. The mass of blocks A and B are 2 kg and 3 kg respectively. The blocks slide over a frictionless surface. The force exerted by block A on block B is :
NCERT | Page-65 | 2024

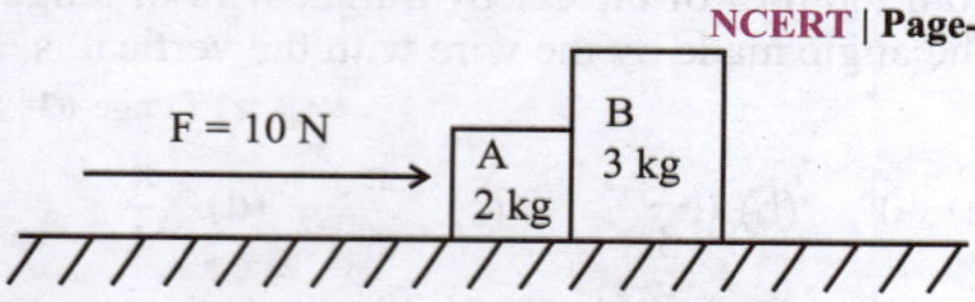

(a) Zero (b) 4 N
(c) 6 N (d) 10 N

64. In the diagram shown, the normal reaction force between 2 kg and 1 kg is (Consider the surface, to be smooth) : Given $g=10 \text{ ms}^{-2}$
NCERT | Page-61 | RS 2022

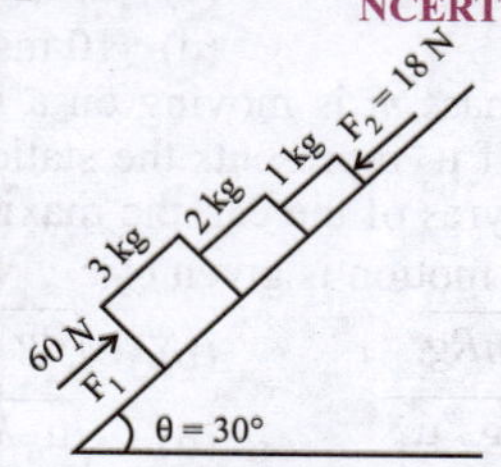

(a) 10 N (b) 25 N (c) 39 N (d) 6N

65. Two bodies of mass 4 kg and 6 kg are tied to the ends of a massless string. The string passes over a pulley which is frictionless (see figure). The acceleration of the system in terms of acceleration due to gravity (g) is:
NCERT | Page-65 | 2020
(a) g/2
(b) g/5
(c) g/10
(d) g

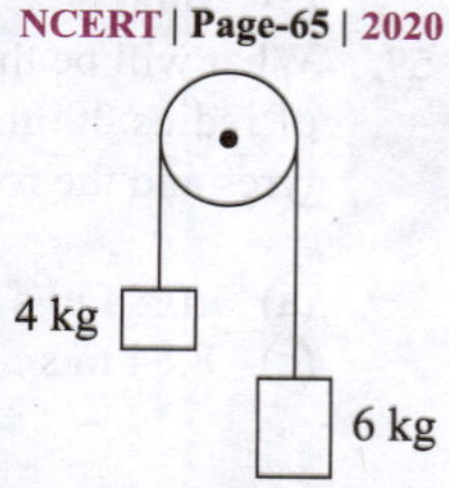

66. A block of mass m is placed on a smooth inclined wedge ABC of inclination θ as shown in the figure. The wedge is given an acceleration 'a' towards the right. The relation between a and θ for the block to remain stationary on the wedge is
NCERT | Page-61 | 2018

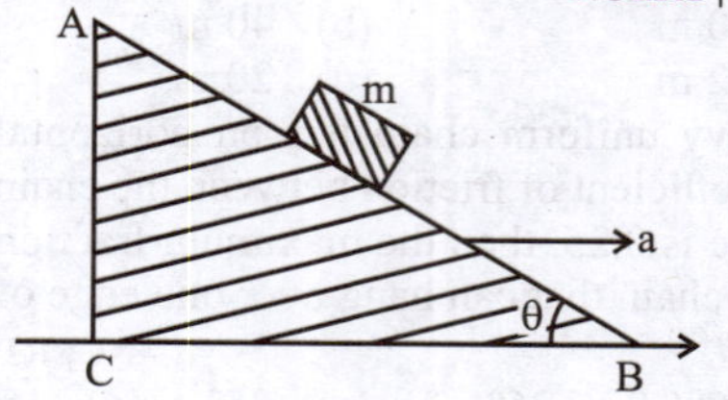

(a) $a = \frac{g}{\operatorname{cosec}\theta}$ (b) $a = \frac{g}{\sin\theta}$
(c) $a = g \tan\theta$ (d) $a = g\cos\theta$

67. Two blocks A and B of masses 3 m and m respectively are connected by a massless and inextensible string. The whole system is suspended by a massless spring as shown in figure. The magnitudes of acceleration of A and B immediately after the string is cut, are respectively :
NCERT | Page-59 | 2017
(a) $\frac{g}{3}, g$
(b) g, g
(c) $\frac{g}{3}, \frac{g}{3}$
(d) $g, \frac{g}{3}$

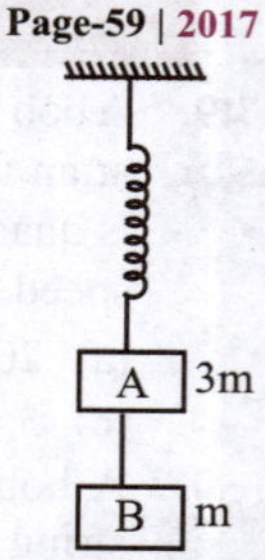

68. Three blocks A, B and C of masses 4 kg, 2 kg and 1 kg respectively, are in contact on a frictionless surface, as shown. If a force of 14 N is applied on the 4 kg block then the contact force between A and B is
NCERT | Page-65 | 2015

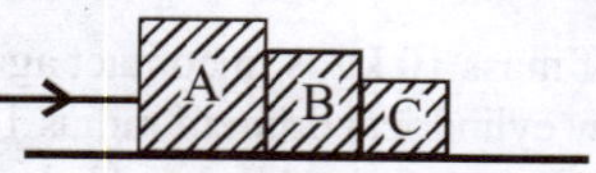

(a) 6 N (b) 8 N
(c) 18 N (d) 2 N

69. A balloon with mass 'm' is descending down with an acceleration 'a' (where a < g). How much mass should be removed from it so that it starts moving up with an acceleration 'a'?
NCERT | Page-62 | 2014
(a) $\frac{2ma}{g+a}$ (b) $\frac{2ma}{g-a}$
(c) $\frac{ma}{g+a}$ (d) $\frac{ma}{g-a}$

70. Three blocks with masses m, 2 m and 3 m are connected by strings as shown in the figure. After an upward force F is applied on block m, the masses move upward at constant speed v. What is the net force on the block of mass 2m? (g is the acceleration due to gravity)
NCERT | Page-65 | 2013

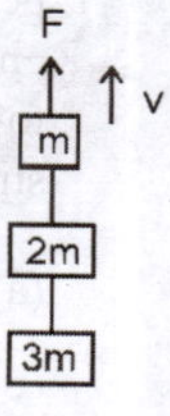

(a) 2 mg (b) 3 mg
(c) 6 mg (d) zero

71. A person of mass 60 kg is inside a lift of mass 940 kg and presses the button on control panel. The lift starts moving upwards with an acceleration 1.0 m/s^2. If g = 10 ms^{-2}, the tension in the supporting cable is
NCERT | Page-58 | 2011

(a) 8600 N (b) 9680 N
(c) 11000 N (d) 1200 N

72. The mass of a lift is 2000 kg. When the tension in the supporting cable is 28000 N, then its acceleration is:
NCERT | Page-58 | 2009

(a) 4 ms^{-2} upwards (b) 4 ms^{-2} downwards
(c) 14 ms^{-2} upwards (d) 30 ms^{-2} downwards

73. A block of mass m is placed on a smooth wedge of inclination θ. The whole system is accelerated horizontally so that the block does not slip on the wedge. The force exerted by the wedge on the block (g is acceleration due to gravity) will be **NCERT | Page-58 | 2004**

(a) mg/cos θ (b) mg cos θ
(c) mg sin θ (d) mg

74. A man weighing 80 kg, stands on a weighing scale in a lift which is moving upwards with a uniform acceleration of 5m/s^2. What would be the reading on the scale? (g = 10 m/s^2)
NCERT | Page-58 | 2003

(a) 1200 N (b) zero
(c) 400 N (d) 800 N

75. A monkey of mass 20 kg is holding a vertical rope. The rope will not break when a mass of 25 kg is suspended from it but will break if the mass exceeds 25 kg. What is the maximum acceleration with which the monkey can climb up along the rope? (g = 10 m/s^2)
NCERT | Page-58 | 2003

(a) 2.5 m/s^2 (b) 5 m/s^2
(c) 10 m/s^2 (d) 25 m/s^2

76. A lift weighing 1000 kg is moving upwards with an accelertion of 1 m/s^2. The tension in the supporting cable is
NCERT | Page-58 | 2002

(a) 980 N (b) 10800 N
(c) 9800 N (d) 8800 N

77. Two blocks m_1 = 5 gm and m_2 = 10 gm are hung vertically over a light frictionless pulley as shown here. What is the acceleration of the masses when they are left free?

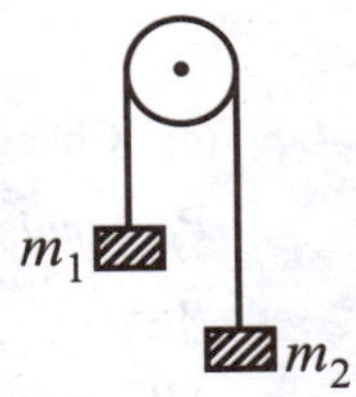

NCERT | Page-59 | 2000

(a) g / 3 (b) g / 2
(c) g (d) g / 5

(where g is acceleration due to gravity)

78. A mass of 1 kg is suspended by a thread. It is
(i) lifted up with an acceleration 4.9 m/s^2,
(ii) lowered with an acceleration 4.9 m/s^2.
The ratio of the tensions is **NCERT | Page-58 | 1998**

(a) 3 : 1 (b) 1 : 2
(c) 1 : 3 (d) 2 : 1

79. A monkey is decending from the branch of a tree with constant acceleration. If the breaking strength is 75% of the weight of the monkey, the minimum acceleration with which monkey can slide down without breaking the branch is
NCERT | Page-58 | 1993

(a) g (b) $\frac{3g}{4}$ (c) $\frac{g}{4}$ (d) $\frac{g}{2}$

ANSWER KEY

1	(c)	9	(c)	17	(b)	25	(a)	33	(d)	41	(d)	49	(a)	57	(d)	65	(b)	73	(a)
2	(a)	10	(c)	18	(c)	26	(d)	34	(c)	42	(a)	50	(b)	58	(c)	66	(c)	74	(a)
3	(a)	11	(a)	19	(c)	27	(a)	35	(b)	43	(c)	51	(c)	59	(a)	67	(a)	75	(a)
4	(c)	12	(a)	20	(b)	28	(a)	36	(a)	44	(a)	52	(d)	60	(c)	68	(a)	76	(b)
5	(c)	13	(b)	21	(c)	29	(d)	37	(c)	45	(d)	53	(b)	61	(b)	69	(a)	77	(a)
6	(c)	14	(c)	22	(d)	30	(b)	38	(b)	46	(b)	54	(d)	62	(d)	70	(d)	78	(a)
7	(c)	15	(d)	23	(b)	31	(d)	39	(d)	47	(a)	55	(d)	63	(c)	71	(c)	79	(c)
8	(b)	16	(b)	24	(b)	32	(c)	40	(c)	48	(b)	56	(b)	64	(b)	72	(a)		

Hints & Solutions

1. **(c)** Newton's first law of motion is related to physical independence of force.

NOTES If no net force acts on a body, then the velocity of the body cannot change i.e., the body cannot accelerate.

2. **(a)** Initial velocity,

$$v_1 = -\sqrt{2gh_1} = -\sqrt{2\times9.8\times10} = -28 \text{ m/s}$$

Final velocity,

$$v_2 = \sqrt{2gh_2} = \sqrt{2\times9.8\times10} = 14 \text{ m/s}$$

∴ Impulse, I = m ($v_2 - v_1$) = 0.5 [14 – (–28)]
= 0.5 × 42 = 21 Ns

3. **(a)** Force is equal to rate of change of momentum

$$\therefore F = \frac{\Delta p}{\Delta t}$$

$\therefore \Delta p = 2\text{ mv}\cos 30°$

$= 2 \times 1 \times \frac{\sqrt{3}}{2} = \sqrt{3}$

$\therefore F = \frac{\Delta p}{\Delta t} = \frac{\sqrt{3}}{0.1} = 10\sqrt{3}\text{ N}$

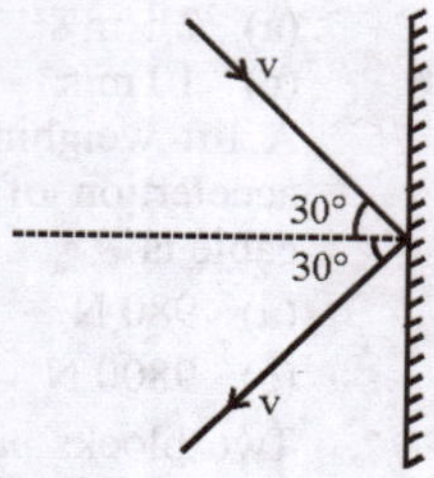

4. **(c)** Change in momentum, $\Delta\vec{P} = \vec{P}_f - \vec{P}_i$

$\vec{P}_f = mu\hat{i}$

$\vec{P}_i = mu(-\hat{j})$

$\Delta\vec{P} = mu\hat{i} - mu(-\hat{j})$

$\Delta\vec{P} = mu(\hat{i} + \hat{j})$

$\vec{F} = \frac{\Delta\vec{P}}{\Delta t}$

Direction of change of momentum and direction of force acting on the player will be same i.e., along north-east.

5. **(c)** Impulse, $\vec{I}$ = change in linear momentum

or, $I = \Delta\vec{P} = \vec{P_f} - \vec{P_i}$

or, $I = m_2v_2 - m_1v_1$

Velocity of ball at ground

$(v) = \sqrt{2gh}\ \left[\because v^2 - u^2 = 2gh\right]$

$= \sqrt{2 \times 10 \times 10} = 10\sqrt{2}\text{ m/s}$

$\therefore \vec{I} = 0.15 \times 10\sqrt{2}\left(-\hat{j}\right) - 0.15 \times 10\sqrt{2}\left(\hat{j}\right)$

$\vec{I} = 2 \times 0.15 \times 10\sqrt{2}\left(-\hat{j}\right) = -4.2\left(\hat{j}\right)$

$\therefore\ I = 4.2\text{ kg m/s}$

> **NOTES** Impulse $\vec{I} = \overrightarrow{\Delta P} = \vec{P}_f - \vec{P}_i$ = change in momentum is called impulse momentum theorem. Impulse of a force is a measure of total effect of force.

6. **(c)** As three forces are forming closed loop in same order, so net force is zero.

i.e., $\vec{F}_{net} = 0$

or $m\frac{d\vec{v}}{dt} = 0$

$\therefore$ Velocity of the particle

$\vec{v}$ = Constant

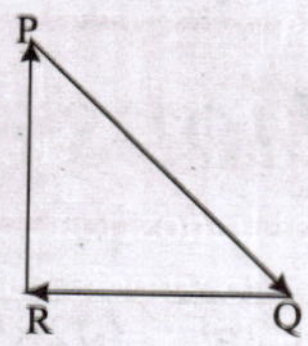

7. **(c)** Change in momentum,

$\Delta p = \int F dt$ = Area of F-t graph

= ar of Δ – *ar* of ▭ + *ar* of ▭

$= \frac{1}{2} \times 2 \times 6 - 3 \times 2 + 4 \times 3 = 12\text{ N-s}$

> **NOTES** According to Newton's 2nd law. $\vec{F} \propto \frac{d\vec{p}}{dt} \Rightarrow \vec{F} = K\frac{d\vec{p}}{dt}$ or, $\vec{F} = \frac{d\vec{p}}{dt}$ [∵ k = 1 in S.I & CGS system]
>
> And $d\vec{p} = \vec{F}dt$

8. **(b)** Momentum $P = mv = m\sqrt{2gh}$

$(\because v^2 = u^2 + 2gh;$ Here $u = 0)$

When stone hits the ground momentum,

$P = m\sqrt{2gh}$

when same stone dropped from $2h$ (100% of initial) then momentum,

$P' = m\sqrt{2g(2h)} = \sqrt{2}P$

$\therefore$ % change in momentum, $\frac{(P' - P)}{P} \times 100\%$

$\Rightarrow \left(\frac{\sqrt{2}P - P}{P}\right) \times 100\% = \left(\sqrt{2} - 1\right) \times 100\%$

$\Rightarrow (1.414 - 1) \times 100\%$

$\Rightarrow 0.414 \times 100\% = 41.4\%$

9. **(c)** Impulse experienced by the body

= change in momentum

= MV – (–MV) = 2MV

10. **(c)** $\vec{F} = 6\hat{i} - 8\hat{j} + 10\hat{k}$,

$|\vec{F}| = \sqrt{36 + 64 + 100} = 10\sqrt{2}\text{ N}$

$\left(\because F = \sqrt{F_x^2 + F_y^2 + F_z^2}\right)$

Given, $a = 1\text{ ms}^{-2}$

$\because F = ma \Rightarrow m = \frac{10\sqrt{2}}{1} = 10\sqrt{2}\text{ kg}$

11. **(a)** $F = \frac{d(Mv)}{dt} = M\frac{dv}{dt} + v\frac{dM}{dt}$

$\therefore v$ is constant,

$\therefore\ F = v\frac{dM}{dt}$ But $\frac{dM}{dt} = M\text{kg/s}$

$\therefore$ F = vM newton.

12. **(a)**

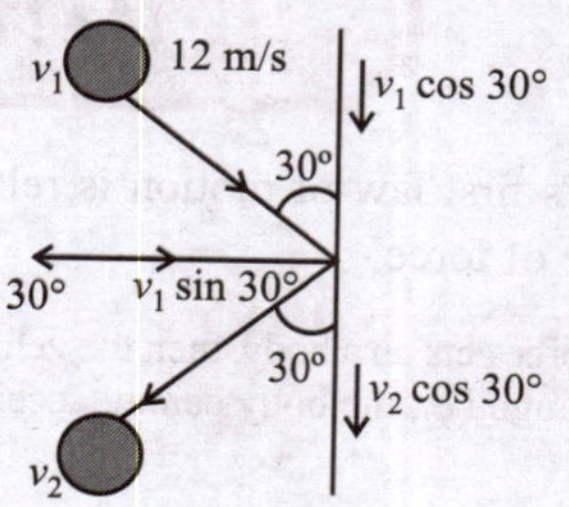

The magnitude of both velocity v_1 and v_2 are equal. Components of velocity along the wall are equal and are in same direction.

But the component of velocity perpendicular to wall are equal but opposite in direction.
So, net change in momentum in perpendicular direction,
$\Delta P = mv_1 \sin 30° - (-mv_2 \sin 30°)$
$\Rightarrow \quad 2mv \sin 30° \qquad$ [As $v_1 = v_2 = v$]
Average force acting on the wall will be given by

$$\Rightarrow \quad F = \frac{\Delta p}{t} = \frac{2mv \sin 30°}{t}$$

$$\Rightarrow \quad F = \frac{2 \times 0.5 \times 12 \times 1}{2 \times 0.25} = 24\text{N}$$

13. **(b)** Net force experienced $= \dfrac{\text{Total Impulse}}{\text{Time taken}}$

$$= \frac{m\Delta v}{t} = 0.15 \times \frac{20}{0.1} = 30N$$

14. **(c)** Change in momentum along the wall
$= mv \cos 60° - mv \cos 60° = 0$
Change in momentum perpendicular to the wall
$= mv \sin 60° - (-mv \sin 60°) = 2mv \sin 60°$

$$\therefore \text{Applied force} = \frac{\text{Change in momentum}}{\text{Time}}$$

$$= \frac{2\,mv \sin 60°}{0.20}$$

$$= \frac{2 \times 3 \times 10 \times \sqrt{3}}{2 \times 0.20} = 50 \times 3\sqrt{3}$$

$= 150\sqrt{3}$ newton

15. **(d)** Given $F = 600 - (2 \times 10^5 t)$
The force is zero at time t, given by
$0 = 600 - 2 \times 10^5\, t$

$$\Rightarrow \quad t = \frac{600}{2 \times 10^5} = 3 \times 10^{-3} \text{ seconds}$$

$$\therefore \text{ Impulse} = \int_0^t F dt = \int_0^{3 \times 10^{-3}} (600 - 2 \times 10^5 t)\, dt$$

$$= \left[600t - \frac{2 \times 10^5 t^2}{2}\right]_0^{3 \times 10^{-3}}$$

$= 600 \times 3 \times 10^{-3} - 10^5 (3 \times 10^{-3})^2$
$= 1.8 - 0.9 = 0.9$ Ns

16. **(b)** Given : Mass of rocket (m) = 5000 kg
Exhaust speed (v) = 800 m/s
Acceleration of rocket (a) = 20 m/s^2
Gravitational acceleration (g) = 10 m/s^2

$$\text{Thrust, F} = \frac{vdm}{dt}$$

We know that upward force,

$F = m(g + a) = 5000\,(10 + 20)$
$= 5000 \times 30 = 150000$ N

This thrust gives upward force, $F = \dfrac{vdm}{dt}$

We also know that amount of gas ejected

$$\left(\frac{dm}{dt}\right) = \frac{F}{v} = \frac{150000}{800} = 187.5 \text{ kg/s}$$

17. **(b)** By Newton's IInd law of motion, $F = ma$
$\Rightarrow 10 = m\,(1) \Rightarrow m = 10$ kg.

18. **(c)** Mass = 150 gm $= \dfrac{150}{1000}$ kg
Force = Mass × acceleration

$$= \frac{150}{1000} \times 20 = 3\,N$$

Impulsive force $= F{\cdot}\Delta t = 3 \times 0.1 = 0.3$ N

19. **(c)** Velocity of the rocket (u) = 300 m/s and force (F) = 345N. According to Newton's second law, thrust (force) = Rate of change of linear momentum.

$$\left(\frac{dm}{dt}\right) = \frac{F}{u} = 1.15 \text{kg / sec}$$

20. **(b)** Thrust on the satellite,

$$F = \frac{-vdM}{dt} = -v(\alpha v) = -\alpha v^2$$

$$\text{Acceleration} = \frac{F}{M} = \frac{-\alpha v^2}{M}$$

21. **(c)** Thrust = weight

$$\Rightarrow \quad \frac{udM}{dt} = mg \Rightarrow \frac{dM}{dt} = \frac{mg}{u}$$

$$= \frac{600 \times 10}{1000} = 6 \text{ kg s}^{-1}$$

22. **(d)** Impulse = final momentum – initial momentum = $m(v_2 - v_1)$

23. **(b)**

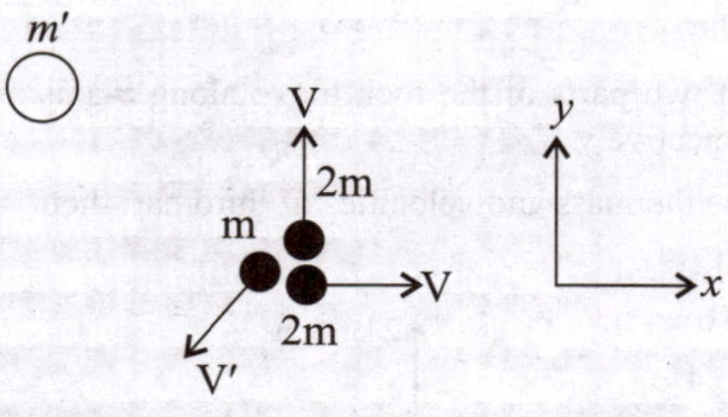

$\vec{P}_i = 0$

$$\vec{P}_f = 2mv\hat{i} + 2mv\hat{j} + m\vec{v}'$$

as, $\quad \vec{P}_i = \vec{P}_f$

$$0 = m\vec{v}' + 2mv\hat{i} + 2mv\hat{j}$$

$$\vec{v} = -2v(\hat{i} + \hat{j}) \Rightarrow |\vec{v}| = 2v\sqrt{1^2 + 1^2} = 2\sqrt{2}v$$

24. (b) $\vec{P}_i = \vec{P}_f$

$$m\vec{v}_i = \left(\frac{m}{6}\vec{v}_1 + \frac{5m}{6}\vec{v}_2\right)$$

$$\vec{v}_i = \left(\frac{\vec{v}_1}{6} + \frac{5}{6}\vec{v}_2\right)$$

$$20\hat{i} + 25\hat{j} - 12\hat{k} = \frac{(100\hat{i} + 35\hat{j} + 8\hat{k})}{6} + \frac{5\vec{v}_2}{6}$$

$$\vec{v}_2 = \frac{20\hat{i} + 115\hat{j} - 80\hat{k}}{5}$$

$$\therefore \quad \vec{v}_2 = 4\hat{i} + 23\hat{j} - 16\hat{k}$$

25. (a)

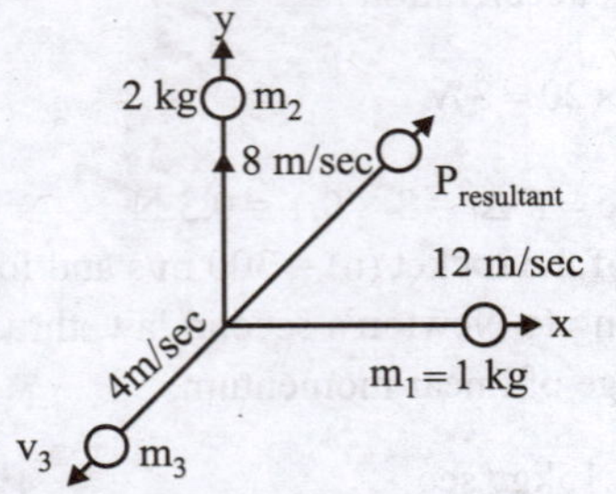

$$P_{resultant} = \sqrt{12^2 + 16^2} = \sqrt{144 + 256} = 20$$

$$m_3 v_3 = 20\ P_{resultant}$$

$$\text{or, } m_3 = \frac{20}{4} = 5 \text{ kg}$$

26. (d) Momentum of first part = 1 × 12 = 12

Momentum of second part = 2 × 8 = 16

Resultant momentum,

$$= \sqrt{(12)^2 + (16)^2} = 20$$

Then M_1 be the mass of third part,

$$M \cdot 4 = 20$$

$$M = \frac{20}{4} = 5 \text{ kg}$$

NOTES Let two parts of the rock move along *x*-axis and *y*-axis respectively.

If M and *v* be the mass and velocities of third part then

$$Mv\cos\theta = 12$$

$$Mv\sin\theta = 16$$

$$\tan\theta = \frac{16}{12} = \frac{4}{3}$$

$$\cos\theta = \frac{3}{5}$$

$$v = 4 \text{ m/s}$$

$$M = \frac{12}{v\cos\theta}$$

$$M = \frac{12 \times 5}{4 \times 3} = \frac{60}{12} = 5\text{kg}$$

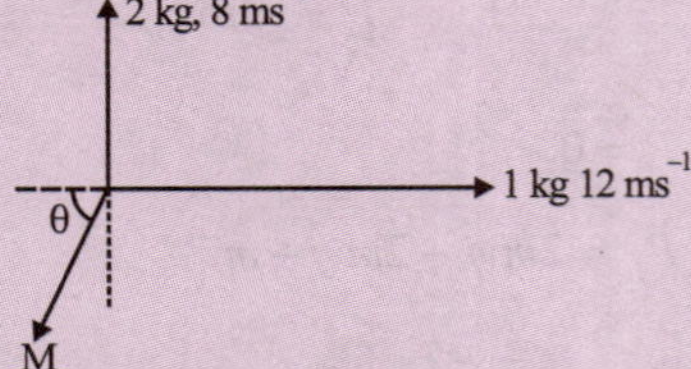

27. (a) Momentum of Ist part = m × 21 = 21m

Momentum of 2nd part = m × 21 = 21m

Resultant momentum, $= \sqrt{(m21)^2 + (m21)^2}$

Resultant momentum = momentum of 3rd part

$$\Rightarrow \quad 21\sqrt{2}\text{ m} = 3mv$$

$$\Rightarrow \quad v = 7\sqrt{2} \text{ m/sec}$$

NOTES Masses of the pieces are 1, 1, 3 kg. Hence $(1 \times 21)^2 + (1 \times 21)^2 = (3 \times V)^2$

That is, $V = 7\sqrt{2}$ m/s

28. (a) The components of 1 N and 2N forces along + *x*-axis = 1 cos 60° + 2 sin 30°

$$= 1 \times \frac{1}{2} + 2 \times \frac{1}{2} = \frac{1}{2} + 1 = \frac{3}{2} = 1.5\text{N}$$

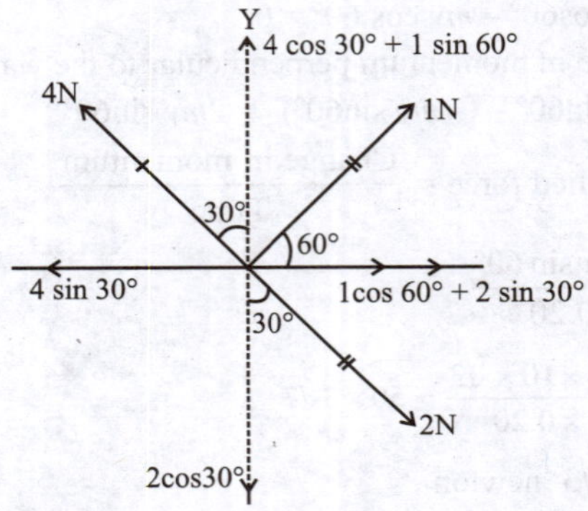

The component of 4 N force along –*x*-axis $= 4\sin 30° = 4 \times \frac{1}{2} = 2\text{N}$.

Therefore, if a force of 0.5N is applied along + *x*-axis, the resultant force along *x*-axis will become zero and the resultant force will be obtained only along *y*-axis.

29. (d)

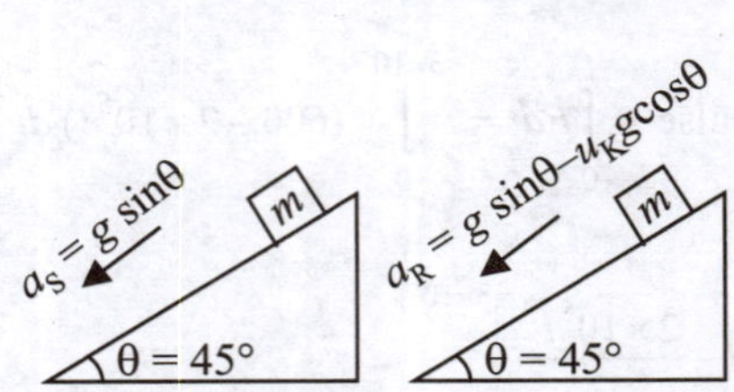

The distance travelled on inclined plane is

$$L = \frac{1}{2}at^2 \Rightarrow a \propto \frac{1}{t^2}$$

$$\Rightarrow \left(\frac{t_S}{t_R}\right)^2 = \frac{a_R}{a_S}$$

$$\Rightarrow \left(\frac{1}{2}\right)^2 = \frac{g\sin\theta - \mu_k g\cos\theta}{g\sin\theta}$$

$$\Rightarrow \frac{1}{4} = \frac{\sin 45° - \mu_k\cos 45°}{\sin 45°}$$

$$\Rightarrow \frac{1}{4} = \frac{\left(\frac{1}{\sqrt{2}}\right)}{\left(\frac{1}{\sqrt{2}}\right)}\left(\frac{1-\mu_k}{1}\right)$$

$$\Rightarrow \frac{1}{4} = 1-\mu_k \Rightarrow \mu_k = 1-\frac{1}{4} = \frac{3}{4} = 0.75$$

30. **(b)** The FBD of the rod is

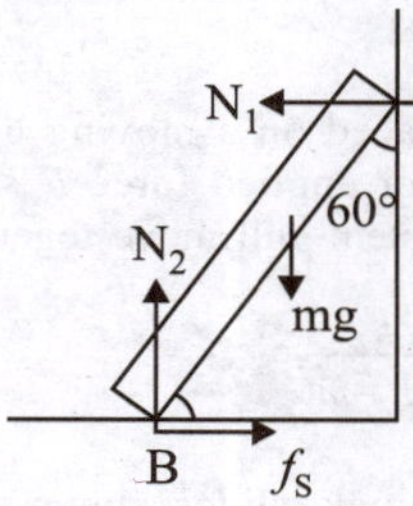

$N_2 = mg = 200$ N

$\Sigma\tau_A = 0$

$$\Rightarrow f_s L\cos 60° + mg\frac{L}{2}\sin 60° = N_2 L \sin 60°$$

$$\Rightarrow f_s\left(\frac{1}{2}\right) + 200\left(\frac{1}{2}\right)\frac{\sqrt{3}}{2} = 200\frac{\sqrt{3}}{2}$$

$$\therefore f_s = 100\sqrt{3}\text{N}$$

31. **(d)**

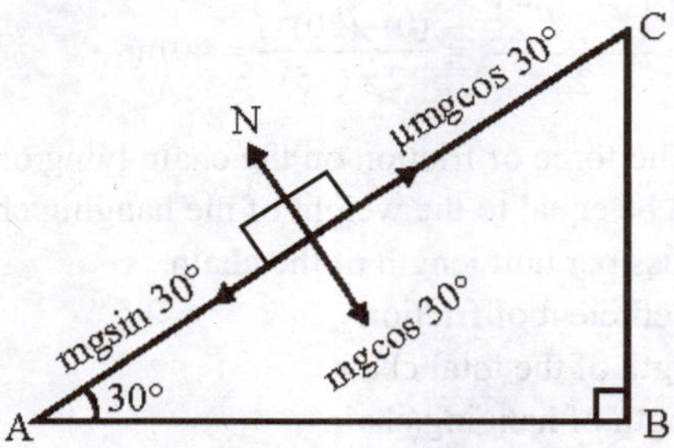

$$\Rightarrow F_{net} = mg\sin 30° - \mu mg\cos 30°$$

$$= \frac{mg}{2} - \frac{1}{\sqrt{3}}mg.\frac{\sqrt{3}}{2} = 0$$

32. **(c)**

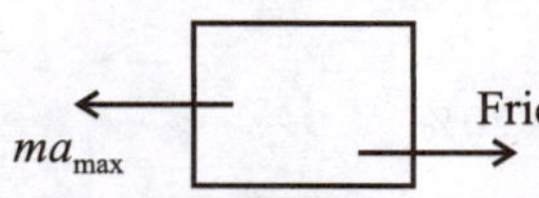

As the body remains stationary,

$\therefore ma_{max} = \mu_s mg$

$\Rightarrow a_{max} = \mu_s g = 0.15 \times 10 = 1.5 \text{ m/s}^{-2}$

33. **(d)** Since body does not move hence it is in equilibrium.

f_r = frictional force which is less than or equal to limiting friction.

Now $N = mg$. If F_{ext} is applied, then

Contact force $\vec{F} = \vec{N} + \vec{f_r}$

$|\vec{F}| \le \sqrt{(mg)^2 + (\mu mg)^2}$ $\quad [\because (f_r)_{max} = \mu mg]$

$|\vec{F}| \le mg\sqrt{1+\mu^2}$

34. **(c)** Coefficient of friction or sliding friction has no dimension.

$$f = \mu_s N \Rightarrow \mu_s = \frac{f}{N} = [M^0L^0T^0]$$

35. **(b)** For the motion of both the blocks

$m_1a = T - \mu_k m_1 g$ ...(i)

$m_2 g - T = m_2 a$ ...(ii)

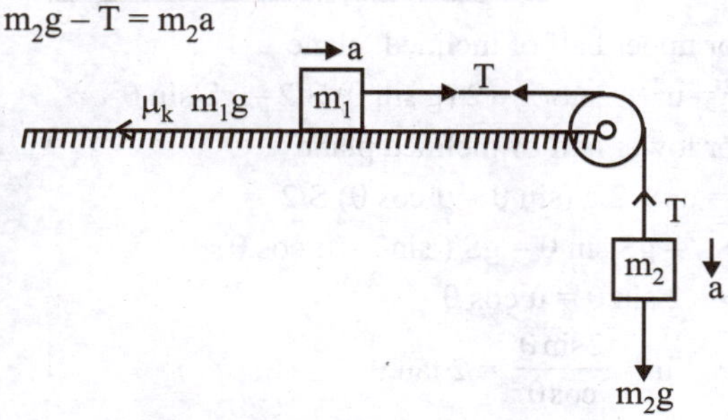

From eq. (i) and (ii), we get

$$a = \frac{m_2 g - \mu_k m_1 g}{m_1 + m_2}$$

From eq. (ii),

$$m_2 g - T = (m_2)\left(\frac{m_2 g - \mu_k m_1 g}{m_1 + m_2}\right)$$

solving we get tension in the string

$$T = \frac{m_1 m_2 (1+\mu_k) g}{m_1 + m_2}$$

36. **(a)** Coefficient of static friction, $\mu_s = \tan\theta$

$$\Rightarrow \mu_s = \tan 30° = \frac{1}{\sqrt{3}} = 0.577 \cong 0.6$$

So, distance covered by plank,

$$S = ut + \frac{1}{2}at^2$$

where, $u = 0$ and $a = g\sin\theta - \mu_k(g)\cos\theta$

$$4 = \frac{1}{2}a(4)^2 \Rightarrow a = \frac{1}{2} = 0.5$$

$[\because \theta = 30°, s = 4\text{m and } t = 4\text{s given}]$

$$\Rightarrow \mu_k = \frac{0.9}{\sqrt{3}} = 0.5$$

37. **(c)** Acceleration

$$= \frac{\text{Net force in the direction of motion}}{\text{Total mass of system}}$$

$$= \frac{m_1 g - \mu(m_2 + m_3)g}{(m_1 + m_2 + m_3)}$$

$$= \frac{g}{3}(1 - 2\mu)$$

$(\because m_1 = m_2 = m_3 = m \text{ given})$

38. (b)

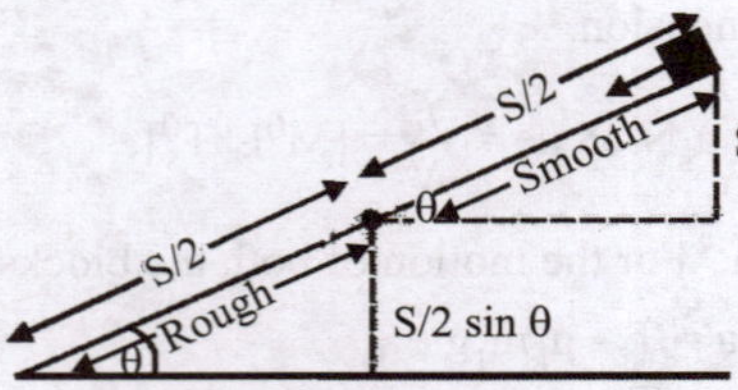

For upper half of inclined plane

$v^2 = u^2 + 2a\ S/2 = 2\ (g \sin\theta)\ S/2 = gS \sin\theta$

For lower half of inclined plane

$0 = u^2 + 2\ g\ (\sin\theta - \mu\cos\theta)\ S/2$

$\Rightarrow\ -gS\sin\theta = gS\ (\sin\theta - \mu\cos\theta)$

$\Rightarrow\ 2\sin\theta = \mu\cos\theta$

$\Rightarrow\ \mu = \dfrac{2\sin\theta}{\cos\theta} = 2\tan\theta$

39. (d) Frictional force on the box $f = \mu mg$

$\therefore$ Acceleration in the box, $a = -\dfrac{f}{m} = -\dfrac{\mu mg}{m}$

$= -\mu g = -5\ ms^{-2}$

$v^2 = u^2 + 2as$

$\Rightarrow\ 0 = 2^2 - 2 \times (5)\ s$

$\Rightarrow\ s = 0.4\ m$

40. (c) Forces acting on the block are as shown in the fig. Normal reaction N equals to ma

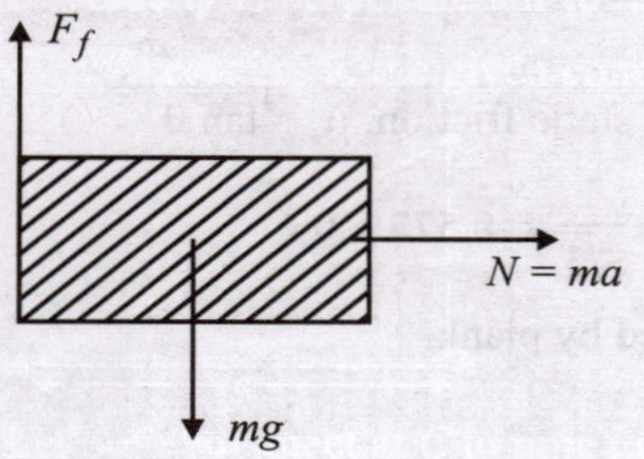

$\therefore\ N = ma$

For the block not to fall, frictional force, $F_f \geq mg$

$\Rightarrow\ \mu N \geq mg\ \Rightarrow\ \mu ma \geq mg$

$\Rightarrow\ a \geq g/\mu$

NOTES When a cart moves with some acceleration towards right then a psendo force (ma) acts on block towards left. This force ma is action force by a block on cart.

41. (d) Friction is the retarding force for the block

$F = ma = \mu R = \mu mg \Rightarrow a = \mu g$

Therefore, from the first equation of motion

$v = u - at$

$0 = V - \mu g \times t \Rightarrow\ t = \dfrac{V}{\mu g}$

42. (a) Condition for the max value of mass of block B so that two blocks do not move is,

$m_B g = \mu_s m_A g$

$\Rightarrow m_B = \mu_s m_A$

or, $m_B = 0.2 \times 2 = 0.4$ kg

43. (c) $a = \dfrac{F - \mu R}{m} = \dfrac{100 - 0.5 \times (10 \times 10)}{10} = 5\,ms^{-2}$

44. (a) Maximum friction force = μmg

$= .6 \times 1 \times 9.8 = 5.88$ N

But here required friction force

$= ma = 1 \times 5 = 5$ N

NOTES For a body placed on a moving body and Ist body not to slide, for applied force $F <$ limiting friction ($= \mu_s\ mg$) combined system will move together with common acceleration $a_1 = a_2 = \dfrac{F}{M + m}$.

45. (d) When the block slides down the plane with a constant speed, then the inclination of the plane is equal to angle of repose (θ).

Coeff. of friction = tan of the angle of repose = tan θ.

46. (b) Here u = 72 km/h = 20 m/s; v = 0;

$a = -\mu g = -0.5 \times 10 = -5\ m/s^2$

As $v^2 = u^2 + 2as$,

$\therefore\ s = \dfrac{(v^2 - u^2)}{2a} = \dfrac{(0 - (20)^2)}{2 \times (-5)} = 40\,m$

47. (a) The force of friction on the chain lying on the table should be equal to the weight of the hanging chain. Let

ρ = mass per unit length of the chain

μ = coefficient of friction

l = length of the total chain

x = length of hanging chain

Now, $\mu(l - x)\rho g = x\rho g$ or $\mu(l - x) = x$

or $\mu l = (\mu + 1)x$ or $x = \mu l/(\mu + 1)$

$\therefore\ x = \dfrac{0.25l}{(0.25 + 1)} = \dfrac{0.25l}{1.25} = 0.2l$

$\dfrac{x}{l} = 0.2 \Rightarrow \dfrac{x}{\ell} \times 100 = 20\%$

48. (b) In presence of friction $a = (g\sin\theta - \mu g\cos\theta)$

$\therefore$ Time taken to slide down the plane

$t_1 = \sqrt{\dfrac{2s}{a}} = \sqrt{\dfrac{2s}{g(\sin\theta - \mu\cos\theta)}}$

In absence of friction, $t_2 = \sqrt{\dfrac{2s}{g\sin\theta}}$

According to the condition,

$t_1 = 2t_2 \qquad \therefore\ t_1^2 = 4t_2^2$

or, $\dfrac{2s}{g(\sin\theta - \mu\cos\theta)} = \dfrac{2s \times 4}{g\sin\theta}$

$\sin\theta = 4\sin\theta - 4\mu\cos\theta$

$\mu = \frac{3}{4}\tan\theta = \frac{3}{4} = 0.75$

49. **(a)** In horizontal circular motion,

Tension, $T = m\omega^2 r$

For constant m and r, $T \propto \omega^2$

$T' = 4T$ (Given)

$\Rightarrow \omega' = 2\omega = 20$ rpm

50. **(b)** Tension, $T = m\ell\omega^2$

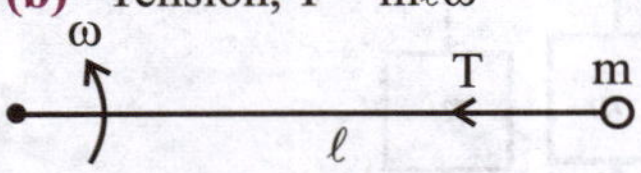

When speed becomes 2ω

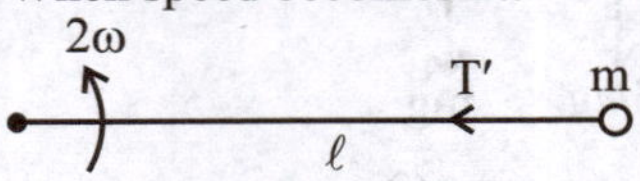

Tension, $T' = m\ell(2\omega)^2$

$\therefore \quad T' = 4T$

51. **(c)** Given mass of block, m = 10 kg; radius of cylindrical drum, r = 1m; coefficient of friction between the block and the inner wall of the cylinder $\mu = 0.1$;

Minimum angular velocity ω_{min}

For equilibrium of the block limiting friction

$f_L \geq mg$

$\Rightarrow \mu N \geq mg \Rightarrow \mu\, mr\omega^2 \geq mg$

Here, $N = mr\omega^2$

or, $\omega \geq \sqrt{\frac{g}{r\mu}}$

or, $\omega_{min} = \sqrt{\frac{g}{r\mu}}$

$\therefore \omega_{min} = \sqrt{\frac{10}{0.1\times 1}} = 10$ rad/s

52. **(d)** Net force on particle in uniform circular motion is centripetal force $\left(\frac{mv^2}{\ell}\right)$ which is provided by tension in string so the net force will be equal to tension *i.e.*, T.

53. **(b)** On a banked road,

$\frac{V_{max}^2}{Rg} = \left(\frac{\mu_s + \tan\theta}{1 - \mu_s\tan\theta}\right)$

Maximum safe velocity of a car on the banked road

$V_{max} = \sqrt{Rg\left[\frac{\mu_s + \tan\theta}{1 - \mu_s\tan\theta}\right]}$

54. **(d)** According to question, two stones experience same centripetal force

i.e., $F_{C_1} = F_{C_2}$

or, $\frac{mv_1^2}{r} = \frac{2mv_2^2}{(r/2)}$ or, $V_1^2 = 4V_2^2$

So, $V_1 = 2V_2$ *i.e.*, $n = 2$

55. **(d)** Given; speed = 10 m/s; radius r = 10 m

Angle made by the wire with the vertical is θ.

Then, $T\cos\theta = mg$

$T\sin\theta = \frac{mv^2}{r}$

So,

$\tan\theta = \frac{v^2}{rg} = \frac{10^2}{10\times 10} = 1$

$\Rightarrow \theta = 45° = \frac{\pi}{4}$

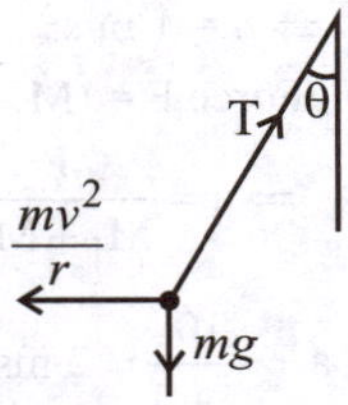

56. **(b)** For banking, $\tan\theta = \frac{V^2}{Rg}$

$\tan 45° = \frac{V^2}{90\times 10} = 1$

$V = 30$ m/s

NOTES If friction is present then maximum safe speed, on a banked frictional road,

$V = \sqrt{\frac{Rg(\mu + \tan\theta)}{1 - \mu\tan\theta}}$

57. **(d)** For smooth driving maximum speed of car v then,

$\frac{mv^2}{R} = \mu_s mg$

$v = \sqrt{\mu_s Rg}$

58. **(c)** Centripetal force $= \frac{mv^2}{r} = \frac{500\times(10)^2}{50}$

= 1000 N $\quad [\because 36 \text{ km/hr} = 10 \text{ m/s}]$

59. **(a)** $r = 30$ *m* and $\mu = 0.4$.

$v_{max} = \sqrt{\mu rg} = \sqrt{0.4\times 30\times 9.8} = 10.84$ m/s

60. **(c)** On the diametrically opposite points, the velocities have same magnitude but opposite directions. Therefore, change in momentum is MV – (– MV) = 2MV

61. **(b)** Cream gets separated from a churned milk due to centrifugal force.

NOTES Centrifugal force is a pseudo force that is equal and opposite to the centripetal force. Centripetal force can be mechanical, electrical or magnetic force.

62. **(d)** $F_{net} = Ma = T - Mg\sin 30$

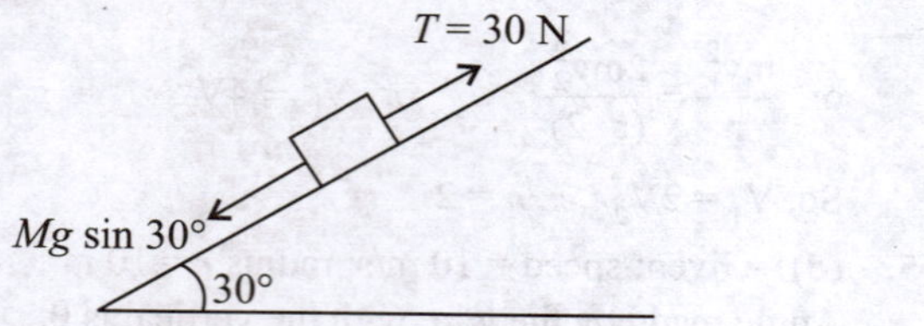

$$\Rightarrow 30 - 5 \times 10 \times \frac{1}{2} = 5a$$

$$\Rightarrow a = 1 \text{ m/s}^2$$

63. (c) Force $F = (M_1 + M_2)a$

$$\Rightarrow a = \frac{F}{M_1 + M_2}$$

or, $a = \frac{10}{2+3} = 2 \text{ ms}^{-2}$

$\therefore$ Force exerted by block A on block B

$F' = M_2(2) = 3 \times 2 = 6$ N

64. (b) $a = \frac{\text{Net pulling force}}{M_{total}}$ (up the inclined)

$$a = \frac{60 - 18 - (3+2+1)g \sin 30°}{3+2+1} = \frac{12}{6} = 2 \text{ m/s}^2$$

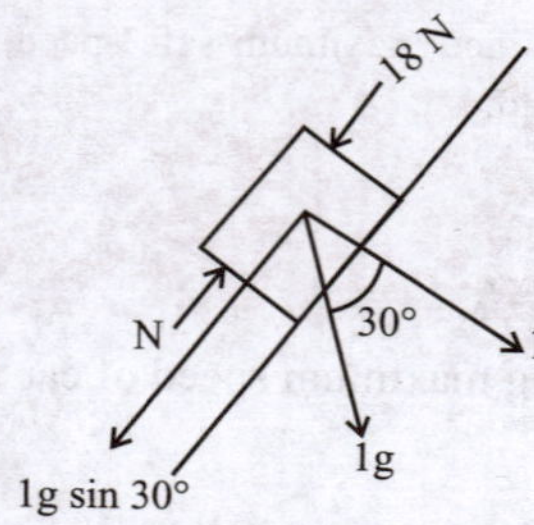

By taking F.B.D. of 1 kg

$$ma = N - 18 - 1 \times 10 \times \frac{1}{2}$$

$$\Rightarrow 2 = N - 23 \Rightarrow N = 25 \text{ N}$$

65. (b) Given : Mass $M_1 = 4$ kg and $M_2 = 6$ kg.

Acceleration of the system,

$$a = \frac{(m_1 - m_2)g}{(m_1 + m_2)} \text{ where } m_1 > m_2$$

$$\therefore a = \frac{(6-4)g}{6+4} = \frac{g}{5}$$

NOTES Here no option is given according to acceleration of COM of the system.

66. (c) Let the mass of block is m. It will remains stationary if forces acting on it are in equilibrium. *i.e.*, $ma \cos \theta = mg \sin \theta$

$\Rightarrow a = g \tan \theta$

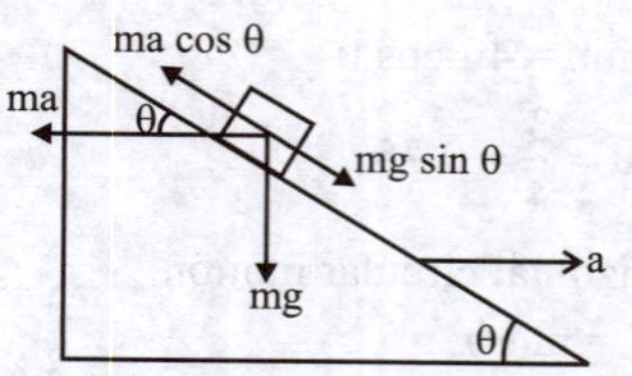

67. (a) Before cutting the string

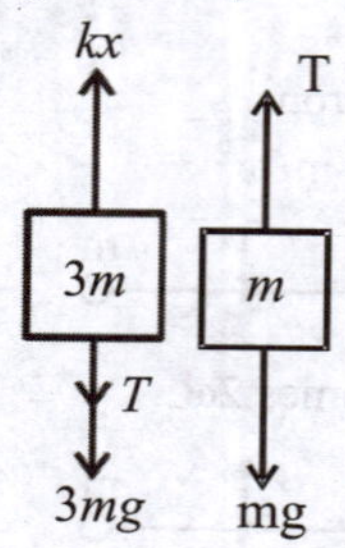

$kx = T + 3$ mg ...(i)

$T = mg$ ...(ii)

Force on spring,

$\Rightarrow kx = 4mg$

After cutting the string T = 0

For block A: $3ma_A = 4mg - 3mg$

$$a_A = \frac{4mg - 3mg}{3m} = \frac{g}{3}$$

$$\therefore \quad a_A = \frac{g}{3}$$

For block B: $ma_B = mg$

$a_B = g$

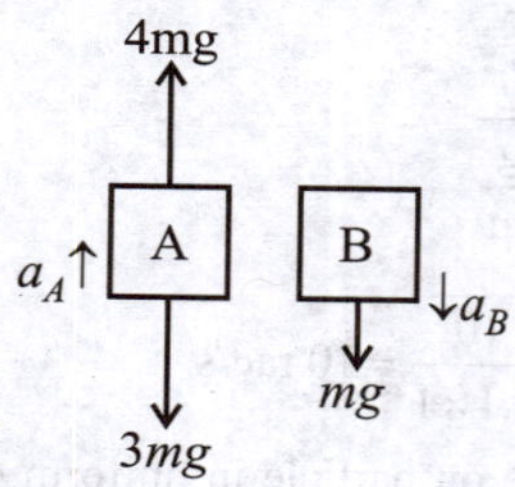

68. (a) Acceleration of system, $a = \frac{F_{net}}{M_{total}}$

$$= \frac{14}{4+2+1} = \frac{14}{7} = 2 \text{ m/s}^2$$

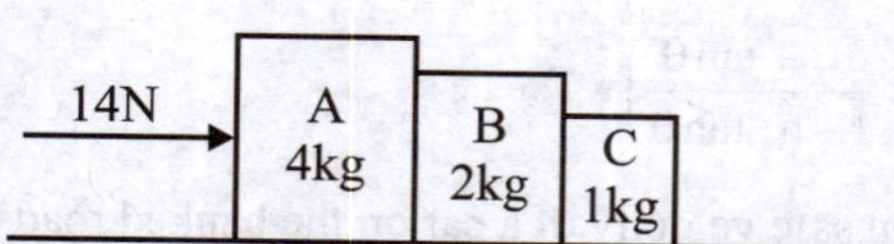

The contact force between A and B

$= (m_B + m_C) \times a = (2 + 1) \times 2 = 6$N

NOTES

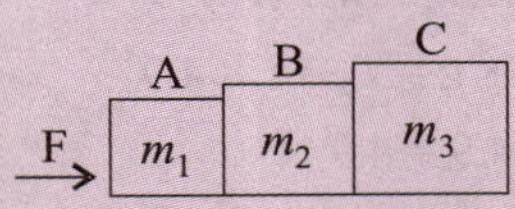

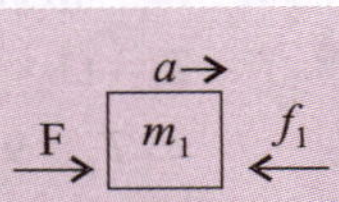	$F - f_1 = m_2 a$	$a = \frac{F}{m_1 + m_2 + m_3}$
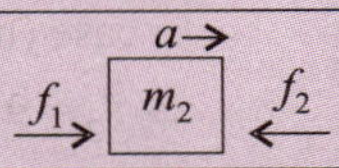	$f_1 - f_2 = m_2 a$	$f_1 = \frac{(m_2 + m_3)F}{m_1 + m_2 + m_3}$
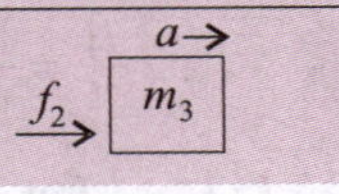	$f_2 = m_2 a$	$f_2 = \frac{m_3 F}{m_1 + m_2 + m_3}$

69. **(a)** Let upthrust of air be F_a then

For downward motion of balloon,

$F_a = mg - ma \Rightarrow F_a = m(g - a)$ ...(i)

For upward motion,

$F_a - (m - \Delta m)g = (m - \Delta m)a$ ...(ii)

From equation (i) and (ii),

Therefore, $\Delta m = \frac{2ma}{(g + a)}$

70. **(d)**

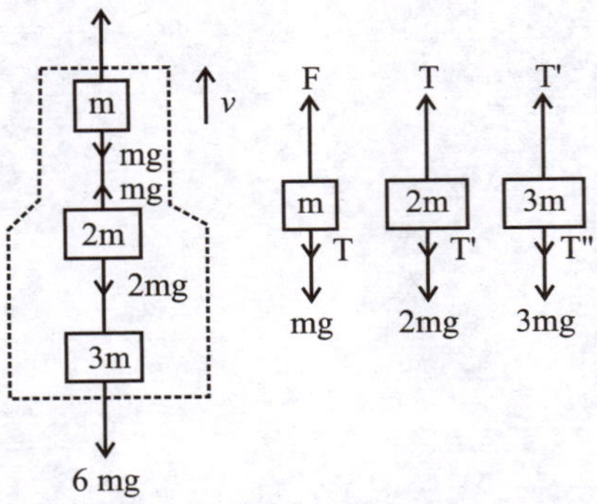

As speed is constant, acceleration $a = 0$

$\therefore$ 6 mg – F = 6ma = 0

$\Rightarrow$ F = 6 mg

$\therefore$ T = 5 mg , T′ = 3 mg

T″ = 0

F_{net} on block of mass 2 m = T – T' – 2 mg = 0

NOTES As all blocks are moving with constant speed, then acceleration is zero, so force is zero.
[$\because$ v = constant so, $a = 0$, Hence, $F_{net} = ma = 0$]

71. **(c)**

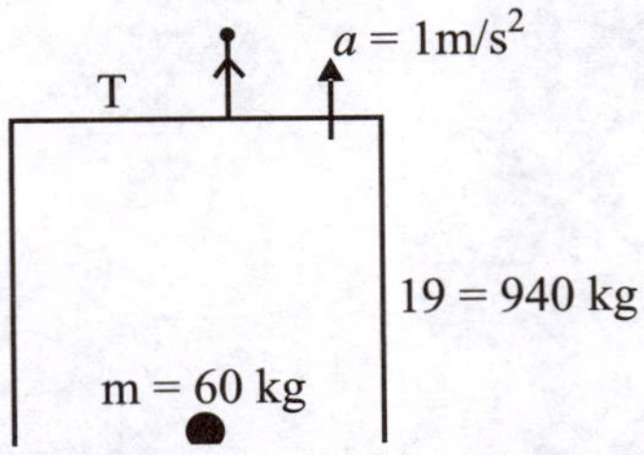

Total mass = (m + m) = (60 + 940) kg = 1000 kg

Let T be the tension in the supporting cable, then

T – 1000g = 1000 × 1 [$\because a = 1\ m/s^2$]

$\Rightarrow$ T = 11000 N

72. **(a)** Net force, $F = T - mg$

$ma = T - mg$

$2000\ a = 28000 - 20000 = 8000$

$a = \frac{8000}{2000} = 4\,ms^{-2} \uparrow$

73. **(a)** According to the condition,

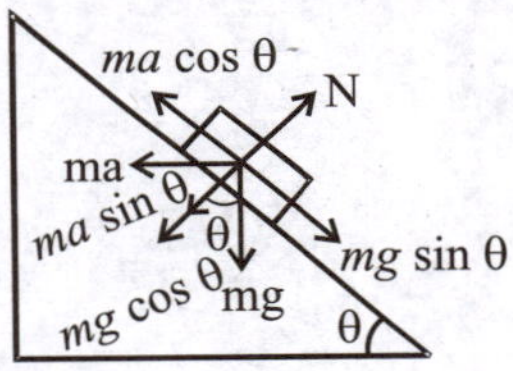

$N = m\,a \sin\theta + mg\cos\theta$ .....(i)

Also, $m\,g\sin\theta = m\,a\cos\theta$ .....(ii)

From (i) & (ii), $a = g\tan\theta$

$\therefore\ N = mg\frac{\sin^2\theta}{\cos\theta} + mg\cos\theta$.

$= \frac{mg}{\cos\theta}(\sin^2\theta + \cos^2\theta) = \frac{mg}{\cos\theta}$

or, $N = \frac{mg}{\cos\theta}$

NOTES The condition for the body to be at rest relative to the inclined plane, a = g sin θ – b cos θ = 0.
Horizontal acceleration, b = g tan θ.

74. **(a)** Reading of the scale

= Apparent wt. of the man = $m(g + a)$

= 80 (10 + 5) = 1200 N

75. **(a)** T = Tension caused in string by monkey

$= m\,(g + a)$

$\therefore\ T \le 25 \times 10 \Rightarrow 20(10 + a) \le 250$

or, $10 + a \le 12.5 \Rightarrow a \le 2.5$

76. **(b)** T – (1000 × 9.8) = 1000 × 1

$\Rightarrow$ T = 10800 N

77. (a) Let T be the tension in the string.

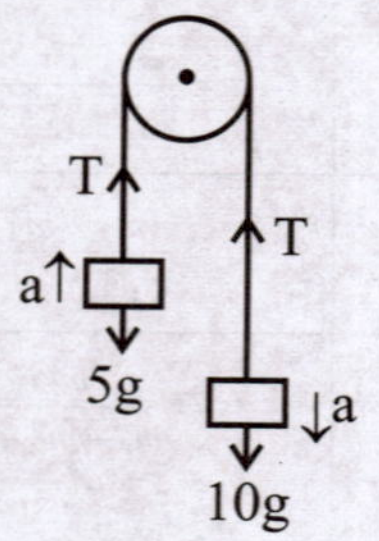

$\therefore$ $10g - T = 10a$ ....(i)

$T - 5g = 5a$ ....(ii)

Adding (i) and (ii),

$5g = 15a \Rightarrow a = \frac{g}{3}$ m/s^2

78. (a) In case (i) we have

$T_1 - (1 \times g) = 1 \times 4.9$

$\Rightarrow T_1 = 9.8 + 4.9 = 14.7$ N

In case (ii), $1 \times g - T_2 = 1 \times 4.9$

$\Rightarrow T_2 = 9.8 - 4.9 = 4.9$ N

$\therefore \frac{T_1}{T_2} = \frac{14.7}{4.9} = \frac{3}{1}$

79. (c) Let T be the tension in the branch of a tree when monkey is descending with acceleration a. Then $mg - T = ma$; and $T = 75\%$ of weight of monkey,

$\therefore ma = mg - \left(\frac{75}{100}\right)mg = \left(\frac{1}{4}\right)mg$

or $a = \frac{g}{4}$.

5 Work, Energy and Power

TREND ANALYSIS

TOPIC	2025		2024		2023		2022		2021		2020	
	Qns.	LOD	Qns.	LOD	Qns.	LOD	Qns.	LOD	Qns.	LOD	Qns.	LOD
Kinetic Energy	1	Avg.										
The Concept of Potential Energy, Conservation of Mechanical Energy	1	Diff.					1	Avg.	1	Avg.		
The Potential Energy of a Spring					1	Easy	1	Avg.				
Power			1	Avg.			2	Avg.	1	Easy		
Collisions			1	Diff.								

How to Code Qns for Better & Faster Revision

Type 1: Able to solve (No code Req)

Type 2: Solved with struggle or by seeing solution **(Use code NR means Needs Revision or orange stickers)**

Type 3: Tricky/ Could not solve on my own. Need help from Teacher/ Friend. **(Use code TQ means Tricky Question or red stickers)**

You can either hand write the code in boxes or else paste coloured stickers given at end of book.

Topic-5.1

Introduction

1. A particle moves so that its position vector is given by $\vec{r} = \cos\omega t\hat{x} + \sin\omega t\hat{y}$. Where ω is a constant. Which of the following is true? NCERT | Page-42 | 2016
(a) Velocity and acceleration both are perpendicular to $\vec{r}$
(b) Velocity and acceleration both are parallel to $\vec{r}$
(c) Velocity is perpendicular to $\vec{r}$ and acceleration is directed towards the origin
(d) Velocity is perpendicular to $\vec{r}$ and acceleration is directed away from the origin

2. If the magnitude of sum of two vectors is equal to the magnitude of difference of the two vectors, the angle between these vectors is : NCERT | Page-72 | 2016
(a) 0° (b) 90° (c) 45° (d) 180°

3. Ifvectors $\vec{A} = \cos\omega t\hat{i} + \sin\omega t\hat{j}$ and $\vec{B} = \cos\frac{\omega t}{2}\hat{i} + \sin\frac{\omega t}{2}\hat{j}$ are functions of time, then the value of t at which they are orthogonal to each other is : NCERT | Page-72 | 2015 RS
(a) $t = \frac{\pi}{2\omega}$ (b) $t = \frac{\pi}{\omega}$
(c) $t = 0$ (d) $t = \frac{\pi}{4\omega}$

4. The vectors $\vec{A}$ and $\vec{B}$ are such that
$|\vec{A} + \vec{B}| = |\vec{A} - \vec{B}|$
The angle between the two vectors is NCERT | Page-72 | 2006, 2001, 1996, 1991
(a) 60° (b) 75° (c) 45° (d) 90°

5. If a vector $2\hat{i} + 3\hat{j} + 8\hat{k}$ is perpendicular to the vector $4\hat{j} - 4\hat{i} + \alpha\hat{k}$, then the value of α is NCERT | Page-72 | 2005
(a) 1/2 (b) –1/2 (c) 1 (d) –1

6. The vector sum of two forces is perpendicular to their vector differences. In that case, the forces NCERT | Page72 | 2003
(a) cannot be predicted
(b) are equal to each other
(c) are equal to each other in magnitude
(d) are not equal to each other in magnitude

7. The angle between the two vectors $\vec{A} = 3\hat{i} + 4\hat{j} + 5\hat{k}$ and $\vec{B} = 3\hat{i} + 4\hat{j} - 5\hat{k}$ will be NCERT | Page-72 | 2001, 1994
(a) zero (b) 45° (c) 90° (d) 180°

8. The magnitudes of vectors $\vec{A}$, $\vec{B}$ and $\vec{C}$ are 3, 4 and 5 units respectively. If $\vec{A} + \vec{B} = \vec{C}$, then the angle between $\vec{A}$ and $\vec{B}$ is NCERT | Page-72 | 1988
(a) π/2 (b) $\cos^{-1} 0.6$
(c) $\tan^{-1} 7/5$ (d) π/4

Topic-5.3

Work

9. A uniform force of $(3\hat{i} + \hat{j})$ newton acts on a particle of mass 2 kg. The particle is displaced from position $(2\hat{i} + \hat{k})$ metre to position $(4\hat{i} + 3\hat{j} - \hat{k})$ metre. The work done by the force on the particle is NCERT | Page-74 | 2017
(a) 6 J (b) 13 J (c) 15 J (d) 9 J

10. A body constrained to move in y-direction, is subjected to a force given by $\vec{F} = (-2\hat{i} + 15\hat{j} + 6\hat{k})N$. What is the work done by this force in moving the body through a distance of 10m along y-axis? **NCERT | Page-75 | 1994**

(a) 190 J (b) 160 J (c) 150 J (d) 20 J

11. A bullet of mass 10g leaves a rifle at an initial velocity of 1000 m/s and strikes the earth at the same level with a velocity of 500 m/s. The work done in joules overcoming the resistance of air will be **NCERT | Page-73 | 1989**

(a) 375 (b) 3750
(c) 5000 (d) 500

Topic-5.4 Kinetic Energy

12. The kinetic energies of two similar cars A and B are 100 J and 225 J respectively. On applying breaks, car A stops after 1000 m and car B stops after 1500 m. If F_A and F_B are the forces applied by the breaks on cars A and B, respectively, then the ratio $\frac{F_A}{F_B}$ is : **NCERT | Page-73, 74 | 2025**

(a) $\frac{3}{2}$ (b) $\frac{2}{3}$ (c) $\frac{1}{3}$ (d) $\frac{1}{2}$

13. An object moving along horizontal x-direction with kinetic energy 10 J is displaced through $x = (3\hat{i})$ m by the force $\vec{F} = (-2\hat{i} + 3\hat{j})$ N. The kinetic energy of the object at the end of the displacement x is **NCERT | Page-73, 74 | Ph-II 2024**

(a) 10 J (b) 16 J
(c) 4 J (d) 6 J

14. A particle of mass 10 g moves along a circle of radius 6.4 cm with a constant tangential acceleration. What is the magnitude of this acceleration if the kinetic energy of the particle becomes equal to 8×10^{-4} J by the end of the second revolution after the beginning of the motion ? **NCERT | Page-79 | 2016**

(a) 0.1 m/s^2 (b) 0.15 m/s^2
(c) 0.18 m/s^2 (d) 0.2 m/s^2

15. A block of mass 10 kg, moving in x direction with a constant speed of 10 ms^{-1}, is subject to a retarding force F = 0.1x J/m during its travel from x = 20 m to 30 m. Its final KE will be: **NCERT | Page-79 | 2015**

(a) 450 J (b) 275 J (c) 250 J (d) 475 J

16. A person holding a rifle (mass of person and rifle together is 100 kg) stands on a smooth surface and fires 10 shots horizontally, in 5 s. Each bullet has a mass of 10 g with a muzzle velocity of 800 ms^{-1}. The final velocity acquired by the person and the average force exerted on the person are **NCERT | Page-80 | Kar. 2013**

(a) –1.6 ms^{-1}; 8 N (b) –0.8 ms^{-1}; 16 N
(c) –0.8 ms^{-1}; 8 N (d) –1.6 ms^{-1}; 16 N

17. An engine pumps water continuously through a hose. Water leaves the hose with a velocity v and m is the mass per unit length of the water jet. What is the rate at which kinetic energy is imparted to water? **NCERT | Page-81 | 2009**

(a) mv^2 (b) $\frac{1}{2}mv^2$ (c) $\frac{1}{2}m^2v^2$ (d) $\frac{1}{2}mv^3$

18. A particle of mass m_1 is moving with a velocity v_1 and another particle of mass m_2 is moving with a velocity v_2. Both of them have the same momentum but their different kinetic energies are E_1 and E_2 respectively. If $m_1 > m_2$ then **NCERT | Page-81 | 2004**

(a) $E_1 = E_2$ (b) $E_1 < E_2$
(c) $\frac{E_1}{E_2} = \frac{m_1}{m_2}$ (d) $E_1 > E_2$

19. A ball of mass 2 kg and another of mass 4 kg are dropped together from a 60 feet tall building. After a fall of 30 feet each towards earth, their respective kinetic energies will be in the ratio of **NCERT | Page-78 | 2004**

(a) 1 : √2 (b) √2 : 1
(c) 1 : 4 (d) 1 : 2

20. If the kinetic energy of a particle is increased by 300%, the momentum of the particle will increase by **NCERT | Page-75 | 2002**

(a) 20% (b) 200% (c) 100% (d) 50%

21. Two bodies with kinetic energies in the ratio 4 : 1 are moving with equal linear momentum. The ratio of their masses is **NCERT | Page-84, 81 | 1999**

(a) 1 : 2 (b) 1 : 1 (c) 4 : 1 (d) 1 : 4

22. Two bodies of masses m and 4 m are moving with equal K.E. The ratio of their linear momenta is **NCERT | Page-81 | 1997**

(a) 4 : 1 (b) 1 : 1 (c) 1 : 2 (d) 1 : 4

23. The kinetic energy acquired by a mass (m) in travelling distance (s) starting from rest under the action of a constant force is directly proportional to **1996, 1994**

(a) $1/\sqrt{m}$ (b) 1/m
(c) $\sqrt{m}$ (d) m^0

24. If the momentum of a body is increased by 50%, then the percentage increase in its kinetic energy is **NCERT | Page-81 | 1995**

(a) 50% (b) 100%
(c) 125% (d) 200%

25. Two masses of 1g and 9g are moving with equal kinetic energies. The ratio of the magnitudes of their respective linear momenta is **NCERT | Page-84, 81 | 1993**

(a) 1 : 9 (b) 9 : 1 (c) 1 : 3 (d) 3 : 1

26. A 4 kg mass and 1 kg are moving with equal kinetic energies. The ratio of the magnitudes of their linear momenta is **NCERT | Page-84, 81 | 1989**

(a) 1 : 2 (b) 1 : 1 (c) 2 : 1 (d) 4 : 1

27. Two bodies of masses m and 4m are moving with equal kinetic energies. The ratio of their linear momenta will be **NCERT | Page-84, 81 | 1988**

(a) 1 : 4 (b) 4 : 1 (c) 1 : 2 (d) 2 : 1

Topic-5.5 Work Done by a Variable Force

28. A force F = 20 + 10y acts on a particle in y-direction where F is in newton and y in metre. Work done by this force to move the particle from y = 0 to y = 1 m is :
NCERT | Page-74 | 2019

(a) 30 J (b) 5 J (c) 25 J (d) 20 J

29. An object of mass 500 g, initially at rest, is acted upon by a variable force whose X-component varies with X in the manner shown. The velocities of the object at the points X = 8 m and X = 12 m, would have the respective values of (nearly)

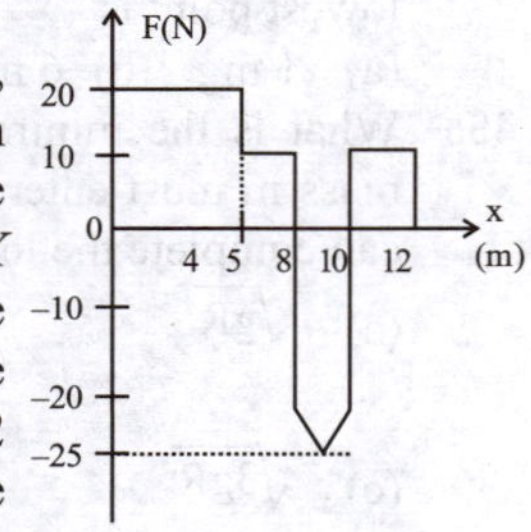

NCERT | Page-73 | Odisha 2019

(a) 18 m/s and 20.6 m/s
(b) 18 m/s and 24.4 m/s
(c) 23 m/s and 24.4 m/s
(d) 23 m/s and 20.6 m/s

30. Force F on a particle moving in a straight line varies with distance d as shown in the figure. The work done on the particle during its displacement of 12 m is
NCERT | Page-75 | 2011

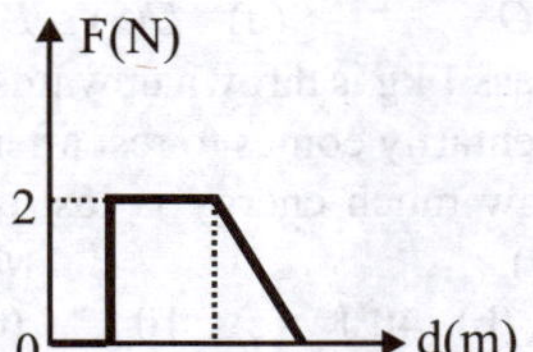

(a) 18 J (b) 21 J (c) 26 J (d) 13 J

Topic-5.6 The Work-Energy Theorem for a Variable Force

31. A body of mass 3 kg is under a constant force which causes a displacement s in metres in it, given by the relation $s = \frac{1}{3}t^2$, where t is in seconds. Work done by the force in 2 seconds is
NCERT | Page-75 | 2006

(a) $\frac{3}{8}$J (b) $\frac{8}{3}$J
(c) $\frac{19}{5}$J (d) $\frac{5}{19}$J

32. A force F acting on an object varies with distance x as shown here. The force is in N and x in m. The work done by the force in moving the object from x = 0 to x = 6 m is
NCERT | Page-75 | 2005

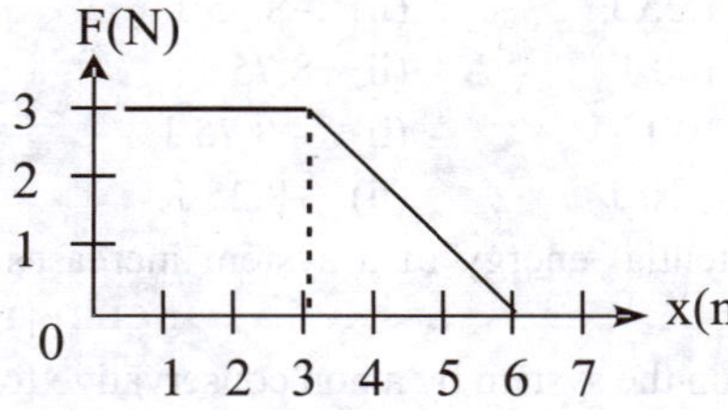

(a) 18.0 J (b) 13.5 J
(c) 9.0 J (d) 4.5 J

33. A force acts on a 30 gm particle in such a way that the position of the particle as a function of time is given by $x = 3t - 4t^2 + t^3$, where x is in metres and t is in seconds. The work done during the first 4 seconds is NCERT | Page-74 | 1998

(a) 576 mJ (b) 450 mJ
(c) 490 mJ (d) 530 mJ

34. A position dependent force, F = $(7 - 2x + 3x^2)$ N acts on a small body of mass 2 kg and displaces it from x = 0 to x = 5 m. Work done in joule is
NCERT | Page-75 | 1992

(a) 35 (b) 70 (c) 135 (d) 270

Topic-5.7 The Concept of Potential Energy

35. A gravitational field is present in a region and a mass is shifted from A to B through different paths as shown. If W_1, W_2 and W_3 represent the work done by the gravitational force along the respective paths, then:
NCERT | Page-77 | RS 2022

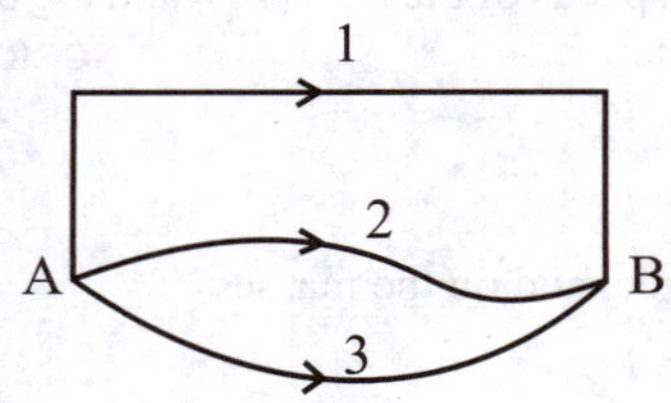

(a) $W_1 < W_2 < W_3$ (b) $W_1 = W_2 = W_3$
(c) $W_1 > W_2 > W_3$ (d) $W_1 > W_3 > W_2$

36. A particle is released from height S from the surface of the Earth. At a certain height its kinetic energy is three times its potential energy. The height from the surface of earth and the speed of the particle at that instant are respectively
NCERT | Page-77 | 2021

(a) $\frac{S}{4}, \sqrt{\frac{3gS}{2}}$ (b) $\frac{S}{4}, \frac{3gS}{2}$
(c) $\frac{S}{4}, \frac{\sqrt{3gS}}{2}$ (d) $\frac{S}{2}, \frac{\sqrt{3gS}}{2}$

37. Consider a drop of rain water having mass 1 g falling from a height of 1 km. It hits the ground with a speed of 50 m/s. Take 'g' constant with a value 10 m/s^2. The work done by the (i) gravitational force and the (ii) resistive force of air is
NCERT | Page-73 | 2017

(a) (i) 1.25 J (ii) –8.25 J
(b) (i) 100 J (ii) 8.75 J
(c) (i) 10 J (ii) – 8.75 J
(d) (i) – 10 J (ii) –8.25 J

38. The potential energy of a system increases if work is done **NCERT | Page-81 | 2011**
(a) upon the system by a non conservative force
(b) by the system against a conservative force
(c) by the system against a non conservative force
(d) upon the system by a conservative force

39. A vertical spring with force constant k is fixed on a table. A ball of mass m at a height h above the free upper end of the spring falls vertically on the spring so that the spring is compressed by a distance d. The net work done in the process is **NCERT | Page-73 | 2007**
(a) $mg(h+d)-\frac{1}{2}kd^2$ (b) $mg(h-d)-\frac{1}{2}kd^2$
(c) $mg(h-d)+\frac{1}{2}kd^2$ (d) $mg(h+d)+\frac{1}{2}kd^2$

40. 300 J of work is done in sliding a 2 kg block up an inclined plane of height 10 m. Taking $g = 10\ m/s^2$, work done against friction is **NCERT | Page-73 | 2006**
(a) 100 J (b) zero
(c) 1000 J (d) 200 J

41. A person slides freely down a frictionless inclined plane while his bag falls down vertically from the same height. The final speeds of the man (V_M) and the bag (V_B) should be such that **NCERT | Page-61 | 2000**
(a) $V_M < V_B$
(b) $V_M = V_B$
(c) they depend on the masses
(d) $V_M > V_B$

Topic-5.8
The Conservation of Mechanical Energy

42. A bob of heavy mass m is suspended by a light string of length l. The bob is given a horizontal velocity v_0 as shown in figure. If the string gets slack at some point P making an angle θ from the horizontal, the ratio of the speed v of the bob at point P to its initial speed v_0 is:

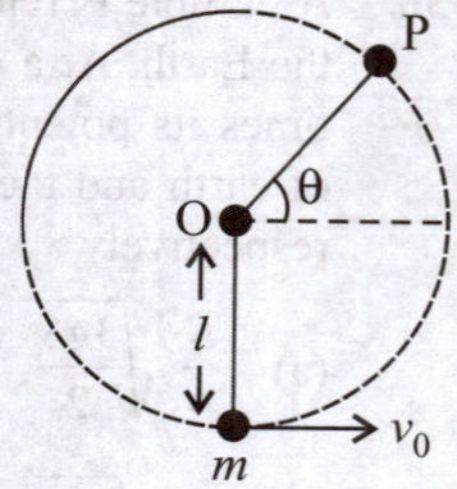

NCERT | Page-79 | 2025
(a) $(\sin\theta)^{\frac{1}{2}}$ (b) $\left(\frac{1}{2+3\sin\theta}\right)^{\frac{1}{2}}$
(c) $\left(\frac{\cos\theta}{2+3\sin\theta}\right)^{\frac{1}{2}}$ (d) $\left(\frac{\sin\theta}{2+3\sin\theta}\right)^{\frac{1}{2}}$

43. An object falls from a height of 10 m above the ground. After striking the ground it loses 50% of its kinetic energy. The height upto which the object can rebounce from the ground is: **NCERT | Page-78, 79 | Ph-II 2024**
(a) 7.5 m (b) 10 m
(c) 2.5 m (d) 5 m

44. A point mass 'm' is moved in a vertical circle of radius 'r' with the help of a string. The velocity of the mass is $\sqrt{7gr}$ at the lowest point. The tension in the string at the lowest point is **NCERT | Page-79 | Ph-II 2020**
(a) 1 mg (b) 6 mg (c) 7 mg (d) 8 mg

45. What is the minimum velocity with which a body of mass m must enter a vertical loop of radius R so that it can complete the loop? **NCERT | Page-79 | 2016**
(a) $\sqrt{gR}$ (b) $\sqrt{2gR}$
(c) $\sqrt{3gR}$ (d) $\sqrt{5gR}$

46. A mass m is attached to a thin wire and whirled in a vertical circle. The wire is most likely to break when: **NCERT | Page-79 | 2014**
(a) the mass is at the highest point
(b) the wire is horizontal
(c) the mass is at the lowest point
(d) inclined at an angle of 60° from vertical

47. A particle with total energy E is moving in a potential energy region $U(x)$. Motion of the particle is restricted to the region when **NCERT | Page-81 | Kar. 2013**
(a) $U(x) > E$ (b) $U(x) < E$
(c) $U(x) = O$ (d) $U(x) \leq E$

48. A body of mass 1 kg is thrown upwards with a velocity 20 m/s. It momentarily comes to rest after attaining a height of 18 m. How much energy is lost due to air friction? ($g = 10\ m/s^2$) **NCERT | Page-78 | 2009**
(a) 30 J (b) 40 J (c) 10 J (d) 20 J

49. In a simple pendulum of length l the bob is pulled aside from its equilibrium position through an angle θ and then released. The bob passes through the equilibrium position with speed **NCERT | Page-81 | 2000**
(a) $\sqrt{2g\ell(1+\cos\theta)}$ (b) $\sqrt{2g\ell\sin\theta}$
(c) $\sqrt{2g\ell}$ (d) $\sqrt{2g\ell(1-\cos\theta)}$

50. A small sphere is attached to a cord and rotates in a vertical circle about a point O. If the average speed of the sphere is increased, the cord is most likely to break at the orientation when the mass is at **NCERT | Page-79 | 2000**

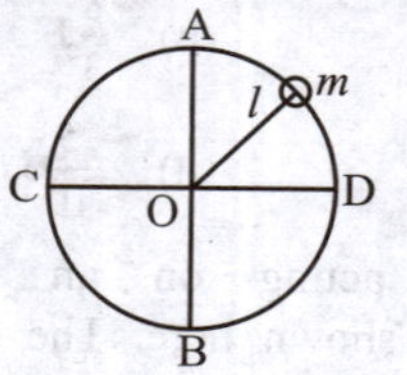

(a) bottom point B (b) the point C
(c) the point D (d) top point A

51. A stone tied with a string, is rotated in a vertical circle. The minimum speed with which the string has to be rotated **NCERT | Page-79 | 1999**

(a) is independent of the mass of the stone
(b) is independent of the length of the string
(c) decreases with increasing mass of the stone
(d) decreases with increasing length of the string

52. A rubber ball is dropped from a height of 5m on a plane, where the acceleration due to gravity is not shown. On bouncing it rises to 1.8 m. The ball loses its velocity on bouncing by a factor of **NCERT | Page-78 | 1998**

(a) $\frac{16}{25}$ (b) $\frac{2}{5}$ (c) $\frac{3}{5}$ (d) $\frac{9}{25}$

53. A body of mass 0.4 kg is whirled in a vertical circle making 2 rev/sec. If the radius of the circle is 1.2 m, then tension in the string when the body is at the top of the circle, is **NCERT | Page-79 | 1999**

(a) 71.88 N (b) 89.86 N
(c) 109.86 N (d) 115.86 N

54. Consider a car moving along a straight horizantal road with a speed of 72 km/h. If the coefficient of static friction between road and tyres is 0.5, the shortest distance in which the car can be stopped is **NCERT | Page-81 | 1994**

(a) 30 m (b) 40 m
(c) 72 m (d) 20 m

Topic-5.9 The Potential Energy of a Spring

55. The potential energy of a long spring when stretched by 2 cm is U. If the spring is stretched by 8 cm, potential energy stored in it will be **NCERT | Page-82 | 2023**

(a) 2 U (b) 4 U (c) 8 U (d) 16 U

56. The restoring force of a spring with a block attached to the free end of the spring is represented by **RS 2022**

(a)
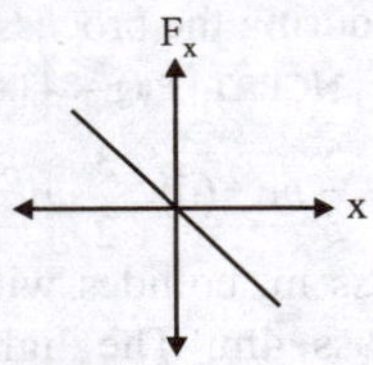

(b)
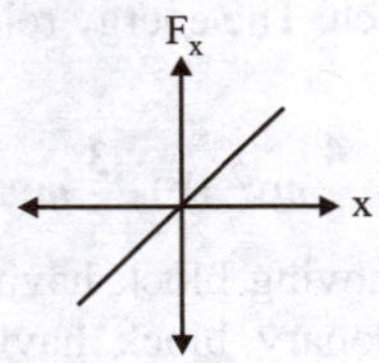

(c)
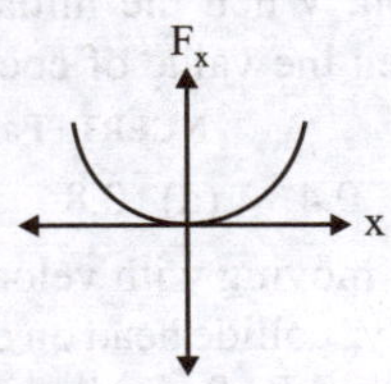

(d)
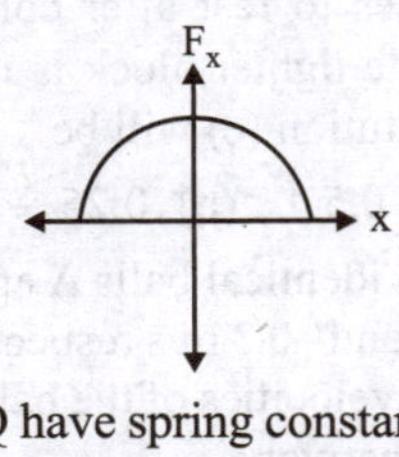

57. Two similar springs P and Q have spring constants K_P and K_Q, such that $K_P > K_Q$. They are stretched, first by the same amount (case a,) then by the same force (case b). The work done by the springs W_P and W_Q are related as, in case (a) and case (b), respectively **NCERT | Page-73 | 2015**

(a) $W_P = W_Q$; $W_P = W_Q$
(b) $W_P > W_Q$; $W_Q > W_P$
(c) $W_P < W_Q$; $W_Q < W_P$
(d) $W_P = W_Q$; $W_P > W_Q$

58. When a long spring is stretched by 2 cm, its potential energy is U. If the spring is stretched by 10 cm, the potential energy stored in it will be **NCERT | Page-81 | 2003**

(a) 25 U (b) U/5 (c) 5 U (d) 10 U

Topic-5.10 Power

59. At any instant of time t, the displacement of any particle is given by 2t – 1 (SI unit) under the influence of force of 5N. The value of instantaneous power is (in SI unit): **NCERT | Page-83 | 2024**

(a) 10 (b) 5 (c) 7 (d) 6

60. The energy that will be ideally rediated by a 100 kW transmitter in 1 hour is: **NCERT | Page-83 | 2022**

(a) 36×10^4 J (b) 36×10^5 J
(c) 1×10^5 J (d) 36×10^7 J

61. An electric lift with a maximum load of 2000 kg (lift + passengers) is moving up with a constant speed of 1.5 ms^{-1}. The frictional force opposing the motion is 3000 N. The minimum power delivered by the motor to the lift in watts is: ($g = 10\ ms^{-2}$) **NCERT | Page-83 | 2022**

(a) 20000 (b) 34500
(c) 23500 (d) 23000

62. Water falls from a height of 60 m at the rate of 15 kg/s to operate a turbine. The losses due to frictional force are 10% of the input energy. How much power is generated by the turbine?
($g = 10\ m/s^2$) **NCERT | Page-83 | 2021**

(a) 7.0 kW (b) 10.2 kW
(c) 8.1 kW (d) 12.3 kW

63. A body of mass 1 kg begins to move under the action of a time dependent force $\vec{F} = (2t\hat{i} + 3t^2\hat{j})$ N, where $\hat{i}$ and $\hat{j}$ are unit vectors alogn *x* and *y*-axis. What power will be developed by the force at the time *t*? **NCERT | Page-83 | 2016**

(a) $(2t^2 + 3t^3)$W (b) $(2t^2 + 4t^4)$W
(c) $(2t^3 + 3t^4)$ W (d) $(2t^3 + 3t^5)$W

64. A particle of mass m is driven by a machine that delivers a constant power of *k* watts. If the particle starts from rest the force on the particle at time *t* is **NCERT | Page-83 | 2015**

(a) $\sqrt{mk}\ t^{-1/2}$ (b) $\sqrt{2mk}\ t^{-1/2}$
(c) $\frac{1}{2}\sqrt{mk}\ t^{-1/2}$ (d) $\sqrt{\frac{mk}{2}}t^{-1/2}$

65. The heart of man pumps 5 litres of blood through the arteries per minute at a pressure of 150 mm of mercury. If the density of mercury be 13.6×10^3 kg/m^3 and g = 10m/s^2 then the power of heart in watt is : NCERT | Page-83 | 2015 RS

(a) 2.35 (b) 3.0 (c) 1.50 (d) 1.70

66. One coolie takes 1 minute to raise a suitcase through a height of 2 m but the second coolie takes 30 s to raise the same suitcase to the same height. The powers of two coolies are in the ratio of NCERT | Page-83 | Kar. 2013

(a) 1 : 2 (b) 1 : 3 (c) 2 : 1 (d) 3 : 1

67. A car of mass m starts from rest and accelerates so that the instantaneous power delivered to the car has a constant magnitude P_0. The instantaneous velocity of this car is proportional to : NCERT | Page-83 | 2012 M

(a) t^2P_0 (b) $t^{1/2}$ (c) $t^{-1/2}$ (d) $\frac{t}{\sqrt{m}}$

68. A body projected vertically from the earth reaches a height equal to earth's radius before returning to the earth. The power exerted by the gravitational force is greatest NCERT | Page-83 | 2011

(a) at the highest position of the body
(b) at the instant just before the body hits the earth
(c) it remains constant all through
(d) at the instant just after the body is projected

69. A force of 250 N is required to lift a 75 kg mass through a pulley system. In order to lift the mass through 3 m, the rope has to be pulled through 12m. The efficiency of system is NCERT | Page-83 | 2011

(a) 50% (b) 75% (c) 33% (d) 90%

70. An engine pumps water through a hose pipe. Water passes through the pipe and leaves it with a velocity of 2 m/s. The mass per unit length of water in the pipe is 100 kg/m. What is the power of the engine? NCERT | Page-83 | 2010

(a) 400 W (b) 200 W(c) 100 W (d) 800 W

71. Water falls from a height of 60 m at the rate of 15 kg/s to operate a turbine. The losses due to frictional force are 10% of energy. How much power is generated by the turbine?
(g = 10 m/s^2) NCERT | Page-83 | 2008

(a) 8.1 kW (b) 10.2 kW
(c) 12.3 kW (d) 7.0 kW

72. A particle moves with a velocity $\vec{v} = (6\hat{i} - 4\hat{j} + 3\hat{k})$ m/s under the influence of a constant force $\vec{F} = (20\hat{i} + 15\hat{j} - 5\hat{k})$ N. The instantaneous power applied to the particle is NCERT | Page-73 | 2000

(a) 45 J/s (b) 35 J/s
(c) 25 J/s (d) 195 J/s

73. How much water, a pump of 2 kW can raise in one minute to a height of 10 m, take g = 10 m/s^2? NCERT | Page-83 | 1990

(a) 1000 litres (b) 1200 litres
(c) 100 litres (d) 2000 litres

Topic-5.11

Collisions

74. Two bodies A and B of same mass undergo completely inelastic one dimensional collision. The body A moves with velocity v_1 while body B is at rest before collision. The velocity of the system after collision is v_2. The ratio $v_1 : v_2$ is NCERT | Page-84 | 2024

(a) 1 : 2 (b) 2 : 1 (c) 4 : 1 (d) 1 : 4

75. A bullet of mass m hits a block of mass M elastically. The transfer of energy is the maximum, when NCERT | Page-84 | Manipur 2023

(a) M << m (b) M >> m
(c) M = m (d) M = 2m

76. Body A of mass 4m moving with speed u collides with another body B of mass 2m, at rest. The collision is head on and elastic in nature. After the collision the fraction of energy lost by the colliding body A is : NCERT | Page-84 | 2019

(a) $\frac{1}{9}$ (b) $\frac{8}{9}$ (c) $\frac{4}{9}$ (d) $\frac{5}{9}$

77. Two bullets are fired horizontally and simultaneously towards each other from roof tops of two buildings 100 m apart of same height of 200 m, with the same velocity of 25 m/s. When and where will the two bullets collide? (g = 10 m/s^2) NCERT | Page-84 | Odisha 2019

(a) They will not collide
(b) After 2 s at a height of 180 m
(c) After 2 s at a height of 20 m
(d) After 4 s at a height of 120 m

78. A particle of mass 5 m at rest suddenly breaks on its own into three fragments. Two fragments of mass m each move along mutually perpendicular direction with speed v each. The energy released during the process is, NCERT | Page-84 | Odisha 2019

(a) $\frac{4}{3}mv^2$ (b) $\frac{3}{5}mv^2$ (c) $\frac{5}{3}mv^2$ (d) $\frac{3}{2}mv^2$

79. A moving block having mass m, collides with another stationary block having mass 4m. The lighter block comes to rest after collision. When the initial velocity of the lighter block is v, then the value of coefficient of restitution (e) will be NCERT | Page-84 | 2018

(a) 0.5 (b) 0.25 (c) 0.4 (d) 0.8

80. Two identical balls A and B moving with velocities +0.5 m/s and –0.3 m/s respectively, collide head on elastically. The velocities of the balls A and B after collision, will be, respectively NCERT | Page-84 | 1991, Ph-II 2016

(a) +0.5 m/s and +0.3 m/s
(b) – 0.3 m/s and +0.5 m/s
(c) +0.3 m/s and 0.5 m/s
(d) –0.5 m/s and +0.3 m/s

81. Two particles A and B, move with constant velocities $\vec{v}_1$ and $\vec{v}_2$. At the initial moment their position vectors are $\vec{r}_1$ and $\vec{r}_2$ respectively. The condition for particles A and B for their collision is: **NCERT | Page-84 | 2015 RS**

(a) $\vec{r}_1 \cdot \vec{v}_1 = \vec{r}_2 \cdot \vec{v}_2$

(b) $\vec{r}_1 \times \vec{v}_1 = \vec{r}_2 \times \vec{v}_2$

(c) $\vec{r}_1 - \vec{r}_2 = \vec{v}_1 - \vec{v}_2$

(d) $\dfrac{\vec{r}_1 - \vec{r}_2}{|\vec{r}_1 - \vec{r}_2|} = \dfrac{\vec{v}_2 - \vec{v}_1}{|\vec{v}_2 - \vec{v}_1|}$

82. On a frictionless surface a block of mass M moving at speed v collides elastically with another block of same mass M which is initially at rest. After collision the first block moves at an angle θ to its initial direction and has a speed $\frac{v}{3}$. The second block's speed after the collision is: **NCERT | Page-84 | 2015 RS**

(a) $\frac{3}{4}v$ (b) $\frac{3}{\sqrt{2}}v$ (c) $\frac{\sqrt{3}}{2}v$ (d) $\frac{2\sqrt{2}}{3}v$

83. A ball is thrown vertically downwards from a height of 20 m with an initial velocity v_0. It collides with the ground loses 50 percent of its energy in collision and rebounds to the same height. The initial velocity v_0 is : (Take $g = 10\ ms^{-2}$) **NCERT | Page-84 | 2015 RS**

(a) $20\ ms^{-1}$ (b) $28\ ms^{-1}$

(c) $10\ ms^{-1}$ (d) $14\ ms^{-1}$

84. A body of mass (4m) is lying in x-y plane at rest. It suddenly explodes into three pieces. Two pieces, each of mass (m) move perpendicular to each other with equal speeds (v). The total kinetic energy generated due to explosion is : **NCERT | Page-84 | 2014**

(a) mv^2 (b) $\frac{3}{2}mv^2$

(c) $2\ mv^2$ (d) $4\ mv^2$

85. Two spheres A and B of masses m_1 and m_2 respectively collide. A is at rest initially and B is moving with velocity v along x-axis. After collision B has a velocity $\frac{v}{2}$ in a direction perpendicular to the original direction. The mass A moves after collision in the direction. **NCERT | Page-84 | 2012**

(a) Same as that of B

(b) Opposite to that of B

(c) $\theta = \tan^{-1}(1/2)$ to the x-axis

(d) $\theta = \tan^{-1}(-1/2)$ to the x-axis

86. A mass m moving horizontally (along the x-axis) with velocity v collides and sticks to mass of 3m moving vertically upward (along the y-axis) with velocity $2v$. The final velocity of the combination is **NCERT | Page-84 | 2011 M**

(a) $\frac{1}{4}v\hat{i} + \frac{3}{2}v\hat{j}$ (b) $\frac{1}{3}v\hat{i} + \frac{2}{3}v\hat{j}$

(c) $\frac{2}{3}v\hat{i} + \frac{1}{3}v\hat{j}$ (d) $\frac{3}{2}v\hat{i} + \frac{1}{4}v\hat{j}$

87. A ball moving with velocity 2 m/s collides head on with another stationary ball of double the mass. If the coefficient of restitution is 0.5, then their velocities (in m/s) after collision will be: **NCERT | Page-84 | 2010**

(a) 0, 1 (b) 1, 1 (c) 1, 0.5 (d) 0, 2

88. A shell of mass 200 gm is ejected from a gun of mass 4 kg by an explosion that generates 1.05 kJ of energy. The initial velocity of the shell is: **NCERT | Page-84 | 2008**

(a) $100\ ms^{-1}$ (b) $80\ ms^{-1}$

(c) $40\ ms^{-1}$ (d) $120\ ms^{-1}$

89. A bomb of mass 30 kg at rest explodes into two pieces of masses 18 kg and 12 kg. The velocity of 18 kg mass is 6 ms^{-1}. The kinetic energy of the other mass is **NCERT | Page-84 | 2005**

(a) 324 J (b) 486 J (c) 256 J (d) 524 J

90. A mass of 0.5 kg moving with a speed of 1.5 m/s on a horizontal smooth surface, collides with a nearly weightless spring of force constant $k = 50$ N/m. The maximum compression of the spring would be **NCERT | Page-84 | 2004**

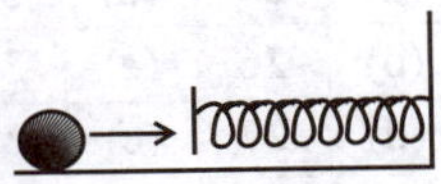

(a) 0.5 m (b) 0.15 m

(c) 0.12 m (d) 1.5 m

91. A stationary particle explodes into two particles of masses m_1 and m_2 which move in opposite directions with velocities v_1 and v_2. The ratio of their kinetic energies E_1/E_2 is **NCERT | Page-84 | 2003**

(a) m_1v_2/m_2v_1 (b) m_2/m_1

(c) m_1/m_2 (d) 1

92. A bomb of mass 1 kg is thrown vertically upwards with a speed of 100 m/s. After 5 seconds it explodes into two fragments. One fragment of mass 400 gm is found to go down with a speed of 25 m/s. What will happen to the second fragment just after the explosion? ($g = 10\ m/s^2$) **NCERT | Page-84 | 2000**

(a) It will go upward with speed 40 m/s

(b) It will go upward with speed 100 m/s

(c) It will go upward with speed 60 m/s

(d) It will also go downward with speed 40m/s

93. Two equal masses m_1 and m_2 moving along the same straight line with velocities + 3 m/s and – 5m/s respectively, collide elastically. Their velocities after the collision will be respectively. **NCERT | Page-84 | 1998**

(a) –3 m/s & +5 m/s

(b) +4 m/s for both

(c) –4 m/s & +4 m/s

(d) –5m/s & +3 m/s

94. A molecule of mass m of an ideal gas collides with the wall of a vessel with a velocity v and returns back with the same velocity. The change in linear momentum of molecule is **NCERT | Page-84 | 1997**

(a) 2 mv (b) 4 mv
(c) 8 mv (d) 10 mv

95. A metal ball of mass 2 kg moving with a velocity of 36 km/h has a head on collision with a stationary ball of mass 3 kg. If after the collision, the two balls move together, the loss in kinetic energy due to collision is **NCERT | Page-84 | 1997**

(a) 140 J (b) 100 J
(c) 60 J (d) 40 J

96. A body of mass m moving with velocity 3 km/h collides with a body of mass 2 m at rest. Now the coalesced mass starts to move with a velocity **NCERT | Page-84 | 1996**

(a) 1 km/h (b) 2 km/h
(c) 3 km/h (d) 4 km/h

97. A shell is fired from a cannon, it explodes in mid air, its total **NCERT | Page-84 | 1994**

(a) momentum increases
(b) momentum decreases
(c) K.E. increases
(d) K.E. decreases

98. The co-efficient of restitution e for a perfectly elastic collision is **NCERT | Page-84 | 1988**

(a) 1 (b) 0
(c) ∞ (d) –1

ANSWER KEY

1	(c)	11	(b)	21	(d)	31	(b)	41	(b)	51	(a)	61	(b)	71	(a)	81	(d)	91	(b)
2	(b)	12	(b)	22	(c)	32	(b)	42	(d)	52	(b)	62	(c)	72	(a)	82	(d)	92	(b)
3	(b)	13	(c)	23	(d)	33	(a)	43	(d)	53	(a)	63	(d)	73	(b)	83	(a)	93	(d)
4	(d)	14	(a)	24	(c)	34	(c)	44	(d)	54	(b)	64	(d)	74	(b)	84	(b)	94	(a)
5	(b)	15	(d)	25	(c)	35	(b)	45	(d)	55	(d)	65	(d)	75	(c)	85	(d)	95	(c)
6	(c)	16	(b)	26	(c)	36	(a)	46	(c)	56	(a)	66	(a)	76	(b)	86	(a)	96	(a)
7	(c)	17	(d)	27	(c)	37	(c)	47	(b)	57	(b)	67	(b)	77	(b)	87	(a)	97	(c)
8	(a)	18	(b)	28	(c)	38	(b)	48	(d)	58	(a)	68	(b)	78	(a)	88	(a)	98	(a)
9	(d)	19	(d)	29	(d)	39	(a)	49	(d)	59	(a)	69	(b)	79	(b)	89	(b)		
10	(c)	20	(c)	30	(d)	40	(a)	50	(a)	60	(d)	70	(a)	80	(b)	90	(b)		

Hints & Solutions

1. **(c) Given:** Position vector

$\vec{r} = \cos \omega t\ \hat{x} + \sin \omega t\ \hat{y}$

$\therefore$ Velocity vector,

$\vec{v} = -\omega \sin \omega t\ \hat{x} + \omega \cos \omega t\ \hat{y}$ acceleration vector,

$\vec{a} = -\omega^2 \cos \omega t\ \hat{x} - \omega^2 \sin \omega t\ \hat{y} = -\omega^2 \vec{r}$

$\vec{r} \cdot \vec{v} = 0$ hence $\vec{r} \perp \vec{v}$ and $\vec{a}$ is directed towards the origin.

NOTES Nothing actually moves in the direction of the angular velocity vector $\vec{\omega}$. The direction of $\vec{\omega}$ simply represents that the circular motion is taking place in a plane.

2. **(b)** $|\vec{A}+\vec{B}| = |\vec{A}-\vec{B}|$

Squaring on both sides

$|\vec{A}+\vec{B}|^2 = |\vec{A}-\vec{B}|^2$

$\Rightarrow \vec{A}\cdot\vec{A} + 2\vec{A}\cdot\vec{B} + \vec{B}\cdot\vec{B}$

$= \vec{A}\cdot\vec{A} - 2\vec{A}\cdot\vec{B} + \vec{B}\cdot\vec{B}$

$\Rightarrow 4\vec{A}\cdot\vec{B} = 0 \Rightarrow 4AB\cos\theta = 0$

$\Rightarrow \cos\theta = 0 \Rightarrow \theta = 90°$

3. **(b)** Two vectors are

$\vec{A} = \cos\omega t\hat{i} + \sin\omega t\hat{j}$

$\vec{B} = \cos\frac{\omega t}{2}\hat{i} + \sin\frac{\omega t}{2}\hat{j}$

For two vectors $\vec{A}$ and $\vec{B}$ to be orthogonal $\vec{A}.\vec{B} = 0$

$\Rightarrow \cos\omega t.\cos\frac{\omega t}{2} + \sin\omega t.\sin\frac{\omega t}{2} = 0$

$\Rightarrow \cos\left(\omega t - \frac{\omega t}{2}\right) = \cos\left(\frac{\omega t}{2}\right) = 0$

So, $\frac{\omega t}{2} = \frac{\pi}{2} \quad \therefore t = \frac{\pi}{\omega}$

4. **(d)** $|\vec{A}+\vec{B}|^2 = |\vec{A}-\vec{B}|^2$

$\Rightarrow A^2 + B^2 + 2AB\cos\theta$

$= A^2 + B^2 - 2AB\cos\theta$

$4AB\cos\theta = 0 \Rightarrow \cos\theta = 0$

$\theta = 90°$

So, angle between A & B is 90°.

5. **(b)** For two vectors to be perpendicular to each other $\vec{A}\cdot\vec{B} = 0$

$(2\hat{i}+3\hat{j}+8\hat{k})\cdot(4\hat{j}-4\hat{i}+\alpha\hat{k})=0$

$-8+12+8\alpha=0$

$\alpha=-\frac{4}{8}=-\frac{1}{2}$

6. **(c)** $\vec{P}=$ vector sum $=\vec{A}+\vec{B}$

$\vec{Q}=$ Vector differences $=\vec{A}-\vec{B}$

Since $\vec{P}$ and $\vec{Q}$ are perpendicular

$\therefore\ \vec{P}.\vec{Q}=0 \Rightarrow (\vec{A}+\vec{B}).(\vec{A}-\vec{B})=0$

$\Rightarrow A^2=B^2 \Rightarrow |\vec{A}|=|\vec{B}|$

7. **(c)** $\vec{A}=3\hat{i}+4\hat{j}+5\hat{k}$, $\vec{B}=3\hat{i}+4\hat{j}-5\hat{k}$

$\therefore\ \vec{A}.\vec{B}=(3\hat{i}+4\hat{j}+5\hat{k}).(3\hat{i}+4\hat{j}-5\hat{k})$

$=9+16-25=0$

$\because\ |\vec{A}|\neq 0,\ |\vec{B}|\neq 0,$

hence, $\cos\theta=0 \Rightarrow \theta=90°$

8. **(a)** $(\vec{A}+\vec{B})^2=(\vec{C})^2$

$\Rightarrow\ A^2+B^2+2\vec{A}\cdot\vec{B}=C^2$

$\Rightarrow\ 3^2+4^2+2\vec{A}\cdot\vec{B}=5^2$

$\Rightarrow\ 2\vec{A}\cdot\vec{B}=0$ or $\Rightarrow \vec{A}\cdot\vec{B}=0\ \therefore\ \vec{A}\perp\vec{B}$

$\therefore\ \theta=\frac{\pi}{2}$

9. **(d)** Given : $\vec{F}=(3\hat{i}+\hat{j})$N

$\vec{r}_1=(2\hat{i}+\hat{k})$m, $\vec{r}_2=(4\hat{i}+3\hat{j}-\hat{k})m$

$\vec{r}=\vec{r}_2-\vec{r}_1=(4\hat{i}+3\hat{j}-\hat{k})-(2\hat{i}+\hat{k})$

or $\vec{r}=(2\hat{i}+3\hat{j}-2\hat{k})m$

So, work done by the given force, $W=\vec{F}\cdot\vec{r}$

$=(3\hat{i}+\hat{j})\cdot(2\hat{i}+3\hat{j}-2\hat{k})=6+3=9$J

10. **(c)** Since displacement is along the y-direction, hence displacement $\vec{s}=10\hat{j}$.

Work done $=\vec{F}\cdot\vec{s}$

$=(-2\hat{i}+15\hat{j}+6\hat{k})\cdot 10\hat{j}=150J$

11. **(b)** $W=\frac{1}{2}m(v_1^2-v_2^2)$

$=\frac{1}{2}\times 0.01[(1000)^2-(500)^2]=3750$J.

12. **(b)** By work-energy theorem,

W.D. $=\Delta$KE

$\frac{(\Delta KE)_A}{(\Delta KE)_B}=\frac{(WD)_A}{(WD)_B}=\frac{-F_AS_A}{-F_BS_B}$

$\Rightarrow \frac{-100}{-225}=\frac{-F_A(1000)}{-F_B(1500)}$

$\Rightarrow \therefore \frac{F_A}{F_B}=\frac{2}{3}$

13. **(c)** $W=\Delta$ K.E

$\Rightarrow K_f-K_i=\vec{F}\cdot\Delta\vec{x}$

$\Rightarrow K_f-10=(-2\hat{i}+3\hat{j})\cdot(3\hat{i})$

$K_f-10=-6$

$K_f=4$ J

14. **(a)** Given: Mass of particle, $M=10g=\frac{10}{1000}$kg

radius of circle R = 6.4 cm

Kinetic energy E of particle $=8\times 10^{-4}$J

acceleration $a_t=?$

$\frac{1}{2}mv^2=E \Rightarrow \frac{1}{2}\left(\frac{10}{1000}\right)v^2=8\times 10^{-4}$

$\Rightarrow\ v^2=16\times 10^{-2}$

$\Rightarrow\ v=4\times 10^{-1}=0.4$ m/s

Now using

$v^2=u^2+2a_ts$ $\qquad (s=4\pi R)$

$(0.4)^2=0^2+2a_t\left(4\times\frac{22}{7}\times\frac{6.4}{100}\right)$

$\Rightarrow\ a_t=(0.4)^2\times\frac{7\times 100}{8\times 22\times 6.4}=0.1$ m/s^2

15. **(d)** Given, m = 10 kg, $u=10$ m/sec

Retarding force, F = 0.1x J/m

Acceleration, $a=\frac{F}{m}$

$a=\frac{0.1x}{10}=0.01x$

From, $v^2=u^2-2\int_{x_1}^{x_2}adx$

$\Rightarrow\ v^2=(10)^2-2\times 0.01\int_{20}^{30}xdx$

$\Rightarrow\ v^2=100-0.02\left[\frac{x^2}{2}\right]_{20}^{30}$

$\Rightarrow\ v^2=100-\frac{0.02}{2}[(30)^2-(20)^2]$

$\Rightarrow\ v^2=100-0.01[900-400]$

$\Rightarrow\ v^2=100-0.01[500]$

$\Rightarrow\ v^2=95$ m/sec

Final K.E. $=\frac{1}{2}mv^2$

$=\frac{1}{2}\times 10\times 95 \Rightarrow 475$ J

NOTES From, F = ma

$$a=\frac{F}{m}=\frac{0.1x}{10}=0.01x=V\frac{dV}{dx}$$

So, $\int_{V_1}^{V_2} VdV=\int_{20}^{30}\frac{x}{100}dx$

$$-\frac{V^2}{2}\Big|_{V_1}^{V_2}=\frac{x^2}{200}\Big|_{20}^{30}=\frac{30\times30}{200}-\frac{20\times20}{200}$$

$= 4.5 - 2 = 2.5$

$\frac{1}{2}m\left(V_2^2-V_1^2\right)=-10\times2.5\,J=-25J$

Final K.E. $=\frac{1}{2}mv_2^2=\frac{1}{2}mv_1^2-25=\frac{1}{2}\times10\times10\times10-25$

$= 500 - 25 = 475$ J

16. (b) According to law of conservation of momentum

MV + mnv = 0

$\Rightarrow V=\frac{-mNv}{M}=\frac{-0.01\text{ kg}\times10\times800\text{ m/s}}{100}$

$\Rightarrow -0.8$ m/s

$F_{av}=\frac{\Delta P}{\Delta t}=\frac{10\times0.01\times800}{5}=16N$

17. (d) m = mass per unit length. If water moves x distance in 't' time.

$\therefore$ Mass flowing per second $=\frac{\text{mx}}{\text{t}}=\text{mv}$

Rate of K.E.$=\frac{1}{2}(mv)\,v^2=\frac{1}{2}mv^3$

18. (b) $E=\frac{p^2}{2m}$

or, $E_1=\frac{p_1^2}{2m_1},E_2=\frac{p_2^2}{2m_2}$

or, $m_1=\frac{p_1^2}{2E_1},m_2=\frac{p_2^2}{2E_2}$

$m_1>m_2\Rightarrow\frac{m_1}{m_2}>1$

$\therefore\ \frac{p_1^2E_2}{E_1p_2^2}>1\Rightarrow\frac{E_2}{E_1}>1$ $[\because p_1=p_2]$

or, $E_2>E_1$

NOTES Kinetic energy of body of mass *m* moving with velocity *v* is

K.E. $=\frac{1}{2}mv^2$ $\Rightarrow$ K.E. $=\frac{1}{2}\frac{mv^2\times m}{m}$

[Multiplying both numerator and denominator by *m*]

$\Rightarrow$ K.E. $=\frac{1}{2}\frac{m^2v^2}{m}$ $\Rightarrow$ K.E. $=\frac{1}{2}\frac{p^2}{m}$ $[\because$ Momentum, $P=mv]$

$\Rightarrow$ $P^2=2m$ K.E.

$\Rightarrow$ $P=\sqrt{2mK.E.}$

19. (d)

NOTES Since height is same for both balls, their velocities on reaching 30 ft above ground will be same.

$$\therefore\ \frac{K.E_1}{K.E_2}=\frac{\frac{1}{2}m_1v_0^2}{\frac{1}{2}m_2v_0^2}=\frac{m_1}{m_2}=\frac{2}{4}=\frac{1}{2}$$

20. (c) New K.E., E′ = E + 3E = 4E

$p=\sqrt{2mE}$ and $p'=\sqrt{2mE'}$

$\frac{p'}{p}=\sqrt{\frac{2m\times4E}{2mE}}=2$

$\frac{p'}{p}-1=2-1$ [on substrating 1 in both sides.]

$\frac{p'-p}{p}\times100=(2-1)\times100=100\%$

21. (d) $\frac{(\text{K.E})_1}{(\text{K.E})_2}=\frac{\frac{1}{2}m_1v_1^2}{\frac{1}{2}m_2v_2^2}=\frac{4}{1}\Rightarrow\frac{m_1v_1^2}{m_2v_2^2}=\frac{4}{1}$

$\Rightarrow\frac{(m_1v_1)^2}{(m_2v_2)^2}\frac{m_2}{m_1}=\frac{4}{1}\Rightarrow\frac{p_1^2}{p_2^2}\times\frac{m_2}{m_1}=\frac{4}{1}$

[Given: $p_1=p_2$]

$\frac{m_2}{m_1}=\frac{4\times p_2^2}{1\times p_1^2}=\frac{4}{1}\times1\Rightarrow\frac{m_1}{m_2}=\frac{1}{4}$

22. (c) $m_1=m,m_2=4$ m

$\text{K.E}_1=\text{K.E}_2$

$\frac{1}{2}m_1v_1^2=\frac{1}{2}m_2v_2^2;\ \frac{1}{2}mv_1^2=\frac{1}{2}4m\,v_2^2$

$\frac{v_1}{v_2}=2\Rightarrow v_1=2v_2$

$\frac{\text{Linear momentum of first body}}{\text{Linear momentum of second body}}$

$=\frac{m_1v_1}{m_2v_2}=\frac{m.2v_2}{4mv_2}=\frac{1}{2}$

23. (d) K.E. $=\frac{1}{2}mv^2$

Further, $v^2=u^2+2as=0+2ad=2ad$

$=2(\text{F/m})d$

Hence, K.E. $=\frac{1}{2}m\times2(F/m)d=Fd$

or, K.E. acquired = Work done

= F × *d* = constant.

i.e., it is independent of mass m.

24. (c) Initial momentum $(p_1)=p$; Final momentum $(p_2)=$ 1.5 *p* and initial kinetic energy (K_1) = K.

Kinetic energy $(K)=\frac{p^2}{2m}\propto p^2$

or, $\frac{K_1}{K_2} = \left(\frac{p_1}{p_2}\right)^2 = \left(\frac{p}{1.5p}\right)^2 = \frac{1}{2.25}$

or, $K_2 = 2.25$ K.

Therefore, increase in kinetic energy is 2.25 K – K = 1.25 K or 125%.

25. (c) $\frac{1}{2}(1)v_1^2 = \frac{1}{2}(9)v_2^2 \Rightarrow \frac{v_1^2}{v_2^2} = 9$ or $\frac{v_1}{v_2} = 3$

Ratio of their linear momenta

$= \frac{m_1 v_1}{m_2 v_2} = \frac{1}{9} \times (3) = \frac{1}{3}$

26. (c) $E = \frac{1}{2}mv^2$. Hence, $mv = (2mE)^{1/2}$.

For same KE, momentum $\propto \sqrt{m}$.

Hence, the ratio is 2 : 1.

27. (c) $\frac{K_1}{K_2} = \frac{p_1^2}{m_1} \times \frac{m_2}{p_2^2}$ $[\because p = mv \Rightarrow K = \frac{p^2}{2m}]$

Hence, $\frac{p_1}{p_2} = \sqrt{\frac{m_1}{m_2}} = \sqrt{\frac{1}{4}} = \frac{1}{2}$

28. (c) Work done by variable force is

$$W = \int_{y_i}^{y_f} F dy$$

Here, F = 20 + 10 y

$$\therefore W = \int_0^1 (20 + 10y)dy = \left[20y + \frac{10y^2}{2}\right]_0^1 = 25\, J$$

29. (d) Using work-energy theorem, ΔK = work = area under F – x graph

From x = 0 to x = 8 m

$\frac{1}{2}mv_1^2 = 100 + 30$

$\therefore v_1 = \sqrt{520} = 23$ m/s

From x = 0 to x = 12 m

$\frac{1}{2}mv_2^2 = 100 + 30 - 47.5 + 20$

$\therefore v_2 = \sqrt{410} = 20.6$ m/s

30. (d) Work done = area under F-d graph

$= [2 \times (7-3)] + \left[\frac{1}{2} \times 2 \times (12-7)\right]$

= 8 + 5 = 13 J.

31. (b) Acceleration $= a = \frac{2}{3}$ m/s²

Force acting on the body

$F = ma = 3 \times \frac{2}{3} = 2$ newton

Displacement in 2 secs, $s = \frac{1}{2}at^2 = \frac{1}{2} \times \frac{2}{3} \times 2^2 = \frac{4}{3}$m

Work done $= F.s = 2 \times \frac{4}{3} = \frac{8}{3}$ J

32. (b)

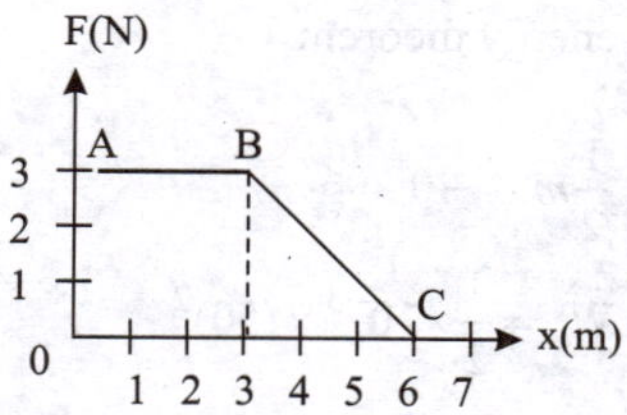

Work done = area under F-x graph = area of trapezium OABC

$= \frac{1}{2}(3+6)(3) = 13.5$ J

33. (a) $x = 3t - 4t^2 + t^3$

$\frac{dx}{dt} = 3 - 8t + 3t^2$

Acceleration $= \frac{d^2x}{dt^2} = -8 + 6t$

Acceleration after 4 sec

$= -8 + 6 \times 4 = 16$ ms^{-2}

Displacement in 4 sec

$= 3 \times 4 - 4 \times 4^2 + 4^3 = 12$ m

∴ Work = Force × displacement

= Mass × acc. × disp.

$= 30 \times 10^{-3} \times 16 \times 12 = 576$ mJ

34. (c) $W = \int_0^s F dx = \int_0^s (7 - 2x + 3x^2)dx$

$= [7x - x^2 + x^3]_0^5 = 135J$

35. (b) Gravitational force is a conservative force and work done by conservative force is path independent

$\therefore W_1 = W_2 = W_3$

36. (a) Let the height from the surface of the earth be x. When body from rest falls through height (S – x) Then from $v^2 = u^2 + 2gs$

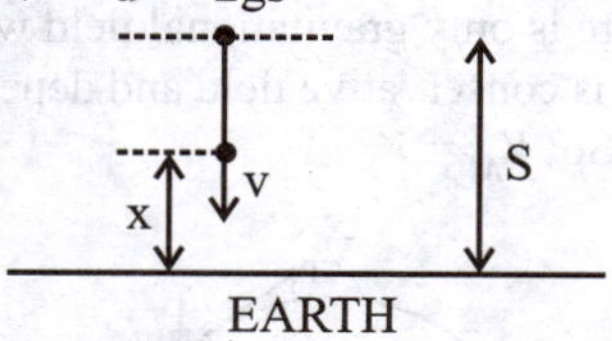

$v^2 = 0^2 + 2g(S - x)$

$v = \sqrt{2g(S-x)}$ ...(i)

At this point, U = mgx

From question, kinetic energy, K = 3U

$\therefore \frac{1}{2}m(v)^2 = 3 \times mg(x)$

$\frac{1}{2} \times m \times 2g(S-x) = 3 \times mgx$ (using (i))

S – x = 3x

$\therefore x = \frac{S}{4}$ ... (ii)

$\therefore v = \sqrt{2 \times g\left(S - \frac{S}{4}\right)} = \sqrt{\frac{3gS}{2}}$ ... (iii)

37. (c) From work-energy theorem,
$W_g + W_a = \Delta K.E.$

or, $mgh + W_a = \frac{1}{2}mv^2 - 0$

$10^{-3} \times 10 \times 10^3 + W_a = \frac{1}{2} \times 10^{-3} \times (50)^2$

$\Rightarrow W_a = -8.75$ J
which is the work done due to air resistance
Work done due to gravity = mgh
$= 10^{-3} \times 10 \times 10^3 = 10$ J

38. (b) When work is done by a system against a conservative force then its potential energy increases.

39. (a)

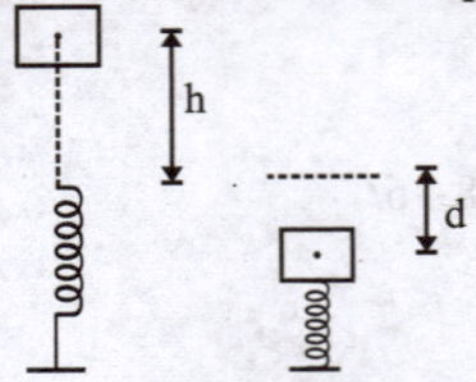

W = Work done by gravity + Work done by spring

$W = mg(h + d) - \frac{1}{2}kd^2$

NOTES Gravitational potential energy of ball gets converted into elastic potential energy of the spring.

$mg(h + d) = \frac{1}{2}kd^2$

Net work done $= mg(h + d) - \frac{1}{2}kd^2 = 0$

40. (a) Work done against gravity = mg sin θ × d
($d \sin\theta = 10$)
$= 2 \times 10 \times 10 = 200$ J
Actual work done = 300 J
Work done against friction = 300 – 200 = 100 J

41. (b) As there is only gravitational field which works. We know it is conservative field and depends only on the end points. So, $V_M = V_B$

42. (d)

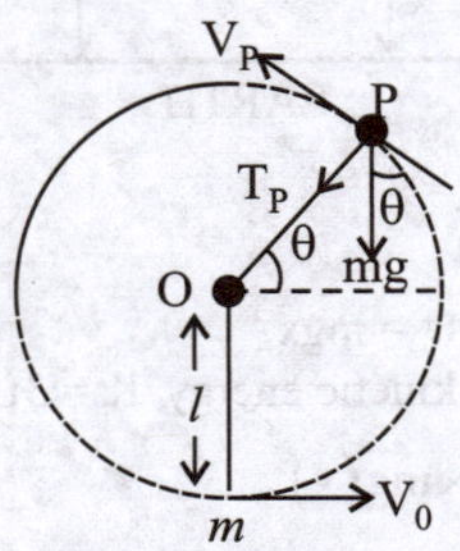

By mechanical energy conservation,

$$\frac{1}{2}mv_0^2 = mg\ell(1+\sin\theta) + \frac{1}{2}mv_p^2 \quad \text{...(i)}$$

At point P,

$$T_p + mg\sin\theta = \frac{mv_p^2}{\ell}$$

Here, $T_p = 0$

$$mg\sin\theta = \frac{mv_p^2}{\ell} \Rightarrow mv_p^2 = mg\ell\sin\theta \quad \text{...(ii)}$$

$\therefore v_p = \sqrt{g\ell\sin\theta}$

From equations (i) & (ii),

$$\frac{1}{2}mv_0^2 = mg\ell(1+\sin\theta+) + \frac{1}{2}mg\ell\sin\theta$$

$$v_0^2 = 2g\ell(1+\sin\theta) + g\ell\sin\theta$$

$$\Rightarrow v_0 = \sqrt{2g\ell + 3g\ell\sin\theta}$$

$$\therefore \frac{v_p}{v_0} = \sqrt{\frac{\sin\theta}{2+3\sin\theta}}$$

43. (d) K.E. before striking the ground is
$K_1 = mgh_1 = mg(10)$
K.E. after striking the ground is

$$= K_1 - \frac{50}{100}K_1 = \frac{50}{100}K_1 = \frac{K_1}{2} = K_2$$

Now, $K_2 = mgh_2 \Rightarrow \frac{K_1}{2} = mgh_2$

$$\Rightarrow \frac{mg(10)}{2} = mgh_2 \Rightarrow h_2 = 5 \text{ m}$$

44. (d) Velocity of point mass in vertical circle at lowest point, $V = \sqrt{7gr}$

$v = \sqrt{7gr} > \sqrt{5gr}$

Hence, point mass will have completed the vertical circular path.

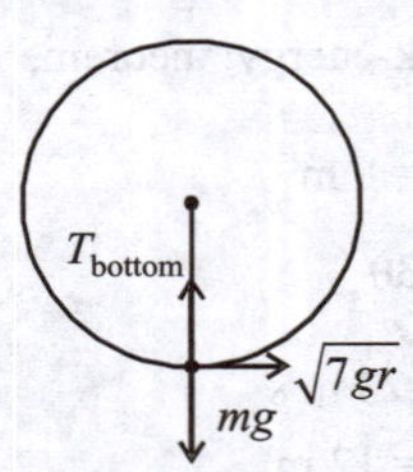

At the lowest point mg will act downward and tension in the upward direction,
$T_{bottom} - mg$ = centripetal force

$$\therefore T_{bottom} - mg = \frac{mv^2}{r} = \frac{m}{r}(\sqrt{7gr})^2$$

$$\Rightarrow T_{bottom} - mg = 7mg$$

$$\Rightarrow T_{bottom} = 8mg$$

45. (d) $\sqrt{5gR}$ is the minimum velocity which the body must possess at the bottom of the circle so as to go round the circle completely.

46. (c) $T - mg = \frac{mv^2}{r}$ [centripetal force]

$$\Rightarrow T = mg + \frac{mv^2}{r}$$

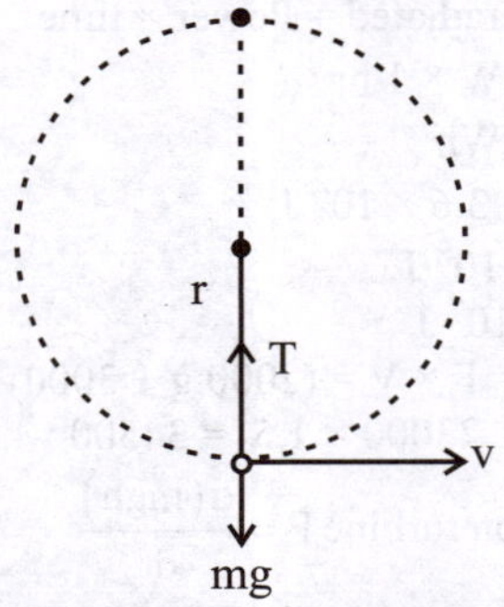

As the velocity is maximum at lowest point so tension is maximum at the lowest position of mass, so the chance of breaking is maximum.

47. **(b)** As the particle is moving in a potential energy region.
And, total energy E = K.E. + P.E.
$\Rightarrow$ P.E = E – K.E.
Kinetic energy > 0
$\Rightarrow U(x) < E$

48. **(d)** When the body is thrown upwards. its K.E is converted into P.E.
The loss of energy due to air friction is the difference of K.E and P.E.

$$\frac{1}{2}mv^2 - mgh = \frac{1}{2}\times1\times400 - 1\times18\times10$$
$$= 200 - 180 = 20 \text{ J}$$

49. **(d)** If l is length of pendulum and θ be angular amplitude then height

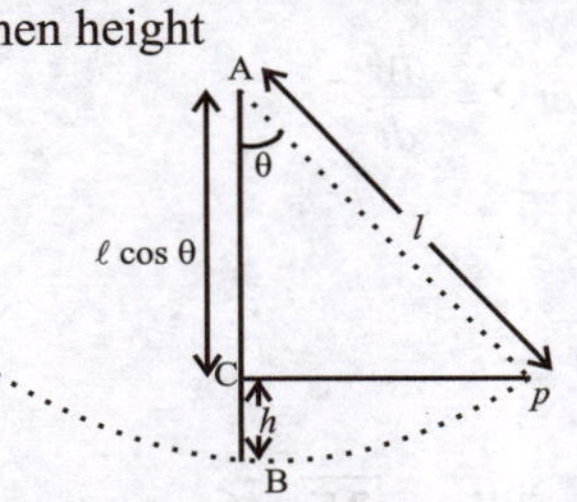

h = AB – AC = $l - l\cos\theta = l(1-\cos\theta)$
At extreme position, potential energy is maximum and kinetic energy is zero; At mean (equilibrium) position potential energy is zero and kinetic energy is maximum, so from principle of conservation of energy.
(KE + PE) at P = (KE + PE) at B

$$0 + mgh = \frac{1}{2}mv^2 + 0$$
$$\Rightarrow v = \sqrt{2gh} = \sqrt{2g\ell(1-\cos\theta)}$$

50. **(a)** In the case of a body describing a vertical circle,

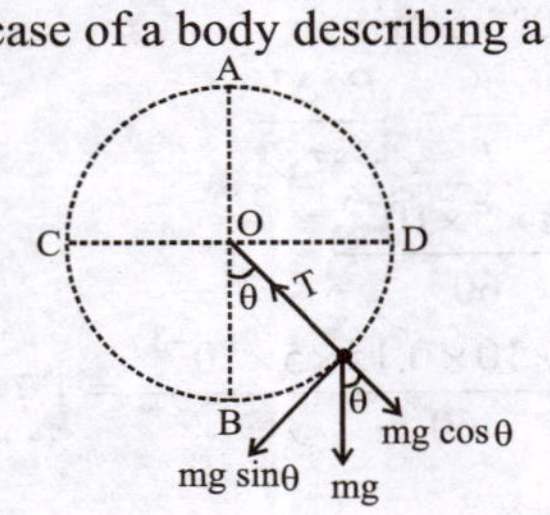

$$T - mg\cos\theta = \frac{mv^2}{l};\ T = mg\cos\theta + \frac{mv^2}{l}$$

Tension is maximum when cos θ = +1 and velocity is maximum.
Both conditions are satisfied at θ = 0º (*i.e.*, at lowest point *B*).

51. **(a)** Minimum speed with which the string is rotating in a vertical circle (v) = $\sqrt{gr}$
The minimum speed of stone is independent of mass of stone.

52. **(b)** According to principle of conservation of energy,
Loss in potential energy = Gain in kinetic energy

$$\Rightarrow mgh = \frac{1}{2}mv^2 \Rightarrow v = \sqrt{2gh}$$

If h_1 and h_2 are initial and final heights, then
$v_1 = \sqrt{2gh_1}$ and $v_2 = \sqrt{2gh_2}$
Loss in velocity,
$\Delta v = v_1 - v_2 = \sqrt{2gh_1} - \sqrt{2gh_2}$
∴ Fractional loss in velocity,

$$= \frac{\Delta v}{v_1} = \frac{\sqrt{2gh_1} - \sqrt{2gh_2}}{\sqrt{2gh_1}} = 1 - \sqrt{\frac{h_2}{h_1}}$$
$$= 1 - \sqrt{\frac{1.8}{5}} = 1 - \sqrt{0.36} = 1 - 0.6$$
$$= 0.4 = \frac{2}{5}$$

53. **(a)** Given : Mass (*m*) = 0.4 kg
Its frequency (*n*) = 2 rev/sec
Radius (*r*) =1.2 m. We know that linear velocity of the body (*v*) = ωr = (2π*n*)*r*
= 2 × 3.14 × 2 × 1.2 = 15.08 m/s.
Therefore, tension in the string when the body is at the top of the circle (T)

$$= \frac{mv^2}{r} - mg$$
$$= \frac{0.4\times(15.08)^2}{1.2} - (0.4\times9.8) = 71.88\,\text{N}$$

54. **(b)** Energy lost due to friction = kinetic energy

$$\mu mg\,s = \frac{1}{2}mv^2$$

[Here, v = 72 km/h $= \frac{72000}{60\times60} = 20\text{m/s}$]

or, $$s = \frac{v^2}{2\mu g} = \frac{20\times20}{2\times0.5\times10} = 40\text{m}$$

NOTES If a body of mass m moves with velocity u on a rough surface and stops after travelling a distance S due to friction. Then, frictional force, $F = ma = \mu R \Rightarrow ma = \mu mg \Rightarrow a = \mu g$

Using $v^2 = u^2 - 2as$

$$\Rightarrow 0 = u^2 - 2\mu g S \Rightarrow S = \frac{u^2}{2\mu g}$$

55. **(d)** Potential energy stored in spring (U) is given by

$$U = \frac{1}{2}Kx^2$$

Initially

$$U_i = \frac{1}{2}K(2)^2 \text{ where } x = 2 \text{ cm}$$

$$\Rightarrow U_i = \frac{1}{2}(K)\cdot(4) = 2K \quad \text{(i)}$$

Finally

$$U_f = \frac{1}{2}K(8)^2 = \frac{1}{2}K \times 64 = 32K \quad \text{(ii)}$$

On dividing (i) by (ii)

$$\frac{U_i}{U_f} = \frac{2K}{32K} = \frac{1}{16}$$

$$\Rightarrow U_f = 16U$$

56. **(a)** Graph (a) represents restoring of a spring, $F = -kx$

57. **(b)** **Case (a):** Suppose the two springs are stretched by the same distance x. Then

$$\frac{W_P}{W_Q} = \frac{\frac{1}{2}k_P x^2}{\frac{1}{2}k_Q x^2} = \frac{k_P}{k_Q}$$

As $k_p > k_Q$ so, $W_p > W_Q$

Case (b): Suppose the two springs stretched by distance x_P and x_Q by the same force F.

Then, $F = k_p x_p = k_Q x_Q$

$$\frac{W_P}{W_Q} = \frac{\frac{1}{2}k_P x_P^2}{\frac{1}{2}k_Q x_Q^2} = \frac{k_P \cdot x_P \cdot x_P}{k_Q \cdot x_Q \cdot x_Q} = \frac{Fx_P}{Fx_Q} = \frac{k_Q}{k_P}$$

As $k_P > k_Q$ $\therefore$ $W_Q > W_P$

58. **(a)** If k be the spring constant, then

$$U = \frac{1}{2}\times k \times (2)^2 = 2k$$

$$U_{final} = \frac{1}{2}\times k \times (10)^2 = 50k$$

$$\Rightarrow \frac{U}{U_{final}} = \frac{2k}{50k} = \frac{1}{25} \Rightarrow U_{final} = 25U$$

59. **(a)** Given, displacement $x = 2t - 1$

$$\therefore \text{ velocity } v = \frac{dx}{dt} = 2\text{ms}^{-1}$$

Therefore power $P = F.\, v = 2 \times 5 = 10$ W

60. **(d)** Energy radiated = Power × time

= 100 kW × 1 hr

= 100 kWh

= $100 \times 3.6 \times 10^6$ J

= 3.6×10^8 J

= 36×10^7 J

61. **(b)** Power = F × V = (2000 g + 3000) × 1.5

= 23000 × 1.5 = 34500 watt

62. **(c)** Power on turbine $P = \frac{d(mgh)}{dt}$

$$= gh\frac{dm}{dt} = 10 \times 60 \times 15 = 9000\text{W}$$

Due to frictional force, losses are 10%

$\therefore$ Power generated

$$= \left(1 - \frac{10}{100}\right) \times 9000 = 8.1 \text{ kW}$$

63. **(d)** Given force $\vec{F} = 2t\hat{i} + 3t^2\hat{j}$

According to Newton's second law of motion,

$$m\frac{d\vec{v}}{dt} = 2t\hat{i} + 3t^2\hat{j} \text{ (m = 1 kg)}$$

$$\Rightarrow \int_0^{\vec{v}} d\vec{v} = \int_0^t \left(2t\hat{i} + 3t^2\hat{j}\right)dt \Rightarrow \vec{v} = t^2\hat{i} + t^3\hat{j}$$

Power $P = \vec{F}\cdot\vec{v}\ (2t\hat{i} + 3t^2\hat{j})\cdot(t^2\hat{i} + t^3\hat{j})$

$= (2t^3 + 3t^5)$W

64. **(d)** As we know power $P = \frac{dw}{dt}$

$$\Rightarrow w = Pt = \frac{1}{2}\text{mV}^2$$

$$\text{So, } v = \sqrt{\frac{2Pt}{m}} = \sqrt{\frac{2kt}{m}}$$

$$\text{Hence, acceleration } a = \frac{dV}{dt} = \sqrt{\frac{2K}{m}}\cdot\frac{1}{2\sqrt{t}}$$

Therefore, force on the particle at time 't'

$$= ma = \sqrt{\frac{2Km^2}{m}}\cdot\frac{1}{2\sqrt{t}} = \sqrt{\frac{Km}{2t}} = \sqrt{\frac{mK}{2}}t^{-1/2}$$

65. **(d)** Work done by the heart, $W = P\Delta V$

where, pressure $P = \rho gh$

and $h = 150$ mm = 0.15 m

$g = 10$ m/sec^2

$\rho = 13.6 \times 10^3$ kg/m^3

$$\text{So, Power} = \frac{W}{t} = \frac{P\cdot\Delta V}{t}$$

$$\text{Power} = \frac{\rho gh \times 5 \times 10^{-3}}{60}$$

$$= \frac{13.6 \times 10^3 \times 10 \times 0.15 \times 5 \times 10^{-3}}{60} = 1.7 \text{ watt}$$

NOTES Power $= F \cdot V = PAV = \rho ghAV$

$$\left[\because P = \frac{F}{A} \text{ and } P = \rho gh\right]$$

$\therefore$ P = 13.6× 10^3 × 10 × 150 × 10^{-3} × 0.5 × 10^{-3}/60

$$= \frac{102}{60} = 1.70 \text{ watt}$$

66. (a) $\because$ Power $P = \frac{W}{t}$

$$\Rightarrow \frac{P_1}{P_2} = \frac{t_2}{t_1} = \frac{30s}{1 \text{ minute}} = \frac{30s}{60s} = \frac{1}{2}$$

(t_1 = 1 minute; t_2 = 30 second given)

67. (b) Constant power of car P_0 = F.V = ma.v

$$P_0 = m\frac{dv}{dt} \cdot v$$

$P_0 dt = mvdv$ [Integrating]

$$P_0 \cdot t = \frac{mv^2}{2} \qquad v = \sqrt{\frac{2P_0 t}{m}}$$

$\because$ P_0, m and 2 are constant $\quad \therefore \quad v \propto \sqrt{t}$

NOTES $P = Fv = \frac{mdv}{dt}v \qquad \left(\because F = \frac{mdv}{dt}\right)$

$$\Rightarrow \quad vdv = \frac{P}{m}dt$$

Integrating both sides, we get

$$\int vdv = \int \frac{P}{m}t$$

$$\Rightarrow \quad \frac{v^2}{2} = \frac{P}{m}t + C_1 \qquad \text{...(i)}$$

At $t = 0, v = 0$

$\therefore \quad C_1 = 0$

From equation (i) we get

$$V^2 = \frac{2Pt}{m} \quad \Rightarrow \quad V = \left(\frac{2Pt}{m}\right)^{1/2}$$

68. (b) Power exerted by a force is given by

P = F.v

When the body is just above the earth's surface, its velocity is greatest. At this instant, gravitational force is also maximum. Hence, the power exerted by the gravitational force is greatest at the instant just before the body hits the earth.

69. (b) Efficiency $= \frac{\text{output work}}{\text{input work}}$

$$= \frac{75g \times 3}{250 \times 12} = \frac{75 \times 10 \times 3}{250 \times 12} = 0.75 = 75\%$$

70. (a) Amount of water flowing per second from the pipe

$$= \frac{m}{\text{time}} = \frac{m}{\ell} \cdot \frac{\ell}{t} = \left(\frac{m}{\ell}\right)v$$

Power = K.E. of water flowing per second

$$= \frac{1}{2}\left(\frac{m}{\ell}\right)v \cdot v^2 = \frac{1}{2}\left(\frac{m}{\ell}\right)v^3$$

$$= \frac{1}{2} \times 100 \times 8 = 400 \text{ W}$$

71. (a) Given, h = 60m, g = 10 ms^{-2},

Rate of flow of water = 15 kg/s

$\therefore$ Power of the falling water

= 15 kgs^{-1} × 10 ms^{-2} × 60 m = 9000 watt.

Loss in energy due to friction

$$= 9000 \times \frac{10}{100} = 900 \text{ watt.}$$

$\therefore$ Power generated by the turbine

= (9000 – 900) watt = 8100 watt = 8.1 kW

72. (a) As we know,

$$P = \vec{F}.\vec{v} = (6\hat{i} - 4\hat{j} + 3\hat{k}).(20\hat{i} + 15\hat{j} - 5\hat{k})$$

= 6 × 20 – 4 × 15 – 3 × 5 = 45 J/s

NOTES Instantanesous power can be calculated by

$P_{ins} = F \cdot \cos\theta v = \vec{F} \cdot \vec{v}$

which is the scalar product of force and velocity vector.

73. (b) $P = \frac{W}{t}$. Here, P = 2kW = 2000 W.

W = Mgh = M × 10 × 10 = 100 M and t = 60 s.

This gives, M = 1200 kg

Its volume = 1200 litre as 1 litre of water contains 1 kg of its mass.

74. (b) (A) → v_1 (B) ⇒ (A)(B) → v_2

(Before collision) rest (After collision)

By conservation of linear momentum,

$P_i = P_f$

$$\Rightarrow \quad mv_1 = mv_2 + mv_2 \ \Rightarrow \ mv_1 = 2mv_2$$

$$\therefore \quad \frac{v_1}{v_2} = \frac{2}{1}$$

75. (c) During elastic collision between two equal masses, the velocity of the two bodies gets interchanged so if one body is at rest, energy transfer will be maximum for M = m

76. (b) $m_1u_1 + m_2u_2 = m_1v_1 + m_2v_2$

Also, $\frac{1}{2}m_1u_1^2 + \frac{1}{2}m_2u_2^2$

$$= \frac{1}{2}m_1v_1^2 + \frac{1}{2}m_2v_2^2$$

For elastic collision

$$v_2 = \frac{2m_1u_1}{m_1 + m_2}$$

as $u_2 = 0; u_1 = u$

m_1 = 4m and m_2 = 2m

so, $v_2 = \frac{4}{3}u$

$$\frac{\Delta K}{K} = \frac{\frac{1}{2}4m\left(u^2 - \frac{u^2}{9}\right)}{\frac{1}{2}4mu^2} = \frac{8}{9}$$

77. (b)

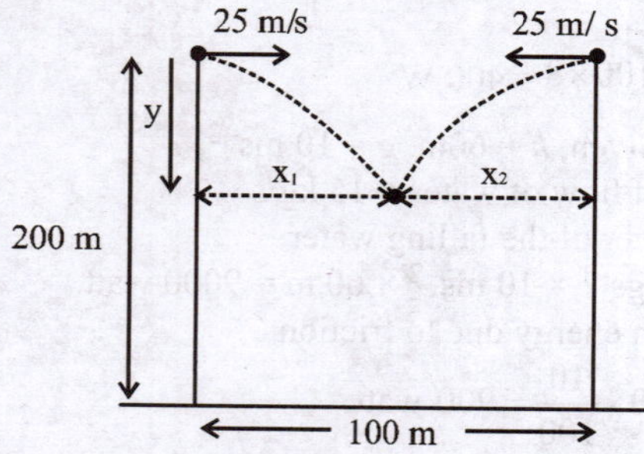

Bullets collide at time t

$x_1 + x_2 = 100$ m

$25t + 25t = 100$

$t = 2s$

$$y = \frac{1}{2}gt^2 = \frac{1}{2} \times 10 \times 2^2 = 20 \text{ m}$$

$h = 200 - 20 = 180$ m

78. (a) From conservation of linear momentum.

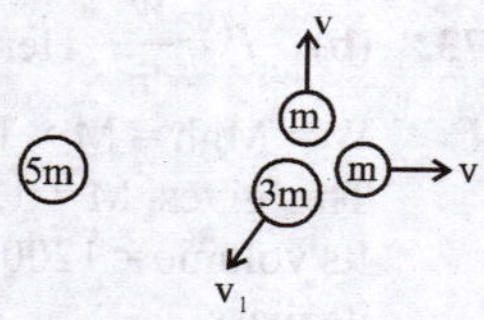

$$0 = mv\hat{j} + mv\hat{i} + 3m\vec{v}_1$$

$$\vec{v}_1 = -\frac{v}{3}(\hat{i} + \hat{j})$$

$$v_1 = \frac{\sqrt{2}}{3}v$$

$$K_f = \frac{1}{2}mv^2 + \frac{1}{2}mv^2 + \frac{1}{2}(3m)\left(\frac{\sqrt{2}}{3}\right)^2 v^2$$

$$= mv^2 + \frac{mv^2}{3} = \frac{4}{3}mv^2$$

$$\Delta KE = K_f - K_i = \frac{4}{3}mv^2$$

79. (b)

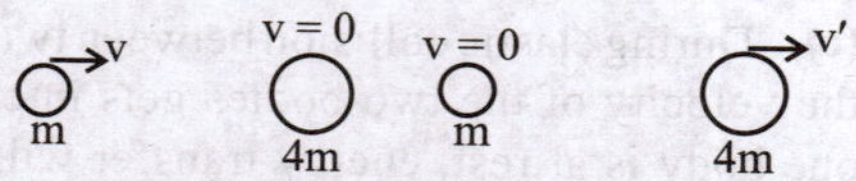

Before Collision After Collision

According to law of conservation of linear momentum,

$$mv + 4m \times 0 = 4\,mv' + 0 \Rightarrow v' = \frac{v}{4}$$

Coefficient of restitution,

$$e = \frac{\text{Relative velocity of separation}}{\text{Relative velocity of approach}}$$

$$= \frac{\frac{v}{4}}{v}$$

$$\text{or, } e = \frac{1}{4} = 0.25$$

80. (b) When the identical balls collide head-on, their velocities are exchanged.

81. (d) For collision $\vec{V}_{B/A}$ should be along $B \rightarrow A$

So,
$$\frac{\vec{V}_2 - \vec{V}_1}{|V_2 - V_1|} = \frac{\vec{r}_1 - \vec{r}_2}{|r_1 - r_2|}$$

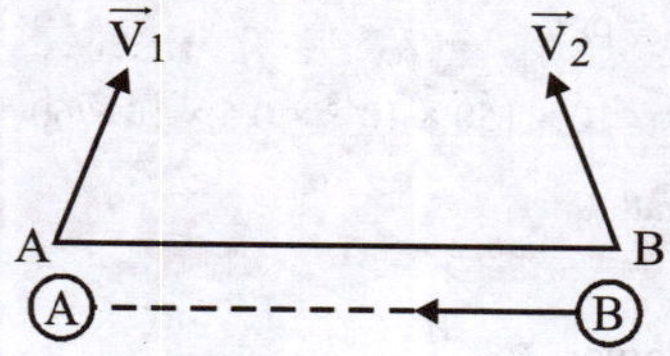

82. (d) Here, $M_1 = M_2$ and $u_2 = 0$

$$u_1 = V, \quad V_1 = \frac{V}{3}; \quad V_2 = ?$$

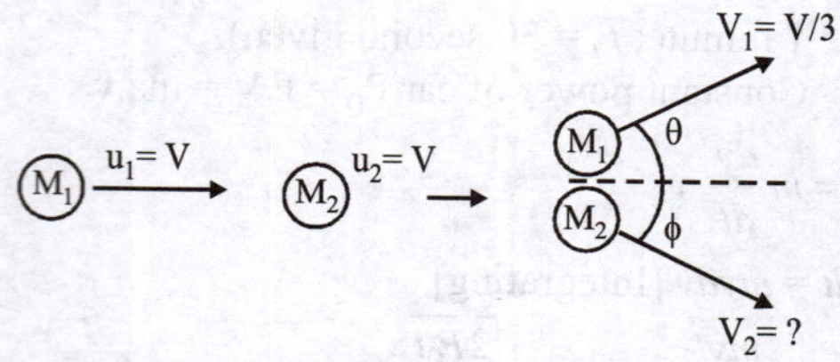

By law of conservation of kinetic energy,

$$\frac{1}{2}M_1u_1^2 + \frac{1}{2}M_2u_2^2 = \frac{1}{2}M_1V_1^2 + \frac{1}{2}M_2V_2^2$$

$$\Rightarrow u_1^2 = V_1^2 + V_2^2$$

$$V^2 = \left(\frac{V}{3}\right)^2 + V_2^2 \qquad \left[\because u_1 = V \text{ and } V_1 = \frac{V}{3}\right]$$

$$\text{or, } V^2 - \left(\frac{V}{3}\right)^2 = V_2^2$$

$$V^2 - \frac{V^2}{9} = V_2^2$$

$$\text{or } V_2^2 = \frac{8}{9}V^2$$

$$\Rightarrow V_2 = \frac{2\sqrt{2}}{3}V$$

83. (a) When ball collides with the ground it loses its 50% of energy

$$\therefore \frac{KE_f}{KE_i} = \frac{1}{2} \Rightarrow \frac{\frac{1}{2}mV_f^2}{\frac{1}{2}mV_i^2} = \frac{1}{2}$$

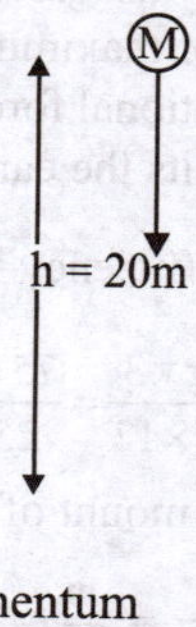

$$\text{or } \frac{V_f}{V_i} = \frac{1}{\sqrt{2}}$$

$$\text{or, } \frac{\sqrt{2gh}}{\sqrt{V_0^2 + 2gh}} = \frac{1}{\sqrt{2}}$$

$$\text{or, } 4gh = V_0^2 + 2gh$$

$$\therefore V_0 = 20 \text{ ms}^{-1}$$

84. (b) By conservation of linear momentum

Magnitude of the momentum of heavier piece of mass (2m) = Magnitude of the vector sum of momentum of each piece of mass (m)

$$(2m)v_1 = \sqrt{(mv)^2 + (mv)^2}$$

$$\Rightarrow \quad 2mv_1 = \sqrt{2}mv \Rightarrow v_1 = \frac{v}{\sqrt{2}}$$

As two masses of each of mass m move perpendicular to each other.

Total KE generated

$$= \frac{1}{2}mv^2 + \frac{1}{2}mv^2 + \frac{1}{2}(2m)v_1^2$$

$$= mv^2 + \frac{mv^2}{2} = \frac{3}{2}mv^2$$

85. (d)

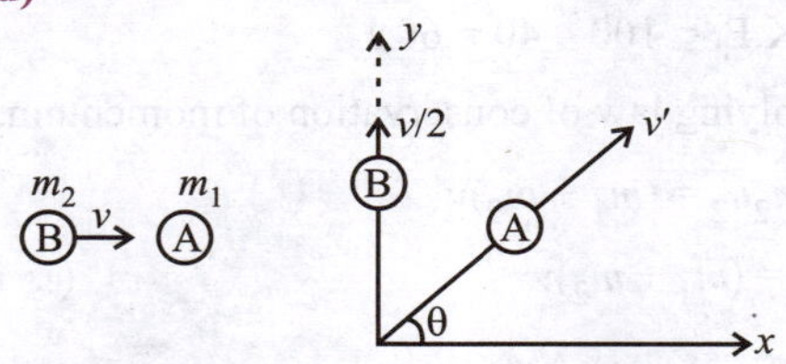

[Before collision] [After collision]

According to law of conservation of linear momentum along x-axis, we get

$$\Rightarrow \quad m_1(0) + m_2v = m_1v' \cos\theta$$

$$\Rightarrow \quad \cos\theta = \frac{m_2v}{m_1v'} \quad \text{...(i)}$$

Similarly, law of conservation of linear momentum along y-axis, we get

$$\Rightarrow \quad m_1(0) + m_2(0) = m_1v' \sin\theta + m_2\left(\frac{v}{2}\right)$$

$$\Rightarrow \quad \sin\theta = -\frac{m_2v}{2m_1v'} \quad \text{...(ii)}$$

From equations, (i) and (ii),

$$\tan\theta = -\left(\frac{m_2v}{2mv'}\right) \times \left(\frac{m_1v'}{m_2v}\right) = -\frac{1}{2}$$

$$\theta = \tan^{-1}\left(\frac{-1}{2}\right) \text{ to the } x\text{-axis.}$$

NOTES Let A moves in the direction, which makes an angle θ with initial direction *i.e.*,

$$\tan\theta = \frac{v_y}{v_x} = -\frac{m_2v}{2m_1} \Big/ \frac{m_2v}{m_1}$$

$$\tan\theta = -\frac{1}{2} \Rightarrow \theta = \tan^{-1}\left(-\frac{1}{2}\right) \text{ to the } x\text{-axis.}$$

86. (a) As the two masses stick together after collision, hence it is inelastic collision. Therefore, only momentum is conserved.

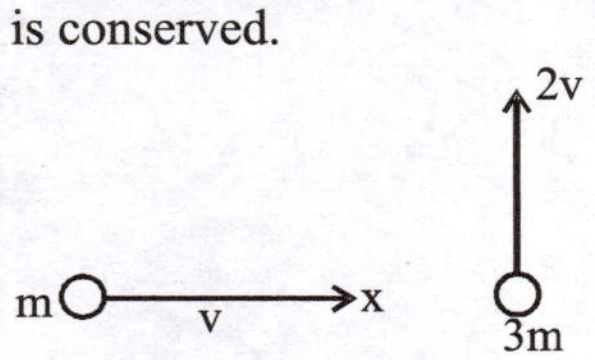

$$\therefore \; mv\hat{i} + 3m(2v)\hat{j} = (4m)\vec{v}$$

$$\vec{v} = \frac{v}{4}\hat{i} + \frac{6}{4}v\hat{j} = \frac{v}{4}\hat{i} + \frac{3}{2}v\hat{j}$$

87. (a) Clearly $v_1 = 2 \text{ ms}^{-1}$, $v_2 = 0$

$m_1 = m$ (say), $m_2 = 2m$

$v_1' = ?$, $v_2' = ?$

$$e = \frac{v_1' - v_2'}{v_2 - v_1} \quad \text{....(i)}$$

By conservation of momentum,

$$2m = mv_1' + 2mv_2' \quad \text{... (ii)}$$

From (i),

$$0.5 = \frac{v_2' - v_1'}{2} \quad \therefore v_2' = 1 + v_1'$$

From (ii),

$$2 = v_1' + 2 + 2v_1' \Rightarrow v_1' = 0 \text{ and } v_2' = 1 \text{ ms}^{-1}$$

88. (a) Let the initial velocity of the shell be v, then by the conservation of momentum $mv = Mv'$

where v' = velocity of gun.

$$\therefore \quad v' = \left(\frac{m}{M}\right)v$$

$$\text{Now, total K.E.} = \frac{1}{2}mv^2 + \frac{1}{2}Mv'^2$$

$$= \frac{1}{2}mv^2 + \frac{1}{2}M\left(\frac{m}{M}\right)^2 v^2$$

$$= \frac{1}{2}mv^2\left[1 + \frac{m}{M}\right]$$

$$= \left(\frac{1}{2} \times 0.2\right)\left(1 + \frac{0.2}{4}\right)v^2 = (0.1 \times 1.05)v^2$$

But total K.E. = 1.05 kJ = 1.05×10^3 J

$\therefore$ $1.05 \times 10^3 = 0.1 \times 1.05 \times v^2$

$$\Rightarrow v^2 = \frac{1.05 \times 10^3}{0.1 \times 1.05} = 10^4$$

$\therefore$ $v = 10^2 = 100 \text{ ms}^{-1}$.

89. (b) From conservation of linear momentum

$$m_1v_1 + m_2v_2 = 0$$

$$v_2 = \left(\frac{-m_1}{m_2}\right)v_1 = \left(\frac{-18}{12}\right)6 = -9\text{ms}^{-1}$$

$$\text{K.E.} = \frac{1}{2}m_2v_2^2 = \frac{1}{2} \times 12 \times 9^2 = 486 \text{ J}$$

90. (b) $\frac{1}{2}mv^2 = \frac{1}{2}kx^2 \Rightarrow mv^2 = kx^2$

or $0.5 \times (1.5)^2 = 50 \times x^2$

$\therefore$ $x = 0.15$ m

91. (b) From law of conservation of momentum, before collision and after collision linear momentum (p) will be same. That is, initial momentum = final momentum.

$$\Rightarrow 0 = m_1v_1 - m_2v_2 \Rightarrow m_1v_1 = m_2v_2$$

$$p_1 = p_2$$

Now, $E = \frac{p^2}{2m}$

$\therefore \frac{E_1}{E_2} = \frac{p_1^2}{2m_1} \times \frac{2m_2}{P_2^2}$

$\Rightarrow \frac{E_1}{E_2} = \frac{m_2}{m_1} [\because p_1 = p_2]$

92. **(b)** Speed of bomb after 5 second,

$v = u - gt = 100 - 10 \times 5 = 50\text{m/s}$

Momentum of 400 g fragment

$= \frac{400}{1000} \times (-25)$ [downward]

Momentum of 600g fragment $= \frac{600}{1000} v$

Momentum of bomb = 1 × 50 = 50

From conservation of momentum

Total momentum before splitting = total momentum after splitting.

$\Rightarrow 50 = -\frac{400}{1000} \times 25 + \frac{600}{1000} v$

$\Rightarrow v = 100$ m/s [upward]

93. **(d)** In elastic collision, the velocities get exchanged if the colliding objects have equal masses.

94. **(a)** 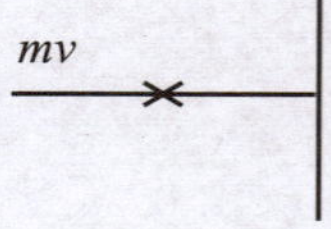

Initial momentum = mv

Final momentum = $m(-v)$

change in momentum

$= mv - m(-v) = 2mv$

95. **(c)** Applying conservation of momentum,

$m_1 v_1 = (m_1 + m_2) v$

$v = \frac{m_1 v_1}{(m_1 + m_2)}$

Here, $v_1 = 36$ km/hr = 10 m/s

$m_1 = 2$ kg, $m_2 = 3$ kg

$v = \frac{10 \times 2}{5} = 4$ m/s

K.E. (initial) $= \frac{1}{2} \times 2 \times (10)^2 = 100\text{ J}$

K.E. (Final) $= \frac{1}{2} \times (3+2) \times (4)^2 = 40$ J

Loss in K.E. = 100 – 40 = 60 J

96. **(a)** Applying law of conservation of momentum,

$m_1 u_1 + m_2 u_2 = (m_1 + m_2) v$

or, $m_1 u_1 = (m_1 + m_2) v$ $(\because u_2 = 0)$

$\Rightarrow m \times 3V = 3m(v) \Rightarrow v = 1\text{ km/hr}$

97. **(c)** When shell explodes in mid air its chemical energy is partly converted into mechanical energy, hence K.E. increases.

98. **(a)** $e = |v_1 - v_2| / |u_1 - u_2|$ which is 1 for a perfectly elastic collision.

6 System of Particles and Rotational Motion

TREND ANALYSIS

TOPIC	2025		2024		2023		2022		2021		2020	
	Qns.	LOD	Qns.	LOD	Qns.	LOD	Qns.	LOD	Qns.	LOD	Qns.	LOD
Centre of Mass							1	Easy			1	Easy
Angular Velocity and its Relation with Linear Velocity					1	Easy						
Torque and Angular Momentum	1	Avg.							1	Avg.		
Moment of Inertia	1	Diff.	1	Avg.			2	Avg.	1	Avg.		
Dynamics of Rotational Motion About a Fixed Axis												
Rolling Motion												

How to Code Qns for Better & Faster Revision

Type 1: Able to solve (No code Req)
Type 2: Solved with struggle or by seeing solution **(Use code NR means Needs Revision or orange stickers)**
Type 3: Tricky/ Could not solve on my own. Need help from Teacher/ Friend. **(Use code TQ means Tricky Question or red stickers)**

You can either hand write the code in boxes or else paste coloured stickers given at end of book.

Topic-6.2 Centre of Mass

1. Two objects of mass 10 kg and 20 kg respectively are connected to the two ends of a rigid rod of length 10 m with negligible mass. The distance of the center of mass of the system from the 10 kg mass is: **NCERT | Page-96 | 2022**

(a) $\frac{20}{3}$ m (b) 10 m (c) 5 m (d) $\frac{10}{3}$ m

2. Two particles of mass 5 kg and 10 kg respectively are attached to the two ends of a rigid rod of length 1 m with negligible mass. The centre of mass of the system from the 5 kg particle is nearly at a distance of : **NCERT | Page-96 | 2020**

(a) 50 cm (b) 67 cm (c) 80 cm (d) 33 cm

3. Three masses are placed on the *x*-axis : 300 g at origin, 500 g at $x = 40$ cm and 400 g at $x = 70$ cm. The distance of the centre of mass from the origin is: **NCERT | Page-96 | 2012 M**

(a) 40 cm (b) 45 cm (c) 50 cm (d) 30 cm

4. Two bodies of mass 1 kg and 3 kg have position vectors $\hat{i}+2\hat{j}+\hat{k}$ and $-3\hat{i}-2\hat{j}+\hat{k}$ respectively. The centre of mass of this system has a position vector: **NCERT | Page-96 | 2009**

(a) $-2\hat{i}-\hat{j}+\hat{k}$ (b) $2\hat{i}-\hat{j}-2\hat{k}$
(c) $-\hat{i}+\hat{j}+\hat{k}$ (d) $-2\hat{i}+2\hat{k}$

5. If the linear density (mass per unit length) of a rod of length 3m is proportional to *x*, where *x* is the distance from one end of the rod, the distance of the centre of gravity of the rod from this end is **NCERT | Page-97 | 2002**

(a) 2.5 m (b) 1 m (c) 1.5 m (d) 2 m

6. The centre of mass of a system of particles does not depend upon **NCERT | Page-96 | 1997**

(a) masses of the particles
(b) forces acting on the particles
(c) position of the particles
(d) relative distances between the particles

7. In carbon monoxide molecule, the carbon and the oxygen atoms are separated by a distance 1.12×10^{-10} m. The distance of the centre of mass, from the carbon atom is **NCERT | Page-96 | 1997**

(a) 0.64×10^{-10} m (b) 0.56×10^{-10} m
(c) 0.51×10^{-10} m (d) 0.48×10^{-10} m

Topic-6.5 Vector Product of Two Vectors

8. Vectors $A\ B$ and $\vec{C}$ are such that $A \cdot B =$ and $\vec{A}\cdot\vec{C}=0$. Then the vector parallel to $\vec{A}$ is **NCERT | Page-102 | Kar. 2013**

(a) $\vec{B}$ and $\vec{C}$ (b) $\vec{A}\times\vec{B}$
(c) $\vec{B}+\vec{C}$ (d) $\vec{B}\times\vec{C}$

9. $\vec{A}$ and $\vec{B}$ are two vectors and θ is the angle between them, if $|\vec{A}\times\vec{B}| = \sqrt{3}(\vec{A}.\vec{B})$, the value of θ is **NCERT | Page-102 | 2007**

(a) 45° (b) 30°
(c) 90° (d) 60°

10. If the angle between the vectors $\vec{A}$ and $\vec{B}$ is θ, the value of the product $(\vec{B}\times\vec{A})\cdot\vec{A}$ is equal to **NCERT | Page-102 | 2005**

(a) $BA^2 \sin\theta$ (b) $BA^2 \cos\theta$
(c) $BA^2 \sin\theta \cos\theta$ (d) zero

11. If $|\vec{A}\times\vec{B}| = \sqrt{3}\vec{A}\cdot\vec{B}$ then the value of $|\vec{A}+\vec{B}|$ is NCERT | Page-102 | 2004

(a) $(A^2+B^2+\sqrt{3}AB)^{1/2}$

(b) $(A^2+B^2+AB)^{1/2}$

(c) $\left(A^2+B^2+\frac{AB}{\sqrt{3}}\right)^{1/2}$

(d) $A+B$

12. The angle between $\vec{A}$ and $\vec{B}$ is θ. The value of the triple product $\vec{A}\cdot(\vec{B}\times\vec{A})$ is NCERT | Page-102 | 1989

(a) A^2B (b) zero

(c) $A^2B\sin\theta$ (d) $A^2B\cos\theta$

Topic-6.6 Angular Velocity and its Relation with Linear Velocity

13. The angular acceleration of a body, moving along the circumference of a circle, is NCERT | Page-105 | 2023

(a) Along the radius, away from centre

(b) Along the radius towards the centre

(c) Along the tangent to its position

(d) Along the axis of rotation

14. Two particles A and B are moving in uniform circular motion in concentric circles of radii r_A and r_B with speed υ_A and υ_B respectively. Their time period of rotation is the same. The ratio of angular speed of A to that of B will be : NCERT | Page-153 | 2019

(a) $r_A : r_B$ (b) $\upsilon_A : \upsilon_B$

(c) $r_B : r_A$ (d) 1 : 1

15. A particle starting from rest, moves in a circle of radius 'r'. It attains a velocity of V_0 m/s in the nth round. Its angular acceleration will be, NCERT | Page-105 | Odisha 2019

(a) $\frac{V_0^2}{4\pi nr}$ rad/s^2 (b) $\frac{V_0}{n}$ rad/s^2

(c) $\frac{V_0^2}{2\pi nr^2}$ rad/s^2 (d) $\frac{V_0^2}{4\pi nr^2}$ rad/s^2

16. A tube of length L is filled completely with an incompressible liquid of mass M and closed at both the ends. The tube is then rotated in a horizontal plane about one of its ends with a uniform angular velocity ω. The force exerted by the liquid at the other end is NCERT | Page-106 | 2006

(a) $\frac{ML^2\omega}{2}$ (b) $ML\omega^2$

(c) $\frac{ML^2\omega^2}{2}$ (d) $\frac{ML\omega^2}{2}$

17. A wheel of radius 1m rolls forward half a revolution on a horizontal ground. The magnitude of the displacement of the point of the wheel initially in contact with the ground is NCERT | Page-153 | 2002

(a) π (b) 2π

(c) $\sqrt{2}\pi$ (d) $\sqrt{\pi^2+4}$

18. What is the linear velocity if angular velocity vector $\vec{\omega} = 3\hat{i}-4\hat{j}+\hat{k}$ and position vector $\vec{r} = 5\hat{i}-6\hat{j}+6\hat{k}$ NCERT | Page-105 | 1999

(a) $6\hat{i}+2\hat{j}-3\hat{k}$ (b) $-18\hat{i}-13\hat{j}+2\hat{k}$

(c) $18\hat{i}+13\hat{j}-2\hat{k}$ (d) $6\hat{i}-2\hat{j}+8\hat{k}$

19. Two racing cars of masses m_1 and m_2 are moving in circles of radii r_1 and r_2 respectively. Their speeds are such that each makes a complete circle in the same time t. The ratio of the angular speeds of the first to the second car is 1999

(a) 1 : 1 (b) $m_1 : m_2$

(c) $r_1 : r_2$ (d) $m_1 m_2 : r_1 r_2$

20. If a flywheel makes 120 revolutions/minute, then its angular speed will be 1996

(a) 8 π rad/sec (b) 6 π rad/sec

(c) 4 π rad/sec (d) 2 π rad/s

21. The angular speed of an engine wheel making 90 revolutions per minute is 1995

(a) 1.5 π rad/s (b) 3π rad/s

(c) 4.5 π rad/s (d) 6π rad/s

22. Two racing cars of masses m and 4m are moving in circles of radii r and 2r respectively. If their speeds are such that each makes a complete circle in the same time, then the ratio of the angular speeds of the first to the second car is 1995

(a) 8 : 1 (b) 4 : 1 (c) 2 : 1 (d) 1 : 1

Topic-6.7 Torque and Angular Momentum

23. The Sun rotates around its centre once in 27 days. What will be the period of revolution if the Sun were to expand to twice its present radius without any external influence? Assume the Sun to be a sphere of uniform density. NCERT | Page-108 | 2025

(a) 100 days (b) 105 days

(c) 115 days (d) 108 days

24. Find the torque about the origin when a force of $3\hat{j}$ N acts on a particle whose position vector is $2\hat{k}$ m . NCERT | Page-106 | 2020

(a) $6\hat{j}$ Nm (b) $-6\hat{i}$ Nm

(c) $6\hat{k}$ Nm (d) $6\hat{i}$ Nm

25. The moment of the force, $\vec{F} = 4\hat{i}+5\hat{j}-6\hat{k}$ at (2, 0, –3), about the point (2, –2, –2), is given by NCERT | Page-106 | 2018

(a) $-8\hat{i}-4\hat{j}-7\hat{k}$ (b) $-4\hat{i}-\hat{j}-8\hat{k}$

(c) $-7\hat{i}-4\hat{j}-8\hat{k}$ (d) $-7\hat{i}-8\hat{j}-4\hat{k}$

26. A solid sphere is rotating freely about its symmetry axis in free space. The radius of the sphere is increased keeping its mass same. Which of the following physical quantities would remain constant for the sphere? NCERT | Page-108 | 2018

(a) Angular velocity

(b) Moment of inertia

(c) Angular momentum

(d) Rotational kinetic energy

27. A force $\vec{F} = \alpha\hat{i} + 3\hat{j} + 6\hat{k}$ is acting at a point $\vec{r} = 2\hat{i} - 6\hat{j} - 12\hat{k}$. The value of α for which angular momentum about origin is conserved is : NCERT | Page-108 | 2015 RS

(a) 2 (b) zero (c) 1 (d) –1

28. When a mass is rotating in a plane about a fixed point, its angular momentum is directed along : NCERT | Page-106 | 2012

(a) a line perpendicular to the plane of rotation
(b) the line making an angle of 45° to the plane of rotation
(c) the radius
(d) the tangent to the orbit

29. If $\vec{F}$ is the force acting on a particle having position vector $\vec{r}$ and $\vec{\tau}$ be the torque of this force about the origin, then: NCERT | Page-107 | 2009

(a) $\vec{r}\cdot\vec{\tau} > 0$ and $\vec{F}\cdot\vec{\tau} < 0$
(b) $\vec{r}\cdot\vec{\tau} = 0$ and $\vec{F}\cdot\vec{\tau} = 0$
(c) $\vec{r}\cdot\vec{\tau} = 0$ and $\vec{F}\cdot\vec{\tau} \neq 0$
(d) $\vec{r}\cdot\vec{\tau} \neq 0$ and $\vec{F}\cdot\vec{\tau} = 0$

30. A particle of mass m moves in the *XY* plane with a velocity *v* along the straight line *AB*. If the angular momentum of the particle with respect to origin *O* is L_A when it is at *A* and L_B when it is at *B*, then NCERT | Page-107 | 2007

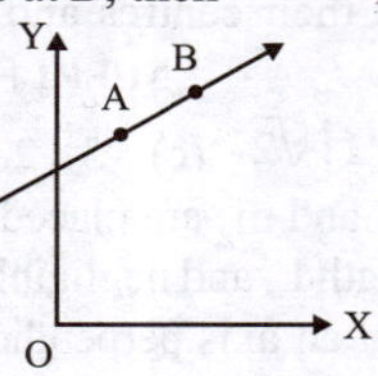

(a) $L_A = L_B$
(b) the relationship between L_A and L_B depends upon the slope of the line *AB*
(c) $L_A < L_B$
(d) $L_A > L_B$

31. A boy suddenly comes and sits on a circular rotating table. What will remain conserved? NCERT | Page-108 | 2002

(a) Angular velocity
(b) Angular momentum
(c) Linear momentum
(d) Kinetic energy

32. A couple produces NCERT | Page-110 | 1997

(a) no motion
(b) purely linear motion
(c) purely rotational motion
(d) linear and rotational

33. Find the torque of a force $\vec{F} = -3\hat{i} + \hat{j} + 5\hat{k}$ acting at the point $\vec{r} = 7\hat{i} + 3\hat{j} + \hat{k}$. NCERT | Page-106 | 1997

(a) $-21\hat{i} + 3\hat{j} + 5\hat{k}$ (b) $-14\hat{i} + 3\hat{j} + 16\hat{k}$
(c) $4\hat{i} + 4\hat{j} + 6\hat{k}$ (d) $14\hat{i} - 38\hat{j} + 16\hat{k}$

34. Angular momentum is NCERT | Page-106 | 1994

(a) vector (axial) (b) vector (polar)
(c) scalar (d) none of the above

35. A particle of mass m = 5 is moving with a uniform speed $v = 3\sqrt{2}$ in the XOY plane along the line $y = x + 4$. The magnitude of the angular momentum of the particle about the origin is NCERT | Page-106 | 1991

(a) 60 units (b) $40\sqrt{2}$ units
(c) zero (d) 7.5 units

Topic-6.8 Equilibrium of a Rigid Body

36. A uniform rod of length 200 cm and mass 500 g is balanced on a wedge placed at 40 cm mark. A mass of 2 kg is suspended from the rod at 20 cm and another unknown mass '*m*' is suspended from the rod at 160 cm mark as shown in the figure. Find the value of '*m*' such that the rod is in equilibrium. ($g = 10\ m/s^2$) NCERT | Page-106 | 2021

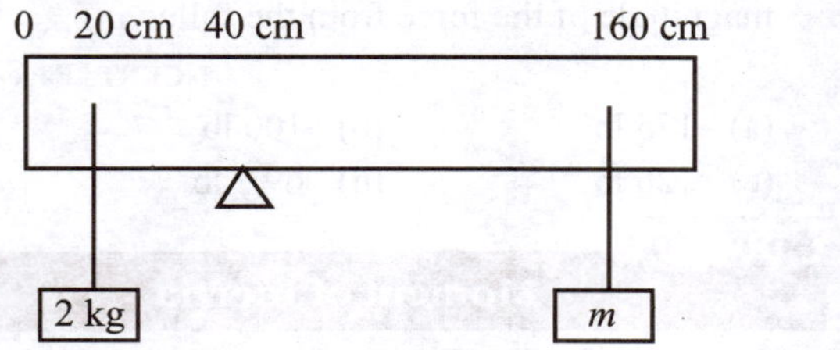

(a) $\frac{1}{12}$kg (b) $\frac{1}{2}$kg (c) $\frac{1}{3}$kg (d) $\frac{1}{6}$kg

37. Which of the following statements are correct?

(A) Centre of mass of a body always coincides with the centre of gravity of the body
(B) Centre of mass of a body is the point at which the total gravitational torque on the body is zero
(C) A couple on a body produce both translational and rotation motion in a body
(D) Mechanical advantage greater than one means that small effort can be used to lift a large load

NCERT | Page-111 | 2017

(a) (A) and (B) (b) (B) and (C)
(c) (C) and (D) (d) (B) and (D)

38. A rod of weight W is supported by two parallel knife edges A and B and is in equilibrium in a horizontal position. The knives are at a distance d from each other. The centre of mass of the rod is at distance x from A. The normal reaction on A is NCERT | Page-109 | 2015

(a) $\frac{Wd}{x}$ (b) $\frac{W(d-x)}{x}$
(c) $\frac{W(d-x)}{d}$ (d) $\frac{Wx}{d}$

39. *ABC* is an equilateral triangle with *O* as its centre. $\vec{F}_1, \vec{F}_2$ and $\vec{F}_3$ represent three forces acting along the sides *AB*, *BC* and *AC* respectively. If the total torque about *O* is zero the magnitude of $\vec{F}_3$ is : NCERT | Page-108 | 2012, 1998

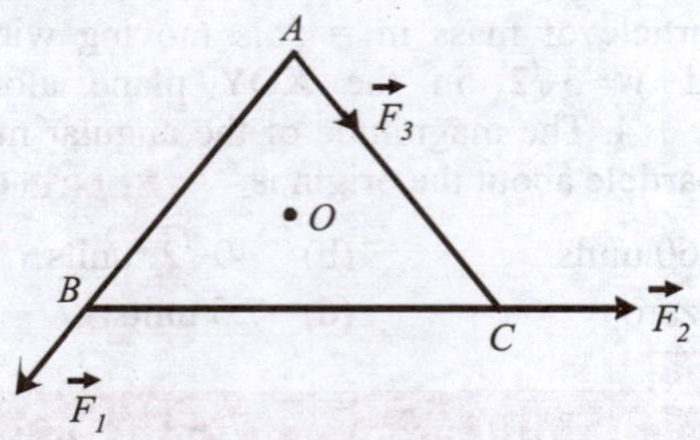

(a) $F_1 + F_2$ (b) $F_1 - F_2$
(c) $\frac{F_1 + F_2}{2}$ (d) $2(F_1 + F_2)$

40. A weightless ladder 20 ft long rests against a frictionless wall at an angle of 60° from the horizontal. A 150 pound man is 4 ft from the top of the ladder. A horizontal force is needed to keep it from slipping. Choose the correct magnitude of the force from the following **NCERT | Page-109 | 1998**

(a) 175 lb (b) 100 lb
(c) 120 lb (d) 69.2 lb

Topic-6.9 Moment of Inertia

41. A sphere of radius R is cut from a larger solid sphere of radius $2R$ as shown in the figure. The ratio of the moment of inertia of the smaller sphere to that of the rest part of the sphere about the Y-axis is : **NCERT | Page-114, 115 | 2025**

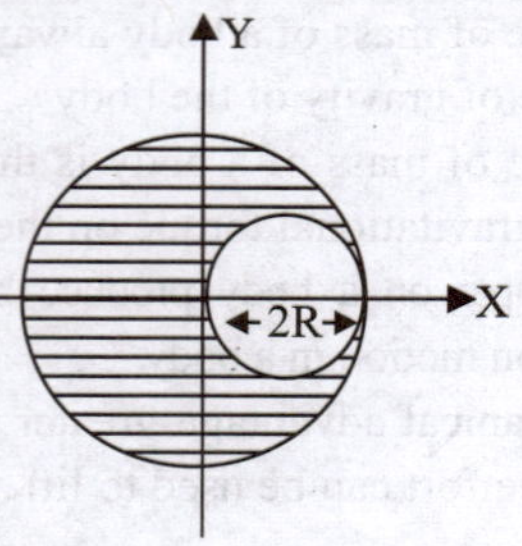

(a) $\frac{7}{8}$ (b) $\frac{7}{40}$ (c) $\frac{7}{57}$ (d) $\frac{7}{64}$

42. The radius of gyration of a solid sphere of mass 5 kg about XY is 5 m as shown in figure. The radius of the sphere is $\frac{5x}{\sqrt{7}}$ m, then the value of x is: **NCERT | Page-115, 116 | Ph-II 2024**

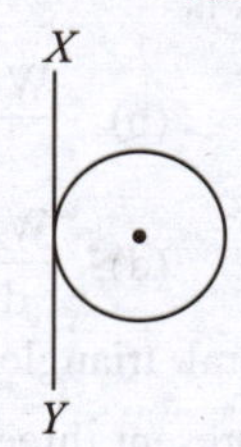

(a) 5 (b) $\sqrt{2}$ (c) $\sqrt{3}$ (d) $\sqrt{5}$

43. The moment of inertia of a thin rod about an axis passing through its mid point and perpendicular to the rod is 2400 g cm^2. The length of the 400 g rod is nearly: **NCERT | Page-116 | 2024**

(a) 8.5 cm (b) 17.5 cm
(c) 20.7 cm (d) 72.0 cm

44. The ratio of radius of gyration of a solid sphere of mass M and radius R about its own axis to the radius of gyration of the thin hollow sphere of same mass and radius about its axis is **NCERT | Page-116 | 2022**

(a) 3 : 5 (b) 5 : 3 (c) 2 : 5 (d) 5 : 2

45. The ratio of the radius of gyration of a thin uniform disc about an axis passing through its centre and normal to its plane to the radius of gyration of the disc about its diameter is: **NCERT | Page-115, 116 | 2022**

(a) $\sqrt{2}:1$ (b) 4 : 1
(c) $1:\sqrt{2}$ (d) 2 : 1

46. From a circular ring of mass 'M' and radius 'R' an arc corresponding to a 90° sector is removed. The moment of inertia of the remaining part of the ring about an axis passing through the centre of the ring and perpendicular to the plane of the ring is 'K' times 'MR^2'. Then the value of 'K' is **NCERT | Page-115 | 2021**

(a) $\frac{1}{8}$ (b) $\frac{3}{4}$ (c) $\frac{7}{8}$ (d) $\frac{1}{4}$

47. The ratio of radii of gyration of a circular ring and a circular disc, of the same mass and radius, about an axis passing through their centres and perpendicular to their planes are **NCERT | Page-115 | Kar. 2013, 2018**

(a) $\sqrt{2}:1$ (b) $1:\sqrt{2}$ (c) 3 : 2 (d) 2 : 1

48. Point masses m_1 and m_2 are placed at the opposite ends of a rigid rod of length L, and negligible mass. The rod is to be set rotating about an axis perpendicular to it. The position of point P on this rod through which the axis should pass so that the work required to set the rod rotating with angular velocity ω_0 is minimum, is given by : **2015 RS**

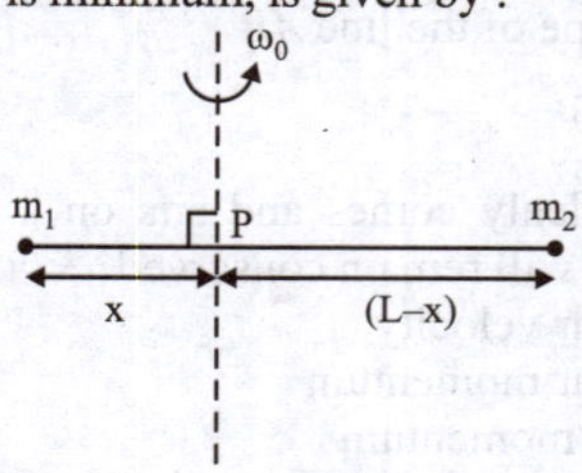

(a) $x = \frac{m_1}{m_2}L$ (b) $x = \frac{m_2}{m_1}L$

(c) $x = \frac{m_2 L}{m_1 + m_2}$ (d) $x = \frac{m_1 L}{m_1 + m_2}$

49. A thin rod of length L and mass M is bent at its midpoint into two halves so that the angle between them is 90°. The moment of inertia of the bent rod about an axis passing through the bending point and perpendicular to the plane defined by the two halves of the rod is: **NCERT | Page116 | 2008**

(a) $\frac{ML^2}{24}$ (b) $\frac{ML^2}{12}$

(c) $\frac{ML^2}{6}$ (d) $\frac{\sqrt{2}ML^2}{24}$

50. The ratio of the radii of gyration of a circular disc about a tangential axis in the plane of the disc and of a circular ring of the same radius about a tangential axis in the plane of the ring is **NCERT | Page-115 | 2004**

(a) $1 : \sqrt{2}$ (b) $1 : 3$

(c) $2 : 1$ (d) $\sqrt{5} : \sqrt{6}$

51. Three particles, each of mass m gram, are situated at the vertices of an equilateral triangle *ABC* of side *l* cm (as shown in the figure). The moment of inertia of the system about a line *AX* perpendicular to AB and in the plane of *ABC*, in gram-cm^2 units will be **NCERT | Page-114 | 2004**

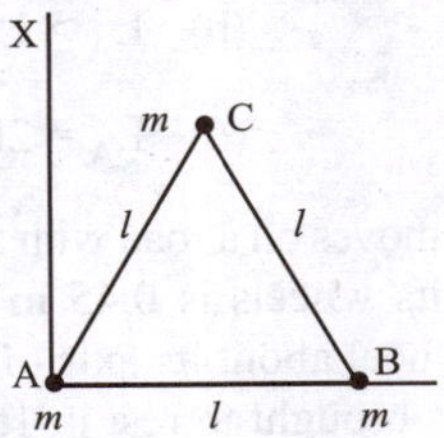

(a) $\frac{3}{2}m\ell^2$ (b) $\frac{3}{4}m\ell^2$

(c) $2\,m\ell^2$ (d) $\frac{5}{4}m\ell^2$

52. A composite disc is to be made using equal masses of aluminium and iron so that it has as high a moment of inertia as possible. This is possible when **NCERT | Page-114 | 2002**

(a) the surfaces of the discs are made of iron with aluminium inside

(b) the whole of aluminium is kept in the core and the iron at the outer rim of the disc

(c) the whole of the iron is kept in the core and the aluminium at the outer rim of the disc

(d) the whole disc is made with thin alternate sheets of iron and aluminium

53. There is a flat uniform triangular plate *ABC* such that *AB* = 4 cm, *BC* = 3 cm and angle *ABC* = 90º. The moment of inertia of the plate about *AB*, *BC* and *CA* as axis is respectively I_1, I_2 and I_3. Which one of the following is true? **NCERT | Page-114 | 2000**

(a) $I_3 > I_2$

(b) $I_2 > I_1$

(c) $I_3 > I_1$

(d) $I_1 > I_2$

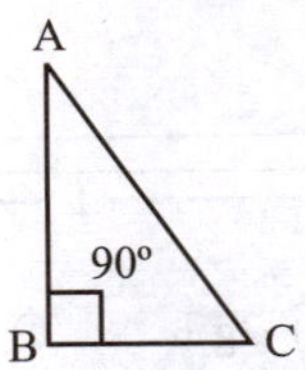

54. ABC is a triangular plate of uniform thickness. The sides are in the ratio shown in the figure. I_{AB}, I_{BC} and I_{CA} are the moments of inertia of the plate about *AB*, *BC* and *CA* as axes respectively. Which one of the following relations is correct? **NCERT | Page-114 | 1995**

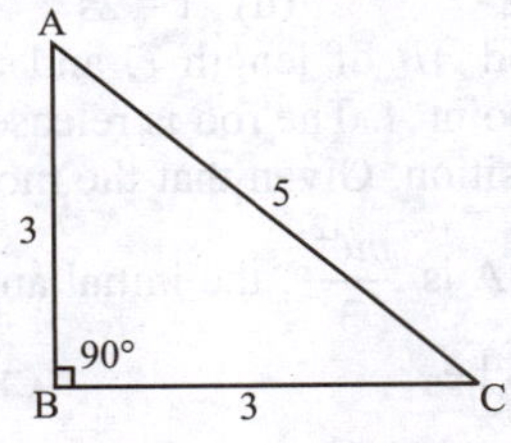

(a) $I_{AB} > I_{BC}$ (b) $I_{BC} > I_{AB}$

(c) $I_{AB} + I_{BC} = I_{CA}$ (d) I_{CA} is maximum

55. In a rectangle *ABCD* (*BC* = 2 *AB*). The moment of inertia is minimum along axis through **NCERT | Page-114 | 1993**

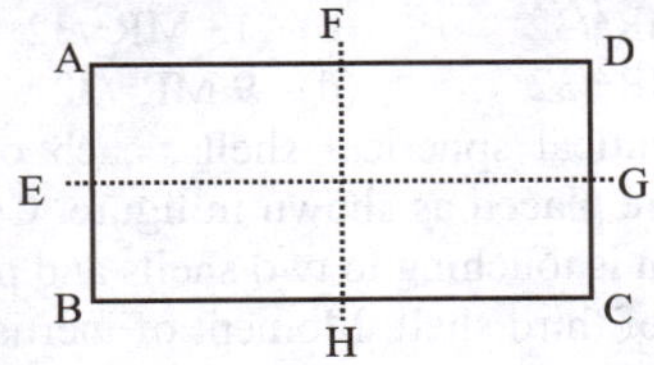

(a) *BC* (b) *BD* (c) *HF* (d) *EG*

Topic-*

Theorems of Perpendicular and Parallel Axes

56. The moment of inertia of a uniform circular disc is maximum about an axis perpendicular to the disc and passing through : **2012 M**

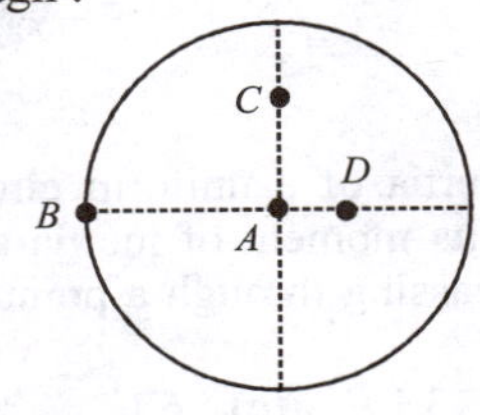

(a) *B* (b) *C* (c) *D* (d) *A*

57. The moment of inertia of a thin uniform rod of mass M and length L about an axis passing through its midpoint and perpendicular to its length is I_0. Its moment of inertia about an axis passing through one of its ends and perpendicular to its length is **2011**

(a) $I_0 + ML^2/2$ (b) $I_0 + ML^2/4$

(c) $I_0 + 2ML^2$ (d) $I_0 + ML^2$

58. The moment of inertia of a uniform circular disc of radius R and mass M about an axis touching the disc at its diameter and normal to the disc is **NCERT | Page-116 | 2006, 2005**

(a) $\frac{2}{5}MR^2$ (b) $\frac{3}{2}MR^2$

(c) $\frac{1}{2}MR^2$ (d) MR^2

59. The moment of inertia of a disc of mass M and radius R about an axis, which is tangential to the circumference of the disc and parallel to its diameter, is **2008**

(a) $\frac{3}{2}MR^2$ (b) $\frac{2}{3}MR^2$

(c) $\frac{5}{4}MR^2$ (d) $\frac{4}{5}MR^2$

60. Four identical thin rods each of mass M and length *l*, form a square frame. Moment of inertia of this frame about an axis through the centre of the square and perpendicular to its plane is : **NCERT | Page-116 | 2009**

(a) $\frac{2}{3}Ml^2$ (b) $\frac{13}{3}Ml^2$

(c) $\frac{1}{3}Ml^2$ (d) $\frac{4}{3}Ml^2$

61. From a disc of radius R and mass M, a circular hole of diameter R, whose rim passes through the centre is cut. What is the moment of inertia of the remaining part of the disc about a perpendicular axis, passing through the centre? **NCERT | Page-116 | 2016**
(a) $15\ MR^2/32$ (b) $13\ MR^2/32$
(c) $11\ MR^2/32$ (d) $9\ MR^2/32$

62. Three identical spherical shells, each of mass m and radius r are placed as shown in figure. Consider an axis XX' which is touching to two shells and passing through diameter of third shell. Moment of inertia of the system consisting of these three spherical shells about XX' axis is **NCERT | Page-157 | 2015**
(a) $3mr^2$
(b) $\frac{16}{5}mr^2$
(c) $4mr^2$
(d) $\frac{11}{5}mr^2$

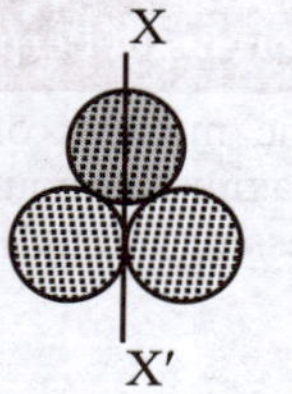

63. Moment of inertia of a uniform circular disc about a diameter is I. Its moment of inertia about an axis ⊥ to its plane and passing through a point on its rim will be **NCERT | Page-167 | 1999**
(a) 5 I (b) 3 I (c) 6 I (d) 4 I

64. The moment of inertia of a disc of mass M and radius R about an axis, which is tangential to the circumference of the disc and parallel to its diameter, is **NCERT | Page-167 | 1999**
(a) $\frac{3}{2}MR^2$ (b) $\frac{2}{3}MR^2$
(c) $\frac{5}{4}MR^2$ (d) $\frac{4}{5}MR^2$

Topic-6.10
Kinematics of Rotational Motion about a Fixed Axis

65. A wheel has angular acceleration of 3.0 rad/sec^2 and an initial angular speed of 2.00 rad/sec. In a time of 2 sec it has rotated through an angle (in radian) of **NCERT | Page-117 | 2007**
(a) 10 (b) 12 (c) 4 (d) 6

66. A uniform circular disc of radius 50 cm at rest is free to turn about an axis which is perpendicular to its plane and passes through its centre. It is subjected to a torque which produces a constant angular acceleration of 2.0 rad s^{-2}. Its net acceleration in ms^{-2} at the end of 2.0s is approximately : **NCERT | Page-120 | 2016**
(a) 8.0 (b) 7.0 (c) 6.0 (d) 3.0

Topic-6.11
Dynamics of Rotational Motion About a Fixed Axis

67. A solid cylinder of mass 2 kg and radius 4 cm is rotating about its axis at the rate of 3 rpm. The torque required to stop after 2π revolutions is: **2019**
(a) 2×10^{-6} N m (b) 2×10^{-3} N m
(c) 12×10^{-4} N m (d) 2×10^{6} N m

68. A rope is wound around a hollow cylinder of mass 3 kg and radius 40 cm. What is the angular acceleration of the cylinder if the rope is pulled with a force of 30 N? **NCERT | Page-120 | 2017**
(a) 0.25 rad/s^2 (b) 25 rad/s^2
(c) 5 m/s^2 (d) 25 m/s^2

69. Two rotating bodies A and B of masses m and 2m with moments of inertia I_A and I_B ($I_B > I_A$) have equal kinetic energy of rotation. If L_A and L_B be their angular momenta respectively, then **NCERT | Page-119 | Ph-II 2016**
(a) $L_B > L_A$ (b) $L_A > L_B$
(c) $L_A = \frac{L_B}{2}$ (d) $L_A = 2L_B$

70. An automobile moves on a road with a speed of 54 km h^{-1}. The radius of its wheels is 0.45 m and the moment of inertia of the wheel about its axis of rotation is 3 kg m^2. If the vehicle is brought to rest in 15s, the magnitude of average torque transmitted by its brakes to the wheel is: **NCERT | Page-120 | 2015 RS**
(a) 8.58 kg m^2 s^{-2} (b) 10.86 kg m^2 s^{-2}
(c) 2.86 kg m^2 s^{-2} (d) 6.66 kg m^2 s^{-2}

71. A solid cylinder of mass 50 kg and radius 0.5 m is free to rotate about the horizontal axis. A massless string is wound round the cylinder with one end attached to it and other hanging freely. Tension in the string required to produce an angular acceleration of 2 revolutions s^{-2} is : **NCERT | Page-120 | 2014**
(a) 25 N (b) 50 N
(c) 78.5 N (d) 157 N

72. A rod PQ of mass M and length L is hinged at end P. The rod is kept horizontal by a massless string tied to point Q as shown in figure. When string is cut, the initial angular acceleration of the rod is **NCERT | Page-120 | 2013**

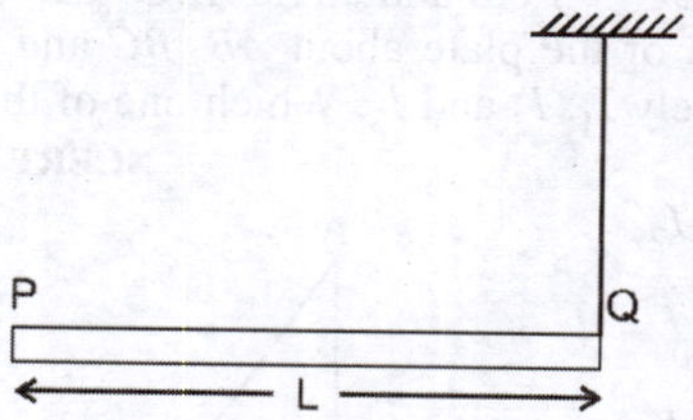

(a) g /L (b) 2g/L (c) $\frac{2g}{3L}$ (d) $\frac{3g}{2L}$

73. The instantaneous angular position of a point on a rotating wheel is given by the equation $\theta(t) = 2t^3 - 6t^2$. The torque on the wheel becomes zero at **NCERT | Page-120 | 2011**
(a) t = 1s (b) t = 0.5 s
(c) t = 0.25 s (d) t = 2s

74. A uniform rod *AB* of length ℓ, and mass m is free to rotate about point *A*. The rod is released from rest in the horizontal position. Given that the moment of inertia of the rod about A is $\frac{m\ell^2}{3}$, the initial angular acceleration of the rod will be **NCERT | Page-120 | 2007**

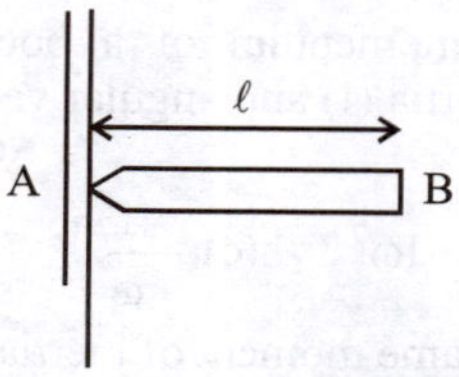

(a) $\frac{mg\ell}{2}$ (b) $\frac{3}{2}g\ell$ (c) $\frac{3g}{2\ell}$ (d) $\frac{2g}{3\ell}$

75. Two bodies have their moments of inertia I and 2I respectively about their axis of rotation. If their kinetic energies of rotation are equal, their angular momenta will be in the ratio **NCERT | Page-120 | 2005**

(a) 2 : 1 (b) 1 : 2
(c) $\sqrt{2}:1$ (d) $1:\sqrt{2}$

76. A wheel having moment of inertia 2 kg-m^2 about its vertical axis, rotates at the rate of 60 rpm about this axis. The torque which can stop the wheel's rotation in one minute would be **NCERT | Page-120 | 2004**

(a) $\frac{\pi}{18}$ *N-m* (b) $\frac{2\pi}{15}$ *N-m*
(c) $\frac{\pi}{12}$ *N-m* (d) $\frac{\pi}{15}$ *N-m*

77. A constant torque of 1000 N-m turns a wheel of moment of inertia 200 kg-m^2 about an axis through its centre. Its angular velocity after 3 seconds is **NCERT | Page-117 | 2001**

(a) 1 rad/s (b) 5 rad/s
(c) 10 rad/s (d) 15 rad/s

78. A constant torque of 100 N m turns a wheel of moment of inertia 300 kg m^2 about an axis passing through its centre. Starting from rest, its angular velocity after 3 s is **NCERT | Page-117 | Manipur 2023**

(a) 10 rad/s (b) 15 rad/s
(c) 1 rad/s (d) 5 rad/s

79. An energy of 484 J is spent in increasing the speed of a flywheel from 60 rpm to 360 rpm.
The moment of inertia of the flywheel is **NCERT | Page-119 | 2022**

(a) 0.07 kg-m^2 (b) 0.7 kg-m^2
(c) 3.22 kg-m^2 (d) 30.8 kg-m^2

80. A small mass attached to a string rotates on frictionless table top as shown. If the tension in the string is increased by pulling the string causing the radius of the circular motion to decrease by a factor of 2, the kinetic energy of the mass will **NCERT | Page-120 | 2011 M**

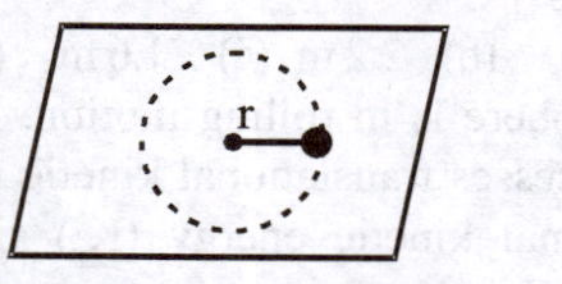

(a) remain constant
(b) increase by a factor of 2
(c) increase by a factor of 4
(d) decrease by a factor of 2

81. The moment of inertia of a body about a given axis is 1.2 kg m^2. Initially, the body is at rest. In order to produce a rotational kinetic energy of 1500 joule, an angular acceleration of 25 radian/sec^2 must be applied about that axis for a duration of **NCERT | Page-120 | 1990**

(a) 4 seconds (b) 2 seconds
(c) 8 seconds (d) 10 seconds

82. A fly wheel rotating about a fixed axis has a kinetic energy of 360 joule when its angular speed is 30 radian/sec. The moment of inertia of the wheel about the axis of rotation is **NCERT | Page-120 | 1990**

(a) 0.6 kg/m^2 (b) 0.15 kg m^2
(c) 0.8 kg m^2 (d) 0.75 kg m^2

83. A ring of mass m and radius r rotates about an axis passing through its centre and perpendicular to its plane with angular velocity ω. Its kinetic energy is **NCERT | Page-170 / N-120 | 1988**

(a) $\frac{1}{2}mr^2\omega^2$ (b) $mr\omega^2$
(c) $mr^2\omega^2$ (d) $\frac{1}{2}mr\omega^2$

Topic-6.12 Angular Momentum in Case of Rotation About a Fixed Axis

84. A mass m moves in a circle on a smooth horizontal plane with velocity v_0 at a radius R_0. The mass is attached to string which passes through a smooth hole in the plane as shown.

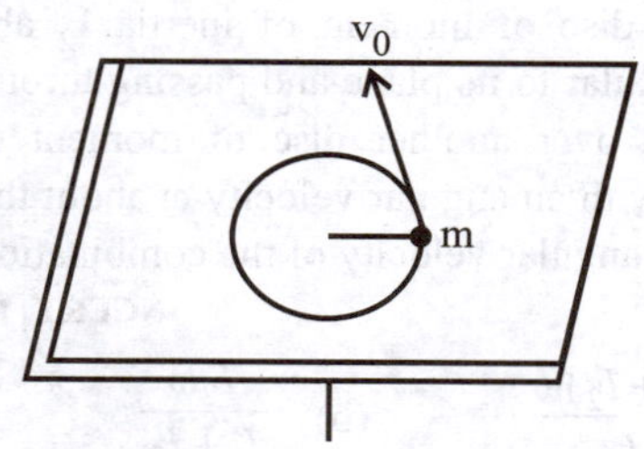

The tension in the string is increased gradually and finally m moves in a circle of radius $\frac{R_0}{2}$. The final value of the kinetic energy is **NCERT | Page-107 | 2015**

(a) $\frac{1}{4}mv_0^2$ (b) $2mv_0^2$ (c) $\frac{1}{2}mv_0^2$ (d) mv_0^2

85. Two discs are rotating about their axes, normal to the discs and passing through the centres of the discs. Disc D_1 has 2 kg mass and 0.2 m radius and initial angular velocity of 50 rad s^{-1}. Disc D_2 has 4kg mass, 0.1 m radius and initial angular velocity of 200 rad s^{-1}. The two discs are brought in contact face to face, with their axes of rotation coincident. The final angular velocity (in rad s^{-1}) of the system is **NCERT | Page-122 | Kar. 2013**

(a) 40 (b) 60 (c) 100 (d) 120

86. A circular platform is mounted on a frictionless vertical axle. Its radius $R = 2$ m and its moment of inertia about the axle is 200 kgm^2. It is initially at rest. A 50 kg man stands on the edge of the platform and begins to walk along the edge at the speed of 1 ms^{-1} relative to the ground. Time taken by the man to complete one revolution is **NCERT | Page-122 | 2012 M**

(a) πs (b) $\frac{3\pi}{2}$s (c) 2πs (d) $\frac{\pi}{2}$s

87. A circular disk of moment of inertia I_t is rotating in a horizontal plane, its symmetry axis, with a constant angular speed ω_i. Another disk of moment of inertia I_b is dropped coaxially onto the rotating disk. Initially the second disk has zero angular speed. Eventually both the disks rotate with a constant angular speed ω_f. The energy lost by the initially rotating disk to friction is: **NCERT | Page-122 | 2010**

(a) $\frac{1}{2}\frac{I_b^2}{(I_t+I_b)}\omega_i^2$ (b) $\frac{I_t^2}{(I_t+I_b)}\omega_i^2$

(c) $\frac{I_b-I_t}{(I_t+I_b)}\omega_i^2$ (d) $\frac{1}{2}\frac{I_bI_t}{(I_t+I_b)}\omega_i^2$

88. A thin circular ring of mass M and radius R is rotating in a horizontal plane about an axis vertical to its plane with a constant angular velocity ω. If two objects each of mass m be attached gently to the opposite ends of a diameter of the ring, the ring will then rotate with an angular velocity: **NCERT | Page-122 | 2009, 1998**

(a) $\frac{\omega M}{M+2m}$ (b) $\frac{\omega(M+2m)}{M}$

(c) $\frac{\omega M}{M+m}$ (d) $\frac{\omega(M-2m)}{M+2m}$

89. A round disc of moment of inertia I_2 about its axis perpendicular to its plane and passing through its centre is placed over another disc of moment of inertia I_1 rotating with an angular velocity ω about the same axis. The final angular velocity of the combination of discs is **NCERT | Page-122 | 2004**

(a) $\frac{(I_1+I_2)\omega}{I_1}$ (b) $\frac{I_2\omega}{I_1+I_2}$

(c) ω (d) $\frac{I_1\omega}{I_1+I_2}$

90. A thin circular ring of mass M and radius r is rotating about its axis with a constant angular velocity ω. Four objects each of mass m, are kept gently to the opposite ends of two perpendicular diameters of the ring. The angular velocity of the ring will be **NCERT | Page-122 | 2003**

(a) $\frac{(M-4m)\omega}{M+4m}$ (b) $\frac{M\omega}{4m}$

(c) $\frac{M\omega}{M+4m}$ (d) $\frac{(M+4m)\omega}{M}$

91. A disc is rotating with angular velocity ω. If a child sits on it, what is conserved ? **NCERT | Page-122 | 2002**

(a) Linear momentum (b) Angular momentum
(c) Kinetic energy (d) Moment of inertia

92. The angular momentum of a body with mass (m), moment of inertia (I) and angular velocity (ω) rad/sec is equal to **NCERT | Page-120 | 1996**

(a) Iω (b) Iω^2 (c) $\frac{I}{\omega}$ (d) $\frac{I}{\omega^2}$

93. Two discs of same moment of inertia rotating about their regular axis passing through centre and perpendicular to the plane of disc with angular velocities ω_1 and ω_2. They are brought into contact face to face coinciding the axis of rotation. The expression for loss of energy during this process is:- **NCERT | Page-122 | 2017**

(a) $\frac{1}{4}I(\omega_1-\omega_2)^2$ (b) $I(\omega_1-\omega_2)^2$

(c) $\frac{1}{8}(\omega_1-\omega_2)^2$ (d) $\frac{1}{2}I(\omega_1+\omega_2)^2$

Topic-* Rolling Motion

94. A wheel of a bullock cart is rolling on a level road as shown in the figure below. If its linear speed is v in the direction shown, which one of the following options is correct (P and Q are any highest and lowest points on the wheel, respectively)? **2024**

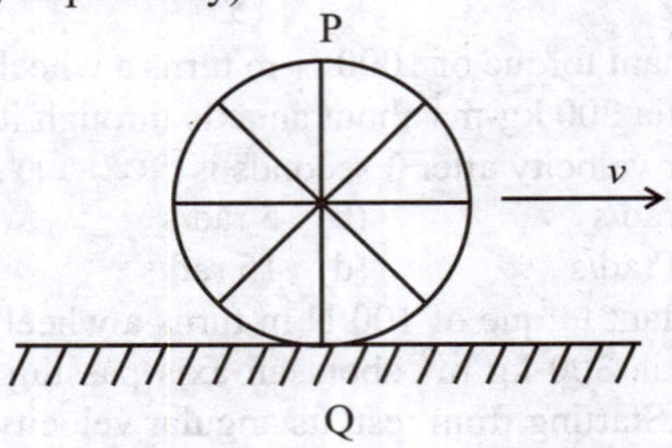

(a) Point P moves slower than point Q
(b) Point P moves faster than point Q
(c) Both the points P and Q move with equal speed
(d) Point P has zero speed

95. A disc of radius 2 m and mass 100 kg rolls on a horizontal floor. Its centre of mass has speed of 20 cm/s. How much work is needed to stop it ? **2019**

(a) 3 J (b) 30 kJ (c) 2 J (d) 1J

96. A solid cylinder of mass 2 kg and radius 50 cm rolls up an inclined plane of angle of inclination 30°. The centre of mass of the cylinder has speed of 4 m/s. The distance travelled by the cylinder on the inclined surface will be, [take g = 10 m/s^2] **Odisha 2019**

(a) 2.4 m (b) 2.2 m (c) 1.6 m (d) 1.2 m

97. A solid sphere is in rolling motion. In rolling motion a body possesses translational kinetic energy (K_t) as well as rotational kinetic energy (K_r) simultaneously. The ratio $K_t : (K_t + K_r)$ for the sphere is **2018**

(a) 7 : 10 (b) 5 : 7
(c) 2 : 5 (d) 10 : 7

98. Three objects, A : (a solid sphere), B : (a thin circular disk) and C : (a circular ring), each have the same mass M and radius R. They all spin with the same angular speed ω about their own symmetry axes. The amounts of work (W) required to bring them to rest, would satisfy the relation **2018**

(a) $W_C > W_B > W_A$ (b) $W_A > W_B > W_C$
(c) $W_A > W_C > W_B$ (d) $W_B > W_A > W_C$

99. A disk and a sphere of same radius but different masses roll off on two inclined planes of the same altitude and length. Which one of the two objects gets to the bottom of the plane first? **2016**

(a) Disk
(b) Sphere
(c) Both reach at the same time
(d) Depends on their masses

100. The ratio of the accelerations for a solid sphere (mass 'm' and radius 'R') rolling down an incline of angle 'θ' without slipping and slipping down the incline without rolling is : **2014**

(a) 5 : 7 (b) 2 : 3 (c) 2 : 5 (d) 7 : 5

101. A small object of uniform density rolls up a curved surface with an initial velocity 'v'. It reaches upto a maximum height of $\frac{3v^2}{4g}$ with respect to the initial position. The object is a **2013**

(a) solid sphere (b) hollow sphere
(c) disc (d) ring

102. A drum of radius R and mass M, rolls down without slipping along an inclined plane of angle θ. The frictional force **2005**

(a) dissipates energy as heat
(b) decreases the rotational motion
(c) decreases the rotational and translational motion
(d) converts translational energy to rotational energy

103. A solid cylinder of mass m and radius R rolls down an inclined plane of height h without slipping. The speed of its centre of mass when it reaches the bottom is **2003, 1989**

(a) $\sqrt{(2gh)}$ (b) $\sqrt{4gh/3}$
(c) $\sqrt{3gh/4}$ (d) $\sqrt{4g/h}$

104. A ball rolls without slipping. The radius of gyration of the ball about an axis passing through its centre of mass is K. If radius of the ball be R, then the fraction of total energy associated with its rotational energy will be **2003**

(a) $\frac{R^2}{K^2+R^2}$ (b) $\frac{K^2+R^2}{R^2}$
(c) $\frac{K^2}{R^2}$ (d) $\frac{K^2}{K^2+R^2}$

105. A solid cylinder and a hollow cylinder both of the same mass and same external diameter are released from the same height at the same time on an inclined plane. Both roll down without slipping. Which one will reach the bottom first? **2000, 2010 M**

(a) Both together
(b) Solid cylinder
(c) One with higher density
(d) Hollow cylinder

106. A spherical ball rolls on a table without slipping. Then the fraction of its total energy associated with rotation is **1994**

(a) 2/5 (b) 2/7 (c) 3/5 (d) 3/7

107. A thin uniform circular ring is rolling down an inclined plane of inclination 30° without slipping. Its linear acceleration along the inclination plane will be **1994**

(a) $\frac{g}{2}$ (b) $\frac{g}{3}$ (c) $\frac{g}{4}$ (d) $\frac{2g}{3}$

108. A solid sphere, disc and solid cylinder all of the same mass and made of the same material are allowed to roll down (from rest) on the inclined plane, then **1993**

(a) solid sphere reaches the bottom first
(b) solid sphere reaches the bottom last
(c) disc will reach the bottom first
(d) all reach the bottom at the same time

109. The speed of a homogenous solid sphere after rolling down an inclined plane of vertical height h from rest without sliding is **1992**

(a) $\sqrt{\frac{10}{7}gh}$ (b) $\sqrt{gh}$
(c) $\sqrt{\frac{6}{5}gh}$ (d) $\sqrt{\frac{4}{3}gh}$

110. If a sphere is rolling, the ratio of the translational energy to total kinetic energy is given by **1991**

(a) 7: 10 (b) 2 : 5
(c) 10 : 7 (d) 5 : 7

111. A solid homogeneous sphere of mass M and radius R is moving on a rough horizontal surface, partly rolling and partly sliding. During this kind of motion of the sphere **NCERT | Page-122 | 1988**

(a) total kinetic energy is conserved
(b) the angular momentum of the sphere about the point of contact with the plane is conserved
(c) only the rotational kinetic energy about the centre of mass is conserved
(d) angular momentum about the centre of mass is conserved.

ANSWER KEY

1	(a)	13	(d)	25	(c)	37	(d)	49	(b)	61	(b)	73	(a)	85	(c)	97	(b)	109	(a)
2	(b)	14	(d)	26	(c)	38	(c)	50	(d)	62	(c)	74	(c)	86	(c)	98	(a)	110	(d)
3	(a)	15	(d)	27	(d)	39	(a)	51	(d)	63	(c)	75	(d)	87	(d)	99	(b)	111	(b)
4	(a)	16	(d)	28	(a)	40	(d)	52	(b)	64	(c)	76	(d)	88	(a)	100	(a)		
5	(d)	17	(d)	29	(b)	41	(c)	53	(b)	65	(a)	77	(d)	89	(d)	101	(c)		
6	(b)	18	(b)	30	(a)	42	(d)	54	(b)	66	(a)	78	(c)	90	(c)	102	(d)		
7	(a)	19	(a)	31	(b)	43	(a)	55	(d)	67	(a)	79	(b)	91	(b)	103	(b)		
8	(d)	20	(c)	32	(c)	44	(N)	56	(a)	68	(b)	80	(c)	92	(a)	104	(d)		
9	(d)	21	(b)	33	(d)	45	(a)	57	(b)	69	(a)	81	(b)	93	(a)	105	(b)		
10	(d)	22	(d)	34	(a)	46	(b)	58	(b)	70	(d)	82	(c)	94	(b)	106	(b)		
11	(b)	23	(d)	35	(a)	47	(a)	59	(c)	71	(d)	83	(a)	95	(a)	107	(c)		
12	(b)	24	(b)	36	(a)	48	(c)	60	(d)	72	(d)	84	(b)	96	(a)	108	(a)		

Hints & Solutions

1. **(a)**

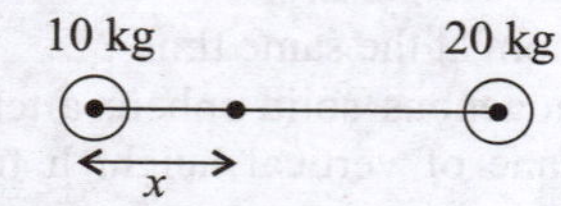

As, distance of COM from

$$m_1 = \frac{m_2 r}{m_1 + m_2}$$

Here, $m_1 = 10$ kg, $m_2 = 20$ kg and $r = 10$ m

So, required distance $= \frac{20 \times 10}{20+10} = \frac{20}{3} m$

2. **(b)**

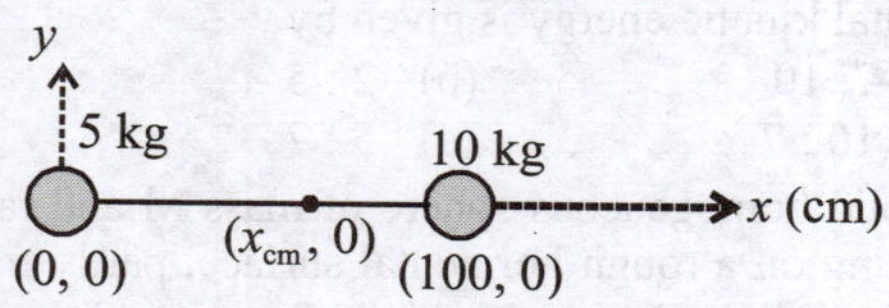

Centre of mass of a system of two particles of masses m_1 and m_2 having position vectors $\vec{x_1}$ and $\vec{x_2}$ is

$$x_{cm} = \frac{m_1 x_1 + m_2 x_2}{m_1 + m_2}$$

Taking particle of mass 5 kg at origin.

$$\Rightarrow x_{cm} = \frac{5 \times 0 + 100 \times 10}{5+10} = \frac{200}{3} = 66.66 \text{ cm}$$

$$\Rightarrow x_{cm} = 67 \text{ cm}$$

3. **(a)** $$X_{cm} = \frac{m_1 x_1 + m_2 x_2 + m_3 x_3}{m_1 + m_2 + m_3}$$

$$X_{cm} = \frac{300 \times (0) + 500(40) + 400 \times 70}{300 + 500 + 400}$$

$$X_{cm} = \frac{500 \times 40 + 400 \times 70}{1200}$$

$$X_{cm} = \frac{50 + 70}{3} = \frac{120}{3} = 40 \text{ cm}$$

4. **(a)** The position vector of the centre of mass of two particle system is given by

$$\vec{R}_{cm} = \frac{m_1 \vec{r_1} + m_2 \vec{r_2}}{(m_1 + m_2)}$$

$$= \frac{1}{4}[-8\hat{i} - 4\hat{j} + 4\hat{k}] = -2\hat{i} - \hat{j} + \hat{k}$$

5. **(d)** Consider an element of length dx at a distance x from end A.

Here, mass per unit length λ of rod

$\lambda \propto x \Rightarrow \lambda = kx$

$\therefore \; dm = \lambda dx = kx\, dx$

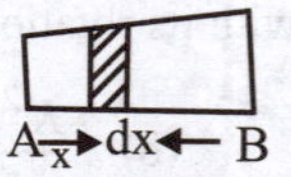

Position of centre of gravity of rod from end A.

$$x_{CG} = \frac{\int_0^L x dm}{\int_0^L dm}$$

$$\therefore\ x_{CG} = \frac{\int_0^3 x(kx\,dx)}{\int_0^3 kx\,dx} = \frac{\left[\frac{x^3}{3}\right]_0^3}{\left[\frac{x^2}{2}\right]_0^3} = \frac{\frac{(3)^3}{3}}{\frac{(3)^3}{2}} = 2m$$

6. **(b)** Centre of mass of system depends upon position and masses of particle. Also, it depends upon relative distance between particles.

7. **(a)**

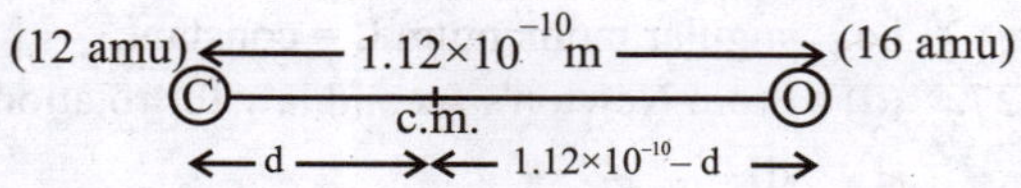

From definition of centre of mass.

$$d = \frac{16\times1.12\times10^{-10} + 12\times0}{16+12}$$

$$= \frac{16\times1.12\times10^{-10}}{28} = 0.64\times10^{-10}\text{ m.}$$

8. **(d)** Vector triple product

$$\vec{A}\times(\vec{B}\times\vec{C}) = \vec{B}(\vec{A}\cdot\vec{C}) - \vec{C}(\vec{A}\cdot\vec{B}) = 0$$

$$\Rightarrow \vec{A}\parallel(\vec{B}\times\vec{C})$$

9. **(d)** $|\vec{A}\times\vec{B}| = \sqrt{3}(\vec{A}.\vec{B})$

$$\Rightarrow AB\sin\theta = \sqrt{3}AB\cos\theta$$

$$\Rightarrow \tan\theta = \sqrt{3} \Rightarrow \theta = 60°$$

10. **(d)** $(\vec{B}\times\vec{A})\cdot\vec{A} = \vec{C}\cdot\vec{A} = CA\cos 90° = 0.$

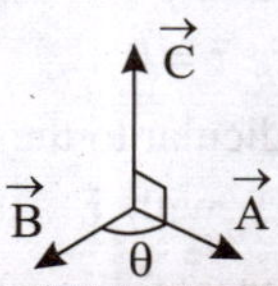

NOTES Dot product of any two perpendicular vector is zero.

11. **(b)** $|\vec{A}\times\vec{B}| = AB\sin\theta$

$\vec{A}.\vec{B} = AB\cos\theta$

$|\vec{A}\times\vec{B}| = \sqrt{3}\ \vec{A}.\vec{B} \Rightarrow AB\sin\theta = \sqrt{3}\,AB\cos\theta$

or, $\tan\theta = \sqrt{3}$, $\quad\therefore\ \theta = 60°$

$$\therefore\ |\vec{A}+\vec{B}| = \sqrt{A^2+B^2+2AB\cos 60°}$$

$$= \sqrt{A^2+B^2+AB}$$

12. **(b)** Note that $(\vec{B}\times\vec{A})\perp\vec{A}$. Hence their dot product is zero.

NOTES Dot product of any two perpendicular vector is zero.

13. **(d)** Using the Right hand screw rule, angular acceleration of a body, moving along the circumference of a cirlce is along the axis of rotation.

14. **(d)** Let T_A and T_B are the time periods of particle A and B respectively. According to question,

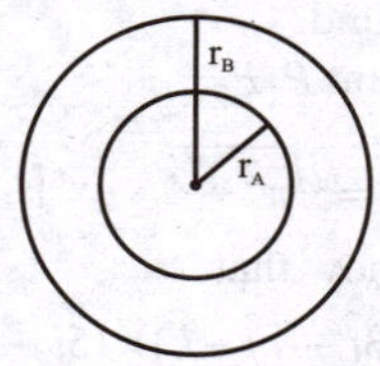

$$T_A = T_B = T$$

If ω_A and ω_B are their angular speeds, then

$$\omega_A = \frac{2\pi}{T_A} \text{ and } \omega_B = \frac{2\pi}{T_B}$$

$$\therefore\ \frac{\omega_A}{\omega_B} = \frac{T_B}{T_A} = \frac{T}{T} = 1:1$$

15. **(d)** $V^2 = u^2 + 2as$

$$V_0^2 = 0 + 2r\alpha(2\pi r)n \Rightarrow \alpha = \frac{V_0^2}{4\pi nr^2}$$

16. **(d)**

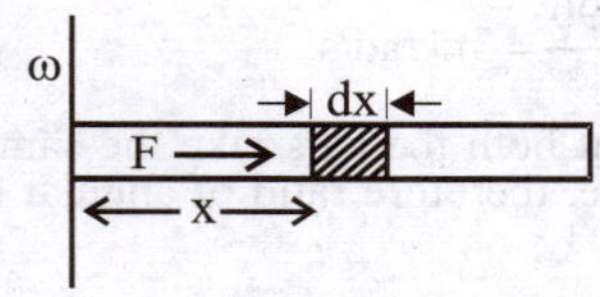

Rotation will create centrifugal force due to which pressure will be created at the other end. Now, this centrifugal force will depend on distance of the liquid from the axis as centrifugal force = $m\omega^2 r$.

Let us consider a small part dx of the liquid column which is at a distance of x.

Mass of this element will be,

$$dm = \frac{M}{L}dx$$

So, force exerted on this element will be,

$\Rightarrow$ $df = dm.\omega^2 x$

As the mass of liquid extends from x = 0 to x = L, the net force on the wall experienced by the mass of the liquid is,

$$\Rightarrow \int_0^f df = \frac{M}{L}\int_0^L \omega^2.x.dx$$

$$\Rightarrow \int_0^f df = \frac{M}{L}\omega^2\int_0^L x.dx$$

$$\Rightarrow f = \frac{M}{L}\omega^2\left[\frac{L^2}{2}\right] = \frac{M\omega^2 L}{2}$$

17. **(d)**

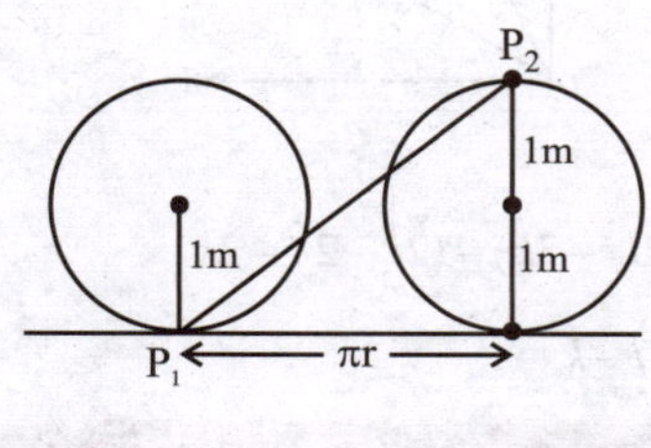

Linear distance moved by wheel in half revolution = πr.
Point P_1 after half revolution reaches at P_2 vertically 2m above the ground.

$\therefore$ Displacement P_1P_2

$= \sqrt{\pi^2 r^2 + 2^2} = \sqrt{\pi^2 + 4}$ $\quad [\because r = 1m]$

18. **(b)** As we know that

$\vec{v} = \vec{\omega} \times \vec{r} = (3\hat{i} - 4\hat{j} + \hat{k}) \times (5\hat{i} - 6\hat{j} + 6\hat{k})$

$= -18\hat{i} - 13\hat{j} + 2\hat{k}$

19. **(a)** As time taken by both car to complete one revolution is same.

As $\omega = \frac{2\pi}{T} \Rightarrow \omega \propto \frac{1}{T}$, as T is same in both cases. Hence 'ω' will also be same.

20. **(c)** Angular speed, $\omega = \frac{120 \times 2\pi}{60} = 4\pi$ rad/sec

21. **(b)** Number of revolutions made by the engine wheel (n) = 90 per minute.

Angular speed of the engine wheel

$(\omega) = \frac{2\pi n}{60} = \frac{2\pi \times 90}{60} = 3\pi$ rad/s.

22. **(d)** We know that both the cars take the same time to complete the circle, therefore ratio of angular speeds of the cars will be 1 : 1.

23. **(d)** The MOI is

$\Rightarrow I \propto r^2 \Rightarrow I_2 = 4 I_1$

$T_1 = 27$ day

By Angular momentum conservation,

$I_1 \omega_1 = I_2 \omega_2$

$\Rightarrow I_1 \omega_1 = (4 I_1)\, \omega_2$

$\therefore \omega_2 = \frac{\omega_1}{4}$

Also, $T = \frac{2\pi}{\omega} \Rightarrow T \propto \frac{1}{\omega}$

$\therefore$ $T_2 = 4T_1 = 4 \times 27 = 108$ days

24. **(b)** Given :

Force, $\vec{F} = 3\hat{j}$ N

Position vector, $\vec{r} = 2\hat{k}$ m

Torque, $\vec{\tau} = \vec{r} \times \vec{F} = 2\hat{k} \times 3\hat{j} = 6(\hat{k} \times \hat{j}) = 6(-\hat{i})$

$\Rightarrow \vec{\tau} = -6\hat{i}$ Nm

25. **(c)** Moment of force, $\vec{\tau} = \vec{r} \times \vec{F}$

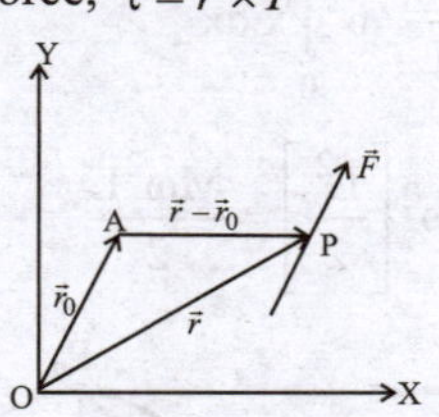

$\vec{\tau} = (\vec{r} - \vec{r}_0) \times \vec{F}$

$\vec{r} - \vec{r}_0 = (2\hat{i} + 0\hat{j} - 3\hat{k}) - (2\hat{i} - 2\hat{j} - 2\hat{k})$

$= 0\hat{i} + 2\hat{j} - \hat{k}$

$\vec{\tau} = (0\hat{i} + 2\hat{j} - \hat{k})(4\hat{i} + 5\hat{j} - 6\hat{k})$

$$\vec{\tau} = \begin{vmatrix} \hat{i} & \hat{j} & \hat{k} \\ 0 & 2 & -1 \\ 4 & 5 & -6 \end{vmatrix} = -7\hat{i} - 4\hat{j} - 8\hat{k}$$

26. **(c)** Angular momentum, remains conserved until the torque acting on sphere remains zero. $\tau_{ex} = 0$

So, $\frac{dL}{dt} = 0$

i.e., angular momentum L = constant.

27. **(d)** From Newton's second law for rotational motion,

$\vec{\tau} = \frac{d\vec{L}}{dt}$, if $\vec{L}$ = constant then $\vec{\tau} = 0$

So, $\vec{\tau} = \vec{r} \times \vec{F} = 0$

$(2\hat{i} - 6\hat{j} - 12\hat{k}) \times (\alpha\hat{i} + 3\hat{j} + 6\hat{k}) = 0$

Solving we get $\alpha = -1$

28. **(a)** $\because \vec{L} = \vec{r} \times \vec{p}$

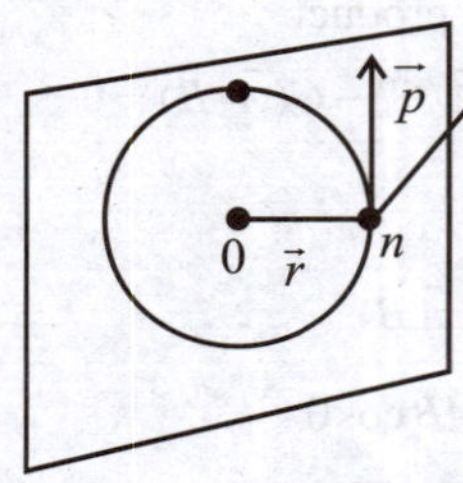

By right hand screw, rule, the direction of $\vec{L}$ is $\perp$ to the plane containing $\vec{r}$ and $\vec{p}$.

29. **(b)** $\vec{\tau} = \vec{r} \times \vec{F} \Rightarrow \vec{r} \cdot \vec{\tau} = 0 \qquad \vec{F} \cdot \vec{\tau} = 0$

Since, $\vec{\tau}$ is perpendicular to the plane of $\vec{r}$ and $\vec{F}$, hence the dot product of $\vec{\tau}$ with $\vec{r}$ and $\vec{F}$ is zero.

NOTES Radial component of force makes no contribution to the torque. Only transverse component of force contributes to the torque.

30. **(a)** Angular momentum = Linear momentum × distance of line of action of linear momentum about the origin.

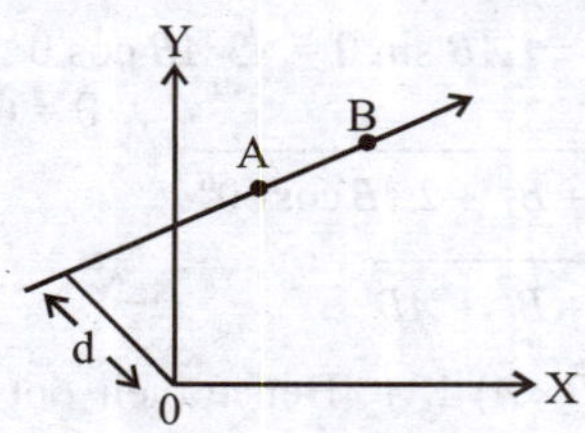

$L_A = p_A \times d$, $L_B = p_B \times d$

As linear momenta are p_A and p_B equal, therefore, $L_A = L_B$.

31. **(b)** As net torque applied is zero.

Hence, $\tau = \frac{dL}{dt}$

$\frac{dL}{dt} = 0$, L = constant.
L (angular momentum) remains conserved.

32. **(c)** A couple is formed of two equal and opposite forces at some separation; so net force is zero. Hence, a couple does not produce translatory motion; but it causes change in rotational motion.

33. **(d)** $\vec{F} = -3\hat{i} + \hat{j} + 5\hat{k}$; $\vec{r} = 7\hat{i} + 3\hat{j} + \hat{k}$

Torque $(\vec{\tau}) = \vec{r} \times \vec{F}$

$= (7\hat{i} + 3\hat{j} + \hat{k}) \times (-3\hat{i} + \hat{j} + 5\hat{k})$

$= 7\hat{k} + 35(-\hat{j}) - 9(-\hat{k}) + 15\hat{i} - 3\hat{j} + (-\hat{i})$

$= 14\hat{i} - 38\hat{j} + 16\hat{k}$

34. **(a)** Angular momentum $\vec{L}$ is defined as $\vec{L} = \vec{r} \times m(\vec{v})$
So, $\vec{L}$ is an axial vector.

35. **(a)**

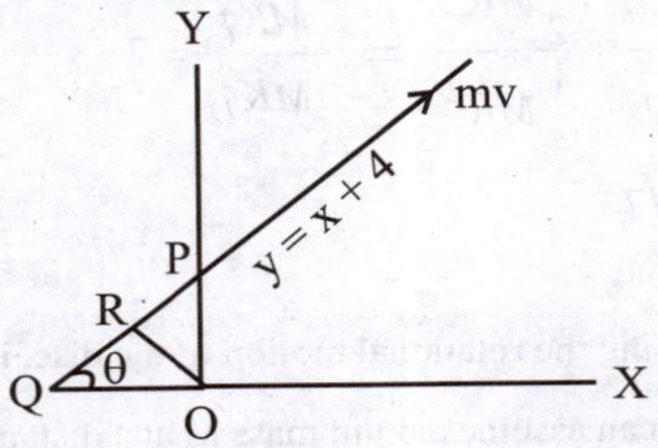

$y = x + 4$ line has been shown in the figure when $x = 0$, $y = 4$. So, $OP = 4$.
The slope of the line can be obtained by comparing with the equation of the straight line
$y = mx + c$
$m = \tan\theta = 1$
$\Rightarrow \theta = 45°$
$\angle OQP = \angle OPQ = 45°$
If we draw a line perpendicular to this line, length of the perpendicular $= OR$

In ΔOPR, $\frac{OR}{OP} = \sin 45°$

$\Rightarrow OR = OP \sin 45°$

$= 4 \times \frac{1}{\sqrt{2}} = \frac{4}{\sqrt{2}} = 2\sqrt{2}$

Angular momentum of particle going along this line $= r \times mv = 2\sqrt{2} \times 5 \times 3\sqrt{2} = 60$ units

36. **(a)**

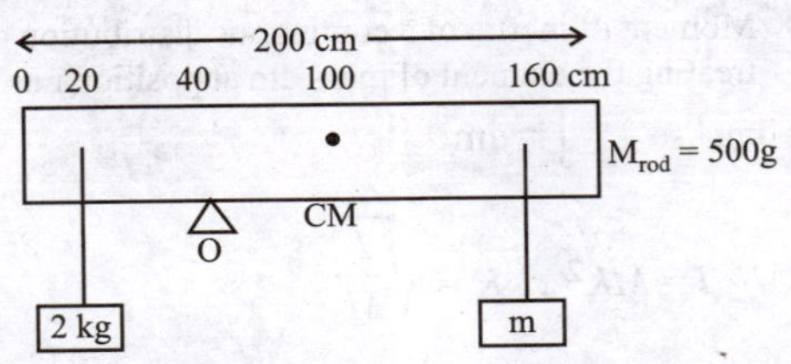

Rod will be in equilibrium, if $\vec{\tau}_{net} = 0$ about point O.
Torque at point O due to 2 kg mass
$\vec{\tau} = \vec{r} \times \vec{F} = rF\sin\theta(\hat{n})$

$\tau_1 = 20 \times 20 \times 10^{-2} \times \sin 90°(\hat{k}) = 4\text{Nm}(\hat{k})$

Torque due to mass of rod

$\tau_2 = 5 \times 60 \times 10^{-2} \times \sin 90°(-\hat{k}) = 3\text{Nm}(-\hat{k})$

Torque due to mass m

$\tau_3 = mg \times 120 \times 10^{-2} \times \sin 90°(-\hat{k})$
$= 12m\ \text{Nm}(-\hat{k})$

So $\vec{\tau_1} + \vec{\tau_2} + \vec{\tau_3} = 0\ (\because \vec{\tau}_{net} = 0)$

$4 - 3 - 12m = 0 \Rightarrow 12m = 1 \quad \therefore m = \frac{1}{12}$ kg

37. **(d)** Mechanical advantage, M. A.$= \frac{\text{Load}}{\text{Effort}}$

If M.A. > 1 $\Rightarrow$ Load > Effort

NOTES Centre of mass may or may not coincide with centre of gravity. Net torque of gravitational pull is zero about centre of mass.

$\tau_g = \Sigma\tau_i = \Sigma r_i \times m_i g = 0$

38. **(c)** By torque balancing about B
N_A (d) = W (d – x)

$N_A = \frac{W(d-x)}{d}$

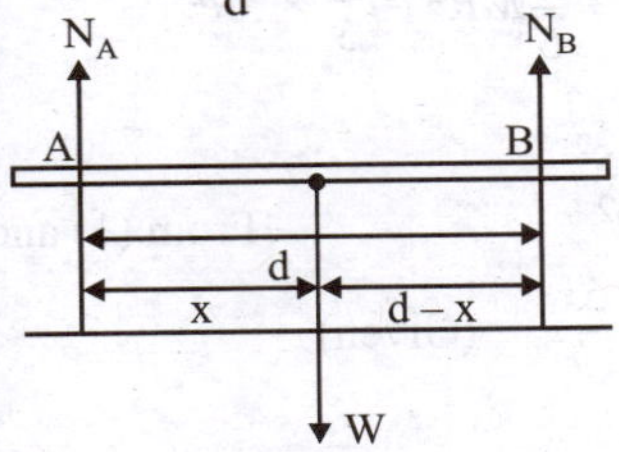

39. **(a)** $F_1 x + F_2 x = F_3 x$
$F_3 = F_1 + F_2$

40. **(d)** AB is the ladder, let F be the horizontal force and W is the weight of man. Let N_1 and N_2 be normal reactions of ground and wall, respectively. Then for vertical equilibrium

$W = N_1$ ...(1)

For horizontal equilibrium, $N_2 = F$ ...(2)

Taking moments about A,

$N_2(AB \sin 60°) - W(AC \cos 60°) = 0$ ......(3)

Using (2) and AB = 20 ft, BC = 4 ft, we get

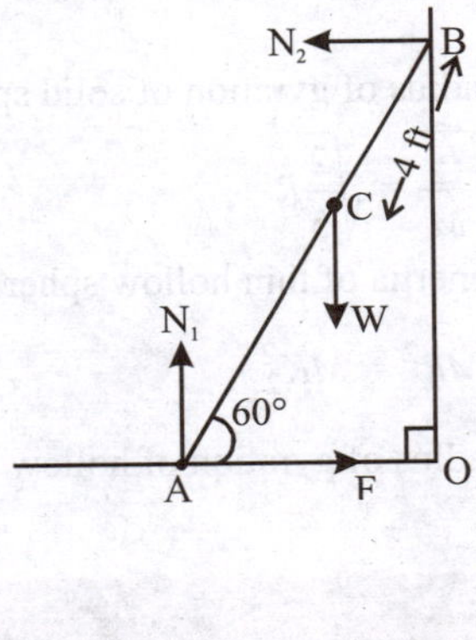

$$F\left(20\times\frac{\sqrt{3}}{2}\right)-W\left(16\times\frac{1}{2}\right)=0$$

$$\Rightarrow F=\frac{8W\times2}{20\sqrt{3}}=\frac{4W}{5\sqrt{3}}=\frac{150\times4}{5\sqrt{3}} \text{ pound}$$

$$=40\sqrt{3}=40\times1.73=69.2 \text{ pound}$$

41. (c) MoI of remaining part = MoI of larger solid sphere - MoI of smaller sphere

(Larger sphere) = $I_1=\frac{2}{5}M(2R)^2$

(Smaller sphere) $I_2=\frac{2}{5}mr^2+md^2$

$$=\frac{2}{5}\left(\frac{M}{8}\right)R^2+\frac{M}{8}(R)^2=\frac{7MR^2}{40}$$

(Remaining part of sphere) $I=I_1-I_2$

$$=\frac{8MR^2}{5}-\frac{7MR^2}{40}=\frac{57}{40}MR^2$$

$$\frac{I_2}{I}=\frac{7}{57}$$

42. (d) $I_{XY}=I_g+MR^2$

$$=\frac{2}{5}MR^2+MR^2=\frac{7}{5}MR^2=\frac{7}{5}\times5R^2$$

$=7R^2$ ...(1)

$I_{XY}=MK^2=5\times5^2$ ...(2)

$\Rightarrow 5\times5^2=7\times R^2$ [From (1) and (2)]

$\Rightarrow R=\sqrt{\frac{5}{7}}\times5=\frac{5x}{\sqrt{7}}$ (Given)

$\Rightarrow x=\sqrt{5}$

43. (a) Moment of inertia of rod = $I=\frac{m\ell^2}{12}$

$$\Rightarrow 2400=400\frac{\ell^2}{12}$$

$$\Rightarrow 72=\ell^2$$

$\therefore \ell=\sqrt{72}=8.48 \text{ cm}\simeq 8.5 \text{ cm}$

44. (N) Moment of inertia of solid sphere about its own axis

$$I_1=\frac{2}{5}MR^2=MK_1^2$$

Here K_1 = radius of gyration of solid sphere

$$\therefore K_1=\sqrt{\frac{I_2}{M}}=\sqrt{\frac{2}{5}}R$$

Moment of inertia of thin hollow sphere,

$$I_2=\frac{2}{3}MR^2=MK_2^2$$

Here K_2 = radius of gyration of hollow sphere

$$\therefore K_2=\sqrt{\frac{I_2}{M}}=\sqrt{\frac{2}{3}}R$$

$$\therefore \frac{K_1}{K_2}=\sqrt{\frac{2}{5}}\times\sqrt{\frac{3}{2}}=\sqrt{\frac{3}{5}}$$

45. (a)

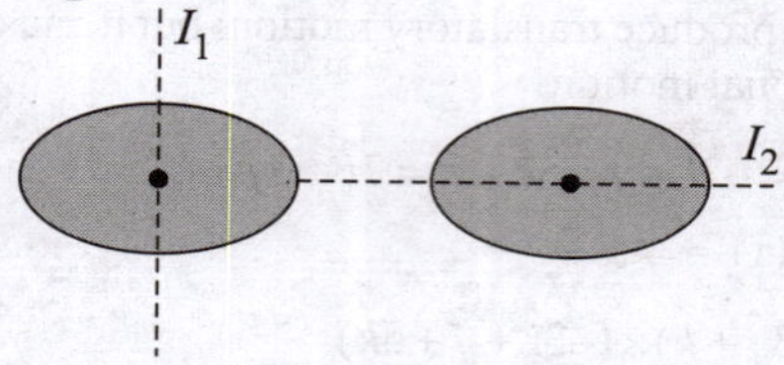

$$I_{centre}=I_1=\frac{1}{2}MR^2$$

$$\Rightarrow I_{Diameter}=I_2=\frac{1}{4}MR^2$$

$$\Rightarrow \frac{I_{centre}}{I_{Diameter}}=\frac{\frac{1}{2}MR^2}{\frac{1}{4}MR^2} \Rightarrow \frac{MK_C^2}{MK_D^2}=2$$

$$\Rightarrow \frac{K_C}{K_D}=\sqrt{2}$$

NOTES For dealing the rotational motion of the disc, instead of disc we can assume a point mass M at a distance $\frac{R}{\sqrt{2}}$.

46. (b) As per question, 90° arc is removed from circular ring,

$\therefore$ Mass removed = $\frac{M}{4}$

$\therefore$ Mass remaining $M-\frac{M}{4}=\frac{3}{4}M$

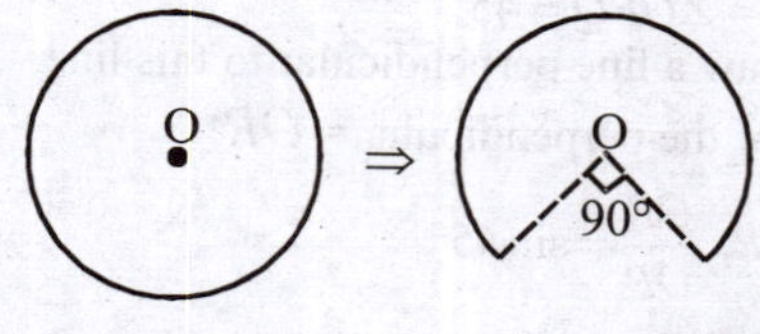

Moment of inertia of remaining $I_{remaining}=\int dmr^2$

$=R^2\int dm \quad (\because r=R)$

$\therefore I_{remaining}=\frac{3MR^2}{4}=KMR^2 \therefore K=\frac{3}{4}M$

NOTES Moment of inertia of a continuous distribution of mass, treating the element of mass dm at position r as particle then $dI=dmr^2$ so I $\int r\ dm$

47. (a) $\because I=MK^2 \therefore K=\sqrt{\frac{I}{M}}$

$$I_{ring}=MR^2 \text{ and } I_{disc}=\frac{1}{2}MR^2$$

$$\frac{K_1}{K_2}=\sqrt{\frac{I_1}{I_2}}=\sqrt{\frac{MR^2}{\left(\frac{MR^2}{2}\right)}}=\sqrt{2}:1$$

48. (c) From figure,
Total moment of inertia of the rod,
$I = m_1x^2 + m_2(L-x)^2$
$I = m_1x^2 + m_2[L^2 + x^2 - 2Lx]$
$I = m_1x^2 + m_2L^2 + m_2x^2 - 2m_2Lx$

As I is minimum, so, $\frac{dI}{dx}=0$ ... (i)

$\Rightarrow \frac{dI}{dx} = 2m_1x + 2m_2x - 2m_2L$

Using equation (i)

$\Rightarrow \quad 2m_1x + 2m_2x - 2m_2L = 0$

$\Rightarrow \quad x = \frac{m_2L}{(m_1+m_2)}$

When I is minimum, the work done by rotating a rod will be minimum.

NOTES Work required to set the rod rotating with angular velocity ω0.

$K.E. = \frac{1}{2}I\omega_0^2$

Work is minimum, when I is minimum. I is minimum about the centre of mass.

$\therefore \quad m1x1 = m2(L - x)$

$\Rightarrow \quad m1x = m2L - M2x$

$\therefore \quad x = \frac{m_2L}{m_1+m_2}$

49. (b) Mass of each part = M/2
Length of each part = L/2

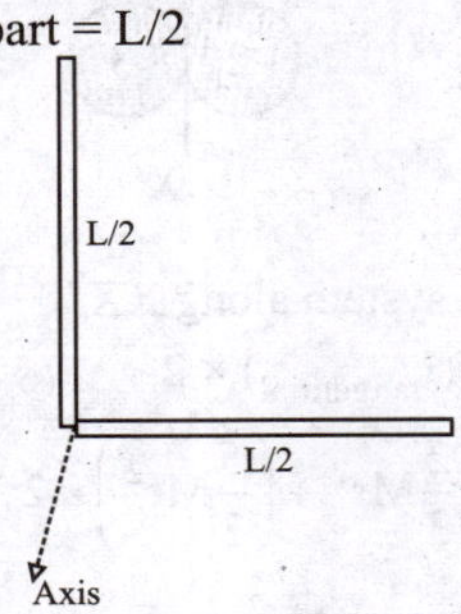

Total M.I. = Sum of M.I.s of both parts

$$=\left(\frac{M}{2}\right)\left(\frac{L}{2}\right)^2\times\frac{1}{3}+\left(\frac{M}{2}\right)\left(\frac{L}{2}\right)^2\times\frac{1}{3}$$

$$=2\times\frac{M}{2}\times\frac{L^2}{4}\times\frac{1}{3}=\frac{ML^2}{12}$$

50. (d)

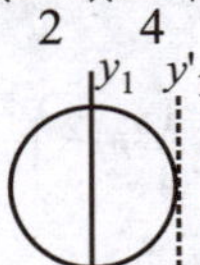

Circular disc [1]

$$I_{y_1}=\frac{MR^2}{4}$$

$$\therefore I'_{y_1}=\frac{MR^2}{4}+MR^2=\frac{5}{4}MR^2$$

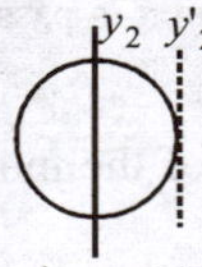

Circular ring [2]

$$I_{y_2}=\frac{MR^2}{2} \qquad \therefore I'_{y_2}=\frac{MR^2}{2}+MR^2=\frac{3}{2}MR^2$$

$I'_{y_1} = MK_1^2,\ I'_{y_2} = MK_2^2$

$$\therefore \frac{K_1^2}{K_2^2}=\frac{I'_{y_1}}{I'_{y_2}} \Rightarrow K_1:K_2=\sqrt{5}:\sqrt{6}$$

51. (d) $I_{AX} = m(AB)^2 + m(OC)^2$
$= m\ell^2 + m(\ell\cos 60^o)^2$
$= m\ell^2 + m\ell^2/4 = 5/4\ m\ell^2$

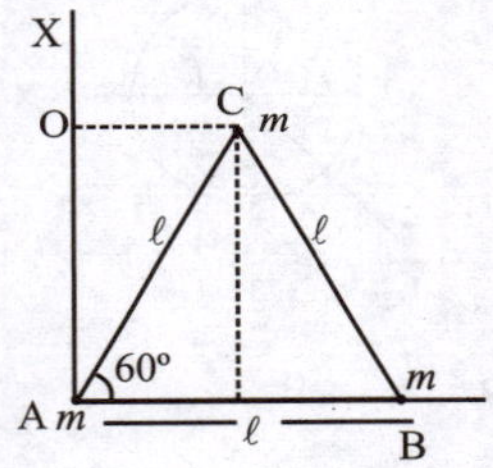

52. (b) Moment of inertia = $\int r^2 dm$.

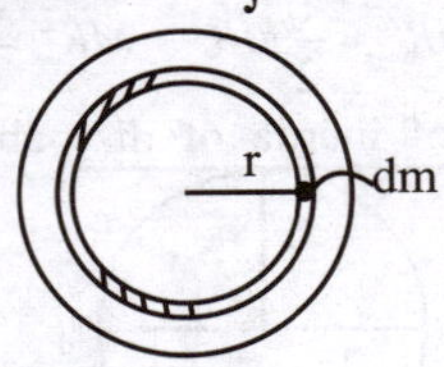

$\therefore$ Since, $\rho_{iron} > \rho_{aluminium}$
So, whole of aluminium is kept in the core and the iron at the outer rim of the disc.

53. (b)

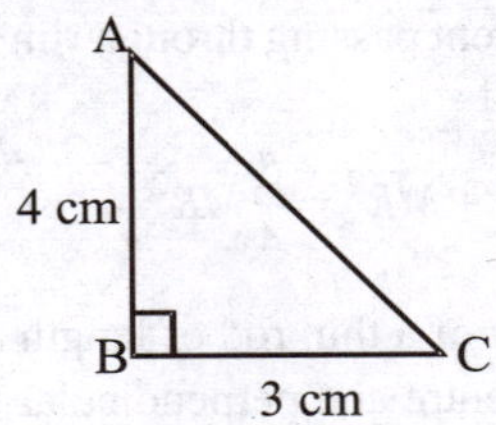

If we choose BC as axis. Distance is maximum. Hence, Moment of Inertia is maximum.

NOTES Moment of Inertia depend upon mass and distribution of masses as I = Σmr2.
Further, as the distance of masses is more, more is the moment of Inertia.

$\therefore I_2 > I_1, I_2 > I_3$

54. (b) Since distances of centre of mass from the sides are related as : $x_{BC} > x_{AB} > x_{AC}$, therefore, $I_{BC} > I_{AB} > I_{AC}$ or $I_{BC} > I_{AB}$.

NOTES The intersection of medians is the centre of mass of the triangle.

55. **(d)** The M.I. is minimum about EG because mass distribution is at minimum distance from EG.

NOTES Among so many parallel axis, moment of inertia of the body is the minimum about on axis which is passing through the centre of mass of the body.

56. **(a)** According to parallel axis theorem of the moment of inertia

$I = I_{cm} + md^2$

d is maximum for point B so I_{max} about B.

57. **(b)** By theorem of parallel axes,

$I = I_{cm} + Md^2$

$I = I_0 + M(L/2)^2 = I_0 + ML^2/4$

58. **(b)** M.I. of a uniform circular disc of radius 'R' and mass 'M' about an axis passing though C.M. and normal to the disc is

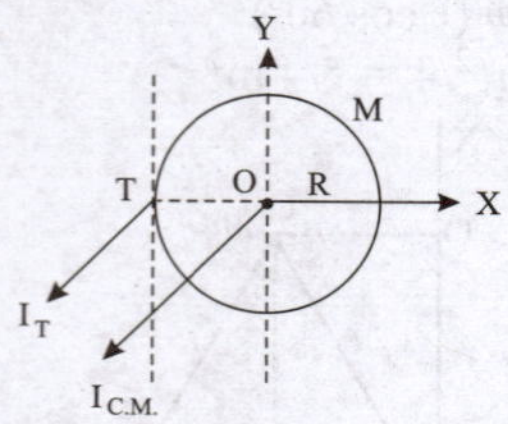

$$I_{C.M.} = \frac{1}{2}MR^2$$

From parallel axis theorem,

$$I_T = I_{C.M.} + MR^2 = \frac{1}{2}MR^2 + MR^2 = \frac{3}{2}MR^2$$

59. **(c)** Moment of inertia of disc about its diameter is

$$I_d = \frac{1}{4}MR^2$$

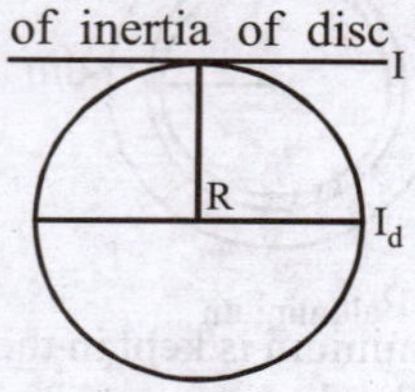

MI of disc about a tangent passing through rim and in the plane of disc is

$$I = I_d + MR^2 = \frac{1}{4}MR^2 + MR^2 = \frac{5}{4}MR^2$$

60. **(d)** Moment of inertia of a thin rod of length l about an axis passing through centre and perpendicular to the rod

$= \frac{1}{12}Ml^2$.

Thus moment of inertia of the frame.

$$\frac{ml^2}{12} + \frac{ml^2}{4} = \frac{4ml^2}{12} = \frac{ml^2}{3}$$

$$\text{Total M.I.} = 4 \times \frac{ml^2}{3}$$

61. **(b)** Moment of inertia of complete disc about point 'O'.

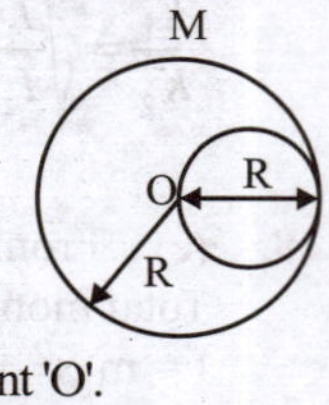

$$I_{\text{Total disc}} = \frac{MR^2}{2}$$

Mass of removed disc

$$M_{\text{Removed}} = \frac{M}{4} \text{ (Mass} \propto \text{area)}$$

Moment of inertia of removed disc about point 'O'.

I_{Removed} (about same perpendicular axis)

$= I_{cm} + mx^2$

$$= \frac{M}{4}\frac{(R/2)^2}{2} + \frac{M}{4}\left(\frac{R}{2}\right)^2 = \frac{3MR^2}{32}$$

Therefore the moment of inertia of the remaining part of the disc about a perpendicular axis passing through the centre,

$$I_{\text{Remaing disc}} = I_{\text{Total}} - I_{\text{Removed}}$$

$$= \frac{MR^2}{2} - \frac{3}{32}MR^2 = \frac{13}{32}MR^2$$

NOTES Moment of inertia of a part of a rigid body (Symmetrically cut from the whose mass) is the same as that of the whole body when whole mass is replaced by mass of that part.

62. **(c)** Moment of inertia of shell 1 along diameter

$$I_{\text{diameter}} = \frac{2}{3}Mr^2$$

Moment of inertia of shell 2 = M.I. of shell 3

$$= I_{\text{tangential}} = \frac{2}{3}Mr^2 + Mr^2 = \frac{5}{3}Mr^2$$

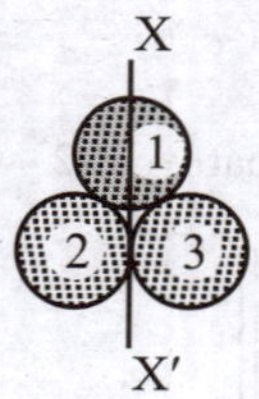

So, I of the system along XX'

$= I_{\text{diameter}} + (I_{\text{tangential}}) \times 2$

$$\text{or, } I_{\text{total}} = \frac{2}{3}Mr^2 + \left(\frac{5}{3}Mr^2\right) \times 2$$

$$= \frac{12}{3}Mr^2 = 4Mr^2$$

63. **(c)** M.I of uniform circular disc about its diameter = I

According to theorem of perpendiclar axes,

$$\text{M.I. of disc about its axis} = \frac{1}{2}mr^2 = 2I$$

$$\text{Applying theorem of } || \text{ axes, } \left(\because I = \frac{1}{4}mr^2\right)$$

M.I of disc about the given axis

$= 2I + mr^2 = 2I + 4I = 6I$

64. **(c)** Moment of inertia of disc about its diameter is $I_d = \frac{1}{4}MR^2$

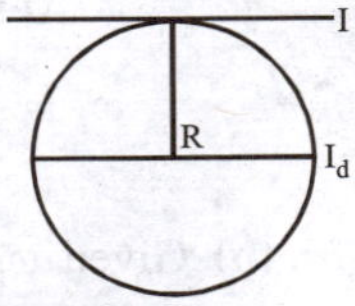

MI of disc about a tangent passing through rim and in the plane of disc is

$$I = I_d + MR^2 = \frac{1}{4}MR^2 + MR^2 = \frac{5}{4}MR^2$$

65. **(a)** Since, $\theta = \omega_0 t + \frac{1}{2}\alpha t^2$

Where α is angular aceleration, ω_0 is the initial angular speed.

$t = 2s$

$$\theta = 2 \times 2 + \frac{1}{2} \times 3(2)^2 = 4 + 6 = 10 \text{ rad}$$

66. **(a)** **Given:** Radius of disc, R = 50 cm

Angular acceleration, $\alpha = 2.0$ rads^{-2}; time t = 2s

Particle at periphery (assume) will have both radial (one) and tangential acceleration

$a_t = R\alpha = 0.5 \times 2 = 1$ m/s^2

Angular speed

$\omega = \omega_0 + \alpha t$

$\omega = 0 + 2 \times 2 = 4$ rad/sec

Centripetal acceleration

$a_c = \omega^2 R = (4)^2 \times 0.5 = 16 \times 0.5 = 8$m/s^2

Linear acceleration, $a_t = \alpha.R = 1$m/sec^2

Net acceleration,

$$a_{total} = \sqrt{a_t^2 + a_c^2} = \sqrt{1^2 + 8^2} \approx 8 \text{ m/s}^2$$

NOTES The acceleration of the particle moving in a circle has two components; tangential acceleration (at) which is along the tangent and radial acceleration (ar) which 15 towards the centre of circle. As the two components are mutually perpendicular therefore, net acceleration of the particle will be

$$a = \sqrt{a_t^2 + a_r^2}$$

67. **(a)** According to the Work-energy theorem,

$$W = \frac{1}{2}I\left(\omega_f^2 - \omega_i^2\right)$$

Given that,

$\theta = 2\pi$ revolution/minute

$\theta = 2\pi \times 2\pi = 4\pi^2$ rad

$\omega_i = 3 \times \frac{2\pi}{60}$ rad / s

$\omega_f = 0$ rad/s

Putting the values of ω_f and ω_i we get

$$\Rightarrow -\tau\theta = \frac{1}{2} \times \frac{1}{2}mr^2\left(0^2 - \omega_i^2\right)$$

$$\Rightarrow -\tau = \frac{\frac{1}{2} \times \frac{1}{2} \times 2 \times \left(4 \times 10^{-2}\right)\left(-3 \times \frac{2\pi}{60}\right)^2}{4\pi^2}$$

$\Rightarrow \tau = 2 \times 10^{-6}$ N-m

68. **(b)** Given, mass of cylinder m = 3kg

R = 40 cm = 0.4 m

F = 30 N ; α = ?

As we know, torque $\tau = I\alpha$

$F \times R = MR^2\alpha$

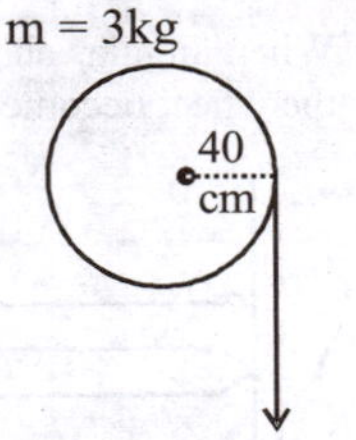

$$\alpha = \frac{F \times R}{MR^2}$$

$$\alpha = \frac{30 \times (0.4)}{3 \times (0.4)^2} \quad \text{or, } \alpha = 25\text{rad / s}^2$$

69. **(a)** Kinetic Energy of rotating body

$$K = \frac{1}{2}I\omega^2 = \frac{L^2}{2I}$$

$$K_A = K_B \Rightarrow \frac{L_A^2}{2I_A} = \frac{L_B^2}{2I_B}$$

$\because I_B > I_A \quad \therefore L_A^2 < L_B^2 \Rightarrow L_A < L_B$

70. **(d)** **Given :** Speed v = 54 kmh^{-1} = 15 ms^{-1}

Moment of inertia, I = 3 kgm^2

Time t = 15s

$$\omega_i = \frac{v}{r} = \frac{15}{0.45} = \frac{100}{3} \qquad [\text{Given } \omega_f = 0]$$

$\omega_f = \omega_i + \alpha t$

$$0 = \frac{100}{3} + (-\alpha)(15) \Rightarrow \alpha = \frac{100}{45}$$

Average torque transmitted by brakes to the wheel,

$$\tau = (I)(\alpha) = 3 \times \frac{100}{45} = 6.66 \text{ kgm}^2\text{s}^{-2}$$

71. **(d)** Here α = 2 revolutions/s^2 = 4π rad/s^2 (given)

$$I_{cylinder} = \frac{1}{2}MR^2 = \frac{1}{2}(50)(0.5)^2 = \frac{25}{4} \text{ Kg-m}^2$$

As $\tau = I\alpha$ so $TR = I\alpha$

$$\Rightarrow T = \frac{I\alpha}{R} = \frac{\left(\frac{25}{4}\right)(4\pi)}{(0.5)} \text{N} = 50\,\pi\text{N} = 157 \text{ N}$$

72. **(d)**

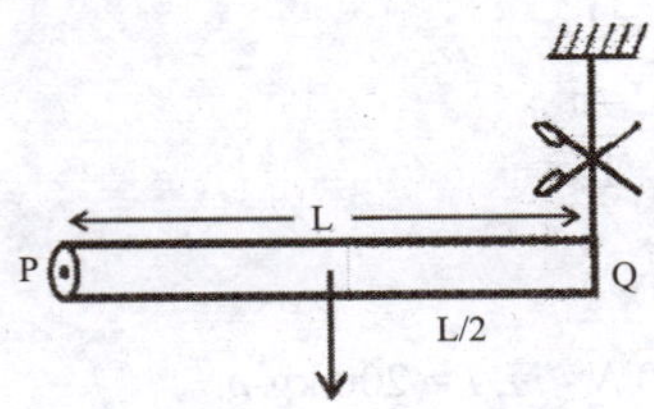

Weight of the rod will produce the torque

$$\tau = mg\frac{L}{2} = I\alpha = \frac{mL^2}{3}\alpha \qquad \left[\because I_{rod} = \frac{ML^2}{3}\right]$$

Hence, angular acceleration, $\alpha = \frac{3g}{2L}$

73. **(a)** $\theta(t) = 2t^3 - 6t^2$

$$\Rightarrow \frac{d\theta}{dt} = 6t^2 - 12t \Rightarrow \alpha = \frac{d^2\theta}{dt^2} = 12t - 12 = 0 \therefore t = 1 \text{ sec.}$$

NOTES When angular acceleration (α) is zero then torque on the wheel becomes zero.

74. (c)

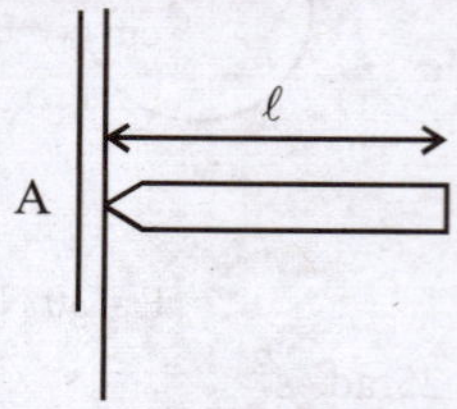

Weight of the rod will produce torque,

$$\tau = mg \times \frac{\ell}{2}$$

Also, $\tau = I\alpha$

where, I is the moment of inertia $= \frac{m\ell^2}{3}$

and α is the angular acceleration

$$\therefore \frac{m\ell^2}{3}\alpha = mg \times \frac{\ell}{2} \Rightarrow \alpha = \frac{3g}{2\ell}$$

75. (d) The relation between K.E. of rotation and angular momentum,

$$K = \frac{L^2}{2I} \Rightarrow L^2 = 2KI \Rightarrow L = \sqrt{2KI}$$

$$\frac{L_1}{L_2} = \sqrt{\left(\frac{K_1}{K_2}\right)\left(\frac{I_1}{I_2}\right)} = \sqrt{\left(\frac{K}{K}\right)\left(\frac{I}{2I}\right)} = \frac{1}{\sqrt{2}}$$

$$L_1 : L_2 = 1 : \sqrt{2}$$

76. (d) $I = 2 kgm^2$, $v_0 = 60 rpm = 1 rps$

$\omega_0 = 2\pi v_0 = 2\pi$ rad/sec

$\omega_f = 0$ and $t = 1$ min $= 60$ sec

$$\text{So, } \alpha = \frac{\omega_f - \omega_0}{t} = \frac{0 - 2\pi}{60}$$

$$\alpha = \frac{-\pi}{30} \text{rad/sec}^2$$

and Torque, $\tau = I.\alpha$

$$\tau = 2.\left(\frac{-\pi}{30}\right) = \frac{-\pi}{15}$$

$$\tau = \frac{\pi}{15} N - m$$

77. (d) $\tau = 1000$ N - m, $I = 200$ kg-m^2

$$\tau = I.\alpha \text{ and } \alpha = \left(\frac{\omega_f - \omega_0}{t}\right)$$

$$\Rightarrow \alpha = \frac{1000}{200} = 5 \text{ rad/sec}^2$$

$$\omega_1 = \omega_0 + \alpha t = 0 + 3 \times 5 = 15 \text{ rad/s}$$

78. (c) About fixed axis, torque = I α

$$\alpha = \frac{100}{300} = \frac{1}{3} \text{rad/s}^2$$

For constant $\alpha \Rightarrow \omega_i = 0$

$$\omega_f = \omega_i + \alpha t$$

$\therefore$ At $t = 3$ s

$$\omega_f = \frac{1}{3} \times 3 = 1 \text{ rad/s}$$

79. (b) Given $\omega_i = 60 rpm = 60 \times \frac{2\pi}{60} = 2\pi \text{ rad/s}$

$$\omega_f = 360 rpm = 360 \times \frac{2\pi}{60} = 12\pi \text{ rad/s}$$

Energy spent $\Delta kE = 484$ J

Energy spent, $\Delta kE = \frac{1}{2} I\left(\omega_f^2 - \omega_i^2\right)$

where I = moment of inertia

$$486 = \frac{1}{2} \times I \times [(12\pi)^2 - (2\pi)^2]$$

$$\therefore I = \frac{2 \times 486}{140\pi^2} \approx 0.7 kg - m^2$$

80. (c) $K.E. = \frac{L^2}{2I}$

The angular momentum L remains conserved about the centre.

That is, L = constant.

$I = mr^2$

$$\therefore K.E. = \frac{L^2}{2mr^2}$$

In 2nd case, $K.E. = \frac{L^2}{2(mr'^2)}$

But $r' = \frac{r}{2}$

$$\therefore K.E' = \frac{L^2}{2m.\frac{r^2}{4}} = \frac{4L^2}{2mr^2} \Rightarrow K.E.' = 4 K.E.$$

$\therefore$ K.E. is increased by a factor of 4.

NOTES When no torque is applied angular momentum remains constant.

$$I_1\omega_1 = I_2\omega_2 \Rightarrow \frac{\omega_2}{\omega_1} = \frac{I_1}{I_2}$$

$$\therefore \frac{E_2}{E_1} = \frac{\frac{1}{2}I_2\omega_2^2}{\frac{1}{2}I_1\omega_1^2} = \frac{I_2}{I_1}\left(\frac{I_1}{I_2}\right)^2 = \frac{I_2}{I_2}$$

So, if moment of inertia (mr^2) decreases K.E. of rotation increases and vice versa.

81. (b) $I = 1.2$ kg m^2, $E_r = 1500$ J, $\alpha = 25$ rad/sec^2, $\omega_1 = 0$, $t = ?$

$$\text{As } E_r = \frac{1}{2}I\omega^2, \Rightarrow \omega = \sqrt{\frac{2E_r}{I}}$$

$$= \sqrt{\frac{2 \times 1500}{1.2}} = 50 \text{ rad/sec}$$

From $\omega_2 = \omega_1 + \alpha t$

$50 = 0 + 25\,t,\ t = 2$ seconds

82. **(c)** $E_r = \frac{1}{2}I\omega^2$

$$I = \frac{2E_r}{\omega^2} = \frac{2\times 360}{30\times 30} = 0.8 \text{ kg m}^2$$

83. **(a)** Kinetic energy $= \frac{1}{2}I\omega^2$

and for ring $I = mr^2$

Hence, $KE = \frac{1}{2}mr^2\omega^2$

84. **(b)** Applying angular momentum conservation

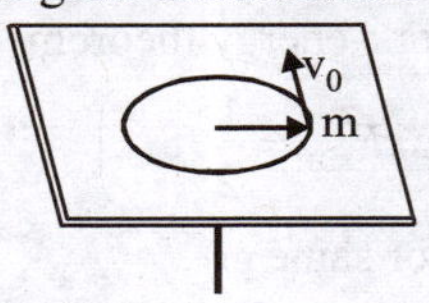

$$mv_0R_0 = (m)(v)\left(\frac{R_0}{2}\right)$$

$\therefore\quad v' = 2v_0$

Therefore, final KE $= \frac{1}{2}m(2v_0)^2 = 2mv_0^2$

85. **(c)** Given: $m_1 = 2$ kg $\quad m_2 = 4$ kg

$r_1 = 0.2$ m $\quad r_2 = 0.1$ m

$\omega_1 = 50$ rad s^{-1} $\quad \omega_2 = 200$ rad s^{-1}

As, angular momentum,

$I_1\omega_1 + I_2\omega_2 = (I_1 + I_2)\omega_1$

$$\therefore\ \omega_f = \frac{I_1\omega_1 + I_2\omega_2}{I_1 + I_2} = \frac{\frac{1}{2}m_1r_1^2\omega_1 + \frac{1}{2}m_2r_2^2\omega_2}{\frac{1}{2}m_1r_1^2 + \frac{1}{2}m_2r_2^2}$$

By putting the value of m_1, m_2, r_1, r_2 and solving we get $= 100$ rad s^{-1}

86. **(c)** According to the conservation of angular momentum, When no external torque acts on system, the angular momentum does not change.

$L_{system} = L_{man} + L_{platform}$ = constant

$L_{man} = -L_{platform}$

$L_{man} = mvR$

where m is mass of man, v is speed of man relative to ground, R is the radius of platform,

$L_{platform} = I\omega$

where I is moment of inertia, ω is angular velocity.

$$\Rightarrow \omega = \frac{L_{platform}}{I} = \frac{L_{man}}{I} = \frac{-mvR}{I}$$

Speed of man relative to platform is

$$v_{platform} = v - \omega R = v + \frac{mvR}{I}R$$

$$= v\left[1 + \frac{mR^2}{I}\right]$$

Time taken by man to complete one revolution is,

$$t = \frac{2\pi R}{v_{platform}} = \frac{2\pi r}{v\left[1+\frac{mR^2}{I}\right]} = 2\times\frac{2}{1(1+1)}\pi$$

$\Rightarrow\quad t = 2\pi$ second

87. **(d)** By conservation of angular momentum,

$I_t\omega_i = (I_t + I_b)\,\omega_f$

where ω_f is the final angular velocity of disks

$$\therefore\quad \omega_f = \left(\frac{I_t}{I_t + I_b}\right)\omega_i$$

Loss in K.E., ΔK= Initial K.E. – Final K.E.

$$= \frac{1}{2}I_t\omega_i^2 - \frac{1}{2}(I_t + I_b)\,\omega_f^2$$

$$= \frac{1}{2}I_t\omega_i^2 - \frac{1}{2}(I_t + I_b)\frac{I_t^2}{(I_t + I_b)^2}\omega_i^2$$

$$= \frac{1}{2}\omega_i^2\frac{I_t}{I_t + I_b}(I_t + I_b - I_t) = \frac{1}{2}\omega_i^2\frac{I_tI_b}{I_t + I_b}$$

88. **(a)** In absence of external torque, $L = I\omega$ = constant

$I_1\omega_1 = I_2\omega_2,\ I_1 = MR^2,\ I_2 = MR^2 + 2mR^2$

(Moment of inertia of a thin circular ring about an axis vertical to its plane = MR^2)

$$\therefore\quad \omega_2 = \frac{I_1}{I_2}\omega = \frac{M}{M + 2m}\omega.$$

89. **(d)** Angular momentum will be conserved

$$I_1\omega = I_1\omega' + I_2\omega' \ \Rightarrow \omega' = \frac{I_1\omega}{I_1 + I_2}$$

90. **(c)** Applying conservation law of angular momentum,

$I_1\omega_1 = I_2\omega_2$

$I_2 = (Mr^2) + 4(m)(r^2) = (M + 4m)r^2$

(Taking $\omega_1 = \omega$ and $\omega_2 = \omega_1$)

$\Rightarrow Mr^2\omega = (M + 4m)r^2\omega_1$

$$\Rightarrow \omega_1 = \frac{M\omega}{M + 4m}$$

91. **(b)** If external torque is zero, angular momentum remains conserved.

$L = I\omega$ = constant.

NOTES External torque is zero because the weight of child acts downward.

92. **(a)** Let body contain m_1, m_2, m_3 ....m_n masses at distance r_1, r_2, r_3 .... r_n from axis OA.

Angular momentum of body

$= m_1v_1r_1 + m_2v_2r_2 \ldots\ldots + m_nv_nr_n$

$= m_1(\omega r_1)r_1 + m_2(\omega r_2)r_2 \ldots\ldots + m_n(\omega r_n)r_n$

$= (m_1r_1^2)\omega + (m_2r_2^2)\omega \ldots\ldots + (m_nr_n^2)\omega$

$$= \left(\sum_{i=1}^{n} m_ir_i^2\right)\omega = I\omega$$

93. **(a)** Here, $I\omega_1 + I\omega_2 = 2I\omega$

$\Rightarrow \quad \omega = \dfrac{\omega_1 + \omega_2}{2}$

$(K.E.)_i = \dfrac{1}{2}I\omega_1^2 + \dfrac{1}{2}I\omega_2^2$

$(K.E.)_f = \dfrac{1}{2} \times 2I\omega^2 = I\left(\dfrac{\omega_1 + \omega_2}{2}\right)^2$

Loss in K.E. = $(K.E)_f - (K.E)_i = \dfrac{1}{4}I(\omega_1 - \omega_2)^2$

94. **(b)** In the case of pure rolling, $v = \omega R$

$\therefore \quad v_P = v + \omega R = 2v$

$v_Q = v - \omega R = 0$

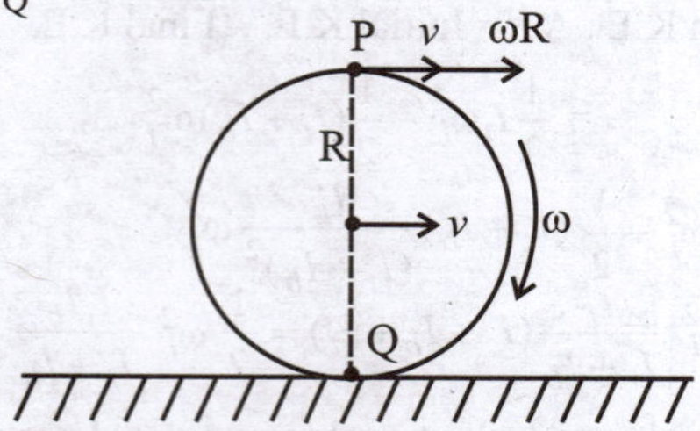

Hence, point P moves faster than point Q.

95. **(a)** Work done to stop the disc = change in total kinetic energy of disc

Final K.E. = 0

Initial K.E. = Translational K.E. + Rotational K.E.

$= \dfrac{1}{2}mv^2 + \dfrac{1}{2}I\omega^2$

$= \dfrac{1}{2}mv^2 + \dfrac{1}{2} \times \dfrac{mR^2}{2} \times \left(\dfrac{v}{R}\right)^2$

$= \dfrac{1}{2}mv^2 + \dfrac{1}{4}mv^2 = \dfrac{3}{4}mv^2$

$= \dfrac{3}{4} \times 100 \times (20 \times 10^{-2})^2 = 3J$

$|\Delta KE| = 3J$

96. **(a)** Kinetic energy at the bottom

$= \dfrac{1}{2}mv_0^2 + \dfrac{1}{2}I\omega^2 = \dfrac{3}{4}mv_0^2$

As $I_{solid\ cylinder} = \dfrac{1}{2}MR^2$ and $V = R\omega$

From work – energy theorem

$\dfrac{3}{4}mv_0^2 = mgh = mgd\sin\theta$

$d = \dfrac{\frac{3}{4}mv_0^2}{mg\sin\theta}$

$\Rightarrow d = \dfrac{3}{4}\dfrac{v_0^2}{g\sin\theta}$

$= \dfrac{3}{4} \cdot \dfrac{16}{10 \times 1/2} = 2.4$ m

97. **(b)** In rolling motion, rotational kinetic energy.

$K_t = \dfrac{1}{2}mv^2$

And, $K_t + K_r = \dfrac{1}{2}mv^2 + \dfrac{1}{2}I\omega^2$

$= \dfrac{1}{2}mv^2 + \dfrac{1}{2}\left(\dfrac{2}{5}mr^2\right)\left(\dfrac{v}{r}\right)^2 = \dfrac{7}{10}mv^2$

$\therefore \quad \dfrac{K_t}{K_t + K_r} = \dfrac{\frac{1}{2}mv^2}{\frac{7}{10}mv^2} = \dfrac{5}{7}$

98. **(a)** Work done required to bring them rest

$\Delta W = \Delta KE$ (work-energy theorem)

$\Delta W = \dfrac{1}{2}I\omega^2 \left(\Delta kE_{rot} = \dfrac{1}{2}I\omega^2\right)$

or, $\quad \Delta W \propto I$ (for same ω)

$I_{solid\ sphere} = \dfrac{2}{5}MR^2$,

$I_{Disk} = \dfrac{1}{2}MR^2$

$I_{Ring} = MR^2$

$\therefore \quad W_C > W_B > W_A$

99. **(b)** Time of descent $\propto \dfrac{K^2}{R^2}$

Order of value of $\dfrac{K^2}{R^2}$

for disc; $\dfrac{K^2}{R^2} = \dfrac{1}{2} = 0.5$

for sphere; $\dfrac{K^2}{R^2} = \dfrac{2}{5} = 0.4$

(sphere) < (disc)

$\therefore$ Sphere reaches first

NOTES

The value $\dfrac{K^2}{R^2}$ is a measure of moment of inertia. Its. value is fixed for a particular shape of the body and independent of mass and readius of the body

100. **(a)** For solid sphere rolling without slipping on inclined plane, acceleration

$a_1 = \dfrac{g\sin\theta}{\left(1 + \dfrac{K^2}{R^2}\right)}$

For solid sphere slipping on inclined plane without rolling, acceleration

$a_2 = g\sin\theta$

Therefore required ratio $=\frac{a_1}{a_2}$

$$=\frac{1}{\left(1+\frac{K^2}{R^2}\right)}=\frac{1}{\left(1+\frac{2}{5}\right)}=\frac{5}{7}$$

101. (c)

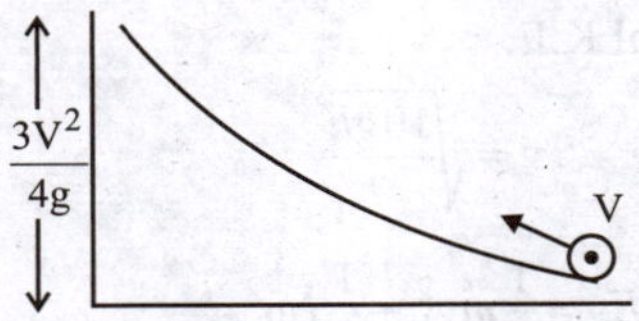

From law of conservation of mechanical energy,

$$\Rightarrow \frac{1}{2}I\omega^2+0+\frac{1}{2}mv^2=mg\times\frac{3v^2}{4g}$$

$$\Rightarrow \frac{1}{2}I\omega^2=\frac{3}{4}mv^2-\frac{1}{2}mv^2$$

$$\Rightarrow \frac{1}{2}I\omega^2=\frac{mv^2}{2}\left(\frac{3}{2}-1\right)$$

or, $\frac{1}{2}I\left(\frac{V^2}{R^2}\right)=\frac{mv^2}{4}$ or, $I=\frac{1}{2}mR^2$

Hence, object is a disc.

102. (d) Net work done by frictional force when drum rolls down without slipping is zero.

$W_{net}=0$

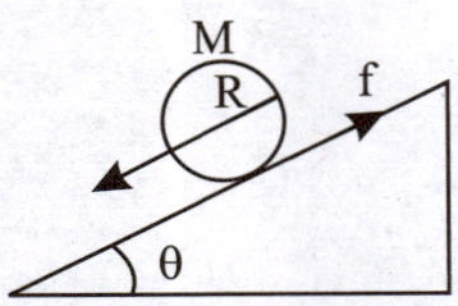

$W_{trans.}+W_{rot.}=0;\ \Delta K_{trans.}+\Delta K_{rot.}=0$

$\Delta K_{trans}=-\Delta K_{rot}.$

i.e., converts translation energy to rotational energy.

103. (b) $K.E.=\frac{1}{2}I\omega^2+\frac{1}{2}mv^2$

$$K.E.=\frac{1}{2}\left(\frac{1}{2}mr^2\right)\omega^2+\frac{1}{2}mv^2$$

$$=\frac{1}{4}mv^2+\frac{1}{2}mv^2=\frac{3}{4}mv^2$$

Now, gain in K.E. = Loss in P.E.

$$\frac{3}{4}mv^2=mgh\Rightarrow v=\sqrt{\left(\frac{4}{3}\right)gh}$$

104. (d) As we know that,

Rotational energy $=\frac{1}{2}I(\omega)^2=\frac{1}{2}(mK^2)\omega^2$ And, Linear kinetic energy $=\frac{1}{2}m\omega^2R^2$

∴ Required fraction,

$$=\frac{\frac{1}{2}(mK^2)\omega^2}{\frac{1}{2}(mK^2)\omega^2+\frac{1}{2}m\omega^2R^2}=\left(\frac{K^2}{K^2+R^2}\right)$$

105. (b)

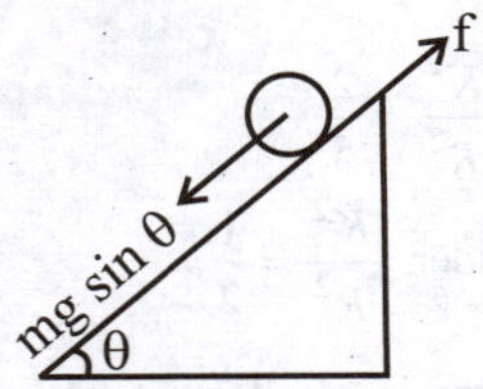

Torque, $I\alpha=f.R.$

Using Newton's IInd law, $mg\sin\theta-f=ma$

∵ pure rolling is there, $a=R\alpha$

$$\Rightarrow mg\sin\theta-\frac{I\alpha}{R}=ma$$

$$\Rightarrow mg\sin\theta-\frac{Ia}{R^2}=ma \qquad \left(\because \alpha=\frac{a}{R}\right)$$

or, acceleration, $a=\frac{mg\sin\theta}{(I/R^2+m)}$

Using, $s=ut+\frac{1}{2}at^2$

or, $s=\frac{1}{2}at^2$

$$\Rightarrow t\ \alpha\ \frac{1}{\sqrt{a}}$$

t minimum means a should be more. This is possible when I is minimum which is the case for solid cylinder. Therefore, solid cylinder will reach the bottom first.

106. (b) $\frac{K_r}{E}=\frac{\frac{1}{2}MK^2\omega^2}{\frac{1}{2}M\omega^2[K^2+R^2]}=\frac{K^2}{K^2+R^2}$

$$=\frac{2/5}{1+2/5}=\frac{2}{7}$$

Here, $K^2=\frac{2}{5}R^2$

NOTES When a body is executing only linear motion, its K.E. is given by $E_t = \frac{1}{2}mv^2$ when a body is rolling without slipping, its centre of mas has linear motion too.
Therefore, its total kinetic energy
$E = E_t + E_r$

$$= \frac{1}{2}mv^2 + \frac{1}{2}I\omega^2 = \frac{1}{2}mv^2 + \frac{1}{2}m(K^2)\frac{v^2}{R^2}$$

$$= \frac{1}{2}mv^2\left(1 + \frac{K^2}{R^2}\right)$$

107. **(c)** $a = \frac{g\sin\theta}{(1 + K^2/R^2)} = \frac{g\sin 30°}{1+1} = \frac{g}{4}$

108. **(a)** For solid sphere, $\frac{K^2}{R^2} = \frac{2}{5}$

For disc and solid cylinder, $\frac{K^2}{R^2} = \frac{1}{2}$

As $\frac{K^2}{R^2}$ for solid sphere is smallest, it takes minimum time to reach the bottom of the incline.

NOTES When bodies of different shapes are allowed to roll down an incline, then velocity at the bottom of incline, acceleration and time taken in reaching the bottom depends on $\frac{K^2}{R^2}$. If the body has less moment of inertia then $\frac{K^2}{R^2}$ will be less, then it will take less time in reaching the bottom.

109. **(a)** P.E. = total K.E.

$$mgh = \frac{7}{10}mv^2, \quad \therefore v = \sqrt{\frac{10gh}{7}}$$

110. **(d)** $E = E_t + E_r = \frac{1}{2}mv^2 + \frac{1}{2}I\omega^2$

$$= \frac{1}{2}mv^2 + \frac{1}{2}\times\left(\frac{2}{5}mr^2\right)\omega^2$$

$$= \frac{1}{2}mv^2 + \frac{1}{5}mv^2 = \frac{7}{10}mv^2$$

$$\therefore \frac{E_t}{E} = \frac{\frac{1}{2}mv^2}{\frac{7}{10}mv^2} = \frac{5}{7}$$

111. **(b)** Angular momentum about the point of contact with the surface includes the angular momentum about the centre. Due to friction linear momentum is conserved.

7 Gravitation

--------------TREND ANALYSIS --------------

TOPIC	2025		2024		2023		2022		2021		2020	
	Qns.	LOD	Qns.	LOD	Qns.	LOD	Qns.	LOD	Qns.	LOD	Qns.	LOD
Kepler's Laws	1	Avg.										
Acceleration Due to Gravity of the Earth			1	Avg.								
Acceleration due to Gravity below and above the Surface of Earth	1	Avg.									1	Avg.
Gravitational Potential Energy					1	Avg.	1	Avg.				
Escape Speed			1	Avg.					2	Easy, Avg.		
Earth Satellites					1	Avg.						

How to Code Qns for Better & Faster Revision

Type 1: Able to solve (No code Req)
Type 2: Solved with struggle or by seeing solution **(Use code NR means Needs Revision or orange stickers)**
Type 3: Tricky/ Could not solve on my own. Need help from Teacher/ Friend. **(Use code TQ means Tricky Question or red stickers)**

You can either hand write the code in boxes or else paste coloured stickers given at end of book.

Topic-7.2

Kepler's Laws

1. The radius of Martian orbit around the Sun is about 4 times the radius of the orbit of Mercury. The Martian year is 687 Earth days. Then which of the following is the length of 1 year on Mercury ? **NCERT | Page-129 | 2025**
(a) 88 earth days (b) 225 earth days
(c) 172 earth days (d) 124 earth days

2. The kinetic energies of a planet in an elliptical orbit about the Sun, at positions A, B and C are K_A, K_B and K_C, respectively. AC is the major axis and SB is perpendicular to AC at the position of the Sun S as shown in the figure. Then **NCERT | Page-129 | 2018**

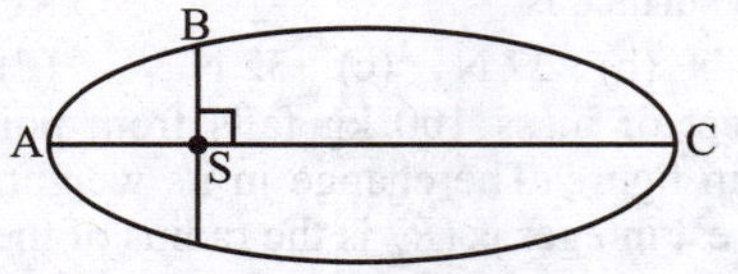

(a) $K_A < K_B < K_C$ (b) $K_A > K_B > K_C$
(c) $K_B > K_A > K_C$ (d) $K_B < K_A < K_C$

3. A planet moving along an elliptical orbit is closest to the sun at a distance r_1 and farthest away at a distance of r_2. If v_1 and v_2 are the linear velocities at these points respectively, then the ratio $\frac{v_1}{v_2}$ is **NCERT | Page-129 | 2011**
(a) $(r_1/r_2)^2$ (b) r_2/r_1 (c) $(r_2/r_1)^2$ (d) r_1/r_2

4. The figure shows elliptical orbit of a planet m about the sun S. The shaded area SCD is twice the shaded area SAB. If t_1 is the time for the planet to move from C to D and t_2 is the time to move from A to B then: **NCERT | Page-129 | 2009**

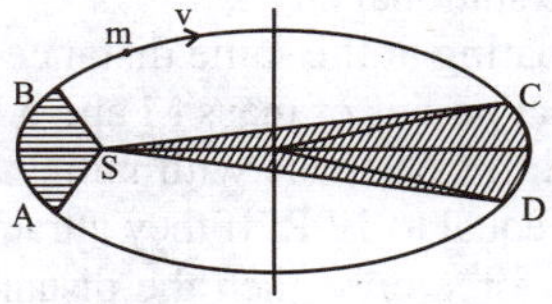

(a) $t_1 = 4t_2$ (b) $t_1 = 2t_2$
(c) $t_1 = t_2$ (d) $t_1 > t_2$

5. The period of revolution of planet A around the Sun is 8 times that of B. The distance of A from the Sun is how many times greater than that of B from the Sun? **NCERT | Page-129 | 1997**
(a) 2 (b) 3 (c) 4 (d) 5

6. The distance of Neptune and Saturn from the sun is nearly 10^{13} and 10^{12} meter respectively. Assuming that they move in circular orbits, their periodic times will be in the ratio **NCERT | Page-129 | 1994**
(a) 10 (b) 100 (c) $10\sqrt{10}$ (d) 1000

7. A satellite A of mass m is at a distance of r from the surface of the earth. Another satellite B of mass 2m is at a distance of 2r from the earth's centre. Their time periods are in the ratio of **NCERT | Page-129 | 1993**
(a) 1 : 2 (b) 1 : 16 (c) 1 : 32 (d) $1 : 2\sqrt{2}$

8. The distance of two planets from the sun are 10^{13} and 10^{12} metres respectively. The ratio of time periods of these two planets is **NCERT | Page-129 | 1988**
(a) $\frac{1}{\sqrt{10}}$ (b) 100 (c) $10\sqrt{10}$ (d) $\sqrt{10}$

9. The largest and the shortest distance of the earth from the sun are r_1 and r_2. Its distance from the sun when it is at perpendicular to the major-axis of the orbit drawn from the sun is **NCERT | Page-128 | 1988**

(a) $\frac{r_1+r_2}{4}$ (b) $\frac{r_1+r_2}{r_1-r_2}$

(c) $\frac{2r_1r_2}{r_1+r_2}$ (d) $\frac{r_1+r_2}{3}$

10. A planet is moving in an elliptical orbit around the sun. If T, V, E and L stand respectively for its kinetic energy, gravitational potential energy, total energy and magnitude of angular momentum about the centre of force, which of the following is correct? **NCERT | Page-138 | 1990**

(a) T is conserved
(b) V is always positive
(c) E is always negative
(d) L is conserved but direction of vector L changes continuously

Topic-7.3
Universal Law of Gravitation

11. Two astronauts are floating in gravitation free space after having lost contact with their spaceship. The two will **NCERT | Page-130 | 2017**

(a) move towards each other.
(b) move away from each other.
(c) become stationary
(d) keep floating at the same distance between them.

12. Two spherical bodies of mass M and 5 M and radii R and 2 R released in free space with initial separation between their centres equal to 12 R. If they attract each other due to gravitational force only, then the distance covered by the smaller body before collision is **NCERT | Page-130 | 2015**

(a) 4.5 R (b) 7.5 R (c) 1.5 R (d) 2.5 R

13. A spherical planet has a mass M_P and diameter D_P. A particle of mass m falling freely near the surface of this planet will experience an acceleration due to gravity, equal to : **NCERT | Page-130 | 2012**

(a) $4GM_P/D_P^2$ (b) GM_Pm/D_P^2

(c) GM_P/D_P^2 (d) $4GM_Pm/D_P^2$

14. Two spheres of masses m and M are situated in air and the gravitational force between them is F. The space around the masses is now filled with a liquid of specific gravity 3. The gravitational force will now be **NCERT | Page-130 | 2003**

(a) $\frac{F}{9}$ (b) $3F$ (c) F (d) $\frac{F}{3}$

Topic-7.5
Acceleration Due to Gravity of the Earth

15. The mass of a planet is $\frac{1}{10}$th that of the earth and its diameter is half that of the earth. The acceleration due to gravity on that planet is: **NCERT | Page-133 | 2024**

(a) 19.6 m s^{-2} (b) 9.8 m s^{-2}
(c) 4.9 m s^{-2} (d) 3.92 m s^{-2}

16. If R is the radius of the earth and g is the acceleration due to gravity on the earth surface. Then the mean density of the earth will be **NCERT | Page-133 | Manipur 2023**

(a) $\frac{3g}{4\pi RG}$ (b) $\frac{4\pi G}{3gR}$ (c) $\frac{\pi RG}{12g}$ (d) $\frac{3\pi R}{4gG}$

17. Imagine a new planet having the same density as that of earth but it is 3 times bigger than the earth in size. If the acceleration due to gravity on the surface of earth is g and that on the surface of the new planet is g', then **NCERT | Page-133 | 2005**

(a) $g' = g/9$ (b) $g' = 27g$
(c) $g' = 9g$ (d) $g' = 3g$

18. The density of a newly discovered planet is twice that of earth. The acceleration due to gravity at the surface of the planet is equal to that at the surface of the earth. If the radius of the earth is R, the radius of the planet would be **NCERT | Page-133 | 2004**

(a) ½ R (b) 2 R (c) 4 R (d) 1/4 R

19. The acceleration due to gravity on the planet A is 9 times the acceleration due to gravity on planet B. A man jumps to a height of 2m on the surface of A. What is the height of jump by the same person on the planet B? **NCERT | Page-133 | 2003**

(a) $\frac{2}{3}$m (b) $\frac{2}{9}$m (c) 18 m (d) 6 m

20. What will be the formula of the mass in terms of g, R and G (R = radius of earth) **NCERT | Page-130 | 1996**

(a) $g^2\frac{R}{G}$ (b) $G\frac{R^2}{g}$ (c) $G\frac{R}{g}$ (d) $g\frac{R^2}{G}$

Topic-7.6
Acceleration due to Gravity below and above the Surface of Earth

21. A body weighs 48 N on the surface of the earth. The gravitational force experienced by the body due to the earth at a height equal to one-third the radius of the earth from its surface is : **NCERT | Page-133 | 2025**

(a) 16 N (b) 27 N (c) 32 N (d) 36 N

22. An object of mass 100 kg falls from point A to B as shown in figure. The change in its weight, corrected to the nearest integer is (R_E is the radius of the earth) **NCERT | Page-133 | 2024**

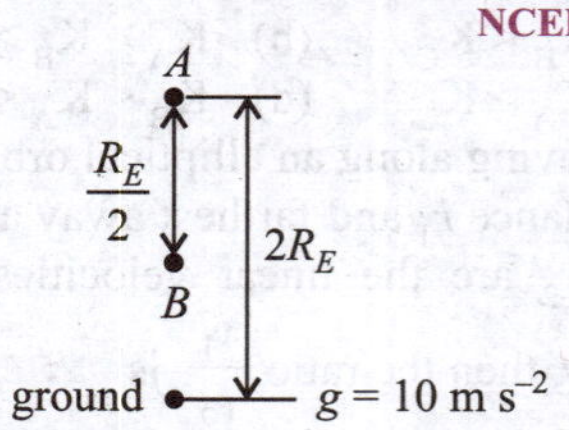

(a) 49 N (b) 89 N (c) 5 N (d) 10 N

23. A body weighs 72 N on the surface of the earth. What is the gravitational force on it, at a height equal to half the radius of the earth? **NCERT | Page-133 | 2020**

(a) 32 N (b) 30 N (c) 24 N (d) 48 N

24. The acceleration due to gravity at a height 1 km above the earth is the same as at a depth d below the surface of earth. Then NCERT | Page-134 | 2017

(a) $d = 1$ km (b) $d = \frac{3}{2}$km

(c) $d = 2$ km (d) $d = \frac{1}{2}$km

25. The height at which the weight of a body becomes 1/16th, its weight on the surface of earth (radius R), is : NCERT | Page-133 | 2012

(a) $5R$ (b) $15R$ (c) $3R$ (d) $4R$

26. Assuming earth to be a sphere of uniform density, what is the value of 'g' in a mine 100 km below the earth's surface? (Given, $R = 6400$ km) NCERT | Page-134 | 2001

(a) 9.65 m/s^2 (b) 7.65 m/s^2

(c) 5.06 m/s^2 (d) 3.10 m/s^2

27. A body weighs 72 N on the surface of the earth. What is the gravitational force on it due to earth at a height equal to half the radius of the earth from the surface? NCERT | Page-133 | 2000

(a) 32 N (b) 28 N (c) 16 N (d) 72 N

Topic-7.7 Gravitational Potential Energy

28. Two bodies of mass m and $9m$ are placed at a distance R. The gravitational potential on the line joining the bodies where the gravitational field equals zero, will be (G = gravitational constant) NCERT | Page-135 | 2023

(a) $-\frac{8Gm}{R}$ (b) $-\frac{12Gm}{R}$

(c) $-\frac{16Gm}{R}$ (d) $-\frac{20Gm}{R}$

29. In a gravitational field, the gravitational potential is given by, $V = -\frac{K}{x}(J/kg)$

The gravitational field intensity at point (2, 0, 3)m is : NCERT | Page-135 | RS 2022

(a) $+\frac{K}{4}$ (b) $+\frac{K}{2}$ (c) $-\frac{K}{2}$ (d) $-\frac{K}{4}$

30. A body of mass 60 g experiences a gravitational force of 3.0 N, when placed at a particular point. The magnitude of the gravitational field intensity at that point is: NCERT | Page-135 | 2022

(a) 50 N/kg (b) 20 N/kg

(c) 180 N/kg (d) 0.05 N/kg

31. The work done to raise a mass m from the surface of the earth to a height h, which is equal to the radius of the earth, is: NCERT | Page-135 | 2019

(a) mgR (b) 2 mgR (c) $\frac{1}{2}$mgR (d) $\frac{3}{2}$mgR

32. Assuming that the gravitational potential energy of an object at infinity is zero, the change in potential energy (final – initial) of an object of mass m, when taken to a height h from the surface of earth (of radius R), is given by, NCERT | Page-135 | Odisha 2019

(a) $\frac{GMm}{R+h}$ (b) $-\frac{GMm}{R+h}$

(c) $\frac{GMmh}{R(R+h)}$ (d) mgh

33. At what height from the surface of earth the gravitational potential and the value of g are -5.4×10^7 J kg^{-1} and 6.0 ms^{-2} respectively?
Take the radius of earth as 6400 km : NCERT | Page-135 | 2016

(a) 2600 km (b) 1600 km

(c) 1400 km (d) 2000 km

34. Dependence of intensity of gravitational field (E) of earth with distance (r) from centre of earth is correctly represented by: 2014

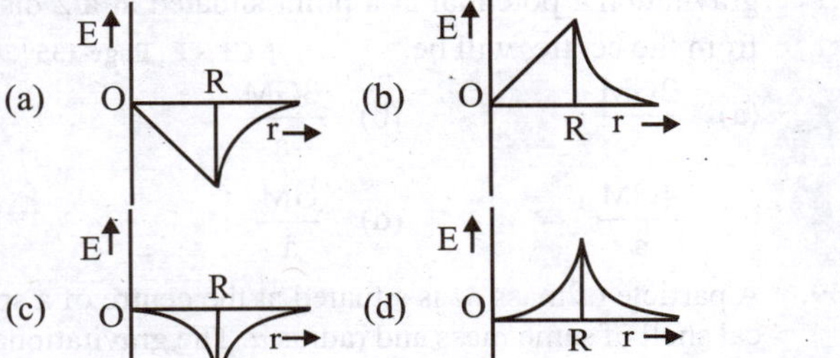

35. A body of mass 'm' is taken from the earth's surface to the height equal to twice the radius (R) of the earth. The change in potential energy of body will be NCERT | Page-135 | 2013

(a) $\frac{2}{3}$ mgR (b) 3 mgR

(c) $\frac{1}{3}$ mgR (d) mg2R

36. Infinite number of bodies, each of mass 2 kg are situated on x-axis at distances 1m, 2m, 4m, 8m, ... respectively, from the origin. The resulting gravitational potential due to this system at the origin will be NCERT | Page-135 | 2013

(a) $-\frac{8}{3}$ G (b) $-\frac{4}{3}$ G

(c) – 4 G (d) – G

37. Which one of the following plots represents the variation of gravitational field on a particle with distance r due to a thin spherical shell of radius R? (r is measured from the centre of the spherical shell) 2012 M

(a)

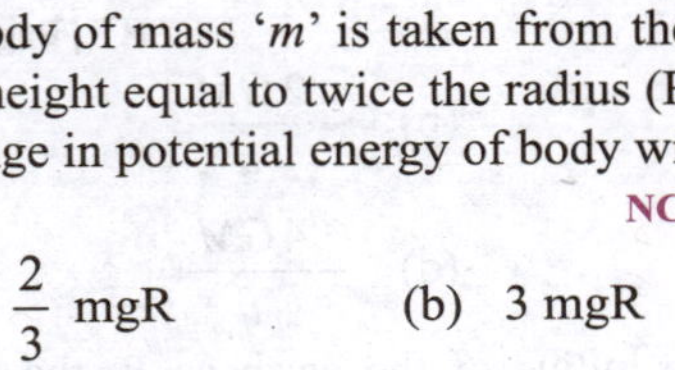

(b)

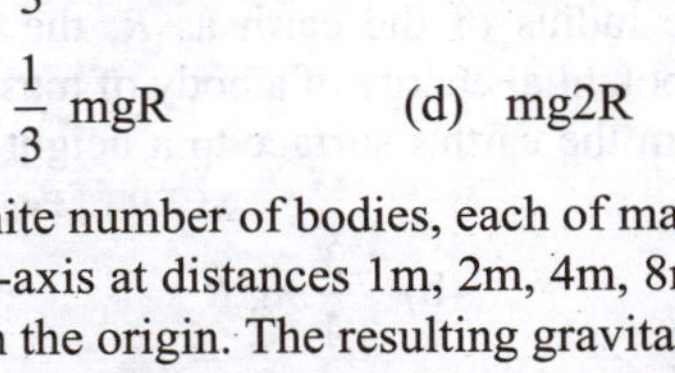

(c)

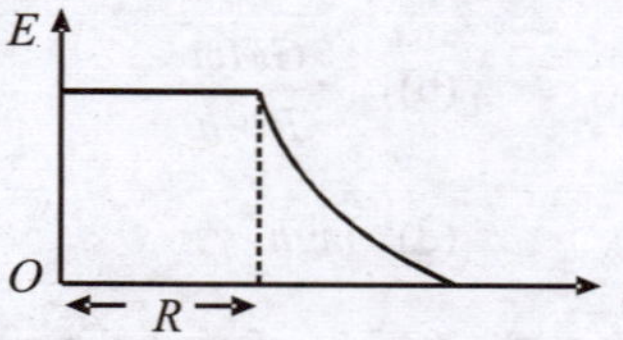

(d)

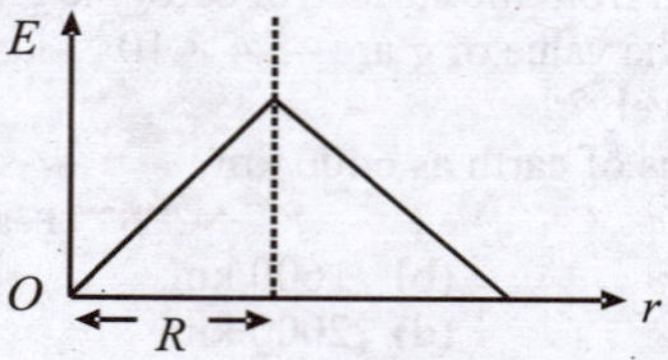

38. A particle of mass M is situated at the centre of spherical shell of mass M and radius a. The magnitude of the gravitational potential at a point situated at a/2 distance from the centre, will be **NCERT | Page-135 | 2011 M**

(a) $\frac{2GM}{a}$ (b) $\frac{3GM}{a}$

(c) $\frac{4GM}{a}$ (d) $\frac{GM}{a}$

39. A particle of mass M is situated at the centre of a spherical shell of same mass and radius a. The gravitational potential at a point situated at $\frac{a}{2}$ distance from the centre, will be: **NCERT | Page-135 | 2010**

(a) $-\frac{3GM}{a}$ (b) $-\frac{2GM}{a}$

(c) $-\frac{GM}{a}$ (d) $-\frac{4GM}{a}$

40. Assuming the radius of the earth as R, the change in gravitational potential energy of a body of mass m, when it is taken from the earth's surface to a height $3R$ above its surface, is **NCERT | Page-135 | 2002**

(a) $3\,mgR$ (b) $\frac{3}{4}mgR$

(c) $1\,mgR$ (d) $\frac{3}{2}mgR$

41. The potential energy of a satellite, having mass m and rotating at a height of 6.4×10^6 m from the earth surface, is **NCERT | Page-135 | 2001**

(a) $-mgR_e$ (b) $-0.67\,mgR_e$

(c) $-0.5\,mgR_e$ (d) $-0.33\,mgR_e$

42. With what velocity should a particle be projected so that its height becomes equal to radius of earth? **NCERT | Page-135 | 2001**

(a) $\left(\frac{GM}{R}\right)^{1/2}$ (b) $\left(\frac{8GM}{R}\right)^{1/2}$

(c) $\left(\frac{2GM}{R}\right)^{1/2}$ (d) $\left(\frac{4GM}{R}\right)^{1/2}$

Topic-7.8

Escape Speed

43. The escape velocity for earth is v. A planet having 9 times mass that of earth and radius, 16 times that of earth, has the escape velocity of : **NCERT | Page-136 | Ph-II 2024**

(a) $\frac{v}{3}$ (b) $\frac{2v}{3}$ (c) $\frac{3v}{4}$ (d) $\frac{9v}{4}$

44. The minimum energy required to launch a satellite of mass m from the surface of earth of mass M and radius R in a circular orbit at an altitude of 2R from the surface of the earth is: **NCERT | Page-135 | 2024**

(a) $\frac{5GmM}{6R}$ (b) $\frac{2GmM}{3R}$

(c) $\frac{GmM}{2R}$ (d) $\frac{GmM}{3R}$

45. The escape velocity of a body on the earth surface is 11.2 km/s. If the same body is projected upward with velocity 22.4 km/s, the velocity of this body at infinite distance from the centre of the earth will be **NCERT | Page-135 | Manipur 2023**

(a) 11.2 km/s (b) $11.2\sqrt{3}$ km/s

(c) $11.2\sqrt{2}$ km/s (d) Zero

46. The escape velocity from the Earth's surface is v. The escape velocity from the surface of another planet having a radius, four times that of Earth and same mass density is **NCERT | Page-136 | 2021**

(a) $4v$ (b) v (c) $2v$ (d) $3v$

47. A particle of mass 'm' is projected with a velocity $v = kV_e$ ($k < 1$) from the surface of the earth. (V_e = escape velocity)

The maximum height above the surface reached by the particle is **NCERT | Page-136 | 2021**

(a) $\frac{Rk^2}{1-k^2}$ (b) $R\left(\frac{k}{1-k}\right)^2$

(c) $R\left(\frac{k}{1+k}\right)^2$ (d) $\frac{R^2k}{1+k}$

48. The ratio of escape velocity at earth (v_e) to the escape velocity at a planet (v_p) whose radius and mean density are twice as that of earth is : **NCERT | Page-136 | 2016**

(a) 1 : 2 (b) $1 : 2\sqrt{2}$ (c) 1 : 4 (d) 1 : 2

49. A black hole is an object whose gravitational field is so strong that even light cannot escape from it. To what approximate radius would earth (mass = 5.98×10^{24} kg) have to be compressed to be a black hole? **NCERT | Page-136 | 2014**

(a) 10^{-9} m (b) 10^{-6} m

(c) 10^{-2} m (d) 100 m

50. The radius of a planet is twice the radius of earth. Both have almost equal average mass-densities. If V_P and V_E are escape velocities of the planet and the earth, respectively, then **NCERT | Page-136 | Kar. 2013**

(a) $V_E = 1.5V_P$ (b) $V_P = 1.5V_E$
(c) $V_P = 2V_E$ (d) $V_E = 3V_P$

51. A particle of mass '*m*' is kept at rest at a height 3*R* from the surface of earth, where '*R*' is radius of earth and '*M*' is mass of earth. The minimum speed with which it should be projected, so that it does not return back, is (*g* is acceleration due to gravity on the surface of earth) **NCERT | Page-136 | Kar. 2013**

(a) $\left(\frac{GM}{R}\right)^{\frac{1}{2}}$ (b) $\left(\frac{GM}{2R}\right)^{\frac{1}{2}}$
(c) $\left(\frac{gR}{4}\right)^{\frac{1}{2}}$ (d) $\left(\frac{2g}{4}\right)^{\frac{1}{2}}$

52. A particle of mass m is thrown upwards from the surface of the earth, with a velocity *u*. The mass and the radius of the earth are, respectively, M and R. G is gravitational constant and g is acceleration due to gravity on the surface of the earth. The minimum value of *u* so that the particle does not return back to earth, is **NCERT | Page-136 | 2011 M**

(a) $\sqrt{\frac{2GM}{R}}$ (b) $\sqrt{\frac{2GM}{R^2}}$
(c) $\sqrt{2gR^2}$ (d) $\sqrt{\frac{2GM}{R^2}}$

53. The Earth is assumed to be a sphere of radius *R*. A platform is arranged at a height *R* from the surface of the Earth. The escape velocity of a body from this platform is *f v*, where *v* is its escape velocity from the surface of the Earth. The value of *f* is **NCERT | Page-135 | 2006**

(a) $\frac{1}{\sqrt{2}}$ (b) $\frac{1}{3}$ (c) $\frac{1}{2}$ (d) $\sqrt{2}$

54. The escape velocity on the surface of earth is 11.2 km/s. What would be the escape velocity on the surface of another planet of the same mass but 1/4 times the radius of the earth? **NCERT | Page-136 | 2000**

(a) 22.4 km/s (b) 44.8 km/s
(c) 5.6 km/s (d) 11.2 km/s

55. The escape velocity of a sphere of mass m is given by (*G* = Universal gravitational constant; *M* = Mass of the earth and R_e = Radius of the earth) **NCERT | Page-136 | 1999**

(a) $\sqrt{\frac{GM}{R_e}}$ (b) $\sqrt{\frac{2GM}{R_e}}$
(c) $\sqrt{\frac{2GMm}{R_e}}$ (d) $\sqrt{\frac{2GM+R_e}{R_e}}$

56. The escape velocity of a body on the surface of the earth is 11.2 km/s. If the earth's mass increases to twice its present value and the radius of the earth becomes half, the escape velocity would become **NCERT | Page-136 | 1997**

(a) 44.8 km/s
(b) 22.4 km/s
(c) 11.2 km/s (remains unchanged)
(d) 5.6 km/s

57. The escape velocity from the surface of the earth is v_e. The escape velocity from the surface of a planet whose mass and radius are three times those of the earth, will be **NCERT | Page-136 | 1995**

(a) v_e (b) $3v_e$ (c) $9v_e$ (d) $1/3v_e$

58. The escape velocity from earth is 11.2 km/s. If a body is to be projected in a direction making an angle 45° to the vertical, then the escape velocity is **NCERT | Page-136 | 1993**

(a) 11.2 × 2 km/s (b) 11.2 km/s
(c) $11.2/\sqrt{2}$ km/s (d) $11.2\sqrt{2}$ km/s

59. For a satellite escape velocity is 11 km/s. If the satellite is launched at an angle of 60° with the vertical, then escape velocity will be **NCERT | Page-136 | 1989**

(a) 11 km/s (b) $11\sqrt{3}$ km/s
(c) $\frac{11}{\sqrt{3}}$ km/s (d) 33 km/s

Topic-7.9 Earth Satellites

60. A satellite is orbiting just above the surface of the earth with period T. If d is the density of the earth and G is the universal constant of gravitation, the quantity $\frac{3\pi}{Gd}$ represents **NCERT | Page-137 | 2023**

(a) T (b) T^2 (c) T^3 (d) $\sqrt{T}$

61. A satellite S is moving in an elliptical orbit around the earth. The mass of the satellite is very small compared to the mass of the earth. Then, **NCERT | Page-137 | 2015 RS**

(a) the total mechanical energy of S varies periodically with time.
(b) the linear momentum of S remains constant in magnitude.
(c) the acceleration of S is always directed towards the centre of the earth.
(d) the angular momentum of S about the centre of the earth changes in direction, but its magnitude remains constant.

62. Kepler's third law states that square of period of revolution (T) of a planet around the sun, is proportional to third power of average distance r between sun and planet *i.e.*, $T^2 = Kr^3$ here K is constant. If the masses of sun and planet are M and m respectively then as per Newton's law of gravitation force of attraction between them is $F = \frac{GMm}{r^2}$, here G is gravitational constant.

The relation between G and K is described as **NCERT | Page-129 | 2015**

(a) $GMK = 4\pi^2$ (b) $K = G$
(c) $K = \frac{1}{G}$ (d) $GK = 4\pi^2$

63. A remote - sensing satellite of earth revolves in a circular orbit at a height of 0.25×10^6 m above the surface of earth. If earth's radius is 6.38×10^6 m and $g = 9.8\ ms^{-2}$, then the orbital speed of the satellite is: **NCERT | Page-137 | 2015 RS**

(a) $8.56\ km\ s^{-1}$ (b) $9.13\ km\ s^{-1}$
(c) $6.67\ km\ s^{-1}$ (d) $7.76\ km\ s^{-1}$

64. If v_e is escape velocity and v_0 is orbital velocity of a satellite for orbit close to the earth's surface, then these are related by : **NCERT | Page-136 | 2012 M**

(a) $v_0 = \sqrt{2}v_e$ (b) $v_0 = v_e$
(c) $v_e = \sqrt{2v_0}$ (d) $v_e = \sqrt{2}v_0$

65. The radii of circular orbits of two satellites A and B of the earth, are 4R and R, respectively. If the speed of satellite A is 3 V, then the speed of satellite B will be: **NCERT | Page-137 | 2010**

(a) 3 V/4 (b) 6 V (c) 12 V (d) 3 V/2

66. Two satellites of earth, S_1 and S_2 are moving in the same orbit. The mass of S_1 is four times the mass of S_2. Which one of the following statements is true? **NCERT | Page-138, 137 | 2007**

(a) The potential energies of earth satellites in the two cases are equal.
(b) S_1 and S_2 are moving with the same speed.
(c) The kinetic energies of the two satellites are equal.
(d) The time period of S_1 is four times that of S_2.

67. A ball is dropped from a satellite revolving around the earth at a height of 120 km. The ball will **NCERT | Page-137 | 1996**

(a) continue to move with same speed along a straight line tangentially to the satellite at that time
(b) continue to move with the same speed along the original orbit of satellite
(c) fall down to earth gradually
(d) go far away in space

68. If the gravitational force between two objects were proportional to $1/R$ (and not as $1/R^2$) where R is separation between them, then a particle in circular orbit under such a force would have its orbital speed v proportional to **NCERT | Page-137 | 1989**

(a) $1/R^2$ (b) R^0 (c) R^1 (d) $1/R$

Topic-7.10

Energy of an Orbiting Satellite

69. For a satellite moving in an orbit around the earth, the ratio of kinetic energy to potential energy is **NCERT | Page-138 | 2005**

(a) $\frac{1}{2}$ (b) $\frac{1}{\sqrt{2}}$
(c) 2 (d) $\sqrt{2}$

70. A satellite of mass m is orbiting around the earth in a circular orbit with a velocity v. What will be its total energy? **NCERT | Page-138 | 1991**

(a) $(3/4)\ mv^2$ (b) $(1/2)\ mv^2$
(c) mv^2 (d) $-(1/2)mv^2$

ANSWER KEY

1	(a)	8	(c)	15	(d)	22	(a)	29	(d)	36	(c)	43	(c)	50	(c)	57	(a)	64	(d)
2	(b)	9	(c)	16	(a)	23	(a)	30	(a)	37	(b)	44	(a)	51	(b)	58	(b)	65	(b)
3	(b)	10	(c)	17	(d)	24	(c)	31	(c)	38	(b)	45	(b)	52	(a)	59	(a)	66	(b)
4	(b)	11	(a)	18	(a)	25	(c)	32	(c)	39	(a)	46	(a)	53	(a)	60	(b)	67	(b)
5	(c)	12	(b)	19	(c)	26	(a)	33	(a)	40	(b)	47	(a)	54	(a)	61	(c)	68	(b)
6	(c)	13	(a)	20	(d)	27	(a)	34	(b)	41	(c)	48	(b)	55	(b)	62	(a)	69	(a)
7	(d)	14	(c)	21	(b)	28	(c)	35	(a)	42	(a)	49	(c)	56	(b)	63	(d)	70	(d)

Hints & Solutions

1. **(a)** Applying Kepler's third law :

Given : $R_{martian} = 4R_{mercury}$

Orbital period of mars = 687 Earth days

$$\frac{T_{Mercury}}{T_{Martian}} = \left(\frac{r_{Mercury}}{r_{Martian}}\right)^{\frac{3}{2}} = \left(\frac{1}{4}\right)^{\frac{3}{2}} = \frac{1}{8}$$

$$T_{Mercury} = \frac{T_{Martian}}{8} = \frac{687}{8} = 85.875 \text{ days} \approx 88 \text{ days}$$

2. **(b)** Speed of the planet will be maximum when its distance from the sun is minimum as mvr = constant.

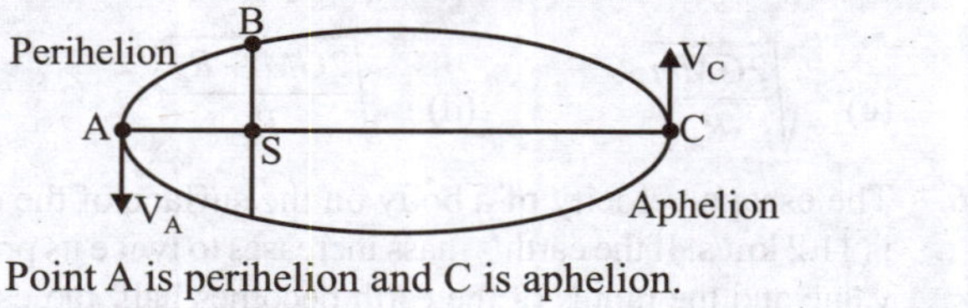

Point A is perihelion and C is aphelion.

Clearly, $V_A > V_B > V_C$

So, $K_A > K_B > K_C$

3. **(b)** Angular momentum is conserved

$\therefore L_1 = L_2$

$\Rightarrow mr_1v_1 = mr_2v_2$

$\Rightarrow r_1v_1 = r_2v_2$

$\Rightarrow \frac{v_1}{v_2} = \frac{r_2}{r_1}$

4. **(b)** According to Kepler's law, the areal velocity of a planet around the sun always remains constant.

SCD : $A_1 - t_1$ (areal velocity constant)

SAB : $A_2 - t_2$

$\frac{A_1}{t_1} = \frac{A_2}{t_2}$,

$t_1 = t_2 \cdot \frac{A_1}{A_2}$ (given $A_1 = 2A_2$)

$= t_2 \cdot \frac{2A_2}{A_2}$

$\therefore t_1 = 2t_2$

NOTES The area covered by radius vector in dt seconds $= \frac{1}{2}r^2 d\theta$

Area velocity $= \frac{1}{2}r^2 \frac{d\theta}{dt} = \frac{1}{2}r^2\omega \quad \left(\because \omega \frac{d\theta}{dt}\right)$

$= \frac{1}{2}rv \quad (\because v = \omega r)$

It follows that speed of the planet is maximum when it is closest to the sun and is minimum when the planet is farthest from the sun.

5. **(c)** Let T_A and T_B be time period of A and B about sun.

$T_A = 8T_B$

$\frac{T_A}{T_B} = 8$ ...(1)

According to Kepler's Law, $T^2 \propto r^3$

$\left(\frac{T_A}{T_B}\right)^2 = \frac{(r_A)^3}{(r_B)^3}$

$\Rightarrow \left(\frac{r_A}{r_B}\right)^3 = 64 \Rightarrow \frac{r_A}{r_B} = 4$

6. **(c)** $T^2 \propto R^3$ (According to Kepler's law)

$T_1^2 \propto (10^{13})^3$ and $T_2^2 \propto (10^{12})^3$

$\therefore \frac{T_1^2}{T_2^2} = (10)^3$

or $\frac{T_1}{T_2} = 10\sqrt{10}$

7. **(d)** Also, $T^2 \propto R^3$.

NOTES Time period does not depend on the mass.

8. **(c)** $T^2 \propto R^3$ (Kepler's law)

$\frac{T_1^2}{T_2^2} = \left(\frac{10^{13}}{10^{12}}\right)^3 \Rightarrow \frac{T_1}{T_2} = 10\sqrt{10}$

9. **(c)** Applying the properties of ellipse, we have

$\frac{2}{R} = \frac{1}{r_1} + \frac{1}{r_2} = \frac{r_1 + r_2}{r_1 r_2}$

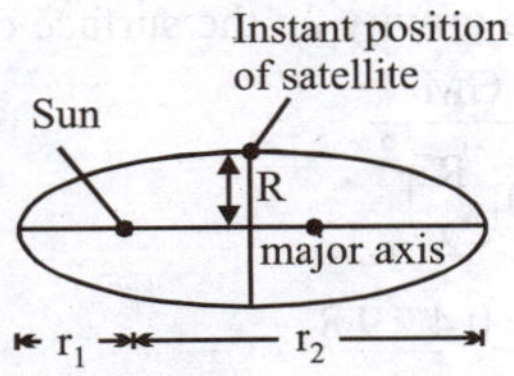

$R = \frac{2r_1 r_2}{r_1 + r_2}$

10. **(c)** In a circular or elliptical orbital motion, $\overline{F}$ and $\overline{r}$ are parallel to each other so torque is zero. So, angular momentum is conserved. In attractive field, potential energy is negative. Kinetic energy changes as velocity increase when distance is less. So, option (c) is correct.

11. **(a)** Both the astronauts are in the condition of weightlessness. Gravitational force between them pulls towards each other. Hence Astronauts move towards each other under mutual gravitional force.

12. **(b)** Let the distance moved by spherical body of mass M is x_1 and by spherical body of mass 5m is x_2

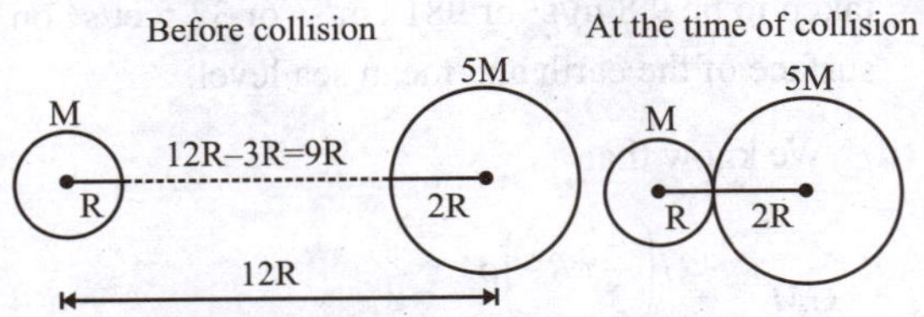

As their C.M. will remain stationary

So, $(M)(x_1) = (5M)(x_2)$ or, $x_1 = 5x_2$

$x_1 + x_2 = 9R$

So, $x_1 = 7.5$ R

13. **(a)** Gravitational attraction force on particle B,

$F_g = \frac{GM_P m}{(D_P/2)^2}$

Acceleration of particle due to gravity

$a = \frac{F_g}{m} = \frac{4GM_P}{D_P^2}$

14. **(c)** Gravitational force is independent of medium, Hence, this will remain same.

NOTES Since, $f = G\frac{M_1M_2}{d^2}$ therefore the affecting factor are:

(i) Mass of the two bodies *i.e.*, $f \alpha M_1M_2$ When mass increases gravitational force increase.

(ii) The distance between thier centers; *i.e.*, $f \alpha \frac{1}{d^2}$ when the distance increases gravitational force decreases.

15. (d) Acceleration due to gravity $g = \frac{GM}{R^2}$

So acceleration due to gravity on the surface of the planet $g' = \frac{GM'}{R'^2} = \frac{GM}{10\left(\frac{R}{2}\right)^2}$

or, $g' = \frac{4}{10}\frac{GM}{R^2} = 0.4 \times g = 0.4 \times 9.8$

$\therefore \quad g' = 3.92 \text{ ms}^{-2}$

16. (a) Acceleration due to gravity on the earth surface

$$g = \frac{GM}{R^2} = \frac{G\frac{4}{3}\pi R^3 \rho}{R^2}$$

$$g = G\frac{4}{3}\pi R\rho \Rightarrow \rho = \frac{3g}{4G\pi R}$$

The average value of acceleration due to gravity 'g' is taken to be 9.8 m/s^2 or 981 cm/s^2 or 32 feet/s^2 on the surface of the earth and mean sea level.

17. (d) We know that

$$g = \frac{GM}{R^2} = \frac{G\left(\frac{4}{3}\pi R^3\right)\rho}{R^2} = \frac{4}{3}\pi G R\rho$$

g depends on radius of the planet

$\frac{g'}{g} = \frac{R'}{R} = \frac{3R}{R} = 3$ [As G & ρ are constants]

$\therefore g' = 3g$

NOTES If the radius of a planet decreases by $n\%$, keeping its mass unchanged, the acceleration due to gravity on its surface increases by $2n\%$

$$\frac{\Delta g}{g} = -\frac{2\Delta R}{R}$$

$= -2(-n) = 2n\%$

18. (a) $g = \frac{GM}{R^2}$. Also, $M = \rho \times \frac{4}{3}\pi R^3$

$\therefore \quad g = G\frac{4}{3}\rho\pi R$.

At the surface of planet, $g_p = \frac{4}{3}G(2\rho)\pi R'$

At the surface of the earth $g_e = \frac{4}{3}G\rho\pi R$

$g_e = g_p \Rightarrow \rho R = 2\rho R' \Rightarrow R' = R/2$

19. (c) Applying conservation of total mechanical energy principle

$\frac{1}{2}mv^2 = mg_Ah_A = mg_Bh_B$

$\Rightarrow g_Ah_A = g_Bh_B$

$\Rightarrow h_B = \left(\frac{g_A}{g_B}\right)h_A = 9 \times 2 = 18 \text{ m}$

NOTES The velocity of the mass while reaching the surface of both the planets will be same.

$\Rightarrow \quad \sqrt{2g'h'} = \sqrt{2gh}$

$\Rightarrow \quad \sqrt{2gh'} = \sqrt{2 \times 9g \times 2}$

$\Rightarrow \quad 2h' = 36 \Rightarrow h' = 18\text{m}$

20. (d) We know that $mg = \frac{GMm}{R^2}$

$\therefore g = \frac{GM}{R^2} \Rightarrow M = \frac{gR^2}{G}$

21. (b) Given : weight of the body on surface of the earth = 48 N

$W' = mg' = mg\left(\frac{R}{R+h}\right)^2 \quad \therefore g' = g\left(\frac{R}{R+h}\right)^2$

$$= 48\left(\frac{R}{R+\frac{R}{3}}\right)^2 = 48 \times \frac{9}{16} = 27\text{N}$$

22. (a) $\therefore$ Weight, $Mg' = Mg\frac{R^2}{(R+h)^2}$

At A, $Mg' = Mg\frac{R^2}{(R+2R)^2} = \frac{Mg}{9}$

At B, $Mg' = Mg\frac{R}{\left(R+\frac{3R}{R}\right)^2} = \frac{Mg \cdot 4}{25}$

Change in weight $= Mg\frac{4}{25} - \frac{Mg}{9} = 49 \text{ N}$

23. (a) Weight of a body on the surface of the earth, $W_S = mgs = 72$ N

Acceleration due to gravity, g varies with height, $h = \frac{R}{2}$ (given)

$$W_h = \frac{mg}{\left(1+\frac{h}{R}\right)^2} = \frac{72}{\left(1+\frac{R/2}{R}\right)^2} = \frac{72}{(3/2)^2}$$

$$= \frac{4}{9} \times 72 = 32 \text{ N}$$

24. **(c)** Above earth surface Below earth surface

$$g_h = g\left(1-\frac{2h}{R_e}\right) \qquad g_d = g\left(1-\frac{d}{R_e}\right)$$

According to question, $g_h = g_d$

$$g\left(1-\frac{2h}{R_e}\right) = g\left(1-\frac{d}{R_e}\right)$$

Clearly,
$d = 2h = 2$ km

NOTES Graph of g, R and d:

(i) g at surface of earth is maximum
(ii) g at centre of earth is zero
(iii) g is maximum at poles and minimum at the equator.

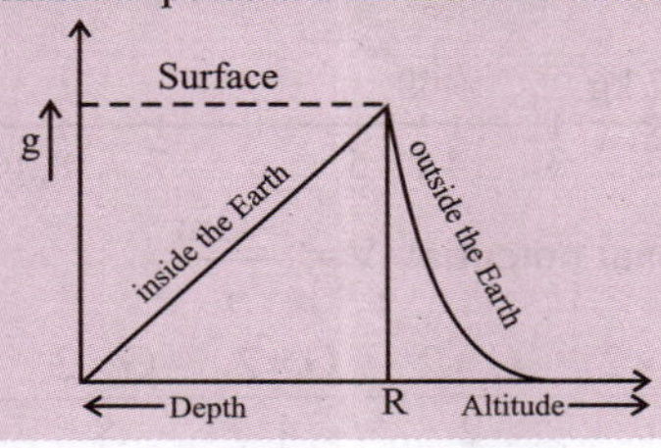

25. **(c)** $\frac{GM_m}{(R+h)^2} = \frac{1}{16}\frac{GM_m}{R^2}$

$$\Rightarrow \frac{1}{(R+h)^2} = \frac{1}{16R^2} \Rightarrow \frac{R}{(R+h)} = \frac{1}{4}$$

$$\Rightarrow \frac{R+h}{R} = 4 \Rightarrow h = 3R$$

26. **(a)** We know that effective gravity g' at depth below earth surface is given by

$$g' = g\left(1-\frac{d}{R}\right).$$

Here, $d = 100$ km, $R = 6400$ km,

$$\therefore \quad g' = 9.8\left(1-\frac{100}{6400}\right) = 9.65 \text{ m/s}^2$$

27. **(a)** $mg = 72$ N (body weight on the surface)

$$g = \frac{GM}{R^2}$$

At a height $H = \frac{R}{2}$,

$$g' = \frac{GM}{\left(R+\frac{R}{2}\right)^2} = \frac{4\,GM}{9\,R^2}$$

Body weight at height $H = \frac{R}{2}$,

$$mg' = m \times \frac{4}{9}\left(\frac{GM}{R^2}\right)$$

$$= m \times \frac{4}{9} \times g = \frac{4}{9}mg$$

$$= \frac{4}{9} \times 72 = 32 \text{ N}$$

28. **(c)** Let gravitational field at point Q be zero

$$\therefore \frac{Gm}{x^2} = \frac{G(9m)}{(R-x)^2}$$

$$\frac{(R-x)^2}{x^2} = 9 \Rightarrow x = \frac{R}{4}$$

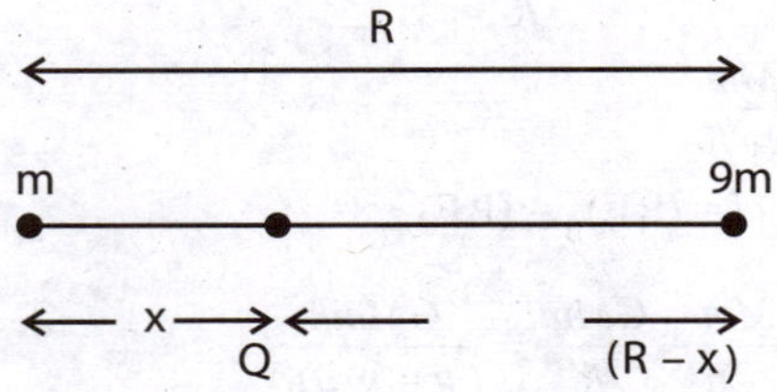

$$V_P = \frac{-Gm}{x} - \frac{G(9m)}{R-x}$$

$$V_P = \frac{-Gm}{\frac{R}{4}} - \frac{G(9m)}{\frac{3R}{4}} = \frac{-4Gm}{R} - \frac{12Gm}{R} = \frac{-16Gm}{R}$$

29. **(d)** Gravitational potential, $v(x) = -\frac{K}{x}$

Gravitational field,

$$E_g = -\frac{dv}{dx} = -\frac{d}{dx}\left(\frac{-k}{x}\right) = -\frac{K}{x^2}\hat{i}$$

Now $\left|\vec{E}_g(2,0,3)\right| = \frac{-K}{(2)^2} = \frac{-K}{4}$

NOTES Negative sign in gravitational field shows that direction of electric field is in the direction of decrease in potential.

30. **(a)** $\vec{F}_m = m\vec{E}$

$$|\vec{E}| = \frac{\vec{F}_m}{m} = \frac{3}{60 \times 10^{-3}} = 50 \text{ N/kg}$$

31. **(c)** Mass to be raised = m

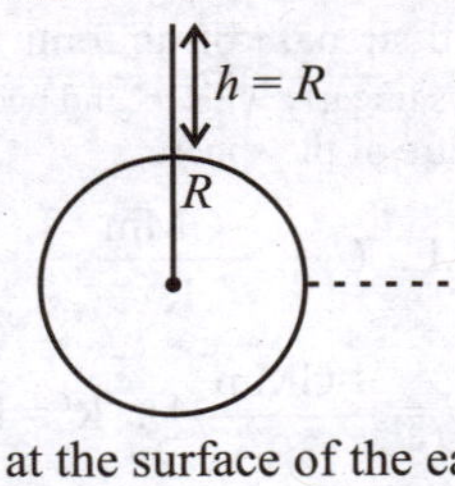

Potential energy at the surface of the earth

$$U_{surface} = \frac{-GMm}{R}$$

Potential energy at a height from the surface of the earth

$h = R$

$U_{\text{height}} = \frac{-GMm}{2R}$

According to work-energy theorem, work done = change in PE

$\therefore\ W = U_{\text{height}} - U_{\text{surface}}$

$\Rightarrow \frac{-GMm}{2R} - \left(-\frac{GMm}{R}\right)$

$= \frac{GMm}{2R} = \frac{gR^2 m}{2R} = \frac{mgR}{2} \quad (\because GM = gR^2)$

32. **(c)** As $U = -\frac{GMm}{r}$

$\therefore\ (P.E)_A = -\frac{GMm}{R}$; $(P.E)_B$

$= -\frac{GMm}{R+h}$

$\therefore\ \Delta U = (P.E)_B - (P.E)_A$

$= -\frac{GMm}{R+h} + \frac{GMm}{R} = \frac{GMmh}{(R)(R+h)}$

33. **(a)** As we know, gravitational potential (V) and acceleration due to gravity (*g*) with height

$V = \frac{-GM}{(R+h)} = -5.4 \times 10^7 \quad ...(1)$

and $g = \frac{GM}{(R+h)^2} = 6 \quad ...(2)$

Dividing (1) by (2)

$\Rightarrow \frac{\frac{-GM}{(R+h)}}{\frac{GM}{(R+h)^2}} = \frac{-5.4\times10^7}{6}$

$\Rightarrow \frac{5.4\times10^7}{(R+h)} = 6$

$\Rightarrow\ (R + h) = 9000$ km so, $h = 2600$ km

34. **(b)** First when $(r < R)$ $E \propto r$ and then when $r \geq R$ $E \propto \frac{1}{r^2}$. Hence graph (b) correctly dipicts.

NOTES The field strength of uniform solid sphere within it decreases linearly with '*r*' and becomes zero as we reach at the centre of the sphere.

35. **(a)** Initial P.E., $U_i = \frac{-GMm}{R}$,

Final P.E., $U_f = \frac{-GMm}{3R}$ $[\because R' = R + 2R = 3R]$

$\therefore$ Change in potential energy,

$\Delta U = \frac{-GMm}{3R} + \frac{GMm}{R}$

$= \frac{GMm}{R}\left(1 - \frac{1}{3}\right) = \frac{2}{3}\frac{GMm}{R} = \frac{2}{3}\, m\, g\, R$

$\left(\because \frac{GMm}{R} = mgR\right)$

NOTES $\Delta U = \frac{mgh}{1+\frac{h}{R}}$

By placing the value of $h = 2R$ we get

$\Delta U = \frac{2}{3}\, mgR.$

Work done against gravitational forces in taking a body of mass *m* from the surface of earth to a height *h* is the change in potential energy of the body and is given by

$\Delta u = \frac{GMmh}{R(R+h)}$

(a) if $h << R$, then $W = \frac{GMmh}{R(R+h)}$

(b) if $h = R$, then $W = \frac{GMm}{2R} = \frac{1}{2} mgh$

36. **(c)**

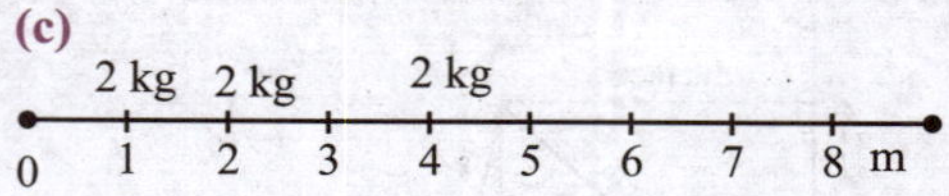

Gravitational potential $V = \frac{-Gm}{r}$

$V_0 = -\frac{G\times2}{1} - \frac{G\times2}{2} - \frac{G\times2}{4} - \frac{G\times2}{8} \text{---} \infty$

$V_0 = -2G\left[1 + \frac{1}{2} + \frac{1}{4} + \frac{1}{8} + ...\infty\right]$

$V_0 = -2G \times \frac{1}{\left(1-\frac{1}{2}\right)} = -2G \times \frac{1}{\frac{1}{2}} = -4\,G.$

37. **(b)** The Gravitational field due to a thin spherical shell of radius *R* at distance *r*.

$E = \frac{GM}{r^2}$ (If $r > R$)

For $r = R$ *i.e.*, on the surface of the shell

$E = \frac{GM}{R^2}$

For $r < R$ *i.e.*, inside the shell.

$E = 0$

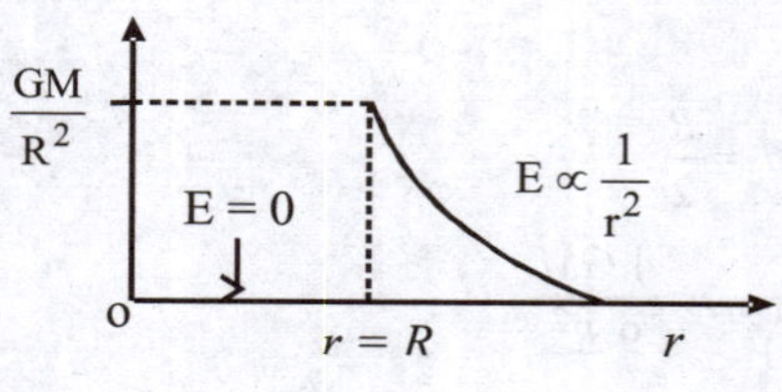

38. **(b)** $V_P = V_{\text{spherical shell}} + V_{\text{particle}}$

$$= \frac{GM}{a} + \frac{GM}{a/2} = \frac{3GM}{a}$$

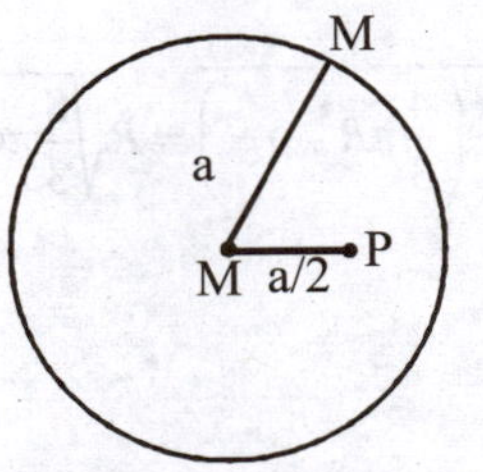

NOTES At every point, inside a spherical shell of radius R, the gravitational potential is same as that on that surface of the shell, *i.e.*,

$$V = -\frac{GM}{R}$$

39. **(a)** Potential at the given point = Potential at the point due to the shell + Potential due to the particle

$$= -\frac{GM}{a} - \frac{2GM}{a} = -\frac{3GM}{a}$$

40. **(b)** Gravitational potential energy (GPE) on the surface of earth,

$$E_1 = -\frac{GMm}{R}$$

GPE at $3R$, $E_2 = -\frac{GMm}{(R+3R)} = -\frac{GMm}{4R}$

$\therefore$ Change in GPE

$$= E_2 - E_1 = -\frac{GMm}{4R} + \frac{GMm}{R} = \frac{3GMm}{4R}$$

$$= \frac{3gR^2m}{4R} \qquad \left(\because g = \frac{GM}{R^2}\right)$$

$$= \frac{3}{4}mgR$$

41. **(c)** Mass of the satellite = m and height of satellite from earth $(h) = 6.4 \times 10^6$ m.

We know that gravitational potential energy of the satellite at height h,

$$\text{GPE} = -\frac{GM_em}{R_e + h} = -\frac{gR_e^2m}{2R_e}$$

$$= -\frac{gR_em}{2} = -0.5\,mgR_e$$

(where, $GM_e = gR_e^2$ and $h = R_e$)

42. **(a)** From conservation of energy

$$\frac{1}{2}mu^2 - \frac{GMm}{R} = \frac{1}{2}m \times (0)^2 - \frac{GMm}{(R+R)}$$

$$\Rightarrow u^2 = \frac{2GM}{R} - \frac{2GM}{2R} = \frac{GM}{R}$$

$$\Rightarrow u = \sqrt{\frac{GM}{R}}$$

NOTES Use $v^2 = \dfrac{2gh}{\left(1+\dfrac{h}{R}\right)}$ where $h = R$

$$v = \sqrt{gR} = \sqrt{\frac{GM}{R}}$$

43. **(c)** Escape velocity of object from planet is

$$(V_e)_P = \sqrt{\frac{2GM_P}{R_P}}$$

$$\therefore (V_e)_P \propto \sqrt{\frac{M_P}{R_P}}$$

Now, $M_p = 9M_e$ and $R_p = 16R_e$

$$\frac{(V_e)_P}{(V_e)_e} = \sqrt{\frac{M_P}{R_P} \times \frac{R_e}{M_e}} = \sqrt{\frac{9M_e}{16R_e} \times \frac{R_e}{M_e}} = \frac{3}{4}$$

$$\Rightarrow (V_e)_P = \frac{3}{4}v$$

44. **(a)** Apply energy conservation,

$$U_i + K_i = U_f + K_f$$

$$\Rightarrow -\frac{GMm}{R} + K_i = -\frac{GMm}{3R} + \frac{1}{2}mv_0^2$$

$$\Rightarrow -\frac{GMm}{R} + K_i = -\frac{GMm}{3R} + \frac{1}{2} \times m \times \frac{GM}{3R}$$

$$\left[\because v_0 = \sqrt{\frac{GM}{r}} \text{ and } r = 3R\right]$$

$$\Rightarrow K_i = -\frac{1}{6}\frac{GMm}{R} + \frac{GMm}{R}$$

$\therefore$ Minimum energy required,

$$K_i = \frac{5}{6} \cdot \frac{GMm}{R}$$

45. **(b)** The velocity of the body at infinite distance from the centre of the earth,

$$V' = \sqrt{V^2 - V_e^2} = \sqrt{(22.4)^2 - (11.2)^2}$$

$$\therefore V' = 11.2\sqrt{3} \text{ km/s}$$

46. (a) Escape velocity from the earth's surface

$$v_e = \sqrt{\frac{2GM}{R}} = \sqrt{\frac{2G\rho \frac{4}{3}\pi R^3}{R}} \quad \left[\because \rho = \frac{M}{V}\right]$$

or, $v_e = \sqrt{\frac{8G\rho\pi}{3}R^2}$ or, $v_e \propto R$ ($\because$ same density)

$\therefore \frac{v}{v_1} = \frac{R}{4R}$ or, $v_1 = 4v$

NOTES Escape velocity is double if the radius of the earth is double keeping the density unchanged. Escape velocity depends on the mass and size of the planet.

47. (a) From law of conservation of mechanical energy,

$$\frac{1}{2}mv^2 - \frac{GmM}{R} = -\frac{GmM}{(R+h)}$$

$$\Rightarrow \frac{v^2}{2} = \frac{GM}{R} - \frac{(GM)}{(R+h)} = \frac{h}{R(R+h)}GM$$

$$\Rightarrow \frac{1}{2}k^2V_e^2 = \frac{GMh}{R(R+h)}$$

We know, escape velocity

$$V_e = \sqrt{\frac{2GM}{R}}$$

$$\Rightarrow \frac{1}{2}k^2\left(\frac{2GM}{R}\right) = \frac{(GM)}{R(R+h)} \text{ or, } k^2 = \frac{h}{(R+h)}$$

$Rk^2 + hk^2 = h \Rightarrow Rk^2 = h(1-k^2)$

$$\therefore h = \frac{Rk^2}{(1-k^2)}$$

48. (b) As we know, escape velocity,

$$v_e = \sqrt{\frac{2GM}{R}} = \sqrt{\frac{2G}{R}\cdot\left(\frac{4}{3}\pi R^3\rho\right)}$$

$v_e \propto R\sqrt{\rho}$

$$\therefore \frac{V_e}{V_p} = \frac{R_e}{R_p}\sqrt{\frac{\rho_e}{\rho_p}} \Rightarrow \frac{V_e}{V_p} = \frac{R_e}{2R_e}\sqrt{\frac{\rho_e}{2\rho_e}}$$

$\therefore$ Ratio $\frac{V_e}{V_p} = 1 : 2\sqrt{2}$

49. (c) From question,

Escape velocity

$$= \sqrt{\frac{2GM}{R}} = c = \text{speed of light}$$

$$\Rightarrow R = \frac{2GM}{C^2}$$

$$= \frac{2\times 6.6\times 10^{-11}\times 5.98\times 10^{24}}{(3\times 10^8)^2}\text{m}$$

$= 10^{-2}$ m

50. (c) Escape velocity,

$$V_e = \sqrt{\frac{2GM_E}{R_E}} = \sqrt{\frac{2G}{R_E}\left(\frac{4}{3}\pi R_E^3\rho_E\right)} = R\sqrt{\frac{8}{3}\pi G\rho_E}$$

$\Rightarrow V_e \propto R$

$$\Rightarrow \frac{V_P}{V_E} = \frac{R_P}{R_E} = 2$$

$\Rightarrow V_P = 2V_E$

51. (b) As we know, the minimum speed with which a body is projected so that it does not return back is called escape speed.

$$V_e = \sqrt{\frac{2GM}{r}} = \sqrt{\frac{2GM}{R+h}} = \sqrt{\frac{2GM}{4R}}$$

$$= \left(\frac{GM}{2R}\right)^{\frac{1}{2}} \quad (\because h = 3R)$$

52. (a) The velocity u should be equal to the escape velocity. That is,

$u = \sqrt{2gR}$

But $g = \frac{GM}{R^2}$

$$\therefore u = \sqrt{2\cdot\frac{GM}{R^2}\cdot R}$$

$$\Rightarrow \sqrt{\frac{2GM}{R}}$$

53. (a) Potential energy at height $R = -\frac{GMm}{2R}$

If m be the mass of a body which is thrown with velocity v_e so that it goes out of gravitational field from distance R, then

$$\frac{1}{2}mv_e^2 = \frac{GM}{2R}.m \Rightarrow v_e = \sqrt{\frac{GM}{R}}$$

or, $v_e = \sqrt{gR}$

Now, $v = \sqrt{2gR}$, So, $v = \sqrt{2}\,v_e$

or, $v_e = \frac{v}{\sqrt{2}}$

Comparing it with given equation, $f = \frac{1}{\sqrt{2}}$.

54. (a) $v_{earth} = \sqrt{\frac{2GM_e}{R_e}}$

$$v_{\text{planet}} = \sqrt{\frac{2GM_p}{R_p}} = \sqrt{\frac{2GM_e}{R_e/4}} = \sqrt{\frac{8GM_e}{R_e}}$$

$$\frac{v_{\text{planet}}}{v_{\text{earth}}} = \sqrt{\frac{8GM_e}{R_e}} \times \sqrt{\frac{R_e}{2GM_e}} = 2$$

$$\therefore v_{\text{planet}} = 2 \times v_{\text{earth}} = 2 \times 11.2 = 22.4 \text{ km/s}$$

55. (b) Escape velocity is the minimum velocity with which a body is projected to escape from earth's gravitational field

$$\frac{1}{2}mv_e^2 = \frac{GMm}{R_e} \Rightarrow v_e = \sqrt{\frac{2GM}{R_e}}$$

56. (b) Escape velcocity

$$v_e = \sqrt{\frac{2GM_e}{R_e}},\ v'_e = \sqrt{\frac{2GM'_e}{R'_e}}$$

$$\therefore \frac{v'_e}{v_e} = \sqrt{\frac{M'_e}{M_e} \times \frac{R_e}{R'_e}}$$

Given $M'_e = 2M_e$ and $R'_e = \frac{R_e}{2}$

$$\therefore \frac{v'_e}{v_e} = \sqrt{\frac{2M_e}{M_e} \times \frac{R_e}{R_e/2}} = \sqrt{4} = 2$$

$$v'_e = 2v_e = 2 \times 11.2 = 22.4 \text{ km/s}$$

57. (a) Escape velocity on surface of earth

$$v_e = \sqrt{\frac{2GM_e}{R_e}} \propto \sqrt{\frac{M_e}{R_e}}.$$

$$\therefore \frac{v_e}{v_P} = \sqrt{\frac{M_e}{R_e}} \times \sqrt{\frac{R_p}{M_p}} = \sqrt{\frac{M_e}{R_e}} \times \sqrt{\frac{3R_e}{3M_e}} = \frac{1}{1} = 1$$

or, $v_P = v_e$.

58. (b) Escape velocity does not depend on the angle of projection.

NOTES The value of escape velocity is given by

$$V_e = \sqrt{\frac{2GM}{R}}$$

Here, M = mass of earth, G = gravitational constant
R = radius of earth

$$V_e = \sqrt{\frac{2GM}{R}}$$

$$= \sqrt{\frac{2 \times G}{R^2} \times \frac{4}{3}\pi R^3 \rho} \quad \left(\because M = \frac{4}{3}\pi R^3 \rho\right)$$

$$V_e = R\sqrt{\frac{8\pi G\rho}{3}}$$

Thus, value of escape velocity depends upon the mass M, density ρ and radius R of earth/planet from which the body is projected and independent of angle of projection.

59. (a) Since, escape velocity $\left(v_e = \sqrt{2gR_e}\right)$ is independent of angle of projection, so it will not change.

60. (b) Time period of satellite just above earth surface (T) is given by

$$T = 2\pi\sqrt{\frac{R}{g}}$$

But $g = \frac{GM}{R^2} = \frac{G}{R^2}\left(\frac{4}{3}\pi R^3 d\right) = Gd\frac{4}{3}\pi R$

$$\therefore T = \sqrt{\frac{R}{G\frac{4}{3}\pi Rd}} = 2\pi\sqrt{\frac{3}{4\pi Gd}}$$

$$\Rightarrow T^2 = \frac{3\pi}{Gd}$$

NOTES Time period of satellite,

$$T = \frac{\text{Circumferance of the orbit of satellite}}{\text{Orbital velocity of satellite}}$$

$$\Rightarrow T = \frac{2\pi r}{v_{\text{Orbital}}} = \frac{2\pi r}{\sqrt{\frac{GM}{r}}} \quad \left(\because v_{\text{Orbital}} = \sqrt{\frac{GM}{r}}\right)$$

$$\Rightarrow T = 2\pi\sqrt{\frac{r^3}{GM}} = 2\pi\sqrt{\frac{r^3}{gR^2}} \left(\because GM = gR^2\right)$$

$$\Rightarrow T = 2\pi\sqrt{\frac{r^3}{gR^2}} = 2\pi\sqrt{\frac{(R+h)^3}{gR^2}} \quad \text{Here, } r = (R+h)$$

When satellite lies close to the earth
$R + h \approx R$

$$T = 2\pi\sqrt{\frac{R^3}{gR^2}} = 2\pi\sqrt{\frac{R}{g}}$$

61. (c) The gravitational force on the satellite will be aiming towards the centre of the earth so acceleration of the satellite will also be aiming towards the centre of the earth.

62. (a) As we know, orbital speed, $v_{\text{orb}} = \sqrt{\frac{GM}{r}}$

Time period $T = \frac{2\pi r}{v_{\text{orb}}} = \frac{2\pi r}{\sqrt{GM}}\sqrt{r}$

Squarring both sides,

$$T^2 = \left(\frac{2\pi r\sqrt{r}}{\sqrt{GM}}\right)^2 = \frac{4\pi^2}{GM}.r^3$$

$$\Rightarrow \frac{T^2}{r^3} = \frac{4\pi^2}{GM} = K$$

$$\Rightarrow \quad GMK = 4\pi^2$$

63. (d) Given: Height of the satellite from the earth's surface $h = 0.25 \times 10^6 m$

Radius of the earth $R = 6.38 \times 10^6 m$

Acceleration due to gravity $g = 9.8\ m/s^2$

Orbital velocity, $v_0 = ?$

$$v_0 = \sqrt{\frac{GM}{(R+h)}} = \sqrt{\frac{GM}{R^2}.\frac{R^2}{(R+h)}}$$

$$= \sqrt{\frac{9.8 \times 6.38 \times 6.38}{6.63 \times 10^6}}$$

$$= 7.76\ km/s \qquad \left[\because \frac{GM}{R^2} = g\right]$$

64. (d) $v_e = \sqrt{\frac{2GM}{R}} \Rightarrow v_0 = \sqrt{\frac{GM}{R}}$

$v_e = \sqrt{2}v_0$

The orbital velocity of a satellite at a height *h* above the surface of earth,

$$v_0 = \sqrt{\frac{GM}{(R+h)}} = \sqrt{\frac{gR^2}{(R+h)}} \quad (\because GM = gR^2)$$

Here, M = mass of earth,

R = radius of earth,

g = acceleration due to If the satellite is very close to the surrface of earth gravity earth on surface of then $h = 0$

$$\therefore \quad v_0 = \sqrt{\frac{gR^2}{R}} = \sqrt{gR} = \sqrt{\frac{GM}{R}}$$

65. (b) Orbital velocity of a satellite in a circular orbit of radius a is given by

$$v = \sqrt{\frac{GM}{a}}$$

$$\Rightarrow \quad v \propto \sqrt{\frac{1}{a}} \Rightarrow \frac{v_2}{v_1} = \sqrt{\frac{a_1}{a_2}}$$

$$\therefore \quad v_2 = v_1\sqrt{\frac{4R}{R}} = 2\,v_1 = 6V$$

66. (b) Since orbital velocity of satellite is

$v = \sqrt{\frac{GM}{r}}$, it does not depend upon the mass of the satellite.

Therefore, both satellites will move with same speed.

67. (b) The orbital speed of satellite is independent of mass of satellite, so the ball will behave as a satellite and will continue to move with the same speed in the original orbit.

68. (b) $F = \frac{k}{R} = \frac{Mv^2}{R}$. Hence $v \propto R^0$

69. (a) K.E. of satellite moving in an orbit around the earth is

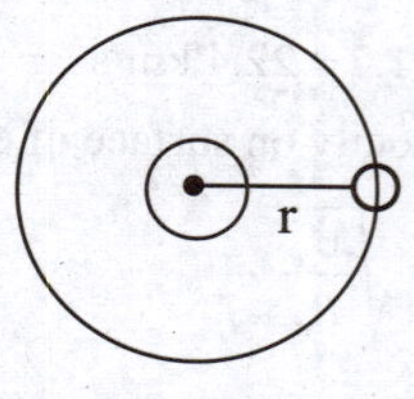

$$K = \frac{1}{2}mv^2 = \frac{1}{2}m\left(\sqrt{\frac{GM}{r}}\right)^2 = \frac{GMm}{2r}$$

P.E. of satellite and earth system is

$$U = \frac{GMm}{r} \Rightarrow \frac{K}{U} = \frac{\frac{GMm}{2r}}{\frac{GMm}{r}} = \frac{1}{2}$$

70. (d) Total energy = – K.E. = $\frac{\text{P.E.}}{2}$

$$\text{K.E.} = \frac{-1}{2}mv^2$$

8 Mechanical Properties of Solids

TREND ANALYSIS

TOPIC	2025		2024		2023		2022		2021		2020	
	Qns.	LOD	Qns.	LOD	Qns.	LOD	Qns.	LOD	Qns.	LOD	Qns.	LOD
Stress and Strain					1	Easy						
Elastic Modulus			1	Avg.			1	Avg.			1	Avg.

How to Code Qns for Better & Faster Revision →

Type 1: Able to solve (No code Req)
Type 2: Solved with struggle or by seeing solution **(Use code NR means Needs Revision or orange stickers)**
Type 3: Tricky/ Could not solve on my own. Need help from Teacher/ Friend. **(Use code TQ means Tricky Question or red stickers)**

You can either hand write the code in boxes or else paste coloured stickers given at end of book.

Topic-8.2 Stress and Strain

1. Let a wire be suspended from the ceiling (rigid support) and stretched by a weight W attached at its free end. The longitudinal stress at any point of cross-sectional area A of the wire is NCERT | Page-168 | 2023

(a) $2W/A$ (b) W/A (c) $W/2A$ (d) Zero

Topic-8.4 Stress-Strain Curve

2. The stress-strain curves are drawn for two different materials *X* and *Y*. It is observed that the ultimate strength point and the fracture point are close to each other for material *X* but are far apart for material *Y*.

We can say that materials *X* and *Y* are likely to be (respectively), NCERT | Page-169 | Odisha 2019

(a) Plastic and ductile (b) Ductile and brittle
(c) Brittle and ductile (d) Brittle and plastic

Topic-8.5 Elastic Modulus

3. The maximum elongation of a steel wire of 1 m length if the elastic limit of steel and its Young's modulus, respectively, are 8×10^8 N m^{-2} and 2×10^{11} N m^{-2}, is: NCERT | Page-170 | 2024

(a) 4 mm (b) 0.4 mm
(c) 40 mm (d) 8 mm

4. Given below are two statements: One is labelled as assertion (A) and the other is labelled as Reason (R). NCERT | Page-174 | 2022

Assertion (A): The stretching of a spring is determined by the shear modulus of the material of the spring.

Reason (R): A coil spring of copper has more tensile strength than a steel spring of same dimensions.

In the light of the above statements, choose the most appropriate answer from the options given below:

(a) Both (A) and (R) are true and (R) is not the correct explanation of (A)
(b) (A) is true but (R) is false
(c) (A) is false but (R) is true
(d) Both (A) and (R) are true and (R) is the correct explanation of (A).

5. A wire of length L, area of cross section A is hanging from a fixed support. The length of the wire changes to L_1 when mass M is suspended from its free end. The expression for Young's modulus is : NCERT | Page-170 | 2020

(a) $\frac{Mg(L_1-L)}{AL}$ (b) $\frac{MgL}{AL_1}$
(c) $\frac{MgL}{A(L_1-L)}$ (d) $\frac{MgL_1}{AL}$

6. Two wires are made of the same material and have the same volume. The first wire has cross-sectional area A and the second wire has cross-sectional area 3A. If the length of the first wire is increased by $\Delta\ell$ on applying a force F, how much force is needed to stretch the second wire by the same amount? NCERT | Page-170 | 2018

(a) 9 F (b) 6 F (c) F (d) 4 F

7. The bulk modulus of a spherical object is 'B'. If it is subjected to uniform pressure 'p', the fractional decrease in radius is NCERT | Page-173 | 2017

(a) $\frac{B}{3p}$ (b) $\frac{3p}{B}$ (c) $\frac{p}{3B}$ (d) $\frac{p}{B}$

8. The Young's modulus of steel is twice that of brass. Two wires of same length and of same area of cross section, one of steel and another of brass are suspended from the same roof. If we want the lower ends of the wires to be at the same level, then the weights added to the steel and brass wires must be in the ratio of : NCERT | Page-170 | 2015 RS

(a) 2 : 1 (b) 4 : 1 (c) 1 : 1 (d) 1 : 2

9. The approximate depth of an ocean is 2700 m. The compressibility of water is 45.4×10^{-11} Pa^{-1} and density of water is 10^3 kg/m^3. What fractional compression of water will be obtained at the bottom of the ocean ? NCERT | Page-183 | 2015
(a) 1.0×10^{-2} (b) 1.2×10^{-2}
(c) 1.4×10^{-2} (d) 0.8×10^{-2}

10. Copper of fixed volume 'V; is drawn into wire of length 'l'. When this wire is subjected to a constant force 'F', the extension produced in the wi re is 'Δl'. Which of the following graphs is a straight line? NCERT | Page-170 | 2014
(a) Δl versus $\frac{1}{l}$ (b) Δl versus l^2
(c) Δl versus $\frac{1}{l^2}$ (d) Δl versus l

11. The following four wires are made of the same material. Which of these will have the largest extension when the same tension is applied ? NCERT | Page-170 | 2013
(a) Length = 100 cm, diameter = 1 mm
(b) Length = 200 cm, diameter = 2 mm
(c) Length = 300 cm, diameter = 3 mm
(d) Length = 50 cm, diameter = 0.5 mm

12. If the ratio of diameters, lengths and Young's modulus of steel and copper wires shown in the figure are p, q and s respectively, then the corresponding ratio of increase in their lengths would be NCERT | Page-170 | Kar. 2013

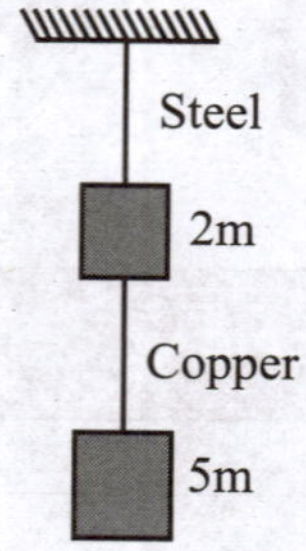

(a) $\frac{7q}{(5sp)}$
(b) $\frac{5q}{(7sp^2)}$
(c) $\frac{7q}{(5sp^2)}$
(d) $\frac{2q}{(5sp)}$

13. The compressibility of water is 4×10^{-5} per unit atmospheric pressure. The decrease in volume of 100 cm^3 of water under a pressure of 100 atmosphere will be NCERT | Page-183 | 1990
(a) 0.4 cm^3 (b) 4×10^{-5} cm^3
(c) 0.025 cm^3 (d) 0.004 cm^3

14. Two wires A and B are of the same material. Their lengths are in the ratio 1 : 2 and the diameter are in the ratio 2 : 1. If they are pulled by the same force, then increase in length will be in the ratio NCERT | Page-170 | 1988
(a) 2 : 1 (b) 1 : 4 (c) 1 : 8 (d) 8 : 1

ANSWER KEY

1	(b)	3	(a)	5	(c)	7	(c)	9	(b)	11	(d)	13	(a)								
2	(c)	4	(b)	6	(a)	8	(a)	10	(b)	12	(c)	14	(c)								

Hints & Solutions

1. **(b)**

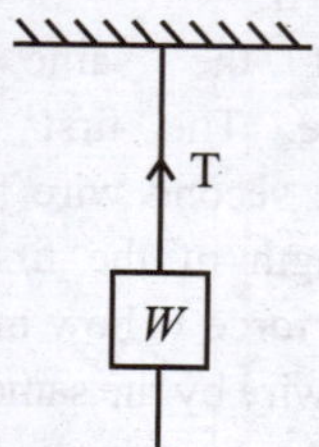

Longitudinal stress

$$= \frac{\text{Internal restoring force}}{\text{Area}} = \frac{F_{ext}}{\text{Area}} \quad \therefore \text{Stress} = \frac{W}{A}$$

2. **(c)** Fracture point and ultimate strength point is close for material X, hence X is brittle in nature and both points are far apart for material Y hence it is ductile.

3. **(a)** For maximum elongation,
Stress = Elastic limit
Maximum elongation

$$= \frac{\sigma_{elastic} \times L}{\text{Young's modulus}} = \frac{8 \times 10^8 \times 1}{2 \times 10^{11}} = 4 \times 10^{-3}$$
= 4 mm

4. **(b)** By stretching coil's, shape changes whereas its length of wire remains same. Due to which shear modules of elasticity is involved so assertion is correct. Elasticity of steel is more than that of copper and also has more tensile stress. So reason is false.

5. **(c)** Stress $= \frac{\text{force}}{\text{cross-section area}} = \frac{Mg}{A}$

Strain $= \frac{\text{change in length}}{\text{original length}} = \frac{\Delta L}{L} = \frac{L_1 - L}{L}$

Young's modulus, $Y = \frac{\text{stress}}{\text{strain}} = \frac{MgL}{A(L_1 - L)}$

6. **(a)** Wire 1: 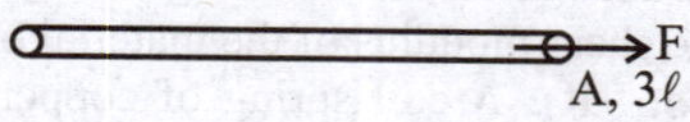

$$\Delta \ell = \left(\frac{F}{AY}\right) 3\ell \quad \text{...(i)}$$

Wire 2: 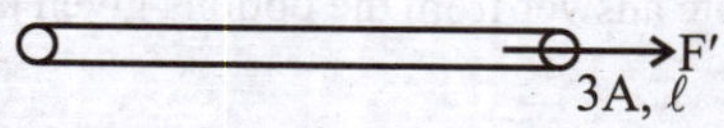

$$\Delta\ell = \left(\frac{F'}{3AY}\right)\ell \quad \text{...(ii)}$$

From equation (i) & (ii),

$$\Delta\ell = \left(\frac{F}{AY}\right)3\ell = \left(\frac{F'}{3AY}\right)\ell \text{ or, } F' = 9F$$

7. **(c)** Bulk modulus is given by

$$B = \frac{p}{\left(\frac{\Delta V}{V}\right)} \quad \text{or} \quad \frac{\Delta V}{V} = \frac{p}{B}$$

$$3\frac{\Delta R}{R} = \frac{p}{B} \text{ (here, } \frac{\Delta R}{R} = \text{fractional decreases in radius)}$$

$$\Rightarrow \frac{\Delta R}{R} = \frac{p}{3B}$$

NOTES If the object of mass m and radius r is surrounded by a liquid in a cylindrical container

$$\frac{\Delta R}{R} = \frac{1}{3}\frac{\Delta V}{V} \quad \text{...(i)}$$

Bulk modulus, $K = -V\frac{\Delta P}{\Delta V}$

$$\therefore \frac{\Delta V}{V} = \frac{\Delta P}{K} = \frac{mg}{AK} \quad \left(\because \Delta P = \frac{mg}{A}\right)$$

$$\therefore \frac{\Delta R}{R} = \frac{1}{3}\frac{mg}{AK} \quad \text{(Here A = area of object)}$$

8. **(a)** Young's modulus $Y = \frac{W}{A}.\frac{l}{\Delta l} \Rightarrow \frac{W_1}{Y_1} = \frac{W_2}{Y_2}$

[$\because$ A, l, Δl same for both brass and steel]

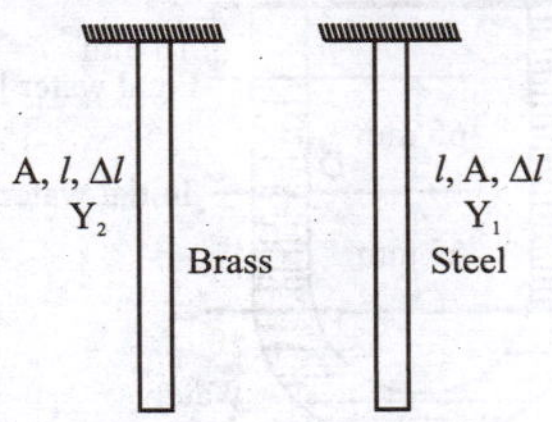

$$\frac{W_1}{W_2} = \frac{Y_1}{Y_2} = 2 \quad [Y_{steel}/Y_{brass} = 2 \text{ given}]$$

9. **(b)** Compressibility of water,
$K = 45.4 \times 10^{-11}\ Pa^{-1}$
density of water $P = 10^3\ kg/m^3$
depth of ocean, h = 2700 m

We have to find $\frac{\Delta V}{V} = ?$

As we know, compressibility,

$$K = \frac{1}{B} = \frac{(\Delta V/V)}{P} \quad [P = \rho gh]$$

So, $(\Delta V/V) = K\rho gh$
$= 45.4 \times 10^{-11} \times 10^3 \times 10 \times 2700 = 1.2258 \times 10^{-2}$

10. **(b)** As $Y = \frac{\frac{F}{A}}{\frac{\Delta l}{l}} \Rightarrow \Delta l = \frac{Fl}{AY}$

But $V = Al$ so $A = \frac{V}{l}$

Therefore $\Delta l = \frac{Fl^2}{VY} \propto l^2$

Hence graph of Δl versus l^2 will give a straight line.

11. **(d)** $F = \frac{YA}{L} \times \Delta L$

So, extension, $\Delta L \propto \frac{L}{A} \propto \frac{L}{D^2}$

[$\because$ F and Y are constant]

$$\Delta L_1 \propto \frac{100}{1^2} \propto 100 \text{ and } \Delta L_2 \propto \frac{200}{2^2} \propto 50$$

$$\Delta L_3 \propto \frac{300}{3^2} \propto \frac{100}{3} \text{ and } \Delta L_4 \propto \frac{50}{\frac{1}{4}} \propto 200$$

The ratio of $\frac{L}{D^2}$ is maximum for case (d).

Hence, option (d) is correct.

12. **(c)** From formula,

Increase in length $\Delta L = \frac{FL}{AY} = \frac{4FL}{\pi D^2 Y}$

$$\frac{\Delta L_S}{\Delta L_C} = \frac{F_S}{F_C}\left(\frac{D_C}{D_S}\right)^2 \frac{Y_C}{Y_S}\frac{L_S}{L_C}$$

$$= \frac{7}{5} \times \left(\frac{1}{p}\right)^2 \left(\frac{1}{s}\right) q = \frac{7q}{(5sp^2)}$$

NOTES If a wire of length l is suspended from a rigid support and A is the area of cross–section of the wire, then mass of the wire, m = Alρ. Then extension in the wire due to its own weight can be find as follows.

Young Modulus, $Y = \frac{mg}{A} \times \frac{(l/2)}{\Delta l}$

Here, L = l/2 because the weight of wire acts at the mid point of wire.

$\therefore$ Extension in the wire due to its own weight $\Delta l = \frac{l^2 \rho g}{2Y}$

13. **(a)** $K = \frac{1}{B} = \frac{\Delta V/V}{P}$. Here, P = 100 atm,
$K = 4 \times 10^{-5}$ and $V = 100\ cm^3$.
Hence, $\Delta V = 0.4\ cm^3$

14. **(c)** We know that Young's modulus $Y = \frac{F}{\pi r^2} \times \frac{L}{\ell}$

Since Y, F are same for both the wires, we have,

$$\ell \propto \frac{L}{r^2}$$

$$\text{or, } \frac{\ell_1}{\ell_2} = \frac{r_2^2 \times L_1}{r_1^2 \times L_2} = \frac{(D_2/2)^2 \times L_1}{(D_1/2)^2 \times L_2}$$

$$\text{or, } \frac{\ell_1}{\ell_2} = \frac{D_2^2 \times L_1}{D_1^2 \times L_2} = \frac{D_2^2}{(2D_2)^2} \times \frac{L_2}{2L_2} = \frac{1}{8}$$

So, $\ell_1 : \ell_2 = 1 : 8$

9 Mechanical Properties of Fluids

--------------TREND ANALYSIS --------------

TOPIC	2025		2024		2023		2022		2021		2020	
	Qns.	LOD	Qns.	LOD	Qns.	LOD	Qns.	LOD	Qns.	LOD	Qns.	LOD
Pressure							1	Avg.				
Bernoulli's Principle					1	Easy						
Viscosity									1	Avg.		
Surface Tension	1	Diff.	1	Avg.	1	Avg.					1	Easy

How to Code Qns for Better & Faster Revision

Type 1: Able to solve (No code Req)

Type 2: Solved with struggle or by seeing solution **(Use code NR means Needs Revision or orange stickers)**

Type 3: Tricky/ Could not solve on my own. Need help from Teacher/ Friend. **(Use code TQ means Tricky Question or red stickers)**

You can either hand write the code in boxes or else paste coloured stickers given at end of book.

Topic-9.2

Pressure

1. Two copper vessels A and B have the same base area but of different shapes. A take twice the volume of water as that B requires to fill upto a particular common height. Then the correct statement among the following is:
NCERT | Page-182, 183 | RS 2022

(a) Vessel B weighs twice that of A.
(b) Pressure on the base area of vessels A and B is same.
(c) Pressure on the base area of A and B is not same.
(d) Both vessels A and B weigh the same.

2. In a u-tube as shown in the fig. water and oil are in the left side and right side of the tube respectively. The heights from the bottom for water and oil columns are 15 cm and 20 cm respectively. The density of the oil is
[take π water = 1000kg/m³]
NCERT | Page-182 | Odisha 2019

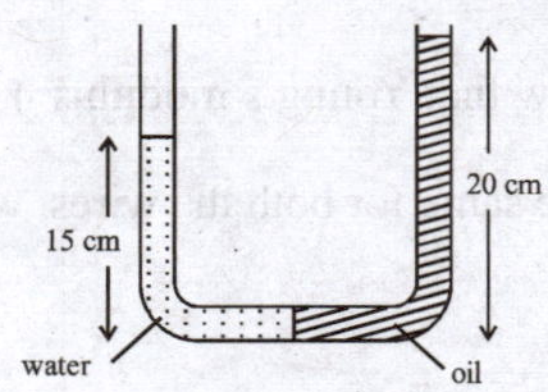

(a) 1333 kg/m³ (b) 1200 kg/m³
(c) 750 kg/m³ (d) 1000 kg/m³

3. A U tube with both ends open to the atmosphere, is partially filled with water. Oil, which is immiscible with water, is poured into one side until it stands at a distance of 10 mm above the water level on the other side. Meanwhile the water rises by 65 mm from its original level (see diagram). The density of the oil is
NCERT | Page-182 | 2017

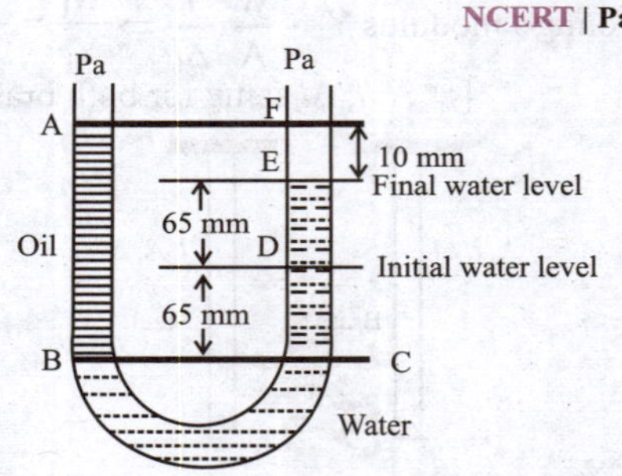

(a) 425 kg m^{-3} (b) 800 kg m^{-3}
(c) 928 kg m^{-3} (d) 650 kg m^{-3}

4. Two non-mixing liquids of densities ρ and nρ (n > 1) are put in a container. The height of each liquid is h. A solid cylinder of length L and density d is put in this container. The cylinder floats with its axis vertical and length pL(p < 1) in the denser liquid. The density d is equal to :
2016

(a) $\{1 + (n + 1)p\}\rho$ (b) $\{2 + (n + 1)p\}\rho$
(c) $\{2 + (n - 1)p\}\rho$ (d) $\{1 + (n - 1)p\}\rho$

5. In rising from the bottom of a lake, to the top, the temperature of an air bubble remains unchanged, but its diameter gets doubled. If *h* is the barometric height (expressed in m of mercury of relative density ρ) at the surface of the lake, the depth of the lake is
NCERT | Page-183 | 1994

(a) 8 ρh m (b) 7ρh m
(c) 9 ρh m (d) 12 ρh m

Topic-9.3

Streamline Flow

6. The cylindrical tube of a spray pump has radius, R, one end of which has n fine holes, each of radius r. If the speed of the liquid in the tube is V, the speed of the ejection of the liquid through the holes is : NCERT | Page-187 | 2015 RS

(a) $\frac{VR^2}{nr^2}$ (b) $\frac{VR^2}{n^3r^2}$ (c) $\frac{V^2R}{nr}$ (d) $\frac{VR^2}{n^2r^2}$

Topic-9.4

Bernoulli's Principle

7. An ideal fluid is flowing in a non-uniform cross-sectional tube XY (as shown in the figure) from end X to end Y. If K_1 and K_2 are the kinetic energy per unit volume of the fluid at X and Y respectively, then the correct option is : NCERT | Page-188 | Ph-II 2024

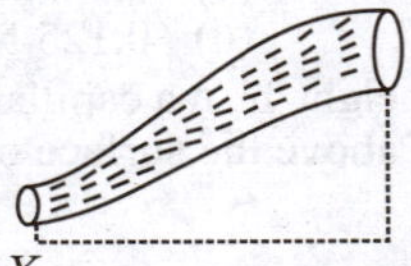

(a) $K_1 = K_2$ (b) $2K_1 = K_2$
(c) $K_1 > K_2$ (d) $K_1 < K_2$

8. The venturi-meter works on 2023
(a) Huygen's principle
(b) Bernoulli's principle
(c) The principle of parallel axes
(d) The principle of perpendicular axes

9. A small hole of area of cross-section 2 mm² is present near the bottom of a fully filled open tank of height 2 m. Taking g = 10 m/s², the rate of flow of water through the open hole would be nearly: NCERT | Page-189 | 2019
(a) 12.6 10^{-6} m³/s (b) 8.9 10^{-6} m³/s
(c) 2.23 10^{-6} m³/s (d) 6.4 10^{-6} m³/s

10. A wind with speed 40 m/s blows parallel to the roof of a house. The area of the roof is 250 m². Assuming that the pressure inside the house is atmosphere pressure, the force exterted by the wind on the roof and the direction of the force will be (ρ_{air} = 1.2 kg/m³) NCERT | Page-188 | 2015
(a) 4.8×10^5 N, upwards
(b) 2.4×10^5 N, upwards
(c) 2.4×10^5 N, downwards
(d) 4.8×10^5 N, downwards

11. A fluid is in streamline flow across a horizontal pipe of variable area of cross section. For this which of the following statements is correct? NCERT | Page-188 | Kar. 2013
(a) The velocity is minimum at the narrowest part of the pipe and the pressure is minimum at the widest part of the pipe
(b) The velocity is maximum at the narrowest part of the pipe and pressure is maximum at the widest part of the pipe
(c) Velocity and pressure both are maximum at the narrowest part of the pipe
(d) Velocity and pressure both are maximum at the widest part of the pipe

Topic-9.5

Viscosity

12. The viscous drag acting on a metal sphere of diameter 1 mm, falling through a fluid of viscosity 0.8 Pa s with a velocity of 2 m s^{-1} is equal to NCERT | Page-191 | Manipur 2023
(a) 1.5×10^{-3} N (b) 20×10^{-3} N
(c) 15×10^{-3} N (d) 30×10^{-3} N

13. The terminal velocity of a copper ball of radius 5 mm falling through a tank of oil at room temperature is 10 cm s^{-1}. If the viscosity of oil at room temperature is 0.9 kg m^{-1}s^{-1}, the viscous drag force is: NCERT | Page-192 | RS 2022
(a) 4.23×10^{-6} N (b) 8.48×10^{-3} N
(c) 8.48×10^{-5} N (d) 4.23×10^{-3} N

14. A spherical ball is dropped in a long column of a highly viscous liquid. The curve in the graph shown, which represents the speed of the ball (v) as a function of time (t) is: NCERT | Page-191 | Odisha 2022

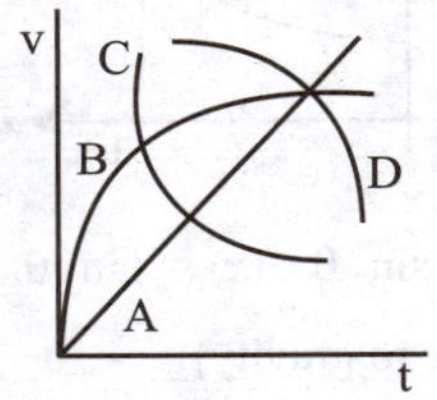

(a) B (b) C (c) D (d) A

15. The velocity of a small ball of mass M and density d, when dropped in a container filled with glycerine becomes constant after some time. If the density of glycerine is $\frac{d}{2}$, then the viscous force acting on the ball will be NCERT | Page-192 | 2021

(a) $2Mg$ (b) $\frac{Mg}{2}$ (c) Mg (d) $\frac{3}{2}Mg$

16. Two small spherical metal balls, having equal masses, are made from materials of densities ρ_1 and ρ_2 ($\rho_1 = 8\rho_2$) and have radii of 1 mm and 2 mm, respectively, they are made to fall vertically (from rest) in a viscous medium whose coefficient of viscosity equals η and whose density is $0.1\rho_2$. The ratio of their terminal velocities would be, NCERT | Page-192 | Odisha 2019

(a) $\frac{79}{36}$ (b) $\frac{79}{72}$ (c) $\frac{19}{36}$ (d) $\frac{39}{72}$

17. A small sphere of radius 'r' falls from rest in a viscous liquid. As a result, heat is produced due to viscous force. The rate of production of heat when the sphere attains its terminal velocity, is proportional to NCERT | Page-192 | 2018

(a) r^3 (b) r^2 (c) r^4 (d) r^5

18. The terminal velocity v_r of a small steel ball of radius r falling under gravity through a column of a viscous liquid of coefficient of viscosity η depends on mass of the ball m, acceleration due to gravity g, coefficient of viscosity η and radius *r*. Which of the following relations is dimensionally correct? **NCERT | Page-192 | 1992**

(a) $v_r \propto \frac{mgr}{\eta}$ (b) $v_r \propto mg\eta r$

(c) $v_r \propto \frac{mg}{r\eta}$ (d) $v_r \propto \frac{\eta mg}{r}$

Topic-9.6

Surface Tension

19. Consider a water tank shown in the figure. It has one wall at $x = L$ and can be taken to be very wide in the z direction. When filled with a liquid of surface tension S and density ρ, the liquid surface makes angle θ_0 ($\theta_0 << 1$) with the x-axis at $x = L$. If $y(x)$ is the height of the surface then the equation for $y(x)$ is: **NCERT | Page-196 | 2025**

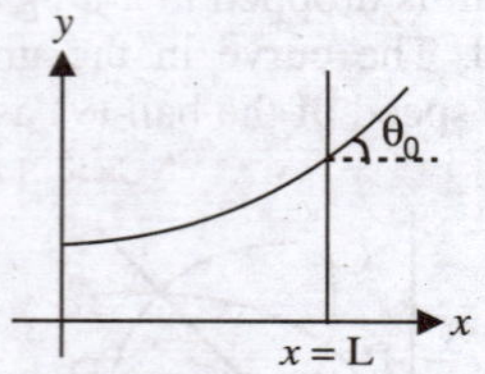

(take $\theta(x) = \sin\theta(x) = \tan\theta(x) = \frac{dy}{dx}$, g is the acceleration due to gravity)

(a) $\frac{d^2y}{dx^2} = \frac{\rho g}{S}x$ (b) $\frac{d^2y}{dx^2} = \frac{\rho g}{S}y$

(c) $\frac{d^2y}{dx^2} = \sqrt{\frac{\rho g}{S}}$ (d) $\frac{dy}{dx} = \sqrt{\frac{\rho g}{S}}x$

20. A thin flat circular disc of radius 4.5 cm is placed gently over the surface of water. If surface tension of water is 0.07 Nm^{-1}, then the excess force required to take it away from the surface is **NCERT | Page-194, 195 | 2024**

(a) 19.8 mN (b) 198 N

(c) 1.98 mN (d) 99 N

21. The amount of energy required to form a soap bubble of radius 2 cm from a soap solution is nearly (surface tension of soap solution = 0.03 N m^{-1}) **NCERT | Page-194 | 2023**

(a) 30.16×10^{-4} J (b) 5.06×10^{-4} J

(c) 3.01×10^{-4} J (d) 50.1×10^{-4} J

22. If a soap bubble expands, the pressure inside the bubble: **NCERT | Page-196 | Odisha 2022**

(a) increases

(b) remains the same

(c) is equal to the atmospheric pressure

(d) decreases

23. A liquid does not wet the solid surface if angle of contact is **NCERT | Page-195 | Ph-II 2020**

(a) zero (b) equal to 45°

(c) equal to 60° (d) greater than 90°

24. A capillary tube of radius r is immersed in water and water rises in it to a height h. The mass of the water in the capillary is 5 g. Another capillary tube of radius 2r is immersed in water. The mass of water that will rise in this tube is : **NCERT | Page-197 | 2020**

(a) 5.0 g (b) 10.0 g

(c) 20.0 g (d) 2.5 g

25. A soap bubble, having radius of 1 mm, is blown from a detergent solution having a surface tension of 2.5×10^{-2} N/m. The pressure inside the bubble equals at a point Z_0 below the free surface of water in a container. Taking $g = 10$ m/s², density of water = 10^3 kg/m³, the value of Z_0 is : **NCERT | Page-196 | 2019**

(a) 100 cm (b) 10 cm

(c) 1 cm (d) 0.5 cm

26. A rectangular film of liquid is extended from (4 cm × 2 cm) to (5 cm × 4 cm). If the work done is 3×10^{-4} J, the value of the surface tension of the liquid is :- **NCERT | Page-194 | Ph-II 2016**

(a) 0.2 Nm^{-1} (b) 8.0 Nm^{-1}

(c) 0.250 Nm^{-1} (d) 0.125 Nm^{-1}

27. Water rises to a height 'h' in a capillary tube. If the length of capillary tube above the surface of water is made less than 'h' then : **NCERT | Page-197 | 2015 RS**

(a) water rises upto the top of capillary tube and stays there without overflowing

(b) water rises upto a point a little below the top and stays there

(c) water does not rise at all.

(d) water rises upto the tip of capillary tube and then starts overflowing like fountain.

28. A certain number of spherical drops of a liquid of radius 'r' coalesce to form a single drop of radius 'R' and volume 'V'. If 'T' is the surface tension of the liquid, then : **NCERT | Page-196 | 2014**

(a) energy $= 4VT\left(\frac{1}{r}-\frac{1}{R}\right)$ is released

(b) energy $= 3VT\left(\frac{1}{r}+\frac{1}{R}\right)$ is absorbed

(c) energy $= 3VT\left(\frac{1}{r}-\frac{1}{R}\right)$ is released

(d) energy is neither released nor absorbed

29. The wetability of a surface by a liquid depends primarily on **NCERT | Page-195 | 2013**

(a) surface tension

(b) density

(c) angle of contact between the surface and the liquid

(d) viscosity

30. The angle of contact between pure water and pure glass, is **NCERT | Page-197 | 1996**

(a) 0° (b) 45°

(c) 90° (d) 135°

ANSWER KEY

1	(b)	4	(d)	7	(c)	10	(b)	13	(b)	16	(a)	19	(b)	22	(d)	25	(c)	28	(c)
2	(c)	5	(b)	8	(b)	11	(b)	14	(a)	17	(d)	20	(a)	23	(d)	26	(d)	29	(c)
3	(c)	6	(a)	9	(a)	12	(c)	15	(b)	18	(c)	21	(c)	24	(b)	27	(a)	30	(a)

Hints & Solutions

1. **(b)** Pressure, P = hρg.

Since water is filled upto same height so pressure at the bottom will be same.

2. **(c)** Pressure at A = Pressure at B.

$P_a + 0.15 \times 10^3 \times g = P_a + 0.20 \times d_0 g$

$d_0 = \frac{0.15 \times 10^3}{0.20} = 0.75 \times 10^3 = 750 \text{ kg/m}^3$

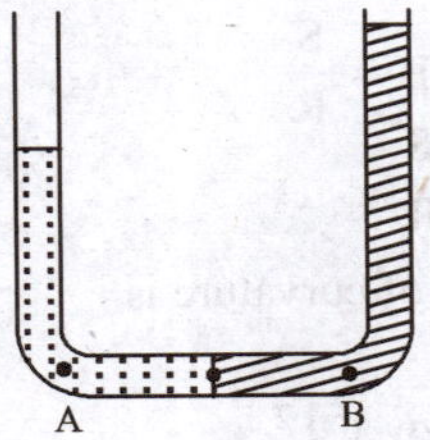

3. **(c)** Here, $h_{oil} \times \rho_{oil} \times g = h_{water} \times \rho_{water} \times g$

$\rho_0 g \times 140 \times 10^{-3} = \rho_w g \times 130 \times 10^{-3}$

$\rho_{oil} = \frac{130}{140} \times 10^3 \approx 928 \text{kg/m}^3 \quad [\because \rho_w = 1 \text{ kgm}^{-3}]$

4. **(d)** As we know,

Pressure P = Vdg

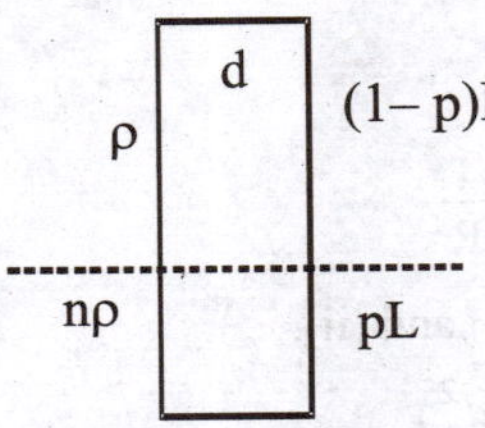

Here, L A d g = (pL) A (nρ)g + (1 – p)L A ρ g

$\Rightarrow \quad d = (1 - p)\rho + pn\,\rho = [1 + (n - 1)p]\rho$

> **NOTES** When a body of volume *V* floats in a liquid then
> Weight of the body = up thrust
> $\Rightarrow V\rho g = V_{in} \sigma g$
> Here, V = volume of body
> ρ = density of body
> σ = density of liquid
> V_{in} = volume of the body immersed in liquid

5. **(b)** $(h\rho g + H \times 1 \times g)\frac{4}{3}\pi r^3 = h\rho g \times \frac{4}{3}\pi(2r)^3$

This gives H = 7hρ

6. **(a)** Inflow rate of volume of the liquid = Outflow rate of volume of the liquid

$\pi R^2 V = n\pi r^2(v) \Rightarrow v = \frac{\pi R^2 V}{n\pi r^2} = \frac{VR^2}{nr^2}$

7. **(c)** By Bernoulli's principle,

$\frac{1}{2}\rho V^2 + \rho gh + P = \text{constant}$

$P + K_1 + \rho g\,(0) = P + K_2 + \rho g\,(h)$

$K_1 = K_2 + \rho gh$

$K_1 > K_2$

8. **(b)** Venturi-meter works on the Bernoulli's principle.

9. **(a)** Volumetric flow rate of liquid

$Q = au = a\sqrt{2gh}$

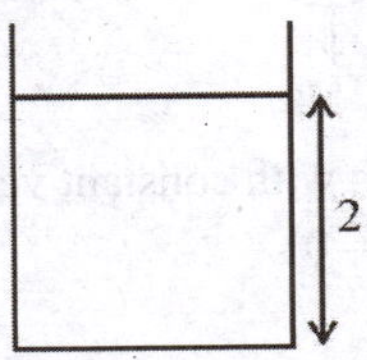

$= 2 \times 10^{-6} \text{m}^2 \times \sqrt{2 \times 10 \times 2} \text{ m/s}$

$= 2 \times 2 \times 3.14 \times 10^{-6} \text{ m}^3/\text{s}$

$= 12.56 \times 10^{-6} \text{ m}^3/\text{s} = 12.6 \times 10^{-6} \text{ m}^3/\text{s}$

10. **(b)** According to Bernoulli's theorem,

$P + \frac{1}{2}\rho v^2 = P_0 + 0$

So, $\Delta P = \frac{1}{2}\rho v^2$

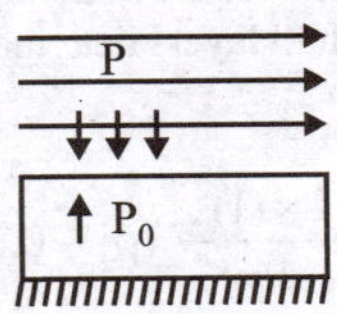

$F = \Delta PA = \frac{1}{2}\rho v^2 A$

$= \frac{1}{2} \times 1.2 \times 40 \times 40 \times 250$

$= 2.4 \times 10^5$ N (upwards)

11. **(b)** According to Bernoulli's theorem,

$P + \frac{1}{2}\rho v^2$ = constant and Av = constant

If *A* is minimum, *v* is maximum, *P* is minimum.

12. **(c)** From stoke's law, Viscous drag acting on sphere

$F_v = 6\pi\eta rv$

$= 6\times\frac{22}{7}\times 0.8\times\frac{10^{-3}}{2}\times 2 = 15.08\times 10^{-3}$ N

$\simeq 15\times 10^{-3}$ N

13. **(b)** From Stoke's formula,

Viscous drag force

$F = 6\pi\eta rV_t$

$F = 6\times\pi\times 0.9\times 5\times 10^{-3}\times 10\times 10^{-2} = 84.78\times 10^{-4}$

$\therefore F = 8.478\times 10^{-3}$ N

14. **(a)** When ball is dropped in highly viscous fluid, the ball's velocity will increase untill it becomes equal to terminal velocity and after attaining terminal velocity it moves downward with constant velocity so, graph will be like as shown below

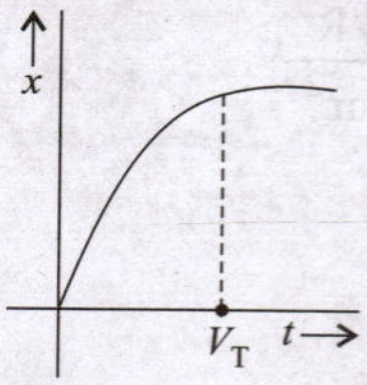

15. **(b)** Let F_v = viscous force and F_B = Bouyant force acting on the ball.

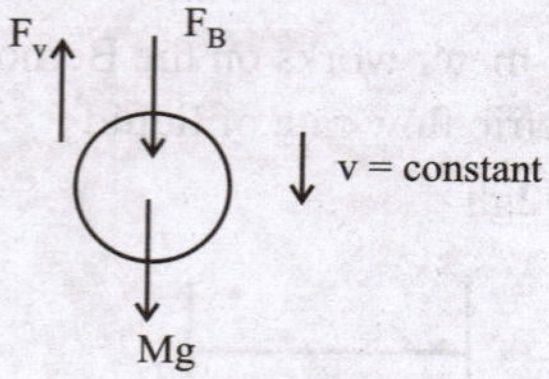

When body moves with constant velocity

$Mg = F_B + F_v$ [a = 0]

$F_v = Mg - F_B$

$= dVg - \frac{d}{2}.Vg$ (M = dVg) V = volume of ball.

or $F_V = \frac{d}{2}Vg = \frac{M}{2}g$ [$\because M = dV$]

NOTES Due to motion of a body through the fluid, the fluid in contact with the body is dragged with it causes relative motion in fluid layers hear the body and hence viscous force starts.

16. **(a)** $V_{T_1} = \frac{2\times(1)^2}{9\eta}(\rho_1 - 0.1\rho_2)g$

$V_{T_1} = \frac{2\times 1}{9\eta}(8\rho_2 - 0.1\rho_2)g$

$V_{T_2} = \frac{2\times(2)^2}{9\eta}(\rho_2 - 0.1\rho_2)g$

$\therefore \frac{V_{T_1}}{V_{T_2}} = \frac{7.9}{4(0.9)} = \frac{79}{36}$

17. **(d)** Power = rate of production of heat = F.V

$= 6\pi\eta r V_T\cdot V_T = 6\pi\eta r V_T^2$

($\because F = 6\pi\eta V_T r$ Stoke's formula)

$V_T \propto r^2$

$\because V_T = \frac{2}{9}\frac{r^2(\rho-\sigma)}{\eta}g \therefore$ Power $\propto r^5$

18. **(c)** Note that according to Stoke's law

$6\pi\eta r v_r = mg$

Hence, the valid relation is $v_r \propto mg/r\eta$

19. **(b)**

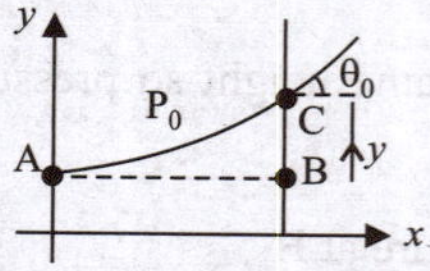

At horizontal level,

$P_A = P_B = P_0$

Along vertical level,

$P_C = P_0 - \rho gy$

Also, $P_C = P_0 - \frac{S}{R}$

$\therefore \rho gy = \frac{S}{R}$ ...(i)

The radius of curvature is

$$R = \frac{\left\{1+\left(\frac{dy}{dx}\right)^2\right\}^{\frac{3}{2}}}{\frac{d^2y}{dx^2}}$$

$\frac{dy}{dx}$ is very small $\therefore \frac{dy}{dx} = 0$

$$R = \frac{1}{\frac{d^2y}{dx^2}}$$

$\therefore \frac{d^2y}{dx^2} = \frac{1}{R}$ ...(ii)

From eq (i) and (ii),

$\rho gy = S\frac{d^2y}{dx^2}$

$\frac{d^2y}{dx^2} = \frac{\rho gy}{S}$

20. **(a)**

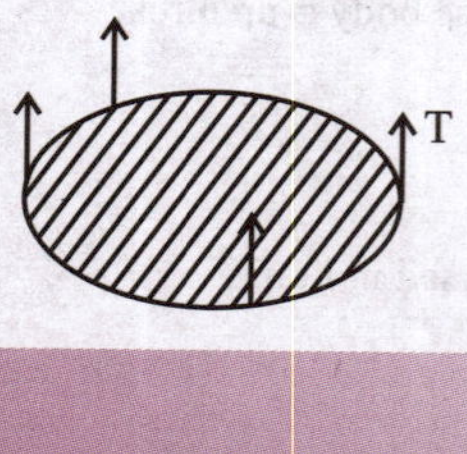

Excess force = T × length = T × 2πR

$$= \frac{7}{100} \times 2 \times 3.14 \times \frac{4.5}{100} = 19.8 \text{ mN}$$

21. (c) Amount of energy required to form a soap bubble

$= [S \times \Delta A] \times 2$

$= [0.03 \times 4 \times \pi \times 4 \times 10^{-4}] \times 2 = 3.015 \times 10^{-4}$ J

22. (d) Pressure inside soap bubble, $P = P_0 + \frac{4T}{R}$

Clearly, $R \uparrow\uparrow \Rightarrow \frac{4T}{R} \downarrow\downarrow \Rightarrow P \downarrow\downarrow$

So, if bubble expands, pressure decreases.

Alternative method :

Here, PV = cons. [∵ T remains same]

$P \propto \frac{1}{V}$, as R $\uparrow\uparrow \Rightarrow V \uparrow\uparrow \Rightarrow P \downarrow\downarrow$

23. (d) If angle of contact is greater than 90°, then liquid does not wet the solid surface.

24. (b) Force of surface tension balances the weight of water in capillary tube.

$F_S = 2\pi r T \cos\theta = mg$

Here, T and θ are constant.

So, $m \propto r$

Let m_1 and m_2 be the mass of water in two capillary tube.

$\therefore \frac{m_2}{m_1} = \frac{r_2}{r_1} \Rightarrow \frac{m_2}{5.0} = \frac{2r}{r} \qquad (\because r_2 = 2r)$

$\Rightarrow m_2 = 10.0$ g

25. (c) Pressure inside the soap bubble

$= P_0 + \frac{4T}{R}$

And pressure at a point below the surface of water

$= \rho g Z_0 + P_0$ [where P_0 = atmospheric pressure]

By equating these pressure we get,

$$P_0 + \frac{4T}{R} = P_0 + \rho g Z_0$$

$$Z_0 = \frac{4T}{R \times \rho g} \Rightarrow Z_0 = \frac{4 \times 2.5 \times 10^{-2}}{10^{-3} \times 1000 \times 10} \text{m}$$

$Z_0 = 1$ cm

26. (d) Work done W = T(2DA) = 3×10^{-4} J

$$\Rightarrow T = \frac{W}{2\Delta A} = \frac{3 \times 10^{-4}}{2 \times 12 \times 10^{-4}} = 0.125 \text{Nm}^{-1}$$

27. (a) Water rises upto the top of capillary tube and stays there without overflowing.

NOTES In case of capillary of insufficient length i.e. $L \angle h$ the liquid will neither overflow from the upper end like a fountain nor it will tickle along the vertical sides of the tube. The liquid after reaching the upper end will increase the radius of its meniscus without changing nature such that

$hr = Lr'$ (Here, r is the initial radius
$\because L < h$ of tube and r' is the new
$\because r' > r$ radius)

28. (c) Volume same $\Rightarrow n\left(\frac{4}{3}\pi r^3\right) = \frac{4}{3}\pi R^3$

$\Rightarrow n = \frac{R^3}{r^3}$

Change in energy = TΔA = T[4πR² – n4πr²]

$$= 4\pi T\left[R^2 - \frac{R^3}{r^3} r^2\right] = 3\left(\frac{4}{3}\pi R^3\right) T\left[\frac{1}{R} - \frac{1}{r}\right]$$

$$= 3VT\left[\frac{1}{R} - \frac{1}{r}\right] (R > r)$$

$= 3VT\left[\frac{1}{r} - \frac{1}{R}\right]$ is released

NOTES Energy released when n drops of radius r coalesec to form a body drop of radius R,

Energy released = $4\pi R^3 T\left[\frac{1}{r} \frac{-1}{R}\right]$

If this energy get converted into kinetic energy of big drop, then

$$\frac{1}{2}mv^2 = 4\pi R^3 T\left[\frac{1}{r} \frac{-1}{R}\right]$$

$$\Rightarrow \frac{1}{2}\left[\frac{4}{3}\pi R^3 d\right] V^2 = 4\pi R^3 T\left[\frac{1}{r} \frac{-1}{R}\right]$$

$$\Rightarrow v^2 = \frac{6T}{d}\left[\frac{1}{r} \frac{-1}{R}\right]$$

$\Rightarrow$ Velocity of big drop $V = \sqrt{\frac{6T}{d}\left(\frac{1}{r} \frac{-1}{R}\right)}$

(Where, d = density of big liquid drop)

29. (c) Wetability of a surface by a liquid primarily depends on angle of contact between the surface and liquid.

If angle of contact is acute liquids wet the solid and vice-versa.

30. (a) We know that angle of contact is the angle between the tangent to liquid surface at the point of contact and solid surface inside the liquid. In case of pure water and pure glass, the angle of contact is zero.

10 Thermal Properties of Matter

TREND ANALYSIS

TOPIC	2025		2024		2023		2022		2021		2020	
	Qns.	LOD	Qns.	LOD	Qns.	LOD	Qns.	LOD	Qns.	LOD	Qns.	LOD
Thermal Expansion			1	Diff.								
Heat Transfer	1	Diff.					1	Avg.				

How to Code Qns for Better & Faster Revision

Type 1: Able to solve (No code Req)
Type 2: Solved with struggle or by seeing solution **(Use code NR means Needs Revision or orange stickers)**
Type 3: Tricky/ Could not solve on my own. Need help from Teacher/ Friend. **(Use code TQ means Tricky Question or red stickers)**

You can either hand write the code in boxes or else paste coloured stickers given at end of book.

Topic-10.3 Measurement of Temperature

1. On a new scale of temperature (which is linear) and called the W scale, the freezing and boiling points of water are 39° W and 239° W respectively. What will be the temperature on the new scale, corresponding to a temperature of 39° C on the Celsius scale? NCERT | Page-203 | 2008
(a) 78° W (b) 117° W (c) 200° W (d) 139° W

2. Mercury thermometer can be used to measure temperature upto NCERT | Page-203 | 1992
(a) 260°C (b) 100°C
(c) 357°C (d) 500°C

3. A centigrade and a Fahrenheit thermometer are dipped in boiling water. The water temperature is lowered until the Fahrenheit thermometer registers 140°. What is the fall in temperature as registered by the Centigrade thermometer? NCERT | Page-203 | 1990
(a) 80° (b) 60°
(c) 40° (d) 30°

Topic-10.5 Thermal Expansion

4. A metallic bar of Young's modulus, 0.5×10^{11} Nm^{-2} and coefficient of linear thermal expansion 10^{-5} °C^{-1}, length 1 m and area of cross-section 10^{-3} m^2 is heated from 0°C to 100°C without expansion or bending. The compressive force developed in it is : NCERT | Page-205 | 2024
(a) 5×10^3 N (b) 50×10^3 N
(c) 100×10^3 N (d) 2×10^3 N

5. A copper rod of 88 cm and an aluminium rod of unknown length have their increase in length independent of increase in temperature. The length of aluminium rod is : ($\alpha_{Cu} = 1.7 \times 10^{-5}$ K^{-1} and $\alpha_{Al} = 2.2 \times 10^{-5}$ K^{-1}) NCERT | Page-205 | 2019
(a) 6.8 cm (b) 113.9 cm
(c) 88 cm (d) 68 cm

6. Coefficient of linear expansion of brass and steel rods are α_1 and α_2. Lengths of brass and steel rods are ℓ_1 and ℓ_2 respectively. If $(\ell_2 - \ell_1)$ is maintained same at all temperatures, which one of the following relations holds good? NCERT | Page-205 | 2016, 1999, 1995
(a) $\alpha_1 \ell_2^2 = \alpha_2 \ell_1^2$ (b) $\alpha_1 \ell_2 = \alpha_2 \ell_1$
(c) $\alpha_1 \ell_2 = \alpha_2 \ell_1$ (d) $\alpha_1 \ell_1 = \alpha_2 \ell_2$

7. The value of coefficient of volume expansion of glycerine is 5×10^{-4} K^{-1}. The fractional change in the density of glycerine for a rise of 40°C in its temperature, is: NCERT | Page-205 | 2015 RS
(a) 0.020 (b) 0.025 (c) 0.010 (d) 0.015

8. The density of water at 20°C is 998 kg/m^3 and at 40°C 992 kg/m^3. The coefficient of volume expansion of water is NCERT | Page-205 | Kar. 2013
(a) 10^{-4}/°C (b) 3×10^{-4}/°C
(c) 2×10^{-4}/°C (d) 6×10^{-4}/°C

Topic-10.6 Specific Heat Capacity

9. An electric kettle takes 4A current at 220 V. How much time will it take to boil 1 kg of water from temperature 20° C? The temperature of boiling water is 100° C. NCERT | Page-208 | 2008
(a) 6.3 min (b) 8.4 min (c) 12.6 min (d) 4.2 min

10. Thermal capacity of 40 g of aluminium ($s = 0.2$ cal /g K) is NCERT | Page-213 | 1990
(a) 168 joule /°C (b) 672 joule/°C
(c) 840 joule/°C (d) 33.6 joule/°C

Topic-10.8 Change of State

11. A piece of ice falls from a height h so that it melts completely. Only one-quarter of the heat produced is absorbed by the ice and all energy of ice gets converted into heat during its fall. The value of h is : NCERT | Page-213 | 2016
[Latent heat of ice is 3.4×10^5 J/kg and g = 10 N/kg]
(a) 34 km (b) 544 km (c) 136 km (d) 68 km

12. Steam at 100°C is passed into 20 g of water at 10°C. When water acquires a temperature of 80°C, the mass of water present will be: [2014]
[Take specific heat of water = 1 cal g^{-1} °C^{-1} and latent heat of steam = 540 cal g^{-1}] **NCERT | Page-209 | 2014**
(a) 24 g (b) 31.5 g
(c) 42.5 g (d) 22.5 g

13. Liquid oxygen at 50 K is heated to 300 K at constant pressure of 1 atm. The rate of heating is constant. Which one of the following graphs represents the variation of temperature with time? **NCERT | Page-213 | 2012**

(a)
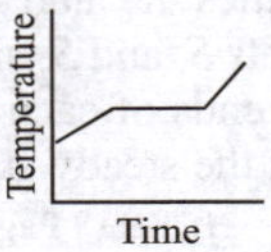

(b)
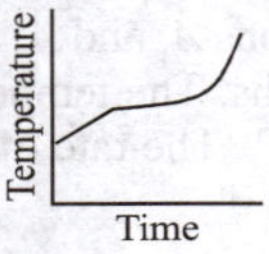

(c)
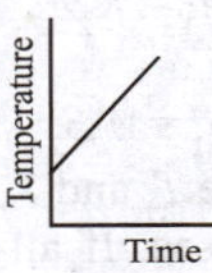

(d)
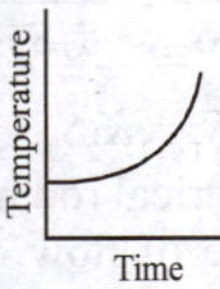

14. If 1 g of steam is mixed with 1 g of ice, then the resultant temperature of the mixture is **NCERT | Page-209 | 1999**
(a) 270ºC (b) 230ºC
(c) 100ºC (d) 50ºC

15. 10 gm of ice cubes at 0°C are released in a tumbler (water equivalent 55 g) at 40°C. Assuming that negligible heat is taken from the surroundings, the temperature of water in the tumbler becomes nearly (L = 80 cal/g) **NCERT | Page-209 | 1988**
(a) 31°C (b) 22°C
(c) 19°C (d) 15°C

Topic-10.9

Heat Transfer

16. Three identical heat conducting rods are connected in series as shown in the figure. The rods on the sides have thermal conductivity 2*K* while that in the middle has thermal conductivity *K*. The left end of the combination is maintained at temperature 3*T* and the right end at *T*. The rods are thermally insulated from outside. In steady state, temperature at the left junction is T_1 and that at the right junction is T_2. The ratio $\frac{T_1}{T_2}$ is **NCERT | Page-216 | 2025**

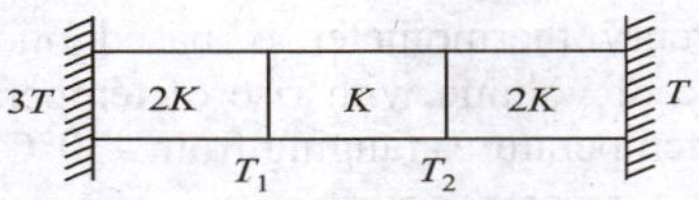

(a) $\frac{3}{2}$ (b) $\frac{4}{3}$ (c) $\frac{5}{3}$ (d) $\frac{5}{4}$

17. Given below are two statements: One is labelled as **Assertion A** and the other is labelled as **Reason R**.
Assertion A: Houses made of concrete roofs overlaid with foam keep the room hotter during summer.
Reason R: The layer of foam insulation prohibits heat transfer, as it contains air pockets.
NCERT | Page-215 | Ph-II 2024
In the light of the above statements, choose the **correct** answer from the options given below.
(a) **A** is true but **R** is false.
(b) **A** is false but **R** is true.
(c) Both **A** and **R** are true and **R** is the correct explanation of **A**.
(d) Both **A** and **R** true but **R** is NOT the correct explanation of **A**.

18. Two rods one made of copper and other made of steel of the same length and same cross sectional area are joined together. The thermal conductivity of copper and steel are 385 J s^{-1} $K^{-1}m^{-1}$ and 50 J $s^{-1}K^{-1}m^{-1}$ respectively. The free ends of copper and steel are held at 100°C and 0°C respectively. The temperature at the junction is, nearly: **NCERT | Page-216 | RS 2022**
(a) 88.5°C (b) 12°C (c) 50°C (d) 73°C

19. A deep rectangular pond of surface area A, containing water (density = ρ), specific heat (capacity = s), is located in a region where the outside air temperature is at a steady value of –26°C. The thickness of the frozen ice layer in this pond, at a certain instant is *x*. Taking the thermal conductivity of ice as K, and its specific latent heat of fusion as L, the rate of increase of the thickness of ice layer, at this instant, would be given by **NCERT | Page-215 | Odisha 2019**
(a) 26K/ρ*x*(L + 4s) (b) 26K/ρ*x*(L – 4s)
(c) 26K/(ρx^2L) (d) 26K/(ρ*x*L)

20. Two rods A and B of different materials are welded together as shown in figure. Their thermal conductivities are K_1 and K_2. The thermal conductivity of the composite rod will be : **NCERT | Page-215 | 2017**

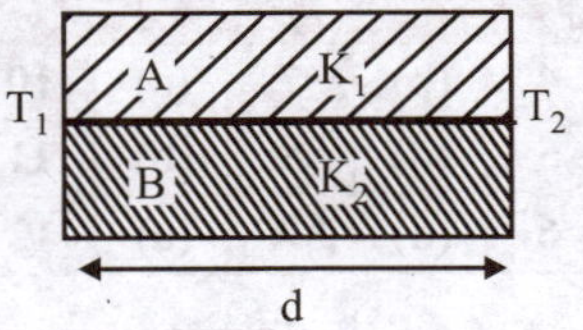

(a) $\frac{3(K_1 + K_2)}{2}$ (b) $K_1 + K_2$
(c) $2(K_1 + K_2)$ (d) $\frac{K_1 + K_2}{2}$

21. The two ends of a metal rod are maintained at temperatures 100°C and 110°C. The rate of heat flow in the rod is found to be 4.0 J/s. If the ends are maintained at temperatures 200°C and 210°C, the rate of heat flow will be **NCERT | Page-215 | 2015**
(a) 16.8 J/s (b) 8.0 J/s (c) 4.0 J/s (d) 44.0 J/s

22. Two metal rods 1 and 2 of same lengths have same temperature difference between their ends. Their thermal conductivities are K_1 and K_2 and cross sectional areas A_1 and A_2, respectively. If the rate of heat conduction in rod 1 is four times that in rod 2, then **NCERT | Page-215 | Kar. 2013**
(a) $K_1A_1 = K_2A_2$ (b) $K_1A_1 = 4K_2A_2$
(c) $K_1A_1 = 2K_2A_2$ (d) $4K_1A_1 = K_2A_2$

23. A slab of stone of area 0.36 m^2 and thickness 0.1 m is exposed on the lower surface to steam at 100°C. A block of ice at 0°C rests on the upper surface of the slab. In one hour 4.8 kg of ice is melted. The thermal conductivity of slab is :
(Given latent heat of fusion of ice = 3.36×10^5 J kg^{-1}) :
NCERT | Page-215, 213 | 2012M
(a) 1.24 J/m/s/°C (b) 1.29 J/m/s/°C
(c) 2.05 J/m/s/°C (d) 1.02 J/m/s/°C

24. A cylindrical metallic rod in thermal contact with two reservoirs of heat at its two ends conducts an amount of heat Q in time t. The metallic rod is melted and the material is formed into a rod of half the radius of the original rod. What is the amount of heat conducted by the new rod, when placed in thermal contact with the two reservoirs in time t ?
NCERT | Page-215 | 2010
(a) $\frac{Q}{4}$ (b) $\frac{Q}{16}$ (c) $2Q$ (d) $\frac{Q}{2}$

25. The two ends of a rod of length L and a uniform cross-sectional area A are kept at two temperatures T_1 and T_2 ($T_1 > T_2$). The rate of heat transfer, $\frac{dQ}{dt}$ through the rod in a steady state is given by: **NCERT | Page-215 | 2009**
(a) $\frac{dQ}{dt} = \frac{k(T_1 - T_2)}{LA}$ (b) $\frac{dQ}{dt} = kLA\,(T_1 - T_2)$
(c) $\frac{dQ}{dt} = \frac{kA\,(T_1 - T_2)}{L}$ (d) $\frac{dQ}{dt} = \frac{kL\,(T_1 - T_2)}{A}$

26. Which of the following circular rods (given radius r and length l), each made of the same material and whose ends are maintained at the same temperature will conduct most heat? **NCERT | Page-215 | 2005**
(a) $r = r_0;\ l = l_0$ (b) $r = 2r_0;\ l = l_0$
(c) $r = r_0;\ l = 2l_0$ (d) $r = 2r_0;\ l = 2l_0$

27. Consider a compound slab consisting of two different materials having equal thicknesses and thermal conductivities K and $2K$, respectively. The equivalent thermal conductivity of the slab is **NCERT | Page-215 | 2003**
(a) $\frac{4}{3}K$ (b) $\frac{2}{3}K$
(c) $\sqrt{3}\,K$ (d) $3\,K$

28. Two rods of thermal conductivities K_1 and K_2, cross-sections A_1 and A_2 and specific heats S_1 and S_2 are of equal lengths. The temperatures of two ends of each rod are T_1 and T_2. The rate of flow of heat at the steady state will be equal if **NCERT | Page-215 | 2002**
(a) $\frac{K_1}{A_1S_1} = \frac{K_2}{A_2S_2}$ (b) $K_1A_1 = K_2A_2$
(c) $K_1S_1 = K_2S_2$ (d) $A_1S_1 = A_2S_2$

29. A cylindrical rod having temperature T_1 and T_2 at its end. The rate of flow of heat is Q_1 cal/sec. If all the linear dimensions are doubled keeping temperature constant, then the rate of flow of heat Q_2 will be
NCERT | Page-215 | 2001
(a) $4Q_1$ (b) $2Q_1$
(c) $Q_1/4$ (d) $Q_1/2$

30. The presence of gravitational field is required for the heat transfer by **NCERT | Page-217 | 2000**
(a) conduction (b) stirring of liquids
(c) natural convection (d) radiation

ANSWER KEY

1	(b)	4	(b)	7	(a)	10	(d)	13	(a)	16	(c)	19	(d)	22	(b)	25	(c)	28	(b)
2	(c)	5	(d)	8	(b)	11	(c)	14	(c)	17	(b)	20	(d)	23	(a)	26	(b)	29	(b)
3	(c)	6	(d)	9	(a)	12	(d)	15	(b)	18	(a)	21	(c)	24	(b)	27	(a)	30	(c)

Hints & Solutions

1. (b) For different temperature scales, we have

$$\frac{x - \text{L.F.P}}{\text{U.F.P.} - \text{L.F.P.}} = \text{constant}$$

Where L.F.P ⇒ Lower Fixed point
U.H.F. ⇒ Upper fixed point
where x is the measurement at that scale. Here, if C and W be the measurements on Celsius and W scale then,

$$\frac{C - 0}{100 - 0} = \frac{W - 39}{239 - 39}$$

$(\therefore C = 39°C)$

$$\Rightarrow W = \frac{39 \times 200}{100} + 39 = 78 + 39 = 117°\text{ W}$$

2. (c) Mercury thermometer is based on the principle of change of volume with rise of temperature and can measure temperatures ranging from –30°C to 357°C.

3. (c) Using $\frac{F - 32}{180} = \frac{C}{100}$

$$\Rightarrow \frac{140 - 32}{180} = \frac{C}{100} \Rightarrow C = 60$$

Temperature of boiling water = 100°C
We get, fall in temperature = 100 – 60 = 40°C

NOTES Centigrade (c) Fahrenheit (F) and Kelvin (K) scales of temperature are related as

$$\frac{C-0}{100}=\frac{F-32}{212-32}=\frac{K-273.15}{373.15-273.15}$$

4. **(b)** The compressive force developed in the metallic bar is F = YA α ΔT

$= 0.5 \times 10^{11} \times 10^{-3} \times 10^{-5} \times 100 = 50 \times 10^3$ N

5. **(d)** $l'_{Cu} = l_{Cu}(1 + \alpha_{Cu}\Delta T) \Rightarrow \Delta l_{Cu} = l_{Cu}\alpha_{Cu}\Delta T$

$l'_{Al} = l_{Al}(1 + \alpha_{Al}\Delta T) \Rightarrow \Delta l_{Al} = l_{Al}\alpha_{Al}\Delta T$

Since, change in length is independent of temperature

$\therefore \alpha_{Cu} l_{Cu} = \alpha_{Al} l_{Al}$

$\Rightarrow 1.7 \times 10^{-5} \times 88 \text{ cm} = 2.2 \times 10^{-5} \times l_{Al}$

$\Rightarrow l_{Al} = \frac{1.7\times 88}{2.2} = 68$ cm

6. **(d)** From question, $(\ell_2 - \ell_1)$ is maintained same at all temperatures hence change in length for both rods should be same

i.e., $\Delta\ell_1 = \Delta\ell_2$

As we know, coefficient of linear expansion,

$$\alpha = \frac{\Delta\ell}{\ell_0\Delta T}$$

$\ell_1\alpha_1\Delta T = \ell_2\alpha_2\Delta T \Rightarrow \ell_1\alpha_1 = \ell_2\alpha_2$

7. **(a)** From question,

Rise in temperature $\Delta t = 40°C$

Coefficient of volume expansion

$\gamma = 5 \times 10^{-4} K^{-1}$

Fractional change in the density $\frac{\Delta\rho}{\rho_0} = \frac{\rho_0 - \rho_T}{\rho_0}$

$\rho_T = \rho_0(1 - \gamma\Delta t)$

$\Rightarrow \frac{\Delta\rho}{\rho_0} = \frac{\rho_0-\rho_T}{\rho_0} = \gamma\Delta T$

$= (5 \times 10^{-4})(40) = 0.020$

NOTES Most substances (solid & liquid) expand when they are heated, so volume of a given mass of a substance increases on heating so density decreases as

$$\rho \propto \frac{1}{V} \Rightarrow \frac{\rho'}{\rho} = \frac{V}{V'} = \frac{V}{V+\Delta V} = \frac{V}{V+\gamma V\Delta\theta} = \frac{1}{1+\gamma\Delta\theta}$$

8. **(b)** Given

$T_1 = 20°$ C, $T_2 = 40°$ C

$\rho_{20} = 998$ kg/m³, $\rho_{40} = 992$ kg/m³

As $\rho_{T2} = \frac{\rho_{T1}}{1+\gamma\Delta T} = \frac{\rho_{T1}}{1+\gamma(T_2-T_1)}$

$\therefore 992 = \frac{998}{1+\gamma(40-20)} \Rightarrow 992 = \frac{998}{1+20\gamma}$

$\Rightarrow 20\gamma = \frac{998}{992} - 1 = \frac{6}{992}$

$\Rightarrow \gamma = \frac{6}{992}\times\frac{1}{20} = 3\times10^{-4}/°C$

9. **(a)** Heat required to raise the temperature of 1kg water from 20°C to 100°C is given by $Q = ms\Delta\theta = (1\times 4200 \times 80)$ J

Power of kettle (P) = VI = (220 × 4)W

$\therefore$ Time taken $= \frac{Q}{P} = \frac{1\times4200\times80}{220\times4}$

= 381.81 sec = 6.36 min

10. **(d)** Thermal capacity = ms = 40 × 0.2 = 8 cal/°C = 4.2 × 8 J = 33.6 joules/°C

11. **(c)** According to question only one-quarter of the heat produced by falling piece of ice is absorbed in the melting of ice.

i.e., $\frac{mgh}{4} = mL$

$\Rightarrow h = \frac{4L}{g} = \frac{4\times3.4\times10^5}{10} = 136$ km.

12. **(d)** According to the principle of calorimetry.

Heat lost = Heat gained

$mL_v + ms_w\Delta\theta = m_w s_w\Delta\theta$

$\Rightarrow$ m × 540 + m × 1 × (100 – 80)

= 20 × 1 × (80 – 10) ⇒ m = 2.5 g

Therefore total mass of water at 80°C

= (20 + 2.5) g = 22.5 g

13. **(a)** Initially liquid oxygen will gain the temperature up to its boiling temperature then it change its state to gas. After this again its temperature will increase, so corresponding graph will be graph (a).

14. **(c)** Heat required by ice at 0°C to reach a temperature of 100°C = $mL + mc\Delta\theta$

= 1 × 80 + 1 × 1 × (100 – 0) = 180 cal

Heat available with 1 g steam to condense into 1 g of water at 100°C = 536 cal.

Obviously the whole steam will not be condensed and ice will attain a temperature of 100°C; so the temperature of mixture = 100°C.

15. **(b)** Let the final temperature be T

Heat gained by ice = mL + m × s × $(T - 0)$

= 10 × 80 + 10 × 1 × T

Heat lost by water = 55 × 1× (40 – T)

By using law of calorimetery,

$800 + 10\,T = 55 \times (40 - T)$

$\Rightarrow T = 21.54°C = 22°C$

16. **(c)** At steady state, $\frac{dQ}{dt} =$ constant

$$\therefore \left(\frac{dQ}{dt}\right)_1 = \left(\frac{dQ}{dt}\right)_2 = \left(\frac{dQ}{dt}\right)_3$$

$$\frac{2KA(3T - T_1)}{L} = \frac{KA(T_1 - T_2)}{L} = \frac{2KA(T_2 - T)}{L}$$

$\Rightarrow 6T - 2T_1 = T_1 - T_2 = 2T_2 - 2T$

$\Rightarrow 6T - 2T_1 = T_1 - T_2$

$\therefore 6T + T_2 = 3T_1$ ...(i)

Also, $T_1 - T_2 = 2T_2 - 2T$

$\therefore 2T - 3T_2 = -T_1$ ...(ii)

From equations (i) and (ii),

$10T_2 = 6T_1$

$$\therefore \frac{T_1}{T_2} = \frac{5}{3}$$

17. **(b)** Layer of foam forms an insulation which prohibits heat transfer, and hence roofs of houses containing foam keep room cooler during summer.

18. **(a)**

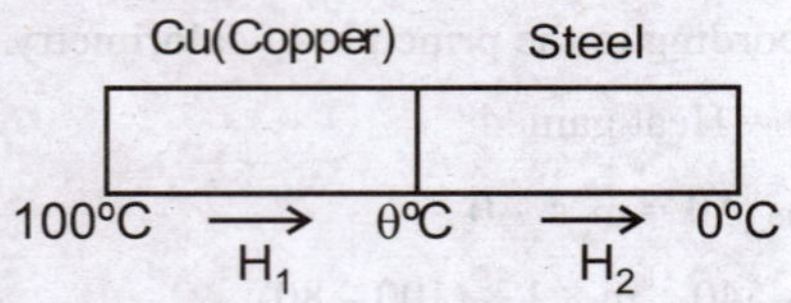

Thermal current, $H_1 = H_2$

$$\frac{K_{Cu}A[100 - \theta]}{\ell} = \frac{K_{steel}A[\theta - 0]}{\ell}$$

$385\,[100 - \theta] = 50\,[\theta - 0]$

$\therefore \theta = 88.5$ °C [junction temperature]

NOTES In series combination of slabs, heat current is same in all the slabs.

19. **(d)** We assume that at any instant thickness of ice is *x*. And time taken to form additional thickness (*dx*) is *dt*.

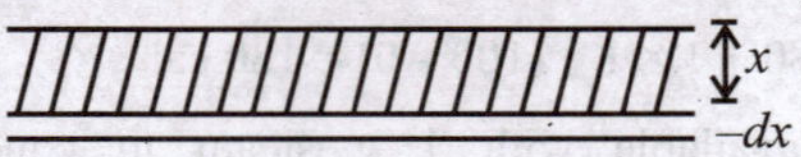

$$mL = \frac{KA[26 - 0]dt}{x}$$

$$\Rightarrow (Adx)\rho L = \frac{KA(26)dt}{x}$$

$$\text{So, } \frac{dx}{dt} = \frac{26K}{x\rho L}$$

20. **(d)** Equivalent thermal conductivity of the composite rod in parallel combination will be,

$$K = \frac{K_1A(T_1 - T_2)}{d} + \frac{K_2A(T_1 - T_2)}{d}$$

$$\frac{K_{EQ}2A(T_1 - T_2)}{d} = \frac{A(T_1 - T_2)}{d}[K_1 + K_2]$$

Hence equivalent thermal conductivities for two rods of equal area is given by

$$K_{EQ} = \frac{k_1 + k_2}{2}$$

21. **(c)** As the temperature difference $\Delta T = 10°C$ as well as the thermal resistance is same for both the cases, so thermal current or rate of heat flow will also be same for both the cases.

22. **(b)** $Q_1 = 4Q_2$ (Given)

$$\Rightarrow \frac{K_1A_1\Delta t}{L} = 4\frac{K_2A_2\Delta t}{L} \Rightarrow K_1A_1 = 4K_2A_2.$$

23. **(a)** According to condition,

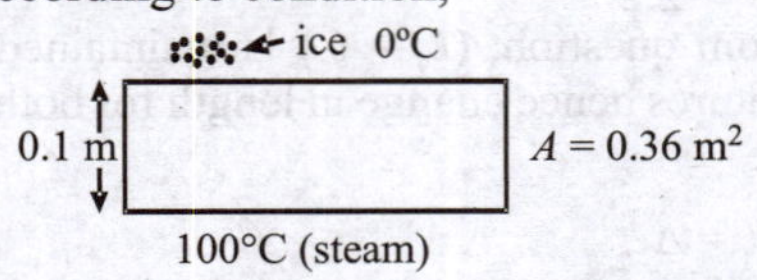

Rate of heat given by steam = Rate of heat taken by ice

where *K* = Thermal conductivity of the slab

m = Mass of the ice

L = Latent heat of melting/fusion

A = Area of the slab

$$\frac{dQ}{dt} = \frac{KA(100 - 0)}{1} = m\frac{dL}{dt}$$

$$\frac{K \times 100 \times 0.36}{0.1} = \frac{4.8 \times 3.36 \times 10^5}{60 \times 60}$$

$K = 1.24$ J/m/s/°C

NOTES In electrical conduction

$$I = \frac{dq}{dt} = \frac{V_1 - V_2}{R} = \frac{\sigma A}{\ell}(V_1 - V_2)$$

In thermal conduction,

$$H = \frac{d\theta}{dt} = \frac{\theta_1 - \theta_2}{R} = \frac{kA}{\ell}(\theta_1 - \theta_2)$$

K = Thermal conductivity of conductor.

24. **(b)** The rate of heat flow is given by

$$\frac{Q}{t} = K.\,A.\,\frac{\Delta T}{\ell}$$

Area of Original rod $A = \pi R^2$;

Area of new rod $A' = \frac{\pi R^2}{4}$

$$A' = \frac{A}{4}$$

Volume of original rod will be equal to the volume of new rod.

So, $A'\ell' = A\ell$

or,

$$\therefore \pi R^2\ell = \pi\left(\frac{R}{2}\right)^2 \ell' \Rightarrow \frac{\ell'}{\ell} = \frac{R^2}{\left(\frac{R^2}{4}\right)} = 4$$

$$\therefore \frac{Q'}{Q} = \frac{A'}{A}\cdot\frac{\ell}{\ell'} = \frac{1}{4}\cdot\frac{1}{4} = \frac{1}{16}$$

$$\therefore Q' = \frac{Q}{16}$$

25. **(c)** $\frac{dQ}{dt} = \frac{kA\,(T_1 - T_2)}{L}$

$[(T_1 - T_2)$ is the temperature difference$]$

NOTES Rate of heat flow is called heat current,

$$H = \frac{dQ}{dt} \text{ and } H = \frac{(T_1 - T_2)}{R}$$

and thermal resistance, $R = \frac{L}{kA}$

26. **(b)** From given option

(i) $r = 2r_0, l = 2l_0$

$$\therefore R = \frac{2\ell_0}{K\pi\,(2r_0)^2} = \frac{\ell_0}{2K\pi r_0^2}$$

(ii) $r = 2r_0, l = l_0$

$$\therefore R = \frac{\ell_0}{K\pi\,(2r_0)^2} = \frac{\ell_0}{4K\pi r_0^2}$$

(iii) $r = r_0, l = 2l_0$

$$\therefore R = \frac{2\ell_0}{K\pi\, r_0^2} = \frac{2\ell_0}{K\pi r_0^2}$$

(iv) $r = r_0, l = l_0$

$$\therefore R = \frac{\ell_0}{K\pi\, r_0^2} = \frac{\ell_0}{K\pi r_0^2}$$

It is clear that for option (b) resistance is minimum, hence, heat flow will be maximum.

(i) Rate of heat flow is directly proportional to area

(ii) inversely proportional to length.

$\therefore$ Heat flow will be maximum when r is maximum and ℓ is minimum.

NOTES We know that $Q = \frac{T_H - T_L}{R}$

Also, Thermal resistance $R = \frac{\ell}{KA} = \frac{\ell}{K\pi r^2}$

Heat flow will be maximum when thermal resistance is minimum.

27. **(a)** In series, equivalent thermal conductivity

$$K_{eq} = \frac{2K_1K_2}{K_1 + K_2} \quad \text{or, } K_{eq} = \frac{2\times K\times 2K}{K + 2K} = \frac{4}{3}K$$

NOTES In electricity equivalent resistance in series grouping $R_s = R_1 + R_2 + R_3 \,.....+ R_n$

In heat-thermal resistance in series $R_s = R_1 + R_2 + R_3 \,..... + R_n$

$$\frac{\ell_1 + \ell_2 + ...\ell_n}{Ks} = \frac{\ell_1}{K_1A} + \frac{\ell_2}{K_2A} + ... + \frac{\ell_n}{K_nA}$$

For n no. of rods of equal length

$$K_s = \frac{n}{\frac{1}{K_1} + \frac{1}{K_2} ..... + \frac{1}{K_n}}$$

28. **(b)** Rate of heat flow for one rod

$$= \frac{K_1A_1\,(T_1 - T_2)}{d} \quad [d \rightarrow \text{Length}]$$

Rate of heat flow for other rod

$$= \frac{K_2A_2\,(T_1 - T_2)}{d}$$

In steady state, $\frac{K_1A_1\,(T_1 - T_2)}{d}$

$$= \frac{K_2A_2\,(T_1 - T_2)}{d} \Rightarrow K_1A_1 = K_2A_2$$

29. **(b)** $Q = \frac{KA(\theta_1 - \theta_2)t}{l}$

Rate of heat flow,

$$H = \frac{Q}{t} = \frac{KA(\theta_1 - \theta_2)}{l} \quad i.e., \; H \propto \frac{A}{l}$$

Dimensions of area $A = [L^2]$, dimensions of distance $l = [L]$

$$\therefore H \propto L \Rightarrow \frac{H_2}{H_1} = \frac{L_2}{L_1} = 2 \Rightarrow H_2 = 2H_1$$

30. **(c)** Natural convection

NOTES In convection, the temperature gradient exists in the vertical direction and not in the horizontal direction. So, up and down movement of particles takes place which depends on the weight or gravity.

11 Thermodynamics

TREND ANALYSIS

TOPIC	2025		2024		2023		2022		2021		2020	
	Qns.	LOD	Qns.	LOD	Qns.	LOD	Qns.	LOD	Qns.	LOD	Qns.	LOD
Specific Heat Capacity	1	Avg.										
Thermodynamic Processes			1	Avg.			1	Avg.			1	Avg.
Carnot Engine					1	Easy						

How to Code Qns for Better & Faster Revision

Type 1: Able to solve (No code Req)
Type 2: Solved with struggle or by seeing solution **(Use code NR means Needs Revision or orange stickers)**
Type 3: Tricky/ Could not solve on my own. Need help from Teacher/ Friend. **(Use code TQ means Tricky Question or red stickers)**

You can either hand write the code in boxes or else paste coloured stickers given at end of book.

Topic-11.5

First Law of Thermodynamics

1. 1 g of water, of volume 1 cm^3 at 100°C, is converted into steam at same temperature under normal atmospheric pressure (= 1×10^5 Pa). The volume of steam formed equals 1671 cm^3. If the specific latent heat of vaporisation of water is 2256 J/g, the change in internal energy is,

NCERT | Page-230 | Odisha 2019

(a) 2256 J (b) 2423 J
(c) 2089 J (d) 167 J

2. A sample of 0.1 g of water at 100°C and normal pressure (1.013×10^5 Nm^{-2}) requires 54 cal of heat energy to convert to steam at 100°C. If the volume of the steam produced is 167.1 cc, the change in internal energy of the sample, is

NCERT | Page-230 | 2019

(a) 104.3 J (b) 208.7 J
(c) 84.5 J (d) 42.2 J

3. A system is taken from state *a* to state *c* by two paths *adc* and *abc* as shown in the figure. The internal energy at *a* is $U_a = 10$ J. Along the path *adc* the amount of heat absorbed $\delta Q_1 = 50$ J and the work done $\delta W_1 = 20$ J whereas along the path abc the heat absorbed $\delta Q_2 = 36$ J. The amount of work done along the path *abc* is

NCERT | Page-230 | Kar. 2013

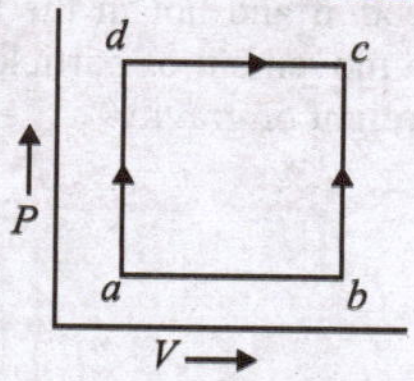

(a) 6 J (b) 10 J (c) 12 J (d) 36 J

4. An ideal gas goes from state *A* to state *B* via three different processes as indicated in the *P-V* diagram :

NCERT | Page-230 | 2012 M

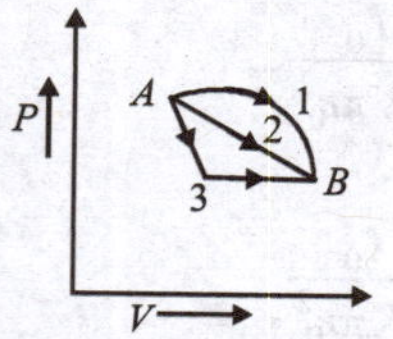

If Q_1, Q_2, Q_3 indicate the heat a absorbed by the gas along the three processes and ΔU_1, ΔU_2, ΔU_3 indicate the change in internal energy along the three processes respectively, then

(a) $Q_1 > Q_2 > Q_3$ and $\Delta U_1 = \Delta U_2 = \Delta U_3$
(b) $Q_3 > Q_2 > Q_1$ and $\Delta U_1 = \Delta U_2 = \Delta U_3$
(c) $Q_1 = Q_2 = Q_3$ and $\Delta U_1 > \Delta U_2 > \Delta U_3$
(d) $Q_3 > Q_2 > Q_1$ and $\Delta U_1 > \Delta U_2 > \Delta U_3$

5. The internal energy change in a system that has absorbed 2 kcals of heat and done 500 J of work is:

NCERT | Page-230 | 2009

(a) 6400 J (b) 5400 J
(c) 7900 J (d) 8900 J

6. 110 joules of heat is added to a gaseous system whose internal energy is 40J. Then the amount of external work done is

NCERT | Page-230 | 1993

(a) 150 J (b) 70 J (c) 110 J (d) 40 J

7. First law of thermodynamics is consequence of conservation of

NCERT | Page-230 | 1988

(a) work (b) energy
(c) heat (d) all of these

Topic-11.6

Specific Heat Capacity

8. Two gases A and B are filled at the same pressure in separate cylinders with movable pistons of radius r_A and r_B, respectively. On supplying an equal amount of heat to both the systems reversibly under constant pressure, the pistons of gas A and B are displaced by 16 cm and 9 cm, respectively. If the change in their internal energy is the same, then the ratio $\frac{r_A}{r_B}$ is equal to

NCERT | Page-231, 232 | 2025

(a) $\frac{4}{3}$ (b) $\frac{3}{4}$ (c) $\frac{2}{\sqrt{3}}$ (d) $\frac{\sqrt{3}}{2}$

Topic-11.7 Thermodynamic State Variables and Equation of State

9. The equilibrium state of a thermodynamic system is described by

NCERT | Page- 232, 233 | Ph-II 2024

A. Pressure B. Total heat
C. Temperature D. Volume
E. Work done

Choose the **most appropriate** answer from the options given below.

(a) A, B and E only (b) B, C and D only
(c) A, B and C only (d) A, C and D only

10. Which of the following is not thermodynamical function ?

NCERT | Page-230 | 1993

(a) Enthalpy (b) Work done
(c) Gibb's energy (d) Internal energy

Topic-11.8 Thermodynamic Processes

11. A thermodynamic system is taken through the cycle abcda. The work done by the gas along the path bc is:

NCERT | Page-235 | 2024

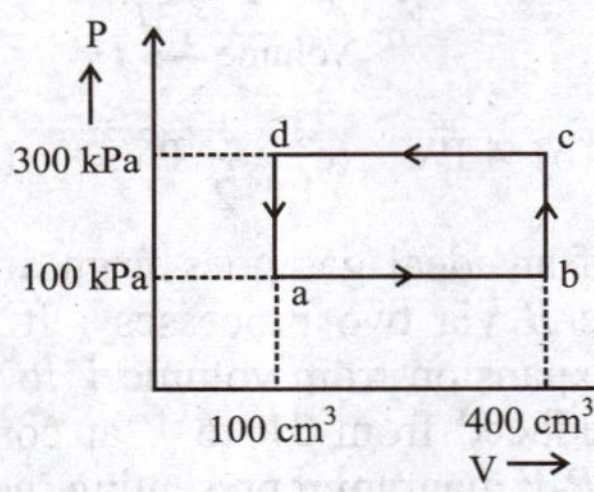

(a) Zero (b) 30 J (c) –90 J (d) –60 J

12. For the given cycle, the work done during isobaric process is:

NCERT | Page-235 | Manipur 2023

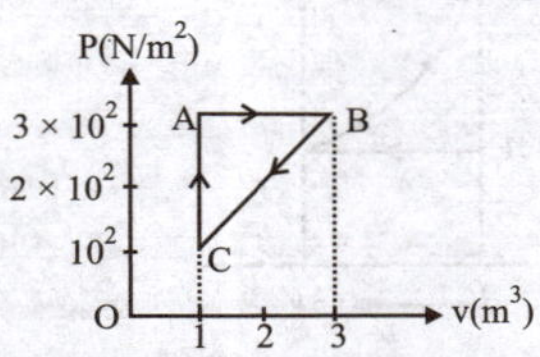

(a) 400 J (b) 600 J (c) 200 J (d) Zero

13. An ideal gas follows a process described by the equation $PV^2 = C$ from the initial (P_1, V_1, T_1) to final (P_2, V_2, T_2) thermodynamic states, where C is a constant. Then :

NCERT | Page-234-235 | RS 2022

(a) If $P_1 > P_2$ then $V_1 > V_2$ (b) If $P_1 > P_2$ then $T_1 < T_2$
(c) If $V_2 > V_1$ then $T_2 > T_1$ (d) If $V_2 > V_1$ then $T_2 < T_1$

14. An ideal gas undergoes four different processes from the same initial state as shown in the figure below. Those processes are adiabatic, isothermal, isobaric and isochoric. The curve which represents the adiabatic process among 1, 2, 3 and 4 is:

NCERT | Page-235 | 2022

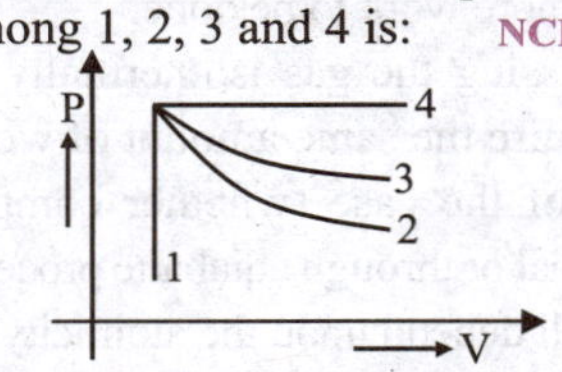

(a) 2 (b) 3 (c) 4 (d) 1

15. Two cylinders A and B of equal capacity are connected to each other via a stop cock. A contains an ideal gas at standard temperature and pressure. B is completely evacuated. The entire system is thermally insulated. The stop cock is suddenly opened. The process is :

NCERT | Page-234 | 2020

(a) adiabatic (b) isochoric
(c) isobaric (d) isothermal

16. In which of the following processes, heat is neither absorbed nor released by a system ?

NCERT | Page-234 | 2019

(a) isothermal (b) adiabatic
(c) isobaric (d) isochoric

17. The volume (V) of a monatomic gas varies with its temperature (T), as shown in the graph. The ratio of work done by the gas, to the heat absorbed by it, when it undergoes a change from state A to state B, is

NCERT | Page-235 | 2018

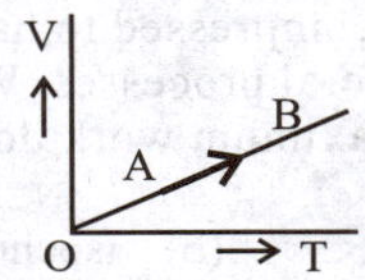

(a) $\frac{2}{5}$ (b) $\frac{2}{3}$ (c) $\frac{2}{7}$ (d) $\frac{1}{3}$

18. Thermodynamic processes are indicated in the following diagram :

NCERT | Page-235 | 2017

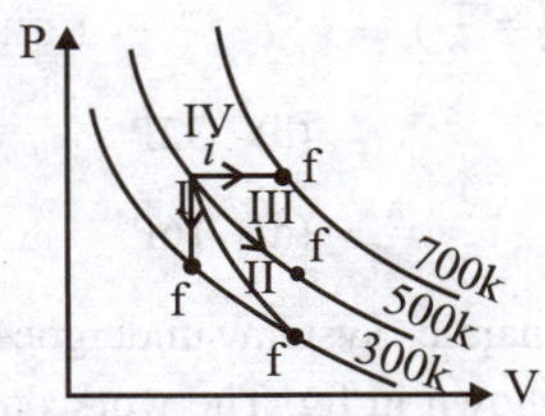

Match the following

	Column-1		Column-2
P.	Process I	A.	Adiabatic
Q.	Process II	B.	Isobaric
R.	Process III	C.	Isochoric
S.	Process IV	D.	Isothermal

(a) P → C, Q → A, R → D, S → B
(b) P → C, Q → D, R → B, S → A
(c) P → D, Q → B, R → A, S → C
(d) P → A, Q → C, R → D, S → B

19. A gas is compressed isothermally to half its initial volume. The same gas is compressed separately through an adiabatic process until its volume is again reduced to half. Then :

NCERT | Page-235 | 2016

(a) Compressing the gas isothermally will require more work to be done.
(b) Compressing the gas through adiabatic process will require more work to be done.
(c) Compressing the gas isothermally or adiabatically will require the same amount of work.
(d) Which of the case (whether compression through isothermal or through adiabatic process) requires more work will depend upon the atomicity of the gas.

20. Figure below shows two paths that may be taken by a gas to go from a state A to a state C.

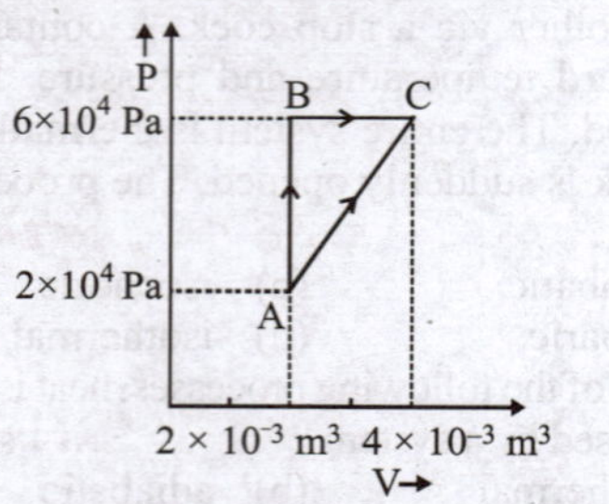

In process AB, 400 J of heat is added to the system and in process BC, 100 J of heat is added to the system. The heat absorbed by the system in the process AC will be NCERT | Page-235 | 2015

(a) 500 J (b) 460 J
(c) 300 J (d) 380 J

21. An ideal gas is compressed to half its initial volume by means of several processes. Which of the process results in the maximum work done on the gas? NCERT | Page-235 | 2015 RS

(a) Isobaric (b) Isochoric
(c) Isothermal (d) Adiabatic

22. A monoatomic gas at a pressure P, having a volume V expands isothermally to a volume 2V and then adiabatically to a volume 16V. The final pressure of the gas is : (take $\gamma = \frac{5}{3}$) NCERT | Page-235 | 2014

(a) 64P (b) 32P
(c) $\frac{P}{64}$ (d) 16P

23. A thermodynamic system undergoes cyclic process ABCDA as shown in fig. The work done by the system in the cycle is : NCERT | Page-235 | 2014

(a) P_0V_0
(b) $2P_0V_0$
(c) $\frac{P_0V_0}{2}$
(d) Zero

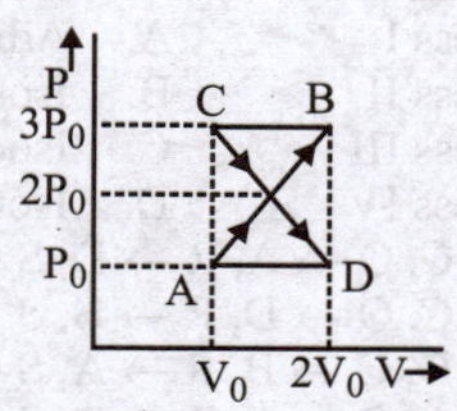

24. A gas is taken through the cycle A → B → C → A, as shown in figure. What is the net work done by the gas? NCERT | Page-235 | 2013

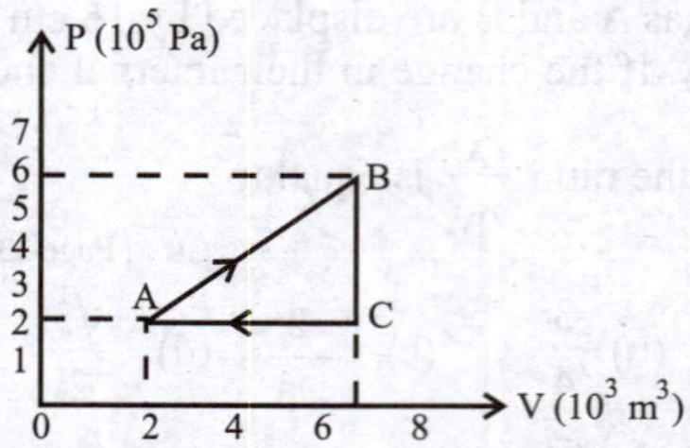

(a) 1000 J (b) zero (c) – 2000 J (d) 2000 J

25. During an adiabatic process, the pressure of a gas is found to be proportional to the cube of its temperature. The ratio of $\frac{C_p}{C_v}$ for the gas is NCERT | Page-235 | 2013

(a) 2 (b) $\frac{5}{3}$ (c) $\frac{3}{2}$ (d) $\frac{4}{3}$

26. Which of the following relations does not give the equation of an adiabatic process, where terms have their usual meaning? NCERT | Page-235 | Kar. 2013

(a) $P^{\gamma}T^{1-\gamma}$ = constant (b) $P^{1-\gamma}T^{\gamma}$ = constant
(c) PV^{γ} = constant (d) $TV^{\gamma-1}$ = constant

27. A thermodynamic system is taken through the cycle *ABCD* as shown in figure. Heat rejected by the gas during the cyclic process is : NCERT | Page-235 | 2012

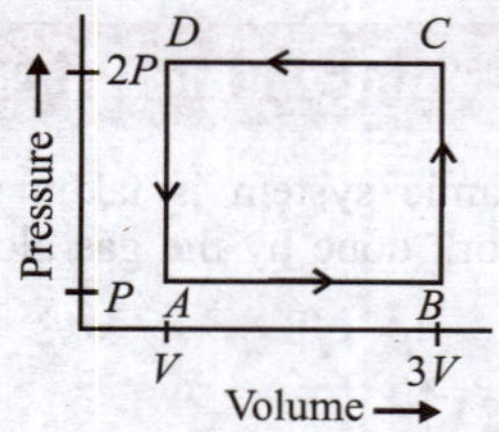

(a) 2 PV (b) 4 PV (c) $\frac{1}{2}$ PV (d) PV

28. One mole of an ideal gas goes from an initial state *A* to final state *B* via two processes : It first undergoes isothermal expansion from volume *V* to 3*V* and then its volume is reduced from 3*V* to *V* at constant pressure. The correct *P-V* diagram representing the two processes is : NCERT | Page-235 | 2012

(a)

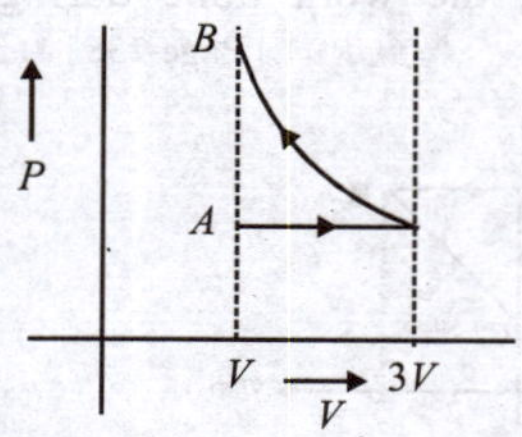

(b)

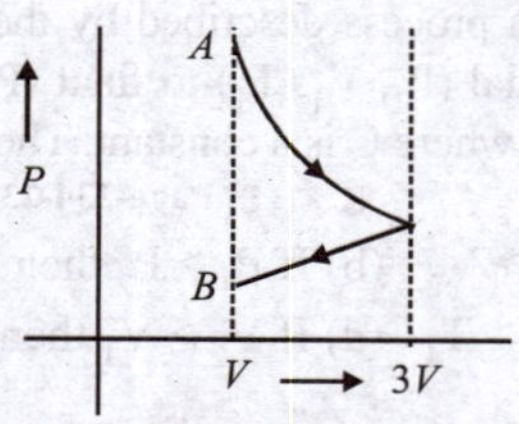

(c) 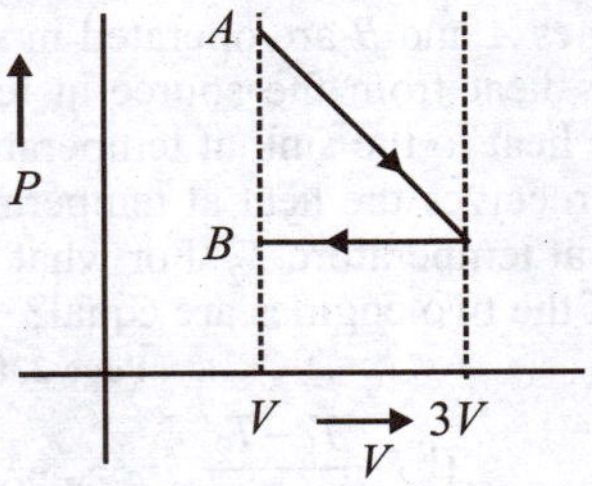

(d) 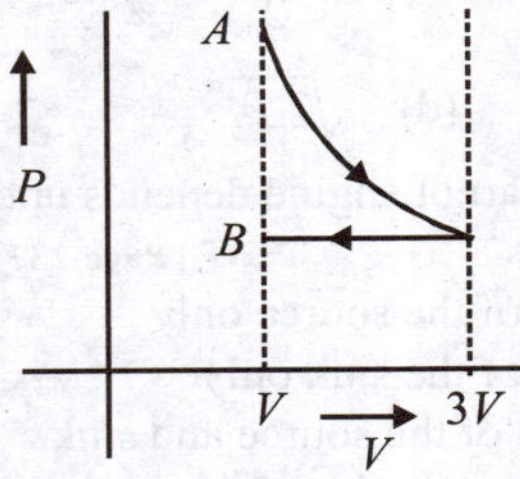

29. During an isothermal expansion, a confined ideal gas does –150 J of work against its surroundings. This implies that NCERT | Page-234 | 2011

(a) 150 J heat has been removed from the gas
(b) 300 J of heat has been added to the gas
(c) no heat is transferred because the process is isothermal
(d) 150 J of heat has been added to the gas

30. A mass of diatomic gas (γ = 1.4) at a pressure of 2 atmospheres is compressed adiabatically so that its temperature rises from 27°C to 927°C. The pressure of the gas in final state is NCERT | Page-235 | 2011 M

(a) 28 atm (b) 68.7 atm
(c) 256 atm (d) 8 atm

31. If ΔU and ΔW represent the increase in internal energy and work done by the system respectively in a thermodynamical process, which of the following is true? NCERT | Page-235 | 2010, 1998

(a) $\Delta U = -\Delta W$, in an adiabatic process
(b) $\Delta U = \Delta W$, in an isothermal process
(c) $\Delta U = \Delta W$, in an adiabatic process
(d) $\Delta U = -\Delta W$, in an isothermal process

32. In thermodynamic processes which of the following statements is not true? NCERT | Page-235 | 2009

(a) In an isochoric process pressure remains constant
(b) In an isothermal process the temperature remains constant
(c) In an adiabatic process PV^{γ} = constant
(d) In an adiabatic process the system is insulated from the surroundings

33. If Q, E and W denote respectively the heat added, change in internal energy and the work done in a closed cyclic process, then: NCERT | Page-235 | 2008

(a) W = 0 (b) Q = W = 0
(c) E = 0 (d) Q = 0

34. One mole of an ideal gas at an initial temperature of T K does 6R joules of work adiabatically. If the ratio of specific heats of this gas at constant pressure and at constant volume is 5/3, the final temperature of gas will be NCERT | Page-235 | 2004

(a) $(T-4)\,K$ (b) $(T+2.4)\,K$
(c) $(T-2.4)\,K$ (d) $(T+4)\,K$

35. An ideal gas at 27°C is compressed adiabatically to $\frac{8}{27}$ of its original volume. The rise in temperature is $\left(\gamma = \frac{5}{3}\right)$ NCERT | Page-235 | 1999

(a) 475°C (b) 402°C (c) 275°C (d) 175°C

36. If the ratio of specific heat of a gas at constant pressure to that at constant volume is γ, the change in internal energy of a mass of gas, when the volume changes from V to $2V$ at constant pressure P, is NCERT | Page-235 | 1998

(a) $\frac{R}{(\gamma-1)}$ (b) PV (c) $\frac{PV}{(\gamma-1)}$ (d) $\frac{\gamma PV}{(\gamma-1)}$

37. A sample of gas expands from volume V_1 to V_2. The amount of work done by the gas is greatest, when the expansion is NCERT | Page-235 | 1997

(a) adiabatic (b) isobaric
(c) isothermal (d) equal in all cases

38. An ideal gas undergoing adiabatic change has the following pressure-temperature relationship NCERT | Page-235 | 1996

(a) $P^{\gamma-1}T^{\gamma}$ = constant (b) $P^{\gamma}T^{\gamma-1}$ = constant
(c) $P^{\gamma}T^{1-\gamma}$ = constant (d) $P^{1-\gamma}T^{\gamma}$ = constant

39. A diatomic gas initially at 18°C is compressed adiabatically to one eighth of its original volume. The temperature after compression will be NCERT | Page-235 | 1996

(a) 18°C (b) 668.4°K
(c) 395.4°C (d) 144°C

40. An ideal gas A and a real gas B have their volumes increased from V to $2V$ under isothermal conditions. The increase in internal energy NCERT | Page-235 | 1993

(a) will be same in both A and B
(b) will be zero in both the gases
(c) of B will be more than that of A
(d) of A will be more than that of B

41. A thermodynamic system is taken from state A to B along ACB and is brought back to A along BDA as shown in the PV diagram. The net work done during the complete cycle is given by the area NCERT | Page-235 | 1992

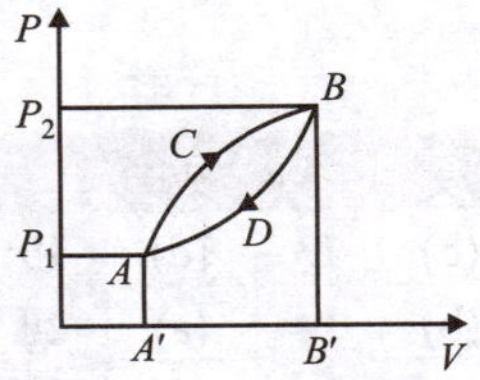

(a) $P_1ACBP_2P_1$ (b) $ACBB'A'A$
(c) $ACBDA$ (d) $ADBB'A'A$

42. A thermodynamic process is shown in the figure. The pressures and volumes corresponding to some points in the figure are

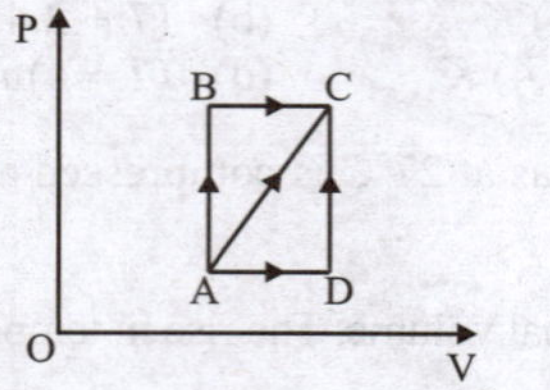

$P_A = 3 \times 10^4$ Pa $\quad V_A = 2 \times 10^{-3}$ m^3
$P_B = 8 \times 10^4$ Pa $\quad V_D = 5 \times 10^{-3}$ m^3

In process *AB*, 600 J of heat is added to the system and in process BC, 200 J of heat is added to the system. The change in internal energy of the system in process *AC* would be **NCERT | Page-235 | 1991**

(a) 560 J (b) 800 J (c) 600 J (d) 640 J

43. At 27° C a gas is compressed suddenly such that its pressure becomes (1/8) of original pressure. Final temperature will be (γ = 5/3) **NCERT | Page-235 | 1989**

(a) 420K (b) 300K (c) – 142°C (d) 327°C

Topic-11.9

Second Law of Thermodynamics

44. When 1 kg of ice at 0°C melts to water at 0°C, the resulting change in its entropy, taking latent heat of ice to be 80 cal/°C, is **NCERT | Page-236 | 2011**

(a) 273 cal/K (b) 8 × 104 cal/K
(c) 80 cal/K (d) 293 cal/K

Topic-11.10

Reversible and Irreversible Processes

45. Which of the following processes is reversible? **NCERT | Page-236 | 2005**

(a) Transfer of heat by conduction
(b) Transfer of heat by radiation
(c) Isothermal compression
(d) Electrical heating of a nichrome wire

Topic-11.11

Carnot Engine

46. A Carnot engine has an efficiency of 50% when its source is at a temperature 327°C. The temperature of the sink is **NCERT | Page-237 | 2023**

(a) 27°C (b) 15°C
(c) 100°C (d) 200°C

47. Two Carnot engines *A* and *B* are operated in series. The engine *A* receives heat from the source at temperature T_1 and rejects the heat to the sink at temperature *T*. The second engine *B* receives the heat at temperature *T* and rejects to its sink at temperature T_2. For what value of *T* the efficiencies of the two engines are equal? **NCERT | Page-238 | Kar. 2013**

(a) $\dfrac{T_1+T_2}{2}$ (b) $\dfrac{T_1-T_2}{2}$
(c) T_1T_2 (d) $\sqrt{T_1T_2}$

48. The efficiency of a Carnot engine depends upon **NCERT | Page-237 | Ph-II 2010**

(a) the temperature of the source only
(b) the temperature of the sink only
(c) the temperatures of the source and sink
(d) the volume of the cylinder of the engine

49. A Carnot engine whose sink is at 300 K has an efficiency of 40%. By how much should the temperature of source be increased so as to increase, its efficiency by 50% of original efficiency ? **NCERT | Page-238 | 2006**

(a) 325 K (b) 250 K
(c) 380 K (d) 275 K

50. A Carnot engine whose efficiency is 50% has an exhaust temperature of 500 K. If the efficiency is to be 60% with the same intake temperature, the exhaust temperature must be (in K) **NCERT | Page-238 | 2002**

(a) 800 (b) 200 (c) 400 (d) 600

51. The efficiency of a Carnot engine operating between the temperatures of 100ºC and –23ºC will be **NCERT | Page-238 | 1997**

(a) $\dfrac{100+23}{100}$ (b) $\dfrac{100-23}{100}$
(c) $\dfrac{373+250}{373}$ (d) $\dfrac{373-250}{373}$

52. An ideal carnot engine, whose efficiency is 40% receives heat at 500 K. If its efficiency is 50%, then the intake temperature for the same exhaust temperature is **NCERT | Page-238 | 1995**

(a) 600 K (b) 700 K
(c) 800 K (d) 900 K

ANSWER KEY

1	(c)	7	(b)	13	(d)	19	(b)	25	(c)	31	(a)	37	(b)	43	(c)	49	(b)		
2	(b)	8	(b)	14	(a)	20	(b)	26	(a)	32	(a)	38	(d)	44	(d)	50	(c)		
3	(a)	9	(d)	15	(a)	21	(d)	27	(a)	33	(c)	39	(b)	45	(c)	51	(d)		
4	(a)	10	(b)	16	(b)	22	(c)	28	(d)	34	(a)	40	(b)	46	(a)	52	(a)		
5	(c)	11	(a)	17	(a)	23	(d)	29	(d)	35	(b)	41	(c)	47	(d)				
6	(b)	12	(b)	18	(a)	24	(a)	30	(c)	36	(c)	42	(a)	48	(c)				

Hints & Solutions

1. (c) $\Delta Q = 2256 \times 1 = 2256$ J
$\Delta W = P[V_{steam} - V_{water}]$ [$\because dw = pdv$]
$= 1 \times 105[1671 - 1] \times 106 = 167$ J
As $\Delta Q = \Delta U + \Delta W$
$2256 = \Delta U + 167$
$\Delta U = 2089$ J

2. (b) Using first law of thermodynamics equation,
$\Delta Q = \Delta U + \Delta W$
$\Rightarrow \quad 54 \times 4.18$
$= \Delta U + 1.013 \times 10^5(167.1 \times 10^{-6} - 0)$ ($\because \Delta W = P\Delta V$)
$\Rightarrow \quad \Delta U = 208.7$ J

3. (a) From first law of thermodynamics
$$Q_{adc} = \Delta U_{adc} + W_{adc}$$
$$50\text{ J} = \Delta U_{adc} + 20\text{ J}$$
$$\Delta U_{adc} = 30\text{ J}$$
Again, $Q_{abc} = \Delta U_{abc} + W_{abc}$
$$W_{abc} = Q_{abc} - \Delta U_{abc} = Q_{abc} - \Delta U_{adc} = 36\text{ J} - 30\text{ J} = 6\text{ J}$$

4. (a) Initial and final condition is same for all process
$\Delta U_1 = \Delta U_2 = \Delta U_3$
from first law of thermodynamics
$\Delta Q = \Delta U + \Delta W$
Work done
$\Delta W_1 > \Delta W_2 > \Delta W_3$ (Area of P.V. graph)
So $\Delta Q_1 > \Delta Q_2 > \Delta Q_3$

5. (c) According to first law of thermodynamics
$Q = \Delta U + W$
$\Delta U = Q - W$
$= 2 \times 4.2 \times 1000 - 500 = 8400 - 500$
$= 7900$ J

6. (b) $\Delta Q = \Delta U + \Delta W$
$\Rightarrow \Delta W = \Delta Q - \Delta U = 110 - 40 = 70$ J

7. (b) The first law of thermodynamics is just a conservation of energy.

8. (b) According to question
Pressure is constant $P_A = P_B$
$\Delta U_A = Q - W_A$
$\Delta U_B = Q - W_B$
Given $\Delta U_A = \Delta U_B$, So $W_A = W_B$
$P_A \Delta V_A = P_B \Delta V_B$
$\left(\pi r_A^2\right) x_B = \left(\pi r_B^2\right) x_A$
$$\frac{r_A}{r_B} = \sqrt{\frac{x_B}{x_A}} = \sqrt{\frac{9}{16}} = \frac{3}{4}$$

9. (d) Equilibrium state of a thermodynamic system is given by state variables P, V and T. Work done and total heat are path variables.

10. (b) Work done is not a thermodynamical function.

NOTES The fundamental thermodynamic equations follow from five primary thermodynamic definitions and describe internal energy, enthalpy, Helmholtz energy and Gibb's energy.

11. (a) Path bc is an isochoric process i.e., volume is constant i.e., $\Delta V = 0$
$\therefore$ Work done by gas along path bc,
$W_{bc} = P. \Delta V = 0$

NOTES In Isobaric process, pressure is kept constant, in isochoric process volume is kept constant and in adiabatic process no exchange of heat $\Delta Q = 0$

12. (b) Work done in isobaric i.e., when pressure is constant (AB) process
$= P\Delta V = 3 \times 10^2$ [2] J = 600 J

In an isobaric process, P is fixed. Work done by the gas is
$W = P(V_2 - V_1) = \mu R(T_2 - T_1)$

13. (d) Given, $PV^2 = C \Rightarrow TV =$ const and $\frac{T^2}{P}$ = const
$P\uparrow \Rightarrow V\downarrow \qquad T\uparrow \Rightarrow V\downarrow \qquad T\uparrow = P\uparrow$
$P_1 > P_2 \Rightarrow V_2 > V_1$ $V_2 > V_1 \Rightarrow T_1 > T_2$ $P_1 > P_2 = T_1 > T_2$

14. (a) Curve 1 and curve 4 are isochoric and isobaric, whereas, slope of curve 3 is less than slope of curve 2. So curve 2 should be adiabatic and curve 3 should be isothermal because (slope) isothermal < (slope) adiabatic.

15. (a) Entire system is thermally insulated. So, no heat exchange with surrounding will take place. Hence, process will be adiabatic.

16. (b) In adiabatic process, there is no exchange of heat, with the surroundings.

17. (a) Gas is monatomic, so $C_p = \frac{5}{2}R$
Given process is isobaric
$\therefore \quad dQ = n\,C_p\,dT$

$\Rightarrow \quad dQ = n\left(\frac{5}{2}R\right)dT$

$dW = P\,dV = n\,RdT$

$\therefore \quad \text{Required ratio} = \frac{dW}{dQ} = \frac{nRdT}{n\left(\frac{5}{2}R\right)dT} = \frac{2}{5}$

18. **(a)** Process I volume is constant hence, it is isochoric. In process IV, pressure is constant hence, it is isobaric.

19. **(b)** W_{ext} = negative of area with volume-axis

W(adiabatic) > W(isothermal)

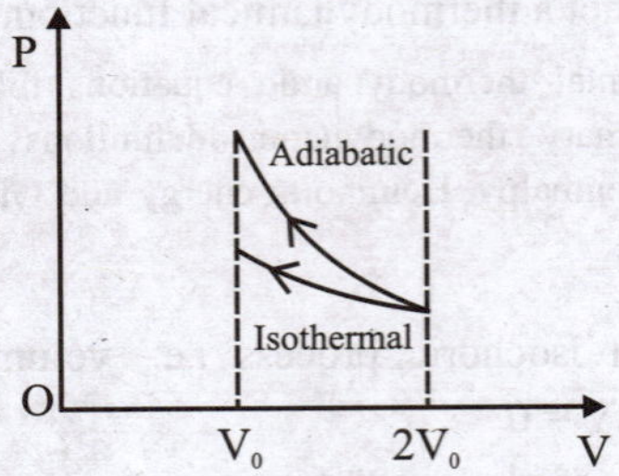

20. **(b)** In cyclic process ABCA

$Q_{cycle} = W_{cycle}$

$Q_{AB} + Q_{BC} + Q_{CA} = \text{ar. of } \Delta ABC$

$+ 400 + 100 + Q_{C\to A} = \frac{1}{2}(2 \times 10^{-3})\,(4 \times 10^{4})$

$\Rightarrow \quad Q_{C\to A} = -460 \text{ J}$

$\Rightarrow Q_{A\to C} = +460 \text{ J}$

21. **(d)**

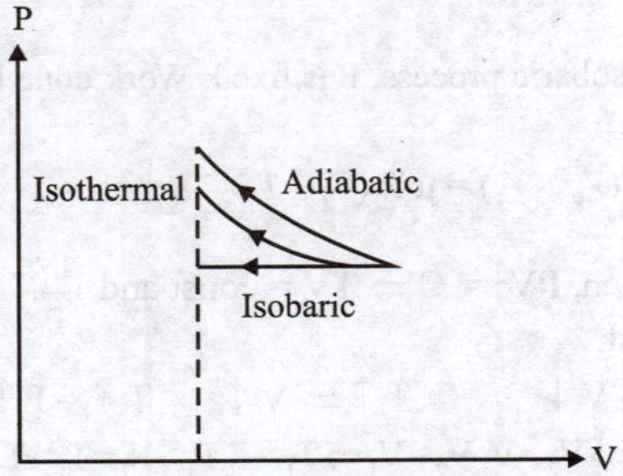

Since area under the curve is maximum for adiabatic process so, work done (W = PdV) on the gas will be maximum for adiabatic process.

22. **(c)** For isothermal process $P_1V_1 = P_2V_2$

$\Rightarrow \quad PV = P_2(2V) \Rightarrow P_2 = \frac{P}{2}$

For adiabatic process

$P_2V_2^{\gamma} = P_3V_3^{\gamma} \Rightarrow \left(\frac{P}{2}\right)(2v)^{\gamma} = P_3 16v)^{\gamma}$

$\Rightarrow P_3 = \frac{3}{2}\left(\frac{1}{8}\right)^{5/3} = \frac{P}{64}$

23. **(d)** Work done by the system in the cycle

= Area under P-V curve and V-axis

$= \frac{1}{2}(2P_0 - P_0)(2V_0 - V_0) +$

$\left[-\left(\frac{1}{2}\right)(3P_0 - 2P_0)(2V_0 - V_0)\right]$

$= \frac{P_0V_0}{2} - \frac{P_0V_0}{2} = 0$

24. **(a)** W_{net} = Area of triangle ABC

$= \frac{1}{2}\ AC \times BC$

$= \frac{1}{2} \times 5 \times 10^{-3} \times 4 \times 10^{5} = 1000 \text{ J}$

25. **(c)** According to question $P \propto T^3$

But as we know for an adiabatic process the pressure $P \propto T^{\frac{\gamma}{\gamma-1}}$.

So, $\frac{\gamma}{\gamma-1} = 3 \Rightarrow \gamma = \frac{3}{2}$ or, $\frac{C_p}{C_v} = \frac{3}{2}$

26. **(a)** Adiabatic equations of state are

PV^{γ} = constant $\quad TV^{\gamma-1}$ = constant

$P^{1-\gamma}T^{\gamma}$ = constant.

NOTES When a thermodynamic system undergoes a change in such a way that to exchange of heat takes place between system and surroundings, then the process is adiabatic process. In this process P, V and T change but $\Delta Q = 0$.

27. **(a)** $\because$ Internal energy is the state function.

$\therefore$ In cyclie process; $\Delta U = 0$

According to 1st law of thermodynamics $\Delta Q = \Delta U + W$

So heat absorbed

$\Delta Q = W$ = Area under the curve

$= -(2V)(P) = -2PV$

So heat rejected = 2PV

28. **(d)** 1st process is isothermal expansion which is only correct shown in option (d).

2nd process is isobaric compression which is correctly shown in option (d).

29. **(d)** According to question work done by gas is given –150J so that according to it answer will be (d).

30. **(c)** $T_1 = 273 + 27 = 300\text{K}$

$T_2 = 273 + 927 = 1200\text{K}$

For adiabatic process,

$P^{1-\gamma}T^{\gamma} = \text{constant} \Rightarrow P_1^{1-\gamma}T_1^{\gamma} = P_2^{1-\gamma}T_2^{\gamma}$

$\Rightarrow \left(\frac{P_2}{P_1}\right)^{1-\gamma} = \left(\frac{T_1}{T_2}\right)^{\gamma} \Rightarrow \left(\frac{P_1}{T_2}\right)^{1-\gamma} = \left(\frac{T_2}{T_1}\right)^{\gamma}$

$$\Rightarrow \left(\frac{P_1}{P_2}\right)^{1-1.4} = \left(\frac{1200}{300}\right)^{1.4} \Rightarrow \left(\frac{P_1}{P_2}\right)^{-0.4} = (4)^{1.4}$$

$$\Rightarrow \left(\frac{P_2}{P_1}\right)^{0.4} = 4^{1.4} \Rightarrow P_2 = P_1\, 4^{\left(\frac{1.4}{0.4}\right)} = P_1\, 4^{\left(\frac{7}{2}\right)}$$

$$\Rightarrow P_1\,(2^7) = 2 \times 128 = 256 \text{ atm}$$

31. **(a)** By first law of thermodynamics,

$\Delta Q = \Delta U + \Delta W$

In adiabatic process, $\Delta Q = 0$

$\therefore \quad \Delta U = -\Delta W$

In isothermal process, $\Delta U = 0 \quad \therefore \Delta Q = \Delta W$

32. **(a)** In an isochoric process volume remains constant whereas pressure remains constant in isobaric process.

33. **(c)** In a cyclic process, the initial state coincides with the final state. Hence, the change in internal energy is zero, as it depends only on the initial and final states.

NOTES In case of cyclic process, $u_f = u_i \therefore \Delta u = u_f - u_i = 0$. And also $\Delta u \propto \Delta T \therefore \Delta T = 0$

34. **(a)** $T_1 = T$, $W = 6R$ joules, $\gamma = \frac{5}{3}$

$$W = \frac{P_1V_1 - P_2V_2}{\gamma - 1} = \frac{nRT_1 - nRT_2}{\gamma - 1}$$

$$= \frac{nR(T_1 - T_2)}{\gamma - 1}$$

$$n = 1,\ T_1 = T \Rightarrow \frac{R(T - T_2)}{5/3 - 1} = 6R$$

$$\Rightarrow T_2 = (T - 4)\text{K}$$

35. **(b)** $T = 27°C = 300$ K

$$\gamma = \frac{5}{3};\quad V_2 = \frac{8}{27}V_1;\quad \frac{V_1}{V_2} = \frac{27}{8}$$

From adiabatic process we know that

$$T_1V_1^{\gamma-1} = T_2V_2^{\gamma-1}$$

$$\frac{T_2}{T_1} = \left(\frac{V_1}{V_2}\right)^{\gamma-1} = \left(\frac{27}{8}\right)^{\frac{5}{3}-1}$$

$$\frac{T_2}{T_1} = \frac{9}{4} \Rightarrow T_2 = \frac{9}{4} \times T_1 = \frac{9}{4} \times 300 = 675\,K$$

$T_2 = 675 - 273°C = 402°C$

36. **(c)** Change in internal energy is equal to work done in adiabatic system

$\Delta W = -\Delta U$ (Expansion in the system)

$$= -\frac{1}{\gamma - 1}(P_1V_1 - P_2V_2)$$

$$\Delta U = \frac{1}{1-\gamma}(P_2V_2 - P_1V_1)$$

Here, $V_1 = V, V_2 = 2V$

$$\therefore \quad \Delta U = \frac{1}{1-\gamma}[P \times 2V - PV] = \frac{PV}{1-\gamma}$$

$$\Rightarrow \Delta U = -\frac{PV}{\gamma - 1}$$

37. **(b)** In thermodynamics for same change in volume, the work done is maximum in isobaric process because in $P - V$ graph, area enclosed by curve and volume axis is maximum in isobaric process.

38. **(d)** We know that in adiabatic process,

$PV^\gamma = \text{constant}$ ....(i)

From ideal gas equation, we know that

$PV = nRT$

$V = \frac{nRT}{P}$ ....(ii)

Putting the value from equation (ii) in equation (i),

$$P\left(\frac{nRT}{P}\right)^\gamma = \text{constant}$$

$$P^{(1-\gamma)}\, T^\gamma = \text{constant}$$

39. **(b)** Initial temperature $(T_1) = 18°C = 291$ K

Let Initial volume $(V_1) = V$

Final volume $(V_2) = \frac{V}{8}$

According to adiabatic process,

$TV^{\gamma - 1} = \text{constant}$

According to question, $T_1V_1^{\gamma-1} = T_2V_2^{\gamma-1}$

$$\Rightarrow T_2 = 293\left(\frac{V_1}{V_2}\right)^{\gamma-1} \Rightarrow T_2 = 293(8)^{\frac{7}{5}-1}$$

$= 293 \times 2.297 = 668.4$K

$\left[\text{For diatomic gas } \gamma = \frac{C_p}{C_v} = \frac{7}{5}\right]$

40. **(b)** Under isothermal conditions, there is no change in internal energy.

41. **(c)** Work done = Area under curve $ACBDA$

42. **(a)** BC is isobaric process,

$\therefore \quad W = P_B \times (V_D - V_A) = 240$ J

$\Delta Q = 600 + 200 = 800$ J

Using $\Delta Q = \Delta U + \Delta W$

$\Rightarrow \Delta U = \Delta Q - \Delta W = 800 - 240 = 560$ J

NOTES Since AB is an isochoric process i.e., $\Delta V = 0$ so, no work is done.

43. (c) $T_1^{\gamma}P_1^{1-\gamma} = T_2^{\gamma}P_2^{1-\gamma}$

$$\Rightarrow \left(\frac{T_2}{T_1}\right)^{\gamma} = \left(\frac{P_1}{P_2}\right)^{-\gamma} \Rightarrow T_2 = T_1.\left(\frac{P_1}{P_2}\right)^{\frac{1-\gamma}{\gamma}}$$

$= 300 \times (8)^{-2/5} = 142°C$

44. (d) Change in entropy is given by

$$\Delta S = \frac{dQ}{T} \text{ or } \Delta S = \frac{\Delta Q}{T} = \frac{mL_f}{273}$$

$$\Delta S = \frac{1000 \times 80}{273} = 293 \text{cal / K.}$$

45. (c) Isothermal process occur very slowly so it is quasi-static and hence it is reversible.

NOTES
- The conditions for reversibility
- Must take place infinitely slowly
- The temperature of the system must not differ appreciably
- There must be complete absence of dissipative forces such as friction, viscosity, electric resistance, etc.

46. (a) Given, efficiency $\eta = \frac{50}{100} = \frac{1}{2}$

Source temperature, $T_1 = 327 + 273 = 600$ K

Efficiency of Carnot engine $\eta = 1 - \frac{T_2}{T_1}$

$$\Rightarrow \eta = 1 - \frac{T_2}{600} \Rightarrow \frac{1}{2} = 1 - \frac{T_2}{600}$$

$$\frac{T_2}{600} = \frac{1}{2} \Rightarrow T_2 = 300 \text{ K}$$

$\therefore$ Sink temperature, $T_2 = 300 - 273 = 27°C$

47. (d) Efficiency of engine A, $\eta_1 = 1 - \frac{T}{T_1}$,

Efficiency of engine B, $\eta_2 = 1 - \frac{T_2}{T}$

Here, $\eta_1 = \eta_2$

$$\therefore \frac{T}{T_1} = \frac{T_2}{T} \Rightarrow T = \sqrt{T_1 T_2}$$

48. (c) Efficiency of Carnot's heat engine is

$\eta = $ ——

where, T_2 = temperature of sink and T_1 = temperature of source.

Hence, η depends upon the temperature of source and sink both.

49. (b) We know that efficiency of Carnot Engine

$$\eta = \frac{T_1 - T_2}{T_1}$$

where, T_1 is temp. of source and T_2 is temp. of sink

$$\therefore \quad 0.40 = \frac{T_1 - 300}{T_1} \Rightarrow T_1 - 300 = 0.40T_1$$

$$0.6T_1 = 300 \Rightarrow T_1 = \frac{300}{.6} = \frac{3000}{6} = 500K$$

Now efficiency to be increased by 50%

$$\therefore \quad 0.60 = \frac{T_1 - 300}{T_1} \Rightarrow T_1 - 300 = 0.6T_1$$

$$0.4T_1 = 300 \Rightarrow T_1 = \frac{300}{.4} = \frac{300 \times 10}{4} = 750$$

Increase in temp = 750 – 500 = 250 K

50. (c) $\eta = 1 - \frac{T_2}{T_1}$ or $\frac{50}{100} = 1 - \frac{500}{T_1}$

$\Rightarrow T_1 = 1000\ K$

Also, $\frac{60}{100} = 1 - \frac{T_2}{1000} \Rightarrow T_2 = 400\ K$

51. (d) $\eta = 1 - \frac{T_1}{T_2}$

$T_1 = -23°C = 250$ K, $T_2 = 100°C = 373$K

$$\eta = 1 - \frac{250}{373} = \frac{373 - 250}{373}$$

52. (a) Efficiency of carnot engine (η_1) = 40% = 0.4; Initial intake temperature (T_1) = 500K and new efficiency (η_2) = 50% = 0.5.

Efficiency $(\eta) = 1 - \frac{T_2}{T_1}$ or $\frac{T_2}{T_1} = 1 - \eta$

Therefore in first case, $\frac{T_2}{500} = 1 - 0.4 = 0.6$

$\Rightarrow T_2 = 0.6 \times 500 = 300$K

And in second case, $\frac{300}{T_1} = 1 - 0.5 = 0.5$

$$\Rightarrow T_1 = \frac{300}{0.5} = 600\ K$$

12 Kinetic Theory

TREND ANALYSIS

TOPIC	2025		2024		2023		2022		2021		2020	
	Qns.	LOD	Qns.	LOD	Qns.	LOD	Qns.	LOD	Qns.	LOD	Qns.	LOD
Behaviour of Gases & Gas Laws	2	Avg.	1	Avg.			1	Avg.			1	Avg.
Kinetic Theory of an Ideal Gas					1	Easy						
Specific Heat Capacity									1	Diffi.		.
Mean Free Path & Gas Laws											2	Avg.

How to Code Qns for Better & Faster Revision

Type 1: Able to solve (No code Req)
Type 2: Solved with struggle or by seeing solution **(Use code NR means Needs Revision or orange stickers)**
Type 3: Tricky/ Could not solve on my own. Need help from Teacher/ Friend. **(Use code TQ means Tricky Question or red stickers)**

You can either hand write the code in boxes or else paste coloured stickers given at end of book.

Topic-12.3

Behaviour of Gases & Gas Laws

1. An oxygen cylinder of volume 30 litre has 18.20 moles of oxygen. After some oxygen is withdrawn from the cylinder, its gauge pressure drops to 11 atmospheric pressure at temperature 27°C. The mass of the oxygen withdrawn from the cylinder is nearly equal to: **NCERT | Page-246 | 2025**

[Given, $R = \frac{100}{12}$ J mol^{-1} K^{-1} and molecular mass of O_2 = 32, 1 atm pressure = 1.01×10^5 N/m]

(a) 0.125 kg (b) 0.144 kg
(c) 0.116 kg (d) 0.156 kg

2. A container has two chambers of volumes V_1 = 2 litres and V_2 = 3 litres separated by a partition made of a thermal insulator. The chambers contains n_1 = 5 and n_2 = 4 moles of ideal gas at pressures p_1 = 1 atm and p_2 = 2 atm, respectively. When the partition is removed, the mixture attains an equilibrium pressure of : **NCERT | Page-246, 247 | 2025**

(a) 1.3 atm (b) 1.6 atm
(c) 1.4 atm (d) 1.8 atm

3. The following graph represents the T-V curves of an ideal gas (where T is the temperature and V the volume) at three pressures P_1, P_2 and P_3 compared with those of Charles's law represented as dotted lines. **NCERT | Page-246 | 2024**

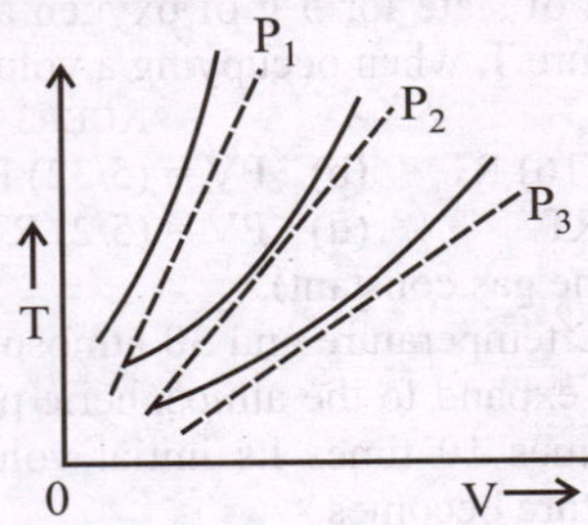

Then the correct relation is:

(a) $P_3 > P_2 > P_1$ (b) $P_1 > P_3 > P_2$
(c) $P_2 > P_1 > P_3$ (d) $P_1 > P_2 > P_3$

4. A container of volume 200 cm^3 contains 0.2 mole of hydrogen gas and 0.3 mole of argon gas. The pressure of the system at temperature 200 K (R = 8.3 J K^{-1} mol^{-1}) will be **NCERT | Page-246 | Manipur 2023**

(a) 4.15×10^5 Pa (b) 4.15×10^6 Pa
(c) 6.15×10^5 Pa (d) 6.15×10^4 Pa

5. The volume occupied by the molecules contained in 4.5 kg water of STP, if the intermolecular forces vanish away is: **NCERT | Page-246 | 2022**

(a) 5.6×10^3 m^3 (b) 5.6×10^{-3} m^3
(c) 5.6 m^3 (d) 5.6×10^6 m^3

6. A cylinder contains hydrogen gas at pressure of 249 kPa and temperature 27°C.
Its density is : (R = 8.3 J mol^{-1} K^{-1}) **NCERT | Page-246 | 2020**

(a) 0.2 kg/m^3 (b) 0.1 kg/m^3
(c) 0.02 kg/m^3 (d) 0.5 kg/m^3

7. Two vessels separately contain two ideal gases A and B at the same temperature. The pressure of A being twice that of B. Under such conditions, the density of A is found to be 1.5 times the density of B. The ratio of molecular weight of A and B is: **NCERT | Page-246 | 2015 RS**

(a) $\frac{3}{4}$ (b) 2 (c) $\frac{1}{2}$ (d) $\frac{2}{3}$

8. In the given (V – T) diagram, what is the relation between pressure P_1 and P_2 ? **NCERT | Page-247 | 2013**

(a) $P_2 > P_1$

(b) $P_2 < P_1$

(c) Cannot be predicted

(d) $P_2 = P_1$

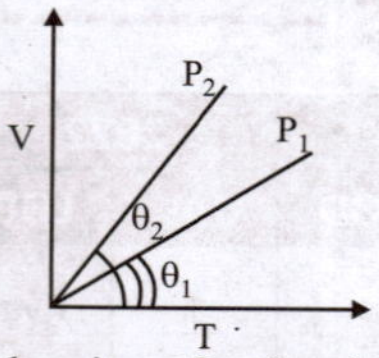

9. At 10° C the value of the density of a fixed mass of an ideal gas divided by its pressure is x. At 110°C this ratio is : **NCERT | Page-247 | 2008**

(a) x (b) $\frac{383}{283}x$ (c) $\frac{10}{110}x$ (d) $\frac{283}{383}x$

10. When volume of system is increased twice and temperature is decreased half of its initial temperature, then pressure becomes **NCERT | Page-246 | 2006**

(a) 2 times (b) 4 times

(c) $\frac{1}{2}$ times (d) $\frac{1}{4}$ times

11. The equation of state for 5 g of oxygen at a pressure P and temperature T, when occupying a volume V, will be **NCERT | Page-246 | 2004**

(a) PV = (5/16) RT (b) PV = (5/32) RT

(c) PV = 5 RT (d) PV = (5/2) RT

(where R is the gas constant).

12. A gas at 27ºC temperature and 30 atmospheric pressure is allowed to expand to the atmospheric pressure. If the volume becomes 10 times its initial volume, then the final temperature becomes **2001**

(a) 100ºC (b) 173ºC

(c) 273ºC (d) –173ºC

13. The equation of state, corresponding to 8g of O_2 is **NCERT | Page-246 | 1994**

(a) $PV = 8\,RT$ (b) $PV = RT/4$

(c) $PV = RT$ (d) $PV = RT/2$

14. Three containers of the same volume contain three different gases. The masses of the molecules are m_1, m_2 and m_3 and the number of molecules in their respective containers are N_1, N_2 and N_3. The gas pressure in the containers are P_1, P_2 and P_3 respectively. All the gases are now mixed and put in one of these containers. The pressure P of the mixture will be **NCERT | Page-250 | 1991**

(a) $P < (P_1 + P_2 + P_3)$ (b) $P = \frac{P_1 + P_2 + P_3}{3}$

(c) $P = P_1 + P_2 + P_3$ (d) $P > (P_1 + P_2 + P_3)$

15. Two containers A and B are partly filled with water and closed. The volume of A is twice that of B and it contains half the amount of water in B. If both are at the same temperature, the water vapour in the containers will have pressure in the ratio of **NCERT | Page-250 | 1988**

(a) 1 : 2 (b) 1 : 1

(c) 2 : 1 (d) 4 : 1

Topic-12.4

Kinetic Theory of an Ideal Gas

16. The temperature of a gas is –50°C. To what temperature the gas should be heated so that the rms speed is increased by 3 times? **NCERT | Page-250 | 2023**

(a) 669°C (b) 3295°C

(c) 3097 K (d) 223 K

17. Three vessels of equal capacity have gases at the same temperature and pressure. The first vessel contains helium (monoatomic), the second contains fluorine (diatomic) and the third contains sulfur hexafluoride (polyatomic). The correct statement, among the following is: **NCERT | Page-250 | RS 2022**

(a) The root mean square speed of sulfur hexafluoride is the largest

(b) All vessels contain unequal number of respective molecules

(c) The root mean square speed of molecules is same in all three cases

(d) The root mean square speed of helium is the largest

18. At what temperature will the rms speed of oxygen molecules become just sufficient for escaping from the Earth's atmosphere?

(Given :

Mass of oxygen molecule (m) = 2.76×10^{-26} kg

Boltzmann's constant $k_B = 1.38 \times 10^{-23}$ JK^{-1}) **NCERT | Page-250 | 2018**

(a) 2.508×10^4 K (b) 8.360×10^4 K

(c) 1.254×10^4 K (d) 5.016×10^4 K

19. The molecules of a given mass of a gas have r.m.s. velocity of 200 ms^{-1} at 27°C and 1.0×10^5 Nm^{-2} pressure. When the temperature and pressure of the gas are respectively, 127°C and 0.05×10^5 Nm^{-2}, the r.m.s. velocity of its molecules in ms^{-1} is : **NCERT | Page-250 | 2016**

(a) $100\sqrt{2}$ (b) $\frac{400}{\sqrt{3}}$

(c) $\frac{100\sqrt{2}}{3}$ (d) $\frac{100}{3}$

20. In a vessel, the gas is at a pressure P. If the mass of all the molecules is halved and their speed is doubled, then the resultant pressure will be **NCERT | Page-249 | Kar. 2013**

(a) $4P$ (b) $2P$ (c) P (d) $P/2$

21. At 0 K, which of the following properties of a gas will be zero? **NCERT | Page-250 | 1996**

(a) Kinetic energy (b) Potential energy

(c) Density (d) Mass

22. The temperature of a gas is raised from 27°C to 927°C. The root mean square speed is **NCERT | Page-250 | 1994**

(a) $\sqrt{(927/27)}$ times the earlier value

(b) remain the same

(c) gets halved

(d) get doubled

23. Relation between pressure (P) and energy (E) density of a gas is **NCERT | Page-250 | 1991**

(a) $P=\frac{2}{3}E$ (b) $P=\frac{1}{3}E$

(c) $P=\frac{1}{2}E$ (d) $P=3E$

24. N molecules each of mass m of a gas A and 2N molecules each of mass 2m of gas B are contained in the same vessel which is maintained at temperature T. The mean square velocity of molecules of B type is v^2 and the mean square rectangular component of the velocity of A type is denoted by ω^2. Then ω^2/v^2 **NCERT | Page-250 | 1991**

(a) 2 (b) 1 (c) 1/3 (d) 2/3

25. According to kinetic theory of gases, at absolute zero temperature **NCERT | Page-247 | 1990**

(a) water freezes

(b) liquid helium freezes

(c) molecular motion stops

(d) liquid hydrogen freezes

26. At constant volume, temperature is increased then **NCERT | Page-247 | 1989**

(a) collision on walls will be less

(b) number of collisions per unit time will increase

(c) collisions will be in straight lines

(d) collisions will not change.

Topic-12.6

Specific Heat Capacity

27. According to the law of equipartition of energy, the number of vibrational modes of a polyatomic gas of constant $\gamma=\frac{C_p}{C_v}$ is (C_p where C_v are the specific heat capacities of the gas at constant pressure and constant volume, respectively): **NCERT | Page-252, 253 | Ph-II 2024**

(a) $\frac{4+3\gamma}{\gamma-1}$ (b) $\frac{3+4\gamma}{\gamma-1}$

(c) $\frac{4-3\gamma}{\gamma-1}$ (d) $\frac{3-4\gamma}{\gamma-1}$

28. Match Column - I and Column - II and choose the correct match from the given choices. **NCERT | Page-249, 250 | 2021**

Column - I	Column - II
(A) Root mean square speed of gas molecules	(i) $\frac{1}{3}nm\bar{v}^2$
(B) Pressure exerted by ideal gas	(ii) $\sqrt{\frac{3RT}{M}}$
(C) Average kinetic energy of a molecule	(iii) $\frac{5}{2}RT$
(D) Total internal energy of 1 mole of a diatomic gas	(iv) $\frac{3}{2}k_BT$

(a) A - (iii), B - (ii), C - (i), D - (iv)

(b) A - (iii), B - (i), C - (iv), D - (ii)

(c) A - (ii), B - (iii), C - (iv), D - (i)

(d) A - (ii), B - (i), C - (iv), D - (iii)

29. Increase in temperature of a gas filled in a container would lead to: **NCERT | Page-250 | 2019**

(a) increase in its mass

(b) increase in its kinetic energy

(c) decrease in its pressure

(d) decrease in intermolecular distance

30. The value of $\gamma=\left(\frac{C_p}{C_v}\right)$, for hydrogen, helium and another ideal diatomic gas X (whose molecules are not rigid but have an additional vibrational mode), are respectively equal to, **NCERT | Page-253 | Odisha 2019**

(a) $\frac{7}{5},\frac{5}{3},\frac{7}{5}$ (b) $\frac{7}{5},\frac{5}{3},\frac{9}{7}$

(c) $\frac{5}{3},\frac{7}{5},\frac{9}{7}$ (d) $\frac{5}{3},\frac{7}{5},\frac{7}{5}$

31. A gas mixture consists of 2 moles of O_2 and 4 moles of Ar at temperature T. Neglecting all vibrational modes, the total internal energy of the system is :- **NCERT | Page-253 | 2017**

(a) 15 RT (b) 9 RT

(c) 11 RT (d) 4 RT

32. One mole of an ideal diatomic gas undergoes a transition from A to B along a path AB as shown in the figure.

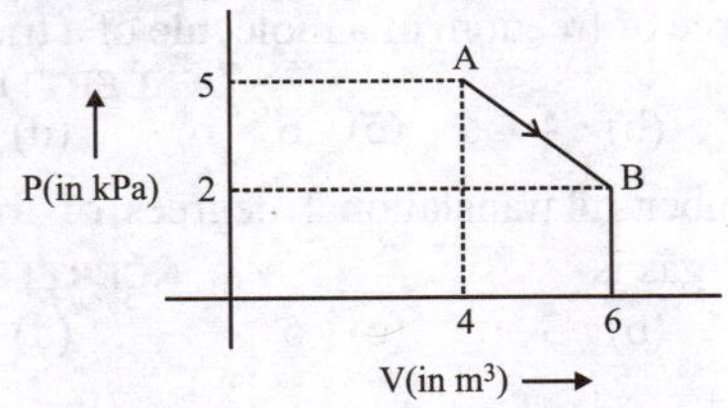

The change in internal energy of the gas during the transition is: **NCERT | Page-253 | 2015**

(a) – 20 kJ (b) 20 J (c) –12 kJ (d) 20 kJ

33. The ratio of the specific heats $\frac{C_p}{C_v}=\gamma$ in terms of degrees of freedom (n) is given by **NCERT | Page-253 | 2015**

(a) $\left(1+\frac{n}{3}\right)$ (b) $\left(1+\frac{2}{n}\right)$

(c) $\left(1+\frac{n}{2}\right)$ (d) $\left(1+\frac{1}{n}\right)$

34. The amount of heat energy required to raise the temperature of 1g of Helium at NTP, from T_1K to T_2K is **NCERT | Page-253 | 2013**

(a) $\frac{3}{2}N_ak_B(T_2-T_1)$ (b) $\frac{3}{4}N_ak_B(T_2-T_1)$

(c) $\frac{3}{4}N_ak_B\frac{T_2}{T_1}$ (d) $\frac{3}{8}N_ak_B(T_2-T_1)$

35. The molar specific heats of an ideal gas at constant pressure and volume are denoted by C_p and C_v, respectively. If $\gamma = \frac{C_p}{C_v}$ and R is the universal gas constant, then C_v is equal to **NCERT | Page-253 | 2013**

(a) $\frac{R}{(\gamma-1)}$ (b) $\frac{(\gamma-1)}{R}$

(c) γR (d) $\frac{1+\gamma}{1-\gamma}$

36. If c_p and c_v denote the specific heats (per unit mass) of an ideal gas of molecular weight M **NCERT | Page-253 | 2010 M**

(a) $C_p - C_v = R/M^2$ (b) $C_p - C_v = R$

(c) $C_p - C_v = R/M$ (d) $C_p - C_v = MR$

Where R is the molar gas constant

37. The molar specific heat at constant pressure of an ideal gas is (7/2) R. The ratio of specific heat at constant pressure to that at constant volume is **NCERT | Page-253 | 2006**

(a) 8/7 (b) 5/7 (c) 9/7 (d) 7/5

38. If γ be the ratio of specific heats of a perfect gas, the number of degrees of freedom of a molecule of the gas is **NCERT | Page-253 | 2000**

(a) $\frac{25}{2}(\gamma-1)$ (b) $\frac{3\gamma-1}{2\gamma-1}$

(c) $\frac{2}{\gamma-1}$ (d) $\frac{9}{2}(\gamma-1)$

39. The degree of freedom of a molecule of a triatomic gas is **NCERT | Page-253 | 1999**

(a) 2 (b) 4 (c) 6 (d) 8

40. The number of translational degrees of freedom for a diatomic gas is **NCERT | Page-253 | 1993**

(a) 2 (b) 3 (c) 5 (d) 6

41. If for a gas, $\frac{R}{C_v} = 0.67$, the gas is made up of molecules which are **NCERT | Page-253 | 1992**

(a) diatomic
(b) mixture of diatomic and polyatomic molecules
(c) monoatomic
(d) polyatomic

42. For hydrogen gas, $C_p - C_v = a$ and for oxygen gas, $C_p - C_v = b$, so the relation between a and b is given by **NCERT | Page-253 | 1991**

(a) $a = 16\,b$ (b) $16\,b = a$

(c) $a = 4\,b$ (d) $a = b$

43. For a certain gas the ratio of specific heats is given to be $\gamma = 1.5$. For this gas **NCERT | Page-253 | 1990**

(a) $C_V = 3R/J$ (b) $C_P = 3R/J$

(c) $C_P = 5R/J$ (d) $C_V = 5R/J$

44. One mole of an ideal monoatomic gas requires 207 J heat to raise the temperature by 10 K when heated at constant pressure. If the same gas is heated at constant volume to raise the temperature by the same 10 K, the heat required is [Given the gas constant $R = 8.3$ J/ mol. K] **NCERT | Page-253 | 1990**

(a) 198.7 J (b) 29 J

(c) 215.3 J (d) 124 J

45. A polyatomic gas with n degrees of freedom has a mean energy per molecule given by **NCERT | Page-253 | 1989**

(a) $\frac{nkT}{N}$ (b) $\frac{nkT}{2N}$ (c) $\frac{nkT}{2}$ (d) $\frac{3kT}{2}$

Topic-12.7 Mean Free Path

46. The Mean Free Path λ for a gas molecule depends upon diameter, d of the molecule as **NCERT | Page-255 | Ph-II 2020**

(a) $\lambda \propto \frac{1}{d}$ (b) $\lambda \propto \frac{1}{d^2}$

(c) $\lambda \propto d$ (d) $\lambda \propto d^2$

47. The mean free path for a gas, with molecular diameter d and number density n can be expressed as : **NCERT | Page-255 | 2020**

(a) $\frac{1}{\sqrt{2}n\pi d^2}$ (b) $\frac{1}{\sqrt{2}n^2\pi d^2}$

(c) $\frac{1}{\sqrt{2}n^2\pi^2 d^2}$ (d) $\frac{1}{\sqrt{2}n\pi d}$

48. The mean free path of molecules of a gas, (radius 'r') is inversely proportional to : **NCERT | Page-255 | 2014**

(a) r^3 (b) r^2 (c) r (d) $\sqrt{r}$

ANSWER KEY

1	(c)	6	(a)	11	(b)	16	(b)	21	(a)	26	(b)	31	(c)	36	(c)	41	(c)	46	(b)
2	(b)	7	(a)	12	(d)	17	(d)	22	(d)	27	(c)	32	(a)	37	(d)	42	(d)	47	(a)
3	(d)	8	(b)	13	(b)	18	(b)	23	(a)	28	(d)	33	(b)	38	(c)	43	(b)	48	(b)
4	(b)	9	(d)	14	(c)	19	(b)	24	(d)	29	(b)	34	(d)	39	(c)	44	(d)		
5	(c)	10	(d)	15	(b)	20	(b)	25	(c)	30	(b)	35	(a)	40	(b)	45	(c)		

Hints & Solutions

1. **(c)** Given, $V = 30L = 30 \times 10^{-3} m^3$,

$n_i = 18.20$ moles,

$P_f = 12$ atm

$T = 27°C = 300k$, $M_{O_2} = 32$ kg

$\therefore$ Final moles of O_2 is

$$n_f = \frac{P_f V}{RT} = \frac{1.2 \times 1.01 \times 10^5 \times 30 \times 10^{-3}}{\left(\frac{100}{12}\right) \times 300}$$

$= 14.544$ moles

$\therefore \Delta n = n_i - n_f = 18.2 - 14.544 = 3.656$ moles

$\therefore$ Mass removed, $\Delta m = (\Delta n)\ (M_{O_2})$

$$= 3.656 \times \frac{32}{1000} \text{ kg} = 0.116 \text{ kg}$$

2. **(b)** Given : $V_1 = 2L$, $n_1 = 5$ mol, $p_1 = 1$ atm

$V_2 = 3L$, $n_2 = 4$ mol, $p_2 = 2$ atm

Using the principle of total pressure after mixing :

$$p_{\text{final}} = \frac{P_1V_1 + P_2V_2}{V_1 + V_2}$$

$$p_{\text{final}} = \frac{1 \times 2 + 2 \times 3}{2+3} = \frac{2+6}{5} = \frac{8}{5}$$

$p_{\text{final}} = 1.6$ ATM

3. **(d)** Using Charle's law, $T \propto V$

$\therefore \quad T = \text{Slope} \times V$ ... (i)

From ideal gas equation,

$P^V = nRT \Rightarrow T = \left(\frac{P}{nR}\right) V$ ... (ii)

From eqs (i) and (ii), we get, slope $\propto P$

As, $(\text{slope})_1 > (\text{slope})_2 > (\text{slope})_3$

$\Rightarrow \quad P_1 > P_2 > P_3$

4. **(b)** We know that

$PV = nRT$

$$P_{net} = P_1 + P_2 = \frac{n_1}{V}RT + = \frac{n_2}{V}RT$$

$$= (n_1 + n_2)\frac{RT}{V}$$

$$= (0.2 + 0.3) \times \frac{8.3 \times 200}{200 \times 10^{-6}}$$

$= 8.3 \times 0.5 \times 10^6 = 4.15 \times 10^6$ Pa

5. **(c)** Number of moles of water molecules (n)

$$= \frac{\text{Mass of water}}{\text{Molecular weight}} = \frac{4.5}{18 \times 10^{-3}} = 250$$

$\because \quad PV = nRT$

$$V = \frac{nRT}{P} = \frac{250 \times 8.314 \times 273}{10^5} = 5.6 \text{ m}^3$$

[at STP, $T = 273K$, $P = 10^5$ Pa]

6. **(a)** Given : Pressure $P = 249\ k\ Pa$

$= 249 \times 10^3$ N/m²

Mass of hydrogen, $M = 2g = 2 \times 10^{-3}$ kg

Temperature, $T = 27 + 273 = 300$ K

Using, ideal gas equation, $PV = nRT$

$$PM = \rho RT \Rightarrow \rho = \frac{PM}{RT} \quad \left(\because \rho = \frac{M}{V}\right)$$

$$\therefore \rho = \frac{(249 \times 10^3)(2 \times 10^{-3})}{8.3 \times 300} = 0.2 \text{ kg/m}^3.$$

7. **(a)** From $PV = nRT$

$$P_A = \frac{\rho_A M_A}{RT} \text{ and } P_B = \frac{\rho_B M_B}{RT}$$

From question,

$$\frac{P_A}{P_B} = \frac{\rho_A}{\rho_B}\frac{M_A}{M_B} = 2\frac{M_A}{M_B} = \frac{3}{2} \text{ So, } \frac{M_A}{M_B} = \frac{3}{4}$$

8. **(b)** $P_1 > P_2$

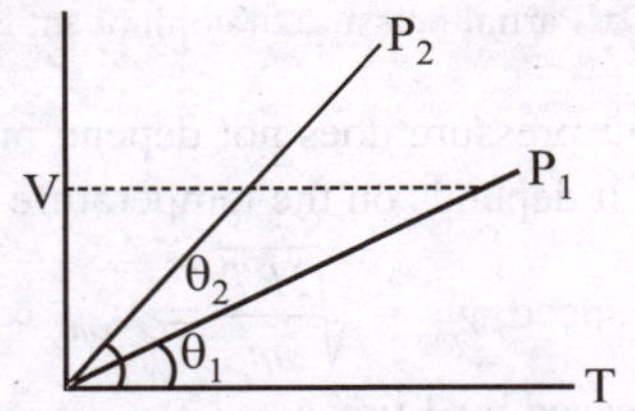

As V = constant $\Rightarrow P \propto T$

Hence from V–T graph $P_1 > P_2$

> **NOTES** From the Gay - Lussac's law, volume remain constant, the pressure of a given mass of a gas is directly proportional to its absolute temperature.
>
> $\therefore \ P \propto T$
>
> $\frac{P}{T} = \text{constant} \Rightarrow \frac{P_1}{T_1} = \frac{P_2}{T_2}$

9. **(d)** Let the mass of the gas be m.

At a fixed temperature and pressure, volume is fixed.

Density of the gas, $r = \frac{m}{V}$

Now $\frac{r}{P} = \frac{m}{PV} = \frac{m}{nRT}$

$\Rightarrow \frac{m}{nRT} = x$ (By question)

$\Rightarrow xT = \text{constant} \Rightarrow x_1T_1 = x_2T_2$

$\Rightarrow x_2 \Rightarrow \frac{x_1T_1}{T_2} = \frac{283}{383}x \quad \begin{bmatrix} \text{Given} & & \\ T_1 & = & 283K \\ T_2 & = & 383K \end{bmatrix}$

10. **(d)** $\frac{PV}{T} = \text{constant}$

11. **(b)** $n = \frac{5}{32} \quad \therefore PV = \frac{5}{32}RT \;[\because PV = nRT]$

12. **(d)** Given : Initial temperature of gas
$(T_1) = 27°C = 300$ K
Initial pressure $(P_1) = 30$ atm
Initial volume $(V_1) = V$
Final pressure $(P_2) = 1$ atm
Final volume $(V_2) = 10\,V$.
We know from the general gas equation that
$\frac{P_1V_1}{T_1} = \frac{P_2V_2}{T_2}$ or, $\frac{30 \times V}{300} = \frac{1 \times 10V}{T_2}$
or, $T_2 = 100\text{ K} = -173°C$.

13. **(b)** 8g of oxygen is equivalent to (1/4) mole.
So, $PV = \frac{1}{4}RT$

14. **(c)** According to Dalton's law of partial pressures, we have $P = P_1 + P_2 + P_3$

NOTES Dalton's law will be valid only if the gases contained are non reacting.
For n non-reacting gases, $P = P_1 + P_2 + P_3 + ... P_n$
Here P = Pressure exerted by mixture
$P_1, P_2, P_3 \,.....\, P_n$ = Partial pressure of component gas.

15. **(b)** Vapour pressure does not depend on the amount of substance. It depends on the temperature alone.

16. **(b)** RMS speed, $v_{rms} = \sqrt{\frac{3RT}{m}} \Rightarrow v_{rms} \propto \sqrt{T}$
v_{rms} is increased by 3 times,
$T_1 = 273 - 50 = 223$ K and $T_2 = ?$
So, final rms speed $= v + 3v = 4v$

$\frac{v_1}{v_2} = \sqrt{\frac{T_1}{T_2}} \Rightarrow \frac{v}{4v} = \sqrt{\frac{223}{T_2}} \Rightarrow \frac{1}{16} = \frac{223}{T_2}$

$\therefore \quad T_2 = 3568$ K or, $T_2 = 3568 - 273 = 3295°C$

17. **(d)** Root mean square speed, $v_{rms} = \sqrt{\frac{3RT}{M}}$
v_{rms} is independent of degree of freedom. M_{Helium} is smallest among all given so, v_{rms} for helium is maximum.

18. **(b)** Let at temperature T rms speed of oxygen molecules become just sufficient for escaping from the Earth's atmosphere

$V_{escape} = 11200$ m/s

Also, $V_{rms} = V_{escape} = \sqrt{\frac{3k_BT}{m_{O_2}}} = 11200$ m/s
Putting value of K_B and m_{O_2} we get,
$T = 8.360 \times 10^4$ K

19. **(b)** Here $v_1 = 200$ m/s;
temperature $T_1 = 27°C = 27 + 273 = 300$ k
temperature $T_2 = 127°C = 127 + 273 = 400$ k, $v_2 = ?$
R.M.S. Velocity, $v \propto \sqrt{T}$

$\Rightarrow \quad \frac{v_2}{200} = \sqrt{\frac{400}{300}}$

$\Rightarrow \quad v_2 = \frac{200 \times 2}{\sqrt{3}}\text{m/s} \Rightarrow v_2 = \frac{400}{\sqrt{3}}\text{m/s}$

20. **(b)** $\because \; P = \frac{1}{3}\frac{mN}{V}v_{rms}^2$
where m = mass of each molecules
N = total number of molecules
V = volume of the gas
When mass is halved and speed is doubled then

Resultant pressure, $P' = \frac{1}{3} \times \left(\frac{m}{2}\right) \times \frac{N}{V}(2v_{rms})^2$
$= 2P$

21. **(a)** At 0 K, molecular motion stops. Hence, kinetic energy of molecules becomes zero.

22. **(d)** $v_{rms} \propto \sqrt{T}$
As temperature increases from 300 K to 1200 K that is four times, so, v_{rms} will be doubled.

23. **(a)** $P = \frac{1}{3}Nmc^2 = \frac{2}{3} \times \left(\frac{1}{2}Nm\right)c^2 = \frac{2}{3}K.E$

24. **(d)** The mean square velocity of A type molecules $= \omega^2 + \omega^2 + \omega^2 = 3\omega^2$

Therefore, $\frac{1}{2}m(3\omega^2) = \frac{1}{2}(2m)v^2$

This gives $\omega^2 / v^2 = 2/3$

NOTES Mean kinetic energy of the two types of molecules should be equal.

25. **(c)** At 0 K velocity of molecules becomes zero.

26. **(b)** As the temperature increases, the average velocity increases. So, the collisions are faster.

27. **(c)** A polygamic gas has 3 translational, 3 rotational and f number of vibration modes

Total internal energy $U = \frac{3}{2}k_BT + \frac{3}{2}k_BT + fk_BT$
$\Rightarrow \; U = (3 + f)k_BT$
$C_V = (3 + f)R$
$C_P = (4 + f)R$
$\therefore \quad \frac{C_P}{C_V} = \frac{4 + f}{3 + f} = \gamma$

$\Rightarrow\ 4 + f = 3\gamma + f\gamma$

$\Rightarrow\ 4 - 3\gamma = f(\gamma - 1)$

$\Rightarrow\ f = \frac{4-3\gamma}{\gamma - 1}$

28. **(d)** Root mean square RMS speed of gas molecule

$V_{RMS} = \sqrt{\frac{3RT}{M}}$

Pressure exerted by ideal gas $P = \frac{1}{3}nm\bar{v}^2$

Average kinetic energy of a molecule $= \frac{3}{2}k_BT$

Total internal energy of a gas is $(U) = \frac{1}{2}nfRT$

Here, for one mole n = 1 and

f = 5 for diatomic

$\therefore U = \frac{5}{2}RT$

29. **(b)** $U = \frac{F}{2}nRT$

As $U \propto T$

$\therefore$ Increase in temperature would lead to the increase in kinetic energy of gas.

30. **(b)** For Hydrogen $f = 5$

For Helium $f = 3$

For gas X

$f = 7$.

$\gamma = 1 + \frac{2}{f} = 1 + \frac{2}{7} = \frac{9}{7}$

So, $\left(\frac{7}{5}, \frac{5}{3}, \frac{9}{7}\right)$

31. **(c)** Internal energy of the system is given by

$U = \frac{f}{2}nRT$

Degree of freedom

$F_{diatomic} = 5$

$f_{monoatomic} = 3$

and, number of moles

$n(O_2) = 2$

$n(Ar) = 4$

$U_{total} = \frac{5}{2}(2)RT + \frac{3}{2}(4)RT = 11RT$

32. **(a)** Change in internal energy from $A \to B$

$\Delta U = \frac{f}{2}nR\Delta T = \frac{f}{2}nR(T_f - T_i)$

$= \frac{5}{2}\{P_fV_f - P_iV_i\}$

$= \frac{5}{2}\{2 \times 10^3 \times 6 - 5 \times 10^3 \times 4\}$

$= \frac{5}{2}\{12 - 20\} \times 10^3\ J = 5 \times (-4) \times 10^3\ J$

$\Delta U = -20\ KJ$

NOTES (As gas is diatomic $\therefore$ f = 5)

33. **(b)** Let 'n' be the degree of freedom

$\gamma = \frac{C_p}{C_v} = \frac{\left(\frac{n}{2}+1\right)R}{\left(\frac{n}{2}\right)R} = \left(1 + \frac{2}{n}\right)$

34. **(d)** From first law of thermodynamics,

$\Delta Q = \Delta U + \Delta W = \frac{3}{2} \cdot \frac{1}{4}R(T_2 - T_1) + 0$

$= \frac{3}{8}N_aK_B(T_2 - T_1)\ [\because K = \frac{R}{N}]$

35. **(a)** $C_p - C_v = R \Rightarrow C_p = C_v + R$

$\because \gamma = \frac{C_p}{C_v} = \frac{C_v + R}{C_v} = \frac{C_v}{C_v} + \frac{R}{C_v}$

$\Rightarrow \gamma = 1 + \frac{R}{C_v} \Rightarrow \frac{R}{C_v} = \gamma - 1$

$\Rightarrow C_v = \frac{R}{\gamma - 1}$

36. **(c)** From Mayer's formula,

$C_p - C_v = R$

$MC_p - MC_v = R$

$C_p - C_v = \frac{R}{M}$

37. **(d)** $C_P = \frac{7}{2}R;\ C_V = C_P - R = \frac{7}{2}R - R = \frac{5}{2}R$

$\frac{C_P}{C_V} = \frac{7/2\,R}{5/2\,R} = \frac{7}{5}$

38. **(c)** We know that ratio of specific heats,

$\gamma = 1 + \frac{2}{n}$ or $n = \frac{2}{\gamma - 1}$

[where n = Degree of freedom]

39. **(c)** No. of degree of freedom $= 3K - N$

where K is no. of atom and N is the number of relations between atoms. For triatomic gas,

$K = 3, N = {}^3C_2 = 3$

No. of degree of freedom = 3 (3) – 3 = 6

40. (b) Number of translational degrees of freedom are same for all types of gases.

41. (c) Since $\frac{R}{C_V} = 0.67 \Rightarrow C_V = \frac{3}{2}R,$ hence gas is monoatomic.

NOTES For a gas having f degree of freedom specific heat at constant volume, $C_v = \frac{1}{2}fR$ specific heat at constant pressure,

$C_p = \left(\frac{f}{2}+1\right)R$

where, R = universal gas constant

42. (d) Both are diatomic gases and $C_p - C_v = R$ for all gases.

43. (b) $\gamma = \frac{C_P}{C_V} = \frac{15}{10} = \frac{3}{2} \Rightarrow C_V = \frac{2}{3}C_P$

$C_P - C_V = \frac{R}{J} \Rightarrow C_P - \frac{2}{3}C_P = \frac{R}{J}$

$\Rightarrow \frac{C_P}{3} = \frac{R}{J} \Rightarrow C_P = \frac{3R}{J}$

44. (d) $C_p = \frac{5}{2}R$ and $C_v = \frac{3}{2}R$

We know that $Q_v = nC_v\Delta T$ and $Q_p = nC_p\Delta T$

$\Rightarrow \frac{Q_v}{Q_p} = \frac{3}{5}$

Given $Q_p = 207\ \text{J} \Rightarrow Q_v \cong 124\ \text{J}$

45. (c) According to law of equipartition of energy, the energy per degree of freedom is $\frac{1}{2}kT$.

For a polyatomic gas with n degrees of freedom, the mean energy per molecule $= \frac{1}{2}nkT$.

46. (b) The mean free path l for a gas molecule is given as

$\lambda = \frac{1}{\sqrt{2}\pi n d^2} \Rightarrow \lambda \propto \frac{1}{d^2}$

where, d = diameter of molecule of gas

n = number of molecules per unit volume.

47. (a) Mean free path for a gas $\lambda_m = \frac{1}{\sqrt{2}n\pi d^2}$

Here, d = diameter of a gas molecule and, n = molecular density.

48. (b) Mean free path $\lambda_m = \frac{1}{\sqrt{2}\pi d^2 n}$

where d = diameter of molecule and d = 2r

$\therefore \quad \lambda_m \propto \frac{1}{r^2}$

13 Oscillations

TREND ANALYSIS

TOPIC	2025		2024		2023		2022		2021		2020	
	Qns.	LOD	Qns.	LOD	Qns.	LOD	Qns.	LOD	Qns.	LOD	Qns.	LOD
Force Law of Simple Harmonic Motion	2	Diff.							1	Avg.		
Simple Harmonic Motion												
Simple Harmonic Motion and Uniform Circular Motion			1	Avg.								
Velocity and Acceleration in Simple Harmonic Motion					1	Avg.					1	Avg.
The Simple Pendulum			1	Avg.			1	Avg.				
Energy in Simple Harmonic Motion									1	Avg.		

How to Code Qns for Better & Faster Revision →

Type 1: Able to solve (No code Req)
Type 2: Solved with struggle or by seeing solution **(Use code NR means Needs Revision or orange stickers)**
Type 3: Tricky/ Could not solve on my own. Need help from Teacher/ Friend. **(Use code TQ means Tricky Question or red stickers)**

You can either hand write the code in boxes or else paste coloured stickers given at end of book.

Topic-13.2 Periodic and Oscillatory Motions

1. Identify the function which represents a non periodic motion. **NCERT | Page-262 | RS 2022**

(a) $\sin(\omega t + \pi/4)$ (b) $e^{-\omega t}$
(c) $\sin \omega t$ (d) $\sin \omega t + \cos \omega t$

Topic-13.3 Simple Harmonic Motion

2. The two-dimensional motion of a particle, described by $\vec{r} = (\hat{i} + 2\hat{j}) A \cos \omega t$ is a/an: **NCERT | Page-265, 266 | Ph-II 2024**

A. parabolic path
B. elliptical path
C. periodic motion
D. simple harmonic motion

Choose the correct answer from the options given below:

(a) B, C and D only (b) A, B and C only
(c) A, C and D only (d) C and D only

3. The displacement of a particle executing simple harmonic motion is given by **NCERT | Page-266 | 2019**

$y = A_0 + A \sin\omega t + B \cos\omega t$.

Then the amplitude of its oscillation is given by:

(a) $A_0 + \sqrt{A^2 + B^2}$ (b) $\sqrt{A^2 + B^2}$
(c) $\sqrt{A_0^2 + (A+B)^2}$ (d) $A + B$

4. The distance covered by a particle undergoing SHM in one time period is (amplitude = A), **NCERT | Page-266 | Odisha 2019**

(a) 4A (b) Zero
(c) A (d) 2A

5. When two displacements represented by $y_1 = a \sin(\omega t)$ and $y_2 = b \cos(\omega t)$ are super-imposed the motion is: **NCERT | Page-265 | 2015**

(a) simple harmonic with amplitude $\frac{a}{b}$
(b) simple harmonic with amplitude $\sqrt{a^2 + b^2}$
(c) simple harmonic with amplitude $\frac{(a+b)}{2}$
(d) not a simple harmonic

6. Out of the following functions, representing motion of a particle, which represents SHM? **NCERT | Page-265 | 2011**

(A) $y = \sin \omega t - \cos \omega t$
(B) $y = \sin^3 \omega t$
(C) $y = 5\cos\left(\frac{3\pi}{4} - 3\omega t\right)$
(D) $y = 1 + \omega t + \omega^2 t^2$

(a) Only (A)
(b) Only (D) does not represent SHM
(c) Only (A) and (C)
(d) Only (A) and (B)

7. Two particles are oscillating along two close parallel straight lines side by side, with the same frequency and amplitudes. They pass each other, moving in opposite directions when their displacement is half of the amplitude. The mean positions of the two particles lie on a straight line perpendicular to the paths of the two particles. The phase difference is NCERT | Page-265 | 2011 M

(a) 0 (b) $2\pi/3$ (c) π (d) $\pi/6$

8. The displacement of a particle along the x-axis is given by $x = a \sin^2 \omega t$. The motion of the particle corresponds to: NCERT | Page-265 | 2010
(a) simple harmonic motion of frequency ω/π
(b) simple harmonic motion of frequency $3\omega/2\pi$
(c) non simple harmonic motion
(d) simple harmonic motion of frequency $\omega/2\pi$

9. A particle executes simple harmonic oscillation with an amplitude a. The period of oscillation is T. The minimum time taken by the particle to travel half of the amplitude from the equilibrium position is NCERT | Page-266 | 2007
(a) $T/8$ (b) $T/12$ (c) $T/2$ (d) $T/4$

10. Two simple harmonic motions act on a particle. These harmonic motions are $x = A\cos(\omega t + \delta)$, $y = A\cos(\omega t + \alpha)$ when $\delta = \alpha + \frac{\pi}{2}$, the resulting motion is NCERT | Page-265 | 2000
(a) a circle and the actual motion is clockwise
(b) an ellipse and the actual motion is counterclockwise
(c) an elllipse and the actual motion is clockwise
(d) a circle and the actual motion is counter clockwise

11. Two simple harmonic motions with the same frequency act on a particle at right angles i.e., along x and y axis. If the two amplitudes are equal and the phase difference is $\pi/2$, the resultant motion will be NCERT | Page-265 | 1997
(a) a circle
(b) an ellipse with the major axis along y-axis
(c) an ellipse with the major axis along x-axis
(d) a straight line inclined at 45° to the x-axis

12. Which one of the following is a simple harmonic motion? NCERT | Page-264 | 1994
(a) Ball bouncing between two rigid vertical walls
(b) Particle moving in a circle with uniform speed
(c) Wave moving through a string fixed at both ends
(d) Earth spinning about its own axis.

13. A particle is subjected to two mutually per-pendicular simple harmonic motions such that its x and y coordinates are given by

$x = 2\sin\omega t;\quad y = 2\sin\left(\omega t + \frac{\pi}{4}\right)$ 1994

The path of the particle will be
(a) a straight line (b) a circle
(c) an ellipse (d) a parabola

14. A simple harmonic oscillator has an amplitude A and time period T. The time required by it to travel from $x = A$ to $x = A/2$ is NCERT | Page-266 | 1992
(a) T/6 (b) T/4 (c) T/3 (d) T/2

15. The composition of two simple harmonic motions of equal periods at right angle to each other and with a phase difference of π results in the displacement of the particle along NCERT | Page-265 | 1990
(a) circle (b) figures of eight
(c) straight line (d) ellipse

Topic-13.4
Simple Harmonic Motion and Uniform Circular Motion

16. If $x = 5\sin\left(\pi t + \frac{\pi}{3}\right)$ m represents the motion of a particle executing simple harmonic motion, the amplitude and time period of motion, respectively, are NCERT | Page-265 | 2024
(a) 5 cm, 2 s (b) 5 m, 2 s
(c) 5 cm, 1 s (d) 5 m, 1 s

17. The radius of circle, the period of revolution, initial position and sense of revolution are indicated in the fig. NCERT | Page-265 | 2019

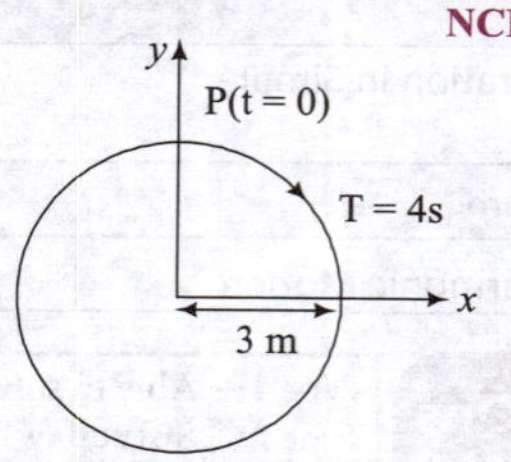

y - projection of the radius vector of rotating particle P is :
(a) $y(t) = -3\cos 2\pi t$, where y in m
(b) $y(t) = 4\sin\left(\frac{\pi t}{2}\right)$, where y in m
(c) $y(t) = 3\cos\left(\frac{3\pi t}{2}\right)$, where y in m
(d) $y(t) = 3\cos\left(\frac{\pi t}{2}\right)$, where y in m

Topic-13.5
Velocity and Acceleration in Simple Harmonic Motion

18. The x-t graph of a particle performing simple harmonic motion is shown in the figure. The acceleration of the particle at t = 2 s is NCERT | Page-266 | 2023

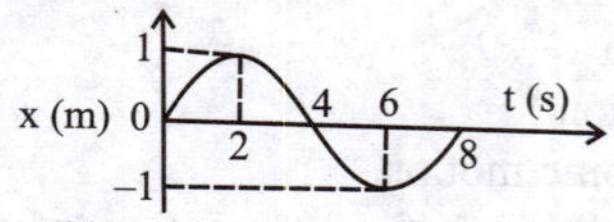

(a) $\frac{\pi^2}{8}\,\text{m s}^{-2}$ (b) $-\frac{\pi^2}{8}\,\text{m s}^{-2}$
(c) $\frac{\pi^2}{16}\,\text{m s}^{-2}$ (d) $-\frac{\pi^2}{16}\,\text{m s}^{-2}$

19. The phase difference between displacement and acceleration of a particle in a simple harmonic motion is : NCERT | Page-266 | 2020
(a) $\frac{3\pi}{2}$ rad (b) $\frac{\pi}{2}$ rad
(c) zero (d) π rad

20. Average velocity of a particle executing SHM in one complete vibration is: NCERT | Page-266 | 2019

(a) $\frac{A\omega}{2}$ (b) $A\omega$ (c) $\frac{A\omega^2}{2}$ (d) zero

21. A particle executes linear simple harmonic motion with an amplitude of 3 cm. When the particle is at 2 cm from the mean position, the magnitude of its velocity is equal to that of its acceleration. Then its time period in seconds is NCERT | Page-266 | 2017

(a) $\frac{\sqrt{5}}{2\pi}$ (b) $\frac{4\pi}{\sqrt{5}}$ (c) $\frac{2\pi}{\sqrt{3}}$ (d) $\frac{\sqrt{5}}{\pi}$

22. A particle is executing SHM along a straight line. Its velocities at distances x_1 and x_2 from the mean position are V_1 and V_2, respectively. Its time period is NCERT | Page-266 | 2015

(a) $2\pi\sqrt{\frac{x_2^2 - x_1^2}{V_1^2 - V_2^2}}$ (b) $2\pi\sqrt{\frac{V_1^2 + V_2^2}{x_1^2 + x_2^2}}$

(c) $2\pi\sqrt{\frac{V_1^2 - V_2^2}{x_1^2 - x_2^2}}$ (d) $2\pi\sqrt{\frac{x_1^2 - x_2^2}{V_1^2 - V_2^2}}$

23. A particle is executing a simple harmonic motion. Its maximum acceleration is a and maximum velocity is b. Then its time period of vibration will be : NCERT | Page-266 | 2015 RS

(a) $\frac{\alpha}{\beta}$ (b) $\frac{\beta^2}{\alpha}$ (c) $\frac{2\pi\beta}{\alpha}$ (d) $\frac{\beta^2}{\alpha^2}$

24. The oscillation of a body on a smooth horizontal surface is represented by the equation,

$X = A\cos(\omega t)$

where X = displacement at time t

ω = frequency of oscillation

Which one of the following graphs shows correctly the variation of 'a' with 't'? NCERT | Page-267 | 2014

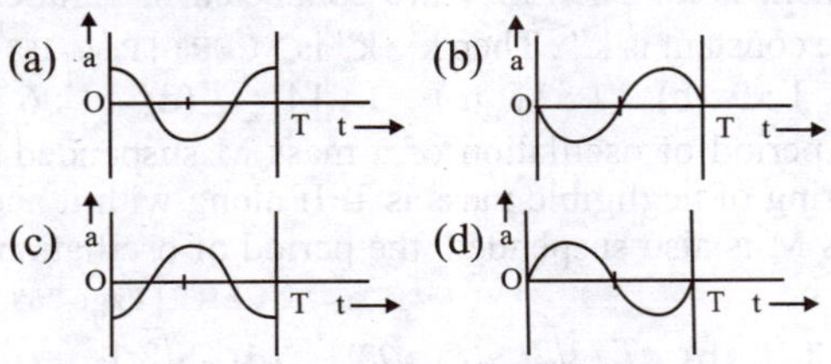

25. A particle of mass *m* oscillates along *x*-axis according to equation $x = a\sin\omega t$. The nature of the graph between momentum and displacement of the particle is NCERT | Page-266 | Kar. 2013

(a) straight line passing through origin
(b) circle
(c) hyperbola
(d) ellipse

26. The equation of a simple harmonic wave is given by

$$y = 3\sin\frac{\pi}{2}(50t - x)$$

Where x and y are in meters and t is in seconds. The ratio of maximum particle velocity to the wave velocity is NCERT | Page-266 | 2012 M

(a) 2π (b) $\frac{3}{2}\pi$ (c) 3π (d) $\frac{2}{3}\pi$

27. A particle of mass m is released from rest and follows a parabolic path as shown. Assuming that the displacement of the mass from the origin is small, which graph correctly depicts the position of the particle as a function of time NCERT | Page-267 | 2011

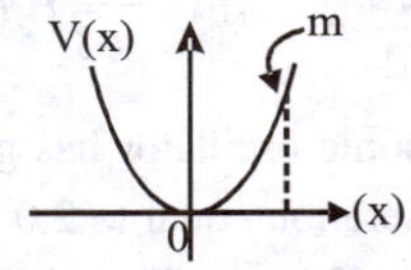

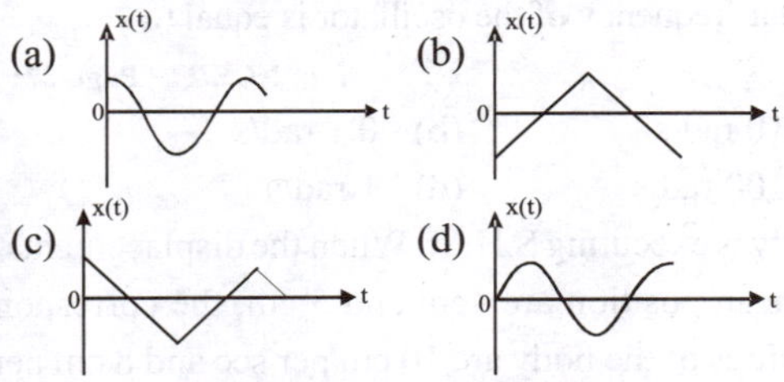

28. Which one of the following equations of motion represents simple harmonic motion? NCERT | Page-266 | 2009

(a) Acceleration = $-k(x + a)$
(b) Acceleration = $k(x + a)$
(c) Acceleration = kx
(d) Acceleration = $-k_0x + k_1x^2$

where k, k_0, k_1 and a are all postive.

29. Two simple harmonic motions of angular frequency 100 and 1000 rad s^{-1} have the same displacement amplitude. The ratio of their maximum accelerations is: NCERT | Page-266 | 2008

(a) 1 : 10 (b) 1 : 10^2 (c) 1 : 10^3 (d) 1 : 10^4

30. A point performs simple harmonic oscillation of period T and the equation of motion is given by $x = a\sin(\omega t + \pi/6)$. After the elapse of what fraction of the time period the velocity of the point will be equal to half of its maximum velocity? NCERT | Page-266 | 2008

(a) T/8 (b) T/6 (c) T/3 (d) T/12

31. The phase difference between the instantaneous velocity and acceleration of a particle executing simple harmonic motion is NCERT | Page-266 | 2007

(a) π (b) 0.707π (c) zero (d) 0.5π

32. A particle executing simple harmonic motion of amplitude 5 cm has maximum speed of 31.4 cm/s. Its oscillation per second NCERT | Page-266 | 2005

(a) 4 (b) 3 (c) 2 (d) 1

33. Which one of the following statements is true for the speed v and the acceleration a of a particle executing simple harmonic motion ? NCERT | Page-266 | 2004

(a) When *v* is maximum, *a* is zero
(b) When *v* is maximum, *a* is maximum
(c) Value of *a* is zero, whatever may be the value of *v*
(d) When *v* is zero, *a* is zero

34. A particle executing S.H.M. has amplitude 0.01m and frequency 60 Hz. The maximum acceleration of the particle is **NCERT | Page-266 | 1999**

(a) $144\ \pi^2\ m/s^2$ (b) $120\ \pi^2\ m/s^2$
(c) $80\ \pi^2\ m/s^2$ (d) $60\ \pi^2\ m/s^2$

35. A particle starts simple harmonic motion from the mean position. Its amplitude is A and time period is T. What is its displacement when its speed is half of its maximum speed **NCERT | Page-266 | 1996**

(a) $\frac{\sqrt{2}}{3}A$ (b) $\frac{\sqrt{3}}{2}A$ (c) $\frac{2}{\sqrt{3}}A$ (d) $\frac{A}{\sqrt{2}}$

36. If a simple harmonic oscillator has got a displacement of 0.02 m and acceleration equal to 2.0 m/s^2 at any time, the angular frequency of the oscillator is equal to

NCERT | Page-271 | 1992

(a) 10 rad/s (b) 0.1 rad/s
(c) 100 rad/s (d) 1 rad/s

37. A body is executing S.H.M. When the displacements from the mean position are 4cm and 5 cm, the corresponding velocities of the body are 10 cm per sec and 8 cm per sec. Then the time period of the body is

NCERT | Page-348 / N-266 | 1991

(a) 2π sec (b) $\pi/2$ sec
(c) π sec (d) $(3\pi/2)$sec

Topic-13.6
Force Law of Simple Harmonic Motion

38. In an oscillating spring mass system, a spring is connected to a box filled with sand. As the box oscillates, sand leaks slowly out of the box vertically so that the average frequency $\omega(t)$ and average amplitude $A(t)$ of the system change with time t. Which one of the following options schematically depicts these changes correctly? **NCERT | Page-268 | 2025**

(a)
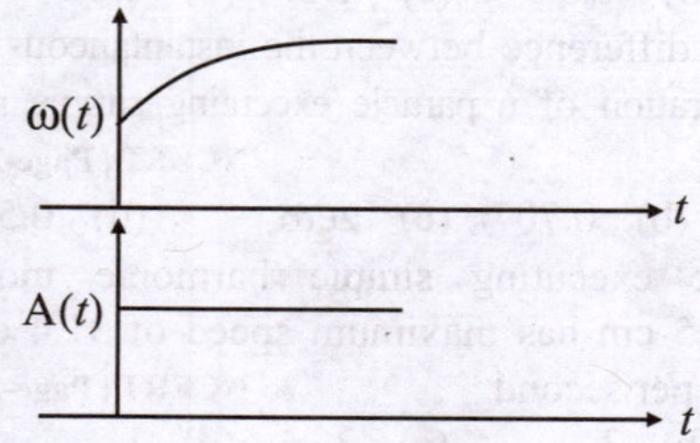

(b)
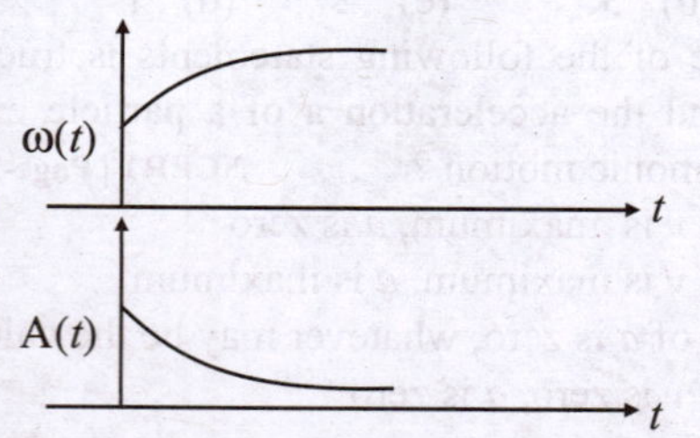

(c)
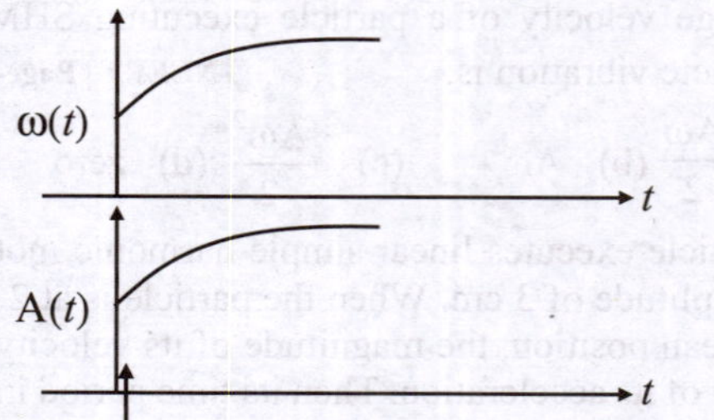

(d)
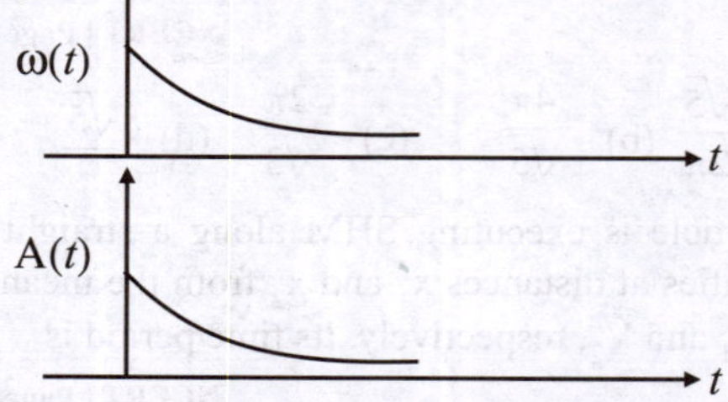

39. Two identical point masses P and Q, suspended from two separate massless springs of spring constants k_1 and k_2, respectively, oscillate vertically. If their maximum speeds are the same, the ratio $\left(\frac{A_Q}{A_P}\right)$ of the amplitude A_Q of mass Q to the amplitude A_P of mass P is : **NCERT | Page-268 | 2025**

(a) $\frac{k_2}{k_1}$ (b) $\frac{k_1}{k_2}$

(c) $\sqrt{\frac{k_2}{k_1}}$ (d) $\sqrt{\frac{k_1}{k_2}}$

40. A spring is stretched by 5 cm by a force 10 *N*. The time period of the oscillations when a mass of 2 kg is suspended by it is **NCERT | Page-268 | 2021**

(a) 0.628 s (b) 0.0628 s
(c) 6.28 s (d) 3.14 s

41. A spring of force constant k is cut into lengths of ratio 1 : 2 : 3. They are connected in series and the new force constant is k′. Then they are connected in parallel and force constant is k″. Then k′ : k″ is **NCERT | Page-268 | 2017**

(a) 1 : 9 (b) 1 : 11 (c) 1 : 14 (d) 1 : 6

42. The period of oscillation of a mass M suspended from a spring of negligible mass is T. If along with it another mass M is also suspended, the period of oscillation will now be **NCERT | Page-268 | 2010**

(a) T (b) $T/\sqrt{2}$ (c) 2T (d) $\sqrt{2}$ T

43. A block of mass M is attached to the lower end of a vertical spring. The spring is hung from a ceiling and has force constant value k. The mass is released from rest with the spring initially unstretched. The maximum extension produced in the length of the spring will be: **NCERT | Page-268 | 2009**

(a) 2 Mg/k (b) 4 Mg/k
(c) Mg/2k (d) Mg/k

44. A mass of 2.0 kg is put on a flat pan attached to a vertical spring fixed on the ground as shown in the figure. The mass of the spring and the pan is negligible. When pressed slightly and released the mass executes a simple harmonic motion. The spring constant is 200 N/m. What should be the minimum amplitude of the motion so that the mass gets detached from the pan (take $g = 10\ m/s^2$)?

NCERT | Page-268 | 2007

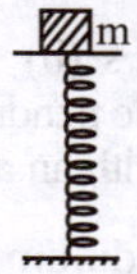

(a) 10.0 cm
(b) any value less than 12.0 cm
(c) 4.0 cm
(d) 8.0 cm

45. Two springs of spring constants k_1 and k_2 are joined in series. The effective spring constant of the combination is given by

NCERT | Page-268 | 2004

(a) $k_1k_2/(k_1 + k_2)$ (b) k_1k_2
(c) $(k_1 + k_2)/2$ (d) $k_1 + k_2$

46. The time period of a mass suspended from a spring is T. If the spring is cut into four equal parts and the same mass is suspended from one of the parts, then the new time period will be NCERT | Page-268 | 2003

(a) $2T$ (b) $\frac{T}{4}$ (c) 2 (d) $\frac{T}{2}$

47. A body of mass M, executes vertical SHM with periods t_1 and t_2, when separately attached to spring A and spring B respectively. The period of SHM, when the body executes SHM, as shown in the figure is t_0. Then

NCERT | Page-268 | 2002

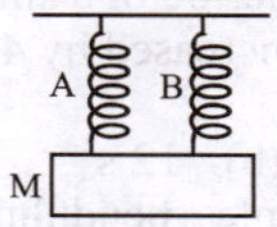

(a) $t_0^{-1} = t_1^{-1} + t_2^{-1}$
(b) $t_0 = t_1 + t_2$
(c) $t_0^2 = t_1^2 + t_2^2$
(d) $t_0^{-2} = t_1^{-2} + t_2^{-2}$

48. A mass m is vertically suspended from a spring of negligible mass; the system oscillates with a frequency n. What will be the frequency of the system, if a mass 4 m is suspended from the same spring? NCERT | Page-268 | 1998

(a) $\frac{n}{4}$ (b) $4n$ (c) $\frac{n}{2}$ (d) $2n$

49. A mass m is suspended from a two coupled springs, connected in series. The force constant for springs are k_1 and k_2. The time period of the suspended mass will be

NCERT | Page-268 | 1990

(a) $T = 2\pi\sqrt{\frac{m}{k_1 - k_2}}$ (b) $T = 2\pi\sqrt{\frac{mk_1k_2}{k_1 + k_2}}$

(c) $T = 2\pi\sqrt{\frac{m}{k_1 + k_2}}$ (d) $T = 2\pi\sqrt{\frac{m(k_1 + k_2)}{k_1k_2}}$

50. A particle moving along the X-axis, executes simple harmonic motion then the force acting on it is given by

NCERT | Page-267 | 1988

(a) $-A\,kx$ (b) $A\cos(kx)$
(c) $A\exp(-kx)$ (d) Akx

where, A and k are positive constants.

Topic-13.7

Energy in Simple Harmonic Motion

51. A particle executing simple harmonic motion with amplitude A has the same potential and kinetic energies at the displacement NCERT | Page-269 | Ph-II 2024

(a) $2\sqrt{A}$ (b) $\frac{A}{2}$
(c) $\frac{A}{\sqrt{2}}$ (d) $A\sqrt{2}$

52. A body is executing simple harmonic motion with frequency 'n', the frequency of its potential energy is 2021

(a) $4n$ (b) n (c) $2n$ (d) $3n$

53. The particle executing simple harmonic motion has a kinetic energy $K_0 \cos^2\omega t$. The maximum values of the potential energy and the total energy are respectively

NCERT | Page-269 | 2007

(a) $K_0/2$ and K_0 (b) K_0 and $2K_0$
(c) K_0 and K_0 (d) 0 and $2K_0$

54. A particle of mass m oscillates with simple harmonic motion between points x_1 and x_2, the equilibrium position being O. Its potential energy is plotted. It will be as given below in the graph. NCERT | Page-268 | 2003

(a) 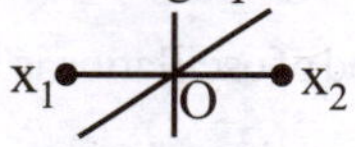

(b) 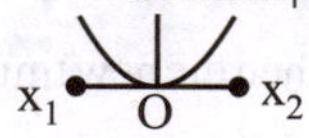

(c) 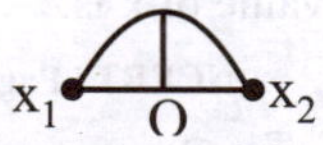

(d) 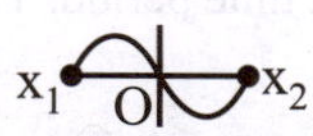

55. The potential energy of a simple harmonic oscillator when the particle is half way to its end point is

NCERT | Page-268 | 2003

(a) $\frac{1}{2}E$ (b) $\frac{2}{3}E$ (c) $\frac{1}{8}E$ (d) $\frac{1}{4}E$

where E is the total energy

56. A particle is executing a simple harmonic motion of amplitude 'a'. Its potential energy is maximum when the displacement from the position of the maximum kinetic energy is NCERT | Page-269 | 2002

(a) 0 (b) $\pm a$ (c) $\pm a/2$ (d) $-a/2$

57. There is a body having mass m and performing S.H.M. with amplitude a. There is a restoring force $F = -kx$. The total energy of body depends upon NCERT | Page-269 | 2001

(a) k, x (b) k, a (c) k, a, x (d) k, a, v

58. In a simple harmonic motion, when the displacement is one-half the amplitude, what fraction of the total energy is kinetic? NCERT | Page-269 | 1996

(a) 0 (b) $\frac{1}{4}$ (c) $\frac{1}{2}$ (d) $\frac{3}{4}$

59. A linear harmonic oscillator of force constant 2×10^6 N/m and amplitude 0.01 m has a total mechanical energy of 160 J. Its NCERT | Page-269 | 1996

(a) potential energy is 160 J
(b) potential energy is 100 J
(c) potential energy is zero
(d) potential energy is 120 J

60. A body executes S.H.M with an amplitude A. At what displacement from the mean position is the potential energy of the body is one fourth of its total energy ?

NCERT | Page-269 | 1993

(a) A/4
(b) A/2
(c) 3A/4
(d) Some other fraction of A.

61. The angular velocity and the amplitude of a simple pendulum is ω and a respectively. At a displacement *x* from the mean position if its kinetic energy is *T* and potential energy is *V*, then the ratio of *T* to *V* is

NCERT | Page-269 | 1991

(a) $\frac{(a^2 - x^2\omega^2)}{x^2\omega^2}$ (b) $\frac{x^2\omega^2}{(a^2 - x^2\omega^2)}$

(c) $\frac{(a^2 - x^2)}{x^2}$ (d) $\frac{x^2}{(a^2 - x^2)}$

Topic-13.8 The Simple Pendulum

62. If the mass of the bob in a simple pendulum is increased to thrice its original mass and its length is made half its original length, then the new time period of oscillation is $\frac{x}{2}$ times its original time period. Then the value of x is:

NCERT | Page-271 | 2024

(a) $\sqrt{3}$ (b) $\sqrt{2}$ (c) $2\sqrt{3}$ (d) 4

63. A simple pendulum oscillating in air has a period of $\sqrt{3}$s. If it is completely immersed in non-viscous liquid, having density $\left(\frac{1}{4}\right)^{th}$ of the material of the bob, the new period will be

NCERT | Page-271 | Manipur 2023

(a) 2s (b) $\frac{\sqrt{3}}{2}$s (c) $2\sqrt{3}$s (d) $\frac{2}{\sqrt{3}}$s

64. Two pendulums of length 121 cm and 100 cm start vibrating in phase. At some instant, the two are at their mean position in the same phase. The minimum number of vibrations of the shorter pendulum after which the two are again in phase at the mean position is:

NCERT | Page-271 | 2022

(a) 9 (b) 10
(c) 8 (d) 11

65. A mass falls from a height '*h*' and its time of fall '*t*' is recorded in terms of time period *T* of a simple pendulum. On the surface of earth it is found that *t* = 2*T*. The entire set up is taken on the surface of another planet whose mass is half of that of earth and radius the same. Same experiment is repeated and corresponding times noted as *t* ' and *T* '.

Then we can say NCERT | Page-271 | Odisha 2019

(a) $t' = 2T'$ (b) $t' = T'$
(c) $t' > 2T'$ (d) $t' < 2T'$

66. A pendulum is hung from the roof of a sufficiently high building and is moving freely to and fro like a simple harmonic oscillator. The acceleration of the bob of the pendulum is 20 m/s^2 at a distance of 5 m from the mean position. The time period of oscillation is

NCERT | Page-271 | 2018

(a) 2π s (b) π s (c) 1 s (d) 2 s

67. A simple pendulum performs simple harmonic motion about x = 0 with an amplitude a and time period T. The speed of the pendulum at $x = \frac{a}{2}$ will be: NCERT | Page-271 | 2009

(a) $\frac{\pi a}{T}$ (b) $\frac{3\pi^2 a}{T}$ (c) $\frac{\pi a\sqrt{3}}{T}$ (d) $\frac{\pi a\sqrt{3}}{2T}$

68. A simple pendulum has a metal bob, which is negatively charged. If it is allowed to oscillate above a positively charged metallic plate, then its time period will

NCERT | Page-271 | 2001

(a) increase (b) decrease
(c) become zero (d) remain the same

69. Masses M_A and M_B hanging from the ends of strings of lengths L_A and L_B are executing simple harmonic motions. If their frequencies are $f_A = 2f_B$, then

NCERT | Page-271 | 2000

(a) $L_A = 2L_B$ and $M_A = M_B/2$
(b) $L_A = 4L_B$ regardless of masses
(c) $L_A = L_B/4$ regardless of masses
(d) $L_A = 2L_B$ and $M_A = 2M_B$

70. The time period of a simple pendulum is 2 seconds. If its length is increased by 4-times, then its period becomes

NCERT | Page-271 | 1999

(a) 16 s (b) 12 s (c) 8 s (d) 4 s

71. Two simple pendulums of length 5m and 20m respectively are given small linear displacement in one direction at the same time. They will again be in the phase when the pendulum of shorter length has completed .... oscillations NCERT | Page-270 | 1998

(a) 5 (b) 1 (c) 2 (d) 3

72. A hollow sphere is filled with water. It is hung by a long thread. As the water flows out of a hole at the bottom, the period of oscillation will NCERT | Page-271 | 1997

(a) first increase and then decrease
(b) first decrease and then increase
(c) go on increasing
(d) go on decreasing

73. If the length of a simple pendulum is increased by 2%, then the time period NCERT | Page-271 | 1997

(a) increases by 2% (b) decreases by 2%
(c) increases by 1% (d) decreases by 1%

74. A simple pendulum is suspended from the roof of a trolley which moves in a horizontal direction with an acceleration a, then the time period is given by $T = 2\pi\sqrt{(l/g)}$, where g is equal to

NCERT | Page-271 | 1991

(a) g (b) $g - a$
(c) $g + a$ (d) $\sqrt{(g^2 + a^2)}$

ANSWER KEY																			
1	(b)	9	(b)	17	(d)	25	(d)	33	(a)	41	(b)	49	(d)	57	(b)	65	(a)	73	(c)
2	(d)	10	(d)	18	(d)	26	(b)	34	(a)	42	(d)	50	(a)	58	(d)	66	(b)	74	(d)
3	(b)	11	(a)	19	(d)	27	(a)	35	(b)	43	(a)	51	(c)	59	(b)	67	(c)		
4	(a)	12	(c)	20	(d)	28	(a)	36	(a)	44	(a)	52	(c)	60	(b)	68	(b)		
5	(b)	13	(c)	21	(b)	29	(b)	37	(c)	45	(a)	53	(c)	61	(c)	69	(c)		
6	(c)	14	(a)	22	(a)	30	(d)	38	(b)	46	(d)	54	(b)	62	(b)	70	(d)		
7	(b)	15	(c)	23	(c)	31	(d)	39	(d)	47	(d)	55	(d)	63	(a)	71	(c)		
8	(a)	16	(b)	24	(c)	32	(d)	40	(a)	48	(c)	56	(b)	64	(d)	72	(a)		

Hints & Solutions

1. **(b)** Periodic motion is represented by sin & cosine (harmonic functions) functions.
So $e^{-\omega t}$ is not harmonic function

2. **(d)** Given, $r = (\hat{i} + 2\hat{j})\, A \cos \omega t$
x component $= A \cos \omega t$
y component $= 2A \cos \omega t$
$\omega y = 2x$
The path is straight line.
The motion is SHM and periodic because,
$$\frac{dr}{dt} = -(\hat{i} + 2\hat{j})\, \omega A \sin \omega T$$
$$\Rightarrow a = \frac{d^2 r}{dt^2} = -(\hat{i} + 2\hat{j})\, \omega^2 A \cos \omega t$$
$$\vec{a} = -\omega^2 \vec{r}$$

3. **(b)**

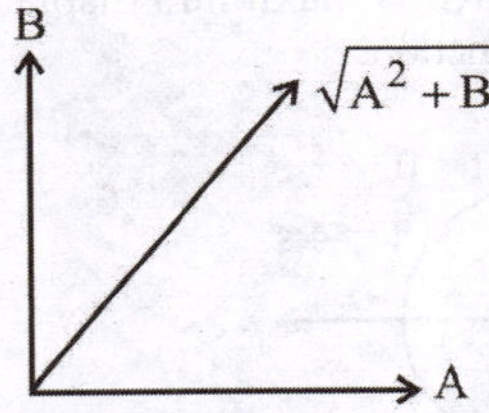

Given equation
$y = A_0 + A \sin \omega t + B \sin \omega t$
Now assume $(y - A_0) = \gamma$
$y - A_0 = A \sin \omega t + B \sin \omega t$
$\gamma = A \sin \omega t + B \cos \omega t$
$= \sqrt{A^2 + B^2} \sin (\omega t + \phi)$
which is S.H.M.
so, resultant amplitude
$= \sqrt{A^2 + B^2}$

4. **(a)** In one time period total distance travelled by the particle is 4A.

5. **(b)** The two displacements equations are $y_1 = a \sin(\omega t)$
and $y_2 = b \cos(\omega t) = b \sin\left(\omega t + \frac{\pi}{2}\right)$
$y_{eq} = y_1 + y_2$
$= a \sin\omega t + b \cos\omega t$
$= a \sin\omega t + b \sin\left(\omega t + \frac{\pi}{2}\right)$

Since the frequencies for both SHMs are same, resultant motion will be SHM.

Now amplitude, $A_{eq} = \sqrt{a^2 + b^2 + 2ab \cos \frac{\pi}{2}}$

$$\Rightarrow A_{eq} = \sqrt{a^2 + b^2}$$

6. **(c)** Only functions given in (A) & (C) represent SHM.

NOTES $y = A \sin\omega t + B \cos\omega t \Rightarrow y = a \sin(\omega t + \phi)$. This is also a equation of S.H.M where $\phi = \sqrt{A^2 + B^2}$ $A = a \cos \phi$, $B = a \sin\phi$ and $\theta = \tan^{-1} \frac{B}{A}$

7. **(b)** Equation of SHM is given by
$x = A \sin (\omega t + \delta)$
$(\omega t + \delta)$ is called phase.
When $x = \frac{A}{2}$, then $\sin (\omega t + \delta) = \frac{1}{2}$
$$\Rightarrow \omega t + \delta = \frac{\pi}{6} \quad \text{or} \quad \phi_1 = \frac{\pi}{6}$$
For second particle,

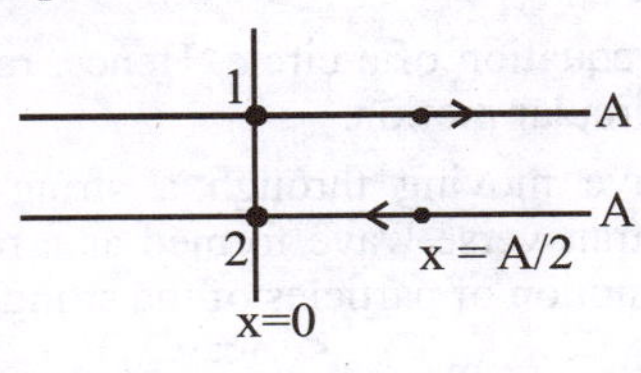

$$\phi_2 = \pi - \frac{\pi}{6} = \frac{5\pi}{6}$$
$$\therefore \phi = \phi_2 - \phi_1 = \frac{4\pi}{6} = \frac{2\pi}{3}$$

8. **(a)** $x = a\sin^2 \omega t = \frac{a}{2}(1 - \cos 2\omega t)$

$$\frac{dx}{dt} = \frac{a}{2} 2\omega \sin 2\omega t$$

$$\Rightarrow \frac{d^2x}{dt^2} = \frac{4\omega^2 a}{2} \cdot \cos 2\omega t$$

This represents an S.H.M. of frequency $= \frac{\omega}{\pi}$

9. **(b)** Displacement from the mean position

$$y = a\sin\left(\frac{2\pi}{T}\right)t$$

According to problem $y = a/2$

$$a/2 = a\sin\left(\frac{2\pi}{T}\right)t$$

$$\Rightarrow \frac{\pi}{6} = \left(\frac{2\pi}{T}\right)t \Rightarrow t = T/12$$

This is the minimum time taken by the particle to travel half of the amplitude from the equilibrium position.

10. **(d)** $x = A\cos(\omega t + \delta)$

$y = A\cos(\omega t + \alpha)$ ....(1)

When $\delta = \alpha + \frac{\pi}{2}$

$$x = A\cos\left(\frac{\pi}{2} + \omega t + \alpha\right)$$

$x = -A\sin(\omega t + \alpha)$ ....(2)

Squaring (1) and (2) and then adding

$x^2 + y^2 = A^2[\cos^2(\omega t + \alpha) + \sin^2(\omega t + \alpha)]$

or $x^2 + y^2 = A^2$, which is the equation of a circle. The present motion is anticlockwise.

11. **(a)** Equation of two simple harmonic motions

$y = A\sin(\omega t + \phi)$ ....(1)

$$x = A\sin\left(\omega t + \phi + \frac{\pi}{2}\right)$$

$\Rightarrow x = A\cos(\omega t + \phi)$ ....(2)

On squaring and adding equations (1) and (2),

$x^2 + y^2 = A^2$

This is an equation of a circle. Hence, resulting motion will be a circular motion.

12. **(c)** A wave moving through a string fixed at both ends, is a transverse wave formed as a result of simple harmonic motion of particles of the string.

13. **(c)** As phase difference $= \frac{\pi}{4}$, the resultant path of particle is an ellipse.

14. **(a)** For S.H.M., $x = A\sin\left(\frac{2\pi}{T}t\right)$

When $x = A$, $A = A\sin\left(\frac{2\pi}{T}.t\right)$

$$\therefore \sin\left(\frac{2\pi}{T}.t\right) = 1 \Rightarrow \sin\left(\frac{2\pi}{T}.t\right) = \sin\left(\frac{\pi}{2}\right)$$

$\Rightarrow t = (T/4)$

When $x = \frac{A}{2}, \frac{A}{2} = A\sin\left(\frac{2\pi}{T}.t\right)$

or, $\sin\frac{\pi}{6} = \sin\left(\frac{2\pi}{T}t\right)$ or $t = (T/12)$

Now, time taken to travel from $x = A$ to $x = A/2$ is $(T/4 - T/12) = T/6$

15. **(c)** $x = a\sin\omega t$

and $y = b\sin(\omega t + \pi) = -b\sin\omega t$

or, $\frac{x}{a} = -\frac{y}{b}$ or $y = -\frac{b}{a}x$

It is an equation of a straight line.

NOTES If two mutually perpendicular S.H.M's of same frequency be $x = 9$, $\sin\omega t$ and $y = a_2\sin(\omega t + \phi)$ then general equation of Lissajous figure is

$$\frac{x^2}{a_1^2} + \frac{y_2}{a_2^2} - \frac{2xy}{a_1 a_2}\cos\phi = \sin^2\phi$$

For $\phi = \pi \frac{x^2}{a_1^2} + \frac{y^2}{a_2^2} - \frac{2xy}{a_1 a_2}\cos(\pi) = \sin 2(\pi)$

$$\Rightarrow \frac{x}{a_1} + \frac{y}{a_2} = 0 \quad \frac{x}{a} + \frac{y}{a^2} = 0 \quad \Rightarrow y = \frac{-a_2}{a_1}x$$

16. **(b)** $x = 5\sin\left(\pi t + \frac{\pi}{3}\right)$ m

Amplitude, A = 5m

Time period, $T = \frac{2\pi}{\omega} = \frac{2\pi}{\pi} = 2\text{ s}$

17. **(d)** At t = 0, y = 3, which is maximum displacement so equation will be cosine function.

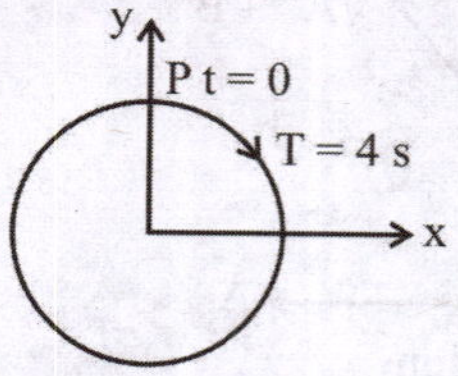

$$\omega = \frac{2\pi}{T} = \frac{2\pi}{4} = \frac{\pi}{2}\text{ rad/s} \quad (\because T = 4\text{ s})$$

$$y = a\cos\omega t \Rightarrow y = 3\cos\frac{\pi}{2}t$$

18. **(d)** Position of particle as function of time

$x = A\sin\omega t$

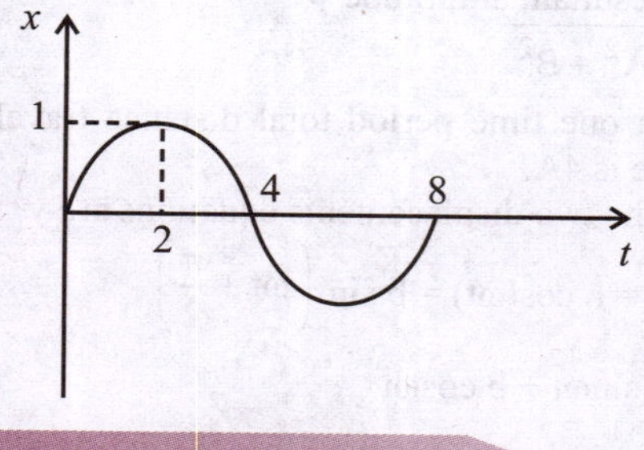

From figure, at $t = 2$ s
$x = 1$
Time period, $T = 8$ s

$$\omega = \frac{2\pi}{T} = \frac{2\pi}{8} = \frac{\pi}{4}$$

$$a = -\omega^2 x = -\left(\frac{\pi}{4}\right)^2 \times 1$$

$$\Rightarrow a = -\frac{\pi^2}{16} \text{m/s}^{-2}$$

NOTES In SHM, we cannot apply equation of translatory motion as acceleration ($a = \omega^2 x$) is not constant

19. (d) Displacement equation of a SHM
$y = A\sin(\omega t + \phi)$

$\therefore$ Velocity, $v = \frac{dy}{dt} = A\omega\cos(\omega t + \phi)$

Acceleration, $a = \frac{dv}{dt}$

or, $a = -A\omega^2 \sin(\omega t + \phi)$

$\therefore a = A\omega^2 \sin(\omega t + \phi + \pi)$

Hence, phase difference between displacement and acceleration is π.

20. (d) Displacement of the particle in one complete vibration is zero, so, average velocity in one complete vibration will be

$$= \frac{\text{Displacement}}{\text{Time interval}} = \frac{y_f - y_i}{T} = 0$$

21. (b) Given, Amplitude A = 3 cm
When particle is at x = 2 cm
According to question, magnitude of velocity = acceleration

$$\omega\sqrt{A^2 - x^2} = x\omega^2$$

$$\sqrt{(3)^2 - (2)^2} = 2\left(\frac{2\pi}{T}\right)$$

$$\sqrt{5} = \frac{4\pi}{T} \Rightarrow T = \frac{4\pi}{\sqrt{5}}$$

22. (a) As we know, for particle undergoing SHM,

$$V = \omega\sqrt{A^2 - X^2}$$

$$V_1^2 = \omega^2 (A^2 - x_1^2)$$

$$V_2^2 = \omega^2 (A^2 - x_2^2)$$

Substracting we get,

$$\frac{V_1^2}{\omega^2} + x_1^2 = \frac{V_2^2}{\omega^2} + x_2^2$$

$$\Rightarrow \frac{V_1^2 - V_2^2}{\omega^2} = x_2^2 - x_1^2$$

$$\Rightarrow \omega = \sqrt{\frac{V_1^2 - V_2^2}{x_2^2 - x_1^2}}$$

$$\Rightarrow T = 2\pi\sqrt{\frac{x_2^2 - x_1^2}{V_1^2 - V_2^2}}$$

23. (c) As, we know, in SHM
Maximum acceleration of the particle, $\alpha = A\omega^2$
Maximum velocity, $\beta = A\omega$

$$\Rightarrow \omega = \frac{\alpha}{\beta}$$

$$\Rightarrow T = \frac{2\pi}{\omega} = \frac{2\pi\beta}{\alpha} \quad \left[\because \omega = \frac{2\pi}{T}\right]$$

24. (c) Displacement, x = A cos (ωt) (given)

Velocity, $v = \frac{dx}{dt} = -A\omega \sin(\omega t)$

Acceleration, $a = \frac{dv}{dt} = -A\omega^2 \cos \omega t$

Hence graph (c) correctly dipicts the variation of a with t.

25. (d) As $\frac{v^2}{a^2\omega^2} + \frac{y^2}{a^2} = 1$ This is the equation of ellipse.

Hence the graph is an ellipse. Momentum (P) versus x graph is similar to v versus x graph.

26. (b) $y = 3\sin\frac{\pi}{2}(50t - x)$

$y = 3\sin\left(25\pi t - \frac{\pi}{2}x\right)$ on comparing with the standard wave equation

$y = a \sin(\omega t - kx)$

Wave velocity $v = \frac{\omega}{k} = \frac{25\pi}{\pi/2} = 50$ m/sec.

The velocity of particle

$$v_p = \frac{\partial y}{\partial t} = 75\pi\cos\left(25\pi t - \frac{\pi}{2}x\right)$$

$$v_{p_{max}} = 75\pi$$

then $\frac{v_{p_{max}}}{v} = \frac{75\pi}{50} = \frac{3\pi}{2}$

27. (a) The given velocity-position graph depicts that the motion of the particle is SHM.
In SHM, at t = 0, v = 0 and $x = x_{max}$
So, option (a) is correct.

28. (a) Acceleration, a = – kX, where, X = x + a.
In simple harmonic motion acceleration is directly proportional to the displacement from the mean position. Also the acceleration is in the opposite direction of displacement.

29. (b) Maximum acceleration of a particle in the simple harmonic motion is directly proportional to the square of angular frequency *i.e.*,
i.e., $a \propto \omega^2$

$$\therefore \frac{a_1}{a_2} = \frac{\omega_1^2}{\omega_2^2} = \frac{(100)^2}{(1000)^2} = \frac{1}{100}$$

$$\Rightarrow a_1 : a_2 = 1 : 10^2.$$

30. (d) We have $x = a\sin\left(\omega t + \frac{\pi}{6}\right)$

$\therefore$ Velocity, $v = \frac{dx}{dt} = a\omega\cos\left(\omega t + \frac{\pi}{6}\right)$

Maximum velocity = aω
According to question,

$$\frac{a\omega}{2} = a\omega\cos\left(\omega t + \frac{\pi}{6}\right)$$

or, $\cos\left(\omega t + \frac{\pi}{6}\right) = \frac{1}{2} = \cos 60°$ or $\cos\frac{\pi}{3}$

$\Rightarrow \omega t + \frac{\pi}{6} = \frac{\pi}{3}$; $\omega t = \frac{\pi}{3} - \frac{\pi}{6}$ or, $\omega t = \frac{\pi}{6}$

or, $\frac{2\pi}{T}.t = \frac{\pi}{6} \Rightarrow t = \frac{T}{12}$

31. (d) Let $y = A\sin\omega t$

$v_{inst} = \frac{dy}{dt} = A\omega\cos\omega t = A\omega\sin(\omega t + \pi/2)$

Acceleration $= -A\omega^2\sin\omega t = A\omega^2\sin(\pi + \omega t)$

$\therefore \phi = \frac{\pi}{2} = 0.5\pi$

32. (d) In S.H.M., $v_{max} = A\omega = A(2\pi f)$

$f = \frac{v_{max}}{2\pi A} = \frac{31.4}{2(3.14)\times 5} = 1$ oscillation per sec.

33. (a) When v is maximum, a is zero

$v = \omega\sqrt{A^2 - x^2}$; $a = \omega^2 x$

v_{max} at $x = 0$, but at $x = 0$, $a = 0$

34. (a) Amplitude (A) = 0.01 m, Frequency = 60 Hz
Maximum acceleration
$= A\omega^2 = 0.01\times(2\pi n)^2$
$= 0.01\times 4\pi^2\times 60\times 60 = 144\pi^2$ m/sec^2

35. (b) $v_{max} = A\omega$ when $v = \frac{v_{max}}{2} = \frac{A\omega}{2}$

$v = \omega\sqrt{A^2 - y^2}$

$\Rightarrow \frac{A\omega}{2} = \omega\sqrt{A^2 - y^2} \Rightarrow y = \pm\frac{\sqrt{3}}{2}A$

NOTES The displacement at which the speed is n times the maximum speed is given by $y = a\sqrt{1 - n^2}$

36. (a) $\omega^2 = \frac{\text{acceleration}}{\text{displacement}} = \frac{2.0}{0.02}$

$\omega^2 = 100$ or $\omega = 10$ rad/s

37. (c) For S.H.M., Velocity,

$v = \omega\sqrt{a^2 - x^2}$ at displacement x.

$\Rightarrow 10 = \omega\sqrt{a^2 - 16}$ .....(1)

and $8 = \omega\sqrt{a^2 - 25}$ .....(2)

Dividing, $\frac{5^2}{4^2} = \frac{a^2 - 16}{a^2 - 25} = \frac{25}{16}$

or, $16a^2 - 256 = 25a^2 - 625 \Rightarrow 9a^2 = 369$

$\therefore a^2 = \frac{369}{9}$

Putting this value in equation (2) mentioned above,

$10 = \omega\sqrt{\frac{369}{9} - 16} \Rightarrow 10 = \omega\sqrt{\frac{225}{9}}$

or, $\omega = \frac{10\times 3}{15} = 2$ radian / sec.

Time period $= \frac{2\pi}{\omega} = \frac{2\pi}{2} = \pi$ sec.

38. (b) The angular frequency of spring-mass system is

$\omega = \sqrt{\frac{k}{m}} \Rightarrow \omega \propto \frac{1}{\sqrt{m}}$

$\therefore m\downarrow \Rightarrow \omega\uparrow$

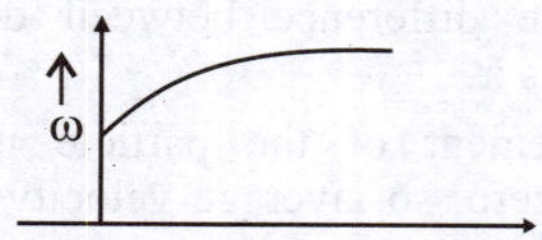

Also, $E = \frac{1}{2}kA^2 = \frac{1}{2}m\omega^2 A^2$

As, the part of sand being removed,
$\therefore$ Energy and amplitude of SHM decreases

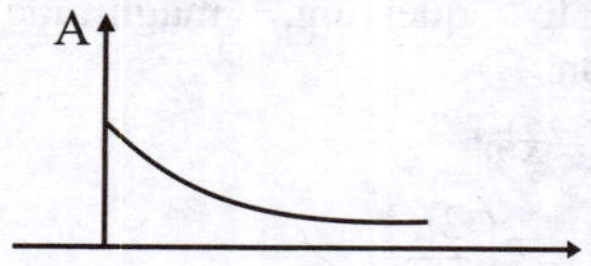

39. (d) Given $m_P = m_Q$
Also $(V_{max})_P = (V_{max})_Q$
Also $(V_{max})_P = A_Q\omega_Q$
$\because v = A\omega$

$\therefore A_P\sqrt{\frac{k_1}{m_P}} = A_Q\sqrt{\frac{k_2}{m_Q}}$ $\left[\because \omega = \sqrt{\frac{k}{m}}\right]$

$\therefore \frac{A_Q}{A_P} = \sqrt{\frac{k_1}{k_2}}$

40. (a) Spring constant

$k = \frac{F}{x} = \frac{10}{5\times 10^{-2}} = 200$ N / m

And for spring-mass system undergoing SHM

$T = 2\pi\sqrt{\frac{m}{k}} = 2\pi\sqrt{\frac{2}{200}} = \frac{2\pi}{10} = 0.628$s

($\because$ m = 2gk given)

41. **(b)** Let ℓ be the complete length of the spring.

Length when cut in ratio, 1 : 2 : 3 are $\frac{\ell}{6}, \frac{\ell}{3}$ and $\frac{\ell}{2}$

Spring constant (k) $\propto \frac{1}{\text{length}(\ell)}$

Spring constant for given segments

$k_1 = 6k, k_2 = 3k$ and $k_3 = 2k$

When they are connected **in series**

$$\frac{1}{k'} = \frac{1}{6k} + \frac{1}{3k} + \frac{1}{2k}$$

$$\Rightarrow \frac{1}{k'} = \frac{6}{6k}$$

$\therefore$ Force constant $k' = k$

And when they are connected in parallel

$k'' = 6k + 3k + 2k \Rightarrow k'' = 11k$

Then the ratios

$\frac{k'}{k''} = \frac{1}{11}$ i.e., $k' : k'' = 1 : 11$

NOTES If a spring of force constant K is divided into n equal parts then spring constant of each part will become nk and if these n parts connected in parallel then $k_{eff} = n^2 k$.

42. **(d)** $T = 2\pi\sqrt{\frac{M}{K}}$

$$\therefore \frac{T_1}{T_2} = \sqrt{\frac{M_1}{M_2}}$$

$$\therefore T_2 = T_1\sqrt{\frac{M_2}{M_1}}$$

$$= T_1\sqrt{\frac{2M}{M}}$$

$T_2 = T_1\sqrt{2} = \sqrt{2}\,T$ (where $T_1 = T$)

NOTES If the spring has a mass M and mass m is suspended from it, effective mass is given by $m_{eff} = m + \frac{M}{3}$

If the two mass m_1 and m_2 are connected by a spring and made to oscillate on horizontal surface, reduced mass $\frac{1}{m_r} = \frac{1}{m_1} + \frac{1}{m_2}$

$\therefore$ Time period $T = 2\pi\sqrt{\frac{m_r}{k}}$

43. **(a)** Loss in gravitation P.E. = gain in spring P.E

$$\Rightarrow mgx = \frac{1}{2}kx^2$$

$$\Rightarrow x = \frac{2mg}{k}$$

44. **(a)** Mass gets detached at the upper extreme position when pan returns to its mean position.

At that point, $R = mg - m\omega^2 a = 0$

i.e. $g = \omega^2 a \Rightarrow a = g/\omega = mg/k$

$$\Rightarrow a = \frac{2\times 10}{200} \quad \left[As = \omega^2 = \frac{k}{m}\right]$$

$$\Rightarrow a = 1/10\,m = 10\text{cm}$$

45. **(a)** $\frac{1}{k_{eq}} = \frac{1}{k_1} + \frac{1}{k_2} \Rightarrow \frac{1}{k_{eq}} = \frac{k_1 + k_2}{k_1 k_2}$

$$\Rightarrow k_{eq} = \frac{k_1 k_2}{k_1 + k_2}$$

46. **(d)** $T = 2\pi\sqrt{\frac{m}{k}}$

When a spring is cut into n parts

Spring constant for each part = nk

Here, $n = 4$; $T_1 = 2\pi\sqrt{\frac{m}{4k}} = \frac{T}{2}$

47. **(d)** The time period of spring mass system,

$$T = 2\pi\sqrt{\frac{m}{K}}$$

$$\therefore \quad t_1 = 2\pi\sqrt{\frac{m}{k_1}} \quad .....(i)$$

$$t_2 = 2\pi\sqrt{\frac{m}{k_2}} \quad .....(ii)$$

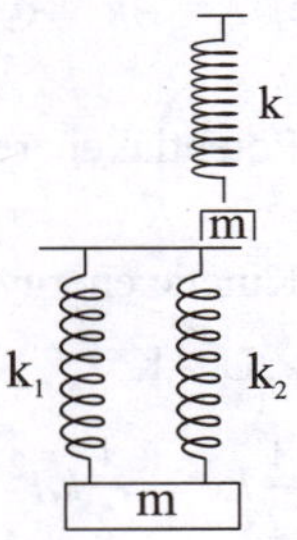

When springs are connected in parallel then

$K_{eff} = k_1 + k_2$

So, $t_0 = 2\pi\sqrt{\frac{m}{K_{eff}}} \Rightarrow 2\pi\sqrt{\frac{m}{(k_1 + k_2)}}$ .... (iii)

from (i), $\frac{1}{t_1^2} = \frac{1}{4\pi^2} \times \frac{k_1}{m}$

from (ii), $\frac{1}{t_2^2} = \frac{1}{4\pi^2} \times \frac{k_2}{m}$

from (iii), $\frac{1}{t_0^2} = \frac{1}{4\pi^2} \times \frac{(k_1 + k_2)}{m}$

from above expressions,

$$\frac{1}{t_0^2} = \frac{1}{t_1^2} + \frac{1}{t_2^2}$$

NOTES

$$t = 2\pi\sqrt{\frac{m}{k}}$$

$\Rightarrow k = \text{Const.}\, t^{-2}$

Here the springs are joined in parallel. So

$k_0 = k_1 + k_2$

where k_0 is resultant force constant

$\therefore$ Const. t_0^{-2} = Const. t_1^{-2} + Const. t_2^{-2}

or, $t_0^{-2} = t_1^{-2} + t_2^{-2}$

48. (c) $n = \frac{1}{2\pi}\sqrt{\frac{k}{m}}$

$$n' = \frac{1}{2\pi}\sqrt{\frac{k}{4m}} = \frac{1}{2}\times\frac{1}{2\pi}\sqrt{\frac{k}{m}}$$

On putting the value of n we get $n' = \frac{n}{2}$

49. (d) The effective spring constant of two springs in series; $K = \frac{k_1k_2}{k_1+k_2}$.

Time period, $T = 2\pi\sqrt{\frac{m}{K}} = 2\pi\sqrt{\frac{m(k_1+k_2)}{k_1k_2}}$

50. (a) For simple harmonic motion, $F = -Kx$.

Here, $K = Ak$.

51. (c) Potential energy, $U = \frac{1}{2}kx^2$

Kinetic energy, $K = \frac{1}{2}kA^2 - \frac{1}{2}kx^2$

$\therefore U = K$

$\frac{1}{2}kx^2 = \frac{1}{2}kA^2 - \frac{1}{2}kx^2 \Rightarrow k^2x^2 = kA^2$

$x^2 = \frac{A^2}{2}$

$x = \frac{A}{\sqrt{2}}$

52. (c) Displacement of particle executing

SHM $y = A\sin(\omega t + \phi)$ ...(i)

Potential energy $U = \frac{1}{2}ky^2$

or, $U = \frac{1}{2}kA^2\sin^2(\omega t + \phi)$ ...(ii)

From eq. (i) and (ii),

Time period of $y = A\sin(\omega t + \phi)$

$T_1 = \frac{2\pi}{\omega}$ $\therefore$ frequency $n_1 = \frac{\omega}{2\pi}$

and time period of $y^2 = A^2\sin^2(\omega t + \phi)$

$T_2 = \frac{\pi}{\omega}$ $\therefore$ frequency $n_2 = \frac{\omega}{\pi}$

$\therefore n_2 = 2n_1 = 2n$

NOTES Potential energy varies periodically with double the frequency of SHM. It is maximum and equal to total energy at extreme position and minimum at mean position.

53. (c) We have, $U + K = E$

where, U = potential energy, K = Kinetic energy, E = Total energy.

Also, we know that, in S.H.M., when potential energy is maximum, K.E. is zero and vice-versa.

$\therefore U_{max} + 0 = E \Rightarrow U_{max} = E$

Further,

$K.E. = \frac{1}{2}m\omega^2a^2\cos^2\omega t$

But by question, $K.E. = K_0\cos^2\omega t$

$\therefore K_0 = \frac{1}{2}m\omega^2a^2$

Hence, total energy, $E = \frac{1}{2}m\omega^2a^2 = K_0$

$\therefore U_{max} = K_0$ & $E = K_0$.

54. (b) Potential energy $\propto$ (Amplitude)^2

$\propto$(Displacement)^2

P.E.is maximum at maximum distance.

P.E. is zero at equilibrium point.

P.E. curve is parabolic.

55. (d) $P.E. = \frac{1}{2}kx^2 = E$

At half way

$$P.E. = \frac{1}{2}k\left(\frac{x}{2}\right)^2 = \frac{\frac{1}{2}kx^2}{4} = \frac{E}{4}$$

56. (b) P.E. of particle executing S.H.M. $= \frac{1}{2}m\omega^2x^2$

At $x = a$, P.E. is maximum i.e. $= \frac{1}{2}m\omega^2a^2$

K.E. $= \frac{1}{2}m\omega^2(a^2 - x^2)$

At $x = 0$, K.E. is maximum. Hence, displacement from position of maximum Kinetic energy $= \pm a$.

57. (b) Total Energy of body performing simple harmonic motion $= \frac{1}{2}m\omega^2a^2 = \frac{1}{2}ka^2$

where $k = m\omega^2$

NOTES

$$T.E. = K.E. + P.E. = \frac{1}{2}m(a^2 - x^2)\omega^2 + \frac{1}{2}m\omega^2x^2$$

Hence energy depends upon amplitude and k (spring constant).

58. **(d)** Total energy of particle executing S.H.M. of amplitude (A).

$$E = \frac{1}{2}m\omega^2 A^2$$

K.E.of the particle

$$= \frac{1}{2}m\omega^2\left(A^2 - \frac{A^2}{4}\right)\left(\text{when } x = \frac{A}{2}\right)$$

$$= \frac{1}{2}m\omega^2 \times \frac{3}{4}A^2 = \frac{1}{2}\times\frac{3}{4}m\omega^2 A^2$$

Clearly, $\frac{KE}{\text{Total Energy}} = \frac{3}{4}$

59. **(b)** Force constant $k = 2 \times 10^6$ N/m

Amplitude $(x) = 0.01$ m

Potential Energy $= \frac{1}{2}kx^2$

$$= \frac{1}{2}\times(2\times10^6)\times(0.01)^2 = 100 \text{ J}$$

60. **(b)** $P.E = \frac{1}{2}m\omega^2 x^2$.

Total energy $E = \frac{1}{2}m\omega^2 A^2$

$$\text{P.E.} = \frac{1}{4}E \Rightarrow \frac{1}{2}m\omega^2 x^2 = \frac{1}{8}m\omega^2 A^2 \quad \therefore x = \frac{1}{2}A.$$

61. **(c)** P.E., $V = \frac{1}{2}m\omega^2 x^2$

and K.E., $T = \frac{1}{2}m\omega^2(a^2 - x^2)$

$$\therefore \frac{T}{V} = \frac{a^2 - x^2}{x^2}$$

62. **(b)** Time period of simple pendulum is

$$T = 2\pi\sqrt{\frac{\ell}{g}} \Rightarrow T \propto \sqrt{\ell}$$

$$\Rightarrow \frac{T_2}{T_1} = \sqrt{\frac{\ell_2}{\ell_1}} = \frac{1}{\sqrt{2}} \quad \left[\because \ell_2 = \frac{\ell_1}{2}\right]$$

$$\Rightarrow T_2 = \frac{1}{\sqrt{2}}T_1 = \frac{x}{2}T_1 \text{ (given)}$$

$$\therefore x = \sqrt{2}$$

NOTES The time period of simple pendulum is independent of mass of the bob

63. **(a)** Time period of simple pendulum $T = 2\pi\sqrt{\frac{1}{g_{eff}}}$

$$\therefore \sqrt{3} = 2\pi\sqrt{\frac{1}{g}}$$

After immersed in liquid $Mg_{eff} = Mg - \frac{Mg}{\frac{\rho_s}{\rho_l}}$

or, $Mg_{eff} = Mg - \frac{Mg}{4} = \frac{3Mg}{4} \quad \therefore g_{eff} = \frac{3g}{4}$

$$\therefore T_1 = 2\pi\sqrt{\frac{1}{\frac{3g}{4}}} = \left(2\pi\sqrt{\frac{1}{g}}\right)\times\frac{2}{\sqrt{3}} = \frac{2T}{\sqrt{3}} = 2\text{s}$$

64. **(d)** $T \propto \sqrt{\ell}$

$$\frac{T_\ell}{T_S} = \frac{\sqrt{121}}{\sqrt{100}} \Rightarrow \frac{T_\ell}{T_S} = \frac{11}{10} \Rightarrow 11\,T_s = 10\,T_c$$

Hence, smaller pendulum will make 1 oscillation more than longer pendulum.

If longer makes n^1 oscillation, then smaller makes $(n + 1)$ oscillation.

For oscillating in same phase

$$(n+1)T_S = nT_e$$

$$\Rightarrow (n+1)\cdot 2\pi\sqrt{\frac{100}{g}} = n\cdot 2\pi\sqrt{\frac{121}{g}}$$

$$\Rightarrow (n+1)10 = n \times 11 \Rightarrow n = 10$$

65. **(a)** At surface of earth time taken in falling h distance.

$$t = \sqrt{\frac{2h}{g}}\text{; and } T = 2\pi\sqrt{\frac{l}{g}}$$

Given $t = 2T$; $\frac{t}{T} = 2$

For surface of other planet

$$g' = \frac{g}{2}$$

Time taken in falling h distance

$$t' = \sqrt{\frac{2h}{g'}} = \sqrt{2}t$$

and $T' = 2\pi\sqrt{\frac{l}{g'}} = \sqrt{2}T$

Here $\frac{t'}{T'} = \frac{\sqrt{2}t}{\sqrt{2}T} = 2$

$$\therefore \quad t' = 2T$$

66. **(b)** From question, acceleration, $a = 20$ m/s^2, and displacement, $y = 5$m

$|a| = \omega^2 y$

$\Rightarrow \quad 20 = \omega^2(5)$

$\Rightarrow \quad \omega = 2$ rad/s

Time period of pendulum,

$$T = \frac{2\pi}{\omega} = \frac{2\pi}{2} = \pi \text{ s}$$

67. (c) Speed $v = \omega\sqrt{a^2 - x^2}$, $x = \frac{a}{2}$

$$\therefore \quad v = \omega\sqrt{a^2 - \frac{a^2}{4}} = \omega\sqrt{\frac{3a^2}{4}}$$

$$= \frac{2\pi}{T}\frac{a\sqrt{3}}{2} = \frac{\pi a\sqrt{3}}{T}$$

68. (b) We know that the time period (T)

$$= 2\pi\sqrt{\frac{l}{g'}} \propto \frac{1}{\sqrt{g'}} \qquad g' = g_{effective}$$

Since the negatively charged bob is attracted by the positively charged plate, therefore acceleration due to gravity will increase and time period will decrease.

69. (c) $f_A = \frac{1}{2\pi}\sqrt{\frac{g}{L_A}}$ and $f_B = \frac{f_A}{2} = \frac{1}{2\pi}\sqrt{\frac{g}{L_B}}$

$$\therefore \quad \frac{f_A}{f_{A/2}} = \frac{1}{2\pi}\sqrt{\frac{g}{L_A}} \times 2\pi\sqrt{\frac{L_B}{g}}$$

$$\Rightarrow 2 = \sqrt{\frac{L_B}{L_A}} \Rightarrow 4 = \frac{L_B}{L_A}, \text{ regardless of mass.}$$

70. (d) $T = 2\pi\sqrt{\frac{\ell}{g}} \qquad T \propto \sqrt{\ell}$

If ℓ is increased by 4 times, time period will increase by two times.

71. (c) Let the pendulums be in phase after t sec of start. Within this time, if the bigger pendulum executes n oscillations, the smaller one will have executed $(n + 1)$ oscillations.

Now, the time of n oscillation $= 2\pi\sqrt{\frac{20}{g}} \times n$

& the time of $(n + 1)$ oscillation

$$= 2\pi\sqrt{\frac{5}{g}} \times (n+1)$$

To be in phase

$$2\pi\sqrt{\frac{20}{g}} \times n = 2\pi\sqrt{\frac{5}{g}} \times (n+1)$$

or, $\quad 2n = n + 1$ or, $\quad n = 1$

Hence, the no. of oscillations executed by shorter pendulum $= n + 1 = 1 + 1 = 2$

72. (a) Time period of simple pendulum $T = 2\pi\sqrt{\left(\frac{\ell}{g}\right)} \propto \sqrt{\ell}$

where l is effective length.

[i.e distance between centre of suspension and centre of gravity of bob]

Initially, centre of gravity is at the centre of sphere. When water leaks the centre of gravity goes down until it is half filled; then it begins to go up and finally it again goes at the centre. That is effective length first increases and then decreases. As $T \propto \sqrt{l}$, so time period first increases and then decreases.

73. (c) We know that $T = 2\pi\sqrt{\frac{l}{g}}$

$$\frac{\Delta T}{T} \times 100 = \frac{1}{2}\frac{\Delta l}{l} \times 100$$

If length is increased by 2%, time period increases by 1%.

NOTES

(i) If g is constant and length varies by n %. Then % change in time period $\frac{\Delta T}{T} \times 100 = \frac{n}{2}\%$

(ii) If l is constant and g varies by n %. Then % change in time period $\frac{\Delta T}{T} \times 100 = \frac{n}{2}\%$

Valid only for percentage less than 10%.

74. (d) The effective value of acceleration due to gravity is $\sqrt{(a^2 + g^2)}$

NOTES

If a simple pendulum is suspended from a trolley that is moving with constant speed v around a circle of radius r

$$T = 2\pi\sqrt{\frac{l}{\sqrt{g^2 + \left(\frac{v^2}{r}\right)^2}}}$$

14 Waves

TREND ANALYSIS

TOPIC	2025		2024		2023		2022		2021		2020	
	Qns.	LOD	Qns.	LOD	Qns.	LOD	Qns.	LOD	Qns.	LOD	Qns.	LOD
The Principle of Superposition of Waves							1	Avg.				
Reflection of Waves	1	Avg.			1	Avg.	1	Avg.			1	Avg.

How to Code Qns for Better & Faster Revision

Type 1: Able to solve (No code Req)

Type 2: Solved with struggle or by seeing solution **(Use code NR means Needs Revision or orange stickers)**

Type 3: Tricky/ Could not solve on my own. Need help from Teacher/ Friend. **(Use code TQ means Tricky Question or red stickers)**

You can either hand write the code in boxes or else paste coloured stickers given at end of book.

Topic-14.2

Transverse and Longitudinal Waves

1. Which one of the following statements is true ? NCERT | Page-280 | 2006

(a) The sound waves in air are longitudinal while the light waves are transverse
(b) Both light and sound waves in air are longitudinal
(c) Both light and sound waves can travel in vacuum
(d) Both light and sound waves in air are transverse

2. With the propagation of a longitudinal wave through a material medium, the quantities transmitted in the propagation direction are NCERT | Page-279 | 1992

(a) Energy, momentum and mass
(b) Energy
(c) Energy and mass
(d) Energy and linear momentum

Topic-14.3

Displacement Relation in a Progressive Wave

3. The displacement of a travelling wave $y = C\sin\frac{2\pi}{\lambda}(at - x)$ where t is time, x is distance and λ is the wavelength, all in S.I. units. Then the frequency of the wave is NCERT | Page- 282, 283 | Ph-II 2024

(a) $\frac{2\pi\lambda}{a}$ (b) $\frac{2\pi a}{\lambda}$
(c) $\frac{\lambda}{a}$ (d) $\frac{a}{\lambda}$

4. A wave travelling in the +ve x-direction having displacement along y-direction as 1m, wavelength 2π m and frequency $\frac{1}{\pi}$ Hz is represented by NCERT | Page-283 | 2013

(a) $y = \sin(2\pi x - 2\pi t)$
(b) $y = \sin(10\pi x - 20\pi t)$
(c) $y = \sin(2\pi x + 2\pi t)$
(d) $y = \sin(x - 2t)$

5. Two waves are represented by the equations $y_1 = a\sin(\omega t + kx + 0.57)$ m and $y_2 = a\cos(\omega t + kx)$ m, where x is in meter and t in sec. The phase difference between them is NCERT | Page-283 | 2011

(a) 1.0 radian (b) 1.25 radian
(c) 1.57 radian (d) 0.57 radian

6. A wave in a string has an amplitude of 2 cm. The wave travels in the + ve direction of x axis with a speed of 128 m/sec and it is noted that 5 complete waves fit in 4 m length of the string. The equation describing the wave is NCERT | Page-283 | 2009

(a) $y = (0.02)$ m $\sin(15.7x - 2010t)$
(b) $y = (0.02)$ m $\sin(15.7x + 2010t)$
(c) $y = (0.02)$ m $\sin(7.85x - 1005t)$
(d) $y = (0.02)$ m $\sin(7.85x + 1005t)$

7. The wave described by $y = 0.25\sin(10\pi x - 2\pi t)$, where x and y are in meters and t in seconds, is a wave travelling along the: NCERT | Page-283 | 2008

(a) –ve x direction with frequency 1 Hz
(b) +ve x direction with frequency π Hz and wavelength $\lambda = 0.2$ m
(c) +ve x direction with frequency 1 Hz and wavelength $\lambda = 0.2$ m
(d) –ve x direction with amplitude 0.25 m and wavelength $\lambda = 0.2$ m

8. The phase difference between two waves, represented by
$y_1 = 10^{-6}\sin\{100\,t + (x/50) + 0.5\}$ m
$y_2 = 10^{-6}\cos\{100\,t + (x/50)\}$ m
where x is expressed in metres and t is expressed in seconds, is approximately NCERT | Page-283 | 2004

(a) 1.5 radians (b) 1.07 radians
(c) 2.07 radians (d) 0.5 radians

9. The equation for a transverse wave travelling along the positive x-axis with amplitude 0.2 m, velocity $v = 360$ ms^{-1} and wavelength $\lambda = 60$ m can be written as

NCERT | Page-283 | 2002

(a) $y = 0.2\sin\left[2\pi\left(6t - \frac{x}{60}\right)\right]$

(b) $y = 0.2\sin\left[\pi\left(6t + \frac{x}{60}\right)\right]$

(c) $y = 0.2\sin\left[\pi\left(6t - \frac{x}{60}\right)\right]$

(d) $y = 0.2\sin\left[2\pi\left(6t + \frac{x}{60}\right)\right]$

10. A transverse wave is represented by the equation $y = y_0 \sin\frac{2\pi}{\lambda}(vt - x)$

For what value of λ is the maximum particle velocity equal to two times the wave velocity?

NCERT | Page-284 | 1998

(a) $\lambda = 2\pi y_0$ (b) $\lambda = \frac{\pi y_0}{3}$
(c) $\lambda = \frac{\pi y_0}{2}$ (d) $\lambda = \pi y_0$

11. In a sinusoidal wave, the time required for a particular point to move from maximum displacement to zero displacement is 0.170 sec. The frequency of the wave is

NCERT | Page-283 | 1998

(a) 1.47 Hz (b) 0.36 Hz
(c) 0.73 Hz (d) 2.94 Hz

12. Two sound waves having a phase difference of 60° have path difference of

NCERT | Page-283 | 1996

(a) 2λ (b) $\frac{\lambda}{2}$ (c) $\frac{\lambda}{3}$ (d) $\frac{\lambda}{6}$

13. The equation of a sound wave is given as: $y = 0.0015 \sin(62.4\,x + 316\,t)$. The wavelength of this wave is

NCERT | Page-283 | 1996

(a) 0.4 unit (b) 0.3 unit
(c) 0.2 unit (d) 0.1 unit

14. From a wave equation: $y = 0.5\sin\frac{2\pi}{3.2}(64t - x)$, the frequency of the wave is

NCERT | Page-283 | 1995

(a) 5 Hz (b) 15 Hz
(c) 20 Hz (d) 25 Hz

15. Which of the following equations represent a wave?

NCERT | Page-283 | 1994

(a) $y = A\sin\omega t$
(b) $y = A\cos kx$
(c) $y = A\sin(at - bx + c)$
(d) $y = A(\omega t - kx)$

16. The frequency of sinusoidal wave $y = 0.40\cos[2000\,t + 0.80]$ would be

NCERT | Page-283 | 1992

(a) $1000\,\pi$ Hz (b) 2000 Hz
(c) 20 Hz (d) $\frac{1000}{\pi}$ Hz

17. Velocity of sound waves in air is 330 m/s. For a particular sound wave in air, a path difference of 40 cm is equivalent to phase difference of $1.6\,\pi$. The frequency of this wave is

NCERT | Page-283 | 1990

(a) 165 Hz (b) 150 Hz
(c) 660 Hz (d) 330 Hz

18. The transverse wave represented by the equation $y = 4\sin\left(\frac{\pi}{6}\right)\sin(3x - 15t)$ has

NCERT | Page-283 | 1990

(a) amplitude = 4

(b) wavelength $= 4\frac{\pi}{3}$

(c) speed of propagation = 5

(d) period $= \frac{\pi}{15}$

19. Equation of a progressive wave is given by

$y = 4\sin\left[\pi\left(\frac{t}{5} - \frac{x}{9}\right) + \frac{\pi}{6}\right]$

Then which of the following is correct?

NCERT | Page-283 | 1988

(a) $v = 5$ cm (b) $\lambda = 18$ cm
(c) $a = 0.04$ cm (d) f = 50 Hz

Topic-14.4

The Speed of a Travelling Wave

20. If the initial tension on a stretched string is doubled, then the ratio of the initial and final speeds of a transverse wave along the string is:

NCERT | Page-285 | 2022

(a) $\sqrt{2}:1$ (b) $1:\sqrt{2}$ (c) 1 : 2 (d) 1 : 1

21. Sound waves travel at 350 m/s through a warm air and at 3500 m/s through brass. The wavelength of a 700 Hz acoustic wave as it enters brass from warm air

NCERT | Page-284 | 2011

(a) decreases by a factor 10
(b) increases by a factor 20
(c) increases by a factor 10
(d) decreases by a factor 20

22. A transverse wave is represented by $y = A\sin(\omega t - kx)$. For what value of the wavelength is the wave velocity equal to the maximum particle velocity?

NCERT | Page-284 | 2010

(a) $\frac{\pi A}{2}$ (b) πA (c) $2\pi A$ (d) A

23. A transverse wave propagating along x-axis is represented by $y(x,t) = 8.0\sin(0.5\pi x - 4\pi t - \frac{\pi}{4})$ where x is in metres and t is in seconds. The speed of the wave is

NCERT | Page-284 | 2006

(a) $0.5\,\pi$ m/s (b) $\frac{\pi}{4}$ m/s
(c) 8 m/s (d) 4π m/s

24. A point source emits sound equally in all directions in a non–absorbing medium. Two points P and Q are at distances of 2 m and 3 m respectively from the source. The ratio of the intensities of the waves at P and Q is **2005**

(a) 3 : 2 (b) 2 : 3
(c) 9 : 4 (d) 4 : 9

25. The equation of a wave is represented by: $y = 10^{-4}\sin\left[100t - \frac{x}{10}\right]$. The velocity of the wave will be **NCERT | Page-284 | 2001**

(a) 100 m/s (b) 250 m/s
(c) 750 m/s (d) 1000 m/s

26. The speed of a wave in a medium is 960 m/s. If 3600 waves are passing through a point in the medium in 1 min., then the wavelength of the wave is **NCERT | Page-283 | 1997**

(a) 8 m (b) 12 m
(c) 16 m (d) 20 m

27. The equation of a travelling wave is $y = 60\cos(180\,t - 6x)$ where y is in microns, t in second and x in metres. The ratio of maximum particle velocity to velocity of wave propagation is **NCERT | Page-284 | 1997**

(a) 3.6 (b) 3.6×10^{-4}
(c) 3.6×10^{-6} (d) 3.6×10^{-11}

28. What is the effect of humidity on sound waves when humidity increases? **NCERT | Page-286 | 1996**

(a) speed of sound waves is more
(b) speed of sound waves is less
(c) speed of sound waves remains same
(d) speed of sound waves becomes zero

29. The speed of a wave in a medium is 760 m/s. If 3600 waves are passing through a point in the medium in 2 min, then their wavelength is **NCERT | Page-284 | 1995**

(a) 13.8 m (b) 25.3 m
(c) 41.5 m (d) 57.2 m

30. A hospital uses an ultrasonic scanner to locate tumours in a tissue. The operating frequency of the scanner is 4.2 MHz. The speed of sound in a tissue is 1.7 km/s. The wavelength of sound in tissue is close to **NCERT | Page-284 | 1995**

(a) 4×10^{-4} m (b) 8×10^{-4} m
(c) 4×10^{-3} m (d) 8×10^{-3} m

31. A standing wave is represented by $y = A\sin(100t)\cos(0.01x)$, where y and A are in millimetre, t in seconds and x is in metre. Velocity of wave is **NCERT | Page-284 | 1994**

(a) 10^4 m/s
(b) 1 m/s
(c) 10^{-4} m/s
(d) not derivable from above data

32. The temperature at which the speed of sound becomes double as was at 27°C is **NCERT | Page-286 | 1993**

(a) 273° C (b) 0° C
(c) 927° C (d) 1027°C

33. When sound waves travel from air to water, which of the following remains constant? **NCERT | Page-283 | 1989**

(a) Velocity (b) Wavelength
(c) Frequency (d) All of the above

34. The velocity of sound in any gas depends upon **NCERT | Page-285 | 1988**

(a) wavelength of sound only
(b) density and elasticity of gas
(c) intensity of sound waves only
(d) amplitude and frequency of sound

35. A uniform rope of length L and mass m_1 hangs vertically from a rigid support. A block of mass m_2 is attached to the free end of the rope. A transverse pulse of wavelength λ_1 is produced at the lower end of the rope. The wavelength of the pulse when it reaches the top of the rope is λ_2 the ratio λ_2/λ_1 is **NCERT | Page-291 | 2016**

(a) $\sqrt{\frac{m_1}{m_2}}$ (b) $\sqrt{\frac{m_1+m_2}{m_2}}$
(c) $\sqrt{\frac{m_2}{m_1}}$ (d) $\sqrt{\frac{m_1+m_2}{m_1}}$

36. A string of 7 m length has a mass of 0.035 kg. If tension in the string is 60.5 N, then speed of a wave on the string is **NCERT | Page-286 | 2001, 1989**

(a) 77 m/s (b) 102 m/s
(c) 110 m/s (d) 165 m/s

37. If the amplitude of sound is doubled and the frequency is reduced to one fourth, the intensity of sound at the same point will be **NCERT | Page-293 | 1989**

(a) increasing by a factor of 2
(b) decreasing by a factor of 2
(c) decreasing by a factor of 4
(d) unchanged

38. 4.0 g of a gas occupies 22.4 litres at NTP. The specific heat capacity of the gas at constant volume is $5.0\,JK^{-1}$. If the speed of sound in this gas at NTP is $952\,ms^{-1}$, then the heat capacity at constant pressure is (Take gas constant $R = 8.3\,JK^{-1}\,mol^{-1}$) **NCERT | Page-253 | 2015 RS**

(a) $7.5\,JK^{-1}\,mol^{-1}$ (b) $7.0\,JK^{-1}\,mol^{-1}$
(c) $8.5\,JK^{-1}\,mol^{-1}$ (d) $8.0\,JK^{-1}\,mol^{-1}$

Topic-14.5
The Principle of Superposition of Waves

39. Two waves of the same frequency and intensity superimpose each other in opposite phases. After the superposition, the intensity and frequency of waves will **NCERT | Page-293 | 1996**

(a) increase (b) decrease
(c) remain constant (d) become zero

Topic-14.6

Reflection of Waves

40. A pipe open at both ends has a fundamental frequency f in air. The pipe is now dipped vertically in a water drum to half of its length. The fundamental frequency of the air column is now equal to : **NCERT | Page-292 | 2025**
(a) $\frac{f}{2}$ (b) f
(c) $\frac{3f}{2}$ (d) $2f$

41. The 4th overtone of a closed organ pipe is same as that of 3rd overtone of an open pipe. The ratio of the length of the closed pipe to the length of the open pipe is: **NCERT | Page-292 | Manipur 2023**
(a) 9 : 8 (b) 7 : 9
(c) 8 : 9 (d) 9 : 7

42. The ratio of frequencies of fundamental harmonic produced by an open pipe to that of closed pipe having the same length is **NCERT | Page-292 | 2023**
(a) 1 : 2 (b) 2 : 1
(c) 1 : 3 (d) 3 : 1

43. An organ pipe filled with a gas at 27°C resonates at 400 Hz in its fundamental mode. If it is filled with the same gas at 90°C, the resonance frequency at the same mode will be: **NCERT | Page-292 | RS 2022**
(a) 512 Hz (b) 420 Hz
(c) 440 Hz (d) 484 Hz

44. In a guitar, two strings A and B made of same material are slightly out of tune and produce beats of frequency 6 Hz. When tension in B is slightly decreased, the beat frequency increases to 7 Hz. If the frequency of A is 530 Hz, the original frequency of B will be **NCERT | Page-293 | 2020**
(a) 524 Hz (b) 536 Hz
(c) 537 Hz (d) 523 Hz

45. A tuning fork with frequency 800 Hz produces resonance in a resonance column tube with upper end open and lower end closed by water surface. Successive resonance are observed at lengths 9.75 cm, 31.25 cm and 52.75 cm. The speed of sound in air is, **NCERT | Page-292 | Odisha 2019**
(a) 172 m/s (b) 500 m/s
(c) 156 m/s (d) 344 m/s

46. A tuning fork is used to produce resonance in a glass tube. The length of the air column in this tube can be adjusted by a variable piston. At room temperature of 27°C two successive resonances are produced at 20 cm and 73 cm of column length. If the frequency of the tuning fork is 320 Hz, the velocity of sound in air at 27°C is **NCERT | Page-292 | 2018**
(a) 330 m/s (b) 339 m/s
(c) 300 m/s (d) 350 m/s

47. The fundamental frequency in an open organ pipe is equal to the third harmonic of a closed organ pipe. If the length of the closed organ pipe is 20 cm, the length of the open organ pipe is **NCERT | Page-292 | 2018**
(a) 13.2 cm (b) 8 cm
(c) 16 cm (d) 12.5 cm

48. The two nearest harmonics of a tube closed at one end and open at other end are 220 Hz and 260 Hz. What is the fundamental frequency of the system? **NCERT | Page-292 | 2017**
(a) 20 Hz (b) 30 Hz
(c) 40 Hz (d) 10 Hz

49. An air column, closed at one end and open at the other, resonates with a tuning fork when the smallest length of the column is 50 cm. The next larger length of the column resonating with the same tuning fork is : **NCERT | Page-292 | 2016**
(a) 66.7 cm (b) 100 cm
(c) 150 cm (d) 200 cm

50. The fundamental frequency of a closed organ pipe of length 20 cm is equal to the second overtone of an organ pipe open at both the ends. The length of organ pipe open at both the ends is **NCERT | Page-292 | 2015**
(a) 100 cm (b) 120 cm
(c) 140 cm (d) 80 cm

51. A string is stretched between two fixed points separated by 75.0 cm. It is observed to have resonant frequencies of 420 Hz and 315 Hz. There are no other resonant frequencies between these two. The lowest resonant frequency for this string is : **NCERT | Page-291 | 2015 RS**
(a) 205 Hz (b) 10.5 Hz
(c) 105 Hz (d) 155 Hz

52. If n_1, n_2 and n_3 are the fundamental frequencies of three segments into which a string is divided, then the original fundamental frequency n of the string is given by : **NCERT | Page-291 | 2014, 2012, 2000**
(a) $\frac{1}{n}=\frac{1}{n_1}+\frac{1}{n_2}+\frac{1}{n_3}$
(b) $\frac{1}{\sqrt{n}}=\frac{1}{\sqrt{n_1}}+\frac{1}{\sqrt{n_2}}+\frac{1}{\sqrt{n_3}}$
(c) $\sqrt{n}=\sqrt{n_1}+\sqrt{n_2}+\sqrt{n_3}$
(d) $n=n_1+n_2+n_3$

53. The number of possible natural oscillation of air column in a pipe closed at one end of length 85 cm whose frequencies lie below 1250 Hz are : (velocity of sound = 340 ms^{-1}) **NCERT | Page-292 | 2014**
(a) 4 (b) 5 (c) 7 (d) 6

54. If we study the vibration of a pipe open at both ends, then which of the following statements is not true ? **NCERT | Page-292 | 2013**
(a) Odd harmonics of the fundamental frequency will be generated
(b) All harmonics of the fundamental frequency will be generated

(c) Pressure change will be maximum at both ends
(d) Antinode will be at open end

55. The length of the wire between two ends of a sonometer is 100 cm. What should be the positions of two bridges below the wire so that the three segments of the wire have their fundamental frequencies in the ratio of 1 : 3 : 5? NCERT | Page-291 | Kar. 2013

(a) $\frac{1500}{23}$ cm, $\frac{2000}{23}$ cm (b) $\frac{1500}{23}$ cm, $\frac{500}{23}$ cm
(c) $\frac{1500}{23}$ cm, $\frac{300}{23}$ cm (d) $\frac{300}{23}$ cm, $\frac{1500}{23}$ cm

56. A standing wave having 3 nodes and 2 antinodes is formed between two atoms having a distance 1.21 Å between them. The wavelength of the standing wave is NCERT | Page-290 | 1998

(a) 1.21 Å (b) 2.42 Å
(c) 6.05 Å (d) 3.63 Å

57. An organ pipe P_1 closed at one end vibrating in its first overtone and another pipe P_2, open at both ends vibrating in its third overtone are in resonance with a given tuning fork. The ratio of lengths of P_1 and P_2 respectively are given by NCERT | Page-292 | 1997

(a) 1 : 2 (b) 1 : 3 (c) 3 : 8 (d) 3 : 4

58. A cylindrical resonance tube open at both ends, has a fundamental frequency, 'f' in air. If half of the length is dipped vertically in water, the fundamental frequency of the air column will be NCERT | Page-292 | 1997

(a) $2f$ (b) $3f/2$ (c) f (d) $f/2$

59. Two waves are approaching each other with a velocity of 20 m/s and frequency n. The distance between two consecutive nodes is NCERT | Page-290 | 1995

(a) $\frac{20}{n}$ (b) $\frac{10}{n}$ (c) $\frac{5}{n}$ (d) $\frac{n}{10}$

60. A wave of frequency 100 Hz is sent along a string towards a fixed end. When this wave travels back after reflection, a node is formed at a distance of 10 cm from the fixed end of the string. The speeds of incident (and reflected) waves are NCERT | Page-290 | 1994

(a) 5 m/s (b) 10 m/s
(c) 20 m/s (d) 40 m/s

61. A stretched string resonates with tuning fork frequency 512 Hz when length of the string is 0.5 m. The length of the string required to vibrate resonantly with a tuning fork of frequency 256 Hz would be NCERT | Page-291 | 1993

(a) 0.25 m (b) 0.5 m
(c) 1 m (d) 2 m

62. A closed organ pipe (closed at one end) is excited to support the third overtone. It is found that air in the pipe has NCERT | Page-291 | 1991

(a) three nodes and three antinodes
(b) three nodes and four antinodes
(c) four nodes and three antinodes
(d) four nodes and four antinodes

Topic-14.7 Beats

63. Three sound waves of equal amplitudes have frequencies (n –1), n, (n +1). They superimpose to give beats. The number of beats produced per second will be: NCERT | Page-293 | Ph-II 2016

(a) 3 (b) 2 (c) 1 (d) 4

64. A source of unknown frequency gives 4 beats/s, when sounded with a source of known frequency 250 Hz. The second harmonic of the source of unknown frequency gives five beats per second, when sounded with a source of frequency 513 Hz. The unknown frequency is NCERT | Page-293 | 2013

(a) 246 Hz (b) 240 Hz
(c) 260 Hz (d) 254 Hz

65. Two sources P and Q produce notes of frequency 660 Hz each. A listener moves from P to Q with a speed of 1 ms^{-1}. If the speed of sound is 330 m/s, then the number of beats heard by the listener per second will be NCERT | Page-293 | Kar. 2013

(a) zero (b) 4 (c) 8 (d) 2

66. Two sources of sound placed close to each other are emitting progressive waves given by $y_1 = 4 \sin 600 \pi t$ and $y_2 = 5 \sin 608 \pi t$. An observer located near these two sources of sound will hear : NCERT | Page-293 | 2012

(a) 4 beats per second with intensity ratio 25 : 16 between waxing and waning.
(b) 8 beats per second with intensity ratio 25 : 16 between waxing and waning
(c) 8 beats per second with intensity ratio 81 : 1 between waxing and waning
(d) 4 beats per second with intensity ratio 81 : 1 between waxing and waning

67. Two identical piano wires kept under the same tension T have a fundamental frequency of 600 Hz. The fractional increase in the tension of one of the wires which will lead to occurrence of 6 beats/s when both the wires oscillate together would be NCERT | Page-293 | 2011M

(a) 0.02 (b) 0.03
(c) 0.04 (d) 0.01

68. A tuning fork of freqqency 512 Hz makes 4 beats per second with the vibrating string of a piano. The beat frequency decreases to 2 beats per sec when the tension in the piano string is slightly increased. The frequency of the piano string before increasing the tension was NCERT | Page-293 | 2010

(a) 510 Hz (b) 514 Hz
(c) 516 Hz (d) 508 Hz

69. Each of the two strings of length 51.6 cm and 49.1 cm are tensioned separately by 20 N force. Mass per unit length of both the strings is same and equal to 1 g/m. When both the strings vibrate simultaneously the number of beats is NCERT | Page-293 | 2009

(a) 7 (b) 8 (c) 3 (d) 5

70. Two sound waves with wavelengths 5.0 m and 5.5m respectively, each propagate in a gas with velocity 330 m/s. We expect the following number of beats per second **NCERT | Page-293 | 2006**
(a) 0 (b) 1 (c) 6 (d) 12

71. Two vibrating tuning forks produce progressive waves given by $y_1 = 4 \sin 500\,\pi t$ and $y_2 = 2 \sin 506\,\pi t$. Number of beats produced per minute is **NCERT | Page-293 | 2005**
(a) 360 (b) 180 (c) 60 (d) 3

72. Two waves of lengths 50 cm and 51 cm produce 12 beats per sec. The velocity of sound is **NCERT | Page-293 | 1999**
(a) 306 m/s (b) 331 m/s
(c) 340 m/s (d) 360 m/s

73. A source of sound gives 5 beats per second, when sounded with another source of frequency 100/sec. The second harmonic of the source, together with a source of frequency 205/sec gives 5 beats per second. What is the frequency of the source? **NCERT | Page-293 | 1995**
(a) 95 sec^{-1} (b) 100 sec^{-1}
(c) 105 sec^{-1} (d) 205 sec^{-1}

74. For production of beats the two sources must have **NCERT | Page-293 | 1992**
(a) different frequencies and same amplitude
(b) different frequencies
(c) different frequencies, same amplitude and same phase
(d) different frequencies and same phase

75. A wave has S.H.M whose period is 4 seconds while another wave which also possess SHM has its period 3 seconds. If both are combined, then the resultant wave will have the period equal to **NCERT | Page-293 | 1993**
(a) 4 seconds (b) 5 seconds
(c) 12 seconds (d) 3 seconds

ANSWER KEY

1	(a)	9	(a)	17	(c)	25	(d)	33	(c)	41	(a)	49	(c)	57	(c)	65	(b)	73	(c)
2	(b)	10	(d)	18	(c)	26	(c)	34	(b)	42	(b)	50	(b)	58	(c)	66	(d)	74	(b)
3	(d)	11	(a)	19	(b)	27	(b)	35	(b)	43	(c)	51	(c)	59	(b)	67	(a)	75	(c)
4	(d)	12	(d)	20	(b)	28	(a)	36	(c)	44	(a)	52	(a)	60	(c)	68	(d)		
5	(a)	13	(d)	21	(c)	29	(b)	37	(c)	45	(d)	53	(d)	61	(d)	69	(a)		
6	(c)	14	(c)	22	(c)	30	(a)	38	(d)	46	(b)	54	(c)	62	(d)	70	(c)		
7	(c)	15	(c)	23	(c)	31	(a)	39	(c)	47	(a)	55	(a)	63	(b)	71	(b)		
8	(b)	16	(d)	24	(c)	32	(c)	40	(b)	48	(a)	56	(a)	64	(d)	72	(a)		

Hints & Solutions

1. **(a)** Sound waves in air are longitudinal and the light waves are transverse.

NOTES Sound waves are longitudinal waves and require medium for their propagation.
Light waves i.e., electromagnetic waves are transverse in nature and do not require medium for their propagation.

2. **(b)** With the propagation of a longitudinal wave, energy alone is propagated.

NOTES Energy is propagated along with the wave motion without any net transport of the matter.

3. **(d)** Given, $y = c\sin\frac{2\pi}{\lambda}(at - x)$

$$\Rightarrow y = c\sin\left(\frac{2\pi}{\lambda}at - \frac{2\pi}{\lambda}x\right)$$

Comparing with $y = A\sin(\omega t - kx)$, we get

$$\omega = 2\pi f = \frac{2\pi a}{\lambda}$$

$$f = \frac{a}{\lambda}$$

4. **(d)** The standard equation of a wave travelling along x-axis (+ve direction) is given by, $Y = A\sin(kx - \omega t)$ where,

angular frequency, $\omega = 2\pi f$

$$\Rightarrow \frac{2\pi}{\pi} = 2 \qquad \left[\because f = \frac{1}{\pi}\right]$$

angular wave number, $k = \frac{2\pi}{\lambda} \Rightarrow \frac{2\pi}{2\pi} = 1$

$[\because \lambda = 2\pi]$

$\therefore Y = 1\sin(x - 2t)$ $[\because$ Amplitude, A = 1 m]

NOTES If the sign between t and x terms is negative the wave is propagating along +(ve) X-axis and if the sign is positive then wave moves in –(ve) X-axis direction.

5. **(a)** Here, $y_1 = a \sin(\omega t + kx + 0.57)$

so, $\phi_1 = 0.57$

and $y_2 = a\cos(\omega t + kx)$

$= a\sin\left[\frac{\pi}{2} + (\omega t + kx)\right]$

so, $\phi_2 = \pi/2$

Phase difference, $\Delta\phi = \phi_2 - \phi_1$

$= \frac{\pi}{2} - 0.57 = \frac{3.14}{2} - 0.57 = 1.57 - 0.57$

= 1 radian

6. **(c)** $A = 2\text{cm}, \frac{\omega}{k} = 128\ \text{ms}^{-1}, 5\lambda = 4, \lambda = \frac{4}{5}\text{m}$

$y = A\sin(kx - \omega t)$,

$k = \frac{2\pi}{\lambda} \Rightarrow \frac{2\pi\times 5}{4} = \frac{31.4}{4} = 7.85$

$y = 0.02\ \text{m}\sin(7.85x - 1005\,t)$

$\omega = 128 \times 7.85 = 1005$

7. **(c)** $y = 0.25\sin(10\pi x - 2\pi t)$

Comparing this equation with the standard wave equation

$y = a\sin(kx - \omega t)$

We get, $k = 10\pi$

$\Rightarrow \frac{2\pi}{\lambda} = 10\pi \Rightarrow \lambda = 0.2\ \text{m}$

And $\omega = 2\pi$ or, $2\pi\nu = 2\pi \Rightarrow \nu = 1\text{Hz}$.

The sign inside the bracket is negative, hence the wave travels in +ve x-direction.

8. **(b)** $y_1 = 10^{-6}\sin(100\,t + x/50 + 0.5)\text{m}$

$= 10^{-6}\cos(100\,t + x/50 - \pi/2 + 0.5)\text{m}$

$y_2 = 10^{-6}\cos(100\,t + x/50)\text{m}$

$\therefore \quad \phi = \pi/2 - 0.5 = 1.07$ rad

9. **(a)** As we know that, velocity $v = n\lambda$

$\Rightarrow$ frequency, $n = \frac{v}{\lambda} = \frac{360}{60} = 6$

Amplitude, $a = 0.2$

For a wave travelling along positive x-axis

$y = a\sin(\omega t - kx)$

$= a\sin\left(2\pi nt - \frac{2\pi x}{\lambda}\right)$

$= a\sin 2\pi\left(nt - \frac{x}{\lambda}\right) = 0.2\sin 2\pi\left(6t - \frac{x}{60}\right)$

10. **(d)** $y = y_0\sin\frac{2\pi}{\lambda}(vt - x)$

Particle velocity,

$\frac{dy}{dt} = y_0 \times \frac{2\pi}{\lambda}v\cos\frac{2\pi}{\lambda}(vt - x)$

Maximum particle velocity $= y_0 \times \frac{2\pi v}{\lambda}$

Wave velocity $= v$ [given]

So, $y_0 \times \frac{2\pi v}{\lambda} = 2v$

$\lambda = \pi \cdot y_0$

11. **(a)** Time taken to move from maximum to zero displacement $= \frac{T}{4}$

$\therefore$ Time period $T = 4 \times 0.170$ second

$\therefore$ Frequency, $n = \frac{1}{T} = \frac{1}{4\times 0.170} = 1.47$ Hz

12. **(d)** Phase difference $= 60° = \frac{\pi}{3}$

Path difference $= \frac{\lambda}{2\pi}$ (phase diff.) $= \frac{\lambda}{2\pi} \times \frac{\pi}{3} = \frac{\lambda}{6}$

13. **(d)** $y = 0.0015\sin(62.4x + 316t)$

On comparing with $y = A\sin(\omega t + kx)$

$\omega = 316, k = 62.4$

$\Rightarrow k = \frac{2\pi}{\lambda} = 62.4 \Rightarrow \lambda = 0.1$ unit

14. **(c)** Given, $y = 0.5\sin\frac{2\pi}{3.2}(64t - x)$

Standard equation of the wave is :

$y = a\sin\frac{2\pi}{\lambda}(vt - x)$

Comparing the given equation with the standard equation, we get $v = 64$ and $\lambda = 3.2$. Therefore, frequency $= \frac{64}{3.2} = 20\,\text{Hz}$.

15. **(c)** $y = A\sin(at - bx + c)$ represents a wave, where a may correspond to ω and b may correspond to K.

16. **(d)** Given $y = 0.40\cos[2000t + 0.80]$

Comparing with the standard equation

$y = A\cos(2\pi\nu t + \phi)$

We get angular frequency, $2\pi\nu = 2000$

$\nu \Rightarrow \frac{2000}{2\pi} = \frac{1000}{\pi}$ Hz

17. **(c)** From $\Delta x = \frac{\lambda}{2\pi}\Delta\phi,$

$$\lambda = 2\pi\frac{\Delta x}{\Delta\phi} = \frac{2\pi(0.4)}{1.6\pi} = 0.5\text{m}$$

$$f = \frac{v}{\lambda} = \frac{330}{0.5} = 660 \text{ Hz}$$

18. **(c)** Given $y = 4\sin\left(\frac{\pi}{6}\right)\sin(3x - 15t)$

Compare the given equation with standard form

$$y = A\sin\left[\frac{2\pi x}{\lambda} - \frac{2\pi t}{T}\right]$$

where, $\frac{2\pi}{\lambda} = 3, \lambda = \frac{2\pi}{3}$ and $\frac{2\pi}{T} = 15$

$$T = \frac{2\pi}{15}$$

Speed of propagation,

$$v = \frac{\lambda}{T} = \frac{2\pi/3}{2\pi/15} = 5$$

19. **(b)** The standard equation of a progressive wave is

$$y = a\sin\left[2\pi\left(\frac{t}{T} - \frac{x}{\lambda}\right) + \phi\right]$$

The given equation can be written as

$$y = 4\sin\left[2\pi\left(\frac{t}{10} - \frac{x}{18}\right) + \frac{\pi}{6}\right]$$

$\therefore$ $a = 4$ cm, $T = 10$ s, $\lambda = 18$ cm and $\phi = \pi/6$

Hence, (b) is correct.

20. **(b)** Speed $V = \sqrt{\frac{T}{\mu}} \Rightarrow V \propto \sqrt{T}$

So, $\frac{V_1}{V_2} = \sqrt{\frac{T_1}{T_2}} = \frac{1}{\sqrt{2}}$

NOTES A transverse wave propagates along a stretched string when plucked and then released, with a speed $v = \sqrt{\frac{T}{M}}$

T = tension and M = mass per unit length.

21. **(c)** We have, $v = n\lambda$

$\Rightarrow v \propto \lambda$ (as n remains constant)

Thus, as v increases 10 times, λ also increases 10 times.

22. **(c)** $y = A\sin(\omega t - kx)$

Particle velocity,

$$v_p = \frac{dy}{dt} = A\,\omega\cos(\omega t - kx)$$

$$\therefore \quad v_{p_{max}} = A\omega$$

wave velocity $= \frac{\omega}{k}$

$$\therefore \quad A\omega = \frac{\omega}{k}$$

i.e., $A = \frac{1}{k}$ But $k = \frac{2\pi}{\lambda}$

$$\therefore \quad \lambda = 2\pi A$$

23. **(c)** Speed of a wave represented by the equation $y(x, t) = A\sin(kx - \omega t + \phi)$ is $v = \frac{\omega}{k}$

By comparison, $\omega = 4\pi$; $k = 0.5\pi$

$$v = \frac{\omega}{k} = \frac{4\pi}{0.5\pi} = 8 \text{ m/sec}$$

24. **(c)** Intensity = Energy/sec/unit area

Area $\propto r^2 \Rightarrow I \propto 1/r^2$

$$\Rightarrow \frac{I_1}{I_2} = \frac{r_2^2}{r_1^2} = \frac{3^2}{2^2} = \frac{9}{4}$$

25. **(d)** $y = 10^{-4}\sin\left[100t - \frac{x}{10}\right]$

frequency, $n = \left(\frac{100}{2\pi}\right)\text{sec}^{-1}, \lambda = 2\pi 10$ m

velocity $v = \lambda n = \left(\frac{100}{2\pi}\right)(2\pi \times 10) = 1000$ m/s

26. **(c)** Given speed of wave (v) = 960m/s

Frequency of wave (f) = 3600/ min

$$= \frac{3600}{60} \text{ rev/sec} = 60 \text{ rev per sec.}$$

Wavelength of waves (λ) $= \frac{v}{f} = \frac{960}{60} = 16$m.

27. **(b)** $y = 60\cos(180t - 6x)$ ....(1)

$\omega = 180, k = 6 \Rightarrow \frac{2\pi}{\lambda} = 6$

Wave velocity,

$$v = \frac{\omega}{k} = \frac{2\pi}{T} \times \frac{\lambda}{2\pi} = \frac{180}{6} = 30 \text{ m/s}$$

Differentiating (1) w.r.t. t,

Particle velocity,

$$v_p = \frac{dy}{dt} = -60 \times 180\sin(180t - 6x)$$

$v_{p\,max} = 60 \times 180$ µm/s

= 10800 µm/s = 0.0108 m/s

$$\frac{v_{p\,max}}{v} = \frac{0.0108}{30} = 3.6 \times 10^{-4}$$

NOTES Wave velocity (v) is the distance travelled by the disturbance in one second. It only depends on the properties of the medium and is independent of time and position.

$$V = n\lambda = \frac{\lambda}{T} = \frac{\omega\lambda}{2\pi} = \frac{\omega}{k}$$

28. **(a)** Velocity of sound $= \sqrt{\frac{\gamma RT}{M}}$

When water vapour are present in air, average molecular weight of air decreases and hence velocity increases.

29. **(b)** Speed of the wave (v) = 760 m/s;

Number of waves = 3600

and time taken (t) = 2 min = 120 sec.

Frequecny of waves

$$(n) = \frac{3600}{120} = 30/\text{sec}.$$

$$\therefore \text{wavelength of waves } (\lambda) = \frac{v}{n} = \frac{760}{30} = 25.3\text{m}$$

30. **(a)** Frequency (n) = 4.2 MHz = 4.2 × 106 Hz and speed of sound (v) = 1.7 km/s = 1.7×10^3 m/s. Wave length of sound in tissue,

$$\lambda = \frac{v}{n} = \frac{1.7\times10^3}{4.2\times10^6} = 4\times10^{-4}\,m$$

31. **(a)** The wave equation is y = Asin (ωt) cos (kx); $c = \omega/k = 100/0.01 = 10^4$ m/s.

32. **(c)** We have $c \propto \sqrt{T}$

$$\therefore \frac{c_1}{c_2} = \sqrt{\frac{T_1}{T_2}}$$

$$\Rightarrow \frac{c_1}{2c_2} = \sqrt{\frac{(27+273)k}{T_2}}$$

$$\Rightarrow T_2 = 1200\,k = 927°\text{C}$$

33. **(c)** When sound wave travel from air to water, frequency remains constant.

34. **(b)** Velocity of sound in any gas depends upon density and elasticity of gas.

35. **(b)** From figure, tension $T_1 = m_2 g$

$T_2 = (m_1 + m_2)g$

As we know

Velocity $\propto \sqrt{T}$

So, $\lambda \propto \sqrt{T}$

$$\Rightarrow \frac{\lambda_1}{\lambda_2} = \frac{\sqrt{T_1}}{\sqrt{T_2}}$$

$$\Rightarrow \frac{\lambda_2}{\lambda_1} = \sqrt{\frac{m_1 + m_2}{m_2}}$$

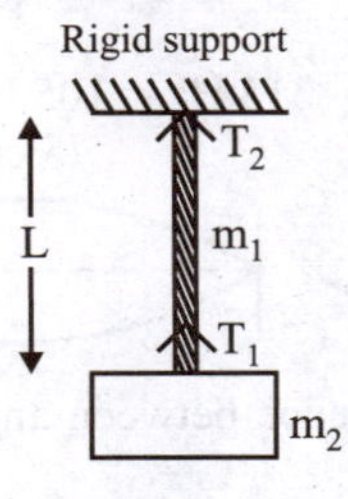

36. **(c)** Given : Length (l) = 7 m

Mass (M) = 0.035 kg and tension (T) = 60.5 N.

Therefore, mass of string per unit length (m) = $\frac{0.035}{7} = 0.005$ kg / m

$$\text{speed of wave, } = \sqrt{\frac{T}{m}} \Rightarrow \sqrt{\frac{60.5}{0.005}} = 110 \text{ m/s}$$

37. **(c)** Intensity is proportional to $(\text{amplitude})^2$ and also intensity α $(\text{frequency})^2$. Therefore, intensity becomes

$$\frac{2^2}{4^2} = \frac{1}{4}\text{th}$$

38. **(d)** Molar mass of the gas = 4g/mol

Speed of sound

$$v = \sqrt{\frac{\gamma RT}{m}} \Rightarrow 952 = \sqrt{\frac{\gamma\times8.3\times273}{4\times10^{-3}}}$$

$$\Rightarrow \quad \gamma = 1.6 = \frac{16}{10} = \frac{8}{5}$$

$$\text{Also, } \gamma = \frac{C_P}{C_V} = \frac{8}{5}$$

$$\text{So, } C_P = \frac{8\times5}{5} = 8\text{JK}^{-1}\text{mol}^{-1}$$

$[C_V = 5.0 \text{ JK}^{-1} \text{ given}]$

39. **(c)** After the superposition, frequency and intensity of waves will remain constant.

40. **(b)** For open pipe, Fundamental frequency

$$f = \frac{V}{2\ell}$$

When half of its length is dipped in water, Now its fundamental frequency

$$f' = \frac{V}{4\left(\frac{\ell}{2}\right)}$$

$$\Rightarrow f' = \frac{V}{2\ell} \Rightarrow f' = f$$

41. **(a)** 4th overtone for closed organ pipe → 9th harmonic and 3rd overtone for open organ pipe → 4th harmonic

$$\therefore 9\left(\frac{V}{4l_c}\right) = 4\left(\frac{V}{2l_o}\right) \Rightarrow \frac{l_c}{l_o} = \frac{9}{8}$$

42. **(b)** Fundamental harmonic produced by open pipe,

$$f_{\text{open pipe}} = \frac{v}{2l}$$

And by closed pipe, $f_{\text{closed pipe}} = \frac{v}{4l}$

$$\therefore \frac{f_0}{f_c} = \frac{v}{2l} \times \frac{4l}{v} = \frac{2}{1} \quad \therefore f_0 : f_c = 2 : 1$$

43. (c) Velocity of sound, V

$\propto \sqrt{T}$ and $f = \frac{V}{\lambda}$

$\therefore \frac{f_1}{f_2} = \sqrt{\frac{T_1}{T_2}}$

$\frac{400}{f_2} = \sqrt{\frac{273+27}{273+90}} = \sqrt{\frac{300}{363}} = \frac{1}{1.1}$

$\therefore f_2 = 440$ Hz

44. (a) Frequency of string, $f = \frac{1}{2l}\sqrt{\frac{T}{m}}$

Frequency $\propto \sqrt{\text{Tension}}$

Difference of f_A and f_B is 6 Hz.

If tension decreases, f_B decreases and becomes f'_B.

Now, difference of f_A and $f'_B = 7$ Hz (increases)

So, $f_A > f_B$

$f_A - f_B = 6$ Hz

$\Rightarrow f_A = 530$ Hz $\Rightarrow f_B = 524$ Hz (original)

45. (d) $l_1 = 9.75$ cm

$l_2 = 31.25$ cm

$l_3 = 52.75$ cm

e = end correction

$\frac{\lambda}{4} + e = 9.75$ cm ...(i)

$\frac{3\lambda}{4} + e = 31.25$ cm ...(ii)

$\frac{3\lambda}{4} - \frac{\lambda}{4} = 31.25 - 9.75$

$\frac{\lambda}{2} = 21.5$

$\lambda = 43$ cm

$v = \lambda \times f$

$v = 34400 \times 10^{-2}$ $\therefore$ $v = 344\ \text{ms}^{-1}$

46. (b) Two successive resonance are produced at 20 cm and 73 cm of column length

$\therefore \frac{\lambda}{2} = (73 - 20) \times 10^{-2}$ m

$\Rightarrow \lambda = 2 \times (73 - 20) \times 10^{-2}$

Velocity of sound, $v = n\lambda$

$= 2 \times 320\,[73 - 20] \times 10^{-2} = 339.2\ \text{ms}^{-1}$

47. (a) For closed organ pipe, third harmonic

$n = \frac{(2N-1)V}{4\ell} = \frac{3V}{4\ell}$ $(\because N = 2)$

For open organ pipe, fundamental frequency

$n = \frac{NV}{2\ell} = \frac{V}{2\ell'}$ $(\because N = 1)$

According to question, $\frac{3V}{4\ell} = \frac{V}{2\ell'}$

$\Rightarrow \ell' = \frac{4\ell}{3 \times 2} = \frac{2\ell}{3} = \frac{2 \times 20}{3} = 13.33$ cm

48. (a) Difference in two successive frequencies of closed pipe,

$\frac{2v}{4l} \Rightarrow 260 - 220 = 40$ Hz

or $\frac{2v}{4l} = 40$ Hz $\Rightarrow \frac{v}{4\ell} = 20$Hz

Which is the fundamental frequency of system of closed organ pipe.

49. (c) For a closed organ pipe first minimum resonating length

$L_1 = \frac{\lambda}{4} = 50$ cm

$\therefore$ Next or second resonating length, L_2

$\Rightarrow \frac{3\lambda}{4} = 150$ cm

NOTES If an open pipe is half submerged in water, it will become a closed organ pipe of length half that of a open pipe. It's fundamental frequency will become

$n' = \frac{V}{4\left(\frac{\ell}{2}\right)} = \frac{V}{2\ell}$ equal to that of open pipe.

50. (b) Fundamental frequency of closed organ pipe,

$v_c = \frac{v}{4l_c}$

Fundamental frequency of open organ pipe

$v_0 = \frac{v}{2l_0}$

Second overtone frequency of open organ pipe $= \frac{3v}{2l_0}$

From question, $\frac{v}{4l_c} = \frac{3v}{2l_0}$

$\Rightarrow l_0 = 6l_c = 6 \times 20 = 120$ cm

51. (c) Resonant frequencies, for a string fixed at both ends, will be

$f_n = \frac{nv}{2L}$ where, $n = 1, 2, 3.....$

The difference between two consecutive resonant frequency is,

$\Delta f_n = f_{n+1} - f_n \Rightarrow \frac{(n+1)v}{2L} - \frac{nv}{2L} = \frac{v}{2L}$

which is also the lowest resonant frequency (n = 1).

Thus, for the given string the lowest resonant frequency will be

= 420 Hz – 315 Hz = 105 Hz

75 cm

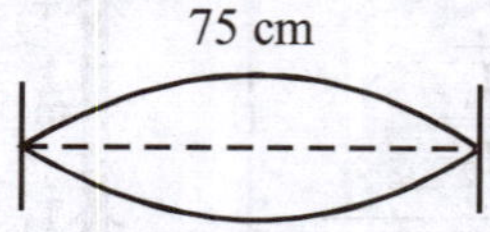

The difference between any two successive frequencies will be 'n'.

According to question, n = 420 – 315 = 105 Hz

So the lowest frequency of the string is 105 Hz.

NOTES In a stretched string all multiples of frequencies can be obtained *i.e.*, if fundamental frequency is n then higher frequencies will be 2n, 3n, 4n ...

52. (a) 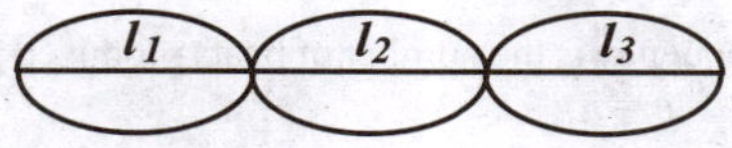

$$n = \frac{1}{2l}\sqrt{\frac{T}{m}}$$

or, $n \propto \frac{1}{l}$ or nl = constant, K

$\therefore n_1 l_1 = K,\ n_2 l_2 = K,\ n_3 l_3 = K$

Also, $l = l_1 + l_2 + l_3$

or, $\frac{K}{n} = \frac{K}{n_1} + \frac{K}{n_2} + \frac{K}{n_3}$ or, $\frac{1}{n} = \frac{1}{n_1} + \frac{1}{n_2} + \frac{1}{n_3}$

53. (d) In case of closed organ pipe frequency,

$$f_n = (2n+1)\frac{v}{4l}$$

for $n = 0,\ f_0 = 100$ Hz

$n = 1,\ f_1 = 300$ Hz

$n = 2,\ f_2 = 500$ Hz

$n = 3,\ f_3 = 700$ Hz

$n = 4,\ f_4 = 900$ Hz

$n = 5,\ f_5 = 1100$ Hz

$n = 6,\ f_6 = 1300$ Hz

Hence possible natural oscillation whose frequencies < 1250 Hz = 6(n = 0, 1, 2, 3, 4, 5)

54. (c) Pressure change will be minimum at both ends. In fact, pressure variation is maximum at $\ell/2$ because the displacement node is pressure antinode.

55. (a) From formula, $f = \frac{1}{x}\sqrt{\frac{T}{m}} \Rightarrow f \propto \frac{1}{l}$

$$\therefore l_1 : l_2 : l_3 = \frac{1}{f_1} : \frac{1}{f_2} : \frac{1}{f_3} = f_2 f_3 : f_1 f_3 : f_1 f_2$$

[Given: $f_1 : f_2 : f_3 = 1 : 3 : 5$]

= 15 : 5 : 3

Therefore the positions of two bridges below the wire are

$$\frac{15 \times 100}{15+5+3}\text{ cm and } \frac{15 \times 100 + 5 \times 100}{15+5+3}\text{ cm}$$

i.e., $\frac{1500}{23}$ cm, $\frac{2000}{23}$ cm

56. (a) Let ℓ be length of string

$$\ell = \left(\frac{\lambda}{2}\right)2 \Rightarrow \lambda = \ell$$

Hence, the wave length of standing wave $\Rightarrow \lambda = \ell$ = 1.21Å

57. (c) We know that the length of pipe closed at one end for first overtone $(l_1) = \frac{3\lambda}{4}$

and length of the open pipe for third overtone (l_2) $= \frac{4\lambda}{2} = 2\lambda.$

Therefore, the ratio of lengths $\frac{l_1}{l_2} = \frac{3\lambda/4}{2\lambda} = \frac{3}{8}$ or $l_1 : l_2$ = 3 : 8.

NOTES In an open pipe all harmonics are present whereas in a closed organ pipe only alternate harmonics of frequencies are present. Hence musical sound produced by an open organ pipe is sweeter than that produced by a closed organ pipe.

58. (c) Fundamental frequency of open pipe, $f = \frac{v}{2l}$

When half of tube is filled with water, then the length of air column becomes half $\left(l' = \frac{l}{2}\right)$ and the pipe becomes closed.

So, new fundamental frequency

$$f' = \frac{v}{4l'} = \frac{v}{4\left(\frac{l}{2}\right)} = \frac{v}{2l}$$

Clearly $f' = f$.

59. (b) Distance between two successive nodes

$$\Rightarrow \frac{\lambda}{2} = \frac{v}{2n} \Rightarrow \frac{20}{2n} = \frac{10}{n} \qquad \left(\text{where } \lambda = \frac{v}{n}\right)$$

60. (c) As fixed end is a node, therefore, distance between two consecutive nodes

$= \frac{\lambda}{2} = 10$ cm.

λ = 20 cm = 0.2 m

As $v = n\lambda \therefore v = 100 \times 0.2 = 20$ m/s

61. (d) $f = \frac{1}{2l}\left[\frac{T}{\mu}\right]^{\frac{1}{2}}$. When f is halved, the length is doubled.

62. (d) Third overtone has a frequency 7 n, which means $L = \frac{7\lambda}{4}$ = three full loops + one half loop, which would make four nodes and four antinodes.

63. (b) Net beat frequency

= LCM of individual beat frequencies

= LCM of [(n, n – 1), (n, n + 1), (n – 1, n + 1)]

= LCM of (1, 1, 2) = 2 Hz

So, no. of beats per second = 2

64. (d) When sounded with a source of known frequency fundamental frequency

= 250 ± 4 Hz = 254 Hz or 246 Hz

2nd harmonic if unknown frequency (suppose) 254 Hz

= 2 × 254 = 508 Hz

As it gives 5 beats

$\therefore$ 508 + 5 = 513 Hz

Hence, unknown frequency is 254 Hz

65. (b) $\frac{\Delta f}{f}=\frac{v}{c} \Rightarrow \frac{(\text{Beats})/2}{f}=\frac{v}{c}$

$\Rightarrow$ Beats $=\frac{2fv}{c}=4$

66. (d) $2\pi f_1 = 600\pi;\ \ f_1 = 300$ ... (1)

$2\pi f_2 = 608\pi; f_2 = 304$ ... (2)

$|f_1 - f_2| = 4$ beats

$\frac{I_{max}}{I_{min}}=\frac{(A_1+A_2)^2}{(A_1+A_2)^2}=\frac{(5+4)^2}{(5-4)^2}=\frac{81}{1}$

where A_1, A_2 are amplitudes of given two sound wave.

67. (a) For fundamental mode, $f=\frac{1}{2\ell}\sqrt{\frac{T}{\mu}}$

Taking logarithm on both sides, we get

$\Rightarrow \log f=\log\left(\frac{1}{2\ell}\right)+\log\left(\sqrt{\frac{T}{\mu}}\right)$

$\Rightarrow \log\left(\frac{1}{2\ell}\right)+\frac{1}{2}\log\left(\frac{T}{\mu}\right)$

or $\log f=\log\left(\frac{1}{2\ell}\right)+\frac{1}{2}[\log T-\log\mu]$

Differentiating both sides, we get

$\frac{df}{f}=\frac{1}{2}\frac{dT}{T}$ (as ℓ and μ are constants)

$\Rightarrow \frac{dT}{T}=2\times\frac{df}{f}$

Here df = 6

f = 600 Hz $\qquad \therefore \frac{dT}{T}=\frac{2\times 6}{600}=0.02$

68. (d) The frequency of the piano string = 512 ± 4 = 516 or 508. When the tension is increased, beat frequency decreases to 2, it means that frequency of the string is 508 as frequency of string increases with tension.

69. (a) The frequency of vibration of a string is given by,

$f=\frac{1}{2l}\sqrt{\frac{T}{m}}$ where m is mass per unit length.

$f_1=\frac{1}{2l_1}\sqrt{\frac{T}{m}},\ f_2=\frac{1}{2l_2}\sqrt{\frac{T}{m}},$

$f_2-f_1=\frac{1}{2}\sqrt{\frac{T}{m}}\frac{(l_1-l_2)}{l_1 l_2}$

$\sqrt{\frac{T}{m}}=\sqrt{\frac{20}{10^{-3}}}=\sqrt{2}\times 10^2=1.414\times 100$

$= 141.4$

$\frac{l_1-l_2}{l_1 l_2}=\frac{(51.6-49.1)\times 10^2}{51.6\times 49.1}=\frac{2.5\times 10^2}{50\times 50}=\frac{1}{10}$

$\therefore \quad f_2-f_1=\frac{1}{2}\times 141.4\times\frac{1}{10}=7$ beats

70. (c) Frequencies of sound waves are $\frac{330}{5}$ and $\frac{330}{5.5}$ *i.e.*, 66 Hz and 60 Hz

Frequency of beat = 66 – 60 = 6 per second

NOTES Beat frequency is the number of beats produced per second.
$n = n_1 \sim n_2$

71. (b) Equation of progressive wave is given by

$y = A\sin 2\pi f t$

Given $y_1 = 4\sin 500\,\pi t$ and $y_2 = 2\sin 506\,\pi t$.

Comparing the given equations with equation of progressive wave, we get

$2f_1 = 500 \Rightarrow f_1 = 250$

$2f_2 = 506 \Rightarrow f_2 = 253$

Beats $= f_2 - f_1 = 253 - 250 = 3$ beats/sec

= 3 × 60 = 180 beats/minute.

72. (a) Given : Wavelength of first wave (λ_1)

= 50 cm = 0.5 m

Wavelength of second wave (λ_2)

= 51 cm = 0.51m

frequency of beats per sec (n) = 12.

We know that the frequency of beats,

$n=12=\frac{v}{\lambda_1}-\frac{v}{\lambda_2} \Rightarrow 12=v\left[\frac{1}{0.5}-\frac{1}{0.51}\right]$

$= v[2-1.9608] = v\times 0.0392$

or, $v=\frac{12}{0.0392}=306$ m/s

[where, v = velocity of sound]

73. (c) Frequency of first source with 5 beats/ sec = 100 Hz and frequency of second source with 5 beats/sec = 205 Hz. The frequency of the first source = 100 ± 5 = 105 or 95 Hz. Therefore, frequency of second harmonic of source = 210 Hz or 190 Hz. As the second harmonic gives 5 beats/ second with the sound of frequency 205 Hz, therefore, frequency of second harmonic source should be 210 Hz or frequency of source = 105 Hz.

74. (b) For production of beats different frequencies are essential. The different amplitudes affect the minimum and maximum amplitude of the beats and different phases affect the time of occurrence of minimum and maximum.

75. (c) Beats are produced. Frequency of beats will be $\frac{1}{3}-\frac{1}{4}=\frac{1}{12}$. Hence time period = 12 s.

NOTES Beat period (T) is the time interval between two successive beats.

$$T=\frac{1}{\text{Beat frequency}}=\frac{1}{n_1-n_2}$$

1 Electric Charges and Fields

TREND ANALYSIS

TOPIC	2025		2024		2023		2022		2021		2020	
	Qns.	LOD	Qns.	LOD	Qns.	LOD	Qns.	LOD	Qns.	LOD	Qns.	LOD
Coulomb's Law	1	Avg.					1	Easy	1	Avg.		
Electric Dipole							1	Easy	1	Avg.		
Dipole in a Uniform External Field					1	Avg.						
Gauss's Law					1	Easy						
Applications of Gauss's Law											1	Avg.

How to Code Qns for Better & Faster Revision

Type 1: Able to solve (No code Req)
Type 2: Solved with struggle or by seeing solution **(Use code NR means Needs Revision or orange stickers)**
Type 3: Tricky/ Could not solve on my own. Need help from Teacher/ Friend. **(Use code TQ means Tricky Question or red stickers)**

You can either hand write the code in boxes or else paste coloured stickers given at end of book.

Topic-1.5 Coulomb's Law

1. Two identical charged conducting spheres A and B have their centres separated by a certain distance. Charge on each sphere is q and the force of repulsion between them is F. A third identical uncharged conducting sphere is brought in contact with sphere A first and then with B and finally removed from both. New force of repulsion between spheres A and B (Radii of A and B are negligible compared to the distance of separation so that for calculating force between them they can be considered as point charges) is best given as :

NCERT | Page-7 | 2025

(a) $\frac{3F}{5}$ (b) $\frac{2F}{3}$
(c) $\frac{F}{2}$ (d) $\frac{3F}{8}$

2. The ratio of Coulomb's electrostatic force to the gravitational force between an electron and a proton separated by some distance is 2.4×10^{39}. The ratio of the proportionality constant, $K = \frac{1}{4\pi\varepsilon_0}$ to the Gravitational constant G is nearly (Given that the charge of the proton and electron each $= 1.6 \times 10^{-19}$ C, the mass of the electron $= 9.11 \times 10^{-31}$ kg, the mass of the proton $= 1.67 \times 10^{-27}$ kg):

NCERT | Page-7 | RS 2022

(a) 10 (b) 10^{20} (c) 10^{30} (d) 10^{40}

3. The acceleration of an electron due to the mutual attraction between the electron and a proton when they are 1.6Å apart is,

$(m_e = 9\times10^{-31}$ kg, $e = 1.6\times10^{-19} C)$

(Take $\frac{1}{4\pi\varepsilon_0} = 9\times10^9$ Nm^{2}C^{-2})

NCERT | Page-7 | Ph-II 2021

(a) 10^{25} m/s^2 (b) 10^{24} m/s^2
(c) 10^{23} m/s^2 (d) 10^{22} m/s^2

4. Two point charges A and B, having charges +Q and –Q respectively, are placed at certain distance apart and force acting between them is F. If 25% charge of A is transferred to B, then force between the charges becomes :

NCERT | Page-7 | 2019

(a) F (b) $\frac{9F}{16}$ (c) $\frac{16F}{9}$ (d) $\frac{4F}{3}$

5. Suppose the charge of a proton and an electron differ slightly. One of them is – e, the other is (e + Δe). If the net of electrostatic force and gravitational force between two hydrogen atoms placed at a distance d (much greater than atomic size) apart is zero, then Δe is of the order of [Given mass of hydrogen $m_h = 1.67 \times 10^{-27}$ kg]

NCERT | Page-7 | 2017

(a) 10^{-23}C (b) 10^{-37} C
(c) 10^{-47} C (d) 10^{-20} C

6. Two identical charged spheres suspended from a common point by two massless strings of lengths l, are initially at a distance d (d << l) apart because of their mutual repulsion. The charges begin to leak from both the spheres at a constant rate. As a result, the spheres approach each other with a velocity v. Then v varies as a function of the distance x between the spheres, as :

NCERT | Page-7 | 2016

(a) $v \propto x^{\frac{1}{2}}$ (b) $v \propto x$
(c) $v \propto x^{-\frac{1}{2}}$ (d) $v \propto x^{-1}$

7. Two pith balls carrying equal charges are suspended from a common point by strings of equal length. The equilibrium separation between them is r. Now the strings are rigidly clamped at half the height. The equilibrium separation between the balls now become

NCERT | Page-7 | 2013

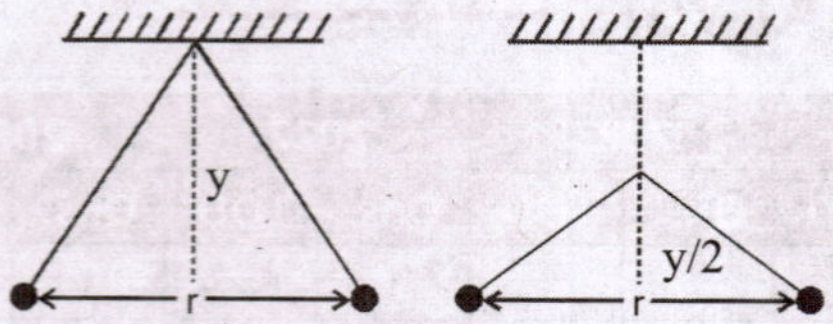

(a) $\left(\frac{r}{\sqrt[3]{2}}\right)$ (b) $\left(\frac{2r}{\sqrt{3}}\right)$

(c) $\left(\frac{2r}{3}\right)$ (d) $\left(\frac{r}{\sqrt{2}}\right)^2$

8. Two positive ions, each carrying a charge q, are separated by a distance d. If F is the force of repulsion between the ions, the number of electrons missing from each ion will be (e being the charge of an electron)

NCERT | Page-7 | 2010

(a) $\frac{4\pi\varepsilon_0 Fd^2}{e^2}$ (b) $\sqrt{\frac{4\pi\varepsilon_0 Fe^2}{d^2}}$

(c) $\sqrt{\frac{4\pi\varepsilon_0 Fd^2}{e^2}}$ (d) $\frac{4\pi\varepsilon_0 Fd^2}{q^2}$

9. An electron is moving around the nucleus of a hydrogen atom in a circular orbit of radius r. The Coulomb force $\vec{F}$ between the two is **NCERT | Page-8 | 2003**

(a) $K\frac{e^2}{r^3}\vec{r}$ (b) $K\frac{e^2}{r^2}\hat{r}$

(c) $-K\frac{e^2}{r^3}\hat{r}$ (d) $-K\frac{e^2}{r^3}\vec{r}$

$\left(\text{where } K = \frac{1}{4\pi\varepsilon_0}\right)$

10. When air is replaced by a dielectric medium of force constant K, the maximum force of attraction between two charges, separated by a distance **NCERT | Page-7 | 1999**

(a) decreases K-times

(b) increases K-times

(c) remains unchanged

(d) becomes $\frac{1}{K^2}$ times

Topic-1.6 Forces between Multiple Charges

11. A charge 'q' is placed at the centre of the line joining two equal charges 'Q'. The system of the three charges will be in equilibrium if 'q' is equal to **NCERT | Page-7 | Kar. 2013, 1995**

(a) $Q/2$ (b) $-Q/4$ (c) $Q/4$ (d) $-Q/2$

12. Point charges $+4q$, $-q$ and $+4q$ are kept on the X-axis at points $x = 0$, $x = a$ and $x = 2a$ respectively. Then

NCERT | Page-8 | 1988

(a) only $-q$ is in stable equilibrium

(b) none of the charges is in equilibrium

(c) all the charges are in unstable equilibrium

(d) all the charges are in stable equilibrium

Topic-1.7 Electric Field

13. A toy car with charge q moves on a frictionless horizontal plane surface under the influence of a uniform electric field $\vec{E}$. Due to the force q $\vec{E}$, its velocity increases from 0 to 6 m/s in one second duration. At that instant the direction of the field is reversed. The car continues to move for two more seconds under the influence of this field. The average velocity and the average speed of the toy car between 0 to 3 seconds are respectively

NCERT | Page-14 | 2018

(a) 2 m/s, 4 m/s (b) 1 m/s, 3 m/s

(c) 1.5 m/s, 3 m/s (d) 1 m/s, 3.5 m/s

14. An electron falls from rest through a vertical distance h in a uniform and vertically upward directed electric field E. The direction of electric field is now reversed, keeping its magnitude the same. A proton is allowed to fall from rest in it through the same vertical distance h. The time of fall of the electron, in comparison to the time of fall of the proton is **NCERT | Page-14 | 2018**

(a) smaller (b) 5 times greater

(c) equal (d) 10 times greater

15. There is an electric field E in x-direction. If the work done on moving a charge of 0.2 C through a distance of 2 m along a line making an angle 60° with x-axis is 4J, then what is the value of E? **1995**

(a) 3 N/C (b) 4 N/C (c) 5 N/C (d) 20 N/C

Topic-1.9 Electric Flux

16. A square surface of side L meter in the plane of the paper is placed in a uniform electric field E (volt/m) acting along the same plane at an angle θ with the horizontal side of the square as shown in Figure. The electric flux linked to the surface, in units of volt. m, is

NCERT | Page-22 | 2010

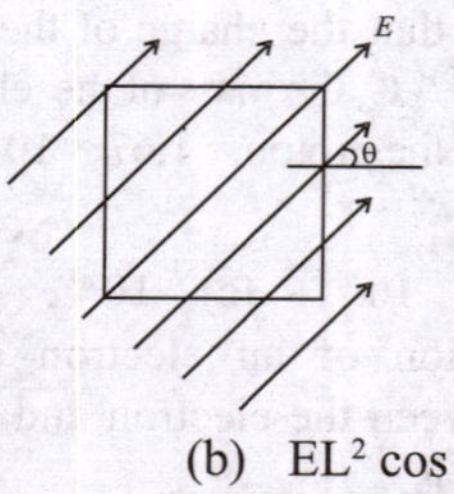

(a) EL^2 (b) $EL^2 \cos\theta$

(c) $EL^2 \sin\theta$ (d) zero

17. A square surface of side L metres is in the plane of the paper. A uniform electric field $\vec{E}$ (volt /m), also in the plane of the paper, is limited only to the lower half of the square surface (see figure). The electric flux in SI units associated with the surface is NCERT | Page-22 | 2006

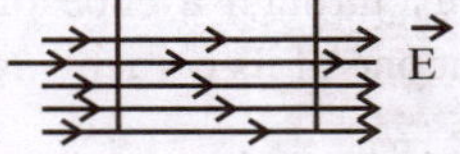

(a) $EL^2/2$ (b) zero
(c) EL^2 (d) $EL^2/(2\varepsilon_0)$

Topic-1.10

Electric Dipole

18. Two point charges –q and +q are placed at a distance of L, as shown in the figure.

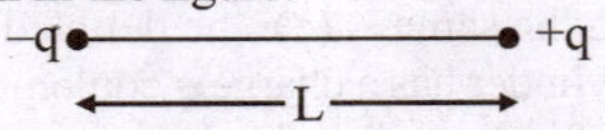

The magnitude of electric field intensity at a distance R(R>>L) varies as: NCERT | Page-14 | 2022

(a) $\frac{1}{R^3}$ (b) $\frac{1}{R^4}$
(c) $\frac{1}{R^6}$ (d) $\frac{1}{R^2}$

19. Polar molecules are the molecules NCERT | Page-23 | 2021
(a) having a permanent electric dipole moment
(b) having zero dipole moment
(c) acquire a dipole moment only in the presence of electric field due to displacement of charges
(d) acquire a dipole moment only when magnetic field is absent

20. Three point charges $+q$, $-q$ and $+q$ are placed at points ($x = 0$, $y = a$, $z = 0$), ($x = 0$, $y = 0$, $z = 0$) and ($x = a$, $y = 0$, $z = 0$) respectively. The magnitude and direction of the electric dipole moment vector of this charge assembly are NCERT | Page-23 | 2007
(a) $\sqrt{2}qa$ along the line joining points ($x = 0, y = 0, z = 0$) and ($x = a, y = a, z = 0$)
(b) qa along the line joining points ($x = 0, y = 0, z = 0$) and ($x = a, y = a, z = 0$)
(c) $\sqrt{2}qa$ along +ve x direction
(d) $\sqrt{2}qa$ along +ve y direction

21. The electric intensity due to a dipole of length 10 cm and having a charge of 500 μC, at a point on the axis at a distance 20 cm from one of the charges in air, is NCERT | Page-24 | 2001
(a) 6.25×10^7 N/C (b) 9.28×10^7 N/C
(c) 13.1×10^{11} N/C (d) 20.5×10^7 N/C

22. A point Q lies on the perpendicular bisector of an electrical dipole of dipole moment p. If the distance of Q from the dipole is r (much larger than the size of the dipole), then the electric field at Q is proportional to NCERT | Page-24 | 1998
(a) p^{-1} and r^{-2} (b) p and r^{-2}
(c) p^2 and r^{-3} (d) p and r^{-3}

23. Intensity of an electric field (E) depends on distance r, due to a dipole, is related as NCERT | Page-24 | 1996
(a) $E \propto \frac{1}{r}$ (b) $E \propto \frac{1}{r^2}$
(c) $E \propto \frac{1}{r^3}$ (d) $E \propto \frac{1}{r^4}$

24. The formation of a dipole is due to two equal and dissimilar point charges placed at a NCERT | Page-23 | 1996
(a) short distance (b) long distance
(c) above each other (d) none of these

Topic-1.11

Dipole in a Uniform External Field

25. An electric dipole is placed at an angle of 30° with an electric field of intensity 2×10^5 NC^{-1}. It experiences a torque equal to 4 N m. Calculate the magnitude of charge on the dipole, if the dipole length is 2 cm. NCERT | Page-27 | 2023
(a) 8 mC (b) 6 mC (c) 4 mC (d) 2 mC

26. If a dipole of dipole moment $\vec{p}$ is placed in a uniform electric field $\vec{E}$, then torque acting on it is given by NCERT | Page-27 | 2001
(a) $\vec{\tau} = \vec{p}.\vec{E}$ (b) $\vec{\tau} = \vec{p} \times \vec{E}$
(c) $\vec{\tau} = \vec{p} + \vec{E}$ (d) $\vec{\tau} = \vec{p} - \vec{E}$

27. An electric dipole, consisting of two opposite charges of 2×10^{-6} C each separated by a distance 3 cm is placed in an electric field of 2×10^5 N/C. Torque acting on the dipole is NCERT | Page-27 | 1995
(a) 12×10^{-1} *Nm* (b) 12×10^{-2} *Nm*
(c) 12×10^{-3} *Nm* (d) 12×10^{-4} *Nm*

Topic-1.12

Continuous Charge Distribution

28. A metal cube of side 5 cm is charged with 6 μC. The surface charge density on the cube is NCERT | Page-14 | Ph-II 2024
(a) 0.125×10^{-3} C m^{-2} (b) 0.25×10^{-3} C m^{-2}
(c) 4×10^{-3} C m^{-2} (d) 0.4×10^{-3} C m^{-2}

29. Two metal spheres, one of radius R and the other of radius $2R$ respectively have the same surface charge density σ. They are brought in contact and separated. What will be the new surface charge densities on them? NCERT | Page-14 | Odisha 2019
(a) $\sigma_1 = \frac{5}{3}\sigma, \sigma_2 = \frac{5}{6}\sigma$ (b) $\sigma_1 = \frac{5}{6}\sigma, \sigma_2 = \frac{5}{2}\sigma$
(c) $\sigma_1 = \frac{5}{2}\sigma, \sigma_2 = \frac{5}{6}\sigma$ (d) $\sigma_1 = \frac{5}{2}\sigma, \sigma_2 = \frac{5}{3}\sigma$

30. A thin conducting ring of radius R is given a charge +Q. The electric field at the centre O of the ring due to the charge on the part AKB of the ring is E. The electric field at the centre due to the charge on the part ACDB of the ring is **NCERT | Page-14 | 2008**

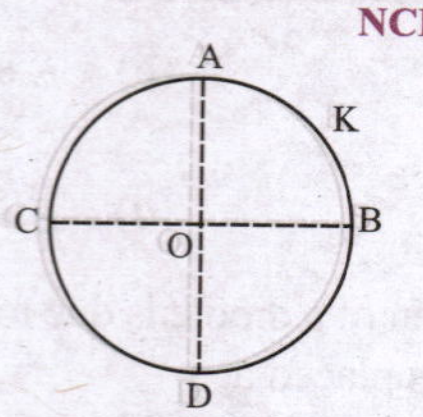

(a) E along KO (b) E along OK
(c) E along KO (d) 3 E along OK

31. A semi-circular arc of radius 'a' is charged uniformly and the charge per unit length is λ. The electric field at the centre of this arc is **NCERT | Page-34 | 2000**

(a) $\dfrac{\lambda}{2\pi\varepsilon_0 a}$ (b) $\dfrac{\lambda}{2\pi\varepsilon_0 a^2}$

(c) $\dfrac{\lambda}{4\pi^2\varepsilon_0 a}$ (d) $\dfrac{\lambda^2}{2\pi\varepsilon_0 a}$

Topic-1.13

Gauss's Law

32. According to Gauss law of electrostatics, electric flux through a closed surface depends on **NCERT | Page-22 | Manipur 2023**

(a) The shape of the surface
(b) The volume enclosed by the surface
(c) The area of the surface
(d) The quantity of charges enclosed by the surface

33. A charge Q μC is placed at the centre of a cube. The flux coming out from any one of its faces will be (in SI unit) **NCERT | Page-22 | Manipur 2023**

(a) $\dfrac{Q}{6\varepsilon_0}\times 10^{-3}$ (b) $\dfrac{Q}{6\varepsilon_0}\times 10^{-6}$

(c) $\dfrac{Q}{\varepsilon_0}\times 10^{-6}$ (d) $\dfrac{2Q}{3\varepsilon_0}\times 10^{-3}$

34. If $\oint_s \vec{E}\cdot\vec{dS} = 0$ over a surface, then **NCERT | Page-29, 30 | 2023**

(a) The number of flux lines entering the surface must be equal to the number of flux lines leaving it
(b) The magnitude of electric field on the surface is constant
(c) All the charges must necessarily be inside the surface
(d) The electric field inside the surface is necessarily uniform

35. A sphere encloses an electric dipole with charges $\pm 3 \times 10^{-6}$ C. What is the total electric flux across the sphere? **NCERT | Page-22 | Odisha 2019**

(a) 6×10^{-6} Nm²/C (b) -3×10^{-6} Nm²/C
(c) Zero (d) 3×10^{-6} Nm²/C

36. The electric field in a certain region is acting radially outward and is given by E = Aa. A charge contained in a sphere of radius 'a' centred at the origin of the field, will be given by **NCERT | Page-22 | 2015**

(a) $A\,\varepsilon_0\,a^2$ (b) $4\,\pi\varepsilon_0\,Aa^3$
(c) $\varepsilon_0\,Aa^3$ (d) $4\,\pi\varepsilon_0\,Aa^2$

37. What is the flux through a cube of side 'a' if a point charge of q is at one of its corner : **NCERT | Page-22 | 2012**

(a) $\dfrac{2q}{\varepsilon_0}$ (b) $\dfrac{q}{8\varepsilon_0}$

(c) $\dfrac{q}{\varepsilon_0}$ (d) $\dfrac{q}{2\varepsilon_0}$

38. A charge Q is enclosed by a Gaussian spherical surface of radius R. If the radius is doubled, then the outward electric flux will **NCERT | Page-30 | 2011**

(a) increase four times (b) be reduced to half
(c) remain the same (d) be doubled

39. A hollow cylinder has a charge q coulomb within it. If ϕ is the electric flux in units of voltmeter associated with the curved surface B, the flux linked with the plane surface A in units of voltmeter will be **NCERT | Page-30 | 2007**

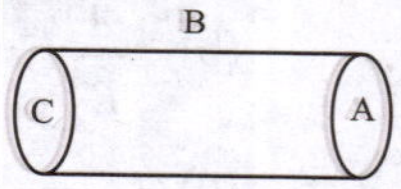

(a) $\dfrac{q}{2\varepsilon_0}$ (b) $\dfrac{\phi}{3}$

(c) $\dfrac{q}{\varepsilon_0}-\phi$ (d) $\dfrac{1}{2}\left(\dfrac{q}{\varepsilon_0}-\phi\right)$

40. A charge q is located at the centre of a cube. The electric flux through any face is **NCERT | Page-30 | 2003**

(a) $\dfrac{q}{6(4\pi\varepsilon_0)}$ (b) $\dfrac{2\pi q}{6(4\pi\varepsilon_0)}$

(c) $\dfrac{4\pi q}{6(4\pi\varepsilon_0)}$ (d) $\dfrac{\pi q}{6(4\pi\varepsilon_0)}$

41. A charge Q μC is placed at the centre of a cube, the flux coming out from any surface will be **NCERT | Page-30 | 2001**

(a) $\dfrac{Q}{6\varepsilon_0}\times 10^{-6}$ (b) $\dfrac{Q}{6\varepsilon_0}\times 10^{-3}$

(c) $\dfrac{Q}{24\varepsilon_0}$ (d) $\dfrac{Q}{8\varepsilon_0}$

42. A charge Q is placed at the corner of a cube. The electric flux through all the six faces of the cube is **NCERT | Page-30 | 2000**

(a) $Q/3\varepsilon_0$ (b) $Q/6\varepsilon_0$
(c) $Q/8\varepsilon_0$ (d) Q/ε_0

43. A point charge $+q$ is placed at mid point of a cube of side 'L'. The electric flux emerging from the cube is **NCERT | Page-30 | 1996**

(a) $\dfrac{q}{\varepsilon_0}$ (b) $\dfrac{6qL^2}{\varepsilon_0}$

(c) $\dfrac{q}{6L^2\varepsilon_0}$ (d) zero

Topic-1.14

Applications of Gauss's Law

44. A spherical conductor of radius 10 cm has a charge of 3.2 × 10^{-7} C distributed uniformly. What is the magnitude of electric field at a point 15 cm from the centre of the sphere? **NCERT | Page-35 | 2020**

$$\left(\frac{1}{4\pi \in_0} = 9\times10^9 \text{ Nm}^2/\text{C}^2\right)$$

(a) 1.28 × 10^5 N/C (b) 1.28 × 10^6 N/C
(c) 1.28 × 10^7 N/C (d) 1.28 × 10^4 N/C

45. Two parallel infinite line charges with linear charge densities +λ C/m and –λ C/m are placed at a distance of 2R in free space. What is the electric field mid-way between the two line charges? **NCERT | Page-34 | 2019**

(a) zero (b) $\frac{2\lambda}{\pi \in_0 R}$ N/C

(c) $\frac{\lambda}{\pi \in_0 R}$ N/C (d) $\frac{\lambda}{2\pi\varepsilon_0 R}$ N/C

46. A hollow metal sphere of radius R is uniformly charged. The electric field due to the sphere at a distance r from the centre : **NCERT | Page-35 | 2019**

(a) increases as r increases for r < R and for r > R
(b) zero as r increases for r < R, decreases as r increases for r > R
(c) zero as r increases for r < R, increases as r increases for r > R
(d) decreases as r increases for r < R and for r > R

47. The electric field at a distance $\frac{3R}{2}$ from the centre of a charged conducting spherical shell of radius R is E. The electric field at a distance $\frac{R}{2}$ from the centre of the sphere is **NCERT | Page-35 | 2010 M**

(a) zero (b) E (c) $\frac{E}{2}$ (d) $\frac{E}{3}$

48. A hollow insulated conduction sphere is given a positive charge of 10 μC. What will be the electric field at the centre of the sphere if its radius is 2 metres? **NCERT | Page-35 | 1998**

(a) zero (b) 5 μCm^{-2}
(c) 20 μCm^{-2} (d) 8 μCm^{-2}

ANSWER KEY

1	(d)	6	(c)	11	(b)	16	(d)	21	(a)	26	(b)	31	(a)	36	(b)	41	(a)	46	(b)
2	(b)	7	(a)	12	(c)	17	(b)	22	(d)	27	(c)	32	(d)	37	(b)	42	(c)	47	(a)
3	(d)	8	(c)	13	(b)	18	(a)	23	(c)	28	(d)	33	(b)	38	(c)	43	(a)	48	(a)
4	(b)	9	(d)	14	(a)	19	(a)	24	(a)	29	(a)	34	(a)	39	(d)	44	(a)		
5	(b)	10	(a)	15	(d)	20	(a)	25	(d)	30	(b)	35	(c)	40	(c)	45	(c)		

Hints & Solutions

1. **(d)** Initial force

$F = \frac{kQ^2}{r^2}$

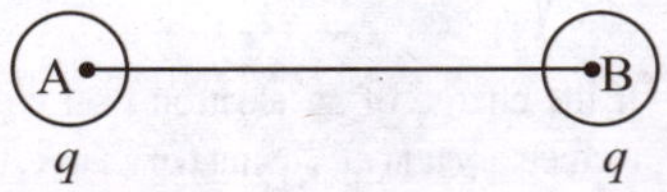

After touching :

Sphere C touches A

Both are identical charge, redistribute equally

$Q'_A = \frac{q}{2}, \quad Q'_C = \frac{q}{2}$

Now C (with $\frac{q}{2}$) touches B (which has q)

Total charge = $q + \frac{q}{2} = \frac{3}{2}q$

Equal distribution :

$Q'_A = \frac{3}{4}q \quad Q'_C = \frac{3}{4}q$

On touching :

$$F' = \frac{k\left(\frac{Q}{2}\right)\left(\frac{3Q}{4}\right)}{r^2} \Rightarrow F' = \frac{3F}{8}$$

2. **(b)** Coulomb's force $F_E = \frac{Kq_1q_2}{r^2}$

Gravitational force, $F_G = \frac{Gm_1m_2}{r^2}$

$$\frac{F_E}{F_G} = \frac{\frac{kq_1q_2}{r^2}}{\frac{Gm_1m_2}{r^2}} = \frac{k}{G}\frac{\left(1.6\times10^{-19}\right)^2}{(1.67\times9.11)\times10^{-58}}$$

$$2.4\times10^{39} = \frac{k}{G}\frac{1.6\times1.6\times10^{-38}}{(1.67\times9.11)\times10^{-58}}$$

$$\therefore \frac{k}{G} \approx 10^{20}$$

3. **(d)** Force of mutual attraction between the electron and proton is

$$F = \frac{Ke^2}{r^2} \qquad \text{(Here, K} = 9 \times 10^9)$$

$$\Rightarrow F = 9 \times 10^9 \times \frac{e^2}{r^2}$$

$$= 9 \times 10^9 \times \frac{(1.6 \times 10^{-19})^2}{(1.6 \times 10^{-10})^2} = 9 \times 10^{-9}\text{N}$$

$$\therefore \text{ Acceleration of electron} = \frac{F}{m_e} = \frac{9 \times 10^{-9}}{9 \times 10^{-31}}$$

$$= 10^{-9} \times 10^{+31} = 10^{22} \text{ m/s}^2$$

4. **(b)** We know that,

$$F = \frac{kQ^2}{r^2}$$

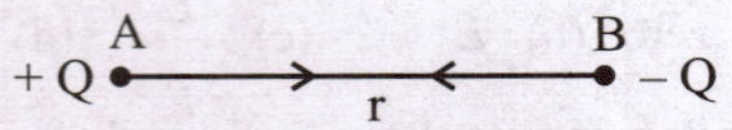

If 25% of charges of A transfer to B then

$$q_A = Q - \frac{Q}{4} = \frac{3Q}{4} \text{ and } q_B = -Q + \frac{Q}{4} = \frac{-3Q}{4}$$

$$F_1 = \frac{kq_A q_B}{r^2}$$

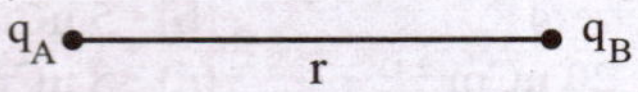

$$\Rightarrow F_1 = \frac{k\left(\frac{3Q}{4}\right)^2}{r^2} = \frac{9}{16}\frac{kQ^2}{r^2} = \frac{9F}{16}$$

5. **(b)** Net charge on one H atom $= -e + (e + \Delta e) = \Delta e$

According to question, the net electrostatic force (F_E) = gravitational force (F_G)

$$F_E = F_G$$

$$\text{or } \frac{1}{4\pi\varepsilon_0}\frac{\Delta e^2}{d^2} = \frac{Gm^2}{d^2}$$

$$\Rightarrow \Delta e = m\sqrt{\frac{G}{k}} \left(\frac{1}{4\pi\varepsilon_0} = k = 9 \times 10^9\right)$$

$$= 1.67 \times 10^{-27}\sqrt{\frac{6.67 \times 10^{-11}}{9 \times 10^9}}$$

$$\Delta e \approx 1.436 \times 10^{-37}\text{ C}$$

6. **(c)** From figure $T \cos\theta = mg$

$$T \sin\theta = \frac{kq^2}{x^2}$$

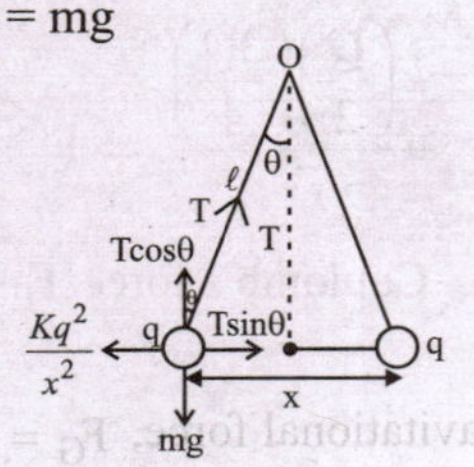

$$\text{so, } \tan\theta = \frac{F_e}{mg} \simeq \theta$$

$$\tan\theta = \frac{kq^2}{x^2 mg}$$

As θ is small $\tan\theta \approx \sin\theta = \frac{x}{2\ell}$

$$\therefore \frac{kq^2}{x^2 mg} = \frac{x}{2\ell}$$

or $x^3 \propto q^2$ ...(1)

or $x^{3/2} \propto q$ ...(2)

Differentiating eq. (1) w.r.t. time

$$3x^2\frac{dx}{dt} \propto 2q\frac{dq}{dt} \text{ but } \frac{dq}{dt} \text{ is constant}$$

so $x^2(v) \propto q$

Replace q from eq. (2)

$x^2(v) \propto x^{3/2}$ or $v \propto x^{-1/2}$

7. **(a)**

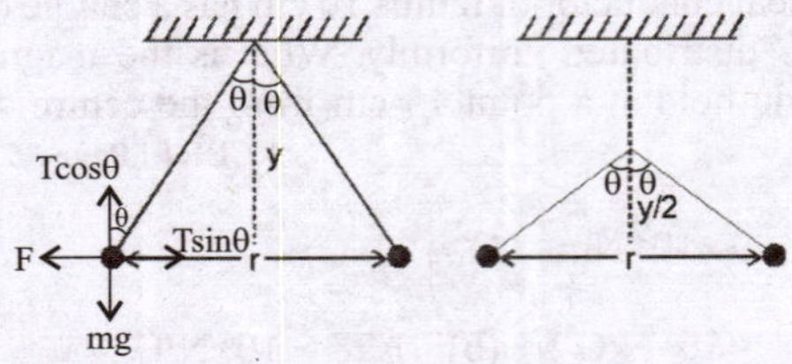

$$T\cos\theta = mg$$

$$T\sin\theta = F$$

$$\therefore \tan\theta = \frac{F}{mg}$$

$$\text{From figure, } \tan\theta = \frac{F_e}{mg}$$

$$\Rightarrow \frac{r/2}{y} = \frac{\frac{kq^2}{r^2}}{mg}$$

$$\left[\because F = \frac{kq^2}{r^2} \text{ from coulomb's law}\right]$$

$$\Rightarrow r^3 \propto y$$

When string is clamped at half the height

$$r'^3 \propto \frac{y}{2}$$

Dividing the above two equations, we have

$$\frac{r'}{r} = \frac{1}{2^{1/3}}$$

$$\Rightarrow r' = \frac{r}{\sqrt[3]{2}}$$

8. **(c)** Let *n* be the number of electrons missing.

$$F = \frac{1}{4\pi\varepsilon_0}\cdot\frac{q^2}{d^2} \Rightarrow q = \sqrt{4\pi\varepsilon_0 d^2 F} = ne$$

$$\therefore \quad n = \sqrt{\frac{4\pi\varepsilon_0 F d^2}{e^2}}$$

NOTES If the charge of an electron ($= 1.6 \times 10^{-19}$ C) is taken as elementary unit *i.e.*, quanta of charge, the charge on any body will be some integral multiple of *e* *i.e.*, $Q = \pm ne$ where $n = 1, 2, 3....$

9. **(d)** Charges (–e) on electron and (e) on proton exert a force of attraction given by

$$\text{Force} = (K)\frac{(-e)(e)}{r^2}\hat{r} = \frac{-Ke^2}{r^3}\vec{r} \qquad \left(\because \hat{r} = \frac{\vec{r}}{|r|}\right)$$

NOTES Magnitude of Coulomb force is given by $\frac{1}{4\pi\varepsilon_0}\frac{q_2 q_2}{r^2}$, but in vector form $\vec{F} = \frac{1}{4\pi\varepsilon_0}\frac{q_1 q_2}{r^3}\vec{r}$

10. **(a)** In air, $F_{air} = \frac{1}{4\pi\varepsilon_0}\frac{q_1 q_2}{r^2}$

In medium, $F_m = \frac{1}{4\pi\varepsilon_0}\frac{q_1 q_2}{Kr^2}$

$\therefore \frac{F_m}{F_{air}} = \frac{1}{K} \Rightarrow F_m = \frac{F_{air}}{K}$ (decreases K-times)

11. **(b)** The system of three charges will be in equilibrium.

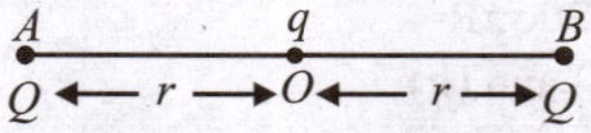

For this, force between charge at A and B + force between charge at point O and either at A or B is zero.

i.e., $\frac{KQ^2}{(2r)^2} + \frac{KQq}{(r)^2} = 0$

By solving we get,

$q = \frac{-Q}{4}$.

12. **(c)** Net force on each of the charge due to the other charges is zero. However, disturbance in any direction other than along the line on which the charges lie, will not make the charges return.

13. **(b)**

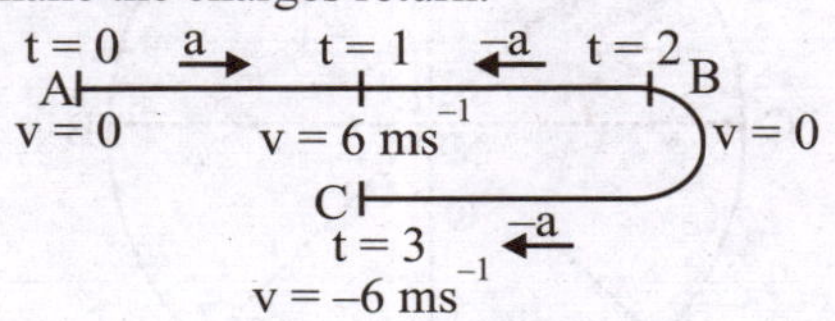

Acceleration, $a = \frac{v-u}{t} = \frac{6-0}{1}$

$= 6 \text{ ms}^{-2}$

For t = 0 to t = 1 s,

$S_1 = \frac{1}{2} \times 6(1)^2 = 3 \text{ m}$ ...(i)

For t = 1 s to t = 2 s,

$S_2 = 6.1 - \frac{1}{2} \times 6(1)^2 = 3\text{m}$ ...(ii)

For t = 2 s to t = 3 s,

$S_3 = 0 - \frac{1}{2} \times 6(1)^2 = -3 \text{ m}$ ...(iii)

Total displacement $S = S_1 + S_2 + S_3 = 3$ m

Average velocity $= \frac{3}{3} = 1 \text{ ms}^{-1}$

Total distance travelled = 9 m

Average speed $= \frac{9}{3} = 3 \text{ ms}^{-1}$

14. **(a)** As we know, F = qE = ma

$\Rightarrow a = \frac{qE}{m}$

$h = \frac{1qE}{2m}t^2 \quad \therefore \quad t = \sqrt{\frac{2hm}{qE}}$

i.e., time $t \propto \sqrt{m}$ as 'q' is same for electron and proton.

Since, electron has smaller mass so it will take smaller time.

15. **(d)** Charge (q) = 0.2 C;

Distance (d) = 2m; Angle θ = 60º and work done (W) = 4J.

Work done in moving the charge (W)

$= F.d\cos\theta = qEd\cos\theta$

or, $E = \frac{W}{qd\cos\theta} = \frac{4}{0.2 \times 2 \times \cos 60^\circ} = \frac{4}{0.4 \times 0.5}$

= 20 N/C.

16. **(d)** Electric flux, $\phi = EA\cos\theta$, where θ = angle between E and normal to the surface.

Here $\theta = \frac{\pi}{2}$

$\Rightarrow \phi = 0$

17. **(b)** Flux $= \vec{E}\cdot\vec{A}$.

$\vec{E}$ is electric field vector & $\vec{A}$ is area vector.

Here, angle between $\vec{E}$ & $\vec{A}$ is 90º.

So, $\vec{E}\cdot\vec{A} = 0$; Flux = 0

18. **(a)** If R >> L, it will be have like a dipole and in dipole we have, $E \propto \frac{1}{r^3}$

NOTES The magnitude of electric field intensity due to dipole at large distances varies as $\frac{1}{r^3}$ and can never be zero unless $r \to \infty$ and $p \to 0$.
Here, p is dipole moment.

19. **(a)** In polar molecules, like HCL → $H^{\delta+}$ $Cl^{\delta-}$ the centre of positive charges does not coincide with the centre of negative charges.
Hence, permanent electric dipole exists.

20. **(a)** Three point charges +*q*, –2*q* and +*q* are placed at points *B* (*x* = 0, *y* = *a*, *z* = 0), *O* (*x* = 0, *y* = 0, *z* = 0) and *A*(*x* = *a*, *y* = 0, *z* = 0)
The system consists of two dipole moment vectors due to (+*q* and –*q*) and again due to (+*q* and –*q*) charges having equal magnitudes qa units – one along $\overrightarrow{OA}$ and other along $\overrightarrow{OB}$. Hence, net dipole moment,

$p_{net} = \sqrt{(qa)^2 + (qa)^2} = \sqrt{2}qa$ along $\overrightarrow{OP}$ at an angle 45° with positive *X*-axis.

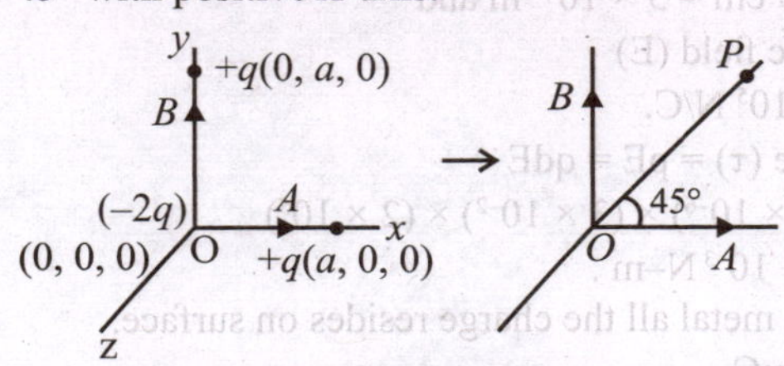

21. **(a)** Given : Length of the dipole (2*l*) = 10cm = 0.1m or *l* = 0.05 m
Charge on the dipole (q) = 500 μC = 500 × 10^{-6} C and distance of the point on the axis from the mid-point of the dipole (*r*) = 20 + 5 = 25 cm = 0.25 m. We know that the electric field intensity due to dipole on the given point

$$(E) = \frac{1}{4\pi\varepsilon_0} \times \frac{2(q.2l)r}{(r^2 - l^2)^2}$$

$$= 9 \times 10^9 \times \frac{2(500 \times 10^{-6} \times 0.1) \times 0.25}{[(0.25)^2 - (0.05)^2]^2}$$

$$= \frac{225 \times 10^3}{3.6 \times 10^{-3}} = 6.25 \times 10^7 \ N/C$$

NOTES

The dipole field $E \propto \frac{1}{r^3}$ decreases much rapidly as compared to the field of a point charge $E \propto \frac{1}{r^2}$

22. **(d)** Electric field at equatorial point

$$E_{eq} = \frac{1p}{4\pi E_o (r^2 + l^2)^{3/2}}$$

$$E = \frac{p}{4\pi\varepsilon_0 . r^3} \qquad [\text{For } r >> l]$$

Apparently, $E \propto p$ and $E \propto \frac{1}{r^3} \propto r^{-3}$.

23. **(c)** Intensity of electric field due to a Dipole

$$E = \frac{p}{4\pi\varepsilon_0 r^3}\sqrt{3\cos^2\theta + 1} \Rightarrow E \propto \frac{1}{r^3}$$

24. **(a)** Dipole is formed when two equal and dissimilar point charges placed at a short distance.

25. **(d)** Given $E = 2 \times 10^5$ N/C, $l = 2$ cm and $\tau = 4$ Nm

$\vec{\tau} = \vec{p} \times \vec{E} = pE\sin\theta$

$4 = p \times 2 \times 10^5 \times \sin 30° \Rightarrow p = 4 \times 10^{-5}$ cm

$\therefore$ Charge on the dipole, $q = \frac{p}{l} = \frac{4 \times 10^{-5}}{0.02} = 2$ mC

26. **(b)** Given : Dipole moment of the dipole = $\vec{P}$ and uniform electric field = $\vec{E}$. We know that dipole moment (p) = q.a (where q is the charge and a is dipole length). And when a dipole of dipole moment $\vec{P}$ is placed in a uniform electric field $\vec{E}$, then Torque (τ) = Either force × Perpendicular distance between the two forces = $qaE\sin\theta$ or $\tau = pE\sin\theta$ or $\vec{\tau} = \vec{p} \times \vec{E}$ (vector form)

27. **(c)** Charges (q) = 2×10^{-6} C,

Distance

(d) = 3 cm = 3×10^{-2} m and

electric field (E)

= 2×10^5 N/C.

Torque (τ) = pE = qdE

T = $(2 \times 10^{-6}) \times (3 \times 10^{-2}) \times (2 \times 10^5)$

= 12×10^{-3} N–m .

28. **(d)** In metal all the charge resides on surface.

$Q = 6\ \mu$C

Total surface area, $A = 6a^2$

$= 6 \times (5 \times 10^{-2})^2 = 150 \times 10^{-4}$ m^2

Surface charge density, $\sigma = \frac{Q}{A}$

$$= \frac{6 \times 10^{-6}}{150 \times 10^{-4}} = 0.4 \times 10^{-3}\ \text{C m}^{-2}$$

29. **(a)** $Q_1 = \sigma 4\pi R_1^{\ 2} = \sigma 4\pi R^2$

$Q_2 = \sigma 4\pi(2R)^2 = \sigma 16\pi R^2$

After redistribution

$$\frac{Q_1'}{Q_2'} = \frac{R}{2R} \Rightarrow Q_2' = 2Q_1' \qquad ...(i)$$

$$Q_1' + Q_2' = 20\sigma\pi R^2 \qquad ...(ii)$$

From eq. (i) and (ii)

$$Q_1' = \frac{20}{3}\sigma\pi R^2 \Rightarrow \sigma_1' = \frac{5}{3}\sigma$$

$$Q_2' = \frac{40}{3}\sigma\pi R^2 \Rightarrow \sigma_2' = \frac{5}{6}\sigma$$

30. **(b)** By the symmetry of the figure, the electric fields at O due to the portions AC and BD are equal in magnitude and opposite in direction. So, they cancel each other.

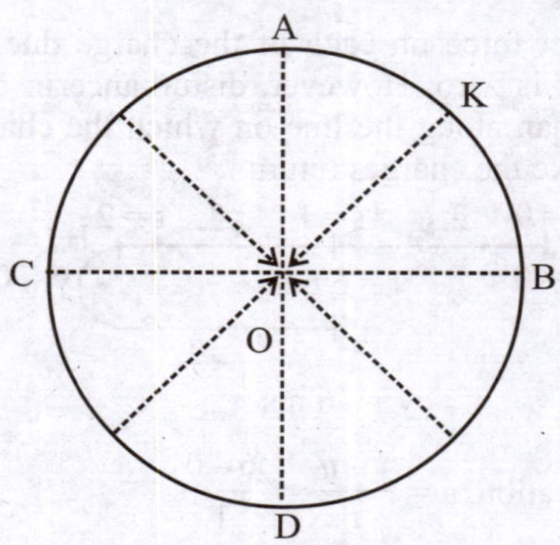

Similarly, the field at O due to CD and AKB are equal in magnitude but opposite in direction. Therefore, the electric field at the centre due to the charge on the part ACDB is E along OK.

31. **(a)** λ = linear charge density;

Charge on elementary portion $dq = \lambda\ dx$.

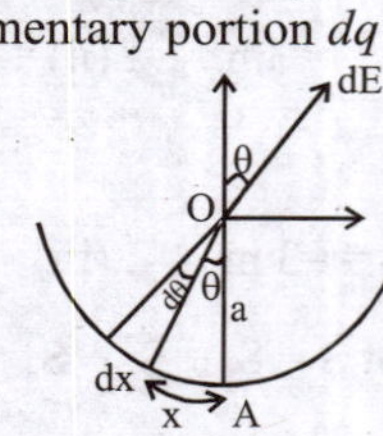

Electric field at O, $dE = \frac{\lambda dx}{4\pi\varepsilon_0 a^2}$

Horizontal electric field, *i.e.*, perpendicular to AO, will be cancelled.

Hence, net electric field = addition of all electrical fields in direction of AO

$= \Sigma dE \cos\theta$

$$\Rightarrow E = \int \frac{\lambda dx}{4\pi\varepsilon_0 a^2}\cos\theta$$

Also, $d\theta = \frac{dx}{a}$ or $dx = ad\theta$

$$E=\int_{-\pi/2}^{\pi/2}\frac{\lambda\cos\theta d\theta}{4\pi\varepsilon_0 a}=\frac{\lambda}{4\pi\varepsilon_0 a}[\sin\theta]_{-\pi/2}^{\pi/2}$$

$$=\frac{\lambda}{4\pi\varepsilon_0 a}[1-(-1)]=\frac{\lambda}{2\pi\varepsilon_0 a}$$

32. **(d)** According to Gauss' law, total electric flux

$$\phi=\oint\vec{E}\cdot\vec{ds}=\frac{q_{enclosed}}{\varepsilon_0}$$

∴ Electric flux depends on quantity of charges enclosed by surface.

Gauss's law is true for any closed surface, no matter what its shape or rize.

33. **(b)** From Gauss's law, total electric flux through a closed surface

$$\phi_{net}=\frac{q}{\varepsilon_0}$$

∴ Total electric flux through the cube

$$\phi_c=\frac{Q\times10^{-6}}{\varepsilon_0}$$

As charge is placed at the centre of the cube, so flux through each face is same.

$$\therefore\phi=\frac{1}{6}\phi_c=\frac{1}{6}\times\frac{Q\times10^{-6}}{\varepsilon_0}$$

34. **(a)** Since, $\phi_{net}=\oint_s\vec{E}\cdot\vec{dS}=0$

i.e., Net flux through surface is zero, so number of flux lines entering the surface must be equal to the number of flux lines leaving it.

> **NOTES** In the expression $\oint\vec{E}.dS=0$ the electric field is the sum of electric field due to charge within the surface and due to charge outside the surface.

35. **(c)** $\phi_{Total}=\frac{q_{enclosed}}{\varepsilon_0}$

$$\phi_{Total}=\frac{0}{\varepsilon_0}=0$$

36. **(b)** Net flux emitted from a spherical surface of radius a according to Gauss's theorem

$$\phi_{net}=\frac{q_{in}}{\varepsilon_0}$$

or, $(Aa)(4\pi a^2)=\frac{q_{in}}{\varepsilon_0}$

So, $q_{in}=4\pi\varepsilon_0 A a^3$

> **NOTES** Electric flux is equal to the product of an area element A and the perpendicular component of electric field $\vec{E}$ integrated over a surface.
> $\phi=E.A\cos\theta$

37. **(b)** Eight identical cubes are required to arrange so that this charge is at centre of the cube formed so flux.

$$\phi=\frac{q}{8\varepsilon_0}$$

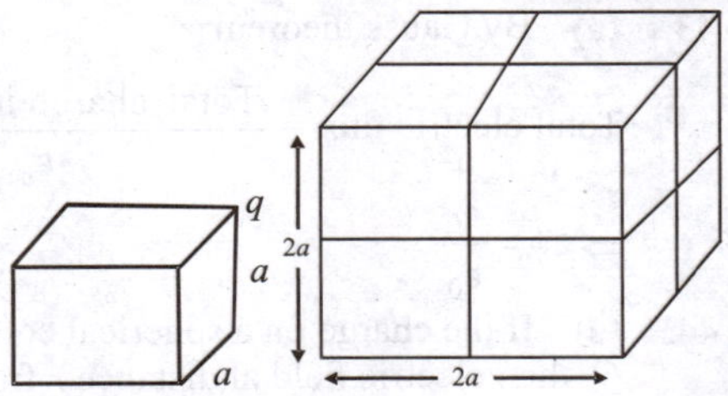

38. **(c)** By Gauss's theorem,

$$\phi=\frac{Q_{in}}{\varepsilon_0}$$

Thus, the net flux depends only on the charge enclosed by the surface. Hence, there will be no effect on the net flux if the radius of the surface is doubled.

39. **(d)** Since $\phi_{total}=\phi_A+\phi_B+\phi_C=\frac{q}{\varepsilon_0}$,

where q is the total charge.

As shown in the figure, flux associated with the curved surface B is $\phi=\phi_B$

Let us assume flux linked with the plane surfaces A and C be

$\phi_A=\phi_C=\phi'$

Therefore,

$$\frac{q}{\varepsilon_0}=2\phi'+\phi_B=2\phi'+\phi\ \Rightarrow\phi'=\frac{1}{2}\left(\frac{q}{\varepsilon_0}-\phi\right)$$

40. **(c)** Cube has 6 faces. Flux through any face is given by

$$\phi=\frac{q}{6\varepsilon_0}=\frac{q4\pi}{6(4\pi\varepsilon_0)}$$

> **NOTES** If charge (Q) is kept at the centre of cube
> $$\phi_{total}=\frac{Q}{\varepsilon_0},\ \phi_{face}=\frac{Q}{6\varepsilon_0}$$
> $$\phi_{corner}=\frac{Q}{8\varepsilon_0},\ \phi_{edge}=\frac{Q}{12\varepsilon_0}$$

41. **(a)** Total flux out of all six faces as per Gauss's theorem should be $\frac{\times10}{\in}$

Therefore, flux coming out of each face

$$=\frac{Q}{6\in_0}\times10^{-6}\ C$$

42. **(c)** According to Gauss's theorem, electric flux through a closed surface $=\frac{Q}{\varepsilon_0}$ where q is charge enclosed by the surface.

The charge is kept at a corner of a cube so it can be visualised as being at the centre of 8 cubes.

∴ The flux through each cube will be $\frac{1}{8}$ of the flux $\frac{\theta}{E_o}$.

∴ Flux through all six faces $=\frac{\theta}{8E_o}$

43. **(a)** By Gauss theorem

Total electric flux = $\frac{\text{Total charge inside cube}}{\varepsilon_0}$

$\Rightarrow \phi = \frac{q}{\varepsilon_0}$

44. **(a)** If the charge on a spherical conductor of radius R is Q, then electric field at distance r from centre is

$E = 0$ (if $r < R$)

$E = \frac{1}{4\pi\varepsilon_0}\frac{Q}{r^2}$ (if $r \geq R$)

Electric field at a distance 15 cm from the centre of sphere will be

$E = \frac{9\times10^9 \times 3.2\times10^{-7}}{225\times10^{-4}}$

$= 0.128\times10^6 = 1.28\times10^5$ N/C

45. **(c)** Electric field due to line charge (1)

$\vec{E}_1 = \frac{\lambda}{2\pi\varepsilon_0 R}\hat{i}$N/C

Electric field due to line charge (2)

$\vec{E}_2 = \frac{\lambda}{2\pi\varepsilon_0 R}\hat{i}$N/C

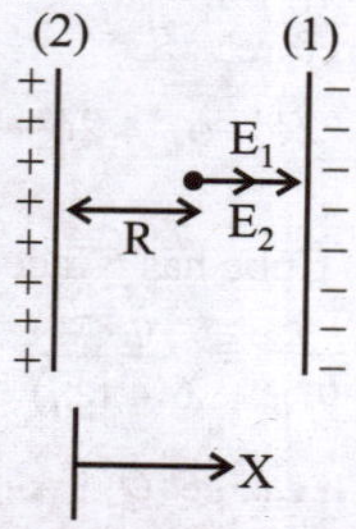

$\vec{E}_{net} = \vec{E}_1 + \vec{E}_2$

$= \frac{\lambda}{2\pi\varepsilon_0 R}\hat{i} + \frac{\lambda}{2\pi\varepsilon_0 R}\hat{i}$

$= \frac{\lambda}{\pi\varepsilon_0 R}\hat{i}$N/C

> **NOTES** Both $\vec{E}_1$ and $\vec{E}_2$ are in the same direction.
> Electric field at a point due to +ve charge (+q) acts away from the charge and due to negative charge (–q) it acts towards the charge.

46. **(b)** Charge Q will be distributed over the surface of hollow metal sphere.

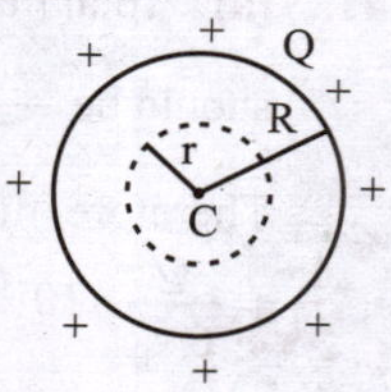

(i) For r < R (inside)

[At a point inside the hollow sphere]

By Gauss's law, $\oint \vec{E}_{in}\cdot\vec{dS} = \frac{q_{en}}{\varepsilon_0} = 0$

As enclosed charge is = 0

So, $E_{in} = 0$ the electric field inside the hollow sphere is always zero.

(ii) For r > R (outside)

[At a point outside hollow sphere]

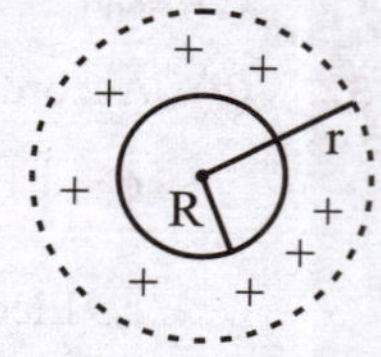

By Gauss's law,

$\oint \vec{E}_0\cdot\vec{dS} = \frac{q_{en}}{\varepsilon_0}$ $(\because q_{en} = Q)$

$\therefore E_0 4\pi r^2 = \frac{Q}{\varepsilon_0}$

$\therefore E_0 \propto \frac{1}{r^2}$

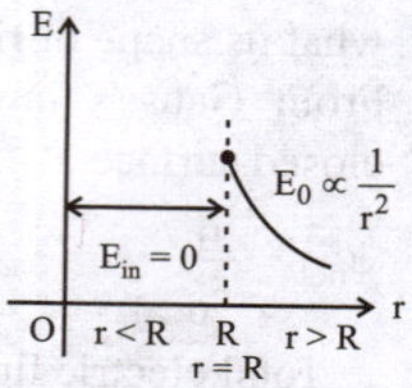

47. **(a)** Electric field inside the shell is zero. So at R/2 from the centre of the sphere is zero.

48. **(a)** Charge resides on the outer surface of a conducting hollow sphere of radius R. We consider a spherical surface of radius $r <$ R.

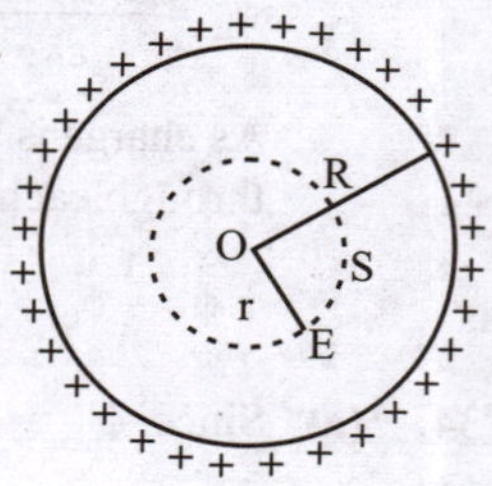

By Gauss's theorem,

$\oint \vec{E}.\vec{ds} = \frac{1}{\varepsilon_0}\times$ charge enclosed

or, $E.4\pi r^2 = \frac{1}{\varepsilon_0}\times 0 \Rightarrow E = 0$

i.e. electric field inside a hollow sphere is zero.

> **NOTES** Electric field inside the charged conducting sphere or shell, $E_{in} = 0$
> And for uniformly charged non-conducting sphere,
> $E_{in} = \frac{1}{4\pi\varepsilon_0}\frac{Q_r}{R^3}$

2 Electrostatic Potential and Capacitance

TREND ANALYSIS

TOPIC	2025		2024		2023		2022		2021		2020	
	Qns.	LOD	Qns.	LOD	Qns.	LOD	Qns.	LOD	Qns.	LOD	Qns.	LOD
Potential due to an Electric Dipole			1	Avg.	1	Avg.					1	Easy
Potential due to a System of Charges			1	Avg.			1	Easy	1	Avg.		
Equipotential Surfaces							1	Easy			1	Avg.
Potential Energy of a System of Charges					1	Avg.			1	Avg.		
The Parallel Plate Capacitor			1	Avg.			1	Easy				
Effect of Dielectric on Capacitance	1	Avg.									1	Avg.
Combination of Capacitors			1	Avg.	1	Avg.	1	Avg.	1	Avg.		
Energy Stored in a Capacitor							1	Diff.	1	Avg.		

How to Code Qns for Better & Faster Revision

Type 1: Able to solve (No code Req)
Type 2: Solved with struggle or by seeing solution **(Use code NR means Needs Revision or orange stickers)**
Type 3: Tricky/ Could not solve on my own. Need help from Teacher/ Friend. **(Use code TQ means Tricky Question or red stickers)**

You can either hand write the code in boxes or else paste coloured stickers given at end of book.

Topic-2.3
Potential due to a Point Charge

1. The value of electric potential at a distance of 9 cm from the point charge 4×10^{-7} C is **NCERT | Page- 48 | Ph-II 2024**

[Given $\frac{1}{4\pi\varepsilon_0} = 9 \times 10^9 \text{ Nm}^2 \text{ C}^{-2}$]:

(a) 4×10^2 V (b) 44.4 V
(c) 4.4×10^5 V (d) 4×10^4 V

Topic-2.4
Potential due to an Electric Dipole

2. Given below are two statements: one is labelled as Assertion A and the other is labelled as Reason R.

Assertion (A): The potential (V) at any axial point, at 2 m distance (r) from the centre of the dipole of dipole moment vector $\vec{P}$ of magnitude, 4×10^{-6} C m, is $\pm 9 \times 10^3$ V. (Take $\frac{1}{4\pi\varepsilon_0} = 9 \times 10^9$ SI units)

Reason (R): $V = \pm\frac{2P}{4\pi\varepsilon_0 r^2}$, where r is the distance of any axial point, situated at 2 m from the centre of the dipole.

In the light of the above statements, choose the correct answer from the options given below: **NCERT | Page-51 | 2024**

(a) Both A and R are true and R is the correct explanation of A.
(b) Both A and R are true and R is NOT the correct explanation of A.
(c) A is true but R is false.
(d) A is false but R is true.

3. An electric dipole is placed as shown in the figure.

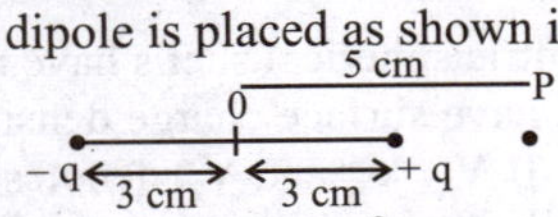

The electric potential (in 10^2 V) at point P due to the dipole is (ϵ_0 = permittivity of free space and $\frac{1}{4\pi\epsilon_0} = K$) **NCERT | Page-50 | Rs 2023**

(a) $\left(\frac{3}{8}\right)qK$ (b) $\left(\frac{5}{8}\right)qK$ (c) $\left(\frac{8}{5}\right)qK$ (d) $\left(\frac{8}{3}\right)qK$

4. A short electric dipole has a dipole moment of 16×10^{-9} C m. The electric potential due to the dipole at a point at a distance of 0.6 m from the centre of the dipole, situated on a line making an angle of 60° with the dipole axis is : **NCERT | Page-51 | 2020**

$\left(\frac{1}{4\pi\epsilon_0} = 9 \times 10^9 \text{ N m}^2/\text{C}^2\right)$

(a) 200 V (b) 400 V (c) zero (d) 50 V

Topic-2.5

Potential due to a System of Charges

5. A thin spherical shell is charged by some source. The potential difference between the two points C and P (in V) shown in the figure is:

(Take $\frac{1}{4\pi\varepsilon_0} = 9 \times 10^9$ SI units) NCERT | Page-52 | 2024

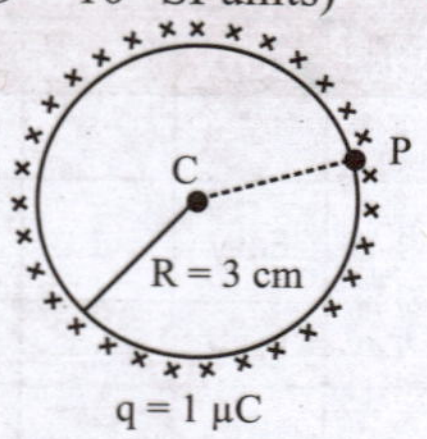

(a) 3×10^5 (b) 1×10^5
(c) 0.5×10^5 (d) Zero

6. Two hollow conducting spheres of radii R_1 and R_2 ($R_1 >> R_2$) have equal charges. The potential would be : NCERT | Page-52 | 2022

(a) more on smaller sphere
(b) equal on both the spheres
(c) dependent on the material property of the sphere
(d) more on bigger sphere

7. Twenty seven drops of same size are charged at 200 *V* each. They combine to form a bigger drop. Calculate the potential of the bigger drop. NCERT | Page-52 | 2021

(a) 1980 *V* (b) 660 *V* (c) 1320 *V* (d) 1520 *V*

8. A conducting sphere of radius R is given a charge Q. The electric potential and the electric field at the centre of the sphere respectively are: NCERT | Page-52 | 2014

(a) Zero and $\frac{Q}{4\pi\varepsilon_0 R^2}$ (b) $\frac{Q}{4\pi\varepsilon_0 R}$ and Zero

(c) $\frac{Q}{4\pi\varepsilon_0 R}$ and $\frac{Q}{4\pi\varepsilon_0 R^2}$ (d) Both are zero

9. Three concentric spherical shells have radii a, b and c (a < b < c) and have surface charge densities σ, – σ and σ respectively. If V_A, V_B and V_C denotes the potentials of the three shells, then for c = a + b, we have NCERT | Page-56 | 2009

(a) $V_C = V_B \neq V_A$ (b) $V_C \neq V_B \neq V_A$
(c) $V_C = V_B = V_A$ (d) $V_C = V_A \neq V_B$

10. A hollow metal sphere of radius 10 cm is charged such that the potential on its surface is 80 V. The potential at the centre of the sphere is NCERT | Page-52 | 1994

(a) zero (b) 80 V (c) 800 V (d) 8 V

11. Four point charges $-Q$, $-q$, $2q$ and $2Q$ are placed, one at each corner of the square. The relation between Q and q for which the potential at the centre of the square is zero is : NCERT | Page-48 | 2012

(a) $Q = -q$ (b) $Q = -\frac{1}{q}$

(c) $Q = q$ (d) $Q = \frac{1}{q}$

12. Four electric charges +q, +q, –q and – q are placed at the corners of a square of side 2L (see figure). The electric potential at point A, midway between the two charges +q and +q, is NCERT | Page-48 | 2011

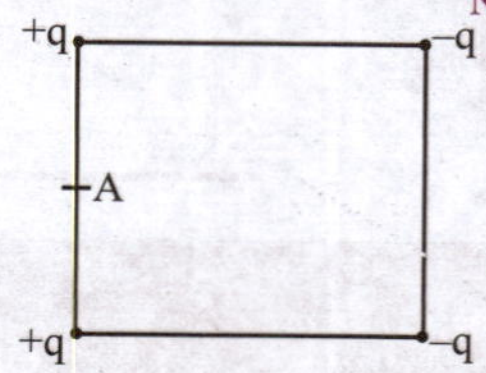

(a) $\frac{1}{4\pi\epsilon_0}\frac{2q}{L}(1+\sqrt{5})$ (b) $\frac{1}{4\pi\epsilon_0}\frac{2q}{L}\left(1+\frac{1}{\sqrt{5}}\right)$

(c) $\frac{1}{4\pi\epsilon_0}\frac{2q}{L}\left(1-\frac{1}{\sqrt{5}}\right)$ (d) zero

Topic-2.6

Equipotential Surfaces

13. The angle between the electric lines of force and the equipotential surface is: NCERT | Page-54 | 2022

(a) 45° (b) 90° (c) 180° (d) 0°

14. In a certain region of space with volume 0.2 m^3, the electric potential is found to be 5 V throughout. The magnitude of electric field in this region is : NCERT | Page-55 | 2020

(a) 0.5 N/C (b) 1 N/C (c) 5 N/C (d) zero

15. The diagrams below show regions of equipotentials. NCERT | Page-54 | 2017

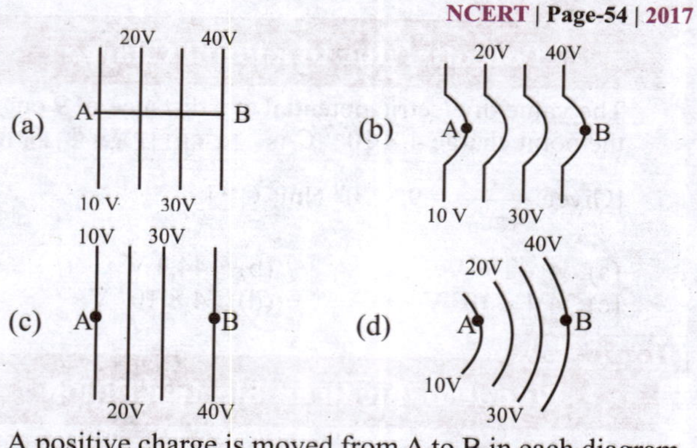

A positive charge is moved from A to B in each diagram.

(a) In all the four cases the work done is the same
(b) Minimum work is required to move q in figure (a)
(c) Maximum work is required to move q in figure (b)
(d) Maximum work is required to move q in figure (c)

16. A, B and C are three points in a uniform electric field. The electric potential is NCERT | Page-55 | 2013

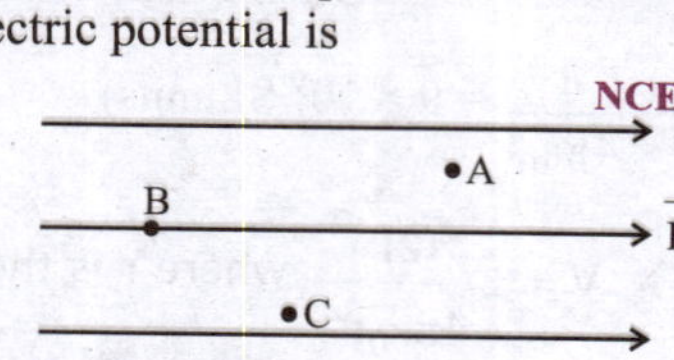

(a) maximum at B
(b) maximum at C
(c) same at all the three points A, B and C
(d) maximum at A

17. The electric potential V at any point (x, y, z), all in meters in space is given by $V = 4x^2$ volt. The electric field at the point (1, 0, 2) in volt/meter is **NCERT | Page-55 | 2011 M**
(a) 8 along positive X-axis
(b) 16 along negative X-axis
(c) 16 along positive X-axis
(d) 8 along negative X-axis

18. The electric potential at a point (x, y, z) is given by $V = -x^2y - xz^3 + 4$. The electric field $\vec{E}$ at that point is **NCERT | Page-55 | 2009**
(a) $\vec{E} = \hat{i}\,2xy + \hat{j}(x^2 + y^2) + \hat{k}(3xz - y^2)$
(b) $\vec{E} = \hat{i}\,z^3 + \hat{j}\,xyz + \hat{k}\,z^2$
(c) $\vec{E} = \hat{i}(2xy - z^3) + \hat{j}\,xy^2 + \hat{k}\,3z^2x$
(d) $\vec{E} = \hat{i}(2xy + z^3) + \hat{j}\,x^2 + \hat{k}\,3xz^2$

19. The electric potential at a point in free space due to a charge Q coulomb is $Q \times 10^{11}$ volts. The electric field at that point is **NCERT | Page-55 | 2008**
(a) $4\pi\varepsilon_0 Q\times10^{22}$ volt/m (b) $12\pi\varepsilon_0 Q\times10^{20}$ volt/m
(c) $4\pi\varepsilon_0 Q\times10^{20}$ volt/m (d) $12\pi\varepsilon_0 Q\times10^{22}$ volt/m

20. From a point charge, there is a fixed point A. At A, there is an electric field of 500 V/m and potential difference of 3000 V. Distance between point charge and A will be **NCERT | Page-55 | 1997**
(a) 6 m (b) 12 m (c) 16 m (d) 24 m

21. If potential (in volts) in a region is expressed as V(x, y, z) = 6 xy − y + 2yz, the electric field (in N/C) at point (1, 1, 0) is : **NCERT | Page-55 | 2015 RS**
(a) $-(6\hat{i}+5\hat{j}+2\hat{k})$ (b) $-(2\hat{i}+3\hat{j}+\hat{k})$
(c) $-(6\hat{i}+9\hat{j}+\hat{k})$ (d) $-(3\hat{i}+5\hat{j}+3\hat{k})$

22. In a region, the potential is represented by V(x, y, z) = 6x − 8xy − 8y + 6yz, where V is in volts and x, y, z are in metres. The electric force experienced by a charge of 2 coulomb situated at point (1, 1, 1) is **NCERT | Page-55 | 2014**
(a) $6\sqrt{5}$ N (b) 30 N
(c) 24 N (d) $4\sqrt{35}$ N

23. The mean free path of electrons in a metal is 4×10^{-8} m. The electric field which can give on an average 2 eV energy to an electron in the metal will be in units of V/m **NCERT | Page-55 2009**
(a) 5×10^{-11} (b) 8×10^{-11}
(c) 5×10^{7} (d) 8×10^{7}

24. If a conducting sphere of radius R is charged. Then the electric field at a distance r (r > R) from the centre of the sphere would be, (V = potential on the surface of the sphere) **NCERT | Page-62 | Manipur 2023**
(a) $\frac{RV}{r^2}$ (b) $\frac{V}{r}$ (c) $\frac{rV}{R^2}$ (d) $\frac{R^2V}{r^3}$

25. Two charged spherical conductors of radius R_1 and R_2 are connected by a wire. Then the ratio of surface charge densities of the spheres (σ_1/σ_2) is **2021**
(a) $\frac{R_1^2}{R_2^2}$ (b) $\frac{R_1}{R_2}$ (c) $\frac{R_2}{R_1}$ (d) $\sqrt{\left(\frac{R_1}{R_2}\right)}$

26. Two metallic spheres of radii 1 cm and 3 cm are given charges of $-1\times10^{-2}\,C$ and $5\times10^{-2}\,C$, respectively. If these are connected by a conducting wire, the final charge on the bigger sphere is : **NCERT | Page-52 | 2012 M**
(a) $2 \times 10^{-2}\,C$ (b) $3 \times 10^{-2}\,C$
(c) $4 \times 10^{-2}\,C$ (d) $1 \times 10^{-2}\,C$

27. A solid spherical conductor is given a charge. The electrostatic potential of the conductor is **NCERT | Page-52 | 2002**
(a) constant throughout the conductor
(b) largest at the centre
(c) largest on the surface
(d) largest somewhere between the centre and the surface

Topic-2.7

Potential Energy of a System of Charges

28. Six charges +q, −q, +q, −q, +q and −q are fixed at the corners of a hexagon of side d as shown in the figure. The work done in bringing a charge q_0 to the centre of the hexagon from infinity is : (ε_0 - permittivity of free space) **NCERT | Page-59 | Rs 2023**

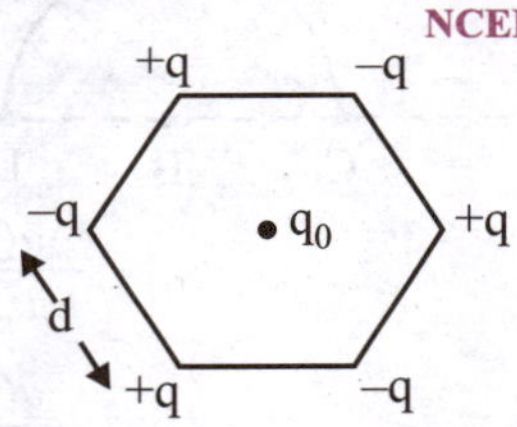

(a) $\frac{-q^2}{4\pi\varepsilon_0 d}\left(6-\frac{1}{\sqrt{2}}\right)$ (b) Zero
(c) $\frac{-q^2}{4\pi\varepsilon_0 d}$ (d) $\frac{-q^2}{4\pi\varepsilon_0 d}\left(3-\frac{1}{\sqrt{2}}\right)$

29. A dipole is placed in an electric field as shown. In which direction will it move? **2021**

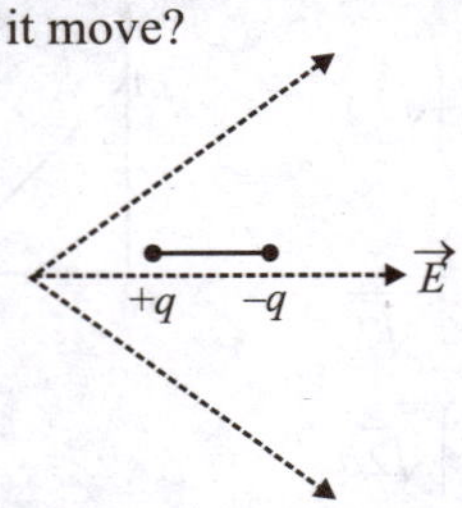

(a) Towards the right as its potential energy will increase.
(b) Towards the left as its potential energy will increase.
(c) Towards the right as its potential energy will decrease.
(d) Towards the left as its potential energy will decrease.

30. Three charges, each +q, are placed at the corners of an isosceles triangle ABC of sides BC, AC and AB. D and E are the mid points of BC and CA. The work done in taking a charge Q from D to E is **NCERT | Page-47 | 2011 M**

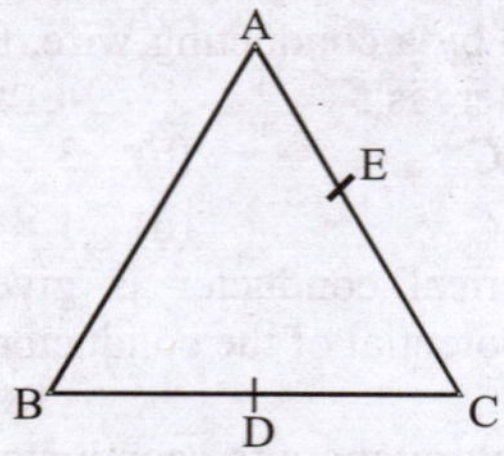

(a) $\frac{3qQ}{8\pi \epsilon_0 a}$ (b) $\frac{qQ}{4\pi \epsilon_0 a}$

(c) zero (d) $\frac{3qQ}{4\pi \epsilon_0 a}$

31. Charges $+q$ and $-q$ are placed at points A and B respectively which are a distance $2L$ apart, C is the midpoint between A and B. The work done in moving a charge $+Q$ along the semicircle CRD is

NCERT | Page-56 | 2007

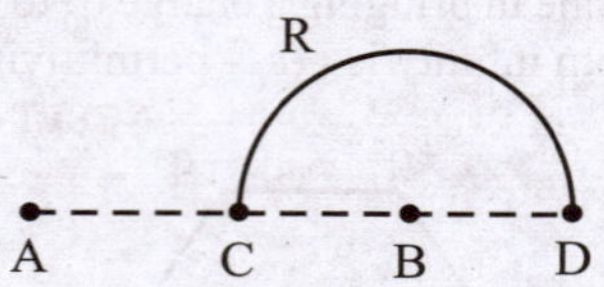

(a) $\frac{qQ}{2\pi\varepsilon_0 L}$ (b) $\frac{qQ}{6\pi\varepsilon_0 L}$

(c) $-\frac{qQ}{6\pi\varepsilon_0 L}$ (d) $\frac{qQ}{4\pi\varepsilon_0 L}$

32. As per the diagram, a point charge $+q$ is placed at the origin O. Work done in taking another point charge $-Q$ from the point A [coordinates $(0, a)$] to another point B [coordinates $(a, 0)$] along the straight path AB is:

NCERT | Page-47 | 2005

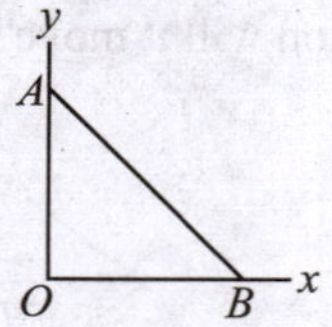

(b) zero

(b) $\left(\frac{-qQ}{4\pi\varepsilon_0}\frac{1}{a^2}\right)\sqrt{2}a$

(c) $\left(\frac{qQ}{4\pi\varepsilon_0}\frac{1}{a^2}\right).\frac{a}{\sqrt{2}}$

(d) $\left(\frac{qQ}{4\pi\varepsilon_0}\frac{1}{a^2}\right).\sqrt{2}a$

33. Two charges q_1 and q_2 are placed 30 cm apart, as shown in the figure. A third charge q_3 is moved along the arc of a circle of radius 40 cm from C to D. The change in the potential energy of the system is $\frac{q_3}{4\pi \epsilon_0}k$, where k is

NCERT | Page-56 | 2005

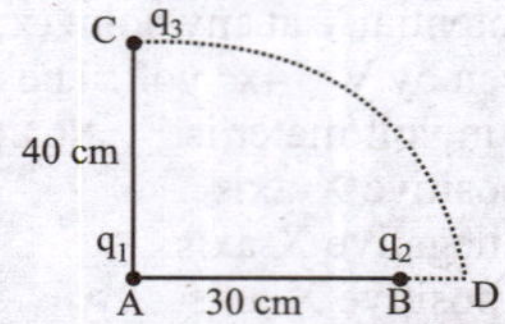

(a) $8q_1$ (b) $6q_1$ (c) $8q_2$ (d) $6q_2$

34. A bullet of mass 2 g is having a charge of 2μC. Through what potential difference must it be accelerated, starting from rest, to acquire a speed of 10 m/s?

NCERT | Page-56 | 2004

(a) 50 V (b) 5 kV (c) 50 kV (d) 5 V

35. Each corner of a cube of side l has a negative charge, $-q$. The electrostatic potential energy of a charge q at the centre of the cube is **NCERT | Page-47 | 2002**

(a) $-\frac{4q^2}{\sqrt{2}\pi\varepsilon_0 l}$ (b) $\frac{\sqrt{3}q^2}{4\pi\varepsilon_0 l}$

(c) $\frac{4q^2}{\sqrt{2}\pi\varepsilon_0 l}$ (d) $-\frac{4q^2}{\sqrt{3}\pi\varepsilon_0 l}$

36. A particle of mass m and charge q is placed at rest in a uniform electric field E and then released. The kinetic energy attained by the particle after moving a distance y is **NCERT | Page-56 | 1998**

(a) qEy^2 (b) $qE^2 y$ (c) qEy (d) $q^2 Ey$

Topic-2.8

Potential Energy in an External Field

37. An electric dipole with dipole moment 5×10^{-6} Cm is aligned with the direction of a uniform electric field of magnitude 4×10^5 N/C. The dipole is then rotated through an angle of 60° with respect to the electric field. The change in the potential energy of the dipole is :

NCERT | Page-60 | 2025

(a) 0.8 J (b) 1.0 J (c) 1.2 J (d) 1.5 J

38. An electric dipole of dipole moment p is aligned parallel to a uniform electric field E. The energy required to rotate the dipole by 90° is **NCERT | Page-60 | Kar. 2013**

(a) pE^2 (b) p^2E (c) pE (d) infinity

39. An electric dipole of moment $'p'$ is placed in an electric field of intensity $'E'$. The dipole acquires a position such that the axis of the dipole makes an angle θ with the direction of the field. Assuming that the potential energy of the dipole to be zero when $\theta = 90°$, the torque and the potential energy of the dipole will respectively be :

NCERT | Page-60 | 2012

(a) $p E \sin\theta, -p E \cos\theta$

(b) $p E \sin\theta, -2 p E \cos\theta$

(c) $p E \sin\theta, 2 p E \cos\theta$

(d) $p E \cos\theta, -p E \cos\theta$

40. An electric dipole has the magnitude of its charge as q and its dipole moment is p. It is placed in uniform electric field E. If its dipole moment is along the direction of the field, the force on it and its potential energy are respectively NCERT | Page-56 | 2004

(a) $q.E$ and max. (b) $2\,q.E$ and min.
(c) $q.E$ and $p.E$ (d) zero and min.

Topic-2.12 The Parallel Plate Capacitor

41. The capacitance of a capacitor with charge q and a potential difference V depends on NCERT | Page-68, 69 | Ph-II 2024

(a) both q and V
(b) the geometry of the capacitor
(c) q only
(d) V only

42. If the plates of a parallel plate capacitor connected to a battery are moved close to each other, then

A. the charge stored in it, increases.
B. the energy stored in it, decreases.
C. its capacitance increases.
D. the ratio of charge to its potential remains the same.
E. the product of charge and voltage increases.

Choose the most appropriate answer from the options given below: NCERT | Page-69 | 2024

(a) A, B and E only (b) A, C and E only
(c) B, D and E only (d) A, B and C only

43. The distance between the two plates of a parallel plate capacitor is doubled and the area of each plate is halved. If C is its initial capacitance, its final capacitance is equal to: NCERT | Page-69 | RS 2022

(a) $\frac{C}{4}$ (b) 2 C (c) $\frac{C}{2}$ (d) 4 C

44. The electrostatic force between the metal plates of an isolated parallel plate capacitor C having a charge Q and area A, is NCERT | Page-68 | 2018

(a) independent of the distance between the plates
(b) linearly proportional to the distance between the plates
(c) inversely proportional to the distance between the plates
(d) proportional to the square root of the distance between the plates

45. A parallel plate air capacitor has capacity 'C' distance of separation between plates is 'd' and potential difference 'V' is applied between the plates. Force of attraction between the plates of the parallel plate air capacitor is : NCERT | Page-68 | 2015 RS

(a) $\frac{CV^2}{2d}$ (b) $\frac{CV^2}{d}$ (c) $\frac{C^2V^2}{2d^2}$ (d) $\frac{C^2V^2}{2d^2}$

46. A parallel plate air capacitor is charged to a potential difference of V volts. After disconnecting the charging battery the distance between the plates of the capacitor is increased using an insulating handle. As a result the potential difference between the plates NCERT | Page-67 | 2006

(a) does not change (b) becomes zero
(c) increases (d) decreases

47. The capacity of a parallel plate condenser is 10 μF when the distance between its plates is 8 cm. If the distance between the plates is reduced to 4 cm then the capacity of this parallel plate condenser will be NCERT | Page-68 | 2001

(a) 5 μF (b) 10 μF (c) 20 μF (d) 40 μF

Topic-2.13 Effect of Dielectric on Capacitance

48. The plates of a parallel plate capacitor are separated by d. Two slabs of different dielectric constant K_1 and K_2 with thickness $\frac{3}{8}d$ and $\frac{d}{2}$, respectively are inserted in the capacitor. Due to this, the capacitance becomes two times larger than when there is nothing between the plates. (If $K_1 = 1.25\,K_2$, the value of K_1 is : NCERT | Page- 70 | 2025

(a) 2.66 (b) 2.33 (c) 1.60 (d) 1.33

49. The capacitance of a parallel plate capacitor with air as medium is 6 μF. With the introduction of a dielectric medium, the capacitance becomes 30 μF. The permittivity of the medium is : NCERT | Page-70 | 2020

($\epsilon_0 = 8.85 \times 10^{-12}\ C^2 N^{-1} m^{-2}$)

(a) $1.77 \times 10^{-12}\ C^2 N^{-1} m^{-2}$
(b) $0.44 \times 10^{-10}\ C^2 N^{-1} m^{-2}$
(c) $5.00\ C^2 N^{-1} m^{-2}$
(d) $0.44 \times 10^{-13}\ C^2 N^{-1} m^{-2}$

50. Two thin dielectric slabs of dielectric constants K_1 and K_2 ($K_1 < K_2$) are inserted between plates of a parallel plate capacitor, as shown in the figure. The variation of electric field 'E' between the plates with distance 'd' as measured from plate P is correctly shown by: NCERT | Page-70 | 2014

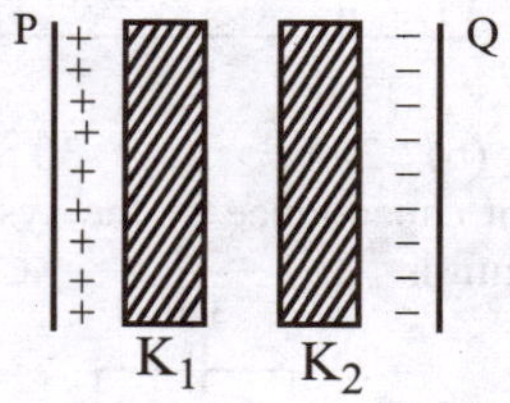

(a) 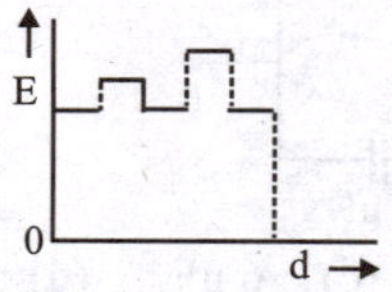

(b) 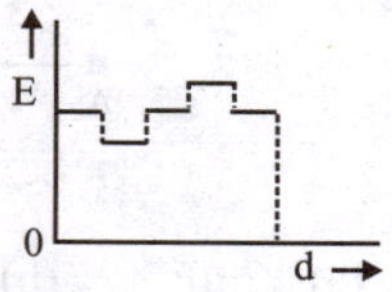

(c) 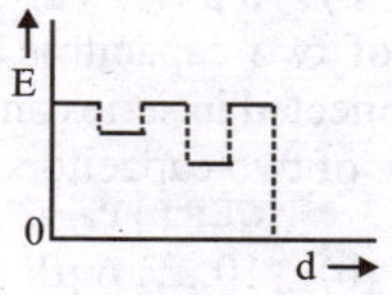

(d) 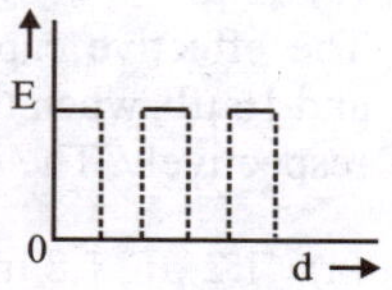

51. Two parallel metal plates having charges $+Q$ and $-Q$ face each other at a certain distance between them. If the plates are now dipped in kerosene oil tank, the electric field between the plates will **NCERT | Page-70 |2010 M**

(a) Become zero (b) Increase
(c) Decrease (d) Remain same

52. Two parallel metal plates having charges + Q and –Q face each other at a certain distance between them. If the plates are now dipped in kerosene oil tank, the electric field between the plates will **NCERT | Page-70 | 2010**

(a) remain same (b) become zero
(c) increases (d) decrease

53. A parallel plate condenser with oil between the plates (dielectric constant of oil $K = 2$) has a capacitance C. If the oil is removed, then capacitance of the capacitor becomes **NCERT | Page-70 | 1999, 97**

(a) $\sqrt{2}\,C$ (b) $2C$ (c) $\frac{C}{\sqrt{2}}$ (d) $\frac{C}{2}$

Topic-2.14 Combination of Capacitors

54. In the following circuit, the equivalent capacitance between terminal A and terminal B is : **NCERT | Page-72 | 2024**

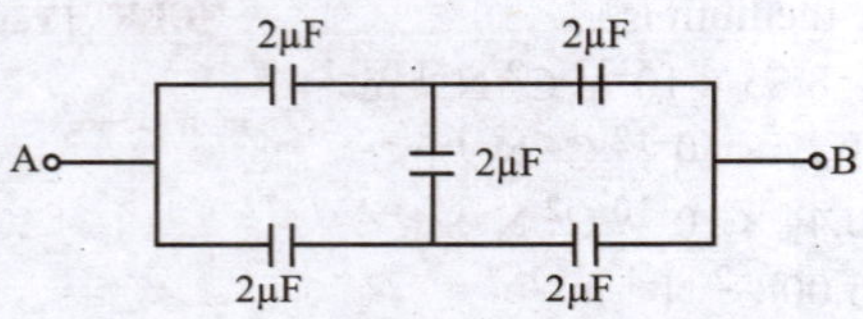

(a) 2 μF (b) 1 μF (c) 0.5 μF (d) 4 μF

55. The equivalent capacitance of the arrangement shown in figure is **NCERT | Page-71 | Manipur 2023**

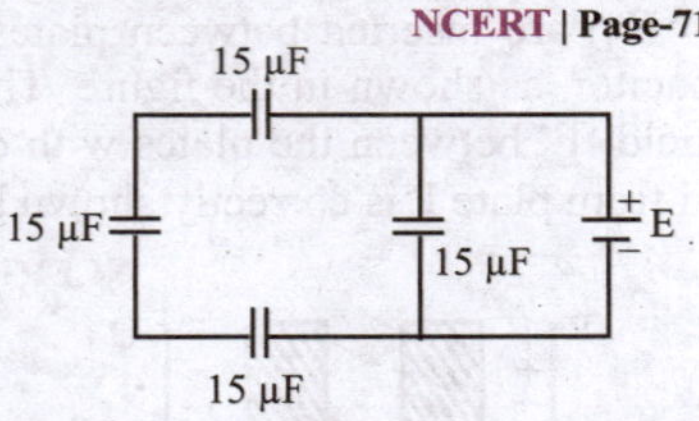

(a) 25 μF (b) 20 μF (c) 30 μF (d) 15 μF

56. The equivalent capacitance of the system shown in the following circuit is **NCERT | Page-72 | 2023**

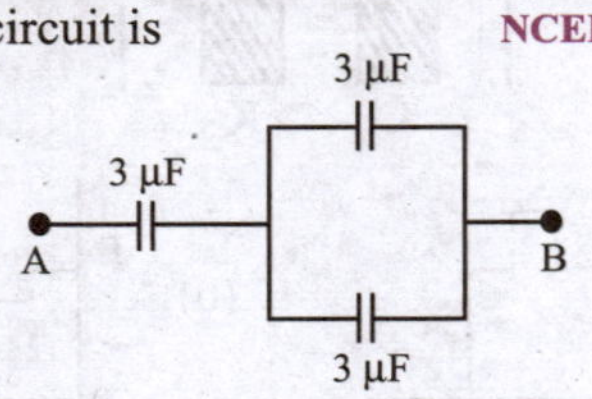

(a) 2 μF (b) 3 μF (c) 6 μF (d) 9 μF

57. The effective capacitances of two capacitors are 3 μF and 16 μF, when they are connected in series and parallel respectively. The capacitance of two capacitors are: **NCERT | Page-72 | RS 2022**

(a) 1.2 μF, 1.8 μF (b) 10 μF, 6 μF
(c) 8 μF, 8 μF (d) 12 μF, 4 μF

58. The equivalent capacitance of the combination shown in the figure is **NCERT | Page-72 | 2021**

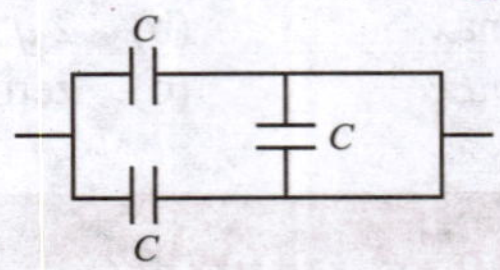

(a) $\frac{3C}{2}$ (b) $3C$ (c) $2C$ (d) $\frac{C}{2}$

59. Three capacitors each of capacitance C and of breakdown voltage V are joined in series. The capacitance and breakdown voltage of the combination will be **NCERT | Page-72 | 2009**

(a) $3C, \frac{V}{3}$ (b) $\frac{C}{3}, 3V$ (c) $3C, 3V$ (d) $\frac{C}{3}, \frac{V}{3}$

60. A network of four capacitors of capacity equal to $C_1 = C$, $C_2 = 2C$, $C_3 = 3C$ and $C_4 = 4C$ are connected to a battery as shown in the figure. The ratio of the charges on C_2 and C_4 is: **NCERT | Page-72 | 2005**

(a) $\frac{4}{7}$

(b) $\frac{3}{22}$

(c) $\frac{7}{4}$

(d) $\frac{22}{3}$

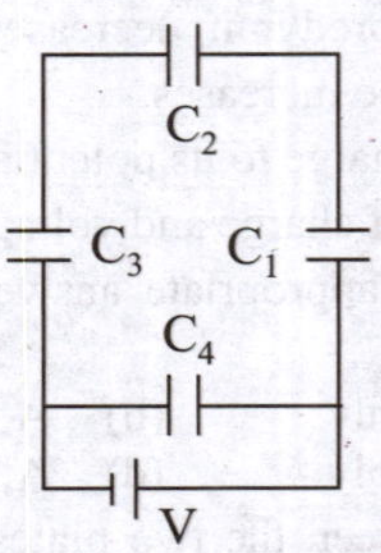

61. Three capacitors each of capacity 4μF are to be connected in such a way that the effective capacitance is 6 μF. This can be done by **NCERT | Page-72 | 2003**

(a) connecting two in parallel and one in series
(b) connecting all of them in series
(c) connecting them in parallel
(d) connecting two in series and one in parallel

62. A capacitor C_1 is charged to a potential difference V. The charging battery is then removed and the capacitor is connected to an uncharged capacitor C_2. The potential difference across the combination is **NCERT | Page-72 | 2002**

(a) $\frac{VC_1}{(C_1 + C_2)}$ (b) $V\left(1+\frac{C_2}{C_1}\right)$

(c) $V\left(1+\frac{C_1}{C_2}\right)$ (d) $\frac{VC_2}{(C_1 + C_2)}$

63. What is the effective capacitance between points X and Y? **NCERT | Page-72 | 1999**

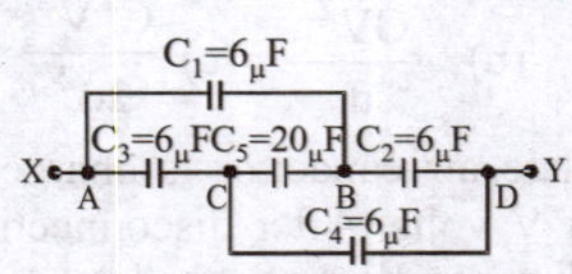

(a) 24 μF (b) 18 μF (c) 12 μF (d) 6 μF

64. The four capacitors, each of 25μ F are connected as shown in fig. The dc voltmeter reads 200 V. The charge on each plate of capacitor is **NCERT | Page-67 | 1994**

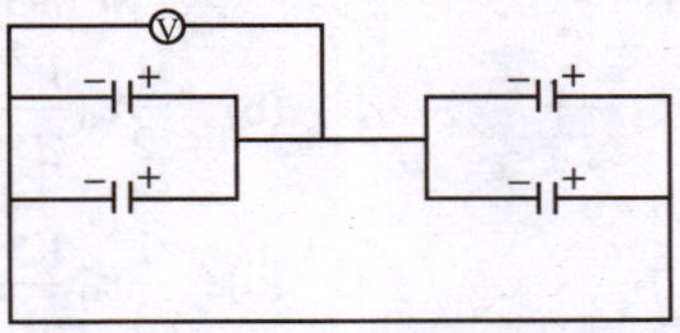

(a) $\pm 2 \times 10^{-3}$ C (b) $\pm 5 \times 10^{-3}$ C
(c) $\pm 2 \times 10^{-2}$ C (d) $\pm 5 \times 10^{-2}$ C

Topic-2.15
Energy Stored in a Capacitor

65. A 12 pF capacitor is connected to a 50 V battery, the electrostatic energy stored in the capacitor in nJ is **NCERT | Page- 74 | Ph-II 2024**

(a) 15 (b) 7.5 (c) 0.3 (d) 150

66. To produce an instantaneous displacement current of 2 mA in the space between the parallel plates of a capacitor of capacitance 4 μF, the rate of change of applied variable potential difference $\left(\frac{dV}{dt}\right)$ must be **NCERT | Page-74 | Manipur 2023**

(a) 200 V/s (b) 400 V/s
(c) 800 V/s (d) 500 V/s

67. A capacitor of capacitance C = 900 pF is charged fully by 100 V battery B as shown in figure (a). Then it is disconnected from the battery and connected to another uncharged capacitor of capacitance C = 900 pF as shown in figure (b). The electrostatic energy stored by the system (b) is **NCERT | Page-74 | 2022**

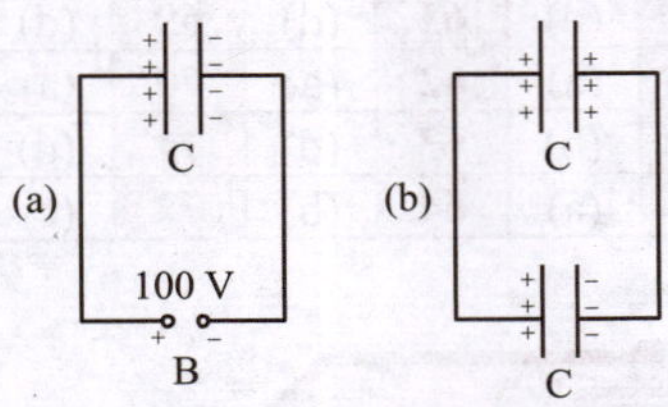

(a) 3.25×10^{-6} J (b) 2.25×10^{-6} J
(c) 1.5×10^{-6} J (d) 4.5×10^{-6} J

68. A parallel plate capacitor has a uniform electric field '$\vec{E}$' in the space between the plates. If the distance between the plates is 'd' and the area of each plate is 'A', the energy stored in the capacitor is (ε_0 = permittivity of free space) **NCERT | Page-74 | 2021**

(a) $\frac{E^2 Ad}{\varepsilon_0}$ (b) $\frac{1}{2}\varepsilon_0 E^2$

(c) $\varepsilon_0 EAd$ (d) $\frac{1}{2}\varepsilon_0 E^2 Ad$

69. Two identical capacitors C_1 and C_2 of equal capacitance are connected as shown in the circuit. Terminals a and b of the key k are connected to charge capacitor C_1 using battery of emf V volt. Now disconnecting a and b the terminals b and c are connected. Due to this, what will be the percentage loss of energy? **NCERT | Page-74 | Odisha 2019**

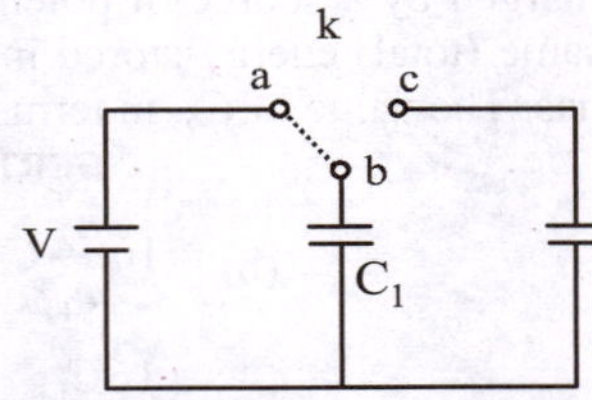

(a) 25% (b) 75% (c) 0% (d) 50%

70. A capacitor is charged by a battery. The battery is removed and another identical uncharged capacitor is connected in parallel. The total electrostatic energy of resulting system: **NCERT | Page-74 | 2017**

(a) decreases by a factor of 2
(b) remains the same
(c) increases by a factor of 2
(d) increases by a factor of 4

71.

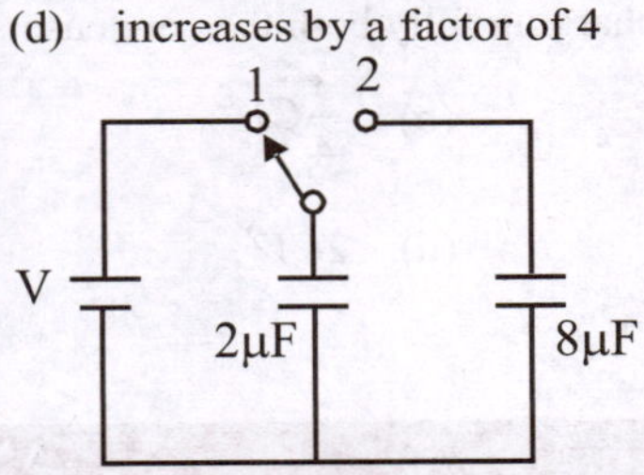

A capacitor of 2μF is charged as shown in the diagram. When the switch S is turned to position 2, the percentage of its stored energy dissipated is : **NCERT | Page-74 | 2016**

(a) 0% (b) 20% (c) 75% (d) 80%

72. A parallel plate air capacitor of capacitance C is connected to a cell of emf V and then disconnected from it. A dielectric slab of dielectric constant K, which can just fill the air gap of the capacitor, is now inserted in it. Which of the following is incorrect ? **NCERT | Page-74 | 2015**

(a) The energy stored in the capacitor decreases K times.
(b) The chance in energy stored is $\frac{1}{2}CV^2\left(\frac{1}{K}-1\right)$
(c) The charge on the capacitor is not conserved.
(d) The potential difference between the plates decreases K times.

73. A parallel plate capacitor has a uniform electric field E in the space between the plates. If the distance between the plates is d and area of each plate is A, the energy stored in the capacitor is : **NCERT | Page-74 | 2012 M, 2011, 2008**

(a) $\frac{1}{2}\varepsilon_0 E^2$ (b) $E^2 Ad/\varepsilon_0$

(c) $\frac{1}{2}\varepsilon_0 E^2 Ad$ (d) $\varepsilon_0 EAd$

74. A series combination of n_1 capacitors, each of value C_1, is charged by a source of potential difference $4V$. When another parallel combination of n_2 capacitors, each of value C_2, is charged by a source of potential difference V, it has the same (total) energy stored in it, as the first combination has. The value of C_2, in terms of C_1, is then **NCERT | Page-74 | 2010**

(a) $\frac{2C_1}{n_1 n_2}$ (b) $16\frac{n_2}{n_1}C_1$

(c) $2\frac{n_2}{n_1}C_1$ (d) $\frac{16C_1}{n_1 n_2}$

75. Two condensers, one of capacity C and other of capacity C/2 are connected to a V-volt battery, as shown. **NCERT | Page-74 | 2007**

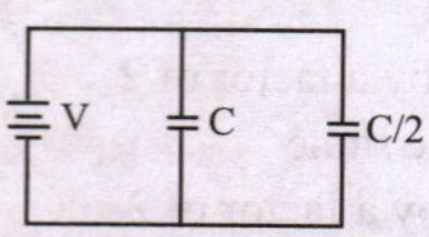

The work done in charging fully both the condensers is

(a) $\frac{1}{4}CV^2$ (b) $\frac{3}{4}CV^2$

(c) $\frac{1}{2}CV^2$ (d) $2CV^2$.

76. In a parallel plate capacitor, the distance between the plates is d and potential difference across the plates is V. Energy stored per unit volume between the plates of capacitor is **NCERT | Page-75 | 2001**

(a) $\frac{Q^2}{2V^2}$ (b) $\frac{1}{2}\varepsilon_0\frac{V^2}{d^2}$

(c) $\frac{1}{2}\frac{V^2}{\varepsilon_0 d^2}$ (d) $\frac{1}{2}\varepsilon_0\frac{V^2}{d}$

77. Energy stored in a capacitor is **NCERT | Page-74 | 2001**

(a) $\frac{1}{2}QV$ (b) QV (c) $\frac{1}{QV}$ (d) $\frac{2}{QV}$

78. A capacitor is charged to store an energy U. The charging battery is disconnected. An identical capacitor is now connected to the first capacitor in parallel. The energy in each of the capacitor is **NCERT | Page-74 | 2000**

(a) $U/2$ (b) $3U/2$

(c) U (d) $U/4$

79. If the potential of a capacitor having capacity 6 μF is increased from 10 V to 20 V, then increase in its energy will be **NCERT | Page-74 | 1995**

(a) 4×10^{-4} J (b) 4×10^{-4} J

(c) 9×10^{-4} J (d) 12×10^{-6} J

80. A 4μF conductor is charged to 400 volts and then its plates are joined through a resistance of 1 kΩ. The heat produced in the resistance is **NCERT | Page-74 | 1989**

(a) 0.16 J (b) 1.28 J (c) 0.64 J (d) 0.32 J

ANSWER KEY

1	(d)	9	(d)	17	(d)	25	(c)	33	(c)	41	(b)	49	(b)	57	(d)	65	(a)	73	(c)
2	(c)	10	(b)	18	(d)	26	(b)	34	(c)	42	(b)	50	(c)	58	(c)	66	(d)	74	(d)
3	(a)	11	(a)	19	(a)	27	(a)	35	(d)	43	(a)	51	(c)	59	(b)	67	(b)	75	(b)
4	(a)	12	(c)	20	(a)	28	(b)	36	(c)	44	(a)	52	(d)	60	(b)	68	(d)	76	(b)
5	(d)	13	(b)	21	(a)	29	(c)	37	(b)	45	(a)	53	(d)	61	(d)	69	(d)	77	(a)
6	(a)	14	(d)	22	(d)	30	(c)	38	(c)	46	(c)	54	(a)	62	(a)	70	(a)	78	(d)
7	(a)	15	(a)	23	(c)	31	(c)	39	(a)	47	(c)	55	(b)	63	(d)	71	(d)	79	(c)
8	(b)	16	(a)	24	(a)	32	(a)	40	(d)	48	(a)	56	(a)	64	(b)	72	(c)	80	(d)

Hints & Solutions

1. **(d)** Electric potential, $V = \frac{kq}{r}$

$= 9\times10^9 \times \frac{4\times10^{-7}}{9\times10^{-2}}$

$= 4 \times 10^4$ V

2. **(c)** The potential V at any point, at distance r from centre of dipole $= \frac{1}{4\pi\varepsilon_0}.\frac{P\cos\theta}{r^2}$

At axial point where $\theta = 0°$, $V = \frac{1}{4\pi\varepsilon_0}.\frac{P}{r^2}$

$= \frac{9\times10^9 \times 4\times10^{-6}}{2^2} = 9\times10^3$ V

At axial point where

$\theta = 180°$, $V = -\frac{1}{4\pi\varepsilon_0}.\frac{P}{r^2} = -9\times10^3$ V

NOTES For point charges, $V \propto \frac{1}{r}$ and for electric dipole, $V \propto \frac{1}{r^2}$

3. **(a)** Electrostatic potential due to a point charge is given by $V = \frac{Kq}{r}$

$$V_{net} = \frac{Kq}{2\times10^{-2}} - \frac{Kq}{8\times10^{-2}}$$

$$\Rightarrow V_{net} = \frac{Kq\times10^2}{2}\left(1-\frac{1}{4}\right) = \left(\frac{3}{8}Kq\right)\times10^2\ V = \frac{3}{8}qK$$

4. **(a)** Given,
Dipole moment of short electric dipole,
$p = 16\times10^{-9}$ C m.
Distance from centre of dipole, $r = 0.6$ m

Electric potential, $V = \frac{kp\cos\theta}{r^2}$

$$\Rightarrow V = \frac{9\times10^9\times16\times10^{-9}\times\cos 60}{0.36} = 200\ V$$

5. **(d)** For uniformly charged spherical shell,

Potential $V = \frac{kq}{R}$ (For $r \le R$)

$\therefore\ V_C = V_P$
or, $V_C - V_P = 0$

6. **(a)** Potential, $V = \frac{KQ}{R}$

$\because\ Q$ = constant

$V \propto \frac{1}{R}$

So, $V_{small\ sphere} > V_{larger\ sphere}$

7. **(a)** Let charge and radius of smaller drop is q and r respectively and R = radius of bigger drop
For smaller drop, electric potential,

$$V = \frac{kq}{r} = 220\ V$$

As volume remains the same

$$\left(\frac{4}{3}\pi r^3\right)\times27 = \frac{4}{3}\pi R^3 \Rightarrow R = \sqrt[3]{27}r = 3r$$

And from conservation of charge,
$Q = 27q$
$\therefore$ Potential of bigger drop,

$$V_{big\ drop} = \frac{kQ}{R} = \frac{k(27q)}{3r} = 9\left(\frac{kq}{r}\right)$$

$= 9\times220 = 1980$ V

8. **(b)** Due to conducting sphere
At centre, electric field E = 0
but electric potential, $V = \frac{Q}{4\pi\epsilon_0 R}$

9. **(d)** $c = a + b$.

$$V_A = \frac{1}{4\pi E_o}\left[\frac{qA}{a}+\frac{qB}{b}+\frac{qC}{a}\right]$$

$$= \frac{1}{4\pi E_o}\left[\frac{4\pi a^2\sigma}{a} - \frac{4\pi b^2\sigma}{b} + \frac{4\pi c^2\sigma}{c}\right]$$

$$= \frac{4\pi}{4\pi E_o}\left[\frac{a^2\sigma}{a} - \frac{b^2\sigma}{b} + \frac{c^2\sigma}{c}\right]$$

$$\Rightarrow V_A = \frac{\sigma a}{\varepsilon_0} - \frac{\sigma b}{\varepsilon_0} + \frac{\sigma c}{\varepsilon_0} = \frac{\sigma}{\varepsilon_0}[c-(b-a)]$$

Similarly

$$V_B = \frac{1}{4\pi E_o}\left[\frac{qA}{a}+\frac{qB}{b}+\frac{qC}{c}\right]$$

$$\Rightarrow V_B = \frac{1}{4\pi E_o}\left[\frac{4\pi\sigma a^2}{b} - \frac{4\pi\sigma b^2}{b} + \frac{4\pi\sigma c^2}{c}\right]$$

$$V_B = \frac{-\sigma b}{\varepsilon_0} + \frac{1}{4\pi\varepsilon_0}\cdot\frac{\sigma\times4\pi a^2}{b} + \frac{\sigma c}{\varepsilon_0}$$

$$= \frac{\sigma}{\varepsilon_0}\left[c - \frac{(b^2-a^2)}{b}\right]$$

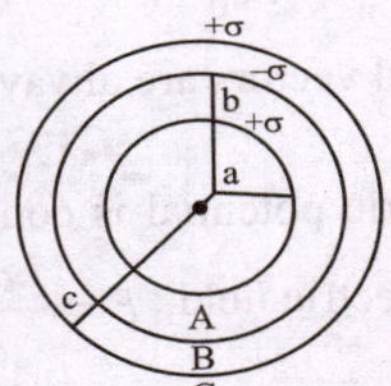

$$V_C = \frac{4\pi}{4\pi E_o}\left[\frac{a^2\sigma}{c} - \frac{b^2\sigma}{c} + \frac{\sigma c^2}{c}\right]$$

$$V_C = \frac{\sigma c}{\varepsilon_0} - \frac{1}{4\pi\varepsilon_0}\cdot\frac{\sigma\times4\pi b^2}{c} + \frac{1}{4\pi\varepsilon_0}\cdot\frac{\sigma\times4\pi a^2}{c}$$

$$= \frac{\sigma}{\varepsilon_0}\left[c - \frac{(b^2-a^2)}{c}\right] = \frac{\sigma}{\varepsilon_0}[c-(b-a)]$$

$V_A = V_C \ne V_B$

10. **(b)** Potential at the centre of the sphere = potential on the surface = 80 V.

11. **(a)** Let the side length of square be '*a*' then potential at centre *O* is

$$AC = BD = \sqrt{a^2+a^2} = a\sqrt{2}$$

$$OA = OB = OC = OD = \frac{a\sqrt{2}}{2} = \frac{a}{\sqrt{2}}$$

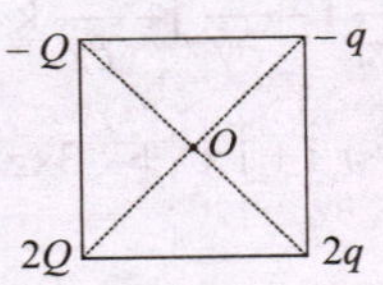

$$V = \frac{k(-Q)}{\left(\frac{a}{\sqrt{2}}\right)} + \frac{k(-q)}{\frac{a}{\sqrt{2}}} + \frac{k(2q)}{\frac{a}{\sqrt{2}}} + \frac{k(2Q)}{\frac{a}{\sqrt{2}}} = 0$$

(Given)

$$= -Q - q + 2q + 2Q = 0 \Rightarrow Q + q = 0$$

$$\Rightarrow Q = -q$$

12. **(c)** Distance of point A from the two +q charges = L.

Distance of point A from the two –q charges

$$= \sqrt{L^2 + (2L)^2} = \sqrt{5}L.$$

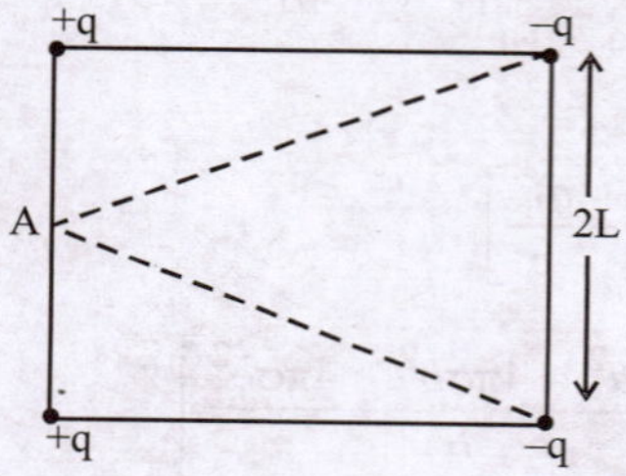

$$\therefore V_A = \left(\frac{Kq}{L} \times 2\right) - \left(\frac{Kq}{\sqrt{5}L} \times 2\right)$$

$$= \frac{2Kq}{L}\left[1 - \frac{1}{\sqrt{5}}\right] = \frac{1}{4\pi \epsilon_0} \cdot \frac{2q}{L}\left(1 - \frac{1}{\sqrt{5}}\right)$$

13. **(b)** Electric field vectors are always ⊥r to equipotential surface.

14. **(d)** Since, electric potential is constant throughout the volume, hence electric field, $E = -\frac{dV}{dr} = 0.$

15. **(a)** As the regions are of equipotentials, so Work done $W = q\Delta V$

ΔV is same in all the cases hence work - done will also be same in all the cases.

16. **(a)** Potential at B, V_B is maximum

$V_B > V_C > V_A$

As ikn the direction of electric field potential decreases.

17. **(d)** $\vec{E} = -\left[\frac{dV}{dx}\hat{i} + \frac{dV}{dy}\hat{j} + \frac{dV}{dz}\hat{k}\right]$

Given, $V = 4x^2$

$$\therefore \quad E = -\hat{i}\frac{d(4x^2)}{dx} = -8x\,\hat{i} \text{ volt/meter}$$

$$\therefore \vec{E}_{(1,0,2)} = -8\hat{i}\,V/m$$

18. **(d)** The electric field at a point is equal to negative of potential gradient at that point.

$$\vec{E} = -\frac{\partial V}{\partial r} = \left[-\frac{\partial V}{\partial x}\hat{i} - \frac{\partial V}{\partial y}\hat{j} - \frac{\partial V}{\partial z}\hat{k}\right]$$

$$= [(2xy + z^3)\,\hat{i} + \hat{j}\,x^2 + \hat{k}\,3xz^2]$$

19. **(a)** Given that, $V = Q \times 10^{11}$ volts

Electric potential at point is given by

$$V = \frac{1}{4\pi \epsilon_0} \cdot \frac{Q}{r} \text{ or, } Q \times 10^{11} = \frac{1}{4\pi \epsilon_0} \cdot \frac{Q}{r}$$

$$\Rightarrow r = \frac{1}{4\pi \epsilon_0} \cdot 10^{-11} \text{m.}$$

$$\text{As } |E| = \frac{|V|}{r} = \frac{Q \times 10^{11}}{\frac{1}{4\pi \epsilon_0} \cdot 10^{-11}}$$

$$= 4\pi \epsilon_0\, Q \times 10^{22} \text{ volt m}^{-1}.$$

20. **(a)** Given : Electric field (E) = 500 V/m and potential difference (V) = 3000 V.

We know that electric field

$$(E) = 500 = \frac{V}{d} \text{ or } d = \frac{3000}{500} = 6\,\text{m}$$

[where d = Distance between point charge and A]

21. **(a)** Potential in a region is given by,

$V = 6xy - y + 2yz$

As we know the relation between electric potential and electric field is

$$\vec{E} = -\left(\frac{\partial V}{\partial x}\hat{i} + \frac{\partial V}{\partial y}\hat{j} + \frac{\partial V}{\partial z}\hat{k}\right)$$

where $V = 6xy - y + 2yz$

$$\vec{E} = -\left[(6y\hat{i} + (6x - 1 + 2z)\hat{j} + (2y)\hat{k}\right]$$

$$\vec{E}_{(1,1,0)} = -(6\hat{i} + 5\hat{j} + 2\hat{k})$$

NOTES

With the help of formula $E = \frac{-dv}{dr}$, potential difference between any two point in an electric field can be determined by knowing the boundary conditions

$$dV = -\int_{r_1}^{r_2} \vec{E}\vec{dr} = -\int_{r_1}^{r_2} E.dr\cos\theta$$

22. **(d)** $\vec{E} = -\frac{\partial V}{\partial x}\hat{i} - \frac{\partial V}{\partial y}\hat{j} - \frac{\partial V}{\partial z}\hat{k}$

where $V(x, y, z) = 6x - 8xy - 8y + 6yz$

$$= -[(6 - 8y)\hat{i} + (-8x - 8 + 6z)\hat{j} + (6y)\hat{k}]$$

At (1, 1, 1), $\vec{E} = 2\hat{i} + 10\hat{j} - 6\hat{k}$

$$\Rightarrow (\vec{E}) = \sqrt{(2)^2 + (10)^2 + (-6)^2} = \sqrt{140} = 2\sqrt{35}$$

$$\therefore \quad F = q\vec{E} = 2 \times 2\sqrt{35} = 4\sqrt{35}$$

23. **(c)** Energy = 2eV

$$\Rightarrow eV_o = 2eV \Rightarrow V_o = 2 \text{ volt}$$

$$E = \frac{V_o}{d} = \frac{2}{4 \times 10^{-8}}$$

$$= 0.5 \times 10^8 = 5 \times 10^7\, Vm^{-1}$$

24. **(a)** For $r > R$,

Electric Field $E = \frac{Kq}{r^2}$ ...(i)

and potential at surface of sphere $V = \frac{Kq}{R}$

$\therefore$ From equation (i)

$V = \frac{Er^2}{R} \Rightarrow E = \frac{RV}{r^2}$

25. **(c)** Charge density $\sigma = \frac{\text{charge}}{\text{area}}$

When two conductors are connected by a conducting wire, then they have same potential

i.e., $V_1 = V_2$

$$\therefore \frac{1}{4\pi\varepsilon_0}\frac{Q_1}{R_1} = \frac{1}{4\pi\varepsilon_0}\frac{Q_2}{R_2}$$

$$\Rightarrow \frac{1}{4\pi\varepsilon_0}\frac{Q_1}{R_1} \times \frac{R_1}{R_1} = \frac{1}{4\pi\varepsilon_0}\frac{Q_2}{R_2} \times \frac{R_2}{R_2}$$

$$\Rightarrow \frac{Q_1R_1}{4\pi R_1^2\varepsilon_0} = \frac{Q_2R_2}{4\pi R_2^2\varepsilon_0} \quad \therefore \frac{Q_1}{4\pi R_1^2} \Big/ \frac{Q_2}{4\pi R_2^2} = \frac{R_2}{R_1}$$

or, $\sigma_1/\sigma_2 = R_2/R_1$

26. **(b)** At equilibrium potential of both sphere becomes same if charge of sphere one x and other sphere $(Q - x)$ then

where $Q = 4 \times 10^{-2}$ C

$V_1 = V_2$

$$\frac{kx}{1\,\text{cm}} = \frac{k(Q-x)}{3\,\text{cm}}$$

$3x = Q - x \Rightarrow 4x = Q$

$$x = \frac{Q}{4} = \frac{4\times10^{-2}}{4}C = 1\times10^{-2}$$

$Q' = Q - x = 3 \times 10^{-2}$C

NOTES Charge given to a conductor always resides on its outer surface. If surface is uniform the charge distributes uniformly on the surface and for irregular surface the distribution of charge i.e., charge density is not uniform. It is maximum where the radius of curvature is maximum and vice-versa

i.e., $\sigma \propto \frac{1}{R}$

27. **(a)** Electric potential is constant (equal to $\frac{kq}{R}$, where $k = \frac{1}{4\pi\epsilon_0}$) within or on the surface of conductor.

28. **(b)** Potential at the centre of hexagon is zero i.e., $V_0 = 0$

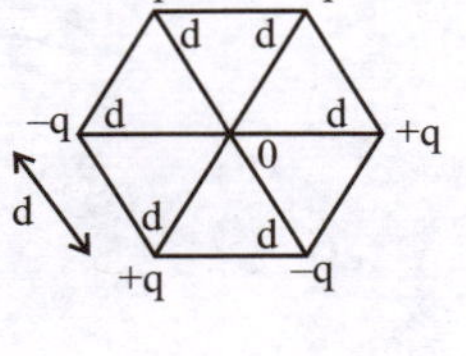

$\therefore W_{\infty\to 0} = q\Delta V = q_0(0) = 0$

29. **(c)** Potential energy of electric dipole in external electric field

$U = -\vec{P}.\vec{E} = -PE\cos\theta = -PE\cos 180°$

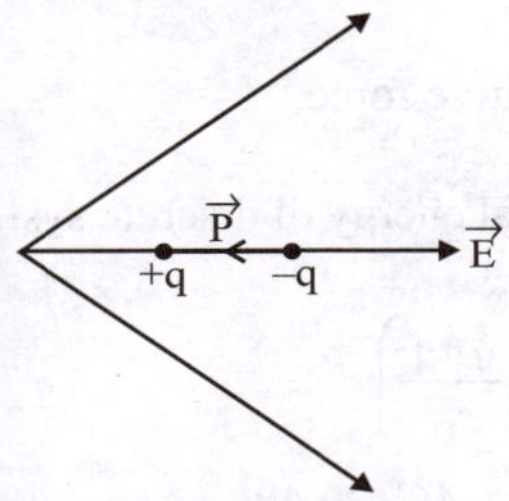

since angle between $\vec{E}$ and $\vec{P}$ is 180°

$\therefore U = +PE$

On moving towards right electric field strength decrease

$\therefore U$ decreases.

Net force on electric dipole is towards right and net torque acting on it is zero so move towards right.

NOTES Potential energy of a dipole $u = -PE\cos\theta$, $\theta = 0° \Rightarrow$ stable equilibrium and $u = -PE$ i.e., minimum value.

For $\theta = 90°$, $u = 0$ and dipole is not in equilibrium. For $\theta = 180°$ $u = PE$ i.e., maximum value and dipole is in unstable equilibrium.

30. **(c)** AC = BC

$V_D = V_E$

We have,

$W = Q(V_D - V_E)$

$\Rightarrow W = 0$

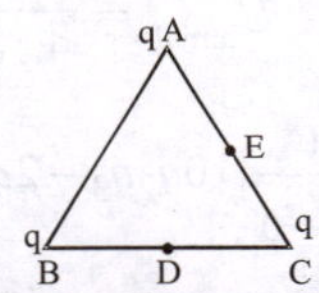

31. **(c)**

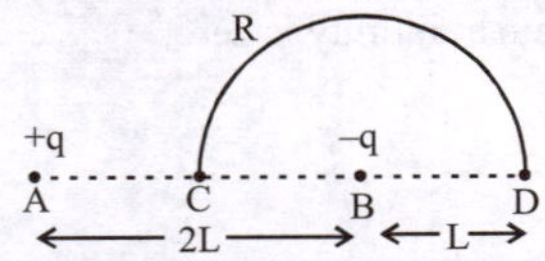

Potential at $C = V_C = 0$

Potential at $D = V_D$

$$= k\left(\frac{-q}{L}\right) + \frac{kq}{3L} = -\frac{2}{3}\frac{kq}{L}$$

Potential difference

$$V_D - V_C = \frac{-2}{3}\frac{kq}{L} = \frac{1}{4\pi\epsilon_0}\left(-\frac{2}{3}\cdot\frac{q}{L}\right)$$

$\Rightarrow$ Work done $= Q(V_D - V_C)$

$$= -\frac{2}{3} \times \frac{1}{4\pi\varepsilon_0}\frac{qQ}{L} = \frac{-qQ}{6\pi\varepsilon_0 L}$$

32. **(a)** We know that potential energy of two charge system is given by

$$U = \frac{1}{4\pi\epsilon_0}\frac{q_1q_2}{r}$$

According to question,

$$U_A = \frac{1}{4\pi \epsilon_0}\frac{(+q)(-Q)}{a} = -\frac{1}{4\pi\varepsilon_0}\frac{Qq}{a}$$

and $$U_B = \frac{1}{4\pi \epsilon_0}\frac{(+q)(-Q)}{a} = -\frac{1}{4\pi\varepsilon_0}\frac{Qq}{a}$$

$\Delta U = U_B - U_A = 0$

We know that for conservative force,

$W = -\Delta U = 0$

33. **(c)** We know that potential energy of discrete system of charges is given by

$$U = \frac{1}{4\pi \epsilon_0}\left(\frac{q_1q_2}{r_{12}} + \frac{q_2q_3}{r_{23}} + \frac{q_3q_1}{r_{31}}\right)$$

Here, $r_{12} = AB = 0.3m$, $r_{31} = AC = 0.4m$

According to question,

$$U_{initial} = \frac{1}{4\pi \epsilon_0}\left(\frac{q_1q_2}{0.3} + \frac{q_2q_3}{0.5} + \frac{q_3q_1}{0.4}\right)$$

$$r_{23} = \sqrt{(0.3)^2 + (0.4)^2} = 0.5m$$

When the charge moved to point D

$r_{12} = 0.3$ m

$r_{23} = 0.1$ m

$r_{31} = 0.4$ m

$$U_{final} = \frac{1}{4\pi \epsilon_0}\left(\frac{q_1q_2}{0.3} + \frac{q_2q_3}{0.1} + \frac{q_3q_1}{0.4}\right)$$

$$U_{final} - U_{initial} = \frac{1}{4\pi \epsilon_0}\left(\frac{q_1q_2}{0.1} - \frac{q_2q_3}{0.5}\right)$$

$$= \frac{1}{4\pi \epsilon_0}[10q_2q_3 - 2q_2q_3] = \frac{q_3}{4\pi \epsilon_0}(8q_2)$$

NOTES When we use the formula of potential energy of system of charges, then we consider that potential energy of the system at infinity is zero.

34. **(c)** $qV = \frac{1}{2}mv^2$

$$\Rightarrow 2\times10^{-6}\times V = \frac{1}{2}\times\frac{2}{1000}\times10\times10$$

$\therefore$ V = 50 kV

35. **(d)** Length of body diagonal = $\sqrt{3}\ell$

$\therefore$ Distance of centre of cube from each corner

$$r = \frac{\sqrt{3}}{2}\ell$$

P.E. at centre

= 8 × Potential Energy of the charge (+q) due to charge (–q) at one corner

$$= 8\times\frac{Kq\times(-q)}{r}$$

$$= 8\times\frac{1}{4\pi\varepsilon_0\sqrt{3}\,l}\times 2\times q\times(-q) = \frac{-4q^2}{\sqrt{3}\pi\varepsilon_0 l}$$

36. **(c)** K.E. = force × distance = $qE.y$

37. **(b)** Given, (Dipole moment) $p = 5 \times 10^{-6}$ Cm

(Electric Field) E = 4×10^5 N/C

$\theta_d = 0°,\ \theta_f = 60°$

$\Delta U = U_f - U_1$ (Change in p.E.)

$= -p\text{E}\cos\theta_1 - (-p\text{E}\cos\theta_q)$

$\Rightarrow p\text{E}(\cos\theta_1 - \cos\theta_f)$

$$\Rightarrow 5\times10^{-6}\times 4\times10^5\left(1-\frac{1}{2}\right) = 1\text{J}$$

38. **(c)** When electric dipole is aligned parallel $\theta = 0°$ and the dipole is rotated by 90° *i.e.*, $\theta = 90°$.

Energy required to rotate the dipole

$W = U_f - U_i = (-pE\cos 90°) - (-pE\cos 0°) = pE$.

39. **(a)** The torque on the dipole is given as

$\tau = pE\sin\theta$

The potential energy of the dipole in the electric field is given as $U = -\ pE\cos\theta$

40. **(d)** When the dipole is in the direction of field then net force is $qE + (-qE) = 0$

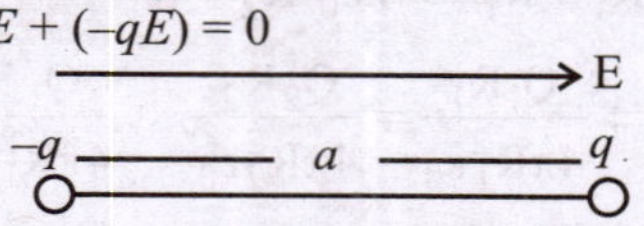

and its potential energy is minimum

= P.E. = $-qaE$

41. **(b)** Capacitance of capacitor is $C = \frac{A\varepsilon_0}{d}$

So capacitance of capacitor is independent of charge (q) and potential (V), it depends on the geometry of the capacitor and medium between plates of the capacitor.

42. **(b)** When the plates of a parallel plate capacitor is connected to a battery, then potential difference remains constant.

i.e., V′ = V = Constant

(i) $C' = \frac{\varepsilon_0 A}{d'}, C = \frac{\varepsilon_0 A}{d}$

$d' < d \Rightarrow C' > C$

Hence, final capacitance is greater than initial capacitance,

(ii) $U' = \frac{1}{2}C'V^2$ and $U = \frac{1}{2}CV^2$

$\Rightarrow U' > U \qquad [\because C' > C]$

Hence, final energy is greater than initial energy

(iii) $\frac{Q'}{V'} = C'$ and $\frac{Q}{V} = C \Rightarrow \frac{Q'}{V'} \neq \frac{Q}{V}$

(iv) Product of charge and voltage

$X' = Q'V = C'V^2$ and $X = QV = CV^2$

$\Rightarrow X' > X \qquad [\because C' > C]$

43. **(a)** Given : $A' = \frac{A}{2}$ and $d' = 2d$

Capacitance $C = \frac{A\varepsilon_0}{d}$

$$\therefore C' = \frac{\frac{A}{2}\varepsilon_0}{2d} = \frac{C}{4}$$

44. **(a)** Electrostatic force between the metal plates

$$F_{plate} = \frac{Q^2}{2A\varepsilon_0}$$

For isolated capacitor Q = constant
Clearly, F is independent of the distance between plates.

45. **(a)** Force of attraction between the plates,

$F = qE$

$= q \times \frac{\sigma}{2\epsilon_0}$

$= q\frac{q}{2A\epsilon_0} \qquad \left[\because E = \frac{\sigma}{2\epsilon_o}\right]$

$= \frac{q^2}{2\left(\frac{\epsilon_0 A}{d}\right)\times d} = \frac{C^2V^2}{2Cd} = \frac{CV^2}{2d}$

Here, $C = \frac{\epsilon_0 A}{d}$, $q = CV$, A = area

NOTES To find the force of attraction between the plates, we use the concept that work done in displacing the plates against the force is equal to the increase in energy of the capacitor.

46. **(c)** If we increase the distance between the plates its capacity decreases resulting in higher potential as we know $Q = CV$. Since Q is constant (battery has been disconnected), on decreasing C, V will increase.

47. **(c)** $C = 10\ \mu F$ $\qquad d = 8$ cm
$C' = ?$ $\qquad d' = 4$ cm

$C = \frac{A\epsilon_0}{d} \Rightarrow C \propto \frac{1}{d}$

If d is halved then C will be doubled.
Hence, $C' = 2C = 2 \times 10\ \mu F = 20\ \mu F$

48. **(a)** The capacitance of capacitor without dielectric is

$C_1 = \frac{\varepsilon_0 A}{d}$

The capacitance of capacitor with dielectric is

$$C_2 = \frac{\varepsilon_0 A}{\frac{3d}{8K_1} + \frac{d}{2K_2} + \left(d - \frac{3d}{8} - \frac{d}{2}\right)}$$

$$= \frac{\frac{\varepsilon_0 A}{d}}{\frac{3}{8K_1} + \frac{1}{2K_2} + \frac{1}{8}} = \frac{\frac{\varepsilon_0 A}{d}}{\frac{3}{8K_1} + \frac{1}{2\left(\frac{4}{5}K_1\right)} + \frac{1}{8}}$$

Given, $C_2 = 2C_1$

$$\Rightarrow \frac{\frac{\varepsilon_0 A}{d}}{\frac{3}{8K_1} + \frac{5}{8K_1} + \frac{1}{8}} = \frac{2\varepsilon_0 A}{d}$$

$$\Rightarrow \frac{1}{\frac{1}{K_1} + \frac{1}{8}} = 2 \Rightarrow \frac{1}{K_1} + \frac{1}{8} = \frac{1}{2} \Rightarrow K_1 = \frac{8}{3} = 2.66$$

49. **(b)** Capacitance of a parallel plate capacitor with air is

$C = \frac{\varepsilon_0 A}{d}$ ... (i)

Here, A = area of plates of capacitor,
d = distance between the plates

Capacitance of a same parallel plate capacitor with introduction of dielectric medium of dielectric constant K is

$C' = \frac{K\varepsilon_0 A}{d}$ ... (ii)

Dividing (ii) by (i)

$\Rightarrow \frac{C'}{C} = K \Rightarrow \frac{30}{6} = K \Rightarrow K = 5 \Rightarrow K = \frac{\varepsilon}{\varepsilon_0}$

$\Rightarrow \varepsilon = K\varepsilon_0 = 5 \times 8.85 \times 10^{-12}$

$= 0.44 \times 10^{-10}\ C^2N^{-1}m^{-2}$

50. **(c)** Electric field, $E \propto \frac{1}{K}$

As $K_1 < K_2$ so $E_1 > E_2$
Hence graph (c) correctly dipicts the variation of electric field E with distance d.

51. **(c)** Electric field between two parallel metal plates in vacuum

$E_0 = \frac{\sigma}{\varepsilon_0}$

In medium of dielectric constant K
Here medium is kerosene

$E = \frac{\sigma}{\varepsilon_0 K}$; $K > 1$

$\therefore$ $E < E_0$ i.e., electric field decreases.

52. **(d)** Electric field between two parallel plates placed in vacuum

$E = \frac{\sigma}{\varepsilon_o} = \frac{Q}{A\varepsilon_o}$

In a medium of dielectric constant K

$E' = \frac{\sigma^-}{E_o K}$

For kerosene $K > 1$, $\therefore E' < E$

53. **(d)** When oil is placed between space of plates

$C = \frac{2A\varepsilon_0}{d}$ ...(1) $\left[\because C = \frac{KA\varepsilon_0}{d}, \text{ if } K = 2\right]$

When oil is removed $C' = \frac{A\varepsilon_0}{d}$ ...(2)

On comparing both equations, we get $C' = C/2$

NOTES When dielectric is partially filled between the plates of a parallel plate capacitor then its capacitance increases but potential difference decreases. To maintain the capacitance and potential difference of capacitor as before separatoin betweeen the plates has to be increased by d. In such case $k = \frac{t}{t-d}$ (Here, t = seperation between plates.)

54. (a) Given circuit is balanced Wheatstone bridge so equivalent circuit is as shown

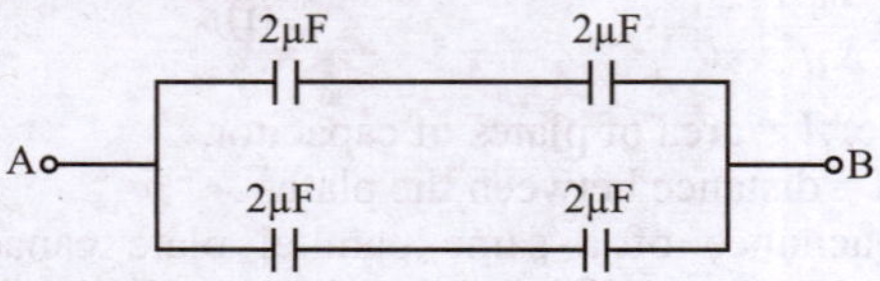

$\therefore$ Equivalent capacitance

$$C_{AB} = \left(\frac{1}{2}+\frac{1}{2}\right)+\left(\frac{1}{2}+\frac{1}{2}\right) = 2\mu F$$

55. (b) The equivalent circuit of the given circuit is as follows

$\therefore C_{net} = 15 + 5 = 20\ \mu F$

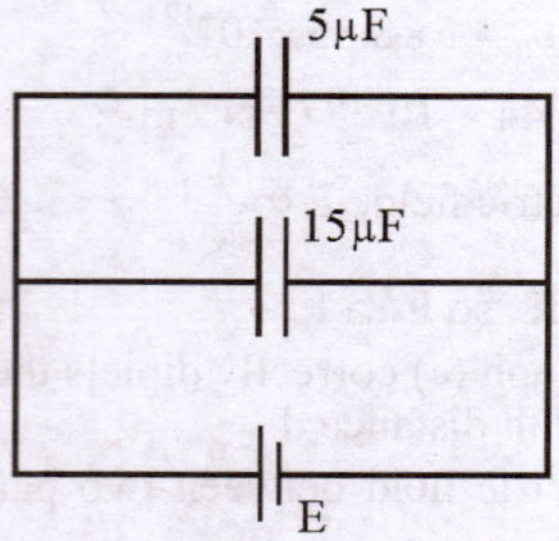

56. (a) For parallel grouping, $C_{eq} = C_1 + C_2 + ...\ C_n$

$C_1 = 3 + 3 = 6\ \mu F$

For series grouping, $\frac{1}{C_{eq}} = \frac{1}{C_1} + \frac{1}{C_2} + ...\ \frac{1}{C_n}$

$$C_{eq} = \frac{C_1C_2}{C_1 + C_2} = \frac{3\times 6}{3+6} = \frac{18}{9} = 2\ \mu F$$

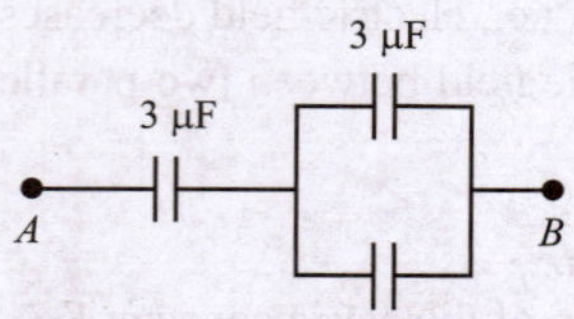

57. (d) In series effective resistance of two capacitors of capacitance C_1 and C_2

$= \frac{C_1C_2}{C_1+C_2} = 3\mu F$ ...(i)

In parallel effective resistance, $= C_1 + C_2$

$= 16\ \mu F$ ...(ii)

From (i) & (ii)

$C_1C_2 = 48\ \mu F$

$(C_1 - C_2)^2 = (C_1 + C_2)^2 - 4C_1\ C_2$

$C_1 - C_2 = 8\ \mu F$ ... (iii)

From (ii) & (iii) we get

$C_1 = 12\ \mu F$ and $C_2 = 4\ \mu F$

58. (c) The circuit is as shown is figure,

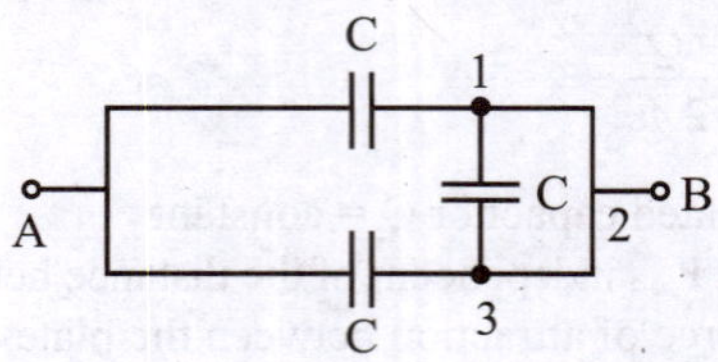

Points 1, 2, 3 are connected by conducting wire so at these points same potential.

So the capacitor is short circuited. So it does not store any charge.

Therefore equivalent capacitance $C_{AB} = C + C = 2C$

59. (b) In series combination of capacitors

$V_{eff} = V + V + V = 3V$

$$\frac{1}{C_{eff}} = \frac{1}{C} + \frac{1}{C} + \frac{1}{C} \Rightarrow C_{eff} = \frac{C}{3}$$

Thus, the capacitance and breakdown voltage of the combination will be $\frac{C}{3}$ and 3V.

60. (b)

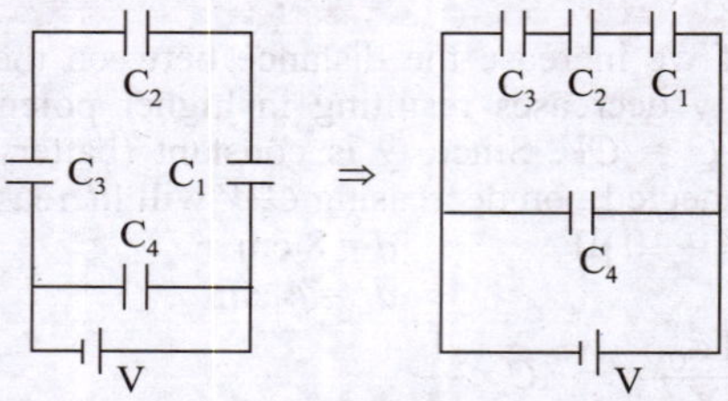

Equivalent capacitance for three capacitors (C_1, C_2 & C_3) in series is given by

$$\frac{1}{C_{eq.}} = \frac{1}{C_1} + \frac{1}{C_2} + \frac{1}{C_3} = \frac{C_2C_3 + C_3C_1 + C_1C_2}{C_1C_2C_3}$$

$$\Rightarrow C_{eq.} = \frac{C_1C_2C_3}{C_1C_2 + C_2C_3 + C_3C_1}$$

$$\Rightarrow C_{eq.} = \frac{C(2C)(3C)}{C(2C)+(2C)(3C)+(3C)C} = \frac{6}{11}C$$

$\Rightarrow$ Charge on capacitors (C_1, C_2 & C_3) in series

$= C_{eq}V = \frac{6C}{11}V$

Charge on capacitor $C_4 = C_4V = 4C\ V$

$$\frac{\text{Charge on } C_2}{\text{Charge on } C_4} = \frac{\frac{6C}{11}V}{4CV} = \frac{6}{11}\times\frac{1}{4} = \frac{3}{22}$$

61. (d) For series, $C' = \frac{C_1 \times C_2}{C_1 + C_2} = \frac{4\times 4}{4+4} = 2\mu F$

For parallel, $C_{eq.} = C' + C_3 = 2 + 4 = 6\mu F$

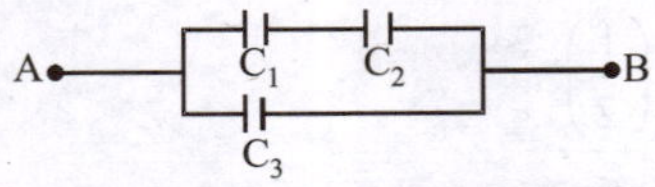

> **NOTES** If CP is the effective capacity when n identical capacitors are connected in parallel and Cs is their effective capacity when connected in series then $\frac{C_p}{C_s} = n^2$

62. **(a)** Charge $Q = C_1V$

Total capacity of combination (parallel)

$C = C_1 + C_2$ P.D. $= \frac{Q}{C} = \frac{C_1V}{C_1 + C_2}$

63. **(d)** From given circuit,

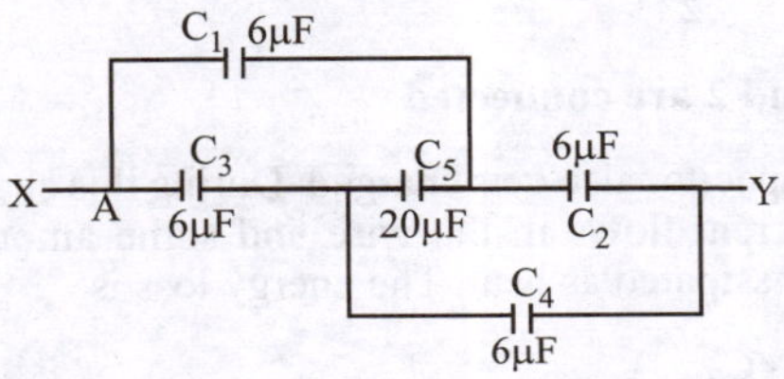

Equivalent circuit

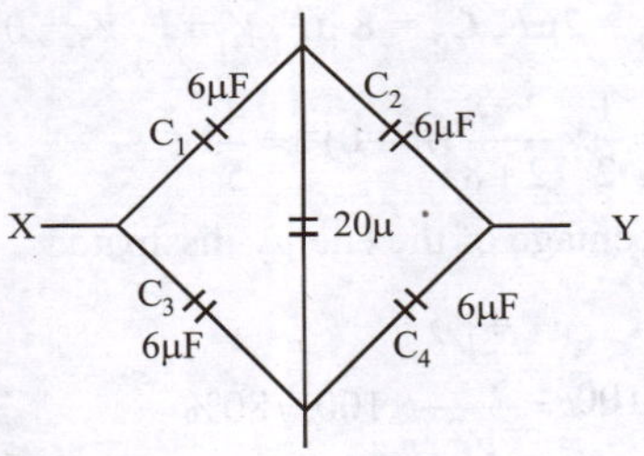

Here, $\frac{C_1}{C_3} = \frac{C_2}{C_4}$

Hence, no charge will flow through 20μF

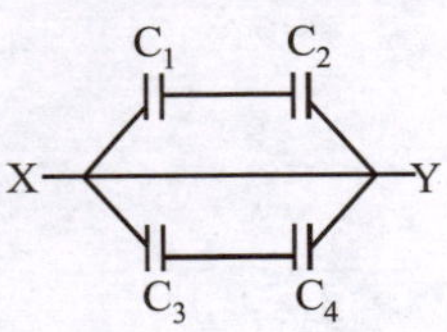

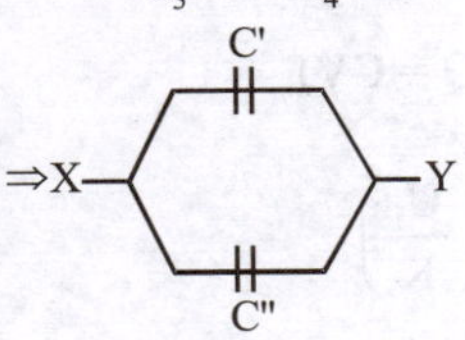

C_1 and C_2 are in series, also C_3 and C_4 are in series.

Hence, $C' = 3\ \mu F$, $C'' = 3\ \mu F$

C' and C'' are in parallel.

Hence net capacitance $= C' + C'' = 3 + 3 = 6\ \mu F$

64. **(b)** Charge on each plate of each capacitor

$$Q = \pm CV = \pm 25 \times 10^{-6} \times 200$$

$$= \pm 5 \times 10^{-3} C$$

65. **(a)** Electrostatic energy, $U = \frac{1}{2}CV^2$

$$= \frac{1}{2} \times 12 \times 10^{-12} \times (50)^2$$

$$= 15 \times 10^{-9}\ J$$

$$= 15\ nJ$$

66. **(d)** Displacement current $i_d = c\frac{dV}{dt}$

$$\Rightarrow 2 \times 10^{-3} = 4 \times 10^{-6} \times \frac{dV}{dt}$$

$$\frac{dV}{dt} = \frac{2 \times 10^{-3}}{4 \times 10^{-6}}$$

$$\therefore \frac{dV}{dt} = 500\ V/s$$

67. **(b)** By conservation of charge

$q_i = q_f$

$\Rightarrow C \times 100 = CV' + CV'$

$\Rightarrow V' = 50\ V$

So, $U_f = \frac{1}{2}CV'^2 \times 2$

$= CV'^2$

$= 900 \times 10^{-12} \times 25 \times 10^2$

$= 225 \times 10^{-8}$

$= 2.25 \times 10^{-6}\ J$

68. **(d)** Energy density i.e. $\frac{\text{energy}}{\text{volume}}$ with electric field

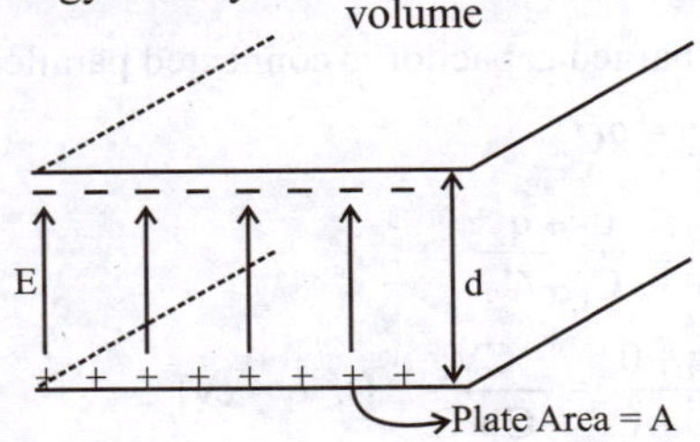

$$u = \frac{dU}{dV} = \frac{1}{2}\varepsilon_0 E^2$$

$$\Rightarrow dU = \frac{1}{2}\varepsilon_0 E^2 dV$$

$\therefore$ Energy stored in the capacitor

$U = \int dU = \int \frac{1}{2}\varepsilon_0 E^2 dV$

$= \frac{1}{2}\varepsilon_0 E^2 \int dV = \frac{1}{2}\varepsilon_0 E^2 V$

$\therefore U = \frac{1}{2}\varepsilon_0 E^2 Ad \qquad [\because V = Ad]$

69. (d)

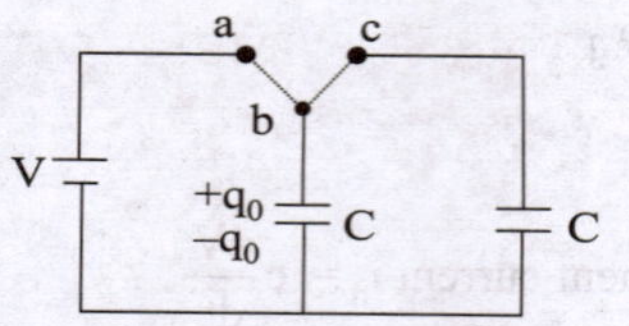

$U_i = \frac{1}{2}CV^2$

On switching key at point *c*

$\frac{q_0 - q}{C} = \frac{q}{C}$

$2q = q_0$

$q = \left(\frac{q_0}{2}\right)$

$U_f = \frac{1}{2}\left(\frac{q_0}{2}\right)^2 \times \frac{1}{C} + \frac{1}{2}\left(\frac{q_0}{2}\right)^2 \times \frac{1}{C}$

$U_f = \frac{1}{4}CV^2$

loss = 50%

70. (a) When battery is replaced by another uncharged capacitor

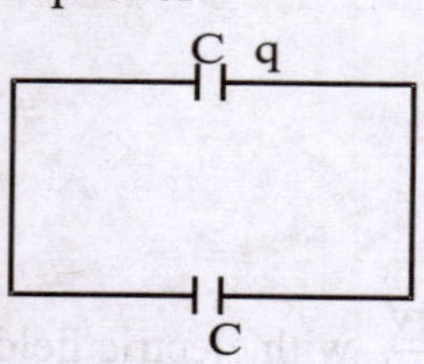

As uncharged capacitor is connected parallel

So, $C' = 2C$

and $V_c = \frac{q_1 + q_2}{C_1 + C_2}$

$V_c = \frac{q + 0}{C + C} = \frac{CV}{C + C} \qquad [\because q = cv]$

$\Rightarrow V_c = \frac{V}{2}$

Initial Energy of system, $U_i = \frac{1}{2}CV^2$ ... (i)

Final energy of system, $U_f = \frac{1}{2}(2C)\left(\frac{V}{2}\right)^2$

$= \frac{1}{2}CV^2\left(\frac{1}{2}\right)$ ...(ii)

From equation (i) and (ii)

$U_f = \frac{1}{2}U_i$

i.e., Total electrostatic energy of resulting system decreases by a factor of 2

71. (d) When S and 1 are connected

The 2μF capacitor gets charged. The potential difference across its plates will be V.

The potential energy stored in 2 μF capacitor

$U_i = \frac{1}{2}CV^2 = \frac{1}{2} \times 2 \times V^2 = V^2$

When S and 2 are connected

The 8μF capacitor also gets charged. During this charging process current flows in the wire and some amount of energy is dissipated as heat. The energy loss is

$\Delta U = \frac{1}{2}\frac{C_1 C_2}{C_1 + C_2}(V_1 - V_2)^2$

Here, $C_1 = 2\mu F$, $C_2 = 8\ \mu F$, $V_1 = V$, $V_2 = 0$

$\therefore \Delta U = \frac{1}{2} \times \frac{2 \times 8}{2 + 8}(V - 0)^2 = \frac{4}{5}V^2$

The percentage of the energy dissipated

$= \frac{\Delta U}{U_i} \times 100 = \frac{\frac{4}{5}V^2}{V^2} \times 100 = 80\%$

72. (c) Capacitance of the capacitor, $C = \frac{Q}{V}$

After inserting the dielectric, new capacitance

$C' = K.C$

New potential difference

$V' = \frac{V}{K}$

$U_i = \frac{1}{2}CV^2 = \frac{Q^2}{2C} \qquad (\because Q = CV)$

$U_f = \frac{Q^2}{2C} = \frac{Q^2}{2KC} = \frac{C^2V^2}{2KC} = \left(\frac{U_i}{K}\right)$

$\Delta U = U_f - U_i = \frac{1}{2}CV^2\left\{\frac{1}{K} - 1\right\}$

As the capacitor is isolated, so charge will remain conserved and p.d. between two plates of the capacitor

$V' = \frac{Q}{KC} = \frac{V}{K}$

NOTES When the battery remains connected, the capacity of capacitor and charge stored is changed. When the battery disconnected, charge remains same but potential and capacitance are changed.

73. (c) The energy stored by a capacitor

$$U=\frac{1}{2}CV^2 \qquad ...(i)$$

V is the p.d. between two plates of the capacitor
potential difference $V = E.d.$
The capacitance of the parallel plate capacitor

$$C=\frac{A\varepsilon_0}{d}$$

Substituting the value of C in equation (i)

$$U=\frac{1}{2}\frac{A\varepsilon_0}{d}(Ed)^2=\frac{1}{2}A\varepsilon_0E^2d$$

74. (d) In series, $C_{eff}=\frac{C_1}{n_1}$

∴ Energy stored,

$$E_S=\frac{1}{2}C_{eff}V_S^2=\frac{1}{2}\frac{C_1}{n_1}16V^2=8V^2\frac{C_1}{n_1}$$

In parallel, $C_{eff}=n_2C_2$

∴ Energy stored, $E_p=\frac{1}{2}n_2C_2V^2$

According to question $E_s = E_p$

$$\therefore \frac{8V^2C_1}{n_1}=\frac{1}{2}n_2C_2V^2$$

$$\Rightarrow C_2=\frac{16C_1}{n_1n_2}$$

75. (b) Work done = Change in energy

$$=\frac{1}{2}\left(C+\frac{C}{2}\right)V^2=\frac{1}{2}\left(\frac{3C}{2}\right)V^2=\frac{3CV^2}{4}$$

76. (b) Energy stored per unit volume

$$=\frac{1}{2}\varepsilon_0E^2=\frac{1}{2}\varepsilon_0\left(\frac{V}{d}\right)^2=\frac{1}{2}\varepsilon_0\frac{V^2}{d^2}\quad\left(\because E=\frac{V}{d}\right)$$

77. (a) Energy stored in capacitor

$$=\frac{1}{2}CV^2=\frac{1}{2}\frac{Q}{V}.V^2 \quad (Q=CV)=\frac{1}{2}QV$$

78. (d) In first case, when capacitor C connected with battery.

Energy stored, $U_1=\frac{1}{2}CV^2=\frac{1}{2}\frac{Q^2}{C}$

When battery is disconnected and another capacitor of same capacity is connected in parallel to first capacitor, then

$C_{eq}=C'=2C$

Voltage across each capacitor, $V'=\frac{Q}{2C}$

∴ Energy stored, $U_2=\frac{1}{2}CV'^2$

$$=\frac{1}{2}C\left(\frac{Q}{2C}\right)^2=\frac{1}{4}\cdot\frac{1}{2}\frac{Q^2}{C}=\frac{U_1}{4}$$

79. (c) Capacitance of capacitor (C) = 6 μF = 6 × 10^{-6} F; Initial potential (V_1) = 10 V and final potential (V_2) = 20 V.

The increase in energy (ΔU)

$$=\frac{1}{2}C(V_2^2-V_1^2)$$

$$=\frac{1}{2}\times(6\times10^{-6})\times[(20)^2-(10)^2]$$

$$=(3\times10^{-6})\times300=9\times10^{-4}\text{ J}$$

80. (d) The energy stored in the capacitor $=\frac{1}{2}CV^2$

$$=\frac{1}{2}\times4\times10^{-6}\times(400)^2=0.32\text{ J}$$

This energy will be converted into heat in the resistor.

3 Current Electricity

TREND ANALYSIS

TOPIC	2025		2024		2023		2022		2021		2020	
	Qns.	LOD	Qns.	LOD	Qns.	LOD	Qns.	LOD	Qns.	LOD	Qns.	LOD
OHM's Law					1	Easy	1	Easy				
Drift of Electrons and the Origin of Resistivity									1	Avg.	1	Avg.
Temperature Dependence of Resistivity					1	Avg.	1	Easy			2	Easy
Electrical Energy, Power			1	Avg.			1	Avg.				
Combination of Resistors – Series and Parallel	2	Avg.	1	Avg.	1	Avg.	1	Avg.	2	Avg.		
Cells, EMF, Internal Resistance			1	Easy								
Cells in Series and in Parallel			1	Avg.								
Kirchoff's Rule	1	Avg.										
Meter Bridge							1	Avg.			1	Avg.

How to Code Qns for Better & Faster Revision

Type 1: Able to solve (No code Req)

Type 2: Solved with struggle or by seeing solution **(Use code NR means Needs Revision or orange stickers)**

Type 3: Tricky/ Could not solve on my own. Need help from Teacher/ Friend. **(Use code TQ means Tricky Question or red stickers)**

You can either hand write the code in boxes or else paste coloured stickers given at end of book.

Topic-3.4

OHM's Law

1. On the basis of electrical conductivity, which one of the following material has the smallest resistivity?
NCERT | Page-88 | Manipur 2023

(a) Glass (b) Silicon
(c) Germanium (d) Silver

2. A certain wire A has resistance 81 Ω. The resistance of another wire B of same material and equal length but of diameter thrice the diameter of A will be
NCERT | Page-88 | 2023

(a) 729 Ω (b) 243 Ω
(c) 81 Ω (d) 9 Ω

3. The reciprocal of resistance is : NCERT | Page-85 | RS 2022

(a) conductance (b) reactance
(c) mobility (d) conductivity

4. A copper wire of length 10 m and radius $(10^{-2}/\pi)$m has electrical resistance of 10 Ω. The current density in the wire for an electric field strength of 10 (V/m) is:
NCERT | Page-84 | 2022

(a) 10^6 A/m^2 (b) 10^{-5} A/m^2
(c) 10^5 A/m^2 (d) 10^4 A/m^2

5. The resistance of a wire is 'R' ohm. If it is melted and stretched to 'n' times its original length, its new resistance will be :-
NCERT | Page-83 | 2017

(a) $\frac{R}{n}$ (b) n^2R (c) $\frac{R}{n^2}$ (d) nR

6. Across a metallic conductor of non-uniform cross section a constant potential difference is applied. The quantity which remains constant along the conductor is
NCERT | Page-86 | 2015

(a) current (b) drift velocity
(c) electric field (d) current density

7. A wire of resistance 4 Ω is stretched to twice its original length. The resistance of stretched wire would be
NCERT | Page-83 | 2013

(a) 4 Ω (b) 8 Ω (c) 16 Ω (d) 2 Ω

8. A wire of a certain material is stretched slowly by ten per cent. Its new resistance and specific resistance become respectively:
NCERT | Page-83 | 2008

(a) 1.2 times, 1.3 times
(b) 1.21 times, same
(c) both remain the same
(d) 1.1 times, 1.1 times

9. The electric resistance of a certain wire of iron is *R*. If its length and radius are both doubled, then
NCERT | Page-83 | 2004

(a) the resistance and the specific resistance, will both remain unchanged
(b) the resistance will be doubled and the specific resistance will be halved

(c) the resistance will be halved and the specific resistance will remain unchanged
(d) the resistance will be halved and the specific resistance will be doubled

10. A 6 volt battery is connected to the terminals of the three metre long wire of uniform thickness and resistance of 100 ohm. The difference of potential between two points on the wire separated by a distance of 50 cm will be NCERT | Page-83 | 2004

(a) 1.5 volt (b) 3 volt
(c) 3 volt (d) 1 volt

11. There are three copper wires of length and cross sectional area (L, A), $(2L, \frac{1}{2}A)$, $(\frac{1}{2}L, 2A)$. In which case is the resistance minimum? NCERT | Page-83 | 1997

(a) It is the same in all three cases
(b) Wire of cross-sectional area $2A$
(c) Wire of cross-sectional area A
(d) Wire of cross-sectional area $\frac{1}{2}A$

12. If a negligibly small current is passed through a wire of length 15 m and of resistance 5Ω having uniform cross-section of 6×10^{-7} m², then coefficient of resistivity of material, is NCERT | Page-83 | 1996

(a) 1×10^{-7} Ω-m (b) 2×10^{-7} Ω-m
(c) 3×10^{-7} Ω-m (d) 4×10^{-7} Ω-m

13. If a wire of resistance R is melted and recasted to half of its length, then the new resistance of the wire will be NCERT | Page-83 | 1995

(a) R / 4 (b) R / 2 (c) R (d) 2R

14. The masses of the three wires of copper are in the ratio of 1 : 3 : 5 and their lengths are in the ratio of 5 : 3 : 1. The ratio of their electrical resistance is NCERT | Page-83 | 1988

(a) 1 : 3 : 5 (b) 5 : 3 : 1
(c) 1 : 25 : 125 (d) 125 : 15 : 1

Topic-3.5
Drift of Electrons and the Origin of Resistivity

15. A uniform wire of diameter d carries a current of 100 mA when the mean drift velocity of electrons in the wire is v. For a wire of diameter $\frac{d}{2}$ of the same material to carry a current of 200 mA, the mean drift velocity of electrons in the wire is NCERT | Page-86 | Ph-II 2024

(a) $4v$ (b) $8v$ (c) v (d) $2v$

16. A copper wire of radius 1 mm contains 10^{22} free electrons per cubic metre. The drift velocity for free electrons when 10 A current flows through the wire will be (Given, charge on electron = 1.6×10^{-19} C)

NCERT | Page-88 | Manipur 2023

(a) $\frac{6.25}{\pi}$ m s^{-1} (b) $\frac{6.25 \times 10^5}{\pi}$ m s^{-1}
(c) $\frac{6.25 \times 10^4}{\pi}$ m s^{-1} (d) $\frac{6.25}{\pi} \times 10^3$ m s^{-1}

17. **Column-I** gives certain physical terms associated with flow of current through a metallic conductor. **Column-II** gives some mathematical relations involving electrical quantities. Match. **Column-I** and **Column-II** with appropriate relations

NCERT | Page-86 | 2021

Column-I	Column-II
(A) Drift velocity	(i) $\frac{m}{ne^2\rho}$
(B) Electrical resistivity	(ii) nev_d
(C) Relaxation period	(iii) $\frac{eE}{m}\tau$
(D) Current density	(iv) $\frac{E}{J}$

(a) A - (iii), B - (ii), C - (iv), D - (i)
(b) A - (iii), B - (iv), C - (i), D - (ii)
(c) A - (iii), B - (iv), C - (ii), D - (i)
(d) A - (iii), B - (i), C - (iv), D - (ii)

18. A charged particle having drift velocity of 7.5×10^{-4} m s^{-1} in an electric field of 3×10^{-10} Vm^{-1}, has a mobility in m² V^{-1} s^{-1} of : NCERT | Page-88 | 2020

(a) 2.5×10^{6} (b) 2.5×10^{-6}
(c) 2.25×10^{-15} (d) 2.25×10^{15}

19. The velocity of charge carriers of current (about 1 amp) in a metal under normal conditions is of the order of NCERT | Page-86 | 1991

(a) a fraction of mm/sec
(b) velocity of light
(c) several thousand metres/second
(d) a few hundred metres per second

Topic-3.6
Limitations of OHM's Law

20. The resistance of a discharge tube is NCERT | Page-83 | 1999

(a) zero (b) ohmic
(c) non-ohmic (d) infinity

Topic-3.8
Temperature Dependence of Resistivity

21. The resistance of platinum wire at 0°C is 2 Ω and 6.8 Ω at 80°C. The temperature coefficient of resistance of the wire is NCERT | Page-90 | 2023

(a) 3×10^{-4} °C^{-1} (b) 3×10^{-3} °C^{-1}
(c) 3×10^{-2} °C^{-1} (d) 3×10^{-1} °C^{-1}

22. As the temperature increase, the electrical resistance: NCERT | Page-90 | 2022

(a) decreases for both conductors and semiconductors
(b) increases for conductors but decreases for semiconductors
(c) decreases for conductors but increase for semiconductors
(d) increases for both conductors and semiconductors.

23. Which of the following graph represents the variation of resistivity (ρ) with temperature (T) for copper? **NCERT | Page-90 | 2020**

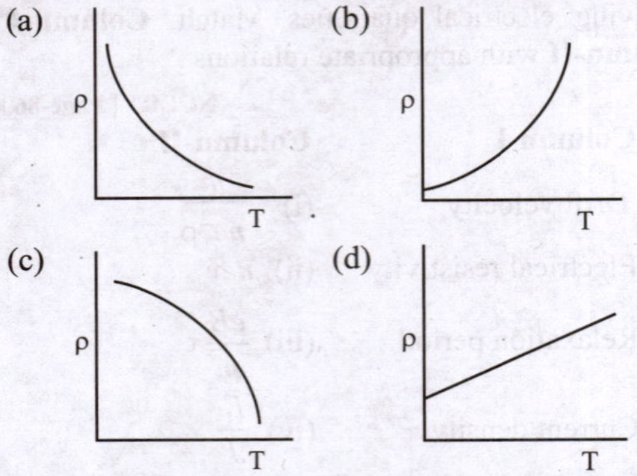

24. The solids which have the negative temperature coefficient of resistance are: **NCERT | Page-90 | 2020**
(a) insulators only
(b) semiconductors only
(c) insulators and semiconductors
(d) metals

25. The resistivity (specific resistance) of a copper wire **NCERT | Page-83 | 2002**
(a) increases with increase in its temperature
(b) decreases with increase in its cross-section
(c) increases with increase in its length
(d) increases with increase in its cross-section

26. Si and Cu are cooled to a temperature of 300 K, then resistivity? **NCERT | Page-90 | 2001**
(a) For Si increases and for Cu decreases
(b) For Cu increases and for Si decreases
(c) Decreases for both Si and Cu
(d) Increases for both Si and Cu

27. A wire has a resistance of 3.1Ω at 30ºC and a resistance 4.5Ω at 100ºC. The temperature coefficient of resistance of the wire **NCERT | Page-90 | 2001**
(a) $0.008\ ^\circ C^{-1}$ (b) $0.0034\ ^\circ C^{-1}$
(c) $0.0025\ ^\circ C^{-1}$ (d) $0.0012\ ^\circ C^{-1}$

28. If the resistance of a conductor is 5 Ω at 50ºC and 7Ω at 100ºC, then the mean temperature coefficient of resistance (of the material) is **NCERT | Page-90 | 1996**
(a) 0.001/ºC (b) 0.004/ºC
(c) 0.006/ºC (d) 0.008/ºC

Topic-3.9 Electrical Energy, Power

29. Two heaters A and B have power rating of 1 kW and 2 kW, respectively. Those two are first connected in series and then in parallel to a fixed power source. The ratio of power outputs for these two cases is: **NCERT | Page-93 | 2024**
(a) 1 : 1 (b) 2 : 9 (c) 1 : 2 (d) 2 : 3

30. Two resistors of resistance, 100 Ω and 200 Ω are connected in parallel in an electrical circuit. The ratio of the thermal energy developed in 100 Ω to that in 200 Ω in a given time is: **NCERT | Page-93 | 2022**
(a) 2 : 1 (b) 1 : 4
(c) 4 : 1 (d) 1 : 2

31. Six similar bulbs are connected as shown in the figure with a DC source of emf E, and zero internal resistance. The ratio of power consumption by the bulbs when (i) all are glowing and (ii) in the situation when two from section A and one from section B are glowing, will be: **NCERT | Page-93 | 2019**

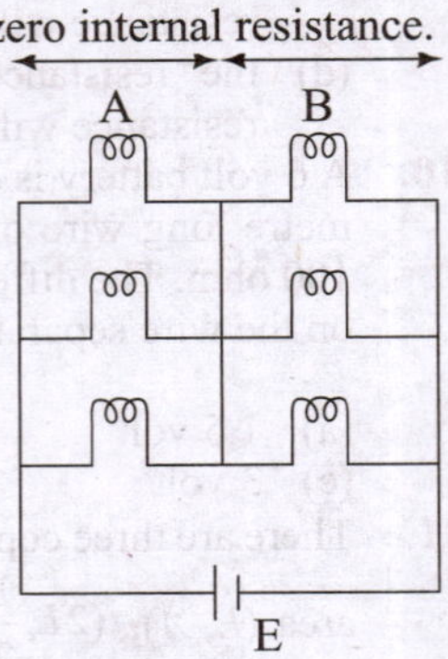

(a) 4 : 9
(b) 9 : 4
(c) 1 : 2
(d) 2 : 1

32. A filament bulb (500 W, 100 V) is to be used in a 230 V main supply. When a resistance R is connected in series, it works perfectly and the bulb consumes 500 W. The value of R is :- **NCERT | Page-93 | Ph-II 2016**
(a) 26Ω (b) 13Ω (c) 230Ω (d) 46Ω

33. The charge flowing through a resistance R varies with time t as $Q = at - bt^2$, where a and b are positive constants. The total heat produ ced in R is: **NCERT | Page-93 | 2016**
(a) $\frac{a^3R}{6b}$ (b) $\frac{a^3R}{3b}$ (c) $\frac{a^3R}{2b}$ (d) $\frac{a^3R}{b}$

34. Two cities are 150 km apart. Electric power is sent from one city to another city through copper wires. The fall of potential per km is 8 volt and the average resistance per km is 0.5 Ω. The power loss in the wires is : **NCERT | Page-93 | 2014**
(a) 19.2 W (b) 19.2 kW
(c) 19.2 J (d) 12.2 kW

35. Ten identical cells connected in series are needed to heat a wire of length one meter and radius '*r*' by 10ºC in time '*t*'. How many cells will be required to heat the wire of length two meter of the same radius by the same temperature in time '*t*'? **NCERT | Page-93 | Kar. 2013**
(a) 10 (b) 20 (c) 30 (d) 40

36. If voltage across a bulb rated 220 Volt-100 Watt drops by 2.5% of its rated value, the percentage of the rated value by which the power would decrease is: **NCERT | Page-93 | 2012**
(a) 20% (b) 2.5% (c) 5% (d) 10%

37. The power dissipated in the circuit shown in the figure is 30 Watts. The value of *R* is: **NCERT | Page-93 | 2012 M**

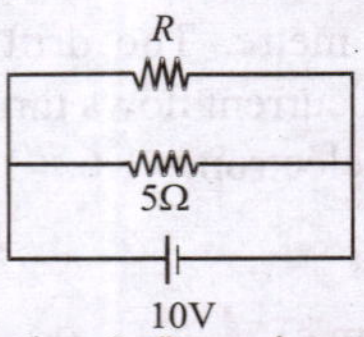

(a) 20 Ω
(b) 15 Ω
(c) 10 Ω
(d) 30 Ω

38. If power dissipated in the 9-Ω resistor in the circuit shown is 36 watt, the potential difference across the 2-Ω resistor is **NCERT | Page-93 | 2011**

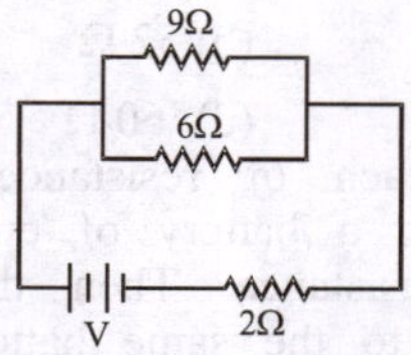

(a) 4 volt (b) 8 volt
(c) 10 volt (d) 2 volt

39. A current of 3 amp flows through the 2Ω resistor shown in the circuit. The power dissipated in the 5Ω resistor is: **NCERT | Page-93 | 2008**

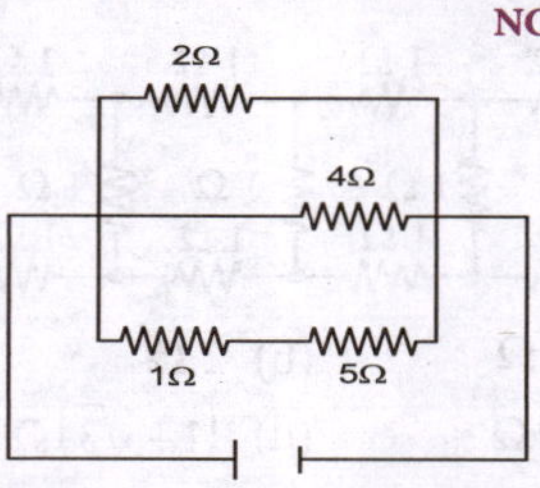

(a) 4 watt (b) 2 watt
(c) 1 watt (d) 5 watt

40. The total power dissipated in watts in the circuit shown here is **NCERT | Page-93 | 2007**

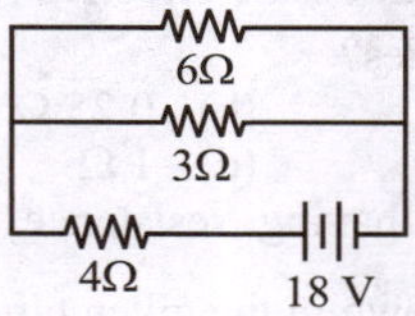

(a) 40 (b) 54 (c) 4 (d) 16

41. Power dissipated across the 8Ω resistor in the circuit shown here is 2 watt. The power dissipated in watt units across the 3Ω resistor is **NCERT | Page-93 | 2006**

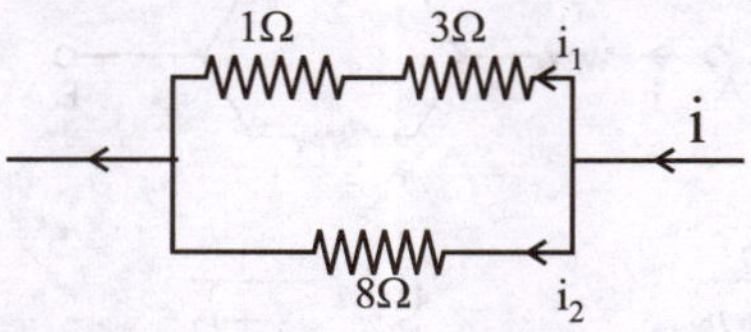

(a) 1.0 (b) 0.5 (c) 3.0 (d) 2.0

42. A 5–ampere fuse wire can withstand a maximum power of 1 watt in the circuit. The resistance of the fuse wire is
(a) 0.04 ohm (b) 0.2 ohm **NCERT | Page-93 | 2005**
(c) 5 ohm (d) 0.4 ohm

43. When three identical bulbs of 60 watt, 200 volt rating are connected in series to a 200 volt supply, the power drawn by them will be **NCERT | Page-93 | 2004**
(a) 20 watt (b) 60 watt
(c) 180 watt (d) 10 watt

44. In India electricity is supplied for domestic use at 220 V. It is supplied at 110 V in USA. If the resistance of a 60 W bulb for use in India is R, the resistance of a 60 W bulb for use in USA will be **NCERT | Page-93 | 2004**
(a) R/2 (b) R (c) 2R (d) R/4

45. An electric kettle has two heating coils. When one of the coils is connected to an a.c. source, the water in the kettle boils in 10 minutes. When the other coil is used, the water boils in 40 minutes. If both the coils are connected in parallel, the time taken by the same quantity of water to boil will be **NCERT | Page-93 | 2003**
(a) 15 min (b) 8 min
(c) 4 min (d) 25 min

46. Two 220 volt, 100 watt bulbs are connected first in series and then in parallel. Each time the combination is connected to a 220 volt a.c. supply line. The power drawn by the combination in each case respectively will be **NCERT | Page-93 | 2003**
(a) 50 watt, 200 watt (b) 50 watt, 100 watt
(c) 100 watt, 50 watt (d) 200 watt, 150 watt

47. Fuse wire is a wire of **NCERT | Page-93 | 2003**
(a) low resistance and high melting point
(b) high resistance and high melting point
(c) high resistance and low melting point
(d) low resistance and low melting point

48. If 25W, 220 V and 100 W, 220 V bulbs are connected in series across a 440 V line, then **NCERT | Page-93 | 2001**
(a) only 25W bulb will fuse
(b) only 100W bulb will fuse
(c) both bulbs will fuse
(d) none of these

49. A battery of 10 V and internal resistance 0.5Ω is connected across a variable resistance R. The value of R for which the power delivered is maximum is equal to **NCERT | Page-93 | 2001, 1992**
(a) 0.25Ω (b) 0.5Ω (c) 1.0Ω (d) 2.0Ω

50. Two electric bulbs, one of 200 V, 40W and other of 200 V, 100W are connected in a domestic circuit. Then **NCERT | Page-93 | 2000**
(a) they have equal resistance
(b) the resistance of 40W bulb is more than 100W bulb
(c) the resistance of 100W bulb is more than 40 W bulb
(d) they have equal current through them

51. Three equal resistors connected across a source of e.m.f. together dissipate 10 watt of power. What will be the power dissipated in watts if the same resistors are connected in parallel across the same source of e.m.f.? **NCERT | Page-93 | 1998**
(a) 10 (b) $\frac{10}{3}$ (c) 30 (d) 90

52. A 5°C rise in temperature is observed in a conductor by passing a current. When the current is doubled the rise in temperature will be approximately **NCERT | Page-93 | 1998**
(a) 10°C (b) 16°C
(c) 20°C (d) 12°C

53. A (100 W, 200 V) bulb is connected to a 160V power supply. The power consumption would be **NCERT | Page-93 | 1997**
(a) 125 W (b) 100 W
(c) 80 W (d) 64 W

54. A heating coil is labelled 100 W, 220 V. The coil is cut in half and the two pieces are joined in parallel to the same source. The energy now liberated per second is **NCERT | Page-93 | 1995**
(a) 25 J (b) 50 J
(c) 200 J (d) 400 J

55. A 4 μF capacitor is charged to 400 volts and then its plates are joined through a resistance of 1k Ω. The heat produced in the resistance is **NCERT | Page-93 | 1994**
(a) 0.16 J (b) 1.28 J
(c) 0.64 J (d) 0.32 J

56. Two identical batteries each of e.m.f 2V and internal resistance 1Ω are available to produce heat in an external resistance by passing a current through it. The maximum power that can be developed across R using these batteries is **NCERT | Page-93 | 1990**

(a) 3.2 W (b) 2.0 W (c) 1.28 W (d) $\frac{8}{9}$W

57. Forty electric bulbs are connected in series across a 220 V supply. After one bulb is fused the remaining 39 are connected again in series across the same supply. The illumination will be **NCERT | Page-93 | 1989**

(a) more with 40 bulbs than with 39
(b) more with 39 bulbs than with 40
(c) equal in both the cases
(d) in the ratio $40^2 : 39^2$

58. A current of 2 A, passing through a conductor produces 80 J of heat in 10 seconds. The resistance of the conductor in ohm is **NCERT | Page-93 | 1989**

(a) 0.5 (b) 2 (c) 4 (d) 20

Topic-*

Combination of Resistors – Series and Parallel

59. The current passing through the battery in the given circuit, is : **2025**

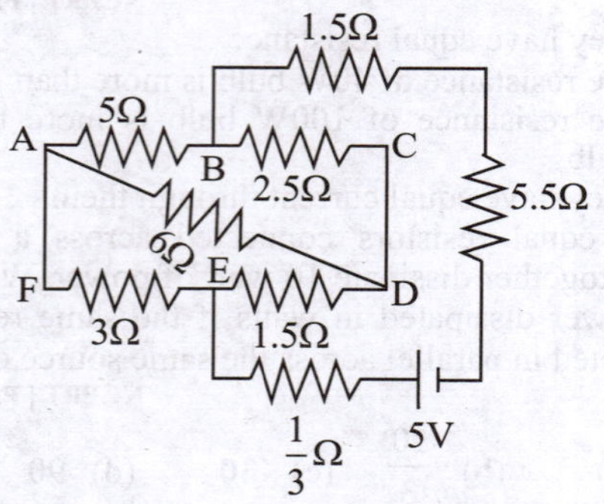

(a) 2.0 A (b) 0.5 A
(c) 2.5 A (d) 1.5 A

60. A wire of resistance R is cut into 8 equal pieces. From these pieces two equivalent resistances are made by adding four of these together in parallel. Then these two sets are added in series. The net effective resistance of the combination is : **2025**

(a) $\frac{R}{64}$ (b) $\frac{R}{32}$ (c) $\frac{R}{16}$ (d) $\frac{R}{8}$

61. A uniform metal wire of length l has 10 Ω resistance. Now this wire is stretched to a length $2l$ and then bent to form a perfect circle. The equivalent resistance across any arbitrary diameter of that circle is **Ph-II 2024**

(a) 10 Ω (b) 5 Ω (c) 40 Ω (d) 20 Ω

62. A wire of length 'l' and resistance 100 Ω is divided into 10 equal parts. The first 5 parts are connected in series while the next 5 parts are connected in parallel. The two combinations are again connected in series. The resistance of this final combination is: **2024**

(a) 26 Ω (b) 52 Ω
(c) 55 Ω (d) 60 Ω

63. 10 resistors, each of resistance R are connected in series to a battery of emf E and negligible internal resistance. Then those are connected in parallel to the same battery, the current is increased n times. The value of n is **2023**

(a) 10 (b) 100 (c) 1 (d) 1000

64. The equivalent resistance of the infinite network given below is : **2022**

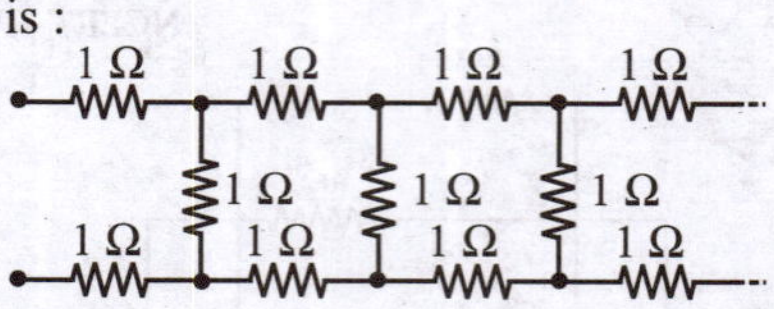

(a) $(1+\sqrt{5})\Omega$ (b) 2 Ω
(c) $(1+\sqrt{2})\Omega$ (d) $(1+\sqrt{3})\Omega$

65. The effective resistance of a parallel connection that consists of four wires of equal length, equal area of cross-section and same material is 0.25 Ω. What will be the effective resistance if they are connected in series? **2021**

(a) 4 Ω (b) 0.25 Ω
(c) 0.5 Ω (d) 1 Ω

66. Three resistors having resistances r_1, r_2 and r_3 are connected as shown in the given circuit. The ratio $\frac{i_3}{i_1}$ of currents in terms of resistances used in the circuit is **2021**

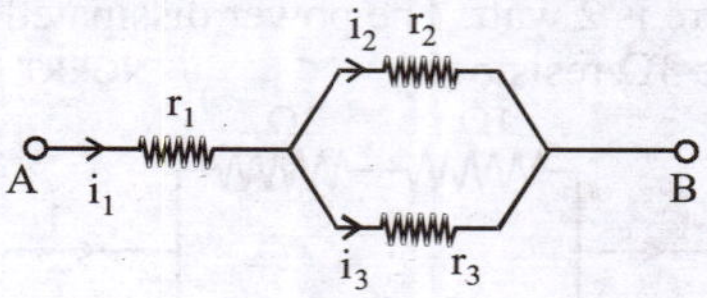

(a) $\frac{r_2}{r_1 + r_3}$ (b) $\frac{r_1}{r_2 + r_3}$
(c) $\frac{r_2}{r_2 + r_3}$ (d) $\frac{r_1}{r_1 + r_2}$

67. A set of 'n' equal resistors, of value 'R' each, are connected in series to a battery of emf 'E' and internal resistance 'R'. The current drawn is I. Now, the 'n' resistors are connected in parallel to the same battery. Then the current drawn from battery becomes 10 I. The value of 'n' is **2018**

(a) 10 (b) 11 (c) 9 (d) 20

68. A, B and C are voltmeters of resistance R, 1.5 R and 3R respectively as shown in the figure. When some potential difference is applied between X and Y, the voltmeter readings are V_A, V_B and V_C respectively. Then **2015**

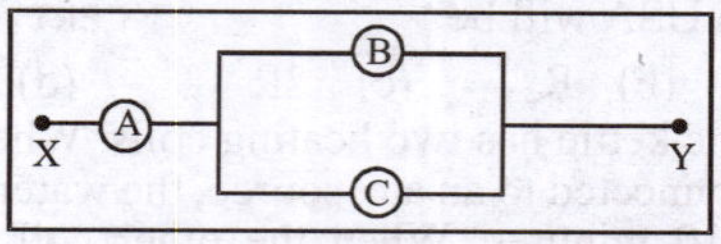

(a) $V_A \neq V_B = V_C$ (b) $V_A = V_B \neq V_C$
(c) $V_A \neq V_B \neq V_C$ (d) $V_A = V_B = V_C$

69. Two metal wires of identical dimension are connected in series. If σ_1 and σ_2 are the conductivities of the metal wires respectively, the effective conductivity of the combination is : **2015 RS**

(a) $\frac{\sigma_1+\sigma_2}{2\sigma_1\sigma_2}$ (b) $\frac{\sigma_1+\sigma_2}{\sigma_1\sigma_2}$

(c) $\frac{\sigma_1\sigma_2}{\sigma_1+\sigma_2}$ (d) $\frac{2\sigma_1\sigma_2}{\sigma_1+\sigma_2}$

70. Two rods are joined end to end, as shown. Both have a cross-sectional area of 0.01 cm^2. Each is 1 meter long. One rod is of copper with a resistivity of 1.7×10^{-6} ohm-centimeter, the other is of iron with a resistivity of 10^{-5} ohm-centimeter.
How much voltage is required to produce a current of 1 ampere in the rods? **2013**

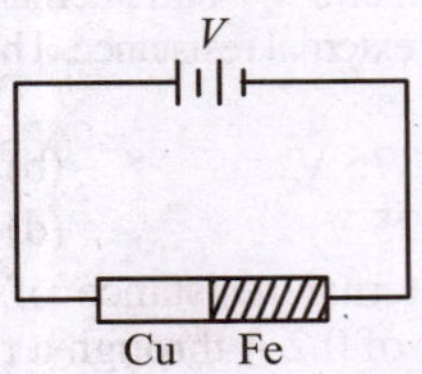

(a) 0.117 V

(b) 0.00145 V

(c) 0.0145 V

(d) 1.7×10^{-6} V

71. A 12 cm wire is given a shape of a right angled triangle *ABC* having sides 3 cm, 4 cm and 5 cm as shown in the figure. The resistance between two ends (*AB*, *BC*, *CA*) of the respective sides are measured one by one by a multi-meter. The resistances will be in the ratio of **Kar. 2013**

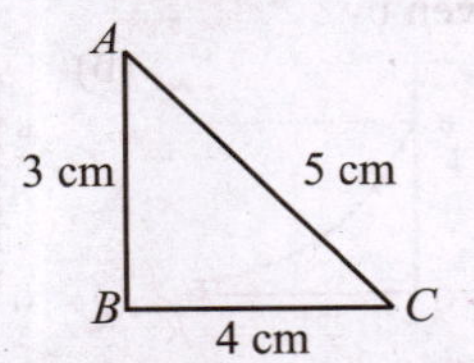

(a) 3 : 4 : 5

(b) 9 : 16 : 25

(c) 27 : 32 : 35

(d) 21 : 24 : 25

72. A ring is made of a wire having a resistance $R_0 = 12\ \Omega$. Find the points *A* and *B* as shown in the figure, at which a current carrying conductor should be connected so that the resistance *R* of the sub-circuit between these points is equal to $\frac{8}{3}\Omega$. **2012**

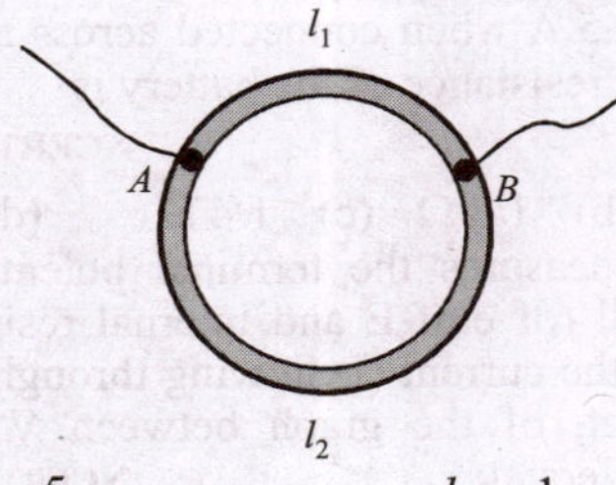

(a) $\frac{l_1}{l_2}=\frac{5}{8}$ (b) $\frac{l_1}{l_2}=\frac{1}{3}$

(c) $\frac{l_1}{l_2}=\frac{3}{8}$ (d) $\frac{l_1}{l_2}=\frac{1}{2}$

73. A wire of resistance 12 ohms per meter is bent to form a complete circle of radius 10 cm. The resistance between its two diametrically opposite points, A and B as shown in the figure, is **2009**

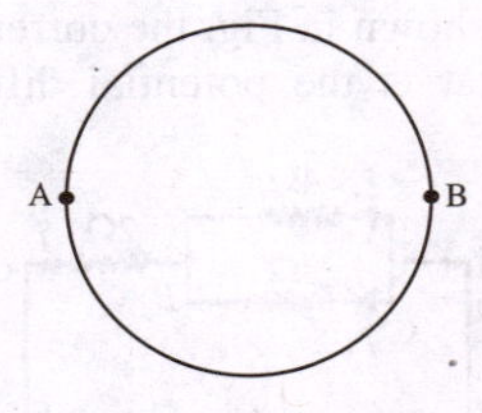

(a) 3Ω (b) 6πΩ

(c) 6Ω (d) 0.6πΩ

74. When a wire of uniform cross–section *a*, length *l* and resistance *R* is bent into a complete circle, resistance between any two of diametrically opposite points will be **2005**

(a) $\frac{R}{4}$ (b) $4R$ (c) $\frac{R}{8}$ (d) $\frac{R}{2}$

75. Resistances n, each of r ohm, when connected in parallel give an equivalent resistance of *R* ohm. If these resistances were connected in series, the combination would have a resistance in ohms, equal to **2004**

(a) nR (b) n^2R

(c) R/n^2 (d) R/n

76. The current (I) in the given circuit is **1999**

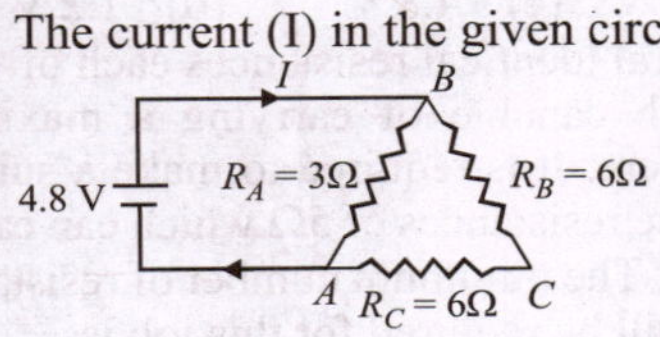

(a) 1.6 A (b) 2.0 A

(c) 0.32 A (d) 3.2 A

77. The current in the following circuit is **1997**

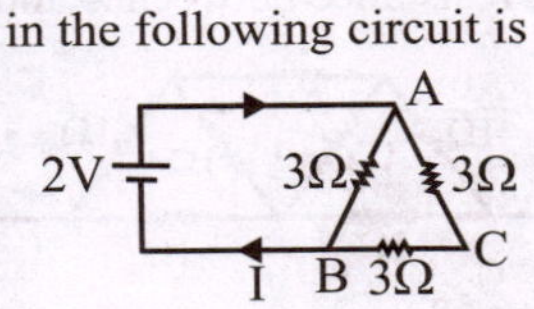

(a) 1*A* (b) $\frac{2}{3}A$ (c) $\frac{2}{9}A$ (d) $\frac{1}{8}A$

78. What will be the equivalent resistance of circuit shown in figure between two points A and D **1996**

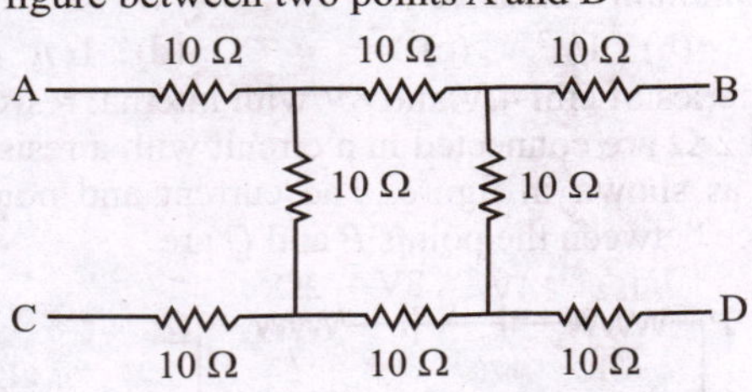

(a) 10Ω (b) 20Ω

(c) 30Ω (d) 40Ω

79. Two wires of the same metal have same length, but their cross-sections are in the ratio 3:1. They are joined in series. The resistance of thicker wire is 10Ω. The total resistance of the combination will be **1995**

(a) 10 Ω (b) 20 Ω

(c) 40 Ω (d) 100 Ω

80. In the circuit shown in Fig, the current in 4 Ω resistance is 1.2 A. What is the potential difference between B and C. 1994

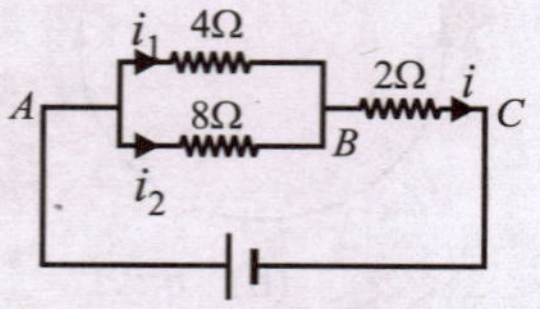

(a) 3.6 volt (b) 6.3 volt
(c) 1.8 volt (d) 2.4 volt

81. Three resistances each of 4 Ω are connected to form a triangle. The resistance between any two terminals is 1993

(a) 12 Ω (b) 2 Ω (c) 6Ω (d) 8/3Ω

82. Current through 3 Ω resistor is

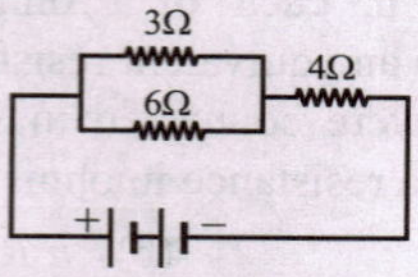

0.8 amp., then potential drop through 4Ω resistor is 1993

(a) 9.6 V (b) 2.6 V (c) 4.8 V (d) 1.2 V

83. You are given several identical resistances each of value $R = 10\Omega$ and each capable of carrying a maximum current of one ampere. It is required to make a suitable combination of these resistances of 5Ω which can carry a current of 4 ampere. The minimum number of resistances of the type R that will be required for this job is 1990

(a) 4 (b) 10 (c) 8 (d) 20

84. In the network shown in the Fig, each resistance is 1Ω. The effective resistance between A and B is 1990

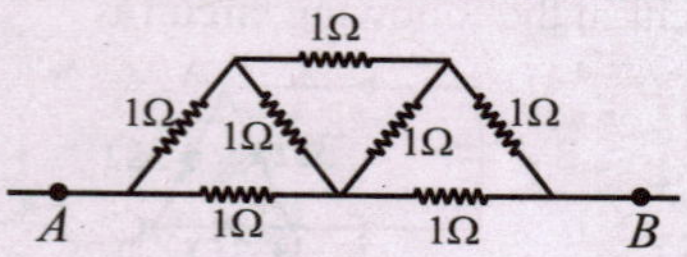

(a) $\frac{4}{3}\Omega$ (b) $\frac{3}{2}\Omega$ (c) 7 Ω (d) $\frac{8}{7}\Omega$

85. n equal resistors are first connected in series and then connected in parallel. What is the ratio of the maximum to the minimum resistance ? 1989

(a) n (b) $1/n^2$ (c) n^2 (d) $1/n$

86. Two batteries of emf 4 V and 8V with internal resistance 1 Ω and 2 Ω are connected in a circuit with a resistance of 9 Ω as shown in figure. The current and potential difference between the points P and Q are 1988

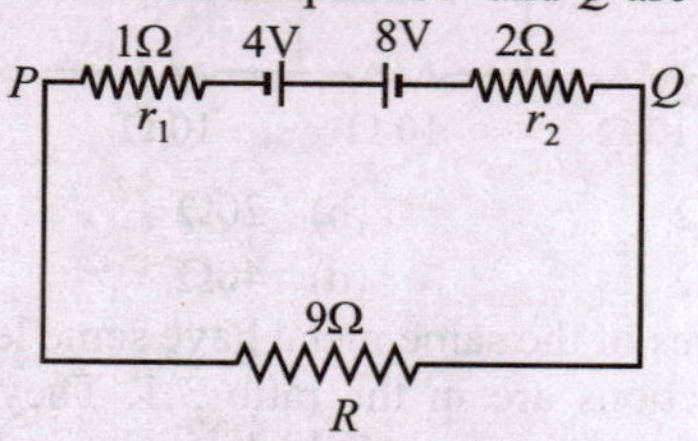

(a) $\frac{1}{3}A$ and 3 V (b) $\frac{1}{6}A$ and 4 V

(c) $\frac{1}{9}A$ and 9 V (d) $\frac{1}{12}A$ and 12 V

Topic-3.10

Cells, EMF, Internal Resistance

87. The terminal voltage of the battery, whose emf is 10 V and internal resistance 1 Ω, when connected through an external resistance of 4 Ω as shown in the figure is: NCERT | Page-94 | 2024

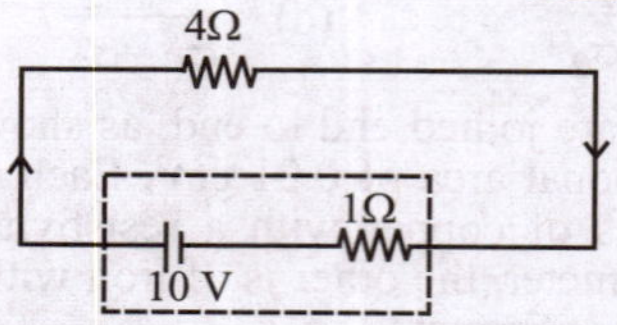

(a) 4 V (b) 6 V (c) 8 V (d) 10 V

88. A cell of emf 4 V and internal resistance 0.5 Ω is connected to a 7.5 Ω external resistance. The terminal potential difference of the cell is. NCERT | Page-94 | RS 2022

(a) 0.375 V (b) 3.75 V
(c) 4.25 V (d) 4 V

89. The internal resistance of a 2.1 V cell which gives a current of 0.2 A through a resistance of 10 Ω is NCERT | Page-94 | 2013

(a) 0.5 Ω (b) 0.8 Ω (c) 1.0 Ω (d) 0.2 Ω

90. Cell having an emf ε and internal resistance r is connected across a variable external resistance R. As the resistance R is increased, the plot of potential difference V across R is given by : NCERT | Page-94 | 2012 M

(a)
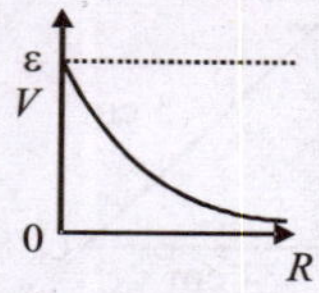

(b)
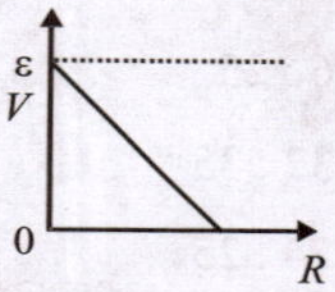

(c)
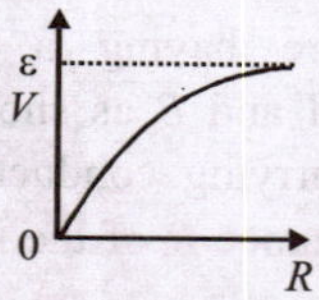

(d)
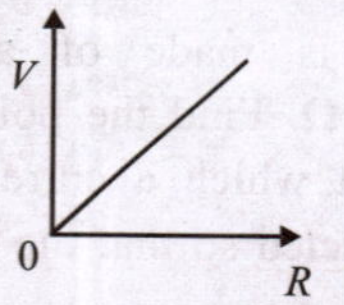

91. A current of 2A flows through a 2Ω resistor when connected across a battery. The same battery supplies a current of 0.5 A when connected across a 9Ω resistor. The internal resistance of the battery is NCERT | Page-94 | 2011

(a) 0.5 Ω (b) 1/3 Ω (c) 1/4 Ω (d) 1 Ω

92. A student measures the terminal potential difference (V) of a cell (of emf E and internal resistance r) as a function of the current (I) flowing through it. The slope and intercept, of the graph between V and I, then, respectively, equal: NCERT | Page-94 | 2009

(a) – r and E (b) r and – E
(c) – E and r (d) E and – r

93. The potential difference between the terminals of a cell in an open circuit is 2.2 V. When a resistor of 5Ω is connected across the terminals of the cell, the potential difference between the terminals of the cell is found to be 1.8 V. The internal resistance of the cell is NCERT | Page-94 | 2002

(a) $\frac{7}{12}\Omega$ (b) $\frac{10}{9}\Omega$ (c) $\frac{9}{10}\Omega$ (d) $\frac{12}{7}\Omega$

94. A car battery has e.m.f. 12 volt and internal resistance 5×10^{-2} ohm. If it draws 60 amp current, the terminal voltage of the battery will be NCERT | Page-94 | 2000

(a) 15 volt (b) 3 volt
(c) 5 volt (d) 9 volt

Topic-3.11
Cells in Series and in Parallel

95. The magnitude and direction of the current in the following circuit is NCERT | Page-98 | 2023

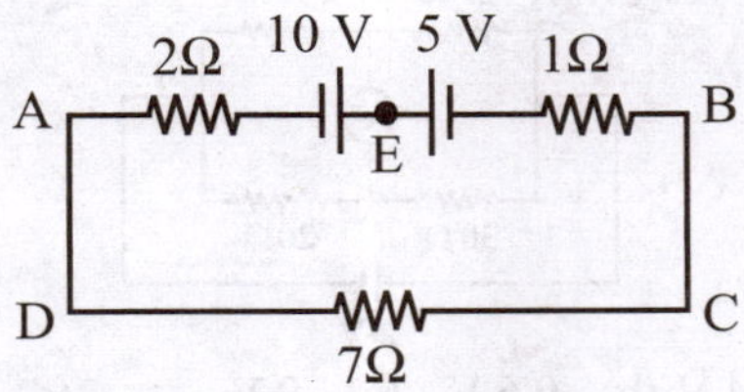

(a) 0.2 *A* from *B* to *A* through *E*
(b) 0.5 *A* from *A* to *B* through *E*
(c) $\frac{5}{9}$ *A* from *A* to *B* through *E*
(d) 1.5 *A* from *B* to *A* through *E*

96. Two cells, having the same e.m.f., are connected in series through an external resistance *R*. Cells have internal resistances r_1 and r_2 ($r_1 > r_2$) respectively. When the circuit is closed, the potential difference across the first cell is zero. The value of R is NCERT | Page-96 | 2006

(a) $\frac{r_1 + r_2}{2}$ (b) $\frac{r_1 - r_2}{2}$
(c) $r_1 + r_2$ (d) $r_1 - r_2$

97. Two batteries, one of emf 18 volt and internal

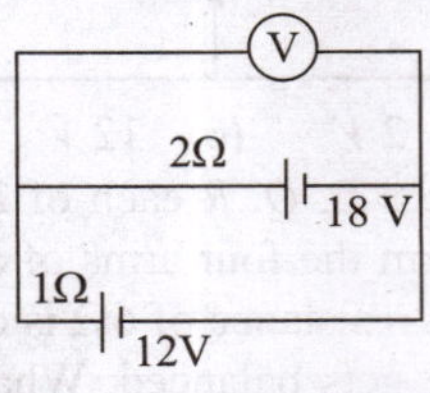

resistance 2Ω and the other of emf 12 volt and internal resistance 1Ω, are connected as shown. The voltmeter V will record a reading of NCERT | Page-97 | 2005

(a) 30 volt (b) 18 volt
(c) 15 volt (d) 14 volt

98. A battery is charged at a potential of 15V for 8 hours when the current flowing is 10A. The battery on discharge supplies a current of 5A for 15 hours. The mean terminal voltage during discharge is 14V. The "watt-hour" efficiency of the battery is 2004

(a) 87.5% (b) 82.5%
(c) 80% (d) 90%

Topic-3.12
Kirchhoff's Rules

99. A constant voltage of 50 V is maintained between the points A and B of the circuit shown in the figure. The current through the branch CD of the circuit is :- NCERT | Page-97, 98 | 2025

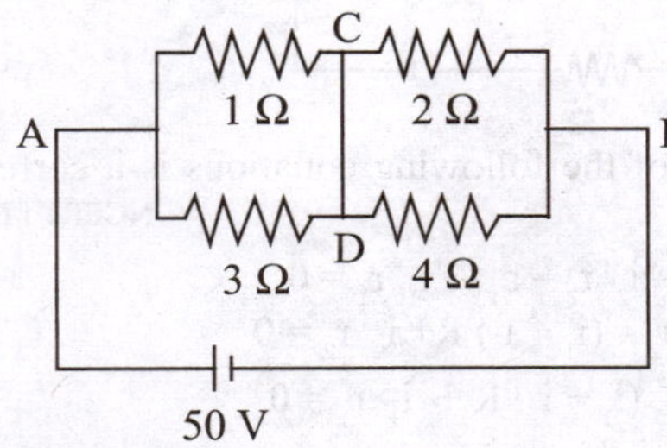

(a) 1.5 A (b) 2.0 A
(c) 2.5 A (d) 3.0 A

100. The steady state current in the circuit shown below is : NCERT | Page-97, 98 | Ph-II 2024

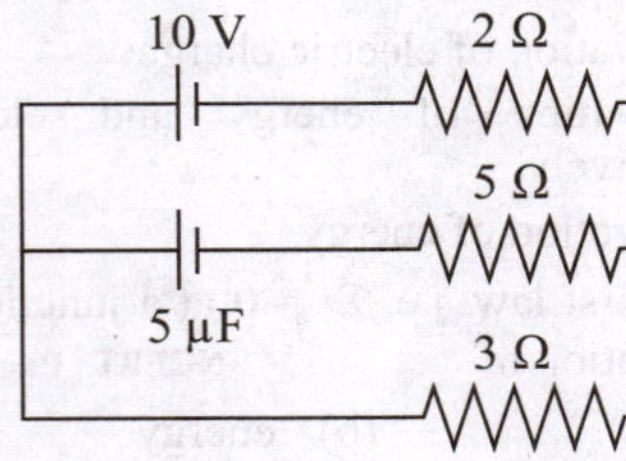

(a) 0.67 A (b) 1.5 A
(c) 2 A (d) 1 A

101. In the circuit shown in the figure, if potential at point A is taken to be zero, the potential at point B is NCERT | Page-97 | 2011 M

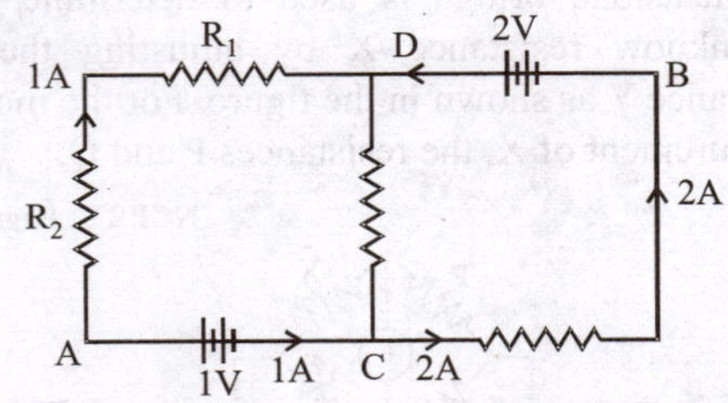

(a) –1V (b) +2V
(c) –2V (d) + 1V

102. Consider the following two statements:

(A) Kirchhoff's junction law follows from the conservation of charge.

(B) Kirchhoff's loop law follows from the conservation of energy.

Which of the following is correct? NCERT | Page-97 | 2010

(a) Both (A) and (B) are wrong
(b) (A) is correct and (B) is wrong
(c) (A) is wrong and (B) is correct
(d) Both (A) and (B) are correct

103. See the electric circuit shown in the figure.

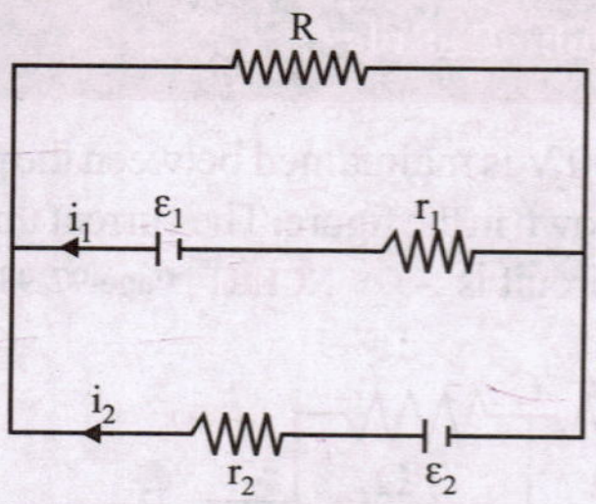

Which of the following equations is a correct equation for it? **NCERT | Page-97 | 2009**

(a) $\varepsilon_2 - i_2 r_2 - \varepsilon_1 - i_1 r_1 = 0$
(b) $-\varepsilon_2 - (i_1 + i_2) R + i_2 r_2 = 0$
(c) $\varepsilon_1 - (i_1 + i_2) R + i_1 r_1 = 0$
(d) $\varepsilon_1 - (i_1 + i_2) R - i_1 r_1 = 0$

104. Kirchhoff's first and second laws for electrical circuits are consequences of **NCERT | Page-97 | 2006**

(a) conservation of electric charge and energy respectively
(b) conservation of electric charge
(c) conservation of energy and electric charge respectively
(d) conservation of energy

105. Kirchoff's first law, i.e. $\Sigma i = 0$ at a junction, deals with the conservation of **NCERT | Page-97 | 1992, 1997**

(a) charge (b) energy
(c) momentum (d) angular momentum

Topic-3.13

Wheatstone Bridge

106. A wheatstone bridge is used to determine the value of unknow resistance X by adjusting the variable resistance Y as shown in the figure. For the most precise measurement of X, the resistances P and Q: **NCERT | Page-101 | 2022**

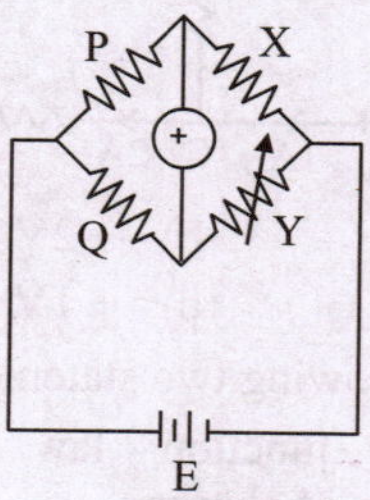

(a) should be approximately equal and are small
(b) should be very large and unequal
(c) do not play any significant role
(d) should be approximately equal to 2X

107. In the circuits shown below, the readings of the voltmeters and the ammeters will be : **2019**

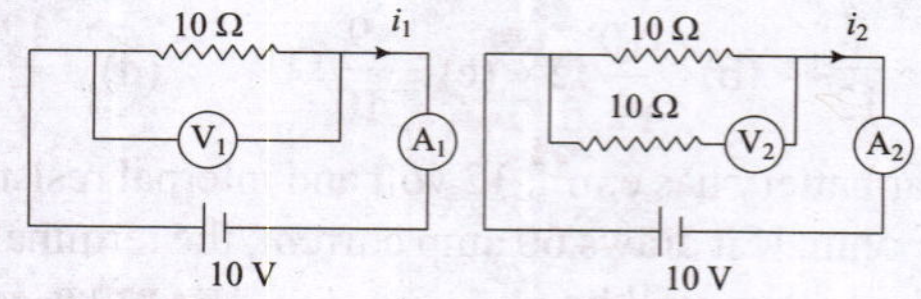

Circuit 1 Circuit 2

(a) $V_2 > V_1$ and $i_1 = i_2$
(b) $V_1 = V_2$ and $i_1 > i_2$
(c) $V_1 = V_2$ and $i_1 = i_2$
(d) $V_2 > V_1$ and $i_1 > i_2$

108. The reading of an ideal voltmeter in the circuit shown is, **NCERT | Page-101 | Odisha 2019**

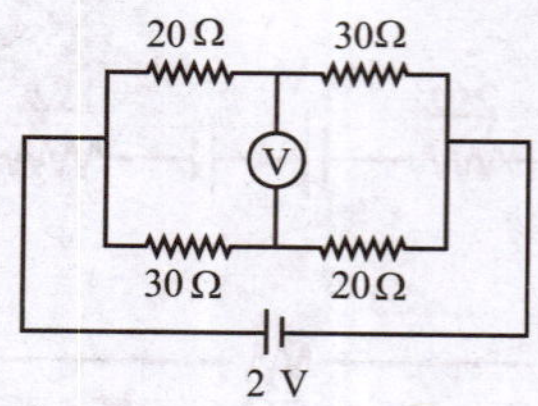

(a) 0.4 V (b) 0.6 V (c) 0 V (d) 0.5 V

109. The resistance of the four arms P, Q, R and S in a Wheatstone's bridge are 10 ohm, 30 ohm, 30 ohm and 90 ohm, respectively. The e.m.f. and internal resistance of the cell are 7 volt and 5 ohm respectively. If the galvanometer resistance is 50 ohm, the current drawn from the cell will be **2013**

(a) 0.2 A (b) 0.1 A
(c) 2.0 A (d) 1.0 A

110. In the circuit shown the cells *A* and *B* have negligible resistances. For $V_A = 12V$, $R_1 = 500\Omega$ and $R = 100\Omega$ the galvanometer (*G*) shows no deflection. The value of V_B is : **2012**

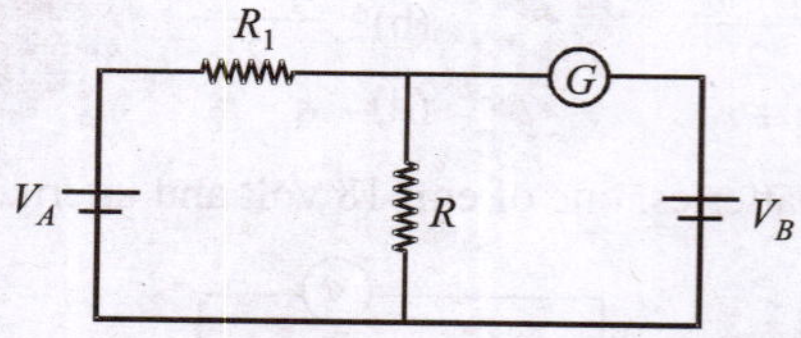

(a) 4 *V* (b) 2 *V* (c) 12 *V* (d) 6 *V*

111. Three resistances *P*, *Q*, *R* each of 2Ω and an unknown resistance S form the four arms of a Wheatstone bridge circuit. When a resistance of 6Ω is connected in parallel to *S* the bridge gets balanced. What is the value of *S*? **2007**

(a) 3Ω (b) 6Ω (c) 1Ω (d) 2Ω

112. In the circuit shown, if a conducting wire is connected between points A and B, the current in this wire will **2006**

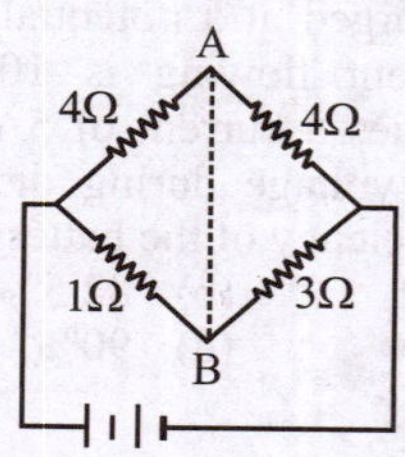

(a) flow in the direction which will be decided by the value of V
(b) be zero
(c) flow from B to A
(d) flow from A to B

113. For the network shown in the Fig. the value of the current i is **2005**

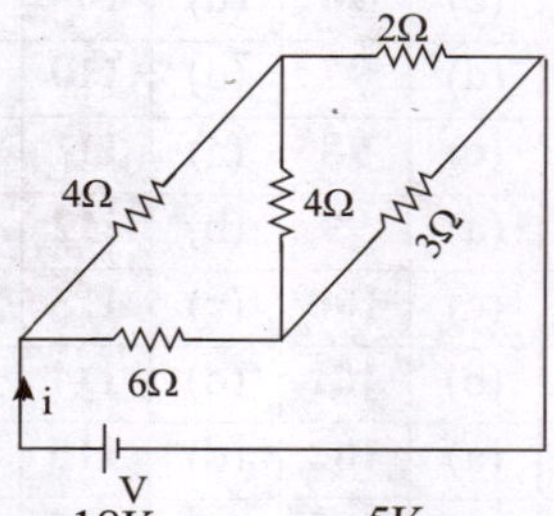

(a) $\frac{9V}{35}$ (b) $\frac{18V}{5}$ (c) $\frac{5V}{9}$ (d) $\frac{5V}{18}$

114. Five equal resistances each of resistance R are connected as shown in the figure. A battery of V volts is connected between A and B. The current flowing in $AFCEB$ will be **2004**

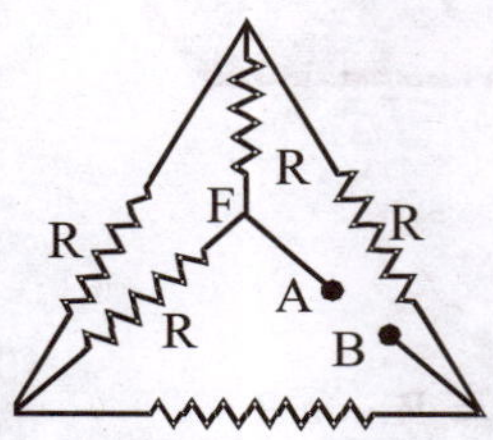

(a) $\frac{2V}{R}$ (b) $\frac{3V}{R}$ (c) $\frac{V}{R}$ (d) $\frac{V}{2R}$

115. In a Wheatstone's bridge all the four arms have equal resistance R. If the resistance of the galvanometer arm is also R, the equivalent resistance of the combination as seen by the battery is **2003**

(a) $2R$ (b) $\frac{R}{4}$ (c) $\frac{R}{2}$ (d) R

116. Five resistances have been connected as shown in the figure. The effective resistance between A and B is **2000**

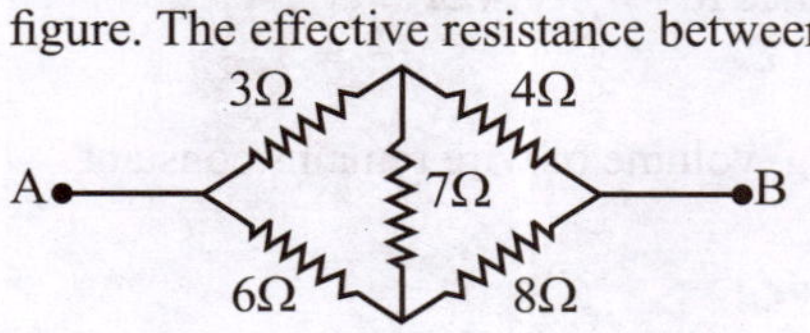

(a) $\frac{14}{3}\Omega$ (b) $\frac{20}{3}\Omega$ (c) 14Ω (d) 21Ω

Topic-*

Meter Bridge

117. The given circuit shows a uniform straight wire AB of 40 cm length fixed at both ends. In order to get zero reading in the galvanometer G, the free end of J is to be placed from B at: **Ph-II 2024**

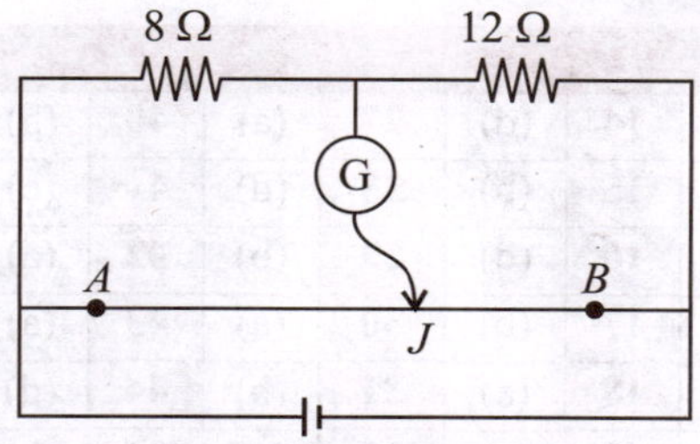

(a) 32 cm (b) 8 cm
(c) 16 cm (d) 24 cm

118. A resistance wire connected in the left gap of a metre bridge balances a 10 Ω resistance in the right gap at a point which divides the bridge wire in the ratio 3 : 2. If the length of the resistance wire is 1.5 m, then the length of 1 Ω of the resistance wire is : **2020**

(a) 1.0×10^{-1} m (b) 1.5×10^{-1} m
(c) 1.5×10^{-2} m (d) 1.0×10^{-2} m

119. The metre bridge shown is in balance position with $\frac{P}{Q} = \frac{l_1}{l_2}$. If we now interchange the position of galvanometer and cell, will the bridge work? If yes, what will be balance condition? **Odisha 2019**

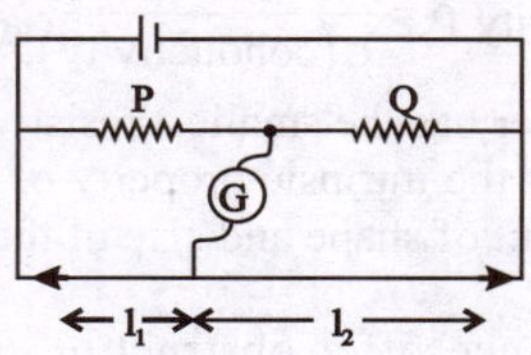

(a) yes, $\frac{P}{Q} = \frac{l_1}{l_2}$ (b) yes, $\frac{P}{Q} = \frac{l_2 - l_1}{l_2 + l_1}$
(c) no, no null point (d) yes, $\frac{P}{Q} = \frac{l_2}{l_1}$

120. The resistances in the two arms of the meter bridge are 5Ω and RΩ, respectively. When the resistance R is shunted with an equal resistance, the new balance point is at 1.6 l_1. The resistance 'R' is : **2014**

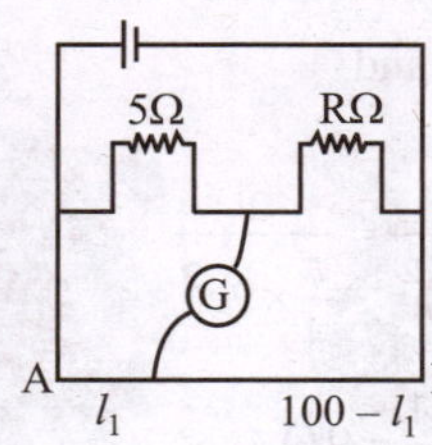

(a) 10Ω
(b) 15Ω
(c) 20Ω
(d) 25Ω

121. In a metre-bridge, the balancing length from the left end when standard resistance of 1 Ω is in right gap is found to be 20 cm. The value of unknown resistance is **1999**

(a) 0.25Ω (b) 0.4Ω
(c) 0.5Ω (d) 4Ω

ANSWER KEY

1	(d)	14	(d)	27	(a)	40	(b)	53	(d)	66	(c)	79	(c)	92	(a)	105	(a)	118	(a)
2	(a)	15	(b)	28	(d)	41	(c)	54	(d)	67	(a)	80	(a)	93	(b)	106	(a)	119	(a)
3	(c)	16	(d)	29	(b)	42	(a)	55	(d)	68	(d)	81	(d)	94	(a)	107	(c)	120	(b)
4	(d)	17	(b)	30	(a)	43	(a)	56	(b)	69	(d)	82	(c)	95	(b)	108	(a)	121	(a)
5	(b)	18	(a)	31	(b)	44	(d)	57	(b)	70	(a)	83	(c)	96	(d)	109	(a)		
6	(a)	19	(a)	32	(a)	45	(b)	58	(b)	71	(c)	84	(d)	97	(d)	110	(b)		
7	(c)	20	(c)	33	(a)	46	(a)	59	(b)	72	(d)	85	(c)	98	(a)	111	(a)		
8	(b)	21	(c)	34	(b)	47	(c)	60	(c)	73	(d)	86	(a)	99	(b)	112	(c)		
9	(c)	22	(b)	35	(b)	48	(a)	61	(a)	74	(a)	87	(c)	100	(c)	113	(d)		
10	(d)	23	(b)	36	(c)	49	(b)	62	(b)	75	(b)	88	(b)	101	(d)	114	(d)		
11	(b)	24	(c)	37	(c)	50	(b)	63	(b)	76	(b)	89	(a)	102	(d)	115	(d)		
12	(b)	25	(a)	38	(c)	51	(d)	64	(d)	77	(a)	90	(c)	103	(d)	116	(a)		
13	(a)	26	(a)	39	(d)	52	(c)	65	(a)	78	(c)	91	(b)	104	(a)	117	(d)		

Hints & Solutions

1. **(d)** Resistivity $\rho = \dfrac{1}{\sigma(\text{conductivity})}$ and σ maximum for silver so silver has the smallest resistivity.
Resistivity is the intrinsic property of the Substance. It is independent of shape and size of the body (i.e., l and A)

2. **(a)** Resistance offer obstruction to current while conductance allow electric current to flow, so conductance is the reciprocal of resistance.

3. **(c)** We have
$J = \sigma E$

$$J=\frac{E}{\rho} \Rightarrow J=\frac{E\ell}{RA} \Rightarrow J=\frac{10\times10}{10\times\pi}\times\left(\frac{10^{-2}}{\sqrt{\pi}}\right)^2=10^5$$

4. **(d)** $\because$ Resistance $R=\dfrac{\rho l}{A}$, where $A=\dfrac{\pi}{4}d^2$

$$\therefore 81=\frac{\rho\times1}{\frac{\pi}{4}d^2}, \text{ and}$$

$$R'=\frac{\rho\times1}{\frac{\pi}{4}\times(3d)^2}=\frac{\rho l}{\frac{\pi}{4}\times9d^2}$$

$$\therefore R'=\frac{R}{9}=\frac{81}{9}=9\Omega$$

5. **(b)** We know that, $R=\dfrac{\rho\ell}{A}$

or $\quad R=\dfrac{\rho\ell^2}{\text{Volume}} \Rightarrow R\propto\ell^2$

According to question $\ell_2 = n\ell_1$

$$\frac{R_2}{R_1}=\frac{n^2l_1^2}{l_1^2}$$

or, $\dfrac{R_2}{R_1}=n^2 \Rightarrow R_2=n^2R_1$

6. **(a)** Here, metallic conductor can be considered as the combination of various conductors connected in series. And in series combination current remains same.

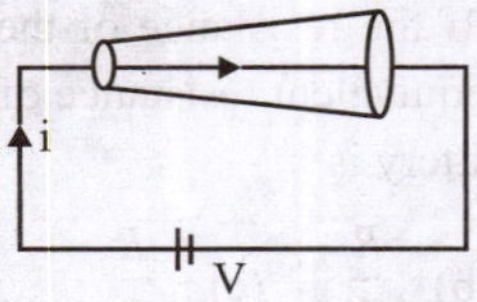

7. **(c)** Resistance $R=\dfrac{\rho\ell}{A}=4\,\Omega$

$\because \ell' = 2\,\ell$

On stretching, volume of wire remains constant

$\therefore \ell A = \ell' A'$

$\therefore A' = \dfrac{A}{2}$

$$\therefore R' = \rho\frac{2\ell}{\frac{A}{2}} = 4R = 4\times4\,\Omega = 16\,\Omega$$

$\Rightarrow \ell A = 2lA'$

Therefore the resistance of new wire becomes 16 Ω.

8. **(b)** Resistance of a wire is given by $R=\rho\dfrac{l}{a}$

If the length is increased by 10% then new

length $l' = l + \frac{1}{10} = \frac{11}{10}l$

In that case, area of cross-section of wire would decrease by 10%

$\therefore$ New area of cross-section

$$A' = A - \frac{A}{10} = \frac{9}{10}A \therefore R' = \rho\frac{\ell'}{A'} = \rho\frac{\frac{11}{10}l}{\frac{9}{10}A}$$

$$R' = \frac{11}{9}\rho\frac{l}{R} \qquad R' = 1.21R$$

The new resistance increases by 1.21 times. The specific resistance (resistivity) remains unchanged as it depends on the nature of the material of the wire.

NOTES After stretching if length of a conductor increases by $x\%$ then resistance will increase by $2x\%$ (Valid only if $x < 10\%$)

9. **(c)** $R = \frac{\rho\ell_1}{A_1}$, now $\ell_2 = 2\ell_1$

$A_2 = \pi(r_2)^2 = \pi(2r_1)^2 = 4\pi r_1^2 = 4A_1$

$$\therefore \quad R_2 = \frac{\rho(2\ell_1)}{4A_1} = \frac{\rho\ell_1}{2A_1} = \frac{R}{2}$$

$\therefore$ Resistance is halved, but specific resistance remains the same.

10. **(d)** $R \propto \ell$

For 300 cm, R = 100 Ω

For 50 cm, $R' = \frac{100}{300} \times 50 = \frac{50}{3}\Omega \therefore IR = 6$

$$\Rightarrow IR' = \frac{6}{R} \times R' = \frac{6}{100} \times \frac{50}{3} = 1 \text{ volt.}$$

11. **(b)** $R = \rho\frac{l}{A}$

$$R_1 = \rho\frac{L}{A} \quad \text{... (i)}$$

$$R_2 = \rho\frac{2L}{A} \times 2 \quad \text{... (ii)}$$

$$R_3 = \rho\frac{L}{2.2A} = \frac{\rho L}{4A} \quad \text{... (iii)}$$

$\Rightarrow R_3 < R_1 < R_2$

12. **(b)** Given : Length of wire (l) = 15m

Area (A) = 6×10^{-7} m²

Resistance (R) = 5Ω.

We know that resistance of the wire material

$$R = \rho\frac{l}{A} \Rightarrow 5 = \rho \times \frac{15}{6 \times 10^{-7}}$$

$\Rightarrow \rho = 2 \times 10^{-7}\ \Omega - m$

[where ρ = coefficient of resistivity]

13. **(a)** Initial resistance (R_1) = R; Initial length is ℓ and final length (ℓ_2) = 0.5 ℓ. Volume of a wire = $\ell \cdot A$. Since the volume of the wire remains the same after recasting, therefore.

$\ell_1 A_1 = \ell_2 A_2$

$$\text{or } \frac{\ell_1}{\ell_2} = \frac{A_2}{A_1} \text{ or } \frac{\ell}{0.5\ell} = \frac{A_2}{A_1} \text{ or } \frac{A_2}{A_1} = 2$$

We also know that resistance of a wire (R)

$$R = \rho \times \frac{\ell}{A}; R \propto \frac{\ell}{A}$$

$$\therefore \frac{R_1}{R_2} = \frac{\ell_1}{\ell_2} \times \frac{A_2}{A_1} = \frac{\ell}{0.5\ell} \times 2 = 4$$

$$\text{or}, R_2 = \frac{R_1}{4} = \frac{R}{4}.$$

NOTES When wires are drawn from same volume but with different area of cross-section, then

$$R \propto \frac{1}{(\text{Area of cross-section})^2}$$

14. **(d)** $R = \frac{\rho l}{\pi r^2}$. But $m = \pi r^2 ld \therefore \pi r^2 = \frac{m}{ld}$

$$\therefore R = \frac{\rho l^2 d}{m}, \; R_1 = \frac{\rho l_1^2 d}{m_1}, \; R_2 = \frac{\rho l_2^2 d}{m_2}$$

$$R_3 = \frac{\rho l_3^2 d}{m_3}$$

$$R_1 : R_2 : R_3 = \frac{l_1^2}{m_1} : \frac{l_2^2}{m_2} : \frac{l_3^2}{m_3}$$

$$R_1 : R_2 : R_3 = \frac{25}{1} : \frac{9}{3} : \frac{1}{5} = 125 : 15 : 1$$

15. **(b)** current, $i = nAv_d e$

$$= n\left(\frac{\pi D^2}{4}\right)v_d e \Rightarrow i \propto D^2 v_d$$

$$\frac{100}{200} = \frac{(d)^2}{\left(\frac{d}{2}\right)^2} \times \frac{v}{v'}$$

$\Rightarrow v' = 2 \times 2^2 v = 8v$

16. **(d)** Drift velocity, $V_d = \frac{i}{neA}$

$$V_d = \frac{10}{10^{22} \times 1.6 \times 10^{-19} \times \pi\left(10^{-3}\right)^2}$$

$$V_d = \frac{10}{1.6\pi \times 10^{-3}} = \frac{6.25}{\pi} \times 10^3 \text{ m/s}$$

Drift velocity per unit electric field i.e., $V_d / \in = \mu$ (mobility of electrons)

17. **(b)** Drift velocity, $v_d = \frac{I}{neA} = \frac{eE\tau}{m}$

Electrical resistivity, $\rho = \frac{1}{\sigma} = \frac{E}{J}$ $\left(\because \sigma = \frac{J}{E}\right)$

Relaxation period, $\tau = \frac{m}{ne^2\rho}$

Current density, $J = \frac{I}{A} = nev_d$

> **NOTES** Relation between drift velocity, v_d, electrical resistivity (ρ) relaxation period (τ) and current density (J)
>
> $$v_d = \frac{i}{neA} = \frac{J}{ne} = \frac{\sigma E}{ne} = \frac{E}{\rho ne} = \frac{v}{\rho l ne};$$
>
> $$R\frac{l}{A} = \frac{m}{ne^2\tau}.\frac{l}{A}$$

18. **(a)** Given,

Drift velocity of charged particle,

$V_d = 7.5 \times 10^{-4}$ m/s

Electric field, $E = 3 \times 10^{-10}$ Vm^{-1}

Mobility,

$$\mu = \frac{V_d}{E} = \frac{7.5 \times 10^{-4}}{3 \times 10^{-10}} = 2.5 \times 10^6 \ m^2V^{-1}s^{-1}$$

19. **(a)** The velocity of charge carriers of current (about 1 amp) in a metal under normal conditions is of the order of a fraction of mm/sec.

20. **(c)** In discharge tube the current is due to flow of positive ions and electrons. Moreover, secondary emission of electrons is also possible. So V-I curve is non-linear; hence resistance is non-ohmic.

21. **(c)** Given,

Resistance at 0°C, $R_0 = 2\ \Omega$

Resistance at 80°C, 6.8 Ω

Using $R = R_0(1 + \alpha\Delta T)$

where α is the thermal coefficient of resistance

$\therefore 6.8 = 2\{1 + \alpha(80 - 0)\}$

$\Rightarrow \frac{6.8}{2} - 1 = \alpha \times 80$

$\Rightarrow \alpha = \frac{3.4 - 1}{80} = \frac{2.4}{80} = 0.03$

$\therefore \alpha = 3 \times 10^{-2}\ °C^{-1}$

22. **(b)** For conductors, $R \propto T$ so, $T\uparrow\uparrow \Rightarrow R\uparrow\uparrow$

For semiconductor, $R \propto \frac{1}{T}$ so $T\uparrow\uparrow \Rightarrow R\downarrow\downarrow$

23. **(b)** For metals like copper, at temperature much lower than 0°C, graph deviates considerably from a straight line.

24. **(c)** For metals temperature coefficient of resistance is positive while for insulators and semiconductors, temperature coefficient of resistance is negative.

25. **(a)** Resistivity of copper wire increases with increase in temprature as $\rho_t = \rho_0(1 + \alpha t)$

> **NOTES** Copper being a metal has positive coefficient of resistivity.

26. **(a)** For metals $P_t = P_o(1 + \alpha.t)$

So resistivity of copper decreases with decrease in temperature. whereas of semiconductor or *si* increases with decreases in temperature.

27. **(a)** $R_1 = 3.1\ \Omega$ at $t = 30°C$

$R_2 = 4.5\ \Omega$ at $t = 100°C$

We have, $R_t = R_0(1 + \alpha t)$

$\therefore R_1 = R_0[1 + \alpha(30)]$

$R_2 = R_0[1 + \alpha(100)]$

$\Rightarrow \frac{R_1}{R_2} = \frac{1 + 30\alpha}{1 + 100\alpha}$

$\Rightarrow \frac{3.1}{4.5} = \frac{1 + 30\alpha}{1 + 100\alpha} \Rightarrow \alpha = 0.008°C^{-1}$

28. **(d)** Resistance of a material at two different temperatures, T_1 and T_2 such that $T_2 > T_1$, are related as

$R_{T_2} = R_{T_1}[1 + \alpha(T_2 - T_1)]$

Using this relation for the present case, we get

$R_{100} = R_{50}[1 + \alpha(100 - 50)]$

$\Rightarrow 7 = 5[1 + \alpha(50)]$

$\Rightarrow \alpha = \frac{(7-5)}{250} = 0.008/°C$

29. **(b)** Rated power, $P = \frac{V^2}{R} \Rightarrow P \propto \frac{1}{R}$

$\frac{P_A}{P_B} = \frac{R_B}{R_A} \Rightarrow R_A = 2R_B$

For Series Combination,

$$P_S = \frac{V^2}{R_A + R_B} = \frac{V^2}{3R_B}$$

For Parallel Combination,

$$P_P = \frac{V^2(R_A + R_B)}{R_A.R_B} = \frac{3V^2}{2R_B}$$

$\therefore \frac{P_S}{P_P} = \frac{2}{9}$

30. **(a)** As power, $P = \frac{V^2}{R}$

$$\frac{P_1}{P_2} = \frac{R_2}{R_1} \Rightarrow \frac{P_{100\Omega}}{P_{200\Omega}} = \frac{200}{100} = \frac{2}{1}$$

> **NOTES** When some potential difference is applied across the conductor then collision of free electrons with ions of the lattice result's in conversion of electrical energy into heat energy.

31. **(b)** When all bulbs are glowing

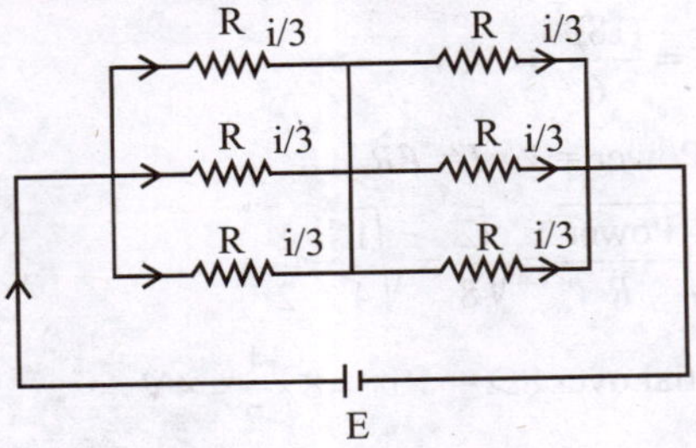

$$R_{eq} = \frac{R}{3} + \frac{R}{3} = \frac{2R}{3}$$

$$\text{Power } (P_i) = \frac{E^2}{R_{eq}} = \frac{3E^2}{2R} \quad \text{...(i)}$$

When two from section A and one from section B are glowing, then

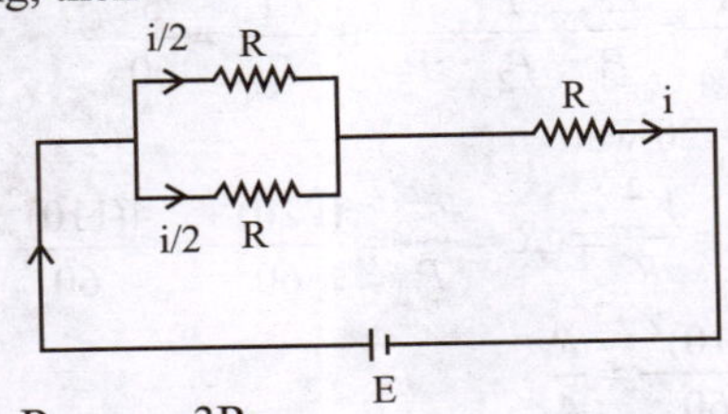

$$R_{eq} = \frac{R}{2} + R = \frac{3R}{2}$$

$$\text{Power } (P_f) = \frac{2E^2}{3R} \quad \text{...(ii)}$$

Dividing equation (i) by (ii) we get

$$\frac{P_i}{P_f} = \frac{3E^2 3R}{2R\, 2E^2} = 9:4$$

32. **(a)**

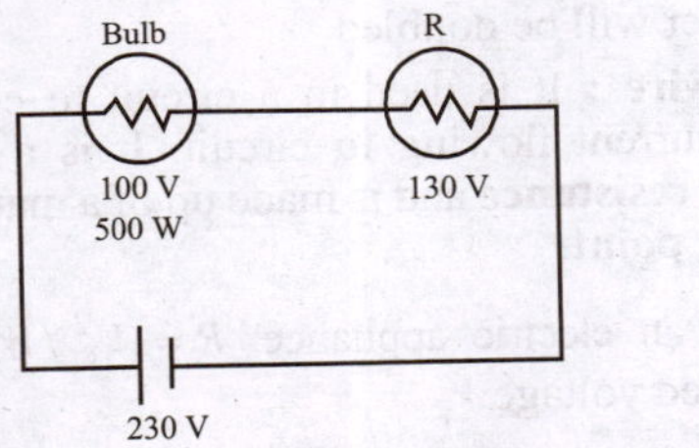

$$R_{bulb} = \frac{V^2}{P} = \frac{100^2}{500} = 20\Omega$$

According to question $\frac{R}{R_{bulb}} = \frac{130}{100}$

$$\Rightarrow \frac{R}{20} = \frac{130}{100} \Rightarrow R = 26\,\Omega$$

33. **(a)** **Given:** Charge $Q = at - bt^2$

$\therefore$ Current $i = \frac{\partial Q}{\partial t} = a - 2bt$

$$\{\text{for } i = 0 \Rightarrow t = \frac{a}{2b}\}$$

From joule's law of heating, heat produced

$dH = i^2Rdt$

$$H = \int_0^{a/2b} (a-2bt)^2 R dt$$

$$H = \left.\frac{(a-2bt)^3 R}{-3\times 2b}\right|_0^{\frac{a}{2b}} = \frac{a^3R}{6b}$$

34. **(b)** Total resistance $R = (0.5\ \Omega/\text{km}) \times (150\ \text{km})$

$= 75\ \Omega$

Total voltage drop = (8 V/km) × (150 km) = 1200 V

$$\text{Power loss} = \frac{(\Delta V)^2}{R} = \frac{(1200)^2}{75}\text{W}$$

$= 19200\ \text{W} = 19.2\ \text{kW}$

35. **(b)** Resistance is directly proportionl to length of the wire. As length is doubled so mass is doubled and resistance is doubled.

We have

$$\frac{(10E)^2}{R}t = mS\Delta T, \text{ Now } \frac{(nE)^2 t}{2R} = (2m)S\Delta T$$

$$\Rightarrow \frac{n^2E^2t}{2R} = 2\frac{10^2E^2t}{R}$$

$$\Rightarrow n = 20$$

36. **(c)** Resistance of bulb is constant

$$P = \frac{V^2}{R} \Rightarrow \frac{\Delta p}{p} = \frac{2\Delta V}{V} + \frac{\Delta R}{R}$$

$$\frac{\Delta p}{p} = 2 \times 2.5 + 0 = 5\%$$

37. **(c)** The power dissipated in the circuit.

$$P = \frac{V^2}{R_{eq}} \quad \text{...(i)}$$

$V = 10$ volt

$$\frac{1}{R_{eq}} = \frac{1}{R} + \frac{1}{5} = \frac{5+R}{5R}$$

$$R_{eq} = \left(\frac{5R}{5+R}\right); P = 30\ \text{W}$$

Substituting the values in equation (i)

$$30 = \frac{(10)^2}{\left(\frac{5R}{5+R}\right)} \Rightarrow \frac{15R}{5+R} = 10$$

$15R = 50 + 10R$

$5R = 50$

$R = 10\ \Omega$

NOTES For a given voltage V, if resistance is changed from R to $\left(\frac{R}{n}\right)$. Power consumed changes from P to nP.

$$P=\frac{V^2}{R}$$

when $R'=\frac{R}{n}$

then $P'=\frac{V^2}{R/n}=\frac{nV^2}{R}=nP$

38. (c) We have, $P=\frac{V^2}{R}$

$\Rightarrow 36=\frac{V^2}{9} \Rightarrow V=18V$

Current passing through the 9Ω resistor is

$i_1=\frac{V}{R}=\frac{18}{9}=2A$

The 9Ω and 6Ω resistors are in parallel, therefore

$i_1=\frac{6}{9+6}\times i$

where i is the current delivered by the battery.

$\therefore i=\frac{2\times 15}{6}=5A$

Thus, potential difference across 2Ω resistor is

$V=iR=5\times 2=10V$

39. (d) Clearly, 2Ω , 4Ω and (1 + 5) Ω resistors are in parallel. Hence, potential difference is same across each of them.

$\therefore I_1\times 2=I_2\times 4=I_3\times 6$

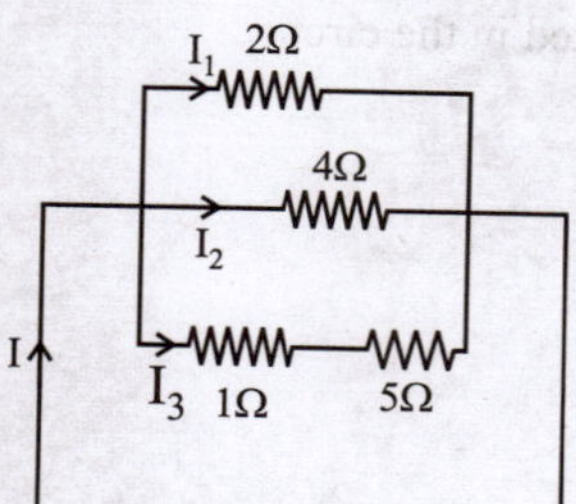

Given $I_1=3A$ $\qquad \therefore I_1\times 2=I_3\times 6$

$\therefore I_1\times 2=I_3\times 6$ provides

$I_3=\frac{I_1\times 2}{6}=\frac{3\times 2}{6}=1A.$

Now, the potential across the 5Ω resistor is

$V=I_3\times 5=1\times 5=5V.$

$\therefore$ the power dissipated in the 5Ω resistor

$P=\frac{V^2}{R}=\frac{5^2}{5}=5\text{watt}.$

40. (b) Power dissipated $=P$

$=\frac{V^2}{R}=\frac{(18)^2}{6}=54\,W$

41. (c) Power $=V.I=I^2R$

$i_2=\sqrt{\frac{\text{Power}}{R}}=\sqrt{\frac{2}{8}}=\sqrt{\frac{1}{4}}=\frac{1}{2}A$

Potential over 8Ω $=Ri_2=8\times\frac{1}{2}=4V$

This is the potential over parallel branch. So,

$i_1=\frac{4}{4}=1A$

Power of 3Ω $=i_1^2R=1\times 1\times 3=3W$

42. (a) $R=\frac{P}{I^2}=\frac{1}{25}=0.04\Omega$

43. (a) $\frac{1}{P_{eq}}=\frac{1}{P_1}+\frac{1}{P_2}+\frac{1}{P_3}$ or $\frac{1}{P_{eq}}=\frac{3}{60}$

$\Rightarrow P_{eq}=20$ watt.

44. (d) $P=\frac{V^2}{R}\Rightarrow R=\frac{V^2}{P}=\frac{(220)^2}{60}=\frac{4(110)^2}{60}$

$R'=\frac{(110)^2}{60}=\frac{R}{4}$

45. (b) Time $=\frac{10\times 40}{10+40}=\frac{400}{50}=8\,\text{min}$

46. (a) Power $\propto\frac{1}{\text{Resistance}}$

In series combination, resistance doubles.
Hence, power will be halved.
In parallel combination, resistance halves.
Hence, power will be doubled.

47. (c) **Fuse wire :** It is used in a circuit to control the maximum current flowing in circuit. It is a thin wire having **high resistance** and is made up of a material with **low melting point**.

48. (a) As for an electric appliance $R=(V_s^2/W)$, so for same specified voltage V_s

$\frac{R_{25}}{R_{100}}=\frac{100}{25}=4$

i.e, $R_{25}=4R$ with $R_{100}=R$

Now in series potential divides in proportion to resistance.

So, $V_1=\frac{R_1}{(R_1+R_2)}V$

i.e., $V_{25}=\frac{4}{5}\times 440=352V$ and $V_2=\frac{R_2}{(R_1+R_2)}V$

i.e., $V_{100}=\frac{1}{5}\times 440=88V$

From this, it is clear that voltage across 100 W bulb (= 88 V) is lesser than specified (220 V) while across 25 W bulb (= 352 V) is greater than specified (220 V), so, 25 W bulb will fuse.

49. **(b)** Power is maximum when $r = R$, $R = r = 0.5\Omega$.

NOTES If a cell of internal resistance r and emf E is connected to an external resistance of radius R, then current given by cell, $I = \frac{E}{R+r}$

Power dissipated in external resistance (load)
$P = VI = (IR)I = I^2R$

$\Rightarrow \quad P = \left(\frac{E}{R+r}\right)^2 R$

when $R = r$

Power, $P = \frac{E^2}{4r}$ (Maximum)

50. **(b)** $R_1 = \frac{V^2}{P_1} = \frac{200 \times 200}{40} = 1000\ \Omega$

$R_2 = \frac{V^2}{P_2} = \frac{200 \times 200}{100} = 400\ \Omega$

$\therefore R_1$ (for 40W) $> R_2$ (for 100 W)

51. **(d)** In series, Equivalent resistance = $3R$

Power $= \frac{V^2}{3R} \Rightarrow 10 = \frac{V^2}{3R} \Rightarrow V^2 = 30R$

In parallel, $\frac{1}{R'} = \frac{1}{R} + \frac{1}{R} + \frac{1}{R} = \frac{3}{R}$

$\therefore$ Equivalent resistance $R' = \frac{R}{3}$

$\therefore$ Power $= \frac{V^2}{R'} = \frac{30R}{R/3} = 90$ W

NOTES If voltage V remains same, then power consumed by a n equal resistors in parallel is n^2 times that of power consumed in series.

52. **(c)**

53. **(d)** Power = 100 W, Voltage = 200 V
Resistance of bulb

$= \frac{V^2}{P} = \frac{200 \times 200}{100} = 400\Omega$

When bulb is applied across 160 V,

Current in bulb $= \frac{160}{400} A$

Power consumption $= VI = 160 \times \frac{160}{400} = 64$W

54. **(d)** Power of heating coil = 100 W and voltage (V) = 220 volts. When the heating coil is cut into two equal parts and these parts are joined in parallel, then the resistance of the coil is reduced to one-fourth of the previous value. Therefore energy liberated per second becomes 4 times i.e., 4 × 100 = 400 J.

55. **(d)** The energy stored in the capacitor

$= \frac{1}{2}CV^2 = \frac{1}{2} \times 4 \times 10^{-4} \times 400 \times 400 = 0.32$ J

This energy will be converted into heat in the resistor.

56. **(b)** For maximum current, the two batteries should be connected in series. The current will be maximum when external resistance is equal to the total internal resistance of cells i.e. 2Ω. Hence power devloped across the resistacne R will be

$I^2R = \left(\frac{2E}{R+2r}\right)^2 R = \left(\frac{2 \times 2}{2+2}\right)^2 \times 2 = 2W$

57. **(b)** Since, the voltage is same for the two combinations, therefore $H \propto \frac{1}{R}$. Hence, the combination of 39 bulbs will glow more.

58. **(b)** $H = I^2Rt$

or $R = \frac{H}{(I^2t)} = \frac{80}{(2^2 \times 10)} = 2\Omega$

59. **(b)** The given circuit can be redrawn as

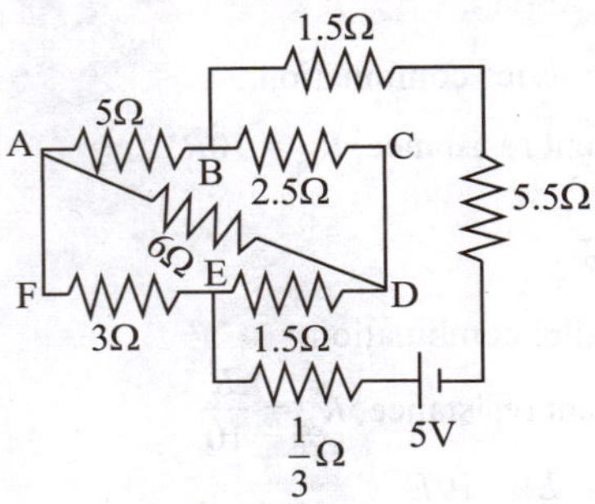

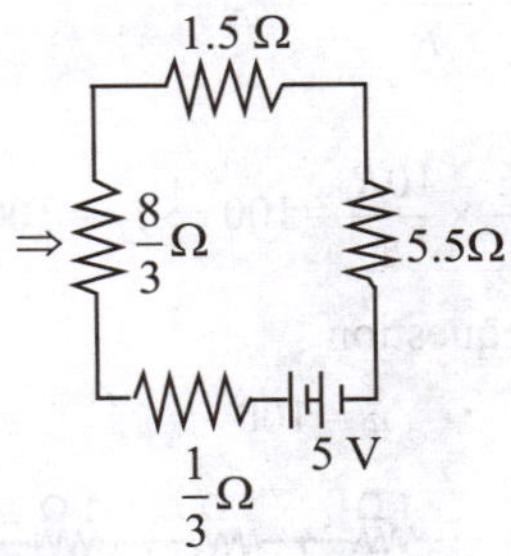

$\therefore$ The current passing through the battery is

$i = \frac{V}{R_{eq}} = \frac{5}{\frac{1}{3}+\frac{8}{3}+1.5+5.5} = \frac{5}{10} = 0.5$A

60. **(c)** Resistance of each part $= \frac{R}{8}$

$R_{eq} = \frac{1}{4}\left(\frac{R}{8}\right) + \frac{1}{4}\left(\frac{R}{8}\right) = \frac{R}{16}$

61. **(a)** $R_0 = 10\ \Omega$, $l_2 = 2l_1 \Rightarrow n = 2$

$R_1 = n^2R_0 = 4R_0 = 40\ \Omega$

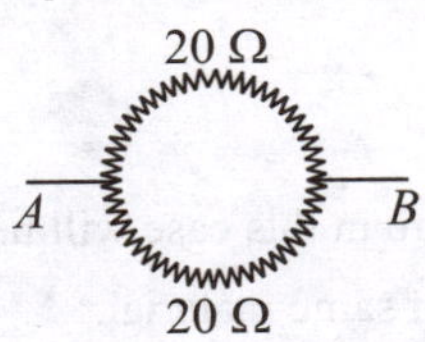

$R_{AB} = \frac{20 \times 20}{20+20} = 10\ \Omega$

62. (b) Resistance $R = \frac{\rho \ell}{A}$

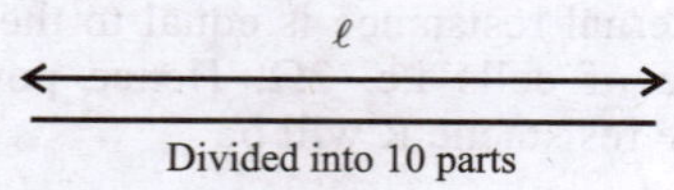

Divided into 10 parts

$\therefore$ resistance of each part = 5Ω

Resistance in series $R_s = 5 \times 10 = 50\Omega$

Resistance in parallel, $R_p = \frac{10}{5} = 2\Omega$

$\therefore \quad R_{eq} = R_S + R_P = 52\ \Omega$

NOTES A wire of resistance R is cut into n equal part and all parts are connected in parallel then $R_{eq} = R/n^2$

63. (b) For series combination,

Equivalent resistance, $R_{eq} = 10R$

$i_1 = \frac{E}{10R}$

For parallel combination,

Equivalent resistance, $R_{eq} = \frac{R}{10}$

$i_2 = \frac{E}{\frac{R}{10}} = \frac{10E}{R}$

$\therefore \quad \frac{i_2}{i_1} = \frac{10E}{R} \times \frac{10R}{E} = 100 \Rightarrow i_2 = 100i$

According to question

$i_2 = ni_1 \quad \therefore \quad n = 100$

64. (d)

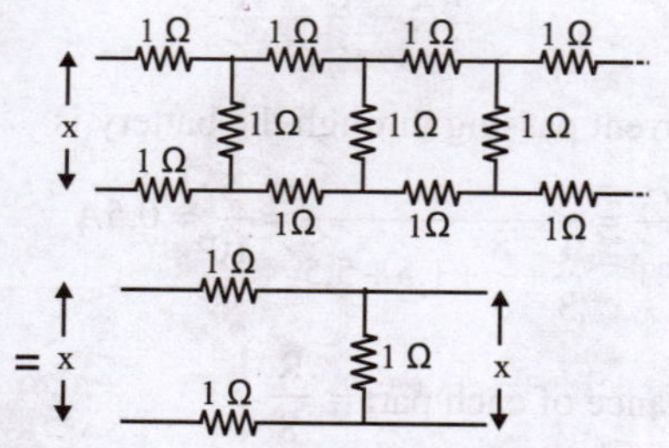

By removing repeated part of circuit,

$\Rightarrow \frac{x \times 1}{x+1} + 2 = x \Rightarrow x^2 - 2x - 2 = 0$

$\Rightarrow x = \frac{2 \pm 2\sqrt{3}}{2}$

$\therefore x = 1 + \sqrt{3}$

65. (a) Resistance of each wire in this case will be same as l & A are same and made of same material.

$\therefore R_1 = R_2 = R_3 = R_4 = R$

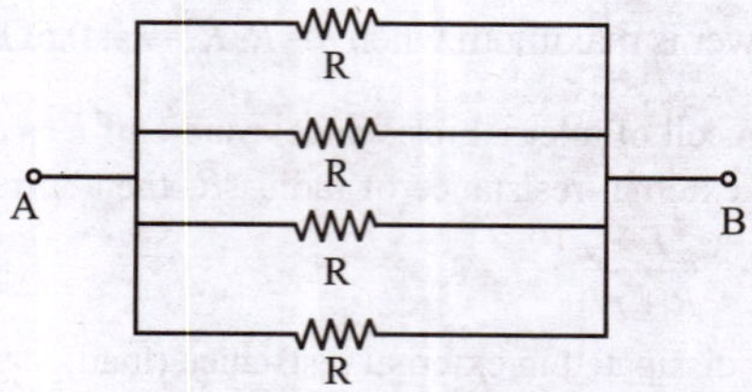

In parallel, $\frac{1}{R_P} = \frac{1}{R_1} + \frac{1}{R_2} + \frac{1}{R_3} + \frac{1}{R_4}$

$\therefore R_P = \frac{R}{4} \Rightarrow 0.25 = \frac{R}{4}$

$\therefore R = 1\ \Omega$

When arranged in series then equivalent resistance:

$R_s = R + R + R + R = 4R$

$\therefore R_s = 4 \times 1 = 4\Omega$

66. (c) Resistance r_2 and r_3 are in parallel so potential difference across $r_2 = r_3 = V$

$\therefore i_2 r_2 = i_3 r_3 \Rightarrow i_2 = \frac{i_3 r_3}{r_2}$ ... (i)

Current $i_1 = i_2 + i_3 \Rightarrow i_1 = \left(\frac{r_3}{r_2} + 1\right) i_3$

$\therefore \frac{i_3}{i_1} = \frac{r_2}{r_2 + r_3}$

67. (a) In series grouping equivalent resistance $R_{series} = nR$

In parallel grouping equivalent resistance $R_{parallel} = \frac{R}{n}$

$I = \frac{E}{nR + R}$ ...(i)

$10\,I = \frac{E}{\frac{R}{n} + R}$ ...(ii)

Dividing eq. (ii) by (i),

$10 = \frac{(n+1)R}{\left(\frac{1}{n} + 1\right)R}$

Solving we get, n = 10

68. (d) Effective resistance of B and C

$= \frac{R_B \cdot R_C}{R_B + R_C} = \frac{1.5R \times 3R}{1.5R + 3R} = \frac{4.5R^2}{4.5R} = R$

i.e., equal to resistance of voltmeter A.

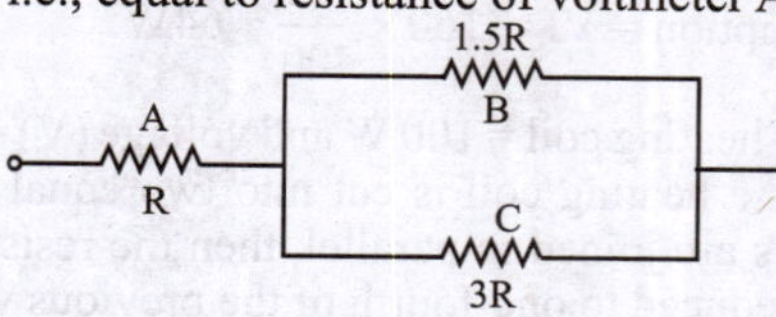

In parallel potential difference is same so, $V_B = V_C$ and in series current is same

So, $V_A = V_B = V_C$

69. (d) In figure, two metal wires of identical dimension are connected in series

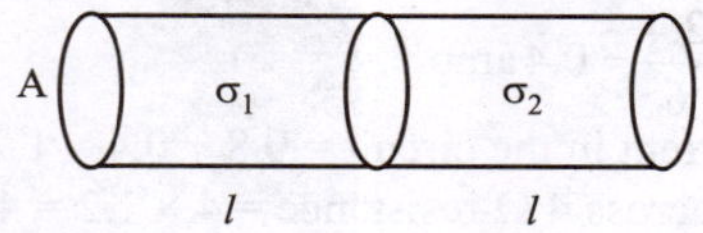

$$R_{eq} = \frac{l}{\sigma_1 A} + \frac{l}{\sigma_2 A} = \frac{l_{eq}}{\sigma_{eq} A_{eq}}$$

$$\frac{2l}{\sigma_{eq} A} = \frac{l}{A}\left(\frac{\sigma_1 + \sigma_2}{\sigma_1 \sigma_2}\right) \quad \therefore \sigma_{eq} = \frac{2\sigma_1\sigma_2}{\sigma_1 + \sigma_2}$$

70. (a) Copper rod and iron rod are joined in series.

$$\therefore R = R_{Cu} + R_{Fe} = (\rho_1 + \rho_2)\frac{\ell}{A} \qquad \left(\because R = \rho\frac{\ell}{A}\right)$$

From ohm's law $V = RI$

$= (1.7 \times 10^{-6} \times 10^{-2} + 10^{-5} \times 10^{-2}) \div 0.01 \times 10^{-4}$ volt

$= 0.117$ volt ($\because$ I = 1A)

71. (c) Resistance is directly proportional to length,

$$\frac{1}{R_{AB}} = \frac{1}{3} + \frac{1}{4+5} = \frac{(4+5)+3}{(3)(4+5)}$$

$$R_{AB} = \frac{3\times(4+5)}{3+(4+5)} = \frac{27}{12}$$

Similarly,

$$R_{BC} = \frac{4\times(3+5)}{4+(3+5)} = \frac{32}{12}$$

$$\Rightarrow R_{AC} = \frac{5\times(3+4)}{5+(3+4)} = \frac{35}{12}$$

$$\therefore R_{AB} : R_{BC} : R_{AC} = 27 : 32 : 35$$

72. (d) Let x is the resistance per unit length then

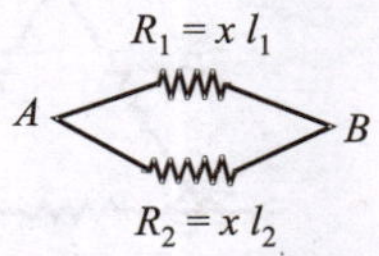

equivalent resistance $R = \dfrac{R_1 R_2}{R_1 + R_2} = \dfrac{(xl_1)(xl_2)}{xl_1 + xl_2}$

$$\Rightarrow \frac{8}{3} = x\frac{l_1 l_2}{l_1 + l_2}$$

$$\frac{8}{3} = x\frac{l_1}{\frac{l_1}{l_2}+1} \qquad \text{...(i)}$$

also $R_0 = xl_1 + x\,l_2$

$12 = x(l_1 + l_2)$

$$12 = xl_2\left(\frac{l_1}{l_2}+1\right) \qquad \text{...(ii)}$$

$$\frac{(i)}{(ii)} \Rightarrow \frac{\frac{8}{3}}{\frac{12}{1}} = \frac{\frac{xl_1}{\left(\frac{l_1}{l_2}+1\right)}}{xl_2\left(\frac{l_1}{l_2}+1\right)} = \frac{l_1}{l_2\left(\frac{l_1}{l_2}+1\right)^2}$$

$$\left(\frac{l_1}{l_2}+1\right)^2 \times \frac{8}{36} = \frac{l_1}{l_2}$$

$$(y^2 + 1 + 2y) \times \frac{8}{36} = y \text{ (where } y = \frac{l_1}{l_2})$$

$8y^2 + 8 + 16y = 36y$

$\Rightarrow \quad 8y^2 - 20y + 8 = 0 \Rightarrow 2y^2 - 5y + 2 = 0$

$\Rightarrow \quad 2y^2 - 4y - y + 2 = 0 \Rightarrow 2y(y-2) - 1(y-2) = 0$

$\Rightarrow \quad (2y-1)(y-2) = 0 \Rightarrow y = \dfrac{l_1}{l_2} = \dfrac{1}{2}$ or 2

73. (d)

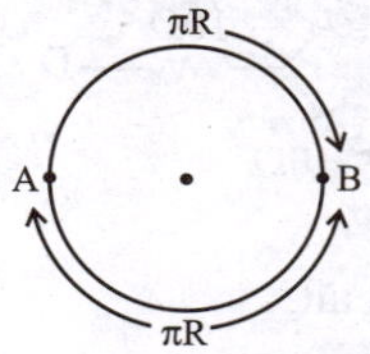

Length of wire $= 2\pi r = 2\pi(0.1) = 0.2\pi$m

Resistances of complete circle/wire $= 12 \times 0.2\pi = 2.4\pi\ \Omega$

$\therefore$ Resistance of each semi-circle $= 1.2\pi\ \Omega$

Hence, equivalent resistance,

$$R_{AB} = \frac{1.2\pi}{2} = 0.6\pi\Omega$$

74. (a)

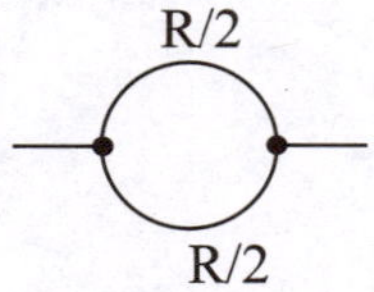

$$R_{eq} = \frac{R_1 R_2}{R_1 + R_2} = \frac{\left(\frac{R}{2}\cdot\frac{R}{2}\right)}{\frac{R}{2}+\frac{R}{2}} \quad \therefore R_{eq} = \frac{R}{4}$$

NOTES If a wire of resistance R is bent in the form of circle, the effective resistance between the ends of diameter is $R/4$

75. (b) $R = \dfrac{r}{n} \Rightarrow r = nR$

When connected in series, $R_{eq} = nr$

$= n(nR) = n^2R$

76. (b) In circuit, R_B and R_C are in series, so, $R_s = 6 + 6 = 12\ \Omega$. This 12 Ω resistance is in parallel with $R_A = 3\ \Omega$, So, equivalent resistance of circuit

$$R = \frac{3\times12}{3+12} = \frac{36}{15} = \frac{12}{5}\Omega$$

$\therefore$ Current in circuit, $I = \dfrac{V}{R}$

$$= \frac{4.8}{\left(\frac{12}{5}\right)} = \frac{4.8\times5}{12} = 2\text{A}$$

77. (a) Resistance of ACB, $R' = 3\Omega + 3\Omega = 6\Omega$.

For net resistance between A and B; $R' = 6\Omega$ and 3Ω are in parallel.

$$R_{AB} = \frac{3\times6}{3+6} = \frac{18}{9} = 2\Omega$$

Current in circuit $(I) = \dfrac{E}{R_{AB}} = \dfrac{2}{2} = 1A$

78. (c)

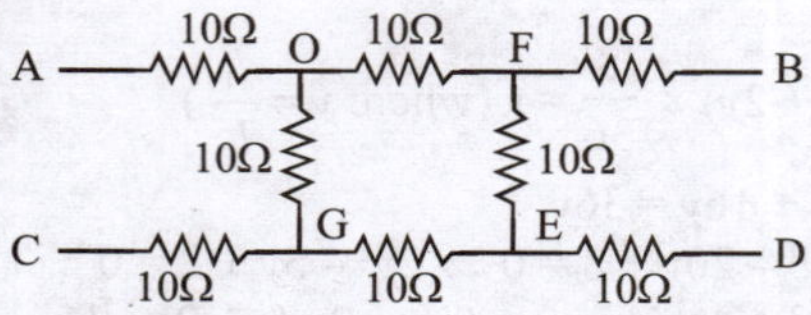

Equivalent Circuit

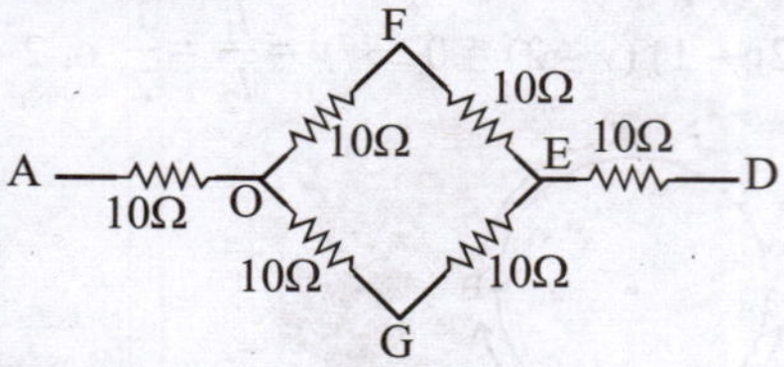

Equivalent Resistance of circuit,

$$= 10\Omega + \frac{20\times20}{20+20} + 10 = 10+10+10 = 30\Omega$$

79. (c) Length of each wire = ℓ; Area of thick wire (A_1) = 3A; Area of thin wire (A_2) = A and resistance of thick wire (R_1) = 10 Ω. Resistance $(R) = \rho\frac{\ell}{A} \propto \frac{1}{A}$ (if ℓ is constant)

$$\therefore \frac{R_1}{R_2} = \frac{A_2}{A_1} = \frac{A}{3A} = \frac{1}{3}$$

or, $R_2 = 3R_1 = 3 \times 10 = 30\ \Omega$

The equivalent resistance of these two resistors in series $= R_1 + R_2 = 30 + 10 = 40\Omega$.

80. (a) The potential difference across 4Ω resistance is given by

$V = 4 \times i_1 = 4 \times 1.2 = 4.8$ volt

So, the potential across 8Ω resistance is also 4.8 volt.

Current $i_2 = \frac{V}{8} = \frac{4.8}{8} = 0.6$ amp

Current in 2Ω resistance $i = i_1 + i_2$

$\therefore i = 1.2 + 0.6 = 1.8$ amp

Potential difference across 2Ω resistance

$V_{BC} = 1.8 \times 2 = 3.6$ volts

81. (d) The two resistances are connected in series and the resultant is connected in parallel with the third resistance.

$\therefore R' = 4\ \Omega + 4\ \Omega = 8\Omega$ and $\frac{1}{R''} = \frac{1}{8} + \frac{1}{4} = \frac{3}{8}$

or $R'' = \frac{8}{3}\Omega$

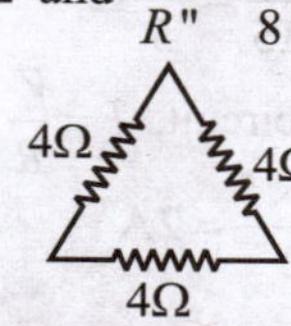

NOTES If three identical resistors each of resistance R are connected in the form of a triangle, the equivalent resistance between the ends of a side is equal to $\left(\frac{2R}{3}\right)$.

82. (c) Voltage across 3 Ω resistance = 3 × 0.8 = 2.4V. This voltage is the same across 6 Ω resistance. Hence current through this resistance

$$i = \frac{V}{R} = \frac{2.4}{6} = 0.4 \text{ amp}$$

Total current in the circuit = 0.8 + 0.4 = 1.2 amp

Voltage across 4 Ω resistance = 4 × 1.2 = 4.8 volts

83. (c) To carry a current of 4 ampere, we need four paths, each carrying a current of one ampere. Let r be the resistance of each path. These are connected in parallel. Hence, their equivalent resistance will be r/4. According to the given problem $\frac{r}{4} = 5$ or $r = 20\Omega$.

For this purpose two resistances should be connected. There are four such combinations. Hence, the total number of resistance = 4 × 2 = 8.

84. (d) At A current is distributed and at B currents are collected. Between A and B, the distribution is symmetrical. It has been shown in the figure. It appears that current in AO and OB remains same. At O, current i_4 returns back without any change. If we detach O from AB there will not be any change in distribution.

Now, CO & OD will be in series hence its total resistance = 2Ω

It is in parallel with CD, so, equivalent resistance = $\frac{2\times1}{2+1} = \frac{2}{3}\Omega$

This equivalent resistance is in series with AC & DB, so, total resistance

$$= \frac{2}{3} + 1 + 1 = \frac{8}{3}\Omega$$

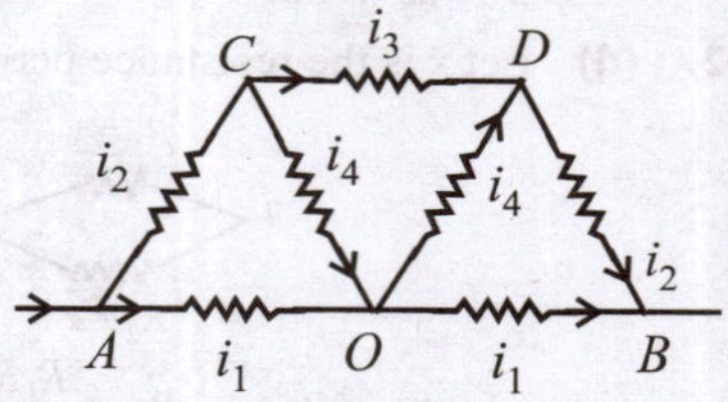

Now $\frac{8}{3}\Omega$ is parallel to AB, that is, 2Ω, so total resistance

$$= \frac{8/3\times2}{8/3+2} = \frac{16/3}{14/3} = \frac{16}{14} = \frac{8}{7}\Omega$$

NOTES

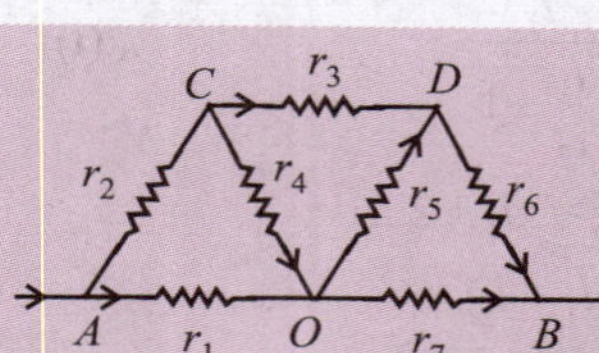

Between C & D, the equivalent resistance is given by

$$1/r = \frac{1}{r_3} + \frac{1}{(r_4+r_5)} = 1 + \frac{1}{2} = \frac{3}{2}$$

Equivalent resistance along

$$ACDB = 1 + \frac{2}{3} + 1 = \frac{8}{3}$$

∴ Effective resistance between A and B is

$$\frac{1}{R} = \frac{3}{8} + \frac{1}{2} = \frac{7}{8} \text{ or } R = \frac{8}{7}\Omega$$

85. (c) In series, $R_s = nR$

In parallel, $\frac{1}{R_p} = \frac{1}{R} + \frac{1}{R} + ...n$ terms

$\therefore R_s/R_p = n^2/1 = n^2$

86. (a) $I = \frac{8-4}{1+2+9} = \frac{4}{12} = \frac{1}{3}A$;

$V_P - V_Q = 4 - \frac{1}{3} \times 3 = 3$ volt

87. (c) Current in circuit $i = \frac{E}{R+r} = \frac{10}{4+1} = 2A$

Terminal voltage $V = E - iR$
$= 10 - 2 \times 1 = 8$ V

88. (b)

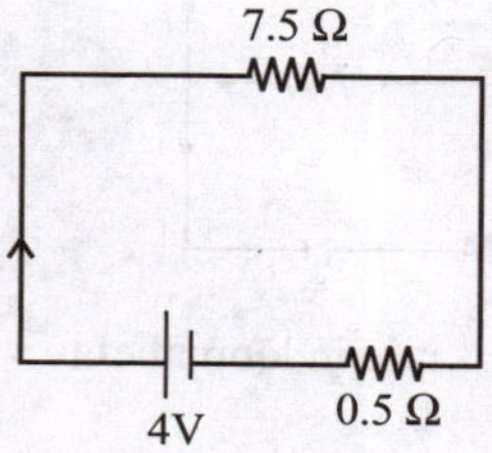

Current $I = \frac{V}{R+r} = \frac{4}{7.5+0.5} = 0.5A$

Terminal voltage $V = E - ir$
$= 4 - 0.5 \times 0.5 = 3.75$ V

NOTES When cell is charging then
$V = E + ir$

89. (a) Given : emf $\varepsilon = 2.1$ V
$I = 0.2$ A, $R = 10\Omega$
Internal resistance $r = ?$
From formula.
$\varepsilon - Ir = V = IR$
$2.1 - 0.2r = 0.2 \times 10$
$2.1 - 0.2r = 2$ or $0.2r = 0.1$
$\Rightarrow r = \frac{0.1}{0.2} = 0.5\ \Omega$

NOTES $i = \frac{\varepsilon}{r+R} \Rightarrow 0.2 = \frac{2.1}{r+10}$

$\Rightarrow 2.1 = 0.2r + 2 \Rightarrow r = \frac{1}{2} = 0.5\ \Omega$

90. (c) The current through the resistance R

$I = \left(\frac{\varepsilon}{R+r}\right)$

The potential difference across R

$V = IR = \left(\frac{\varepsilon}{R+r}\right)R$

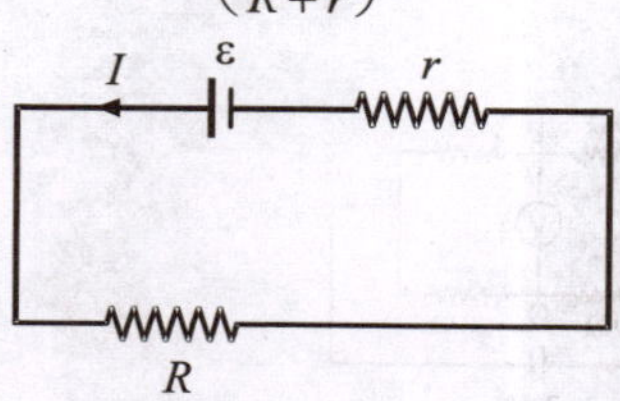

$V = \frac{\varepsilon}{\left(1+\frac{r}{R}\right)}$

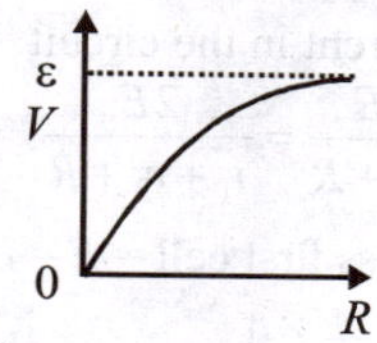

when $R = 0$, $V = 0$,
$R = \infty$, $V = \varepsilon$

Thus V increases as R increases upto certain limit, but it does not increase further.

91. (b) Let the internal resistance of the battery be r. Then the current flowing through the circuit is given by

$i = \frac{E}{R+r}$

In first case,

$2 = \frac{E}{2+r}$ ...(i)

In second case,

$0.5 = \frac{E}{9+r}$ ...(ii)

From (i) & (ii),

$4 + 2r = 4.5 + 0.5r \Rightarrow 1.5r = 0.5 \Rightarrow r = \frac{1}{3}\ \Omega$.

NOTES If I_1 be the current in a circuit with an external resistance R_1 and I_2 be the current in the circuit with external resistance R, then internal resistance of cell can be find using

$r = \frac{I_2R_2 - I_1R_1}{I_1 - I_2}$

92. (a) The terminal potential difference of a cell is given by $V + Ir = E$

$V = V_A - V_B$

or $V = E - Ir$

$\Rightarrow \frac{dV}{dI} = -r,$

Also for, $i = 0$ then $V = E$

$\therefore$ slope $= -r$, intercept $= E$

93. (b) $E = V + ir$

$2.2 = 1.8 + \frac{1.8}{5} \times r \Rightarrow r = \frac{10}{9}\Omega$

94. (a) $E = V - Ir$

$12 = V - 60 \times 5 \times 10^{-2}$

$12 = V - 3$

$\Rightarrow V = 15$ volt

95. (b) Net resistance of the circuit,

$R_{eq} = 2\ \Omega + 1\ \Omega + 7\ \Omega = 10\ \Omega$

Net Voltage, $V_{net} = 10 - 5 = 5$ V

Using Kirchhoff's law

$i = \frac{V_{net}}{R_{eq}} = \frac{5}{10} = 0.5$ A

96. **(d)** Current in the circuit

$$= \frac{E+E}{r_1+r_2+R} = \frac{2E}{r_1+r_2+R}$$

P.D. across first cell $= E - ir_1$

$$= E - \frac{2E \times r_1}{(r_1+r_2)+R}$$

Now, $E = \dfrac{2Er_1}{(r_1+r_2)+R} = 0$

$$\Rightarrow E = \frac{2Er_1}{r_1+r_2+R} \Rightarrow 2r_1 = r_1+r_2+R$$

$R = r_1 - r_2$

NOTES In series grouping of cells their emf's are additive or subtractive while their internal resistances are always additive. If dissimilar plates of cells are connected together their emf's are added to each other while if their similar plates are connected together their emf's are subtractive.

E_1 E_2

$E_{eq} = E_1 + E_2$ | E_1 E_2

$r_{eq} = r_1 + r_2$ | $E_{eq} = E_1 - E_2\ (E_1 > E_2)$

$r_{eq} = r_1 + r_2$

97. **(d)** $V = \dfrac{\dfrac{E_1}{r_1} + \dfrac{E_2}{r_2}}{\dfrac{1}{r_1} + \dfrac{1}{r_2}} = \dfrac{\dfrac{18}{2} + \dfrac{12}{1}}{\dfrac{1}{2} + \dfrac{1}{1}} = 14\,V$

(Since the cells are in parallel).

98. **(a)** Efficiency is given by $\eta = \dfrac{\text{output}}{\text{input}}$

$$= \frac{5 \times 15 \times 14}{10 \times 8 \times 15} = 0.875 \text{ or } 87.5\ \%$$

99. **(b)**

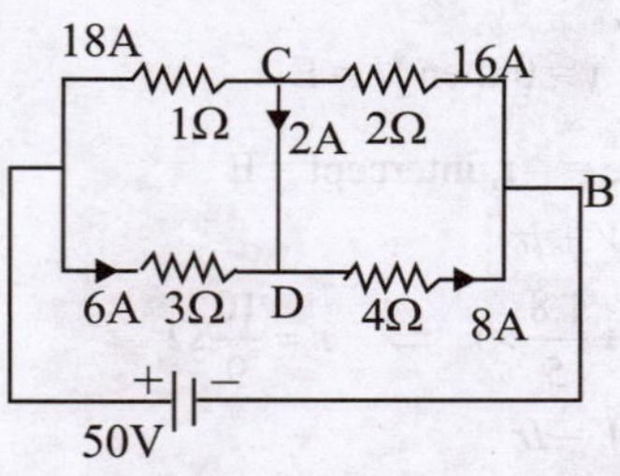

Voltage across 1Ω = 18V, voltage across 2Ω = 32V

Applying Kirchoff's current law

Current in branch CD = 18 – 16 = 2A

100. **(c)** In steady state current will not pass through capacitor,

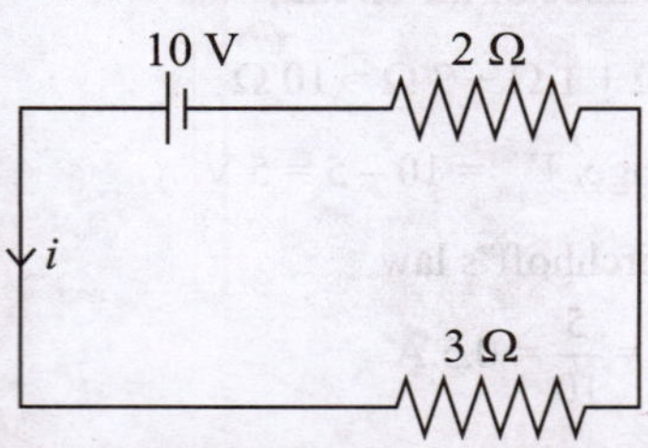

Current is given by $i = \dfrac{V}{R} \Rightarrow i = \dfrac{10}{2+3} = 2A$

101. **(d)** Current from D to C = 1A

$\therefore V_D - V_C = 2 \times 1 = 2V$

$V_A = 0 \qquad \therefore V_C = 1V \therefore V_D - V_C = 2$

$\Rightarrow V_D - 1 = 2 \qquad \therefore V_D = 3V$

$\therefore V_D - V_B = 2 \qquad \therefore 3 - V_B = 2 \therefore V_B = 1V$

102. **(d)** Junction law follows from conservation of charge and loop law is the conservation of energy.

103. **(d)**

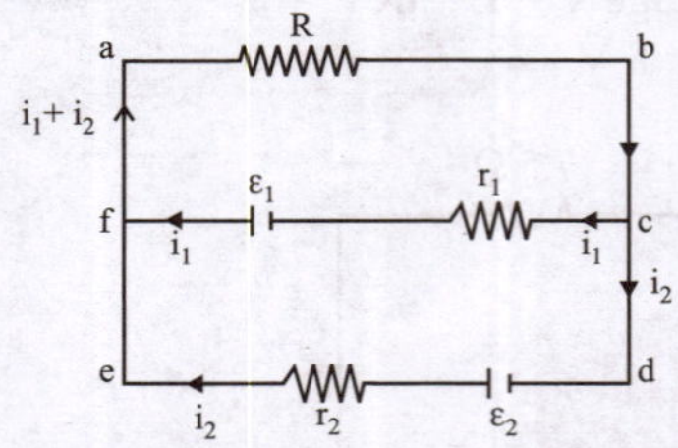

Applying Kirchhoff 's rule in loop **abcfa**

$\varepsilon_1 - (i_1 + i_2) R - i_1 r_1 = 0.$

104. **(a)** Kirchhoff's first law deals with conservation of electrical charge and the second law deals with conservation of electrical energy.

105. **(a)** We know from the Kirchhoff 's first law that the algebraic sum of the current meeting at any junction in the circuit is zero (i.e. $\Sigma i = 0$) or the total charge remains constant. Therefore, Kirchhoff's first law at a junction deals with the conservation of charge.

106. **(a)** You precise measurement, we always take P and Q approx equal and small.

107. **(c)** Resistance for ideal voltmeter = ∞ Resistance for ideal ammeter = 0

For I[st] circuit,

$$V_1 = i_1 \times 10 = \frac{10}{10} \times 10 = 10 \text{ volt}$$

For IInd circuit,

$$V_2 = i_2 \times 10 = \frac{10}{10} \times 10 = 10 \text{ volt}$$

$V_1 = V_2$ and, $i_1 = i_2 = \dfrac{10V}{10\Omega} = 1A$

108. **(a)** Current in first branch $= \dfrac{\quad}{50}$

Current in second branch $= \dfrac{2}{50} A$

ΔV from A to P

$$\Delta V_1 = 2 - \frac{2}{50} \times 20$$

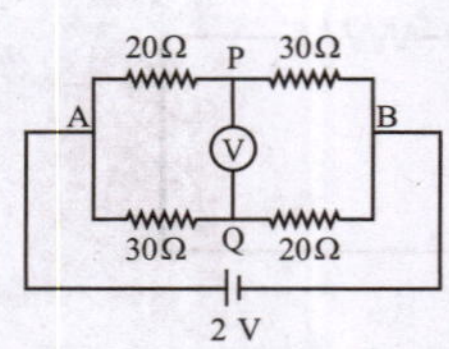

ΔV from A to Q

$$\Delta V_2 = 2 - \frac{2}{50} \times 30$$

$\Rightarrow$ Voltage difference = 0.4 V

109. (a) Given : V = 7 V; r = 5Ω

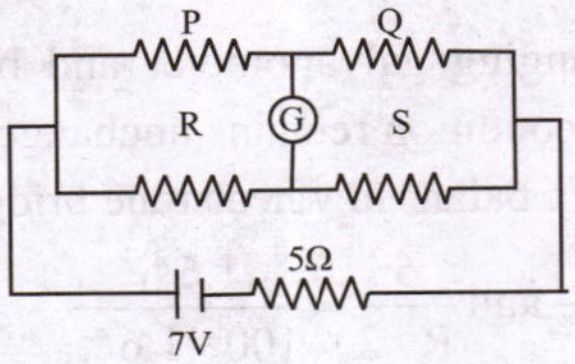

$$R_{eq} = \frac{40 \times 120}{40 + 120}\,\Omega$$

$$i = \frac{V}{R} = \frac{7}{5 + \frac{40 \times 120}{40 + 120}} = \frac{7}{5 + 30} = \frac{1}{5} = 0.2\text{ A.}$$

110. (b)

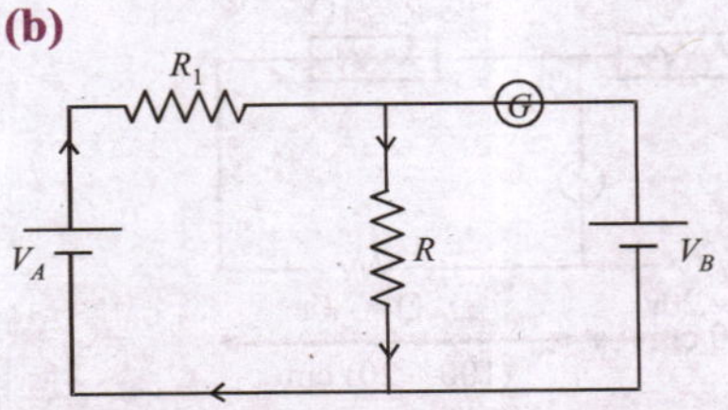

Since deflection in galvanometer is zero so current will flow as shown in the above diagram.

$$\text{current } I = \frac{V_A}{R_1 + R} = \frac{12}{500 + 100} = \frac{12}{600}$$

$$\text{So} \quad V_B = IR = \frac{12}{600} \times 100 = 2V$$

111. (a) A balanced wheatstone bridge simply requires

$$\frac{P}{Q} = \frac{R}{S} \Rightarrow \frac{2}{2} = \frac{2}{S}$$

Therefore, S should be 2Ω.

A resistance of 6Ω is connected in parallel.

In parallel combination,

$$\frac{1}{R} = \frac{1}{R_1} + \frac{1}{R_2}; \ \frac{1}{2} = \frac{1}{6} + \frac{1}{S} \quad \Rightarrow S = 3\Omega$$

> **NOTES** If the resistance is connected in parallel to the right gap resistor in the metre bridge, then balancing length increases and hence jockey moves towards right.

112. (c) Current will flow from *B* to *A*

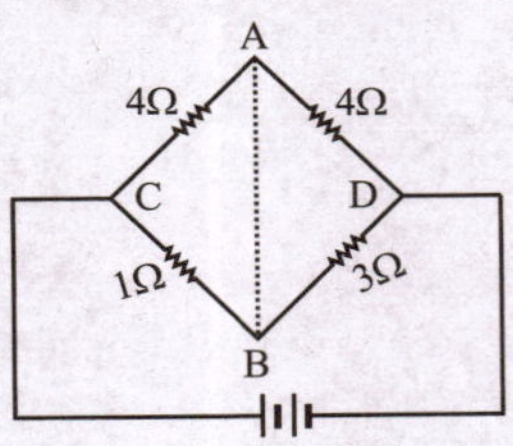

Potential drop over the resistance *CA* will be more due to higher value of resistance. So potential at A will be less as compared with at B. Hence, current will flow from *B* to *A*.

113. (d) It is a balanced Wheatstone bridge. Hence resistance 4Ω can be eliminated.

$$\therefore R_{eq} = \frac{6 \times 9}{6 + 9} = \frac{18}{5} \qquad \therefore i = \frac{V}{R_{eq}} = \frac{5V}{18}$$

114. (d) A balanced Wheatstone's bridge exists between *A* & *B*.

$\therefore R_{eq} = R$

Current through circuit = V/R

Current through AFCEB = V/2R

115. (d) Since, Wheatstone's bridge is balanced, then resistance of galvanometer will be uneffective.

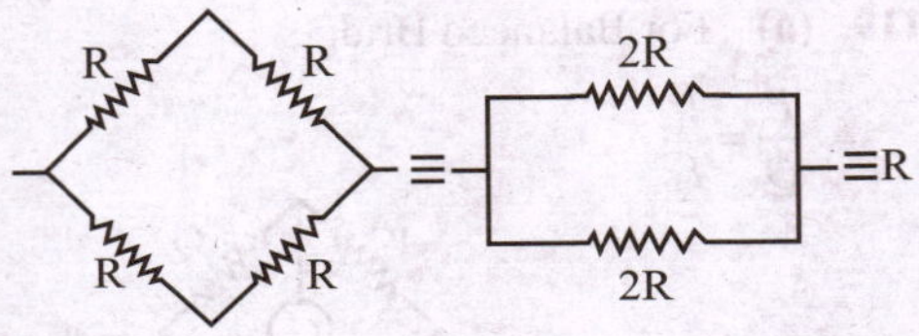

> **NOTES** Wheatstone bridge is most sensitive if all the arms of bridge have equal resistances.

116. (a) Potential gradient = Potential fall per unit length. In this case resistance of unit length.

$$R = \frac{\rho l}{A} = \frac{10^{-7} \times 1}{10^{-6}} = 10^{-1}\Omega$$

Potential fall across R is

$$V = I.R = 0.1 \times 10^{-1} = 0.01 \text{ volt/m.}$$

$$= 10^{-2} \text{ volt / m}$$

117. (d) From the balancing condition of meter bridge.

$$\frac{8}{\ell} = \frac{12}{40 - \ell}$$

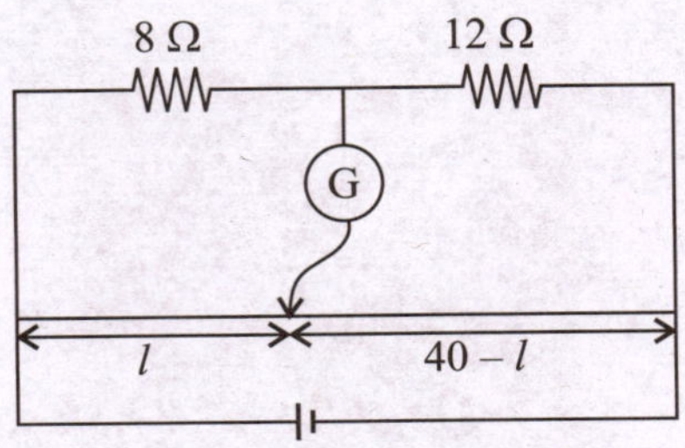

$$\frac{2}{\ell} = \frac{3}{40 - \ell}$$

$\Rightarrow 80 - 2\ell = 3\ell$

$\Rightarrow 16 = \ell$

from B

= 40 – 16 = 24 cm

118. (a) Let R_1 be the resistance of resistance wire.

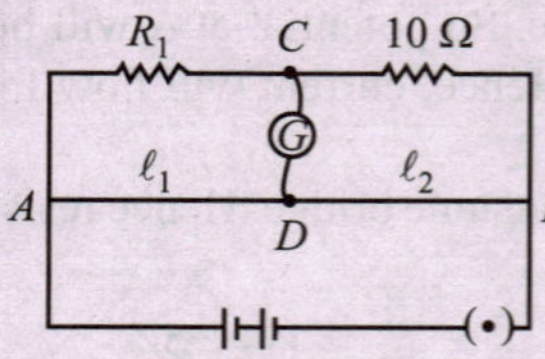

From the balancing condition of metre bridge,

$$\frac{R_1}{10} = \frac{\ell_1}{\ell_2} = \frac{3}{2} \Rightarrow R_1 = \frac{30}{2} = 15\Omega$$

Length of 15Ω resistance wire is 1.5 m.

∴ Length of 1 Ω resistance wire

$$= \frac{1.5}{15} = 0.1 \text{ m} = 1.0 \times 10^{-1} \text{ m}$$

119. (a) For Balanced Bridge

$$\frac{P}{Q} = \frac{l_1}{l_2}$$

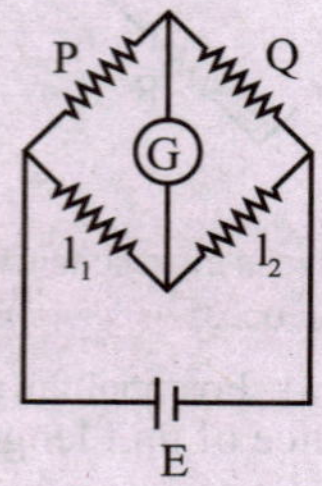

For balanced bridge (finally)

$$\frac{P}{l_1} = \frac{Q}{l_2}$$

$$\frac{P}{Q} = \frac{l_1}{l_2}$$

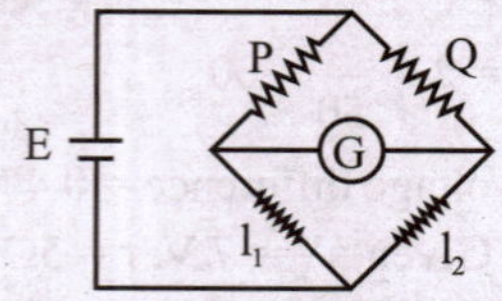

On interchanging galvanometer and battery positions, the balance condition remains unchanged.

120. (b) This is a balanced wheatstone bridge condition,

$$\frac{5}{R} = \frac{\ell_1}{100 - \ell_1} \text{ and } \frac{5}{R/2} = \frac{1.6\ell_1}{100 - 1.6\ell_1}$$

$\Rightarrow$ R = 15 Ω

121. (a) Let unknown resistance be X. Then condition of Wheatstone's bridge gives $\frac{X}{R} = \frac{20r}{80r}$, where r is resistance of wire per cm.

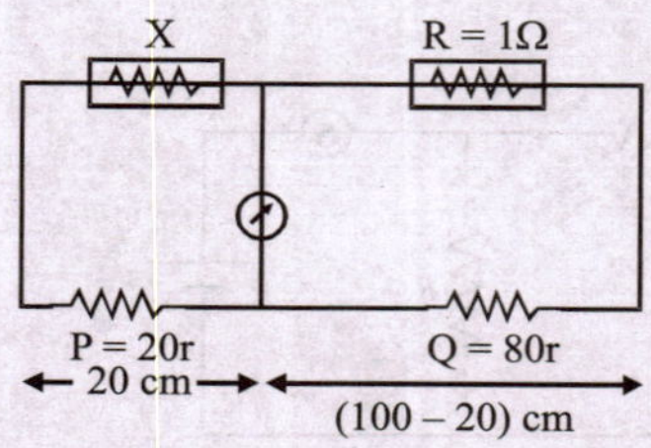

$$\therefore X = \frac{20}{80} \times R = \frac{1}{4} \times 1 = 0.25\Omega$$

4

Moving Charges and Magnetism

TREND ANALYSIS

TOPIC	2025		2024		2023		2022		2021		2020	
	Qns.	LOD	Qns.	LOD	Qns.	LOD	Qns.	LOD	Qns.	LOD	Qns.	LOD
Magnetic Force					1	Easy			1	Avg.		
Motion in Combined Electric and Magnetic Field	1	Avg.										
Magnetic Field on the Axis of a Circular Current Loop	1	Easy	1	Avg.	1	Easy	2	Avg.	1	Avg.		
Ampere's Circuital Law							2	Easy/ Avg.	2	Avg.		
The Solenoid							1	Avg.			1	Avg.
Force between Two Parallel Currents, the Ampere							1	Avg.				
Torque on Current Loop, Magnetic Dipole												
The Moving Coil Galvanometer					1	Easy						

How to Code Qns for Better & Faster Revision

Type 1: Able to solve (No code Req)

Type 2: Solved with struggle or by seeing solution **(Use code NR means Needs Revision or orange stickers)**

Type 3: Tricky/ Could not solve on my own. Need help from Teacher/ Friend. **(Use code TQ means Tricky Question or red stickers)**

You can either hand write the code in boxes or else paste coloured stickers given at end of book.

Topic-4.2

Magnetic Force

1. A long straight wire of length 2 m and mass 250 g is suspended horizontally in a uniform horizontal magnetic field of 0.7 T. The amount of current flowing through the wire will be ($g = 9.8\ ms^{-2}$) **NCERT | Page-110 | Manipur 2023**

(a) 2.75 A (b) 1.75 A
(c) 2.45 A (d) 2.25 A

2. A uniform electric field and a uniform magnetic field are acting along the same direction in a certain region. If an electron is projected in the region such that its velocity is pointed along the direction of fields, then the electron **NCERT | Page-108 | Manipur 2023**

(a) Speed will decrease
(b) Speed will increase
(c) Will turn towards right of direction of motion
(d) Will turn towards left of direction of motion

3. A wire carrying a current I along the positive x-axis has length L. It is kept in a magnetic field $\vec{B} = (2\hat{i} + 3\hat{j} - 4\hat{k})$ T. The magnitude of the magnetic force acting on the wire is **NCERT | Page-110 | 2023**

(a) $3\ IL$ (b) $\sqrt{5}\ IL$ (c) $5\ IL$ (d) $\sqrt{3}\ IL$

4. In the product $\vec{F} = q(\vec{v} \times \vec{B})$
$= q\vec{v} \times (B\hat{i} + B\hat{j} + B_0\hat{k})$
For $q = 1$ and $\vec{v} = 2\hat{i} + 4\hat{j} + 6\hat{k}$
and $\vec{F} = 4\hat{i} - 20\hat{j} + 12\hat{k}$ **NCERT | Page-109 | 2021**

What will be the complete expression for $\vec{B}$

(a) $6\hat{i} + 6\hat{j} - 8\hat{k}$ (b) $-8\hat{i} - 8\hat{j} - 6\hat{k}$
(c) $-6\hat{i} - 6\hat{j} - 8\hat{k}$ (d) $8\hat{i} + 8\hat{j} - 6\hat{k}$

5. Two identical bar magnets are fixed with their centres at a distance d apart. A stationary charge Q is placed at P in between the gap of the two magnets at a distance D from the centre O as shown in the Figure **NCERT | Page-109 | 2010**

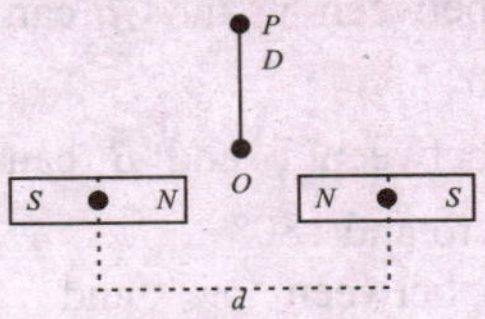

The force on the charge Q is

(a) directed perpendicular to the plane of paper
(b) zero
(c) directed along OP
(d) directed along PO

6. A metallic rod of mass per unit length 0.5 kg m^{-1} is lying horizontally on a smooth inclined plane which makes an angle of 30° with the horizontal. The rod is not allowed to slide down by flowing a current through it when a magnetic field of induction 0.25 T is acting on it in the vertical direction. The current flowing in the rod to keep it stationary is **NCERT | Page-110 | 2018**
(a) 7.14 A (b) 5.98 A
(c) 11.32 A (d) 14.76 A

7. A long straight wire carries a certain current and produces a magnetic field of $2 \times 10^{-4} \frac{\text{weber}}{m^2}$ at a perpendicular distance of 5 cm from the wire. An electron situated at 5 cm from the wire moves with a velocity 10^7 m/s towards the wire along perpendicular to it. The force experienced by the electron will be **NCERT | Page-109 | Kar. 2013**
(charge on electron $=1.6 \times 10^{-19}$ C)
(a) Zero (b) 3.2 N
(c) 3.2×10^{-16} N (d) 1.6×10^{-16} N

8. A current carrying loop in the form of a right angle isosceles triangle ABC is placed in a uniform magnetic field acting along AB. If the magnetic force on the arm BC is F, what is the force on the arm AC? **NCERT | Page-110 | 2011**

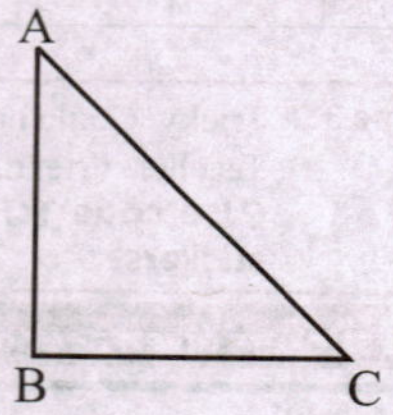

(a) $-\sqrt{2}\,\vec{F}$
(b) $-\vec{F}$
(c) $\vec{F}$
(d) $\sqrt{2}\,\vec{F}$

9. A circular coil *ABCD* carrying a current *i* is placed in a uniform magnetic field. If the magnetic force on the segment *AB* is $\vec{F}$, the force on the remaining segment *BCDA* is **Kar. 2013, 2010**

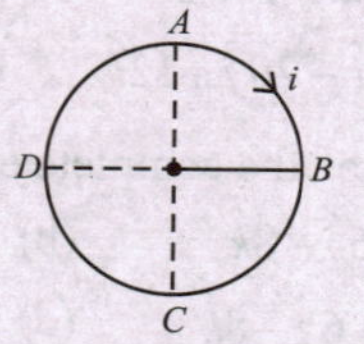

(a) $\vec{F}$
(b) $-\vec{F}$
(c) $3\vec{F}$
(d) $-3\vec{F}$

10. When a charged particle moving with velocity $\vec{v}$ is subjected to a magnetic field of induction $\vec{B}$, the force on it is non-zero. This implies that **NCERT | Page-109 | 2006**
(a) angle between $\vec{v}$ and $\vec{B}$ can have any value other than 90°
(b) angle between $\vec{v}$ and $\vec{B}$ can have any value other than zero and 180°
(c) angle between $\vec{v}$ and $\vec{B}$ is either zero or 180°
(d) angle between $\vec{v}$ and $\vec{B}$ is necessarily 90°

11. A very long straight wire carries a current I. At the instant when a charge $+Q$ at point *P* has velocity $\vec{v}$, as shown, the force on the charge is **NCERT | Page-109 | 2005**

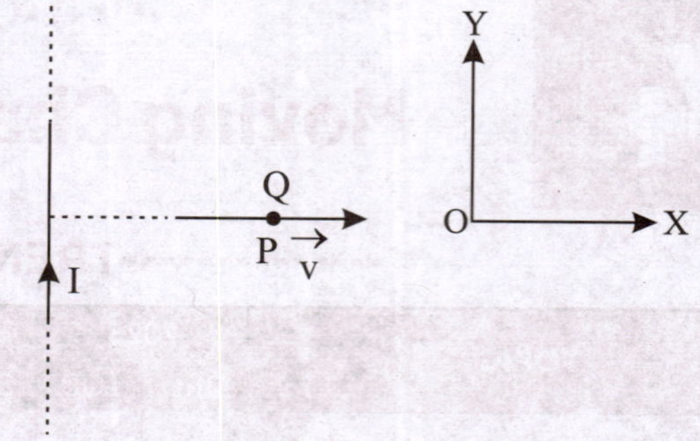

(a) along *OY* (b) opposite to *OY*
(c) along *OX* (d) opposite to *OX*

12. A particle having charge *q* moves with a velocity $\vec{v}$ through a region in which both an electric field $\vec{E}$ and a magnetic field $\vec{B}$ are present. The force on the particle is **NCERT | Page-109 | 2002**
(a) $q\vec{E}+q(\vec{B}\times\vec{v})$ (b) $q\vec{E}.(\vec{B}\times\vec{v})$
(c) $q\vec{v}+q(\vec{E}\times\vec{B})$ (d) $q\vec{E}+q(\vec{v}\times\vec{B})$

13. Two long parallel wires are at a distance of 1 metre. Both of them carry one ampere of current. The force of attraction per unit length between the two wires is **NCERT | Page-123 | 1998**
(a) 2×10^{-7} N/m (b) 2×10^{-8} N/m
(c) 5×10^{-8} N/m (d) 10^{-7} N/m

14. A straight wire of length 0.5 metre and carrying a current of 1.2 ampere is placed in uniform magnetic field of induction 2 tesla. The magnetic field is perpendicular to the length of the wire. The force on the wire is **NCERT | Page-110 | 1992**
(a) 2.4 N (b) 1.2 N (c) 3.0 N (d) 2.0 N

15. Tesla is the unit of **NCERT | Page-109 | 1988**
(a) magnetic flux (b) magnetic field
(c) magnetic induction (d) magnetic moment

Topic-4.3 Motion in a Magnetic Field

16. Ionized hydrogen atoms and α-particles with same momenta enters perpendicular to a constant magnetic field, B. The ratio of their radii of their paths $r_H : r_\alpha$ will be : **NCERT | Page-112 | 2019**
(a) 2 : 1 (b) 1 : 2 (c) 4 : 1 (d) 1 : 4

17. A proton and an alpha particle both enter a region of uniform magnetic field B, moving at right angles to field B. If the radius of circular orbits for both the particles is equal and the kinetic energy acquired by proton is 1 MeV the energy acquired by the alpha particle will be: **NCERT | Page-112 | 2015 RS**
(a) 0.5 MeV (b) 1.5 MeV
(c) 1 MeV (d) 4 MeV

18. An α-particle moves in a circular path of radius 0.83 cm in the presence of a magnetic field of 0.25 Wb/m^2. The wavelength associated with the particle will be : **2012**
(a) 1 Å (b) 0.1 Å (c) 10 Å (d) 0.01 Å

19. A proton carrying 1 MeV kinetic energy is moving in a circular path of radius R in uniform magnetic field. What should be the energy of an α-particle to describe a circle of same radius in the same field? **NCERT | Page-112 | 2012M**

(a) 2 MeV (b) 1 MeV
(c) 0.5 MeV (d) 4 MeV

20. The magnetic force acting on a charged particle of charge – 2 μC in a magnetic field of 2T acting in y direction, when the particle velocity is $\left(2\hat{i}+3\hat{j}\right)\times10^6\,ms^{-1}$, is **NCERT | Page-109 | 2009**

(a) 4 N in z direction (b) 8 N in y direction
(c) 8 N in z direction (d) 8 N in $-z$ direction

21. Under the influence of a uniform magnetic field, a charged particle moves with constant speed v in a circle of radius R. The time period of rotation of the particle **NCERT | Page-112 | 2009**

(a) depends on R and not on v
(b) is independent of both v and R
(c) depends on both v and R
(d) depends on v and not on R

22. A particle of mass m, charge Q and kinetic energy T enters a transverse uniform magnetic field of induction $\vec{B}$. After 3 seconds, the kinetic energy of the particle will be: **NCERT | Page-112 | 2008**

(a) 3T (b) 2T (c) T (d) 4T

23. In a mass spectrometer used for measuring the masses of ions, the ions are initially accelerated by an electric potential V and then made to describe semicircular path of radius R using a magnetic field B. If V and B are kept constant, the ratio $\left(\frac{\text{charge on the ion}}{\text{mass of the ion}}\right)$ will be proportional to **NCERT | Page-112 | 2007**

(a) $1/R^2$ (b) R^2 (c) R (d) 1/R

24. Under the influence of a uniform magnetic field a charged particle is moving in a circle of radius R with constant speed v. The time period of the motion **NCERT | Page-112 | 2007**

(a) depends on both R and v
(b) is independent of both R and v
(c) depends on R and not on v
(d) depends on v and not on R

25. An electron moves in a circular orbit with a uniform speed v. It produces a magnetic field B at the centre of the circle. The radius of the circle is proportional to **NCERT | Page-112 | 2005**

(a) $\sqrt{\frac{B}{v}}$ (b) $\frac{B}{v}$ (c) $\sqrt{\frac{v}{B}}$ (d) $\frac{v}{B}$

26. A charged particle moves through a magnetic field in a direction perpendicular to it. Then the **NCERT | Page-112 | 2003**

(a) velocity remains unchanged
(b) speed of the particle remains unchanged
(c) direction of the particle remains unchanged
(d) acceleration remains unchanged

27. A charged particle of charge q and mass m enters perpendicularly in a magnetic field $\vec{B}$. Kinetic energy of the particle is E; then frequency of rotation is **NCERT | Page-112 | 2001**

(a) $\frac{qB}{m\pi}$ (b) $\frac{qB}{2\pi m}$ (c) $\frac{qBE}{2\pi m}$ (d) $\frac{qB}{2\pi E}$

28. A proton moving with a velocity 3×10^5 m/s enters a magnetic field of 0.3 tesla at an angle of 30° with the field. The radius of curvature of its path will be (e/m for proton = 10^8 C/kg) **NCERT | Page-112 | 2000**

(a) 2 cm (b) 0.5 cm
(c) 0.02 cm (d) 1.25 cm

29. When a proton is accelerated through 1 V, then its kinetic energy will be **NCERT | Page-112 | 1999**

(a) 1840 eV (b) 13.6 eV
(c) 1 eV (d) 0.54 eV

30. A positively charged particle moving due east enters a region of uniform magnetic field directed vertically upwards. The particle will **NCERT | Page-112 | 1997**

(a) continue to move due east
(b) move in a circular orbit with its speed unchanged
(c) move in a circular orbit with its speed increased
(d) gets deflected vertically upwards.

31. A 10 eV electron is circulating in a plane at right angles to a uniform field at magnetic induction 10^{-4} Wb/m^2 (= 1.0 gauss). The orbital radius of the electron is **NCERT | Page-112 | 1996**

(a) 12 cm (b) 16 cm
(c) 11 cm (d) 18 cm

32. A charge moving with velocity v in X-direction is subjected to a field of magnetic induction in negative X-direction. As a result, the charge will **NCERT | Page-109 | 1993**

(a) remain unaffected
(b) start moving in a circular path Y–Z plane
(c) retard along X-axis
(d) move along a helical path around X-axis

33. A deuteron of kinetic energy 50 keV is describing a circular orbit of radius 0.5 metre in a plane perpendicular to the magnetic field B. The kinetic energy of the proton that describes a circular orbit of radius 0.5 metre in the same plane with the same B is **NCERT | Page-112 | 1991**

(a) 25 keV (b) 50 keV
(c) 200 keV (d) 100 keV

34. A uniform magnetic field acts at right angles to the direction of motion of electron. As a result, the electron moves in a circular path of radius 2cm. If the speed of electron is doubled, then the radius of the circular path will be **NCERT | Page-112 | 1991**

(a) 2.0 cm (b) 0.5 cm
(c) 4.0 cm (d) 1.0 cm

Topic-* Motion in Combined Electric and Magnetic Field

35. An electron (mass 9×10^{-31} kg and charge 1.6×10^{-19} C) moving with speed $\frac{c}{100}$ (c = speed of light) is injected into a magnetic field $\vec{B}$ of magnitude 9×10^{-4} T perpendicular to its direction of motion. We wish to apply an uniform electric field $\vec{E}$ together with the magnetic field so that the electron does not deflect from its path. Then (speed of light $c = 3 \times 10^8\ ms^{-1}$) 2025

(a) $\vec{E}$ is perpendicular to $\vec{B}$ and its magnitude is $27 \times 10^4\ Vm^{-1}$

(b) $\vec{E}$ is perpendicular to $\overrightarrow{\ }$ and its magnitude is $27 \times 10^2\ Vm^{-1}$

(c) $\vec{E}$ is parallel to $\vec{B}$ and its magnitude is $27 \times 10^2\ Vm^{-1}$

(d) $\vec{E}$ is parallel to $\vec{B}$ and its magnitude is $27 \times 10^4\ Vm^{-1}$

36. In a certain region of space electric field $\vec{E}$ and magnetic field $\vec{B}$ are perpendicular to each other and an electron enters in region perpendicular to the direction of $\vec{B}$ and $\vec{E}$ both and moves undeflected, then velocity of electron is 2001

(a) $\frac{|\vec{E}|}{|\vec{B}|}$ (b) $\vec{E} \times \vec{B}$ (c) $\frac{|\vec{B}|}{|\vec{E}|}$ (d) $\vec{E} \cdot \vec{B}$

37. A beam of electrons is moving with constant velocity in a region having simultaneous perpendicular electric and magnetic fields of strength $20\ Vm^{-1}$ and 0.5 T respectively at right angles to the direction of motion of the electrons. Then the velocity of electrons must be

NCERT | Page-109 | 1996

(a) 8 m/s (b) 20 m/s

(c) 40 m/s (d) $\frac{1}{40}$ m/s

38. An electron enters a region where magnetic field (B) and electric field (E) are mutually perpendicular, then

NCERT | Page-109 | 1994

(a) it will always move in the direction of B
(b) it will always move in the direction of E
(c) it always possesses circular motion
(d) it can go undeflected also

39. When a proton is released from rest in a room, it starts with an initial acceleration a_0 towards west. When it is projected towards north with a speed v_0 it moves with an initial acceleration $3a_0$ towards west. The electric and magnetic fields in the room are respectively

NCERT | Page-109 | 2013

(a) $\frac{ma_0}{e}$ west, $\frac{2ma_0}{ev_0}$ down

(b) $\frac{ma_0}{e}$ east, $\frac{3ma_0}{ev_0}$ up

(c) $\frac{ma_0}{e}$ east, $\frac{3ma_0}{ev_0}$ down

(d) $\frac{ma_0}{e}$ west, $\frac{2ma_0}{ev_0}$ up

40. A particle having a mass of 10^{-2} kg carries a charge of 5×10^{-8}C. The particle is given an initial horozontal velocity of $10^5\ ms^{-1}$ in the presence of electric field $\vec{E}$ and magnetic field $\vec{B}$. To keep the particle moving in a horizontal direction, it is necessary that

NCERT | Page-109 | 2010

(1) $\vec{B}$ should be perpendicular to the direction of velocity and $\vec{E}$ should be along the direction of velocity.

(2) Both $\vec{B}$ and $\vec{E}$ should be along the direction of velocity.

(3) Both $\vec{B}$ and $\vec{E}$ are mutually perpendicular and perpendicular to the direction of velocity.

(4) $\vec{B}$ should be along the direction of velocity and $\vec{E}$ should be perpendicular to the direction of velocity.

Which one of the following pairs of statements is possible?

(a) (2) and (4)
(b) (1) and (3)
(c) (3) and (4)
(d) (2) and (3)

41. A uniform electric field and uniform magnetic field are acting along the same direction in a certain region. If an electron is projected in the region such that its velocity is pointed along the direction of fields, then the electron

NCERT | Page-112 | 2011

(a) will turn towards right of direction of motion
(b) speed will decrease
(c) speed will increase
(d) will turn towards left direction of motion

42. A beam of electron passes undeflected through mutually perpendicular electric and magnetic fields. If the electric field is switched off, and the same magnetic field is maintained, the electrons move NCERT | Page-112 | 2007

(a) in a circular orbit
(b) along a parabolic path
(c) along a straight line
(d) in an elliptical orbit.

Topic-4.4 Magnetic Field due to a Current Element, Biot-Savart's Law

43. Given below are two statements:

Statement I: Biot-Savart's law gives us the expression for the magnetic field strength of an infinitesimal current element(Idl) of a current carrying conductor only.

Statement II: Biot-Savart's law is analogous to Coulomb's inverse square law of change q, with the former being related to the field produced by a scalar source, Idl while the latter being produced by a vector source, q.

In light of above statement choose the most **appropriate** answer from the options given below:

NCERT | Page-113 | 2022

(a) Both statement I and Statement II are incorrect
(b) Statement I is correct and Statement II is incorrect
(c) Statement I is incorrect and Statement II is correct
(d) Both statement I and Statement II are correct.

44. The magnetic field $(d\vec{B})$ due to a small element $(d\vec{\ell})$ at a distance $(\vec{r})$ and element carrying current i is

NCERT | Page-113 | 1996

(a) $d\vec{B} = \frac{\mu_0}{4\pi} i \left(\frac{d\vec{\ell} \times \vec{r}}{r} \right)$

(b) $d\vec{B} = \frac{\mu_0}{4\pi} i^2 \left(\frac{d\vec{\ell} \times \vec{r}}{r^2} \right)$

(c) $d\vec{B} = \frac{\mu_0}{4\pi} i^2 \left(\frac{d\vec{\ell} \times \vec{r}}{r} \right)$

(d) $d\vec{B} = \frac{\mu_0}{4\pi} i \left(\frac{d\vec{\ell} \times \vec{r}}{r^3} \right)$

Topic-4.5 Magnetic Field on the Axis of a Circular Current Loop

45. A 2 amp current is flowing through two different small circular copper coils having radii ratio 1 : 2. The ratio of their respective magnetic moments will be NCERT | Page-116 | 2025

(a) 1 : 4 (b) 1 : 2 (c) 2 : 1 (d) 4 : 1

46. A tightly wound 100 turns coil of radius 10 cm carries a current of 7A. The magnitude of the magnetic field at the centre of the coil is (Take permeability of free space as $4\pi \times 10^{-7}$ SI units): NCERT | Page-116 | 2024

(a) 44 mT (b) 4.4 T
(c) 4.4 mT (d) 44 T

47. A very long conducting wire is bent in a semi-circular shape from A to B as shown in figure. The magnetic field at point P for steady current configuration is given by NCERT | Page-116 | 2023

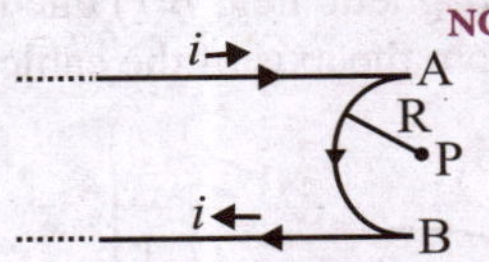

(a) $\frac{\mu_0 i}{4R}$ pointed into the page

(b) $\frac{\mu_0 i}{4R}$ pointed away from the page

(c) $\frac{\mu_0 i}{4R}\left[1 - \frac{2}{\pi}\right]$ pointed away from page

(d) $\frac{\mu_0 i}{4R}\left[1 - \frac{2}{\pi}\right]$ pointed into the page

48. The magnetic field on the axis of a circular loop of radius 100 cm carrying current $I = \sqrt{2}$A, at point 1 m away from the centre of the loop is given by: NCERT | Page-116 | RS 2022

(a) 6.28×10^{-4} T (b) 3.14×10^{-7} T
(c) 6.28×10^{-7} T (d) 3.14×10^{-4} T

49. A closely packed coil having 1000 turns has an average radius of 62.8 cm. If current carried by 62.8 cm the wire of the coil is 1 A the value of magnetic field produced at the centre of the coil will be (permeability of free space $= 4\pi \times 10^{-7}$ H/m) nearly. NCERT | Page-116 | RS 2022

(a) 10^{-3} T (b) 10^{-1} T
(c) 10^{-2} T (d) 10^{2} T

50. A uniform conducting wire of length $12a$ and resistance 'R' is wound up as a current carrying coil in the shape of, NCERT | Page-126 | 2021

(i) an equilateral triangle of side 'a'.
(ii) a square of side 'a'.

The magnetic dipole moments of the coil in each case respectively are

(a) $4 Ia^2$ and $3 Ia^2$ (b) $\sqrt{3} Ia^2$ and $3 Ia^2$
(c) $3 Ia^2$ and Ia^2 (d) $3 Ia^2$ and $4 Ia^2$

51. A straight conductor carrying current i splits into two parts as shown in the figure. The radius of the circular loop is R. The total magnetic field at the centre P of the loop is, NCERT | Page-116 | Odisha 2019

(a) $\frac{\mu_0 i}{2R}$, inward (b) Zero
(c) $3\mu_0 i/32R$, outward (d) $3\mu_0 i/32R$, inward

52. An electron moving in a circular orbit of radius r makes n rotations per second. The magnetic field produced at the centre has magnitude: NCERT | Page-116 | 2015

(a) Zero (b) $\frac{\mu_0 n^2 e}{r}$ (c) $\frac{\mu_0 ne}{2r}$ (d) $\frac{\mu_0 ne}{2\pi r}$

53. A wire carrying current I has the shape as shown in adjoining figure. Linear parts of the wire are very long and parallel to X-axis while semicircular portion of radius R is lying in Y-Z plane. Magnetic field at point O is : NCERT | Page-120 | 2015

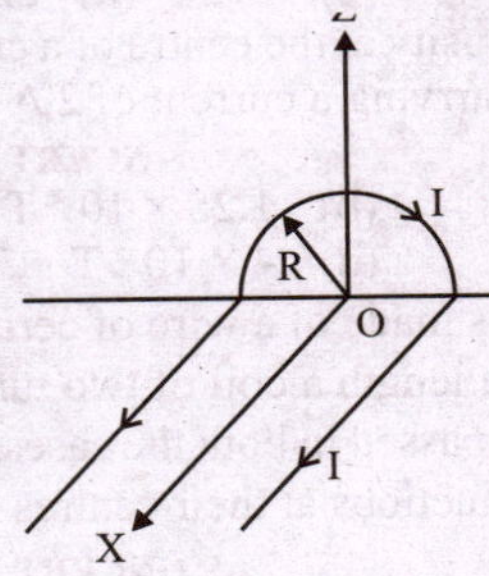

(a) $\vec{B} = -\frac{\mu_0}{4\pi}\frac{I}{R}\left(\mu\hat{i} \times 2\hat{k}\right)$

(b) $\vec{B} = -\frac{\mu_0}{4\pi}\frac{I}{R}\left(\pi\hat{i} + 2\hat{k}\right)$

(c) $\vec{B} = \frac{\mu_0}{4\pi}\frac{I}{R}\left(\pi\hat{i} - 2\hat{k}\right)$

(d) $\vec{B} = \frac{\mu_0}{4\pi}\frac{I}{R}\left(\pi\hat{i} + 2\hat{k}\right)$

54. Two similar coils of radius R are lying concentrically with their planes at right angles to each other. The currents flowing in them are I and $2\ I$, respectively. The resultant magnetic field induction at the centre will be: NCERT | Page-116 | 2012

(a) $\frac{\sqrt{5}\mu_0 I}{2R}$ (b) $\frac{3\mu_0 I}{2R}$ (c) $\frac{\mu_0 I}{2R}$ (d) $\frac{\mu_0 I}{R}$

55. Charge q is uniformly spread on a thin ring of radius R. The ring rotates about its axis with a uniform frequency f Hz. The magnitude of magnetic induction at the centre of the ring is **NCERT | Page-116 | 2011M, 2010**

(a) $\frac{\mu_0 qf}{2R}$ (b) $\frac{\mu_0 q}{2fR}$ (c) $\frac{\mu_0 q}{2\pi f R}$ (d) $\frac{\mu_0 qf}{2\pi R}$

56. A current loop consists of two identical semicircular parts each of radius R, one lying in the x-y plane and the other in x-z plane. If the current in the loop is i., the resultant magnetic field due to the two semicircular parts at their common centre is **NCERT | Page-116 | 2010**

(a) $\frac{\mu_0 i}{\sqrt{2}R}$ (b) $\frac{\mu_0 i}{2\sqrt{2}R}$ (c) $\frac{\mu_0 i}{2R}$ (d) $\frac{\mu_0 i}{4R}$

57. Two circular coils 1 and 2 are made from the same wire but the radius of the 1st coil is twice that of the 2nd coil. What potential difference in volts should be applied across them so that the magnetic field at their centres is the same **NCERT | Page-116 | 2006**

(a) 4 (b) 6 (c) 2 (d) 3

58. A wire carries a current. Maintaining the same current it is bent first to form a circular plane coil of one turn which produces a magnetic field B at the centre of the coil. The same length is now bent more sharply to give a double loop of smaller radius. The magnetic field at the centre of the double loop, caused by the same current is **NCERT | Page-116 | 2002**

(a) $4B$ (b) $B/4$ (c) $B/2$ (d) $2B$

59. Magnetic field intensity at the centre of a coil of 50 turns, radius 0.5 m and carrying a current of 2 A is **NCERT | Page-116 | 1999**

(a) 0.5×10^{-5} T (b) 1.25×10^{-4} T
(c) 3×10^{-5} T (d) 4×10^{-5} T

60. A coil of one turn is made of a wire of certain length and then from the same length a coil of two turns is made. If the same current is passed in both the cases, then the ratio of the magnetic inductions at their centres will be **NCERT | Page-116 | 1998**

(a) 2 : 1 (b) 1 : 4 (c) 4 : 1 (d) 1 : 2

61. Energy in a current carrying coil is stored in the form of **NCERT | Page-116 | 1989**

(a) electric field (b) magnetic field
(c) dielectric strength (d) heat

Topic-4.6
Ampere's Circuital Law

62. From Ampere's circuital law for a long straight wire of circular cross-section carrying a steady current, the variation of magnetic field in the inside and outside region of the wire is: **NCERT | Page-118 | 2022**

(a) a linearly increasing function of distance upto the boundary of the wire and then linearly decreasing for the outside region.
(b) a linearly increasing function of distance r upto the boundary of the wire and then decreasing one with 1/r dependence for the outside region.
(c) a linearly decreasing function of distance upto the boundary of the wire and then a linearly increasing one for the outside region.
(d) uniform and remains constant for both the regions.

63. The shape of the magnetic field lines due to an infinite long, straight current carrying conductor is : **NCERT | Page-117 | RS 2022**

(a) a plane (b) a straight line
(c) circular (d) elliptical

64. An infinitely long straight conductor carries a current of 5 A as shown. An electron is moving with a speed of 10^5 m/s parallel to the conductor. The perpendicular distance between the electron and the conductor is 20 cm at an instant. Calculate the magnitude of the force experienced by the electron at that instant. **NCERT | Page-120 | 2021**

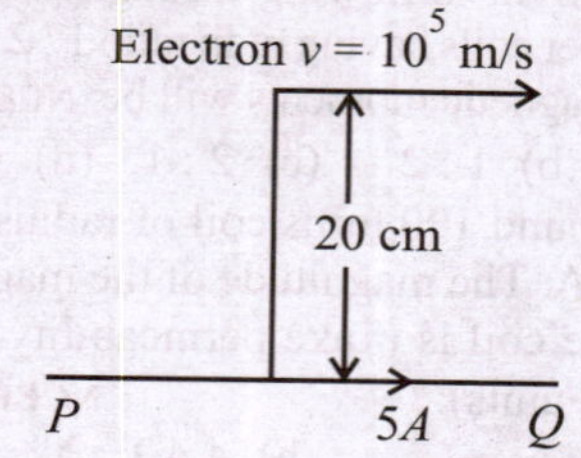

(a) 8×10^{-20} N (b) 4×10^{-20} N
(c) $8\pi \times 10^{-20}$ N (d) $4\pi \times 10^{-20}$ N

65. A thick current carrying cable of radius 'R' carries current 'I' uniformly distributed across its cross-section. The variation of magnetic field $B(r)$ due to the cable with the distance 'r' from the axis of the cable is represented by **2021**

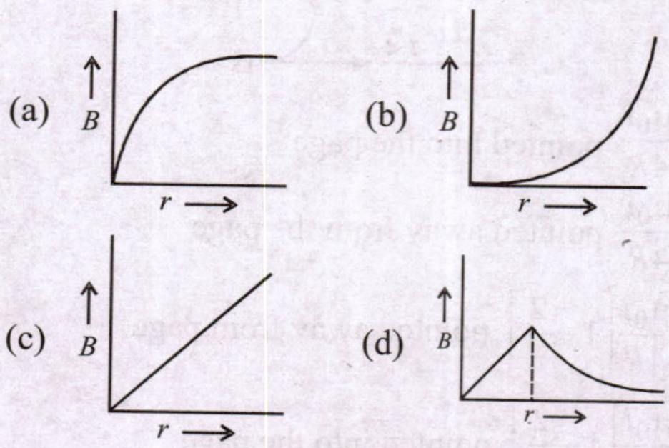

66. A cylindrical conductor of radius R is carrying a constant current. The plot of the magnitude of the magnetic field, B with the distance, d, from the centre of the conductor, is **correctly** represented by the figure : **NCERT | Page-118 | 2019**

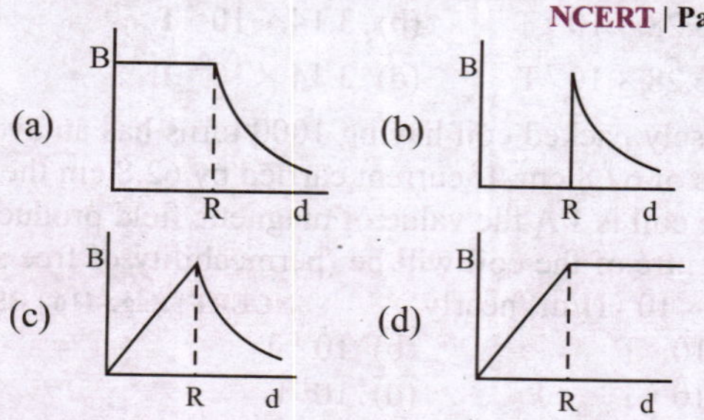

67. A long straight wire of radius a carries a steady current I. The current is uniformly distributed over its cross-section. The ratio of the magnetic fields B and B′, at radial distances $\frac{a}{2}$ and $2a$ respectively, from the axis of the wire is : **NCERT | Page-116| 2016**

(a) $\frac{1}{4}$ (b) $\frac{1}{2}$ (c) 1 (d) 4

68. Two identical long conducting wires AOB and COD are placed at right angle to each other, with one above other such that 'O' is their common point for the two. The wires carry I_1 and I_2 currents respectively. Point 'P' is lying at distance 'd' from 'O' along a direction perpendicular to the plane containing the wires. The magnetic field at the point 'P' will be : **NCERT | Page-120 | 2014**

(a) $\frac{\mu_0}{2\pi d}\left(\frac{I_1}{I_2}\right)$ (b) $\frac{\mu_0}{2\pi d}(I_1 + I_2)$

(c) $\frac{\mu_0}{2\pi d}(I_1^2 - I_2^2)$ (d) $\frac{\mu_0}{2\pi d}(I_1^2 + I_2^2)^{1/2}$

69. Two long parallel wires P and Q are both perpendicular to the plane of the paper with distance of 5 m between them. If P and Q carry currents of 2.5 amp and 5 amp respectively in the same direction, then the magnetic field at a point half-way between the wires is **NCERT | Page-116 | 2000**

(a) $\frac{3\mu_0}{2\pi}$ (b) $\frac{\mu_0}{\pi}$ (c) $\frac{\sqrt{3}\mu_0}{2\pi}$ (d) $\frac{\mu_0}{2\pi}$

70. If a long hollow copper pipe carries a current, then magnetic field is produced **NCERT | Page-118 | 1999**

(a) inside the pipe only
(b) outside the pipe only
(c) both inside and outside the pipe
(b) no where

71. Two equal electric currents are flowing perpendicular to each other as shown in the figure. *AB* and *CD* are perpendicular to each other and symmetrically placed with respect to the current flow. Where do we expect the resultant magnetic field to be zero? **1996**

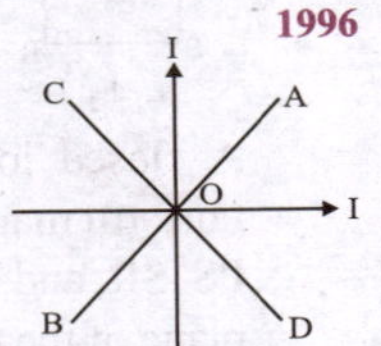

(a) on *AB*
(b) on *CD*
(c) on both *AB* and *CD*
(d) on both *OD* and *BO*

72. A straight wire of diameter 0.5 mm carrying a current of 1 A is replaced by another wire of 1 mm diameter carrying same current. The strength of magnetic field far away is **NCERT | Page-120 | 1995, 97, 99**

(a) twice the earlier value
(b) same as the earlier value
(c) one-half of the earlier value
(d) one-quarter of the earlier value

73. At what distance from a long straight wire carrying a current of 12 A will the magnetic field be equal to 3×10^{-5} Wb/m²? **NCERT | Page-120 | 1995**

(a) 8×10^{-2} m (b) 12×10^{-2} m
(c) 18×10^{-2} m (d) 24×10^{-2} m

74. The magnetic field at a distance r from a long wire carrying current i is 0.4 tesla. The magnetic field at a distance $2r$ is **NCERT | Page-120 | 1992**

(a) 0.2 tesla (b) 0.8 tesla
(c) 0.1 tesla (d) 1.6 tesla

75. The magnetic induction at a point P which is at a distance of 4 cm from a long current carrying wire is 10^{-3} T. The field of induction at a distance 12 cm from the current will be **NCERT | Page-120 | 1990**

(a) 3.33×10^{-4} T (b) 1.11×10^{-4} T
(c) 3×10^{-3} T (d) 9×10^{-3} T

Topic-4.7 The Solenoid

76. A long solenoid of radius 1 mm has 100 turns per mm. If 1A current flows in the solenoid, the magnetic field strength at the centre of the solenoid is **NCERT | Page-122 | 2022**

(a) 12.56×10^{-2} T (b) 12.56×10^{-4} T
(c) 6.28×10^{-4} T (d) 6.28×10^{-2} T

77. A long solenoid of 50 cm length having 100 turns carries a current of 2.5 A. The magnetic field at the centre of the solenoid is : **NCERT | Page-122 | 2020**

($\mu_0 = 4\pi \times 10^{-7}$ T m A^{-1})

(a) 3.14×10^{-4} T (b) 6.28×10^{-5} T
(c) 3.14×10^{-5} T (d) 6.28×10^{-4} T

78. A long solenoid carrying a current produces a magnetic field B along its axis. If the current is double and the number of turns per cm is halved, the new value of the magnetic field is **NCERT | Page-122 | 2003**

(a) $4B$ (b) $B/2$ (c) B (d) $2B$

Topic-4.8 Force between Two Parallel Currents, the Ampere

79. Two very long, straight, parallel conductors A and B carry current of 5 A and 10 A respectively and are at a distance of 10 cm from each other. The direction of current in two conductors is same. The force acting per unit length between two conductors is: **NCERT | Page-123 | RS 2022**

($\mu_0 = 4\pi \times 10^{-7}$ SI unit)

(a) 1×10^{-4} Nm^{-1} and is repulsive
(b) 2×10^{-4} Nm^{-1} and is attractive
(c) 2×10^{-4} Nm^{-1} and is repulsive
(d) 1×10^{-4} Nm^{-1} and is attractive

80. An arrangement of three parallel straight wires placed perpendicular to plane of paper carrying same current 'I along the same direction is shown in fig. Magnitude of force per unit length on the middle wire 'B' is given by **NCERT | Page-123 | 2017**

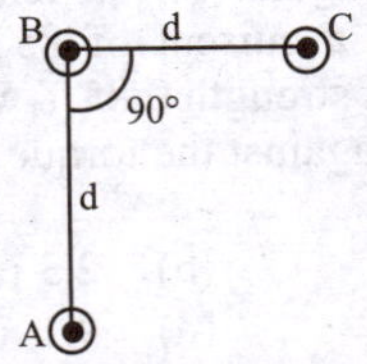

(a) $\frac{2\mu_0 i^2}{\pi d}$ (b) $\frac{\sqrt{2}\mu_0 i^2}{\pi d}$

(c) $\frac{\mu_0 i^2}{\sqrt{2}\pi d}$ (d) $\frac{\mu_0 i^2}{2\pi d}$

81. A square loop ABCD carrying a current i, is placed near and coplanar with a long straight conductor XY carrying a current I, the net force on the loop will be :

NCERT | Page-123 | 2016

(a) $\frac{2\mu_0 Ii}{3\pi}$

(b) $\frac{\mu_0 Ii}{2\pi}$

(c) $\frac{2\mu_0 IiL}{3\pi}$

(d) $\frac{\mu_0 IiL}{2\pi}$

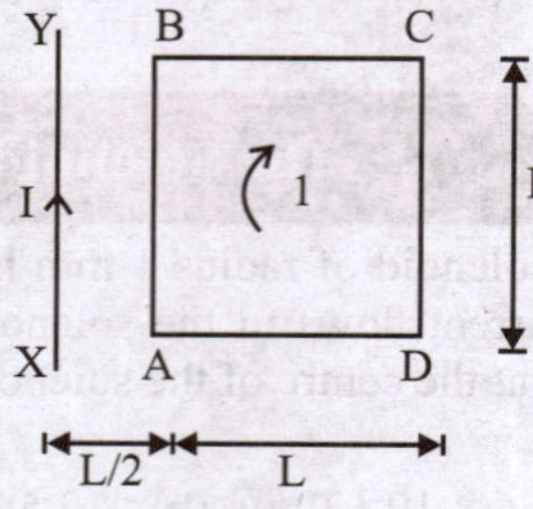

82. A square loop, carrying a steady current I, is placed in a horizontal plane near a long straight conductor carrying a steady current I_1 at a distance d from the conductor as shown in figure. The loop will experience

NCERT | Page-123 | 2011M

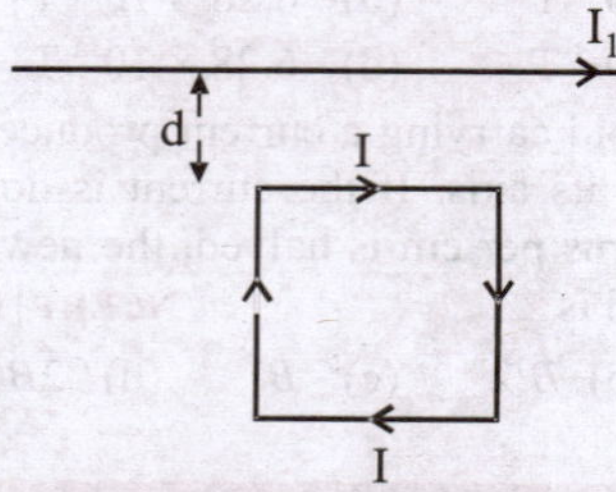

(a) a net repulsive force away from the conductor
(b) a net torque acting upward perpendicular to the horizontal plane
(c) a net torque acting downward normal to the horizontal plane
(d) a net attractive force towards the conductor

Topic-4.9 Torque on Current Loop, Magnetic Dipole

83. A current carrying coil is subjected to a uniform magnetic field. The coil will orient so that its plane becomes

NCERT | Page-125 | 1988

(a) inclined at 45° to the magnetic field
(b) inclined at any arbitrary angle to the magnetic field
(c) parallel to the magnetic field
(d) perpendicular to the magnetic field

84. A 250-turn rectangular coil of length 2.1 cm and width 1.25 cm carries a current of 85 μA and subjected to magnetic field of strength 0.85 T. Work done for rotating the coil by 180° against the torque is

NCERT | Page-126 | 2017

(a) 4.55 μJ (b) 2.3 μJ
(c) 1.15 μJ (d) 9.1 μJ

85. A rectangular coil of length 0.12 m and width 0.1 m having 50 turns of wire is suspended vertically in a uniform magnetic field of strength 0.2 weber/m². The coil carries a current of 2A. If the plane of the coil is inclined at an angle of 30° with the direction of the field, the torque required to keep the coil in stable equilibrium will be :

NCERT | Page-126 | 2015RS

(a) 0.20 Nm (b) 0.24 Nm
(c) 0.12 Nm (d) 0.15 Nm

86. A current loop in a magnetic field **NCERT | Page-126 | 2013**

(a) can be in equilibrium in one orientation
(b) can be in equilibrium in two orientations, both the equilibrium states are unstable
(c) can be in equilibrium in two orientations, one stable while the other is unstable
(d) experiences a torque whether the field is uniform or non-uniform in all orientations

87. A closely wound solenoid of 2000 turns and area of cross-section 1.5×10^{-4} m² carries a current of 2.0 A. It is suspended through its centre and perpendicular to its length, allowing it to turn in a horizontal plane in a uniform magnetic field 5×10^{-2} tesla making an angle of 30° with the axis of the solenoid. The torque on the solenoid will be **NCERT | Page-126 | 2010M**

(a) 3×10^{-3} N.m (b) 1.5×10^{-3} N.m
(c) 1.5×10^{-2} N.m (d) 3×10^{-2} N.m

88. A closely wound solenoid of 2000 turns and area of cross-section 1.5×10^{-4} m² carries a current of 2.0 *A*. It suspended through its centre and perpendicular to its length, allowing it to turn in a horizontal plane in a uniform magnetic field 5×10^{-2} tesla making an angle of 30° with the axis of the solenoid. The torque on the solenoid will be: **NCERT | Page-126 | 2010**

(a) 3×10^{-2} N-m (b) 3×10^{-3} N-m
(c) 1.5×10^{-3} N-m (d) 1.5×10^{-2} N-m

89.

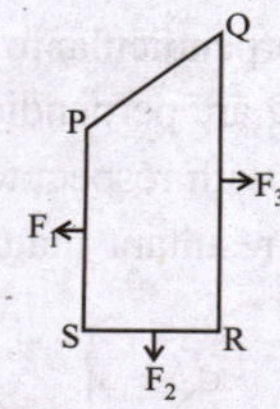

A closed loop PQRS carrying a current is placed in a uniform magnetic field. If the magnetic forces on segments PS, SR, and RQ are F_1, F_2 and F_3 respectively and are in the plane of the paper and along the directions shown, the force on the segment QP is **NCERT | Page-126 | 2008**

(a) $F_3 - F_1 - F_2$ (b) $\sqrt{(F_3 - F_1)^2 + F_2^2}$

(c) $\sqrt{(F_3 - F_1)^2 - F_2^2}$ (d) $F_3 - F_1 + F_2$

90. A coil carrying electric current is placed in uniform magnetic field, then **NCERT | Page-126 | 1993**

(a) torque is formed
(b) e.m.f is induced
(c) both (a) and (b) are correct
(d) none of the above

Topic-4.10 The Moving Coil Galvanometer

91. If the galvanometer G does not show any deflection in the circuit shown, the value of R is given by **NCERT | Page-129 | 2023**

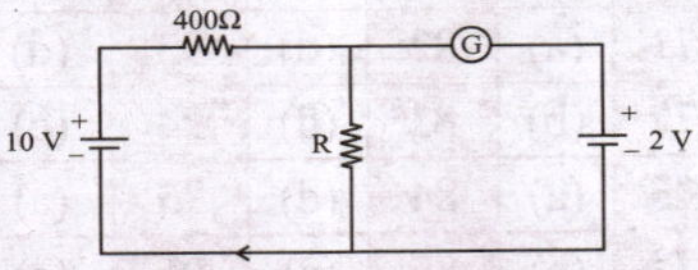

(a) 200 Ω (b) 50 Ω
(c) 100 Ω (d) 400 Ω

92. Current sensitivity of a moving coil galvanometer is 5 div/mA and its voltage sensitivity (angular deflection per unit voltage applied) is 20 div/V. The resistance of the galvanometer is **NCERT | Page-130 | 2018**

(a) 40 Ω (b) 25 Ω (c) 500 Ω (d) 250 Ω

93. A circuit contains an ammeter, a battery of 30V and a resistance 40.8Ω all connected in series. If the ammeter has a coil of resistance 480Ω and a shunt of 20Ω, the reading in the ammeter will be: **NCERT | Page-131 | 2015RS**

(a) 0.25 A (b) 2A (c) 1 A (d) 0.5 A

94. In an ammeter 0.2% of main current passes through the galvanometer. If resistance of galvanometer is G, the resistance of ammeter will be : **NCERT | Page-131 | 2014**

(a) $\frac{1}{499}G$ (b) $\frac{499}{500}G$
(c) $\frac{1}{500}G$ (d) $\frac{500}{499}G$

95. A milli voltmeter of 25 milli volt range is to be converted into an ammeter of 25 ampere range. The value (in ohm) of necessary shunt will be : **NCERT | Page-131 | 2012**

(a) 0.001 (b) 0.01 (c) 1 (d) 0.05

96. A galvanometer of resistance, G is shunted by a resistance S ohm. To keep the main current in the circuit unchanged, the resistance to be put in series with the galvanometer is **NCERT | Page-131 | 2011M**

(a) $\frac{S^2}{(S+G)}$ (b) $\frac{SG}{(S+G)}$
(c) $\frac{G^2}{(S+G)}$ (d) $\frac{G}{(S+G)}$

97. A galvanometer has a coil of resistance 100 ohm and gives a full-scale deflection for 30 *mA* current. It is to work as a voltmeter of 30 volt range, the resistance required to be added will be **NCERT | Page-131 | 2010**

(a) 900 Ω (b) 1800 Ω (c) 500 Ω (d) 1000 Ω

98. A galvanometer having a coil resistance of 60 Ω shows full scale deflection when a current of 1.0 amp passes through it. It can be converted into an ammeter to read currents upto 5.0 amp by **NCERT | Page-131 | 2009**

(a) putting in series a resistance of 15Ω
(b) putting in series a resistance of 240Ω
(c) putting in parallel a resistance of 15Ω
(d) putting in parallel a resistance of 240Ω

99. A galvanometer of resistance 50 Ω is connected to battery of 3V along with a resistance of 2950 Ω in series. A full scale deflection of 30 divisions is obtained in the galvanometer. In order to reduce this deflection to 20 divisions, the resistance in series should be **NCERT | Page-131 | 2008**

(a) 5050 Ω (b) 5550 Ω
(c) 6050 Ω (d) 4450 Ω

100. The resistance of an ammeter is 13 Ω and its scale is graduated for a current upto 100 amps. After an additional shunt has been connected to this ammeter it becomes possible to measure currents upto 750 amperes by this meter. The value of shunt-resistance is **NCERT | Page-131 | 2007**

(a) 2 Ω (b) 0.2 Ω
(c) 2 kΩ (d) 20 Ω

101. A galvanometer acting as a voltmeter will have **NCERT | Page-131 | 2004, 2002**

(a) a low resistance in series with its coil
(b) a high resistance in parallel with its coil
(c) a high resistance in series with its coil
(d) a low resistance in parallel with its coil

102. A galvanometer of 50 ohm resistance has 25 divisions. A current of 4×10^{-4} ampere gives a deflection of one per division. To convert this galvanometer into a voltmeter having a range of 25 volts, it should be connected with a resistance of **NCERT | Page-131 | 2004, 2002**

(a) 2450 Ω in series (b) 2500 Ω in series.
(c) 245 Ω in series. (d) 2550 Ω in series.

103. A galvanometer having a resistance of 8 ohms is shunted by a wire of resistance 2 ohms. If the total current is 1 amp, the part of it passing through the shunt will be **NCERT | Page-131 | 1998**

(a) 0.25 amp (b) 0.8 amp
(c) 0.2 amp (d) 0.5 amp

104. A galvanometer of resistance 20 Ω gives full scale deflection with a current of 0.004 A. To convert it into an ammeter of range 1 A, the required shunt resistance should be **NCERT | Page-131 | 1996**

(a) 0.38Ω (b) 0.21Ω
(c) 0.08Ω (d) 0.05Ω

105. To convert a galvanometer into an ammeter, one needs to connect a **NCERT | Page-131 | 1992**

(a) low resistance in parallel
(b) high resistance in parallel
(c) low resistance in series
(d) high resistance in series

ANSWER KEY

1	(b)	12	(d)	23	(a)	34	(c)	45	(a)	56	(b)	67	(c)	78	(c)	89	(b)	100	(a)
2	(a)	13	(a)	24	(b)	35	(b)	46	(c)	57	(None)	68	(d)	79	(d)	90	(a)	101	(c)
3	(c)	14	(b)	25	(c)	36	(a)	47	(c)	58	(a)	69	(d)	80	(c)	91	(c)	102	(a)
4	(c)	15	(b)	26	(b)	37	(c)	48	(b)	59	(b)	70	(b)	81	(a)	92	(d)	103	(b)
5	(b)	16	(a)	27	(b)	38	(d)	49	(a)	60	(b)	71	(a)	82	(d)	93	(d)	104	(c)
6	(c)	17	(c)	28	(b)	39	(a)	50	(b)	61	(b)	72	(b)	83	(d)	94	(c)	105	(a)
7	(c)	18	(d)	29	(c)	40	(d)	51	(b)	62	(b)	73	(a)	84	(d)	95	(a)		
8	(b)	19	(b)	30	(b)	41	(b)	52	(c)	63	(c)	74	(a)	85	(a)	96	(c)		
9	(b)	20	(d)	31	(c)	42	(a)	53	(b)	64	(a)	75	(a)	86	(c)	97	(a)		
10	(b)	21	(b)	32	(a)	43	(b)	54	(a)	65	(d)	76	(a)	87	(c)	98	(c)		
11	(a)	22	(c)	33	(d)	44	(d)	55	(a)	66	(c)	77	(d)	88	(d)	99	(d)		

Hints & Solutions

1. **(b)** From FBD of the given diagram,

F = mg or, I × ℓ × B = mg

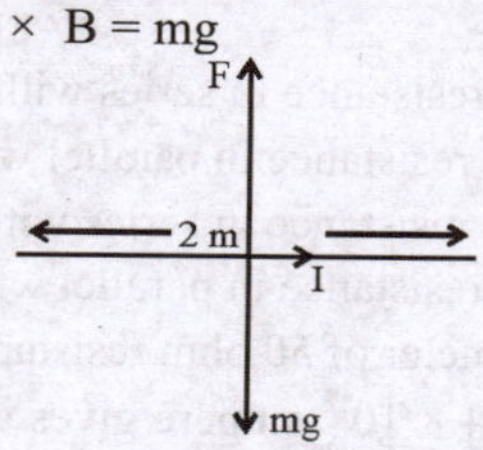

$$\Rightarrow I = \frac{mg}{\ell B} = \frac{250\times10^{-3}\times9.8}{(2)(0.7)} = 1750\times10^{-3}\text{ A}$$

$= 1.75$ A

2. **(a)** Force due to magnetic field $F_m = qvB\sin\theta$

Here $\theta = 0 \Rightarrow F_m = 0$

Force due to electric field $\vec{F}_e = q\vec{E}$

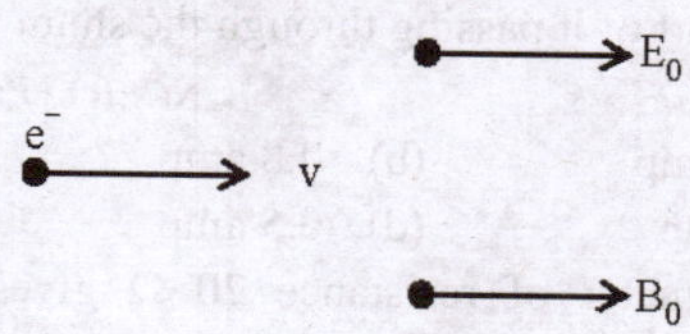

Here, $\vec{F} = -e\vec{E}$

Clearly force due to electric field is opposite to the direction of motion, so its speed will decrease.

When the value of **E** and **B** such that magnitudes of the two forces are equal. Then, total force on the charge is zero and the charge will move in the fields undeflected.

This happens when, $qE = qvB$ or $v = \dfrac{E}{B}$

3. **(c)** Given,

Magnetic Field, B = $(2\hat{i} + 3\hat{j} - 4\hat{k})T$

Magnetic force acting on a current carrying wire is

$$\vec{F} = I\vec{l}\times\vec{B}$$

$$= IL\hat{i}\times(2\hat{i}+3\hat{j}-4\hat{k}) = 3IL\hat{k} + 4IL\hat{j}$$

Magnitude of force,

$$|\vec{F}| = \sqrt{(3IL)^2 + (4IL)^2} = 5IL$$

4. **(c)** Lorentz force, $\vec{F} = q(\vec{v}\times\vec{B})$

$$= q\vec{v}\times(B\hat{i} + B\hat{j} + B_0\hat{k})$$

$\therefore q = 1, \vec{v} = 2\hat{i} + 4\hat{j} + 6\hat{k}$ and

$\vec{F} = 4\hat{i} - 20\hat{j} + 12\hat{k}$ (given)

$$\therefore (4\hat{i} - 20\hat{j} + 12\hat{k})$$

$$= -1\times\left[(2\hat{i}+4\hat{j}+6\hat{k})\times(B\hat{i}+B\hat{j}+B_0\hat{k})\right]$$

or $(4\hat{i} - 20\hat{j} + 12\hat{k})$

$$= i(4B_0 - 6B) - j(2B_0 - 6B) + \hat{k}(2B - 4B)$$

Comparing L.H.S and R.H.S, we get

$4B_0 - 6B = 4 \Rightarrow 2B_0 - 3B = 2$ ...(i)

$-(2B_0 - 6B) = -20 \Rightarrow B_0 - 3B = 10$ ...(ii)

$2B - 4B = 12 \Rightarrow B = -6$ ...(iii)

From eq. (ii) and (iii)

$B = -6$ and $B_0 = -8$

$\therefore$ Complete expression for $\vec{B} = -6\hat{i} - 6\hat{j} - 8\hat{k}$

> **NOTES** In a magnetic field a moving charged particle experiences magnetic force $\vec{F}_m = q(\vec{v}\times\vec{B})$
>
> $\vec{F}_m = q(\vec{v}\times\vec{B})$ is called Lorentz force.

5. **(b)** Force on a charged particle is given by $F = qvB$. Here $v = 0$ and also resultant B is zero.

$\therefore$ Force = 0

6. **(c)** From figure, for equilibrium,

mg sin 30° = I/B cos 30°

$\Rightarrow \quad I = \frac{mg}{\ell B} \tan 30°$

$= \frac{0.5 \times 9.8}{0.25 \times \sqrt{3}} = 11.32 \text{ A}$

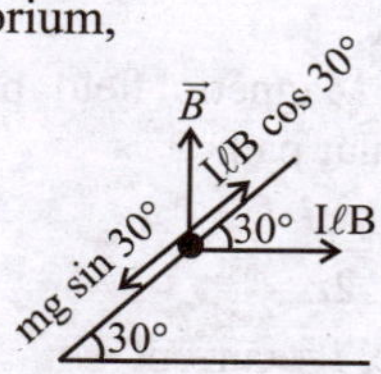

7. **(c)** Given:

Magnetic field B = 2 × 10^{-4} weber/m²

Velocity of electron, $v = 10^7$ m/s

Lorentz force $F = qvB \sin\theta$

$= 1.6 \times 10^{-19} \times 10^7 \times 2 \times 10^{-4} (\because \theta = 90°)$

$= 3.2 \times 10^{-16}$ N

8. **(b)** Let a current i be flowing in the loop ABC in the direction shown in the figure. If the length of each of the sides AB and BC be x then

$|\vec{F}| = i \times B$

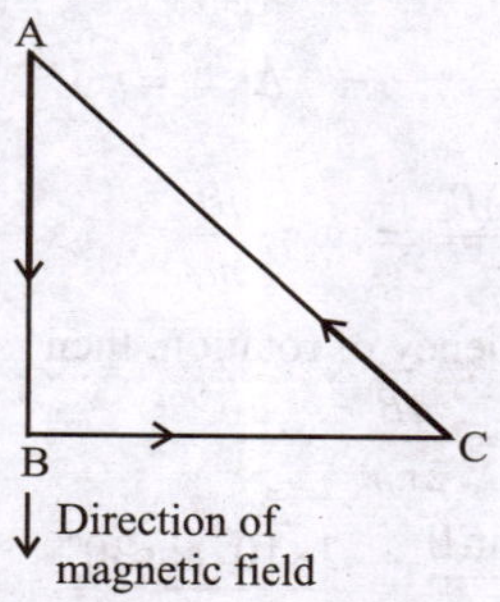

where B is the magnitude of the magnetic force.

The direction of $\vec{F}$ will be in the direction perpendicular to the plane of the paper and going into it.

By Pythagorus theorem,

$AC = \sqrt{x^2 + x^2} = \sqrt{2}x$

$\therefore$ Magnitude of force on AC

$= i\sqrt{2} \times B \sin 45° = i\sqrt{2} x B \times \frac{1}{\sqrt{2}} = ixB = |\vec{F}|$

The direction of the force on AC is perpendicular to the plane of the paper and going out of it. Hence, force on AC = $-\vec{F}$

9. **(b)** Here, $\vec{F}_{AB} + \vec{F}_{BCDA} = \vec{0}$

$\Rightarrow \vec{F}_{BCDA} = -\vec{F}_{AB} = -\vec{F} \quad (\because F_{AB} = \vec{F})$

10. **(b)** Force on a particle moving with velocity v in a magnetic field B is $F = q(\vec{v} \times \vec{B})$

If angle between $\vec{v}$ & $\vec{B}$ is either zero or 180°, then value of F will be zero as cross product of $\vec{v}$ and $\vec{B}$ will be zero.

So option (b) is correct.

11. **(a)** The direction of $\vec{B}$ is along $(-\hat{k})$

$\therefore$ The magnetic force

$\vec{F} = Q(\vec{v} \times \vec{B}) = Q(v\hat{i}) \times B(-\hat{k}) = QvB\hat{j}$

$\Rightarrow \vec{F}$ is along OY.

12. **(d)** Force due to electric field = $q\vec{E}$

Force due to magnetic field = $q(\vec{v} \times \vec{B})$

Net force experienced = $q\vec{E} + q(\vec{v} \times \vec{B})$

13. **(a)** $F = \frac{\mu_0}{4\pi} \times \frac{2 i_1 i_2 l}{r} = 10^{-7} \times \frac{2 \times 1 \times 1 \times 1}{1}$

$= 2 \times 10^{-7}$ N/m.

[This relates to the definition of ampere]

14. **(b)** $F = Bi\ell = 2 \times 1.2 \times 0.5 = 1.2$ N

15. **(b)** Tesla is the unit of magnetic field.

16. **(a)** Radius of the path $= r = \frac{mv}{qB} = \frac{P}{qB}$

For H⁺ion, $r_H = \frac{p_H}{eB}$

For α particle $r_\alpha = \frac{p_\alpha}{2eB}$

$\frac{r_H}{r_\alpha} = \frac{\frac{p}{eB}}{\frac{p}{2eB}}$ [as given $p_H = p_\alpha = p$]

$\Rightarrow \frac{r_H}{r_\alpha} = \frac{2}{1}$

17. **(c)** As we know, $F = qvB = \frac{mv^2}{R}$

$\therefore R = \frac{mv}{qB} = \frac{\sqrt{2m(KE)}}{qB}$

Since R is same so, KE $\propto \frac{q^2}{m}$

Therefore KE of α particle

$= \frac{q^2}{m} = \frac{(2)^2}{4} = 1$ MeV

18. **(d)** Wavelength

$\lambda = \frac{h}{p} \Rightarrow \lambda = \frac{h}{mv},$

h = Plank's constant = 6.63 × 10^{-34} J.S

For circular motion = $F_c = qvB$

$\Rightarrow \frac{mv^2}{r} = qvB \quad \Rightarrow \frac{mv}{qB} = r$

$r = \frac{mv}{qB} \Rightarrow mv = qrB$

$\Rightarrow (2e)(0.83 \times 10^{-2}) \left(\frac{1}{4}\right)$

$$\lambda = \frac{6.6\times10^{-34}\times4}{2\times1.6\times10^{-19}\times0.83\times10^{-2}}$$

$\lambda = 9.93\times10^{-34+21}\approx 0.01$ Å

19. **(b)** According to the principal of circular motion in a magnetic field,

$$F_c = F_m \Rightarrow \frac{mv^2}{R} = qVB \Rightarrow R = \frac{mv}{qB} = \frac{P}{qB} = \frac{\sqrt{2m\cdot K}}{qB}$$

$$R_\alpha = \frac{\sqrt{2(4m)K'}}{2qB}$$

$$\frac{R}{R_\alpha} = \sqrt{\frac{K}{K'}}$$

but $R = R_\alpha$ (given)

Thus $K = K' = 1$ MeV

20. **(d)** The magnetic force acting on the charged paraticle is given by

$$\vec{F} = q\left(\vec{v}\times\vec{B}\right)$$

$= (-2\times10^{-6})[\{2\hat{i}+3\hat{j})\times10^6\}\times(2\hat{j})] = -4(2\hat{k}) = -8\hat{k}$

$\therefore$ Force is of 8N along negative z-axis.

21. **(b)** The time period of the charged particle is given by,

$$T = \frac{2\pi m}{qB}$$

Thus, time period is independent of both v and R.

22. **(c)** When a charged particle enters a transverse magnetic field it traverse a circular path. Its kinetic energy remains constant.

23. **(a)** In mass spectrometer, when ions are accelerated through potential V

$$\frac{1}{2}mv^2 = qV \quad \text{..........(i)}$$

As the magnetic field curves the path of the ions in a semicircular orbit

$$Bqv = \frac{mv^2}{R} \Rightarrow v = \frac{BqR}{m} \quad \text{.......... (ii)}$$

Substituting (ii) in (i)

$$\frac{1}{2}m\left[\frac{BqR}{m}\right]^2 = qV \quad \text{or} \quad \frac{q}{m} = \frac{2V}{B^2R^2}$$

Since V and B are constants,

$$\therefore \frac{q}{m} \propto \frac{1}{R^2}$$

24. **(b)**

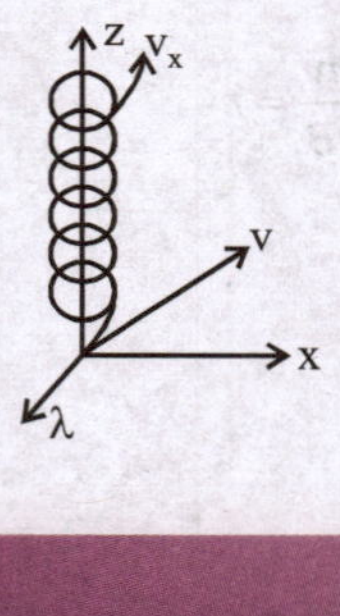

When a test charge q_0 enters a magnetic field $\vec{B}$ directed along z-axis, with a velocity $\vec{v}$ making angles d with the z-axis. The time period of the motion is independent of R and v.

25. **(c)** Magnetic field produced by moving electron in circular path

$$B = \frac{\mu_o i}{2r}$$

Here, i = current

r = radius of circular path

But $i = \frac{q}{t} = \frac{q}{2\pi r}v \quad \left[\because t = \frac{\text{Distance}}{\text{Velocity}} = \frac{2\pi r}{v}\right]$

$\therefore$ Manetic field at centre, $B = \frac{\mu_o}{2r}\times\frac{qv}{2\pi r}$

$$\Rightarrow B = \frac{\mu_o qv}{4\pi r^2} \Rightarrow r \propto \sqrt{\frac{V}{B}}$$

26. **(b)** Magnetic force acts perpendicular to the velocity. Hence speed remains constant.

27. **(b)** For circular path in magnetic field,

$mr\omega^2 = qvB$

$$\Rightarrow \omega^2 = \frac{qvB}{mr} \qquad \text{As } v = r\omega$$

$$\therefore \omega^2 = \frac{q(r\omega)B}{mr} \Rightarrow \omega = \frac{qB}{m}$$

$\therefore$ If ν is frequency of rotation, then

$$\nu = \frac{\omega}{2\pi} \Rightarrow \nu = \frac{qB}{2\pi m}$$

28. **(b)** $r = \frac{mv\sin\theta}{Be} = \frac{3\times10^5\sin30°}{0.3\times10^8}$

$$\frac{3\times10^5\times\frac{1}{2}}{3\times10^7} = 0.5\times10^{-2}m = 0.5 \text{ cm.}$$

NOTES If the change particle is moving perpendicular to the magnetic field then $qVB = \frac{mv^2}{r}$

$$\Rightarrow r = \frac{mv}{qB}$$

If the charge particle is moving at an, angle (other than 0°, 90°, 180°) to the field, then

$$q(v\sin\theta)B = \frac{m(v\sin\theta)^2}{r}$$

$$\Rightarrow r = \frac{mv\sin\theta}{qB}$$

29. **(c)** Potential difference (V) = 1V,

K.E. acquired = qV

$= 1.6\times10^{-19}\times1 = 1.6\times10^{-19}$ joules $= 1\,eV$

30. **(b)** In a perpendicular magnetic field, the path of a charged particle is a circle, and the magnetic field does not cause any change in energy.

31. (c) K.E. of electron = 10 eV

$\Rightarrow \frac{1}{2}mv^2 = 10\ eV$

$\Rightarrow \frac{1}{2}(9.1\times10^{-31})v^2 = 10\times1.6\times10^{-19}$

$\Rightarrow v^2 = \frac{2\times10\times1.6\times10^{-19}}{9.1\times10^{-31}}$

$\Rightarrow v^2 = 3.52 \times 10^{12} \Rightarrow v = 1.88 \times 10^6$ m

Also, we know that for circular motion

$\frac{mv^2}{r} = Bev \Rightarrow r = \frac{mv}{Be} = 11$ cm

32. (a) The force acting on a charged particle in magnetic field is given by

$\vec{F} = q(\vec{v}\times\vec{B})$ or $F = qvB\sin\theta$,

When angle between v and B is 180°,

$F = 0$

Force, $\vec{F} = \vec{q}(\vec{v}\times\vec{B}) = qvB\sin\theta$

When θ = 0°, 180° then particle will be moving in a direction parallel or antiparallel to the field. In such cases, the trajectory of the particle is a straight line.

33. (d) For a charged particle orbiting in a circular path in a magnetic field

$\frac{mv^2}{r} = Bqv \Rightarrow v = \frac{Bqr}{m}$

or, $mv^2 = Bqvr$

Also,

$E_K = \frac{1}{2}mv^2 = \frac{1}{2}Bqvr = Bq\frac{r}{2}\cdot\frac{Bqr}{m} = \frac{B^2q^2r^2}{2m}$

For deuteron, $E_1 = \frac{B^2q^2r^2}{2\times2m}$

For proton, $E_2 = \frac{B^2q^2r^2}{2m}$

$\frac{E_1}{E_2} = \frac{1}{2} \Rightarrow \frac{50keV}{E_2} = \frac{1}{2} \Rightarrow E_2 = 100keV$

34. (c) $r = \frac{mv}{qB}$ or $r \propto v$

As v is doubled, the radius also becomes double. Hence, radius = 2 × 2 = 4 cm

35. (b) For electron to be undeflected

$\vec{F}_e = -\vec{F}_m$

$\Rightarrow$ eE = e v B sin θ

$\therefore E = \frac{c}{100}\times9\times10^{-4}\sin90°$

= 27 × 10^2 V/m

36. (a) Electron moves undeflected if force exerted due to electric field is equal to force due to magnetic field.

$q|\vec{v}||\vec{B}| = q|\vec{E}| \Rightarrow |\vec{v}| = \frac{|\vec{E}|}{|\vec{B}|}$

37. (c) The electron moves with constant velocity without deflection. Hence, force due to magnetic field is equal and opposite to force due to electric field.

$qvB = qE \Rightarrow v = \frac{E}{B} = \frac{20}{0.5} = 40$ m/s

38. (d) When the deflection produced by electric field is equal to the deflection produced by magnetic field, then the electron can go undeflected.

39. (a)

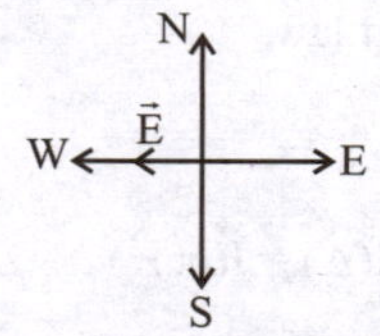

When moves with an acceleration a_0 towards west, electric field

$E = \frac{F}{q} = \frac{ma_0}{e}$ (West)

When moves with an acceleration $3a_0$ towards east, magnetic field

$B = \frac{2ma_0}{ev_0}$ (downward)

40. (d) Force due to electric field acts along the direction of the electric field but force due to the magnetic field acts along a direction perpendicular to both the velocity of the charged particle and the magnetic field. Hence both statements (2) and (3) are true. In statement (2), magnetic force is zero, so, electric force will keep the particle continue to move in horizontal direction. In statement (3), both electric and magnetic forces will be opposite to each other. If their magnitudes will be equal then the particle will continue horizontal motion.

41. (b) $\vec{v}$ and $\vec{B}$ are in same direction so that magnetic force on electron becomes zero, only electric force acts. But force on electron due to electric field is opposite to the direction of velocity.

NOTES

(i) If $\vec{v}, \vec{E}$ and $\vec{B}$ are all collinear. In this case the magnetic force acting on the electron is zero and only electric force will act. So, acceleration $\vec{a} = \frac{\vec{F}}{m}$. The electr on will undeflected in a straight line path with change in speed.

(ii) When $\vec{v}, \vec{E}$ and $\vec{B}$ are mutually perpendicular. In this case $\vec{F} = \vec{F}_e + \vec{F}_m = 0$

Here, F_e = electric force

F_m = magnetic force

Here, $\vec{a} = \frac{\vec{F}}{m} = 0$

The particle (electron) will pass with same velocity without any deviation in path.

42. (a) If the electric field is switched off, and the same magnetic field is maintained, the electrons move in a circular orbit and electron will travel a magnetic field perpendicular to its velocity.

43. (b) Coulomb's law and universal law of gravitation are analogous to each other, whereas Biot-savart's law is different from the above two laws as it is used to calculate magnetic produced by vector source, Idl. so statement II is incorrect and statement I is correct.

44. (d) According to Biot Savart law,

$$d\vec{B} = \frac{\mu_0}{4\pi}\frac{i(d\vec{\ell}\times\vec{r})}{r^3}$$

45. (a) Magnetic moment, $M = I(A) = I(\pi r^2)$

$\Rightarrow M \propto r^2$

$$\frac{M_1}{M_2} = \left(\frac{r_1}{r_2}\right)^2 = \left(\frac{1}{2}\right)^2 = \frac{1}{4}$$

46. (c) The magnitude of magnetic field due to circular coil of N turns

$$B_C = \frac{\mu_0 iN}{2R} = \frac{4\pi\times10^{-7}\times7\times100}{2\times0.1}$$

$= 4.4 \times 10^{-3}\,T = 4.4\,mT$

47. (c)

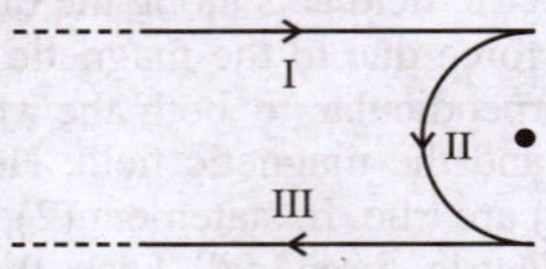

Magnetic field due to wire I $= \frac{\mu_0 i}{4\pi R}$.

Magnetic field due to wire III $= \frac{\mu_0 i}{4\pi R}$.

Magnetic field due to semiconductor arc II $= \frac{\mu_0 i}{4R}$.

Magnetic field due to wire I and II

$$= \frac{\mu_0 i}{4\pi R}\otimes + \frac{\mu_0 i}{4\pi R} = \frac{\mu_0 i}{2\pi R}\otimes$$

$$B_{total} = B_{arc} - B_{wires} = -\frac{\mu_0 i}{2\pi R} + \frac{\mu_0 i}{4R}$$

$$= \frac{\mu_0 i}{4R}\left[-\frac{2}{\pi}+1\right] = \frac{\mu_0 i}{4R}\left[1-\frac{2}{\pi}\right]$$

Pointed away from page.

48. (b) Magnetic field, on the axis of wire,

$$B = \frac{\mu_0 IR^2}{2(R^2+x^2)^{3/2}} = \frac{4\pi\times10^{-7}\times\sqrt{2}\times1^2}{2(1^2+1^2)^{3/2}}$$

$$= \frac{4\pi\times10^{-7}\times\sqrt{2}}{2\times2\sqrt{2}} = \pi\times10^{-7}\,T$$

$\therefore B = 3.14 \times 10^{-7}\,T$

NOTES If a circular loop of radius r is carrying current i, then the ratio of magnetic field at its centre and on it axis is given by

$$\frac{B_{centre}}{B_{axis}} = \left(\frac{1+x^2}{r^2}\right)^{3/2}$$ where, x = distance of point on axis from the centre of coil

49. (a) Magnetic field at the centre of the coil

$$B = \frac{\mu_0 nI}{2R}$$

$$= \frac{4\pi\times10^{-7}\times1000\times1}{2\times62.8\times10^{-2}} = 10^{-3}\,T$$

50. (b) Current in the loop $I = \frac{V}{R}$ equal for both loops.

Magnetic moment of equilateral triangle of side a

$$M_T = \left(\frac{12a}{3a}\right)\cdot I\cdot\frac{\sqrt{3}}{4}a^2 = \sqrt{3}Ia^2\;[\because M = NIA]$$

Similarly magnetic moment of square of side a

$$= \left(\frac{12a}{4a}\right)\cdot I\cdot a^2\;[\because M = NIA]$$

$M_s = 3Ia^2$

NOTES A current carrying coil behaves as a bar magnet and its magnetic moment M = NiA, N = no, of turns, i = current through the coil, A = area of the coil. For a given perimeter circular shape has maximum area hence maximum magnetic moment.

51. (b) Net magnetic field at point 'P'

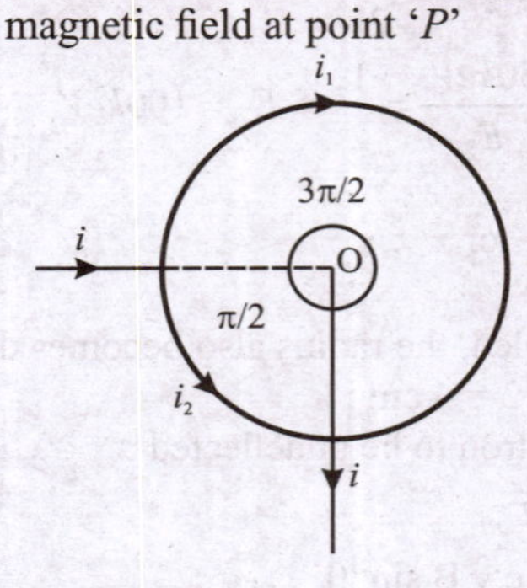

$$B_{net} = \vec{B}_1 + \vec{B}_2$$

Hence, $B_{net} = B_1 - B_2$

$$i_1 = i\left(\frac{\theta}{2\pi}\right) \Rightarrow B_1 = \frac{\mu_0 i_1}{2R}\left(\frac{2\pi-\theta}{2\pi}\right)$$

$$i_2 = i\left(\frac{2\pi-\theta}{2\pi}\right) \Rightarrow B_2 = \frac{\mu_0 i_2}{2R}\left(\frac{\theta}{2\pi}\right)$$

Here $\theta = \frac{\pi}{2}$

$B_{net} = B_1 - B_2 = 0$

52. **(c)** Radius of circular orbit $= r$

No. of rotations per second $= n$

i.e., $T = \frac{1}{n}$

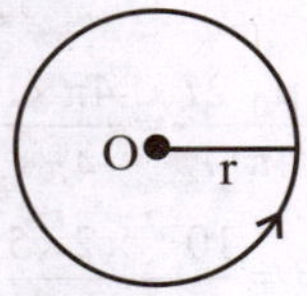

Magnetic field at its centre,

$B_c = ?$

As we know, current

$i = \frac{e}{T} = \frac{e}{(1/n)} = en =$ equivalent current

Magnetic field at the centre of circular orbit,

$$B_c = \frac{\mu_0 i}{2r} = \frac{\mu_0 ne}{2r}$$

53. **(b)** Magnetic field due to segment '1'

$$\vec{B_1} = \frac{\mu_0 I}{4\pi R}[\sin 90° + \sin 0°](-\hat{k})$$

$$= \frac{-\mu_0 I}{4\pi R}(\hat{k}) = \vec{B_3}$$

Magnetic field due to segment 2

$$B_2 = \frac{\mu_0 I}{4R}(-\hat{i}) = \frac{-\mu_0 I}{4\pi R}(\pi\hat{i})$$

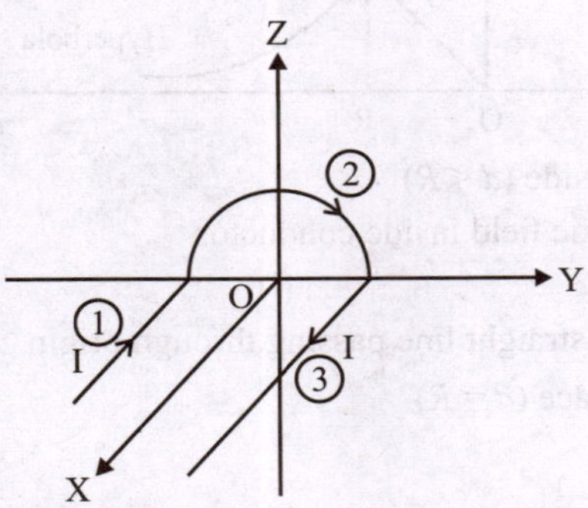

$\therefore$ $\vec{B}$ at centre

$$\vec{B}c = \vec{B}1 + \vec{B}2 + \vec{B}3 = \frac{-\mu_0 I}{4\pi R}(\pi\hat{i} + 2\hat{k})$$

54. **(a)**

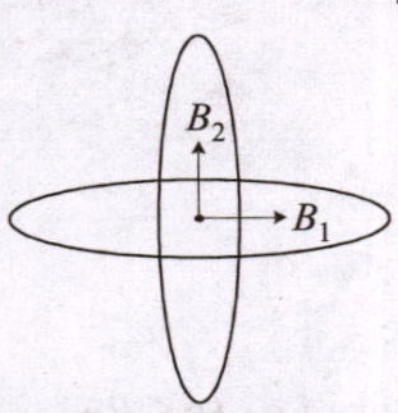

The magnetic field, due the coil, carrying current I Ampere

$$B_1 = \frac{\mu_0 I}{2R}$$

The magnetic field due to the coil, carrying current $2I$ Ampere

$$B_2 = \frac{\mu_0(2I)}{2R}$$

The resultant B

$$B_{net} = \sqrt{B_1^2 + B_2^2 + 2B_1B_2\cos\theta},\ \theta = 90°$$

$$B_{net} = \sqrt{B_1^2 + B_2^2} = \frac{\mu_0(2I)}{2R}\sqrt{1+4} = \frac{\sqrt{5}\mu_0 I}{2R}$$

55. **(a)** When the ring rotates about its axis with a uniform frequency f Hz, the current flowing in the ring is

$$I = \frac{q}{T} = qf$$

Magnetic field at the centre of the ring is

$$B = \frac{\mu_0 I}{2R} = \frac{\mu_0 qf}{2R}$$

56. **(b)** Magnetic fields due to the two parts at their common centre are respectively,

$$B_y = \frac{\mu_0 i}{4R} \text{ and } B_z = \frac{\mu_0 i}{4R}$$

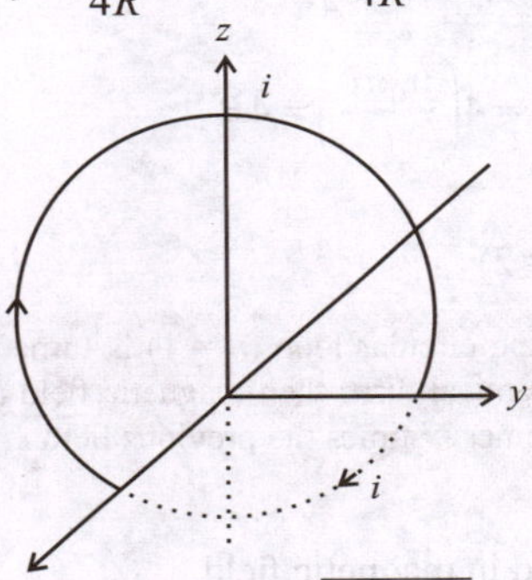

Resultant field $= \sqrt{B_y^2 + B_z^2}$

$$= \sqrt{\left(\frac{\mu_0 i}{4R}\right)^2 + \left(\frac{\mu_0 i}{4R}\right)^2} = \sqrt{2}\cdot\frac{\mu_0 i}{4R} = \frac{\mu_0 i}{2\sqrt{2}R}$$

57. **(None)** If R_1 & R_2 be the radius of the circular wires, $\frac{R_1}{R_2} = \frac{2}{1}$. If same potential is applied on them, current in Ist will be half that in the later. If V potential is applied on them, current in them $= \frac{V}{2R} \& \frac{V}{R}$.

Now magnetic field at the centre of circular coil, $= \frac{\mu_0 I}{2r}$

For first wire, field $B_1 = \frac{\mu_0 V}{2R \times 2R}$

For second wire, field $B_2 = \frac{\mu_0 V}{2(R/2) \times R}$

Given $B_1 = B_2$

The given data do not provide any required result. There is a mistake in the framing of the question.

58. **(a)** Let I be current and l be the length of the wire.

For Ist case : $B = \frac{\mu_0 In}{2r} = \frac{\mu_0 I \times \pi}{l}$

where $2\pi r = l$ and $n = 2$

For IInd Case : $l = 2(2\pi r') \Rightarrow r' = \frac{l}{4\pi}$

$$B' = \frac{\mu_0 nI}{2r'} = \frac{\mu_0 2I}{2\frac{l}{4\pi}} = \frac{4\mu_0 I\pi}{l} = 4B$$

59. (b) We know that magnetic field at the centre of circular coil,

$$B = \frac{\mu_0 In}{2r} = \frac{4\pi\times10^{-7}\times2\times50}{2\times0.5} = 1.25\times10^{-4} N$$

60. (b) Let ℓ be length of wire.

1st case : $\ell = 2\pi r \Rightarrow r = \frac{\ell}{2\pi}$

$$B = \frac{\mu_0 In}{2r} = \frac{\mu_0 I\times2\pi}{2\ell} = \frac{\mu_0\pi I}{\ell} [\because n = 1] \quad ...(1)$$

2nd Case : $\ell = 2(2\pi r') \Rightarrow r' = \frac{\ell}{4\pi}$

$$B' = \frac{\mu_0 In}{2\frac{\ell}{4\pi}} = \frac{2\mu_0 I\pi}{\frac{\ell}{2}} = 4\left(\frac{\mu_0\pi I}{\ell}\right) = 4B,$$

using (1) (where $n = 2$)

NOTES If a current carrying circular loop ($n = 1$) is turned into a coil having n identical turns then magnetic field at the centre of the coil becomes n^2 times the previous field *i.e.*, $B_{(n\ turn)} = n^2 B_{(single\ turn)}$

61. (b) Energy is stored in magnetic field.

62. (b) For long straight wire

$$B = \frac{\mu_0\, i\, r}{2\pi R^2} \Rightarrow B \propto r, r < R$$

$$B = \frac{\mu_0 I}{2\pi r} \Rightarrow B \propto \frac{1}{r}$$

63. (c) For infinite long current carrying conductor magnetic field lines are concentric circles.

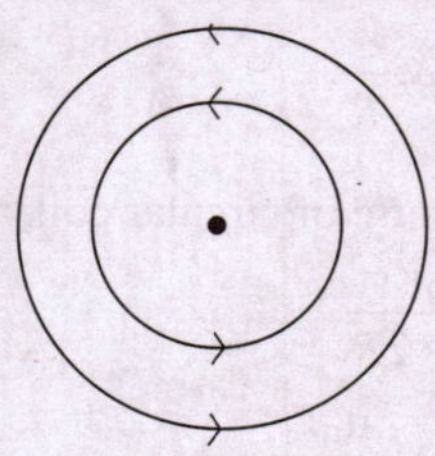

64. (a) Magnetic field due to a current carrying wire at a distance r from the wire.

or, $B = \frac{10^{-7}\times2\times5}{20\times10^{-2}} = \frac{1}{2}\times10^{-5}$

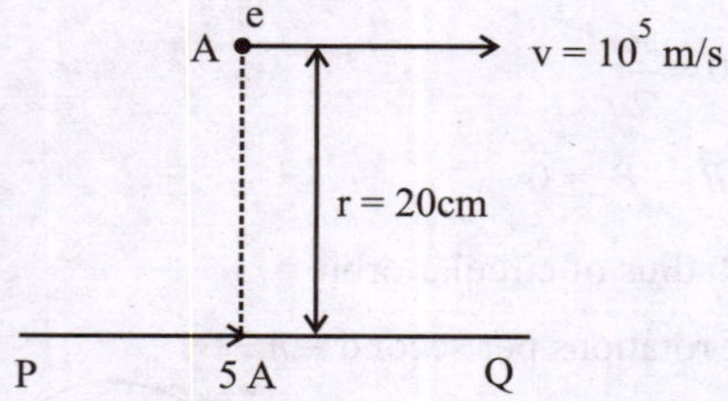

$$B = \frac{\mu_0}{4\pi}\frac{2I}{r} = \frac{4\pi\times10^{-7}\times2\times5}{4\pi\times20\times10^{-2}}$$

or, $B = \frac{10^{-7}\times2\times5}{20\times10^{-2}} = \frac{1}{2}\times10^{-5}$ T

upward to the plane of paper

Therefore, force acting on electron

$$\vec{F} = q(\vec{v}\times\vec{B})$$

$$|\vec{F}| = 1.6\times10^{-19}\times10^5\times\frac{1}{2}\times10^{-5} = 8\times10^{-20} N$$

65. (d) From Ampere's circuital law, magnetic field

$B = \frac{\mu_0 l}{2\pi R^2}\cdot r$ If $r < R \Rightarrow B_{inside} \propto r$

$B = \frac{\mu_0 l}{2\pi r}$ If $r \geq R \Rightarrow B_{outside} \propto \frac{1}{r}$

Hence the correct magnetic field B versus distance r graph is as shown below.

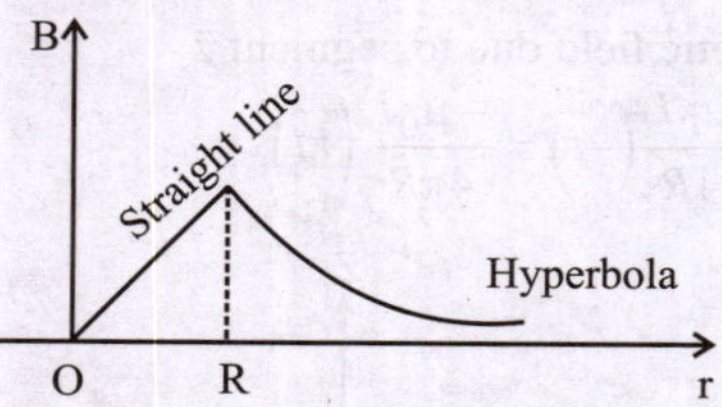

66. (c) Inside ($d < R$)

Magnetic field inside conductor

$B = Kd$ ..... (i)

This is straight line passing through origin

At surface ($d = R$)

$$B = \frac{\mu_0}{2\pi}\frac{1}{d}$$

Maximum at surface

Outside ($d > R$)

$$B = \frac{\mu_0}{2\pi}\frac{1}{d}$$

or $B \propto \frac{1}{d}$ $\therefore$ Hyperbolic

67. (c) For points inside the wire *i.e.*, ($r \leq R$)

Magnetic field $B = \frac{\mu_0 Ir}{2\pi R^2}$

For points outside the wire ($r \geq R$)

Magnetic field, $B' = \dfrac{\mu_0 I}{2\pi R}$

$$\therefore \quad \frac{B}{B'} = \frac{\dfrac{\mu_0 I(a/2)}{2\pi a^2}}{\dfrac{\mu_0 I}{2\pi(2a)}} = 1:1$$

68. (d) Net magnetic field, $B = \sqrt{B_1^2 + B_2^2}$

$$= \sqrt{\left(\frac{\mu_0 I_1}{2\pi d}\right)^2 + \left(\frac{\mu_0 I_2}{2\pi d}\right)^2}$$

$$\left(\because B_1 = \frac{\mu_0 I_1}{2\pi d} \text{ and } B_2 = \frac{\mu_0 I_2}{2\pi d}\right)$$

$$= \frac{\mu_0}{2\pi d}\sqrt{I_1^2 + I_2^2}$$

69. (d) When the current flows in both wires in the same direction then magnetic field at half way due to the wire P,

$$\vec{B}_P = \frac{\mu_0 I_1}{2\pi \frac{5}{2}} = \frac{\mu_0 I_1}{\pi \cdot 5} = \frac{\mu_0}{2\pi} \quad (\text{where } I_1 = 2.5 \text{ amp})$$

The direction of $\vec{B}_P$ is downward ⊙

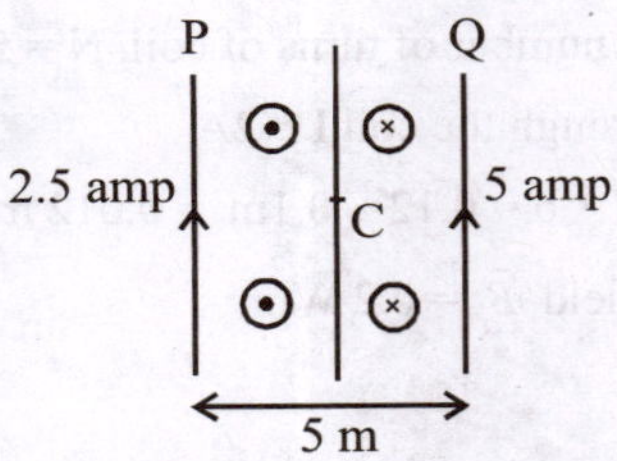

Magnetic field at half way due to wire Q

$$\vec{B}_Q = \frac{\mu_0 I_2}{2\pi \frac{5}{2}} = \frac{\mu_0}{\pi} \quad [\text{upward } \otimes]$$

[where $I_2 = 2.5$ amp.]

Net magnetic field at half way

$$\vec{B} = \vec{B}_P + \vec{B}_Q$$

$$= -\frac{\mu_0}{2\pi} + \frac{\mu_0}{\pi} = \frac{\mu_0}{2\pi} \quad (\text{upward})$$

Hence, net magnetic field at midpoint $= \dfrac{\mu_0}{2\pi}$

70. (b) Inside a hollow pipe carrying current, the magnetic field is zero, since according to Ampere's law, $B_i \cdot 2\pi r = \mu_0 \times 0 \Rightarrow B_i = 0$.

But for external points, the current behaves as if it was concentrated at the axis only; so, outside, $B_0 = \dfrac{\mu_0 i}{2\pi r}$.

Thus, the magnetic field is produced outside the pipe only.

71. (a)

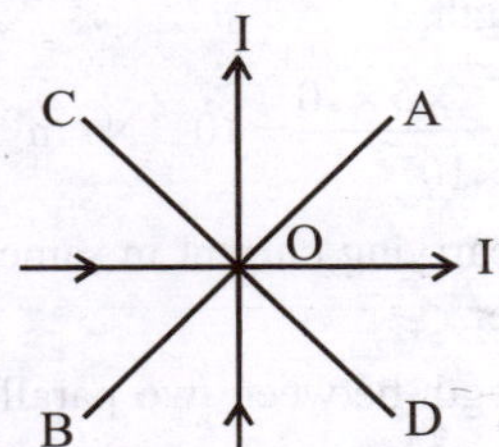

Net magnetic field on AB is zero because magnetic field due to both current carrying wires is equal in magnitude but opposite in direction.

72. (b) $B = \dfrac{\mu_0 i}{2\pi r}$ and so it is independent of thickness.

The current is same in both the wires, hence magnetic field induced will be same.

73. (a) Current (I) = 12 A and magnetic field (B) = 3×10^{-5} Wb/m². Consider magnetic field $\vec{B}$ at distance r.

Magnetic field, $B = \dfrac{\mu_0 I}{2\pi r}$

$$\Rightarrow r = \frac{\mu_0 I}{2\pi B} = \frac{(4\pi \times 10^{-7}) \times 12}{2 \times \pi \times (3 \times 10^{-5})} = 8 \times 10^{-2} \text{ m}$$

74. (a) $B = \dfrac{\mu_0 i}{2\pi r}$ or $B \propto \dfrac{1}{r}$

When r is doubled, the magnetic field becomes half, *i.e.*, now the magnetic field will be 0.2 T.

75. (a) $B = \dfrac{\mu_0 I}{2\pi r} \Rightarrow B \propto \dfrac{1}{r}$

As the distance is increased to three times, the magnetic induction reduces to one third. Hence, $B = \dfrac{1}{3} \times 10^{-3}$ tesla $= 3.33 \times 10^{-4}$ tesla

76. (a) Magnetic field $B_{centre} = \mu_0 nI$

$$= 4\pi \times 10^{-7} \times \frac{100}{10^{-3}} \times 1$$

$$= 4\pi \times 10^{-2} = 12.56 \times 10^{-2} T$$

77. (d) Magnetic field at the centre of solenoid,

$B_{solenoid} = \mu_0 ni$

Given : No. of turns / length,

$$n = \frac{N}{L} = \frac{100}{50 \times 10^{-2}} = 200 \text{ turns/ m}$$

Current, $I = 2.5$ A

$\therefore B_{solenoid} = \mu_0 nI = 4\pi \times 10^{-7} \times 200 \times 2.5 = 6.28 \times 10^{-4}$ T

78. (c) $B = \mu_0 ni$

$$B_1 = (\mu_0)\left(\frac{n}{2}\right)(2i) = \mu_0 ni = B$$

$$\Rightarrow B_1 = B$$

79. (d) Force per unit length,

$$\frac{F}{\ell}=\frac{\mu_0 i_1 i_2}{2\pi d}=\frac{2\times10^{-7}\times5\times10}{10\times10^{-2}}=10^{-4}\,N/m$$

Two parallel wires carrying current in same direction will attract each other.

80. (c) Force per unit length between two parallel current carrying conductors,

$$F=\frac{\mu_0 i_1 i_2}{2\pi d}$$

Since same current flowing through both the wires

$i_1 = i_2 = i$

so $F_1=\frac{\mu_0 i^2}{2\pi d}=F_2$

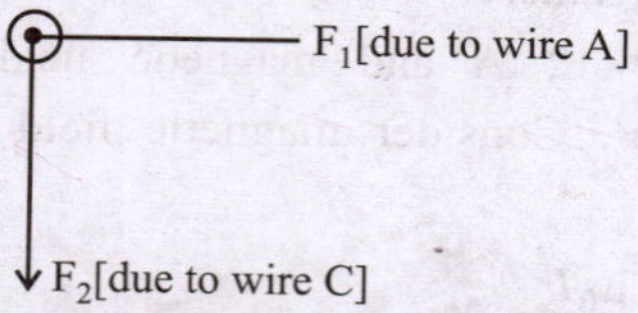

∴ Magnitude of force per unit length on the middle wire 'B'

$$F_{net}=\sqrt{F_1^2+F_2^2}=\frac{\mu_0 i^2}{\sqrt{2}\pi d}$$

81. (a) The direction of current in conductor

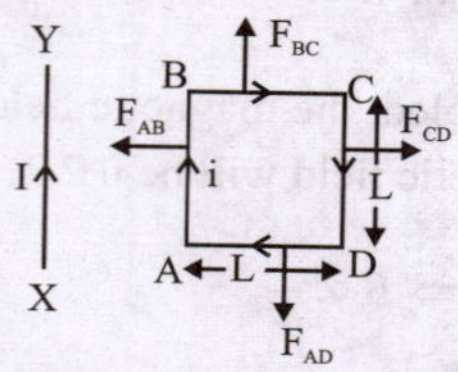

XY and AB is same

∴ $F_{AB} = i\ell B$ (attractive)

$$F_{AB}=i(L).\frac{\mu_0 I}{2\pi\left(\frac{L}{2}\right)}(\leftarrow)=\frac{\mu_0 iI}{\pi}(\leftarrow)$$

F_{BC} opposite to F_{AD}

F_{BC} (↑) and F_{AD} (↓) ⇒ cancels each other

$F_{CD} = i\ell B$ (repulsive)

$$F_{CD}=i(L)\frac{\mu_0 I}{2\pi\left(\frac{3L}{2}\right)}(\rightarrow)=\frac{\mu_0 iI}{3\pi}(\rightarrow)$$

Therefore the net force on the loop

$$F_{net}=F_{AB}+F_{BC}+F_{CD}+F_{AD}$$

$$\Rightarrow\quad F_{net}=\frac{\mu_o iI}{\pi}-\frac{\mu_o iI}{3\pi}=\frac{2\mu_o iI}{3\pi}$$

82. (d)

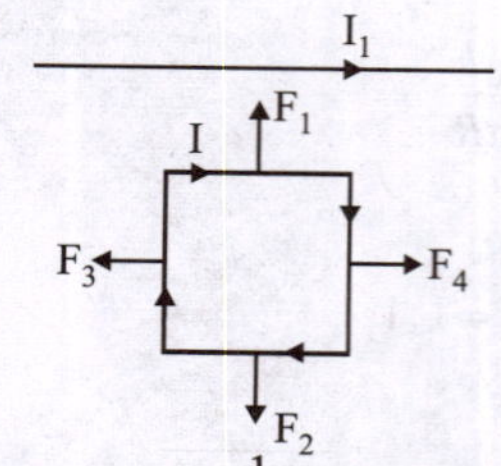

$F_1 > F_2$ as $F\propto\frac{1}{d}$, and F_3 and F_4 are equal and opposite.

Hence, the net attraction force will be towards the conductor.

83. (d) The plane of coil will orient itself so that area vector aligns itself along the magnetic field.

So, the plane will orient perpendicular to the magnetic field.

84. (d) Work done, $W = MB(\cos\theta_1 - \cos\theta_2)$

When it is rotated by angle 180° then

$W = MB(\cos 0° - \cos 180°) = MB(1 + 1)$

$W = 2MB$

$W = 2(NIA)B$

$= 2\times250\times85\times10^{-6}[1.25\times2.1\times10^{-4}]\times85\times10^{-2}$

$= 9.1\ \mu J$

85. (a) Here, number of turns of coil, N = 50

Current through the coil I = 2A

Area $A = l\times b = 0.12\times0.1 m^2 = 0.012\ m^2$

Magnetic field $\vec{B} = 0.2\ W/m^2$

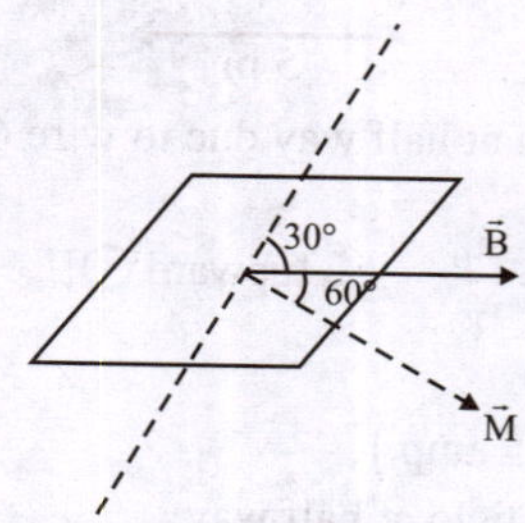

Torque required to keep the coil in stable equilibrium.

$\tau=\vec{M}\times\vec{B}=MB\sin60°=NiAB\sin60°$

$=50\times2\times0.12\times0.1\times0.2\times\frac{\sqrt{3}}{2}$

$=12\sqrt{3}\times10^{-2}=0.20784\ Nm$

> **NOTES** Torque $\tau = nIBA\sin\theta$
> Here, θ is the angle which a normal drawn on the plane of the coil makes with the direction of magnetic field.

86. (c) A current loop in a magnetic field is in equilibrium in two orientations one is stable and another unstable.

NOTES $\because \quad \vec{\tau} = \vec{M} \times \vec{B} = MB \sin\theta$

If $\theta = 0° \Rightarrow \tau = 0$ (stable)

If $\theta = \pi \Rightarrow \tau = 0$ (unstable)

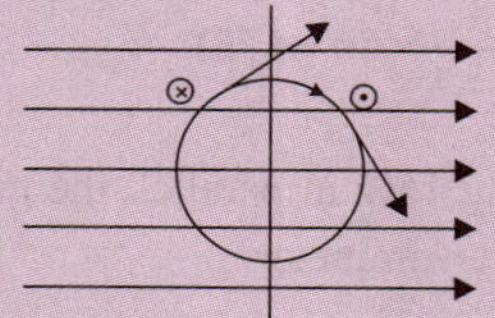

87. **(c)** Torque $\tau = MB \sin\theta$

$= NIAB \sin\theta$

$= 2000 \times 2 \times 1.5 \times 10^{-4} \times 5 \times 10^{-2} \times \frac{1}{2}$

$= 0.6 \times 5 \times 10^{-2} \times \frac{1}{2} = 1.5 \times 10^{-2}$ Nm

88. **(d)** Torque on the solenoid is given by

$\tau = MB \sin\theta$

where θ is the angle between the magnetic field and the axis of solenoid.

$M = niA$

$\therefore \quad \tau = niA B \sin 30°$

$= 2000 \times 2 \times 1.5 \times 10^{-4} \times 5 \times 10^{-2} \times \frac{1}{2}$

$= 1.5 \times 10^{-2}\ N-m$

89. **(b)** According to the figure the magnitude of force on the segment QM is $F_3 - F_1$ and PM is F_2.

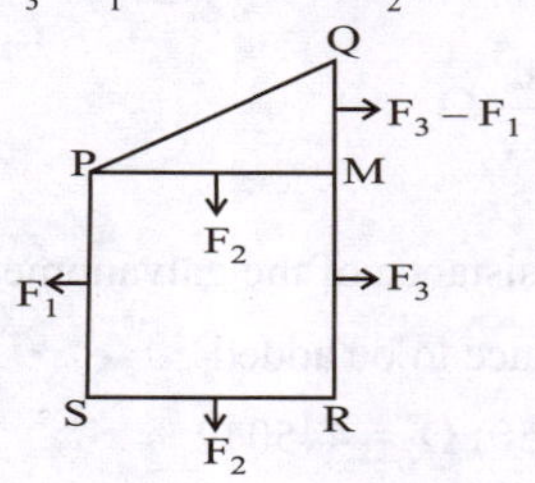

Therefore, the magnitude of the force on segment PQ is $\sqrt{(F_3 - F_1)^2 + F_2^2}$

90. **(a)** A current carrying coil has magnetic dipole moment. Hence, a torque acts on it in magnetic field.

91. **(c)** Since galvanometer does not show any deflection

$\therefore \ i_g = 0$

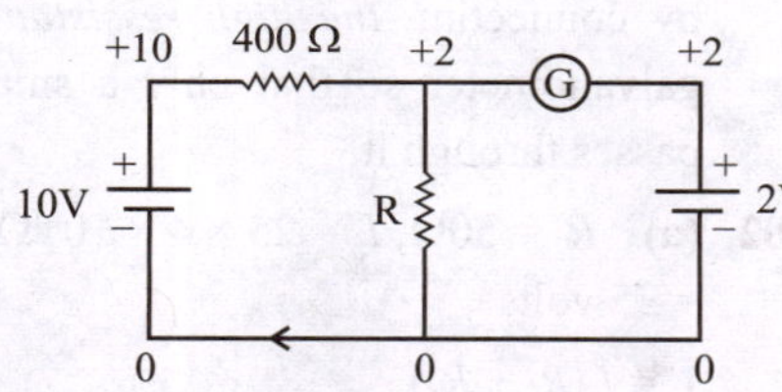

From the circuit diagram,

$$\frac{10-2}{400} = \frac{2}{R} \therefore R = \frac{2 \times 400}{8} = 100\ \Omega$$

92. **(d)** Current sensitivity of moving coil galvanometer

$$I_S = \frac{NBA}{C} \quad ...(i)$$

Voltage sensitivity of moving coil galvanometer,

$$V_S = \frac{NBA}{CR_G} \quad ...(ii)$$

Dividing eqn. (i) by (ii)

Resistance of galvanometer

$$R_G = \frac{I_s}{V_s} = \frac{5 \times 1}{20 \times 10^{-3}} = \frac{5000}{20} = 250\ \Omega$$

93. **(d)** From circuit diagram

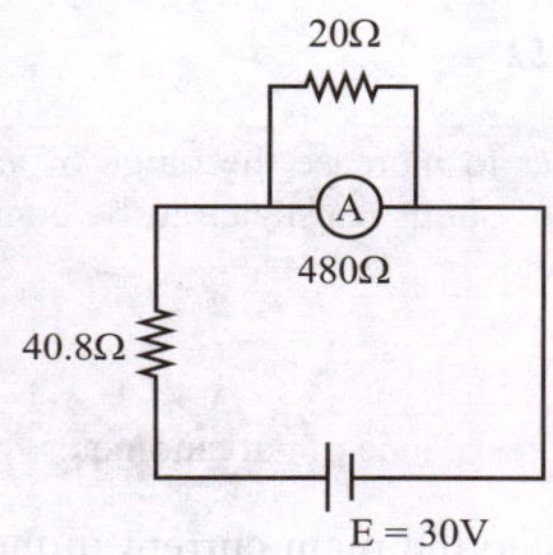

Resistance of ammeter $= \frac{480 \times 20}{480 + 20} = 19.2\Omega$.

Total resistance R = 40.8 + 19.2 = 60Ω

Reading in the ammeter $i = \frac{V}{R}$

$= \frac{30}{40.8 + 19.2} = 0.5$A

94. **(c)** As 0.2% of main current passes through the galvanometer hence $\frac{998}{1000} I$ current through the shunt.

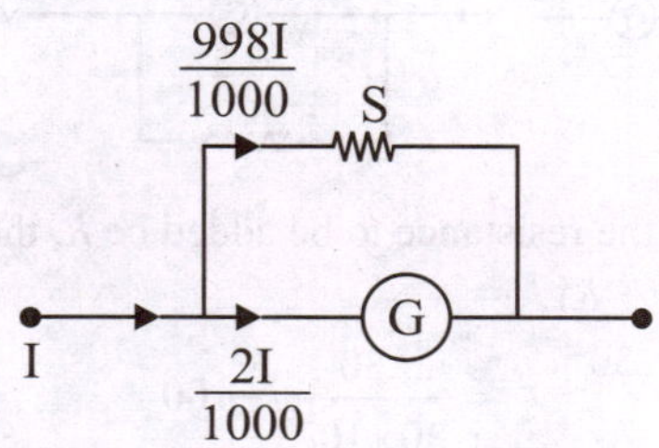

$$\left(\frac{2I}{1000}\right) G = \left(\frac{998I}{1000}\right) S \Rightarrow S = \frac{G}{499}$$

Total resistance of Ammeter

$$R = \frac{SG}{S+G} = \frac{\left(\frac{G}{499}\right) G}{\left(\frac{G}{499}\right) + G} = \frac{G}{500}$$

95. (a) Galvanometer is converted into ammeter, by connected a shunt, in parallel with it.

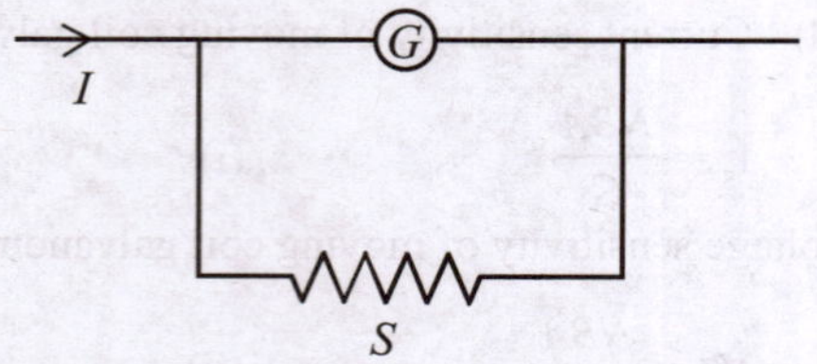

$$\frac{GS}{G+S}=\frac{V_G}{I}=\frac{25\times10^{-3}}{25}$$

$$\frac{GS}{G+S}=0.001\Omega$$

Here $S << G$ so

$S = 0.001\ \Omega$

> **NOTES** In order to increase the range of ammeter *n* times, the value of shunt resistance to be connected in parallel to galvanometer is
>
> $S=\frac{G}{n-1}$
>
> Here, G = resistance of galvanometer

96. (c) To keep the main current in the circuit unchanged, the resistance of the galvanometer should be equal to the net resistance.

$$\therefore G=\left(\frac{GS}{G+S}\right)+S'$$

$$\Rightarrow G-\frac{GS}{G+S}=S'$$

$$\therefore S'=\frac{G^2}{G+S}.$$

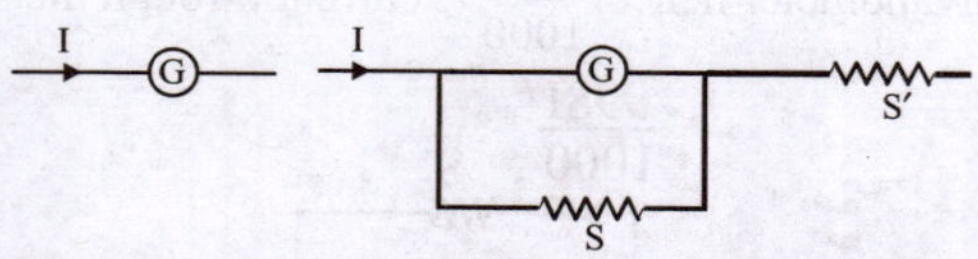

97. (a) Let the resistance to be added be R, then

$30 = I_g\,(r + R)$

$$\therefore\quad R=\frac{30}{I_g}-r=\frac{30}{30\times10^{-3}}-100$$

$= 1000 – 100 = 900\ \Omega$

98. (c) $G = 60\Omega$, $I_g = 1.0$A, $I = 5$A.

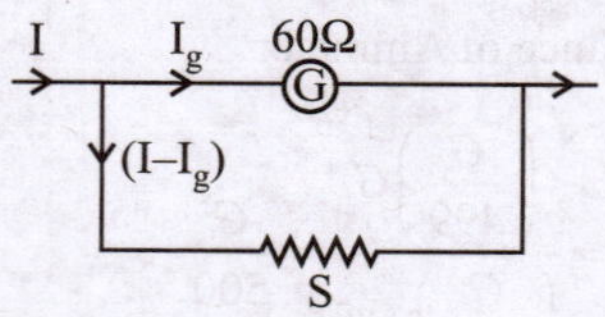

Let S be the shunt resistance connected in parallel to galvanometer

$I_g\,G = (I - I_g)\,S,$

$$S=\frac{I_gG}{I-I_g}=\frac{1}{5-1}\times60=15\Omega$$

Thus by putting 15Ω in parallel, the galvanometer can be converted into an ammeter.

99. (d) Total internal resistance = (50 + 2950) Ω

= 3000 Ω

Emf of the cell, ε = 3V

$$\therefore \text{Current}=\frac{\varepsilon}{R}=\frac{3}{3000}=1\times10^{-3}\text{ A}=1.0\text{ mA}$$

∴ Current for full scale deflection of 30 divisions is 1.0 mA.

∴ Current for a deflection of 20 divisions,

$$I=\left(\frac{20}{30}\times1\right)\text{mA or } I=\frac{2}{3}\text{mA}$$

Let the resistance be x Ω. Then

$$x=\frac{\varepsilon}{I}=\frac{3V}{\left(\frac{2}{3}\times10^{-3}A\right)}$$

$$=\frac{3\times3\times10^3}{2}\Omega$$

= 4500 Ω

But the resistance of the galvanometer is 50Ω.

∴ Resistance to be added

= (4500 –50) Ω = 4450 Ω

100. (a) We know

$$\frac{I}{I_S}=1+\frac{G}{S}$$

$$\frac{750}{100}=1+\frac{13}{S}$$

$S \Rightarrow 2\Omega$

101. (c) A galvanometer can be converted into a voltmeter by connecting *the high ressistance in series* with the galvanometer so that only a small amount of current passes through it.

102. (a) $R_g = 50\Omega$, $I_g = 25 \times 4 \times 10^{-4}\Omega = 10^{-4}$ A Range of V = 25 volts

$V = I_g(R_e + R_g)$

$$\therefore R_e = \frac{V}{I_g} - R_g = 2450\Omega$$

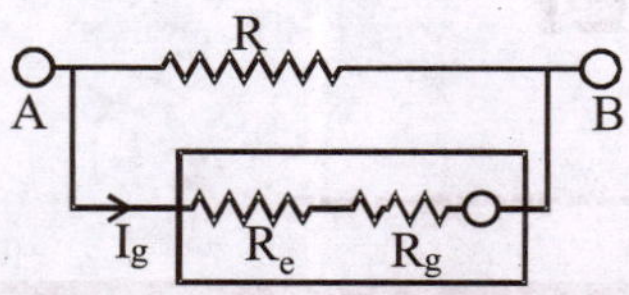

NOTES In order to increase the range of voltmeter n times the value of resistance to be connected in series with galvanometer is
$R = (n-1)G$
Here, G = resistance of galvanometer

103. (b) $I_s = I \times \frac{G}{S+G} = 1 \times \frac{8}{2+8}$

$= \frac{8}{10} = 0.8$ amp.

104. (c) Maximum current which can pass through galvanometer, $I_g = 0.004$A

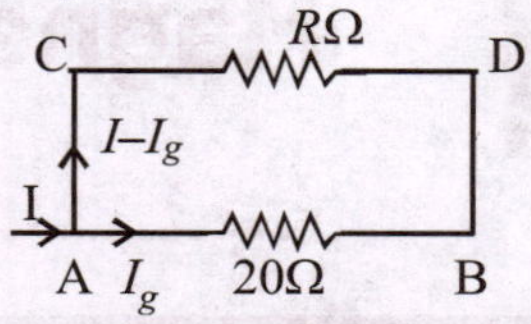

Let R be the resistance of shunt.
We know potential drop across AB
= Potential drop across CD

$R(I - I_g) = I_g(20)$

$\Rightarrow R(1 - 0.004) = 0.004 \times 20 \Rightarrow R = 0.08\ \Omega$

105. (a) To convert a galvanometer into an ammeter, one needs to connect a low resistance in parallel so that maximum current passes through the shunt wire and ammeter remains protected.

5 Magnetism and Matter

TREND ANALYSIS

TOPIC	2025		2024		2023		2022		2021		2020	
	Qns.	LOD	Qns.	LOD	Qns.	LOD	Qns.	LOD	Qns.	LOD	Qns.	LOD
The Bar Magnet			2	Easy								
Magnetisation and Magnetic Intensity											1	Avg.
Magnetic Properties of Materials			1	Avg.								

How to Code Qns for Better & Faster Revision

Type 1: Able to solve (No code Req)
Type 2: Solved with struggle or by seeing solution **(Use code NR means Needs Revision or orange stickers)**
Type 3: Tricky/ Could not solve on my own. Need help from Teacher/ Friend. **(Use code TQ means Tricky Question or red stickers)**

You can either hand write the code in boxes or else paste coloured stickers given at end of book.

Topic-5.2

The Bar Magnet

1. The magnetic potential energy, when a magnetic bar of magnetic moment $\vec{m}$ is placed perpendicular to the magnetic field $\vec{B}$ is **NCERT | Page-139 | Ph-II 2024**

(a) $-\frac{mB}{2}$ (b) Zero (c) $-mB$ (d) mB

2. The magnetic moment and moment of inertia of a magnetic needle as shown are, respectively, 1.0×10^{-2} A m^2 and $\frac{10^{-6}}{\pi^2}$ kg m^2. If it completes 10 oscillations in 10 s, the magnitude of the magnetic field is **Ph-II 2024**

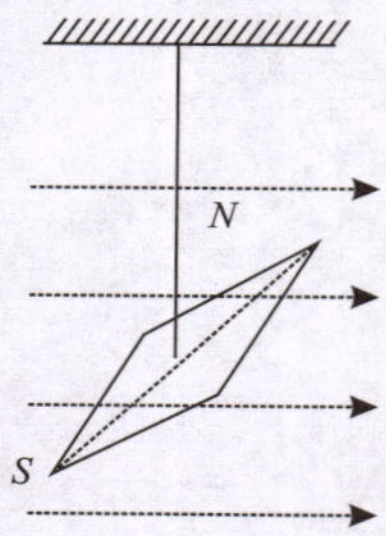

(a) 0.4 T (b) 4 T
(c) 0.4 mT (d) 4 mT

3. The magnetic moment of an iron bar is M. It is now bent in such a way that it forms an arc section of a circle subtending an angle of 60° at the centre. The magnetic moment of this arc section is **NCERT | Page-139 | Ph-II 2024**

(a) $\frac{3M}{\pi}$ (b) $\frac{4M}{\pi}$ (c) $\frac{M}{\pi}$ (d) $\frac{2M}{\pi}$

4. In a uniform magnetic field of 0.049 T, a magnetic needle performs 20 complete oscillations in 5 seconds as shown. The moment of inertia of the needle is 9.8×10^{-6} kg m^2. If the magnitude of magnetic moment of the needle is $x \times 10^{-5}$ Am^2, then the value of 'x' is : **NCERT | Page-139 | 2024**

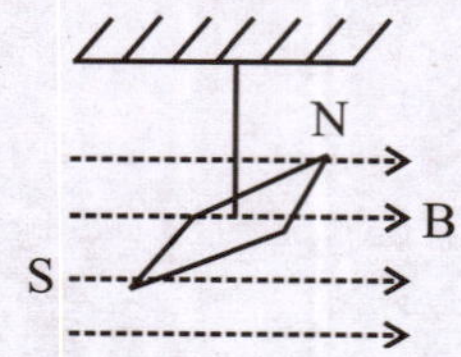

(a) $5\pi^2$ (b) $128\pi^2$ (c) $50\pi^2$ (d) $1280\pi^2$

5. An iron bar of length L has magnetic moment M. It is bent at the middle of its length such that the two arms make an angle 60° with each other. The magnetic moment of this new magnet is : **NCERT | Page-139 | 2024**

(a) M (b) $\frac{M}{2}$ (c) 2M (d) $\frac{M}{\sqrt{3}}$

6. Following figures show the arrangement of bar magnets in different configurations. Each magnet has magnetic dipole moment $\vec{m}$. Which configuration has highest net magnetic dipole moment ? **2014**

A.
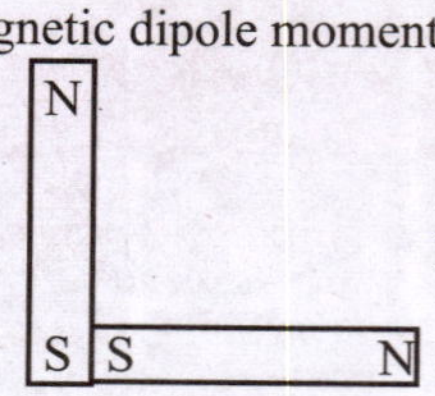

B.

C.
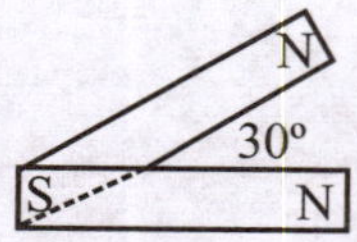

D.
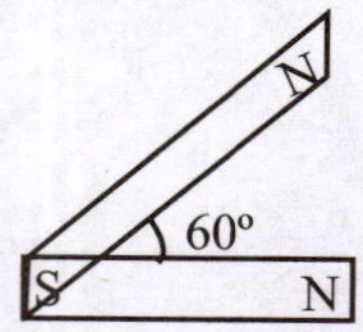

(a) A (b) B (c) C (d) D

7. A bar magnet of length 'ℓ' and magnetic dipole moment 'M' is bent in the form of an arc as shown in figure. The new magnetic dipole moment will be **2013**

(a) $\frac{3}{\pi}M$

(b) $\frac{2}{\pi}M$

(c) $\frac{M}{2}$

(d) M

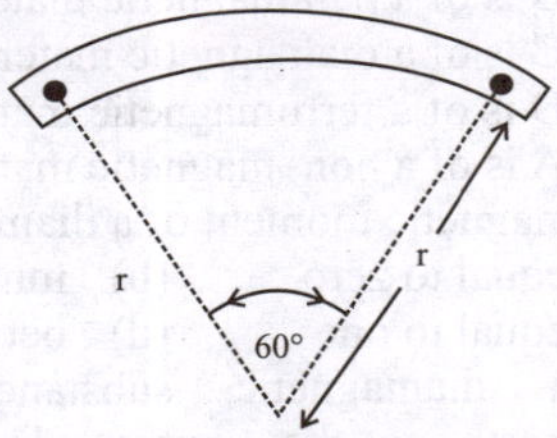

8. A bar magnet of magnetic moment M is placed at right angles to a magnetic induction B. If a force F is experienced by each pole of the magnet, the length of the magnet will be **Kar. 2013**

(a) F/MB (b) MB/F

(c) BF/M (d) MF/B

9. A magnetic needle suspended parallel to a magnetic field requires $\sqrt{3}$ J of work to turn it through 60°. The torque needed to maintain the needle in this position will be : **NCERT | Page-139 | 2012 M**

(a) $2\sqrt{3}$J (b) 3J (c) $\sqrt{3}$J (d) $\frac{3}{2}$J

10. A short bar magnet of magnetic moment 0.4J T^{-1} is placed in a uniform magnetic field of 0.16 T. The magnet is in stable equilibrium when the potential energy is **NCERT | Page-139 | 2011 M**

(a) –0.064 J (b) zero

(c) –0.082 J (d) 0.064 J

11. Two identical bar magnets are fixed with their centres at a distance d apart. A stationary charge Q is placed at P in between the gap of the two magnets at a distance D from the centre O as shown in the figure

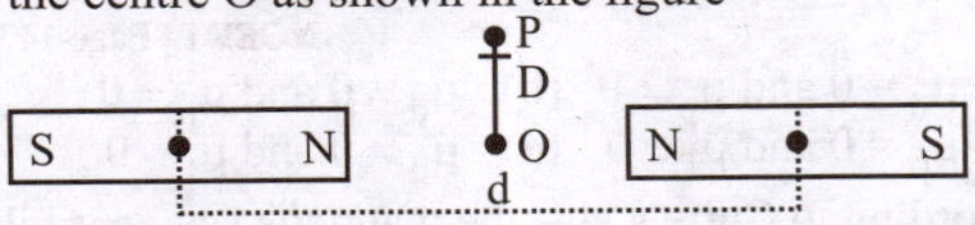

NCERT | Page-139 | 2010 M

The force on the charge Q is

(a) Zero

(b) Directed along OP

(c) Directed along PO

(d) Directed perpendicular to the plane of paper

12. A bar magnet having a magnetic moment of 2×10^4 JT^{-1} is free to rotate in a horizontal plane. A horizontal magnetic field B = 6×10^{-4} T exists in the space. The work done in taking the magnet slowly from a direction parallel to the field to a direction 60° from the field is **NCERT | Page-139 | 2009**

(a) 12 J (b) 6 J (c) 2 J (d) 0.6 J

13. A coil in the shape of an equilateral triangle of side l is suspended between the pole pieces of a permanent magnet such that $\vec{B}$ is in the plane of the coil. If due to a current i in the triangle a torque τ acts on it, the side l of the triangle is **NCERT | Page-139 | 2005**

(a) $\frac{2}{\sqrt{3}}\left(\frac{\tau}{B.i}\right)^{\frac{1}{2}}$ (b) $2\left(\frac{\tau}{\sqrt{3}B.i}\right)^{\frac{1}{2}}$

(c) $\frac{2}{\sqrt{3}}\left(\frac{\tau}{B.i}\right)$ (d) $\frac{2}{\sqrt{3}}\left(\frac{\tau}{B.i}\right)$

14. Current i is flowing in a coil of area A and number of turns N, then magnetic moment of the coil, M is **NCERT | Page-139 | 2001**

(a) NiA (b) $\frac{Ni}{A}$ (c) $\frac{Ni}{\sqrt{A}}$ (d) N^2Ai

15. A bar magnet, of magnetic moment $\vec{M}$, is placed in a magnetic field of induction $\vec{B}$. The torque exerted on it is **NCERT | Page-139 | 1999**

(a) $\vec{M}.\vec{B}$ (b) $-\vec{M}.\vec{B}$

(c) $\vec{M} \times \vec{B}$ (d) $\vec{B} \times \vec{M}$

16. The work done in turning a magnet of magnetic moment M by an angle of 90° from the meridian, is n times the corresponding work done to turn it through an angle of 60°. The value of n is given by **NCERT | Page-139 | 1995**

(a) 2 (b) 1 (c) 0.5 (d) 0.25

Topic-5.4 Magnetisation and Magnetic Intensity

17. An iron rod of susceptibility 599 is subjected to a magnetising field of 1200 A m^{-1}. The permeability of the material of the rod is :
($\mu_0 = 4\pi \times 10^{-7}$ T m A^{-1}) **NCERT | Page-146 | 2020**

(a) 8.0×10^{-5} T m A^{-1} (b) $2.4\pi \times 10^{-5}$ T m A^{-1}

(c) $2.4\pi \times 10^{-7}$ T m A^{-1} (d) $2.4\pi \times 10^{-4}$ T m A^{-1}

Topic-5.5 Magnetic Properties of Materials

18. The incorrect relation for a diamagnetic material (all the symbols carry their usual meaning and ε is a small positive number) is **NCERT | Page-147 | Ph-II 2024**

(a) $\mu < \mu_0$ (b) $0 \le \mu_r < 1$

(c) $-1 \le \chi < 0$ (d) $1 < \mu_r < 1 + \varepsilon$

19. Match List-I with List-II.

List-I (Material)	List-II (Susceptibility (χ))
A. Diamagnetic	I. $\chi = 0$
B. Ferromagnetic	II. $0 > \chi \ge -1$
C. Paramagnetic	III. $\chi >> 1$
D. Non-magnetic	IV. $0 < \chi < \varepsilon$ (a small positive number)

Choose the correct answer from the options given below **NCERT | Page-147 | 2024**

(a) A-II, B-III, C-IV, D-I

(b) A-II, B-I, C-III, D-IV

(c) A-III, B-II, C-I, D-IV

(d) A-IV, B-III, C-II, D-I

20. The variation of susceptibility (χ) with absolute temperature (T) for a paramagnetic material is represented as:

NCERT | Page-147 | Manipur 2023

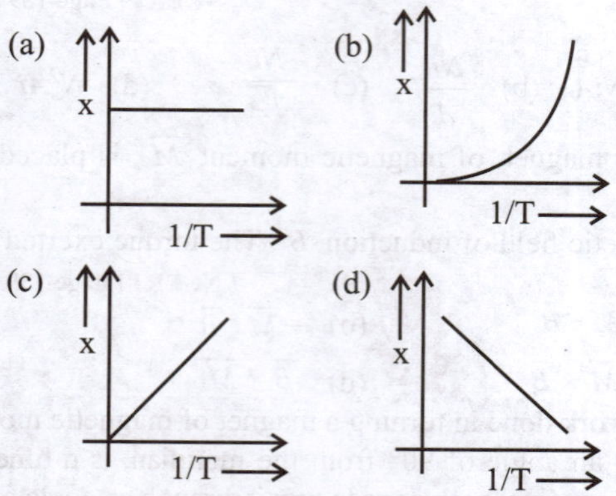

21. A thin diamagnetic rod is placed vertically between the poles of an electromagnet. When the current in the electromagnet is switched on, then the diamagnetic rod is pushed up, out of the horizontal magnetic field. Hence the rod gains gravitational potential energy. The work required to do this comes from

2018

(a) the current source
(b) the magnetic field
(c) the induced electric field due to the changing magnetic field
(d) the lattice structure of the material of the rod

22. The magnetic susceptibility is negative for :

NCERT | Page-146 | 2016

(a) diamagnetic material only
(b) paramagnetic material only
(c) ferromagnetic material only
(d) paramagnetic and ferromagnetic materials

23. There are four light–weight–rod samples A,B,C,D separately suspended by threads. A bar magnet is slowly brought near each sample and the following observations are noted

NCERT | Page-148 | 2011

(i) A is feebly repelled
(ii) B is feebly attracted
(iii) C is strongly attracted
(iv) D remains unaffected

Which one of the following is true ?

(a) B is of a paramagnetic material
(b) C is of a diamagnetic material
(c) D is of a ferromagnetic material
(d) A is of a non–magnetic material

24. The magnetic moment of a diamagnetic atom is **2010**

(a) equal to zero (b) much greater than one
(c) equal to one (d) between zero and one

25. If a diamagnetic substance is brought near the north or the south pole of a bar magnet, it is:

NCERT | Page-147 | 2009, 1999

(a) repelled by the north pole and attracted by the south pole
(b) attracted by the north pole and repelled by the south pole
(c) attracted by both the poles
(d) repelled by both the poles

26. Curie temperature is the temperature above which

2008, 2006

(a) ferromagnetic material becomes paramagnetic material
(b) paramagnetic material becomes diamagnetic material
(c) paramagnetic material becomes ferromagnetic material
(d) ferromagnetic material becomes diamagnetic material.

27. Nickel shows ferromagnetic property at room temperature. If the temperature is increased beyond Curie temperature, then it will show **2007**

(a) anti ferromagnetism
(b) no magnetic property
(c) diamagnetism
(d) paramagnetism

28. If the magnetic dipole moment of an atom of diamagnetic material, paramagnetic material and ferromagnetic material are denoted by μ_d, μ_p and μ_f respectively, then

NCERT | Page-147 | 2005

(a) $\mu_d = 0$ and $\mu_p \neq 0$ (b) $\mu_d \neq 0$ and $\mu_p = 0$
(c) $\mu_p = 0$ and $\mu_f \neq 0$ (d) $\mu_d \neq 0$ and $\mu_f \neq 0$

29. According to Curie's law, the magnetic sus-ceptibility of a substance at an absolute temperature T is proportional to **2003**

(a) T^2 (b) $1/T$ (c) T (d) $1/T^2$

30. A diamagnetic material in a magnetic field moves

NCERT | Page-148 | 2003

(a) perpendicular to the field
(b) from stronger to the weaker parts of the field
(c) from weaker to the stronger parts of the field
(d) in none of the above directions

ANSWER KEY

1	(b)	4	(d)	7	(a)	10	(a)	13	(b)	16	(a)	19	(a)	22	(a)	25	(d)	28	(a)
2	(c)	5	(b)	8	(b)	11	(a)	14	(a)	17	(d)	20	(c)	23	(a)	26	(a)	29	(b)
3	(a)	6	(c)	9	(b)	12	(b)	15	(c)	18	(d)	21	(a)	24	(a)	27	(d)	30	(b)

Hints & Solutions

1. **(b)** Potential energy stored in external magnetic field is
$U = -\vec{M}\cdot\vec{B}$
$\because$ Angle between $\vec{M}$ and $\vec{B}$ is 90º
$\therefore U = -MB\cos 90° = 0$

2. **(c)** Time period of oscillation of magnet inside the magnetic field is $T = 2\pi\sqrt{\frac{I}{MB}}$
$T = \frac{t}{n} = \frac{10}{10} = 1\text{s}$
$1 = 2\pi\sqrt{\frac{10^{-6}}{\pi^2 \times 1.0 \times 10^{-2} \times B}} \Rightarrow B = 0.4 \text{ mT}$

3. **(a)** Magnetic moment $M = m \cdot L$

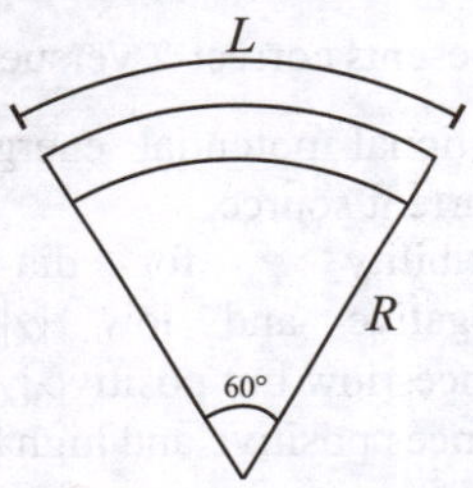

From figure, $R\theta = L \Rightarrow \frac{R\pi}{3} = L$
$\Rightarrow R = \frac{3L}{\pi}$
New magnetic moment $M' = m(2R)\sin 30°$
$= m(2)\frac{3L}{\pi} \times \frac{1}{2} = \frac{3}{\pi}mL = \frac{3M}{\pi}$

4. **(d)** Time period of oscillation, $T = 2\pi\sqrt{\frac{I}{MB}}$
$\Rightarrow \frac{1}{4} = 2\pi\sqrt{\frac{9.8\times10^{-6}}{M\times0.049}}$
$\Rightarrow \frac{1}{16} = 4\pi^2 \times \frac{9.8\times10^{-6}}{M\times49\times10^{-3}}$
$\Rightarrow M = \frac{4\pi^2\times9.8\times10^{-6}}{49\times10^{-3}} \times 16$
$= \frac{4\pi^2\times9.8\times16\times10^{-3}}{49}$
$= 12.8\pi^2 \times 10^{-3} \times 10^{-2} \times 10^2$
$= 1280\pi^2 \times 10^{-5}\ \text{Am}^2$
$\therefore\ x = 1280\,\pi^2$

5. **(b)**

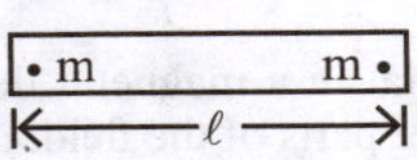

$M = m\ell$

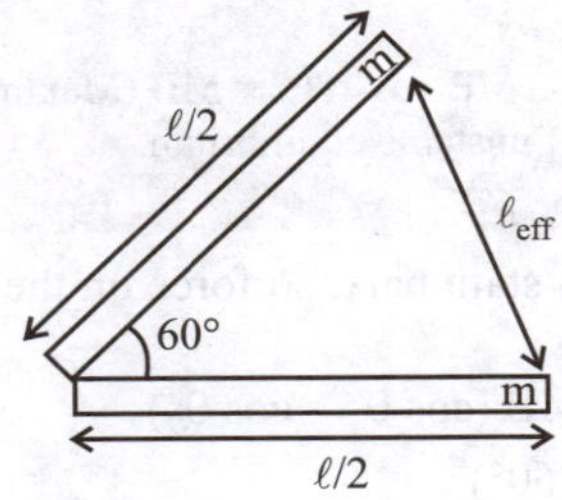

Effective length, $\ell_{eff} = 2\frac{\ell}{2}\sin 30° = \frac{\ell}{2}$
$\therefore$ New magnetic moment, $M' = m\ell_{eff} = m\ell/2 = M/2$

> **NOTES** When a magnet is cut along perpendicular to its length, then its pole strength remains unchanged, whereas by cutting it along its axis, its pole strength becomes half.

6. **(c)** Net magnetic dipole moment $= 2\,M\cos\frac{\theta}{2}$
As value of $\cos\frac{\theta}{2}$ is maximum in case (c) hence net magnetic dipole moment is maximum for option (c).

7. **(a)** Magnetic dipole moment
$M = m \times \ell \qquad M' = m \times r$
From figure
$\ell = \frac{\pi r}{3}$ or $r = \frac{3\ell}{\pi}$

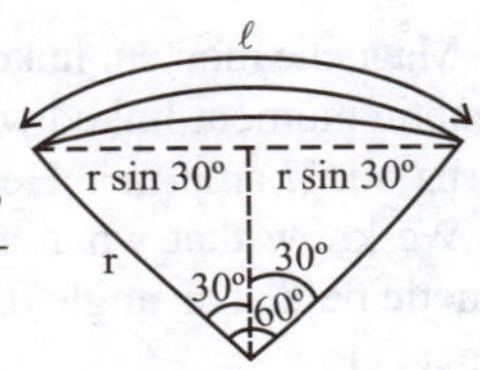

so, $M' = m \times r = \frac{m\times3\ell}{\pi}$
$= \frac{3}{\pi}M$

8. **(b)** Torque $FL = MB \Rightarrow L = \frac{MB}{F}$

9. **(b)** According to work energy theorem
$W = U_{final} - U_{initial} = MB(\cos 0 - \cos 60°)$
$W = \frac{MB}{2} = \sqrt{3}\text{J}$ ...(i)
$\tau = \vec{M}\times\vec{B} = MB\sin 60° = \left(\frac{MB\sqrt{3}}{2}\right)$ ...(ii)
From equation (i) and (ii)
$\tau = \frac{2\sqrt{3}\times\sqrt{3}}{2} = 3\text{J}$

10. **(a)** For stable equilibrium
$U = -MB = -(0.4)(0.16) = -0.064\text{ J}$

> **NOTES** When the magnetic dipole is aligned along the magnetic field it is stable equilibrium having minimum potential energy
> $U = MB \cos\theta = -MB \cos 0°$
> $= -MB$ i.e., $\theta = 0°$
> When $\theta = 180°$
> $U = -MB\cos\theta = -MB \cos 180° = MB$ (Maximum)
> This is the position of unstable equilibrium.

11. **(a)** As charge Q is stationary, so force on the charge Q is zero.

12. **(b)** Work done = MB $(\cos\theta_1 - \cos\theta_2)$

$= MB(\cos 0° - \cos 60°)$

$= MB\left(1-\frac{1}{2}\right) = \frac{2\times10^4\times6\times10^{-4}}{2} = 6\text{ J}$

13. **(b)** $\tau = MB\sin\theta$

$\tau = iAB \sin 90°$

$\therefore\ A = \frac{\tau}{iB}$

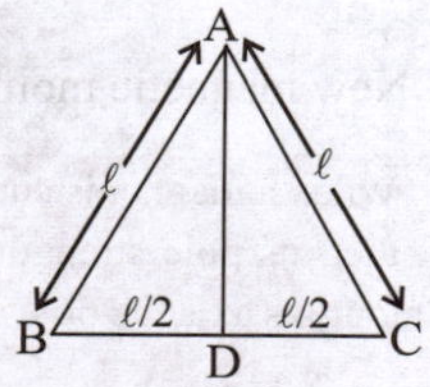

Also, $A = 1/2\,(BC)(AD)$

But $\frac{1}{2}(BC)(AD) = \frac{1}{2}(l)\sqrt{l^2-\left(\frac{l}{2}\right)^2}$

$= \frac{\sqrt{3}}{4}l^2 \Rightarrow \frac{\sqrt{3}}{4}(l)^2 = \frac{\tau}{Bi}$

$\therefore\quad l = 2\left(\frac{\tau}{\sqrt{3}\,B.i}\right)^{\frac{1}{2}}$

14. **(a)** Magnetic moment linked with one turn = iA
Magnetic moment linked with,
N turns = iNA amp-m^2. Here, A = Area of current loop.

15. **(c)** We know that when a bar magnet is placed in the magnetic field at an angle θ, then torque acting on the bar magnet (τ)

$= MB\sin\theta = \vec{M}\times\vec{B}$.

> **NOTES** The direction of torque (τ) is perpendicular to the plane containing M and B and is given by right hand screw rule.

16. **(a)** Magnetic moment = M; Initial angle through which magnet is turned $(\theta_1) = 90°$ and final angle through which magnet is turned $(\theta_2) = 60°$. Work done in turning the magnet through $90°(W_1)$ = MB (cos 0° – cos 90°)

$= MB(1-0) = MB$.

Similarly, $W_2 = MB(\cos 0° - \cos 60°)$

$= MB\left(1-\frac{1}{2}\right) = \frac{MB}{2}$

$\therefore\ W_1 = 2W_2$ or $n = 2$.

17. **(d)** Given : Magnetic susceptibility of iron, $\chi_m = 599$

Using, $\mu_r = 1 + \chi_m = 600$

$\mu = \mu_r\mu_0 = 600\times4\pi\times10^{-7} = 2400\pi\times10^{-7}$

Hence, permeability of the material of the rod,

$\mu = 2.4\pi\times10^{-4}\ \text{T m A}^{-1}$

18. **(d)** For diamagnetic material,

$0 \le \mu_r < 1$

$\chi = \mu_r - 1 \Rightarrow -1 \le \chi < 0$

$\mu_r = \frac{\mu}{\mu_0} < 1 \Rightarrow \mu < \mu_0$

19. **(a)** The susceptibility of different magnetic materials are given below

Material	Susceptibility
Diamagnetic	(II) $0 > \chi \ge -1$
Ferromagnetic	(III) $\chi >> 1$
Paramagnetic	(IV) $0 < \chi < \varepsilon$
Non-magnetic	(I) $\chi = 0$

20. **(c)** As per Curie law $\chi_m \propto \frac{1}{T}$ (Paramagnetic)

Susceptibility χ varies inversely with absolute temperature.

Therefore option (c) represents correct χ versue $\frac{1}{T}$ graph.

21. **(a)** Rod gains gravitational potential energy which comes from energy of current source.

22. **(a)** Magnetic susceptibility χ for dia-magnetic materials only is negative and low $|\chi| = -1$; for paramagnetic substances low but positive $|\chi| = 1$ and for ferromagnetic substances positive and high $|\chi| = 10^2$.

23. **(a)** A → diamagnetic
B → paramagnetic
C → Ferromagnetic
D → Non magnetic

24. **(a)** The magnetic moment of a diamagnetic atom is equal to zero.

25. **(d)** Diamagnetic substances do not have any unpaired electron. And they are magnetised in direction opposite to that of magnetic field. Hence, when they are brought to north or south pole of a bar magnet, they are repelled by poles.

26. **(a)** Curie temperature is the temperature above which ferromagnetic material becomes paramagnetic material.

27. **(d)** Beyond Curie temperature, ferromagnetic substances behaves like a paramagnetic substance.

28. **(a)** The magnetic dipole moment of diamag-netic material is zero as each of its pair of electrons have opposite spins, i.e., $\mu_d = 0$.
Paramagnetic substances have dipole moment > 0, i.e. $\mu_p \ne 0$, because of excess of electrons in its molecules spinning in the same direction.
Ferro-magnetic substances are very strong magnets and they also have permanent magnetic moment, i.e. $\mu_p \ne 0$.

29. **(b)** According to Curie's law, $\chi_m \propto \frac{1}{T}$

30. **(b)** A diamagnetic material in a magnetic field moves from stronger to the weaker parts of the field.

6 Electromagnetic Induction

TREND ANALYSIS

TOPIC	2025		2024		2023		2022		2021		2020	
	Qns.	LOD	Qns.	LOD	Qns.	LOD	Qns.	LOD	Qns.	LOD	Qns.	LOD
Magnetic Flux							1	Easy				
Faraday's Law of Induction							2	Avg.				
Lenz's Law and Conservation of Energy			1	Avg.								
Motional Electromotive Force			1	Diff.								
Inductance	1	Avg.			1	Avg.			1	Avg.		
Eddy Current												

How to Code Qns for Better & Faster Revision

Type 1: Able to solve (No code Req)
Type 2: Solved with struggle or by seeing solution **(Use code NR means Needs Revision or orange stickers)**
Type 3: Tricky/ Could not solve on my own. Need help from Teacher/ Friend. **(Use code TQ means Tricky Question or red stickers)**

You can either hand write the code in boxes or else paste coloured stickers given at end of book.

Topic-6.3 Magnetic Flux

1. A square loop of side 1 m and resistance 1 Ω is placed in a magnetic field of 0.5 T. If the plane of loop is perpendicular to the direction of magnetic field, the magnetic flux through the loop is: NCERT | Page-155 | 2022

(a) 0.5 weber (b) 1 weber
(c) Zero weber (d) 2 weber

Topic-6.4 Faraday's Law of Induction

2. The magnetic flux linked to a circular coil of radius R is: NCERT | Page-158 | RS 2022

$\phi = 2t^3 + 4t^2 + 2t + 5$ Wb

The magnitude of induced emf in the coil at t = 5 s is:

(a) 192 V (b) 108 V
(c) 197 V (d) 150 V

3. A big circular coil of 1000 turns and average radius 10 m is rotating about its horizontal diameter at 2 rad s^{-1}. If the vertical component of earth's magnetic field at that place is 2×10^{-5} T and electrical resistance of the coil is 12.56 Ω, then the maximum induced current in the coil will be: NCERT | Page-158 | 2022

(a) 1.5 A (b) 1 A
(c) 2 A (d) 0.25 A

4. A 800 turn coil of effective area 0.05 m^2 is kept perpendicular to a magnetic field 5×10^{-5} T. When the plane of the coil is rotated by 90° around any of its coplanar axis in 0.1 s, the emf induced in the coil will be : NCERT | Page-158 | 2019

(a) 2 V (b) 0.2 V
(c) 2×10^{-3} V (d) 0.02 V

5. A long solenoid of diameter 0.1 m has 2×10^4 turns per meter. At the centre of the solenoid, a coil of 100 turns and radius 0.01 m is placed with its axis coinciding with the solenoid axis. The current in the solenoid reduces at a constant rate to 0A from 4 A in 0.05 s. If the resistance of the coil is $10\pi^2\Omega$. The total charge flowing through the coil during this time is :- NCERT | Page-168 | 2017

(a) 16 μC (b) 32 μC
(c) 16 π μC (d) 32 π μC

6. A thin semicircular conducting ring (PQR) of radius 'r' is falling with its plane vertical in a horizontal magnetic field B, as shown in figure. The potential difference developed across the ring when its speed is v, is : NCERT | Page-158 | 2014

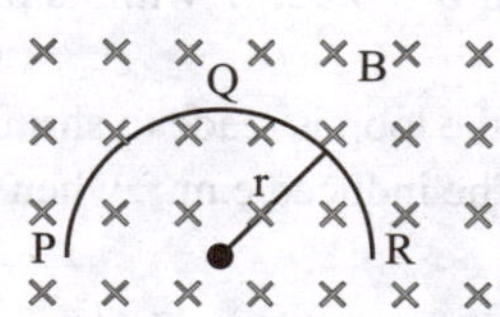

(a) Zero
(b) $B v \pi r^2/2$ and P is at higher potnetial
(c) $\pi r B v$ and R is at higher potnetial
(d) 2rBv and R is at higher potential

7. A wire loop is rotated in a magnetic field. The frequency of change of direction of the induced e.m.f. is NCERT | Page-158 | 2013

(a) twice per revolution
(b) four times per revolution
(c) six times per revolution
(d) once per revolution

8. A coil of resistance 400Ω is placed in a magnetic field. If the magnetic flux ϕ (wb) linked with the coil varies with time t (sec) as $\phi = 50t^2 + 4$. The current in the coil at $t = 2$ sec is : **NCERT | Page-158 | 2012**

(a) 0.5 A (b) 0.1 A
(c) 2 A (d) 1 A

9. In a coil of resistance 10 Ω, the induced current developed by changing magnetic flux through it, is shown in figure as a function of time. The magnitude of change in flux through the coil in Weber is : **NCERT | Page-158 | 2012 M**

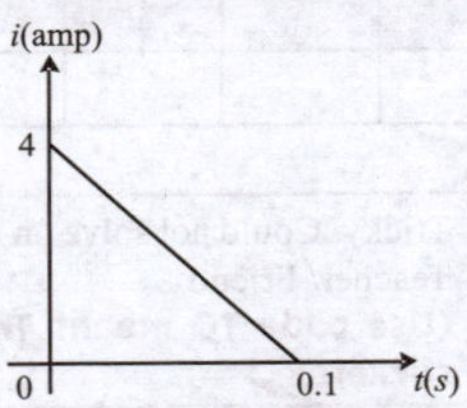

(a) 8 (b) 2 (c) 6 (d) 4

10. The current i in a coil varies with time as shown in the figure. The variation of induced emf with time would be **NCERT | Page-158 | 2011**

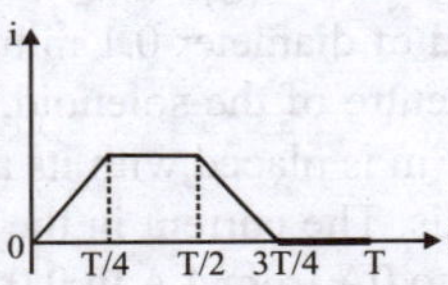

(a)
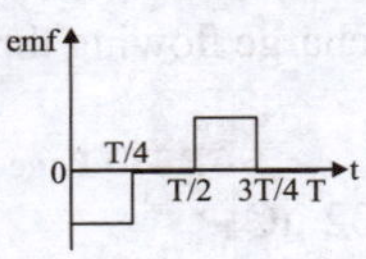

(b)
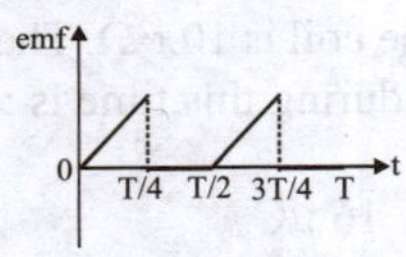

(c)
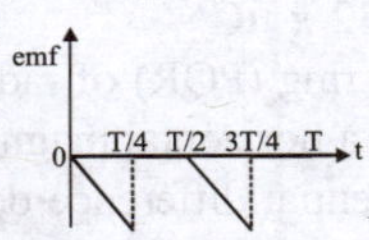

(d)
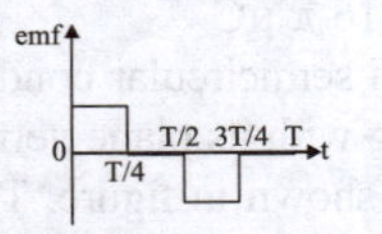

11. A conducting circular loop is placed in a uniform magnetic field, $B = 0.025\ T$ with its plane perpendicular to the loop. **NCERT | Page-158 | 2010**
The radius of the loop is made to shrink at a constant rate of 1 mm s^{-1}. The induced e.m.f. when the radius is 2 cm, is

(a) $2\pi\mu V$ (b) $\pi\mu V$
(c) $\frac{\pi}{2}\mu V$ (d) $2\mu V$

12. A rectangular, a square, a circular and an elliptical loop, all in the (x – y) plane, are moving out of a uniform magnetic field with a constant velocity, $\vec{V} = v\hat{i}$. The magnetic field is directed along the negative z axis direction. The induced emf, during the passage of these loops, out of the field region, will not remain constant for **NCERT | Page-158 | 2009**

(a) the circular and the elliptical loops.
(b) only the elliptical loop.
(c) any of the four loops.
(d) the rectangular, circular and elliptical loops.

13. A conducting circular loop is placed in a uniform magnetic field of 0.04 T with its plane perpendicular to the magnetic field. The radius of the loop starts shrinking at 2 mm/s. The induced emf in the loop when the radius is 2 cm is **NCERT | Page-158 | 2009**

(a) $4.8\ \pi\ \mu V$ (b) $0.8\ \pi\ \mu V$
(c) $1.6\ \pi\ \mu V$ (d) $3.2\ \pi\ \mu V$

14. A circular disc of radius 0.2 meter is placed in a uniform magnetic field of induction $\frac{1}{\pi}\left(Wb/m^2\right)$ in such a way that its axis makes an angle of 60° with $\vec{B}$. The magnetic flux linked with the disc is: **NCERT | Page-158 | 2008**

(a) 0.02 Wb (b) 0.06 Wb
(c) 0.08 Wb (d) 0.01 Wb

15. The magnetic flux through a circuit of resistance R changes by an amount $\Delta\phi$ in a time Δt. Then the total quantity of electric charge Q that passes any point in the circuit during the time Δt is represented by **NCERT | Page-158 | 2004**

(a) $Q = R.\frac{\Delta\phi}{\Delta t}$ (b) $Q = \frac{1}{R}.\frac{\Delta\phi}{\Delta t}$
(c) $Q = \frac{\Delta\phi}{R}$ (d) $Q = \frac{\Delta\phi}{\Delta t}$

16. A rectangular coil of 20 turns and area of cross-section 25 sq. cm has a resistance of 100Ω. If a magnetic field which is perpendicular to the plane of coil changes at a rate of 1000 tesla per second, the current in the coil is **NCERT | Page-158 | 1992**

(a) 1 A (b) 50 A (c) 0.5 A (d) 5 A

17. The total charge induced in a conducting loop when it is moved in a magnetic field depend on **NCERT | Page-158 | 1992**

(a) the rate of change of magnetic flux
(b) initial magnetic flux only
(c) the total change in magnetic flux
(d) final magnetic flux only

18. A magnetic field of 2×10^{-2} T acts at right angles to a coil of area 100 cm^2, with 50 turns. The average e.m.f. induced in the coil is 0.1 V, when it is removed from the field in t sec. The value of t is **NCERT | Page-158 | 1991**

(a) 10 s (b) 0.1 s (c) 0.01 s (d) 1 s

19. A cycle wheel of radius 0.5 m is rotated with constant angular velocity of 10 rad/s in a region of magnetic field of 0.1 T which is perpendicular to the plane of the wheel. The EMF generated between its centre and the rim is **NCERT | Page-158 | Odisha 2019**

(a) zero (b) 0.25 V (c) 0.125 V (d) 0.5 V

20. The total charge induced in a conducting loop when it is moved in a magnetic field depends on

NCERT | Page-158 | 1990

(a) the rate of change of magnetic flux
(b) initial magnetic flux only
(c) the total change in magnetic flux
(d) final magnetic flux only

21. In the circuit of Fig, the bulb will become suddenly bright if

NCERT | Page-158 | 1989

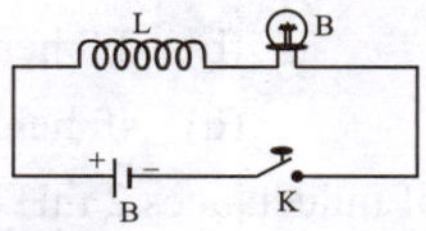

(a) contact is made or broken
(b) contact is made
(c) contact is broken
(d) won't become bright at all

Topic-6.5
Lenz's Law and Conservation of Energy

22.

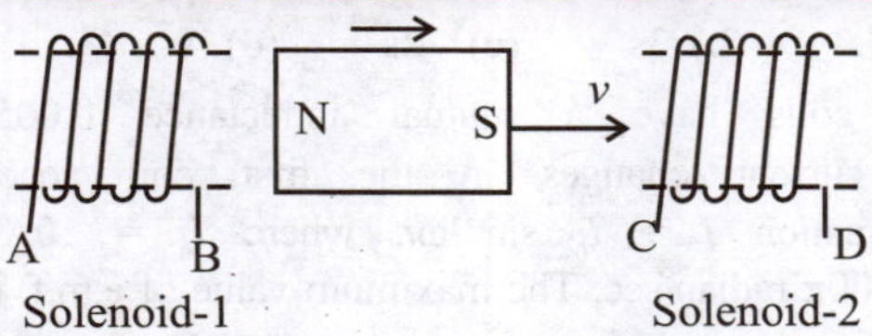

In the above diagram, a strong bar magnet is moving towards solenoid-2 from solenoid-1. The direction of induced current in solenoid-1 and that in solenoid-2, respectively, are through the directions:

NCERT | Page-160 | 2024

(a) AB and DC
(b) BA and CD
(c) AB and CD
(d) BA and DC

23. An electron moves on a straight line path XY as shown. The abcd is a coil adjacent to the path of electron. What will be the direction of current if any, induced in the coil?

NCERT | Page-160 | 2015 RS

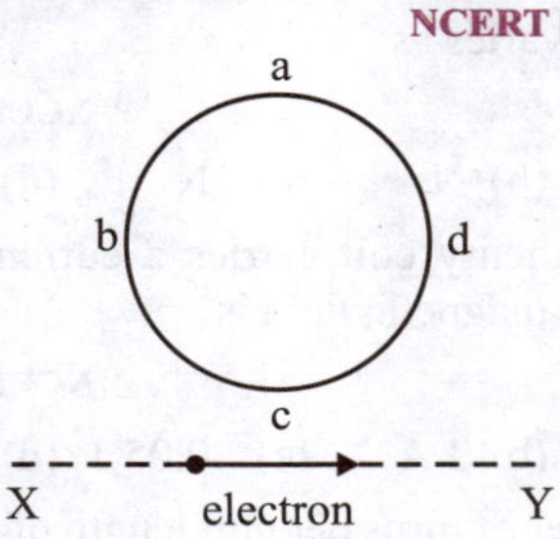

(a) adcb
(b) The current will reverse its direction as the electron goes past the coil
(c) No current induced
(d) abcd

Topic-6.6
Motional Electromotive Force

24. A conducting square frame of side 'a' and a long staight wire carrying current I are located in the same plane as shown in the figure. The frame moves to the right with a constant velocity 'V'. The emf induced in the frame will be proportional to

NCERT | Page-162 | 2015

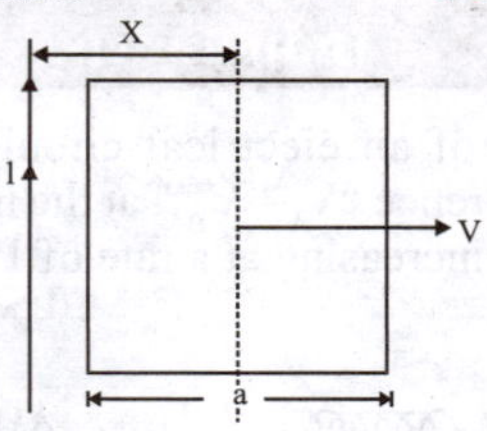

(a) $\frac{1}{(2x-a)^2}$ (b) $\frac{1}{(2x+a)^2}$

(c) $\frac{1}{(2x-a)(2x+a)}$ (d) $\frac{1}{x^2}$

25. As a result of change in the magnetic flux linked to the closed loop shown in the Fig,

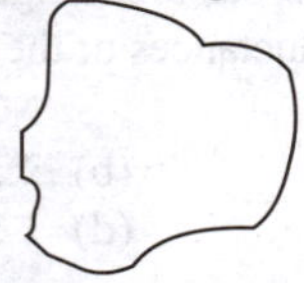

an e.m.f. V volt is induced in the loop. The work done (joules) in taking a charge Q coulomb once along the loop is

NCERT | Page-163 | 2005

(a) QV (b) 2QV (c) QV/2 (d) Zero

26. A conductor of length 0.4 m is moving with a speed of 7 m/s perpendicular to a magnetic field of intensity 0.9 Wb/m^2. The induced e.m.f. across the conductor is

NCERT | Page-162 | 1995

(a) 1.26 V (b) 2.52 V
(c) 5.04 V (d) 25.2 V

Topic-*
Eddy Current

27. A sheet is placed on a horizontal surface in front of a strong magnetic pole. A force is needed to:

A. hold the sheet there if it is magnetic.
B. hold the sheet there if it is non-magnetic.
C. move the sheet away from the pole with uniform velocity if it is conducting.
D. move the sheet away from the pole with uniform velocity if it is both, non-conducting and non-polar.

Choose the correct statement(s) from the options given below:

NCERT | Page-N-218 | 2024

(a) B and D only (b) A and C only
(c) A, C and D only (d) C only

28. In which of the following devices, the eddy current effect is not used ? 2019

(a) induction furnace
(b) magnetic braking in train
(c) electromagnet
(d) electric heater

Topic-6.7 Inductance

29. AB is a part of an electrical circuit (see figure). The potential difference "$V_A - V_B$", at the instant when current $i = 2$ A and is increasing at a rate of 1 amp/second is: **NCERT | Page-168 | 2025**

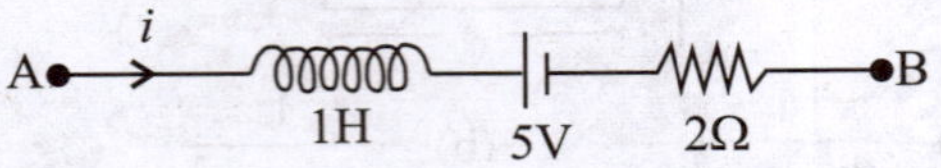

(a) 5 volt (b) 6 volt
(c) 9 volt (d) 10 volt

30. Let us consider two solenoids *A* and *B*, made from same magnetic material of relative permeability μ_r and equal area of cross-section. Length of *A* is twice that of *B* and the number of turns per unit length in *A* is half that of *B*. The ratio of self inductances of the two solenoids, $L_A : L_B$ is **NCERT | Page-168 | Ph-II 2024**

(a) 1 : 2 (b) 2 : 1
(c) 8 : 1 (d) 1 : 8

31. The magnetic energy stored in an inductor of inductance 4 μH carrying a current of 2 A is **NCERT | Page-169 | 2023**

(a) 4 μJ (b) 4 mJ (c) 8 mJ (d) 8 μJ

32. Two conducting circular loops of radii R_1 and R_2 are placed in the same plane with their centres coinciding. If $R_1 >> R_2$, the mutual inductance *M* between them will be directly proportional to **NCERT | Page-166 | 2021**

(a) $\frac{R_2^2}{R_1}$ (b) $\frac{R_1}{R_2}$
(c) $\frac{R_2}{R_1}$ (d) $\frac{R_1^2}{R_2}$

33. The magnetic potential energy stored in a certain inductor is 25 mJ, when the current in the inductor is 60 mA. This inductor is of inductance **NCERT | Page-169 | 2018**

(a) 0.138 H (b) 138.88 H
(c) 13.89 H (d) 1.389 H

34. A long solenoid has 1000 turns. When a current of 4A flows through it, the magnetic flux linked with each turn of the solenoid is 4×10^{-3} Wb. The self inductance of the solenoid is : **NCERT | Page-168 | 2016**

(a) 4H (b) 3H
(c) 2H (d) 1H

35. A current of 2.5 A flows through a coil of inductance 5 H. The magnetic flux linked with the coil is **NCERT | Page-168 | Kar. 2013**

(a) 2 Wb (b) 0.5 Wb
(c) 12.5 Wb (d) Zero

36. A long solenoid has 500 turns. When a current of 2 ampere is passed through it, the resulting magnetic flux linked with each turn of the solenoid is 4×10^{-3} Wb. The self- inductance of the solenoid is **NCERT | Page-168 | 2008**

(a) 2.5 henry (b) 2.0 henry
(c) 1.0 henry (d) 40 henry

37. Two coils of self inductances 2 mH and 8 mH are placed so close together that the effective flux in one coil is completely linked with the other. The mutual inductance between these coils is **NCERT | Page-166 | 2006**

(a) 6 mH (b) 4 mH (c) 16 mH (d) 10 mH

38. In an inductor of self-inductance $L = 2$ mH, current changes with time according to relation $i = t^2e^{-t}$. At what time emf is zero? **NCERT | Page-168 | 2001**

(a) 4s (b) 3s (c) 2s (d) 1s

39. Two coils have a mutual inductance 0.005 H. The current changes in the first coil according to equation $I = I_0 \sin \omega t$, where $I_0 = 10$A and $\omega = 100\pi$ radian/sec. The maximum value of e.m.f. in the second coil is **NCERT | Page-166 | 1998**

(a) 2π (b) 5π (c) π (d) 4π

40. A varying current in a coil changes from 10A to zero in 0.5 sec. If the average e.m.f induced in the coil is 220V, the self-inductance of the coil is **NCERT | Page-168 | 1995**

(a) 5 H (b) 6 H (c) 11 H (d) 12 H

41. What is the self-inductance of a coil which produces 5V when the current changes from 3 ampere to 2 ampere in one millisecond? **NCERT | Page-168 | 1993**

(a) 5000 henry (b) 5 milli-henry
(c) 50 henry (d) 5 henry

42. If N is the number of turns in a coil, the value of self inductance varies as **NCERT | Page-168 | 1993**

(a) N^0 (b) N (c) N^2 (d) N^{-2}

43. A 100 millihenry coil carries a current of 1A. Energy stored in its magnetic field is **NCERT | Page-169 | 1991**

(a) 0.5 J (b) 1 A (c) 0.05 J (d) 0.1 J

44. If the number of turns per unit length of a coil of solenoid is doubled, the self-inductance of the solenoid will **NCERT | Page-168 | 1991**

(a) remain unchanged (b) be halved
(c) be doubled (d) become four times

45. An inductor may store energy in NCERT | Page-169 | 1990

(a) its electric field
(b) its coils
(c) its magnetic field
(d) both in electric and magnetic fields

46. The current in self inductance L = 40 mH is to be increased uniformly from 1 amp to 11 amp in 4 milliseconds. The e.m.f. induced in the inductor during the process is

NCERT | Page-168 | 1990

(a) 100 volt (b) 0.4 volt
(c) 4.0 volt (d) 440 volt

Topic-6.8 AC Generator

47. An emf is generated by an ac generator having 100 turn coil, of loop area 1 m^2. The coil rotates at a speed of one revolution per second and placed in a uniform magnetic field of 0.05 T perpendicular to the axis of rotation of the coil. The maximum value of emf is

NCERT | Page-162 | Manipur 2023

(a) 62.8 V (b) 6.28 V
(c) 3.14 V (d) 31.4 V

48. The variation of EMF with time for four types of generators are shown in the figures. Which amongst them can be called AC?

NCERT | Page-171 | Odisha 2019

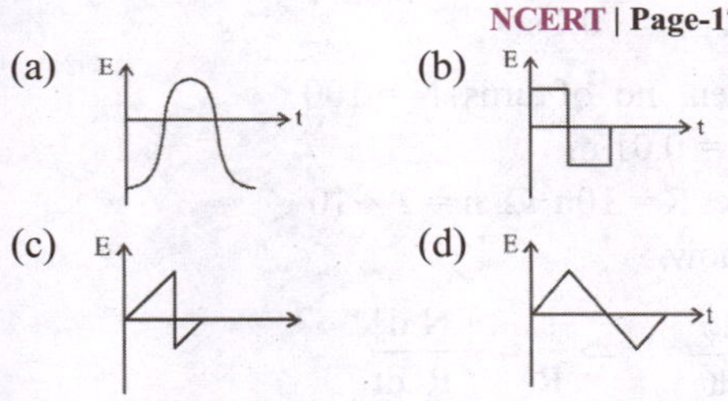

(a) Only (a)
(b) (a) and (d)
(c) (a), (b), (c), (d)
(d) (a) and (b)

ANSWER KEY

1	(a)	6	(d)	11	(b)	16	(c)	21	(c)	26	(b)	31	(d)	36	(c)	41	(b)	46	(a)
2	(a)	7	(a)	12	(a)	17	(c)	22	(a)	27	(b)	32	(a)	37	(b)	42	(c)	47	(d)
3	(b)	8	(a)	13	(d)	18	(b)	23	(b)	28	(d)	33	(c)	38	(c)	43	(c)	48	(c)
4	(d)	9	(b)	14	(a)	19	(c)	24	(c)	29	(d)	34	(d)	39	(b)	44	(d)		
5	(b)	10	(a)	15	(c)	20	(c)	25	(a)	30	(a)	35	(c)	40	(c)	45	(c)		

Hints & Solutions

1. **(a)**

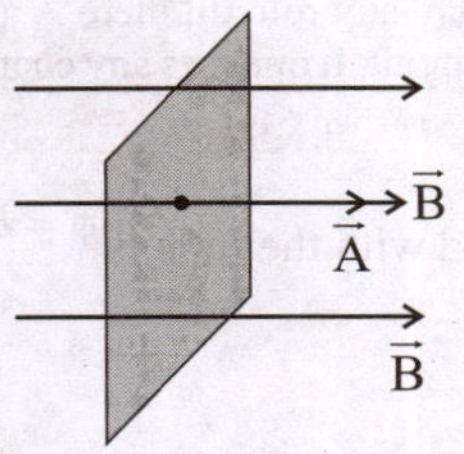

Clearly angle between $\vec{B}$ and $\vec{A}$ is 0°

so, ϕ (flux) $= \vec{B}.\vec{A}$

$= BA\cos 0°$

$= BA = 0.5 \times 1^2 = 0.5$ weber

2. **(a)** Given magnetic flux, $\phi = 2t^3 + 4t^2 + 2t + 5$

Induced emf, $e = \dfrac{-d\phi}{dt}$

$\therefore |e| = \left|\dfrac{d\phi}{dt}\right| = \dfrac{d}{dt}\,[2t^3 + 4t^2 + 2t + 5]$

$= 6t^2 + 8t + 2$

$e(t) = 6t^2 + 8t + 2$

$e(t = 5) = 6(5)^2 + 8(5) + 2 = 150 + 40 + 2$

$\therefore |e| = 192$ V

3. **(b)** $e = \dfrac{Nd\phi}{dt} = \dfrac{Nd}{dt}(BA\cos\omega t) = NBA\omega\sin\omega t$

$$i_0 = \frac{Ne_0}{R} = \frac{NBA\omega}{R}$$

$$= \left(\frac{2\times10^{-5}\times\pi\,100\times2}{4\pi}\right)\times1000 = 1A$$

4. **(d)** Given :

Magnetic field B = 5 × 10^{-5} T
Number of turns in coil N = 800
Area of coil A = 0.05 m^2
Time taken to rotate = Δt = 0.1 s
Initial angle $\theta_1 = 0°$

Final angle $\theta_2 = 90°$

Change in magnetic flux $\Delta\phi$

$= NBA \cos 90° - NBA \cos 0° = -NBA$

$= -800 \times 5 \times 10^{-5} \times 0.05 = -2 \times 10^{-2}$ weber

$$E = -\frac{\Delta\phi}{\Delta t} = \frac{2\times10^{-3}\,Wb}{0.1\,s} = 0.02\ V$$

5. **(b)** Given, no. of turns N = 100

radius, r = 0.01 m

resistance, $R = 10\pi^2\Omega$, $n = 2 \times 10^4$

As we know,

$$\varepsilon = -N\frac{d\phi}{dt} \quad \Rightarrow \frac{\varepsilon}{R} = -\frac{N}{R}\frac{d\phi}{dt}$$

$$\Delta I = -\frac{N}{R}\frac{d\phi}{dt} \quad \Rightarrow \frac{\Delta q}{\Delta t} = -\frac{N}{R}\frac{\Delta\phi}{\Delta t}$$

$$\Delta q = -\left[\frac{N}{R}\left(\frac{\Delta\phi}{\Delta t}\right)\right]\Delta t$$

'–' ve sign shows that induced emf opposes the change of flux.

$$\Delta q = \left[\mu_0 nN\pi r^2\left(\frac{\Delta i}{\Delta t}\right)\right]\frac{1}{R}\Delta t = \frac{\mu_0 nN\pi r^2 \Delta i}{R}$$

$$\Delta q = \frac{4\pi\times10^{-7}\times100\times4\times\pi\times(0.01)^2\times2\times10^4}{10\pi^2}$$

$\Delta q = 32\mu C$

6. **(d)** Rate of decrease of area of semicircular ring $= \frac{dA}{dt} = (2r)V$

From Faraday's law of electromagnetic induction

$$e = -\frac{d\phi}{dt} = -B\frac{dA}{dt} = -B(2rV)$$

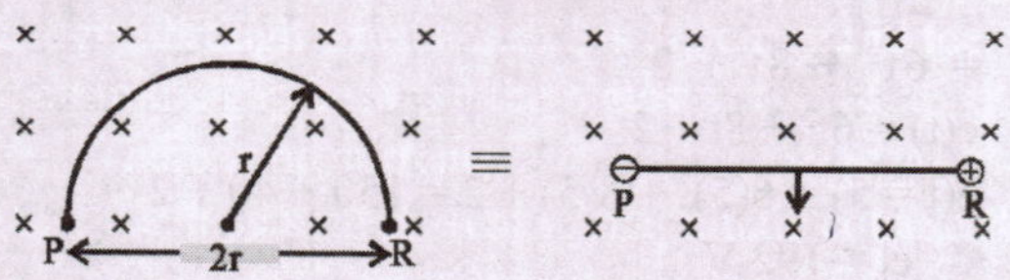

As induced current in ring produces magnetic field in upward direction hence R is at higher potential.

7. **(a)** This is the case of periodic EMI

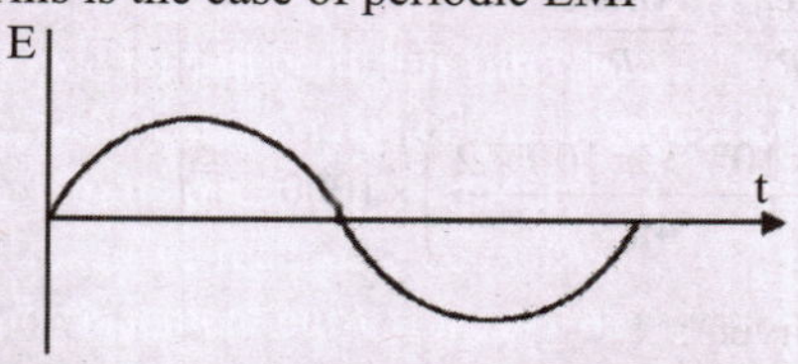

From graph, it is clear that direction is changing once in $\frac{1}{2}$ cycle.

8. **(a)** According, to Faraday's law of induction

Induced e.m.f. $\varepsilon = -\frac{d\phi}{dt} = -(100t)$

Induced current i at $t = 2$ sec.

$$= \left|\frac{\varepsilon}{R}\right| = +\frac{100\times2}{400} = +0.5\,A$$

NOTES The magnetic flux linked with a loop does not change when magnet and loop are moving with the same velocity.

9. **(b)** The charge through the coil = area of current-time $(i - t)$ graph

$$q = \frac{1}{2}\times0.1\times4 = 0.2\ C$$

$q = \frac{\Delta\phi}{R}$ ∵ Change in flux $(\Delta\phi) = q \times R$

$$q = 0.2 = \frac{\Delta\phi}{10}$$

$\Delta\phi = 2$ Weber

10. **(a)** $e = -L\frac{di}{dt}$

During 0 to $\frac{T}{4}, \frac{di}{dt} =$ const.

∴ e = – ve

During $\frac{T}{4}$ to $\frac{T}{2}, \frac{di}{dt} = 0$

∴ e = 0

During $\frac{T}{2}$ to $\frac{3T}{4}, \frac{di}{dt} =$ const.

∴ e = +ve

Thus graph given in option (a) represents the variation of induced emf with time.

NOTES Inductance does not play any role till there is a constant current flowing in the circuit. It opposes any change in the current in the circuit.

11. **(b)** Magnetic flux linked with the loop is $\phi = B\pi r^2$

$$|e| = \frac{d\phi}{dt} = B\pi\cdot2r\frac{dr}{dt}$$

When $r = 2$ cm, $\frac{dr}{dt} = 1\ mm\ s^{-1}$

$e = 0.025\times \pi \times2 \times2 \times10^{-2}\times10^{-3}$

$= 0.100 \times \pi \times 10^{-5} = \pi \times 10^{-6}\ V = \pi\mu V$

12. **(a)** The induced emf will remain constant only in the case of rectangular and square loops. In case of the circular and the elliptical loops, the rate of change of area of the loops during their passage out of the field is not constant, hence induced emf will not remain constant for them.

13. (d) Induced emf in the loop is given by

$$e = -B.\frac{dA}{dt}\text{, where A is the area of the loop.}$$

$$e = -B.\frac{d}{dt}(\pi r^2) = -B\pi 2r\frac{dr}{dt}$$

$r = 2cm = 2\times 10^{-2}$ m

$dr = 2$ mm $= 2\times 10^{-3}$ m

$dt = 1s$

$$e = -0.04\times 3.14\times 2\times 2\times 10^{-2}\times\frac{2\times 10^{-3}}{1}\text{ V}$$

$$= 0.32\pi\times 10^{-5}\text{V} = 3.2\pi\times 10^{-6}\text{V} = 3.2\pi\ \mu\text{V}$$

14. (a) Here, $B = \frac{1}{\pi}$ (Wb/m²)

$\theta = 60°$

Area normal to the plane of the disc

$$= \pi r^2\cos 60° = \frac{\pi r^2}{2}$$

$$\text{Flux} = B\times\text{normal area} = \frac{0.2\times 0.2}{2} = 0.02\text{Wb}$$

15. (c) $\frac{\Delta\phi}{\Delta t} = \varepsilon = iR \Rightarrow \Delta\phi = (i\Delta t)R = QR$

$$\Rightarrow Q = \frac{\Delta\phi}{R}$$

16. (c) $i = \frac{e}{R} = \frac{nAdB}{\frac{dt}{R}}$

$$= \frac{20\times(25\times 10^{-4})\times 1000}{100} = 0.5\text{A}$$

17. (c) $q = \int idt = \frac{1}{R}\int edt = \frac{1}{R}\int\left(\frac{-d\phi}{dt}\right)dt$

$= \frac{1}{R}\int d\phi$ (taking only magnitude of e)

Hence, total charge induced in the conducting loop depends upon the total change in magnetic flux.

18. (b) $e = \frac{-(\phi_2-\phi_1)}{t} = \frac{-(0-NBA)}{t} = \frac{NBA}{t}$

$$t = \frac{NBA}{e} = \frac{50\times 2\times 10^{-2}\times 10^{-2}}{0.1} = 0.1\,s$$

19. (c) Using Faraday's law,

$$e = \frac{Bl^2\omega}{2} = \frac{1}{2}\times 0.1\times\left(\frac{1}{2}\right)^2\times 10 = \frac{1}{8} = 0.125\text{ V}$$

20. (c) $e = \frac{d\phi}{dt};\ i = \frac{e}{R} = \frac{1}{R}\frac{d\phi}{dt}$

$$\text{Total charge induced} = \int i\,dt = \int\frac{1}{R}\frac{d\phi}{dt}.dt$$

$$= \frac{1}{R}\int_{\phi_1}^{\phi_2}d\phi = \frac{1}{R}(\phi_2-\phi_1)$$

21. (c) When a circuit is broken, the induced e.m.f. is largest. So the answer is (c).

22. (a) North of magnet is moving away from solenoid 1 so end B of solenoid 1 is South and as south of magnet is approaching solenoid 2 so end C of solenoid 2 is South.

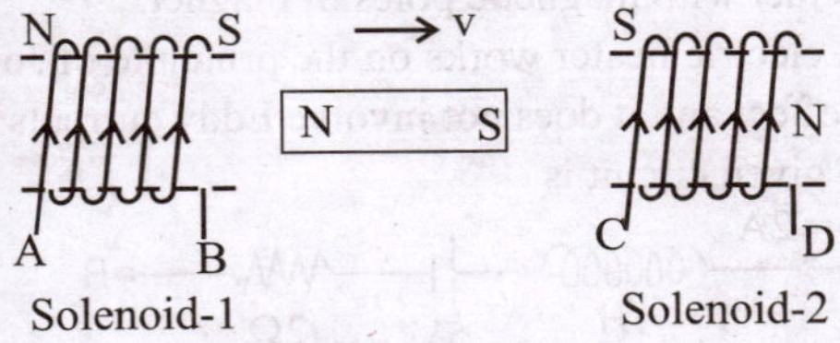

23. (b) Current will be induced,

when e⁻ comes closer the induced current will be anticlockwise

when e⁻ comes further induced current will be clockwise

NOTES Presence of magnetic flux is not enough. The amount of magnetic flux linked with the coil must change in order to produce any induced emf in the coil.

24. (c) Emf induced in side 1 of frame $e_1 = B_1 V\ell$

$$B_1 = \frac{\mu_0 I}{2\pi(x-a/2)}$$

Emf induced in side 2 of frame $e_2 = B_2 V\ell$

$$B_2 = \frac{\mu_0 I}{2\pi(x+a/2)}$$

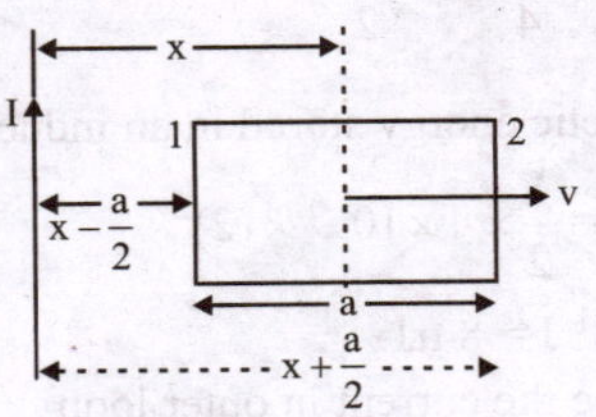

Emf induced in square frame

$e = B_1 V\ell - B_2 V\ell$

$$= \frac{\mu_0 I}{2\pi(x-a/2)}\ell V - \frac{\mu_0 I}{2\pi(x+a/2)}\ell V$$

$$\text{or, } e\propto\frac{1}{(2x-a)(2x+a)}$$

25. (a) $V = \frac{W}{Q}\Rightarrow W = QV$

26. (b) Length of conductor (l) = 0.4 m; Speed (v) = 7 m/s and magnetic field (B) = 0.9 Wb/ m². Induced e.m.f. (V) = Blv sin θ = 0.9 × 0.4 × 7 × sin 90° = 2.52 V.

27. (b) A. A magnetic pole will repel or attract magnetic sheet, so force is need to hold the sheet.

B. If sheet is non-magnetic, then magnetic pole will not exert force on it.

C. If it is conducting, then there will be eddy current in sheet, which opposes the motion. So forces is needed to move sheet with uniform speed.

D. The non-conducting and non-polar sheet do not interact with magnetic poles of magnet.

28. (d) An electric heater works on the principle of Joule's heating effect and it does not involve Eddy currents.

29. (d) The given circuit is

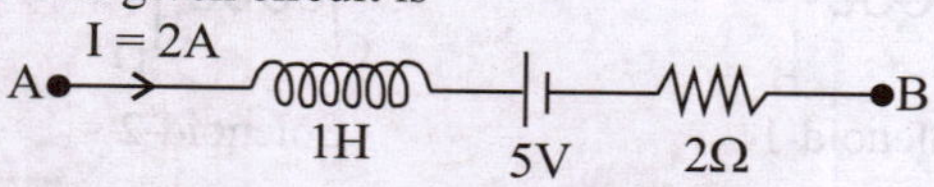

By KVL between points A and B,

$$V_A - \frac{LdI}{dt} - 5 - IR = V_B$$

$$V_A - V_B = \frac{LdI}{dt} + 5 + IR$$

Here, $I = 2A, \frac{dI}{dt} = 1\frac{A}{S}$

$\therefore V_A - V_B = 1 \times 1 + 5 + 2 \times 2 = 10$ V

30. (a) Self inductance $L = \mu_0\mu_r \times n \times A \times N$

$$\Rightarrow L = \mu_0\mu_r n \times A \times \frac{N}{l} \times l$$

$$\Rightarrow L = \mu_0\mu_r \times n^2 \times A \times l \Rightarrow L \propto n^2 l$$

$$\therefore \frac{L_A}{L_B} = \frac{n_A^2}{n_B^2} \times \frac{l_A}{l_B}$$

$$\Rightarrow \frac{L_A}{L_B} = \frac{1}{4} \times 2 = \frac{1}{2}$$

31. (d) Magnetic energy stored in an inductor

$$U = \frac{1}{2}Li^2 = \frac{1}{2} \times 4 \times 10^{-6} \times (2)^2$$

$= 8 \times 10^{-6}$ J $= 8$ μJ

32. (a) Let i be the current in outer loop

Magnetic field at centre $B = \frac{\mu_0 i}{2R_1}$

Magnetic flux through inner coil $\phi_i = B \times \pi R_2^2$

$$\phi_i = \frac{\mu_0 i}{2R_1} \times \pi R_2^2$$

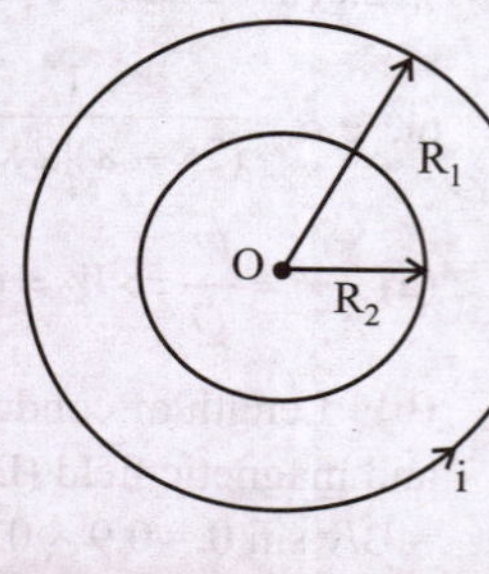

$$\phi_i = \frac{\mu_0 i}{2} \times \frac{\pi R_2^2}{R_1}$$

$\therefore \phi = Mi$

$$\therefore M = \left(\frac{\mu_0 \pi}{2}\right)\frac{R_2^2}{R_1} \text{ or, } M \propto \frac{R_2^2}{R_1}$$

33. (c) From question energy stored in inductor, $U = 25 \times 10^{-3}$ J

Current, I = 60 mA

Energy stored in inductor $U = \frac{1}{2}LI^2$

$$25 \times 10^{-3} = \frac{1}{2} \times L \times (60 \times 10^{-3})^2$$

$$L = \frac{25 \times 2 \times 10^6 \times 10^{-3}}{3600} = 13.89 \text{ Hm}$$

34. (d) Here, number of turns n = 1000; current through the solenoid i = 4A; flux linked with each turn $= 4 \times 10^{-3}$ Wb

∴ Total flux linked,

$= 1000[4 \times 10^{-3}] = 4$ Wb

$\phi_{total} = 4 \Rightarrow Li = 4 \Rightarrow L = 1$ H

NOTES Number of flux linkages with the coil ∝ current (I)
$N\phi \propto I \Rightarrow N\phi = L_I \Rightarrow L = \frac{N\phi}{I}$.

35. (c) Given: current I = 2.5 A

Inductance, L = 5H

Magnatic flux, φ = ?

We know, $\phi = LI \Rightarrow 5 \times 2.5$ Wb $= 12.5$ Wb

36. (c) Total number of turns in the solenoid, N = 500

Current, I = 2A.

Magnetic flux linked with each turn

$= 4 \times 10^{-3}$ Wb

As, $\phi = LI$ or $N\phi = LI \Rightarrow L = \frac{N\phi}{1}$

$$= \frac{500 \times 4 \times 10^{-3}}{2} \text{ henry} = 1 \text{ H.}$$

37. (b) Mutual Inductance of two coils

$$M = \sqrt{M_1 M_2} = \sqrt{2mH \times 8mH} = 4mH$$

NOTES It is not possible to have mutual inductance without self inductance but if may or may not be possible self inductance without mutual inductance.

38. (c) $L = 2$mH, $i = t^2e^{-t}$

$$E = -L\frac{di}{dt} = -L[-t^2e^{-t} + 2te^{-t}]$$

when $E = 0$,

$-e^{-t}t^2 + 2te^{-t} = 0$

or, $2t\,e^{-t} = e^{-t}t^2$

$\Rightarrow t = 2$ sec.

39. (b) $e = -M\frac{di}{dt} = -0.005 \times \frac{d(i_0 \sin \omega t)}{dt}$

$= -0.005 \times i_0 \times (\omega \cos \omega t)$

$e_{max.} = 0.005 \times i_0 \times \omega$ (when $\cos \omega t = -1$)

$= 0.005 \times 10 \times 100\pi = 5\pi$ V

40. (c) Initial current (I_1) = 10 A; Final current (I_2) = 0; Time (t) = 0.5 sec and induced e.m.f. (ε) = 220 V.

Induced e.m.f. (ε)

$= -L\frac{dI}{dt} = -L\frac{(I_2 - I_1)}{t} = -L\frac{(0-10)}{0.5} = 20L$

or, $L = \frac{220}{20} = 11H$

[where L = Self inductance of coil]

41. (b) $L = \frac{e}{\frac{di}{dt}} = \frac{e\,dt}{di} = \frac{5 \times 10^{-3}}{(3-2)} H = 5mH$

42. (c) $L = \frac{N\phi}{i}$; $\phi = BA$; $B = \frac{\mu_0 Ni}{2R}$

$\Rightarrow L = \frac{N}{i}\left(\frac{\mu_0 Ni}{2R}\right)A = \frac{\mu_0 N^2}{2R}A \Rightarrow L \propto N^2$

43. (c) $E = \frac{1}{2}Li^2 = \frac{1}{2} \times (100 \times 10^{-3}) \times 1^2 = 0.05$ J

NOTES In building a steady current in the circuit, the source emf has to do work against self inductance of coil and whatever energy consumed for tis work stored in magnetic field of coil-called magnetic potential energy (u) of coil $u = \frac{1}{2}LI^2 = \frac{N\phi i}{2}$

44. (d) Self inductance of a solenoid = $\frac{\mu n^2 A}{\ell}$

So, self inductance $\propto n^2$

So, inductance becomes 4 times when n is doubled.

45. (c) An inductor stores energy in its magnetic field.

46. (a) $e = L\frac{di}{dt}$

Given that L = 40 × 10^{-3} H,

di = 11 A – 1 A = 10 A

and dt = 4 × 10^{-3} s

$\therefore$ e = 40 × 10^{-3} × $\left(\frac{10}{4 \times 10^{-3}}\right) = 100V$

47. (d) Maximum value of emf generated by an ac generator

$\varepsilon_0 = NBA_\omega$

= 100 × 0.05 × 1 × 2π = 31.415 V

$\therefore \varepsilon_0 = 31.4$ V

48. (c) A current which changes its direction periodically is called alternating current.

7 Alternating Current

TREND ANALYSIS

TOPIC	2025		2024		2023		2022		2021		2020	
	Qns.	LOD	Qns.	LOD	Qns.	LOD	Qns.	LOD	Qns.	LOD	Qns.	LOD
AC Voltage Applied to a Resistor							1	Avg.			1	Easy
AC Voltage Applied to a Capacitor			1	Avg.			1	Avg.				
AC Voltage Applied to a Series LCR Circuit	1	Avg.			2	Avg.	1	Avg.	2	Avg.		
Power in AC Circuit: The Power Factor											1	Avg.
Transformers			1	Easy					1	Easy		

How to Code Qns for Better & Faster Revision

Type 1: Able to solve (No code Req)
Type 2: Solved with struggle or by seeing solution **(Use code NR means Needs Revision or orange stickers)**
Type 3: Tricky/ Could not solve on my own. Need help from Teacher/ Friend. **(Use code TQ means Tricky Question or red stickers)**

You can either hand write the code in boxes or else paste coloured stickers given at end of book.

Topic-7.2
AC Voltage Applied to a Resistor

1. A standard filament lamp consumes 100 W when connected to 200 V ac mains supply. The peak current through the bulb will be: **NCERT | Page-180 | RS 2022**

(a) 2 A (b) 0.707 A
(c) 1A (d) 1.414

2. The peak voltage of the ac source is equal to: **NCERT | Page-180 | 2022**

(a) the rms value of the ac source
(b) $\sqrt{2}$ times the rms values of the ac source
(c) $\frac{1}{\sqrt{2}}$ time the rms value of the ac source
(d) the value of voltage supplied to the circuit.

3. A 40 μF capacitor is connected to a 200 V, 50 Hz ac supply. The rms value of the current in the circuit is, nearly : **NCERT | Page-184 | 2020**

(a) 2.05 A (b) 2.5 A
(c) 25.1 A (d) 1.7 A

4. In an a.c. circuit, the r.m.s. value of current, i_{rms} is related to the peak current, i_0 by the relation **NCERT | Page-180 | 1994**

(a) $I_{rms} = \sqrt{2}\, I_0$ (b) $I_{rms} = \pi I_0$
(c) $I_{rms} = \frac{1}{\pi} I_0$ (d) $I_{rms} = \frac{1}{\sqrt{2}} I_0$

5. In a region of uniform magnetic induction B = 10^{-2} tesla, a circular coil of radius 30 cm and resistance π^2 ohm is rotated about an axis which is perpendicular to the direction of B and which forms a diameter of the coil. If the coil rotates at 200 rpm the amplitude of the alternating current induced in the coil is **NCERT | Page-179 | 1988**

(a) $4\pi^2$ mA (b) 30 mA
(c) 6 mA (d) 200 mA

Topic-7.4
AC Voltage Applied to an Inductor

6. In the circuit shown below, the inductance L is connected to an ac source. The current flowing in the circuit is $I = I_0 \sin\omega t$. The voltage drop (V_L) across L is **NCERT | Page- 182, 183 | Ph-II 2024**

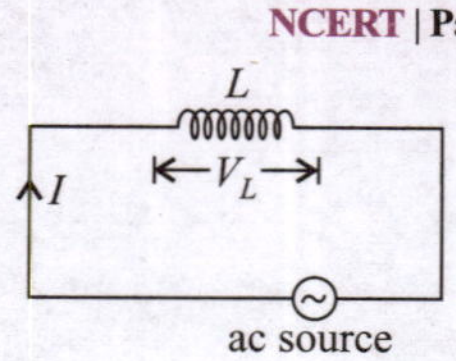

(a) $\omega L I_0 \sin \omega t$ (b) $\frac{I_0}{\omega L} \sin \omega t$
(c) $\frac{I_0}{\omega L} \cos \omega t$ (d) $\omega L I_0 \cos \omega t$

7. An ac source is connected in the given circuit. The value of ϕ will be **NCERT | Page-182 | Manipur 2023**

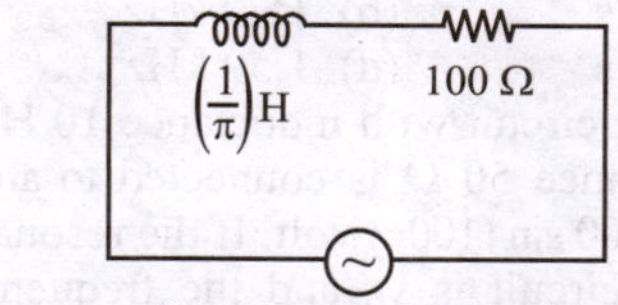

V = 220 sin(100 πt + ϕ) volt

$V = 220 \sin(100\pi t + \phi)$ volt

(a) 30° (b) 45° (c) 60° (d) 90°

8. An inductor of inductance 2 mH is connected to a 220 V, 50 Hz a.c. source. Let the inductive reactance in the circuit is X_1. If a 220 V dc source replaces the ac source in the circuit, then the inductive reactance in the circuit is X_2. X_1 and X_2 respectively are **NCERT | Page-182 | RS 2022**
(a) 0.628 Ω, infinity (b) 6.28 Ω, zero
(c) 6.28 Ω, infinity (d) 0.628 Ω, zero

9. A coil of self-inductance L is connected in series with a bulb B and an AC source. Brightness of the bulb decreases when **2013**
(a) number of turns in the coil is reduced
(b) a capacitance of reactance $X_C = X_L$ is included in the same circuit
(c) an iron rod is inserted in the coil
(d) frequency of the AC source is decreased

10. The current (*I*) in the inductance is varying with time according to the plot shown in figure. **NCERT | Page-182 | 2012**

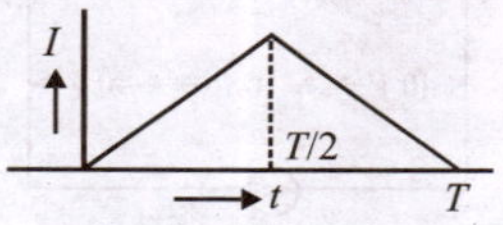

Which one of the following is the correct variation of voltage with time in the coil?

(a) 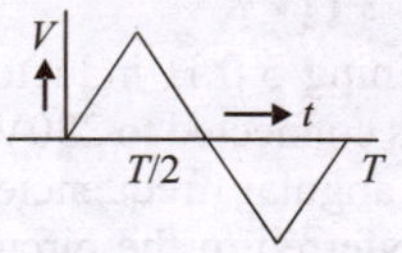
(b) 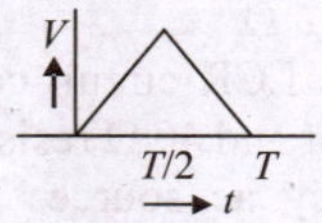
(c) 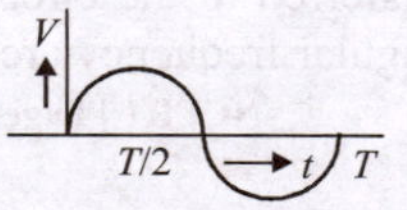
(d) 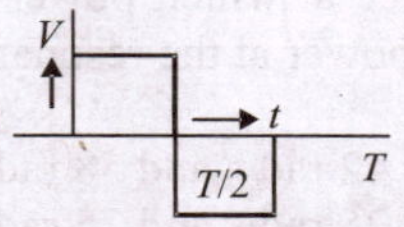

11. An ac voltage is applied to a resistance R and an inductor L in series. If R and the inductive reactance are both equal to 3Ω, the phase difference between the applied voltage and the current in the circuit is **2011**
(a) π/6 (b) π/4 (c) π/2 (d) zero

12. A coil has resistance 30 ohm and inductive reactance 20 ohm at 50 Hz frequency. If an ac source, of 200 volt, 100 Hz, is connected across the coil, the current in the coil will be **2011 M**
(a) 4.0 A (b) 8.0 A (c) $\frac{20}{\sqrt{13}}$ A (d) 2.0 A

Topic-7.5
AC Voltage Applied to a Capacitor

13. A 10 μF capacitor is connected to a 210 V, 50 Hz source as shown in figure. The peak current in the circuit is nearly (π = 3.14): **NCERT | Page-184 | 2024**

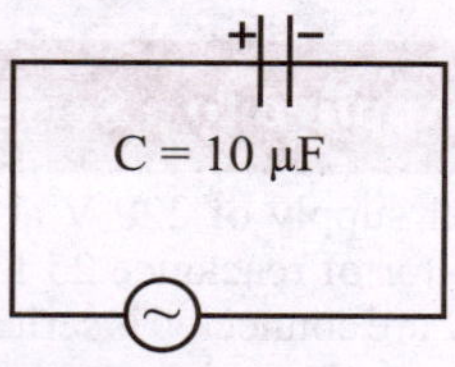

210V

(a) 0.58 A (b) 0.93 A
(c) 1.20 A (d) 0.35 A

14. Given below are two statements.
Statement I : In an ac circuit, the current through a capacitor leads the voltage across it.
Statement II : In a.c circuits containing pure capacitance only, the phase difference between the current and the voltage is π.In the light of the above statements, choose the most appropriate answer from the options given below: **NCERT | Page-184 | RS 2022**
(a) Statement I is incorrect but Statement II is correct
(b) Both Statement I and Statement II are correct
(c) Both Statement I and Statement II are incorrect
(d) Statement I is correct but Statement II is incorrect

15. A small signal voltage $V(t) = V_0 \sin \omega t$ is applied across an ideal capacitor C : **NCERT | Page-184 | 2016**
(a) Current I (t), lags voltage V(t) by 90°.
(b) Over a full cycle the capacitor C does not consume any energy from the voltage source.
(c) Current I (t) is in phase with voltage V(t).
(d) Current I (t) leads voltage V(t) by 180°.

16. In an ac circuit an alternating voltage $e = 200\sqrt{2} \sin 100$ t volts is connected to a capacitor of capacity 1 μF. The r.m.s. value of the current in the circuit is **NCERT | Page-184 | 2011**
(a) 10 mA (b) 100 mA
(c) 200 mA (d) 20 mA

17. A series R-C circuit is connected to an alternating voltage source. Consider two situations: **2015 RS**
(A) When capacitor is air filled.
(B) When capacitor is mica filled.
Current through resistor is i and voltage across capacitor is V then :
(a) $V_a > V_b$ (b) $i_a > i_b$
(c) $V_a = V_b$ (d) $V_a < V_b$

18. The r.m.s. value of potential difference V shown in the figure is **2011 M**

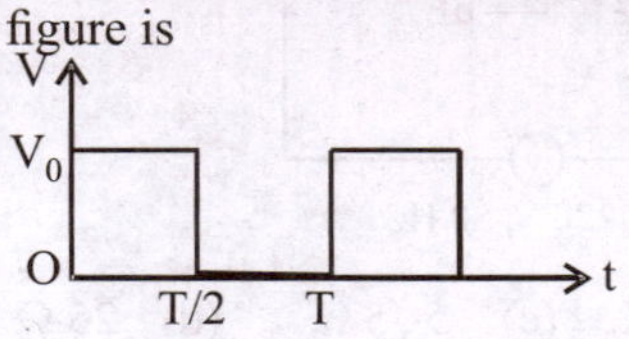

(a) V_0 (b) $V_0/\sqrt{2}$
(c) $V_0/2$ (d) $V_0/\sqrt{3}$

19. A capacitor has capacity *C* and reactance *X*. If capacitance and frequency become double, then reactance will be **NCERT | Page-184 | 2001**
(a) 4*X* (b) *X*/2 (c) *X*/4 (d) 2*X*

Topic-7.6

AC Voltage Applied to a Series LCR Circuit

20. To an ac power supply of 220 V at 50 Hz, a resistor of 20 Ω, a capacitor of reactance 25 Ω and an inductor of reactance 45 Ω are connected is series. The corresponding current in the circuit and the phase angle between the current and the voltage is, respectively- **NCERT | Page-187, 188 | 2025**

(a) 7.8 A and 30° (b) 7.8 A and 45°
(c) 15.6 and 30° (d) 15.6 and 45°

21. The amplitude of the charge oscillating in a circuit decreases exponentially as $Q = Q_0 e^{-Rt/2L}$, where Q_0 is the charge at $t = 0$ s. The time at which charge amplitude decreases to 0.50 Q_0 is nearly: **Ph-II 2024**
[Given that $R = 1.5\ \Omega$, $L = 12$ mH, ln (2) = 0.693]

(a) 19.01 ms (b) 11.09 ms
(c) 19.01 s (d) 11.09 s

22. If Z_1 and Z_2 are the impedances of the given circuits (a) and (b) as shown in figures, then choose the correct option. **NCERT | Page-187 | Manipur 2023**

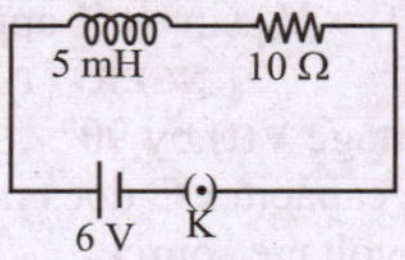

figure (a)

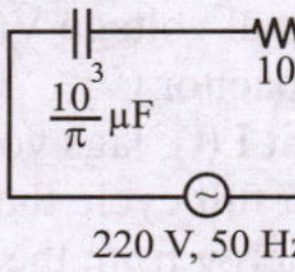

figure (b)

(a) $Z_1 = Z_2$ (b) $Z_1 > Z_2$
(c) $Z_1 < Z_2$ (d) $Z_1 + Z_2 = 20\ \Omega$

23. For very high frequencies, the effective impedance of the circuit (shown in the figure) will be **NCERT | Page-186| Manipur 2023**

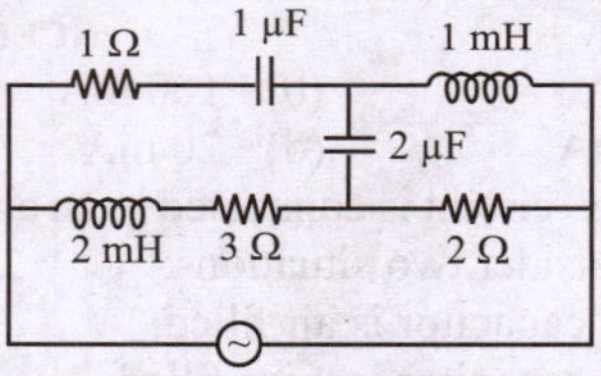

(a) 1 Ω (b) 3 Ω (c) 4 Ω (d) 6 Ω

24. The net impedance of circuit (as shown in figure) will be **NCERT | Page-187 | 2023**

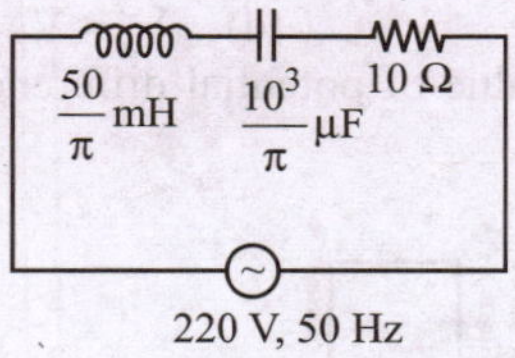

(a) $10\sqrt{2}\ \Omega$ (d) 15 Ω (c) $5\sqrt{5}\ \Omega$ (d) 25 Ω

25. In a series LCR circuit, the inductance L is 10 mH, capacitance C is 1 μF and resistance R is 100 Ω. The frequency at which resonance occurs is **NCERT | Page-189 | 2023**

(a) 15.9 rad/s (b) 15.9 kHz
(c) 1.59 rad/s (d) 1.59 kHz

26. A series LCR circuit with inductance 10 H, capacitance 10 μF, resistance 50 Ω is connected to an ac source of voltage, V = 200 sin (100 t) volt. If the resonant frequency of the LCR circuit is v_0 and the frequency of the ac source is v, then **NCERT | Page-181 | 2022**

(a) $v_0 = v = \frac{50}{\pi}$ Hz

(b) $v_0 = \frac{50}{\pi}$ Hz, $v = 50$Hz

(c) $v = 100$ Hz; $v_0 = \frac{100}{\pi}$ Hz

(d) $v_0 = v = 50$ Hz

27. An inductor of inductance L, a capacitor of capacitance C and a resistor of resistance 'R' are connected in series to an ac source of potential difference 'V' volts as shown in figure.
Potential difference across L, C and R is 40 V, 10 V and 40 V, respectively. The amplitude of current flowing through LCR series circuit is $10\sqrt{2}$ A. The impedance of the circuit is **NCERT | Page-187 | 2021**

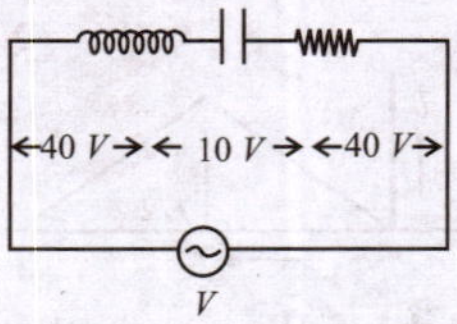

(a) 5 Ω (b) $4\sqrt{2}\ \Omega$
(c) $5\sqrt{2}\ \Omega$ (d) 4 Ω

28. A series LCR circuit containing 5.0 H inductor, 80μ F capacitor and 40 Ω resistor is connected to 230V variable frequency ac source. The angular frequencies of the source at which power transferred to the circuit is half the power at the resonant angular frequency are likely to be **NCERT | Page-189 | 2021**

(a) 42 rad/s and 58 rad/s
(b) 25 rad/s and 75 rad/s
(c) 50 rad/s and 25 rad/s
(d) 46 rad/s and 54 rad/s

29. A circuit when connected to an AC source of 12 V gives a current of 0.2 A. The same circuit when connected to a DC source of 12 V, gives a current of 0.4 A. The circuit is **NCERT | Page-187 | Odisha 2019**

(a) Series LCR (b) Series LR
(c) Series RC (d) Series LC

30. In the given circuit the reading of voltmeter V_1 and V_2 are 300 volts each. The reading of the voltmeter V_3 and ammeter A are respectively **2010**

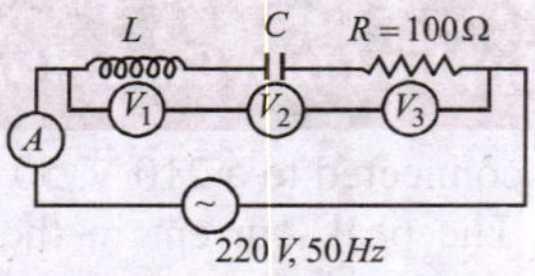

(a) 150 V, 2.2 A (b) 220 V, 2.2 A
(c) 220 V, 2.0 A (d) 100 V, 2.0 A

31. A condenser of capacity C is charged to a potential difference of V_1. The plates of the condenser are then connected to an ideal inductor of inductance L. The current through the inductor when the potential difference across the condenser reduces to V_2 is **2010**

(a) $\left(\frac{C(V_1^2-V_2^2)}{L}\right)^{1/2}$ (b) $\left(\frac{C(V_1-V_2)^2}{L}\right)^{1/2}$

(c) $\frac{C(V_1^2-V_2^2)}{L}$ (d) $\frac{C(V_1-V_2)}{L}$

32. What is the value of inductance L for which the current is maximum in a series LCR circuit with C = 10 μF and ω = $1000s^{-1}$? **2007**

(a) 1 mH
(b) cannot be calculated unless R is known
(c) 10 mH
(d) 100 mH

33. In a circuit, L, C and R are connected in series with an alternating voltage source of frequency f. The current leads the voltage by 45°. The value of C is **2005**

(a) $\frac{1}{\pi f(2\pi fL-R)}$ (b) $\frac{1}{2\pi f(2\pi fL-R)}$

(c) $\frac{1}{\pi f(2\pi fL+R)}$ (d) $\frac{1}{2\pi f(2\pi fL+R)}$

34. A coil of 40 henry inductance is connected in series with a resistance of 8 ohm and the combination is joined to the terminals of a 2 volt battery. The time constant of the circuit is **2004**

(a) 20 seconds (b) 5 seconds
(c) 1/5 seconds (d) 40 seconds

35. In an experiment, 200 V A.C. is applied at the ends of an LCR circuit. The circuit consists of an inductive reactance $(X_L) = 50\ \Omega$, capacitive reactance $(X_C) = 50\ \Omega$ and ohmic resistance (R) = 10 Ω. The impedance of the circuit is **NCERT | Page-186 | 1996**

(a) 10Ω (b) 20Ω (c) 30Ω (d) 40Ω

36. An LCR series circuit is connected to a source of alternating current. At resonance, the applied voltage and the current flowing through the circuit will have a phase difference of **1994**

(a) π (b) $\frac{\pi}{2}$

(c) $\frac{\pi}{4}$ (d) 0

37. The time constant of C–R circuit is **1992**

(a) 1/CR (b) C/R
(c) CR (d) R/C

Topic-7.7

Power in AC Circuit: The Power Factor

38. The maximum power is dissipated for an ac in a/an **NCERT | Page-190| Manipur 2023**

(a) Inductive circuit (b) Capacitive circuit
(c) Resistive circuit (d) LC circuit

39. The instantaneous values of alternating current and voltages in a circuit are given as

$i = \frac{1}{\sqrt{2}}\sin(100\pi t)$ amper

$e = \frac{1}{\sqrt{2}}\sin(100\pi t + \pi/3)$ Volt

The average power in Watts consumed in the circuit is : **NCERT | Page-191 | 2012 M**

(a) $\frac{1}{4}$ (b) $\frac{\sqrt{3}}{4}$ (c) $\frac{1}{2}$ (d) $\frac{1}{8}$

40. A series LCR circuit is connected to an ac voltage source. When L is removed from the circuit, the phase difference between current and voltage is $\frac{\pi}{3}$. If instead C is removed from the circuit, the phase difference is again $\frac{\pi}{3}$ between current and voltage. The power factor of the circuit is: **NCERT | Page-191 | 2020, 2012**

(a) 0.5 (b) 1.0 (c) –1.0 (d) zero

41. An inductor 20 mH, a capacitor 100 μF and a resistor 50Ω are connected in series across a source of emf, V = 10 sin 314 t. The power loss in the circuit is **NCERT | Page-191 | 2018**

(a) 0.79 W (b) 0.43 W
(c) 1.13 W (d) 2.74 W

42. Figure shows a circuit that contains three identical resistors with resistance R = 9.0 Ω each, two identical inductors with inductance L = 2.0 mH each, and an ideal battery with emf ε = 18 V. The current 'i' through the battery just after the switch closed is **2017**

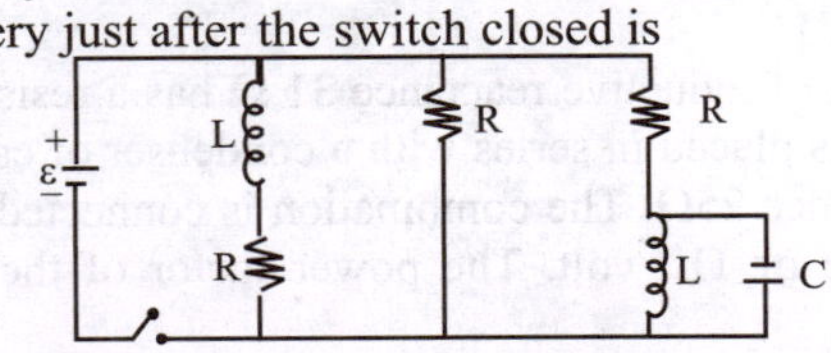

(a) 0.2 A (b) 2 A (c) 0 (d) 2 mA

43. An inductor 20 mH, a capacitor 50 μF and a resistor 40Ω are connected in series across a source of emf V = 10 sin 340 t. The power loss in A.C. circuit is : **NCERT | Page-191 | 2016**

(a) 0.51 W (b) 0.67 W
(c) 0.76 W (d) 0.89 W

44. A resistance 'R' draws power 'P' when connected to an AC source. If an inductance is now placed in series with the resistance, such that the impedance of the circuit becomes 'Z', the power drawn will be **NCERT | Page-191 | 2015**

(a) $P\sqrt{\frac{R}{Z}}$ (b) $P\left(\frac{R}{Z}\right)$ (c) P (d) $P\left(\frac{R}{Z}\right)^2$

45. Power dissipated in an LCR series circuit connected to an a.c source of emf ε is 2009

(a) $\dfrac{\varepsilon^2\sqrt{R^2+\left(L\omega-\dfrac{1}{C\omega}\right)^2}}{R}$

(b) $\dfrac{\varepsilon^2\left[R^2+\left(L\omega-\dfrac{1}{C\omega}\right)^2\right]}{R}$

(c) $\dfrac{\varepsilon^2 R}{\sqrt{R^2+\left(L\omega-\dfrac{1}{C\omega}\right)^2}}$

(d) $\dfrac{\varepsilon^2 R}{\left[R^2+\left(L\omega-\dfrac{1}{C\omega}\right)^2\right]}$

46. In an a.c circuit the e.m.f. (e) and the current (i) at any instant are given respectively by
$e = E_0 \sin \omega t$ NCERT | Page-190 | 2008
$i = I_0 \sin(\omega t - \phi)$
The average power in the circuit over one cycle of a.c. is

(a) $\dfrac{E_0 I_0}{2}$ (b) $\dfrac{E_0 I_0}{2}\sin\phi$
(c) $\dfrac{E_0 I_0}{2}\cos\phi$ (d) $E_0 I_0$

47. In an A.C. circuit with voltage V and current I the power dissipated is NCERT | Page-190 | 1997
(a) dependent on the phase between V and I
(b) $\dfrac{1}{\sqrt{2}}VI$
(c) $\dfrac{1}{2}VI$
(d) VI

48. A coil of inductive reactance 31 Ω has a resistance of 8 Ω. It is placed in series with a condenser of capacitative reactance 25Ω. The combination is connected to an a.c. source of 110 volt. The power factor of the circuit is 2006
(a) 0.64 (b) 0.80
(c) 0.33 (d) 0.56

49. In a series resonant circuit, having *L*, *C* and *R* as its elements, the resonant current is *i*. The power dissipated in the circuit at resonance is 2002

(a) $\dfrac{i^2R}{\left(\omega L-\dfrac{1}{\omega C}\right)}$ (b) zero

(c) $i^2\omega L$ (d) i^2R

where ω is the angular resonance frequency.

Topic-7.8
Transformers

50. A step up transformer is connected to an ac mains supply of 220 V to operate at 11000 V, 88 watt. The current in the secondary circuit, ignoring the power loss in the transformer, is NCERT | Page-195 | Ph-II 2024
(a) 8 mA (b) 4 mA (c) 0.4 A (d) 4 A

51. In an ideal transformer, the turns ratio is $\dfrac{N_P}{N_S}=\dfrac{1}{2}$. The ratio $V_S : V_P$ is equal to (the symbols carry their usual meaning) : NCERT | Page-195 | 2024
(a) 1 : 2 (b) 2 : 1 (c) 1 : 1 (d) 1 : 4

52. A step down transformer connected to an ac mains supply of 220 *V* is made to operate at 11 *V*, 44 *W* lamp. Ignoring power losses in the transformer, what is the current in the primary circuit? NCERT | Page-194 | 2021
(a) 4 A (b) 0.2 A (c) 0.4 A (d) 2 A

53. A 12 V, 60 W lamp is connected to the se ondary of a step-down transformer, whoseprmary is connected to ac mains of 220 V. Assuming the transformer to be ideal, what is the current in the primary winding? NCERT | Page-194, 195 | 2015
(a) 0.27 A (b) 2.7 A
(c) 3.7 A (d) 0.37 A

54. A transformer having efficiency of 90% is working on 200V and 3kW power supply. If the current in the secondary coil is 6A, the voltage across the secondary coil and the current in the primary coil respectively are : NCERT | Page-194 | 2014
(a) 300V, 15A (b) 450V, 15A
(c) 450V, 13.5A (d) 600V, 15A

55. The primary of a transformer when connected to a dc battery of 10 volt draws a current of 1 mA. The number of turns of the primary and secondary windings are 50 and 100 respectively. The voltage in the secondary and the current drawn by the circuit in the secondary are respectively NCERT | Page-194 | Kar. 2013
(a) 20 V and 0.5 mA
(b) 20 V and 2.0 mA
(c) 10 V and 0.5 mA
(d) Zero and therefore no current

56. A 220 volts input is supplied to a transformer. The output circuit draws a current of 2.0 ampere at 440 volts. If the efficiency of the transformer is 80%, the current drawn by the primary windings of the transformer is NCERT | Page-194 | 2010
(a) 3.6 ampere (b) 2.8 ampere
(c) 2.5 ampere (d) 5.0 ampere

57. The primary and secondary coil of a transformer have 50 and 1500 turns respectively. If the magnetic flux φ linked with the primary coil is given by $\phi = \phi_0 + 4t$, where φ is in webers, *t* is time in seconds and ϕ_0 is a constant, the output voltage across the secondary coil is NCERT | Page-195 | 2007
(a) 120 volts (b) 220 volts
(c) 30 volts (d) 90 volts

58. A transformer is used to light a 100 W and 110 V lamp from a 220 V mains. If the main current is 0.5 amp, the efficiency of the transformer is approximately **NCERT | Page-195 | 2007**

(a) 50% (b) 90% (c) 10% (d) 30%

59. The core of a transformer is laminated because **NCERT | Page-195 | 2006**

(a) the weight of the transformer may be reduced
(b) rusting of the core may be prevented
(c) ratio of voltage in primary and secondary may be increased
(d) energy losses due to eddy currents may be minimised

60. A step-up transformer operates on a 230 V line and supplies a load of 2 ampere. The ratio of the primary and secondary windings is 1 : 25. The current in the primary is **NCERT | Page-195 | 1998**

(a) 25 A (b) 50 A
(c) 15 A (d) 12.5 A

61. The primary winding of a transformer has 500 turns whereas its secondary has 5000 turns. The primary is connected to an A.C. supply of 20 V, 50 Hz. The secondary will have an output of **NCERT | Page-195 | 1997**

(a) 2 V, 5 Hz (b) 200 V, 500 Hz
(c) 2V, 50 Hz (d) 200 V, 50 Hz

62. Eddy currents are produced when **1988**

(a) a metal is kept in varying magnetic field
(b) a metal is kept in steady magnetic field
(c) a circular coil is placed in a magnetic field
(d) through a circular coil, current is passed

ANSWER KEY

1	(b)	8	(d)	15	(b)	22	(c)	29	(b)	36	(d)	43	(a)	50	(a)	57	(a)
2	(b)	9	(c)	16	(d)	23	(b)	30	(b)	37	(c)	44	(d)	51	(b)	58	(b)
3	(b)	10	(d)	17	(a)	24	(c)	31	(a)	38	(c)	45	(d)	52	(b)	59	(d)
4	(d)	11	(b)	18	(b)	25	(d)	32	(d)	39	(d)	46	(c)	53	(a)	60	(b)
5	(c)	12	(a)	19	(c)	26	(a)	33	(d)	40	(b)	47	(a)	54	(b)	61	(d)
6	(d)	13	(b)	20	(b)	27	(a)	34	(b)	41	(a)	48	(b)	55	(d)	62	(a)
7	(b)	14	(d)	21	(b)	28	(d)	35	(a)	42	(b)	49	(d)	56	(d)		

Hints & Solutions

1. (b) Power, $P = V_{rms} I_{rms}$

$$\Rightarrow i_{rms} = \frac{P}{V_{rms}} = \frac{1}{2}$$

Peak current,

$$i_0 = i_{rms} \times \sqrt{2} = \frac{1}{\sqrt{2}} = 0.707 A$$

2. (b) We know that

$$V_{rms} = \frac{V_0}{\sqrt{2}} \Rightarrow V_0 = \sqrt{2}\, V_{rms}$$

3. (b) Given :

Capacitance, $C = 40\ \mu F = 40 \times 10^{-6}$ F

Frequency, $f = 50$ Hz $\quad \therefore \omega = 2\pi f = 100\pi$

$$\varepsilon_{rms} = 200 \text{ V} \quad \therefore I_{rms} = \frac{\varepsilon_{rms}}{X_C} = \frac{\varepsilon_{rms}}{\frac{1}{C\omega}}$$

$= 200 \times 40 \times 10^{-6} \times 2\pi \times 50 = 2.5$ A.

4. (d) $I_{rms} = \dfrac{I_0}{\sqrt{2}}$

5. (c) $I_0 = \dfrac{E_0}{R} = \dfrac{nBA\omega}{R}$

Given, $n = 1$, $B = 10^{-2}$ T, $A = \pi(0.3)^2 m^2$, $R = \pi^2$

$f = (200/60)$ and $\omega = 2\pi(200/60)$

Substituting these values and solving, we get

$I_0 = 6 \times 10^{-3}$ A = 6mA

6. (d) V_L leads current I by phase angle $\dfrac{\pi}{2}$

$$\therefore V_L = V_0 \sin\left(\omega t + \frac{\pi}{2}\right) = V_0 \cos \omega t \ (\because I = I_0 \sin \omega t)$$

$V_0 = I_0 X_L$

$\Rightarrow V_L = I_0 X_L \cos(\omega t) = I_0 \omega L \cos(\omega t)$

7. (b) Inductive reactance, $X_L = \omega L$

$$X_L = 100\,\pi \times \frac{1}{\pi} = 100\ \Omega$$

$\therefore \tan\phi = \frac{X_L}{R} = \frac{100}{100} = 1 \;\therefore \phi = 45°$

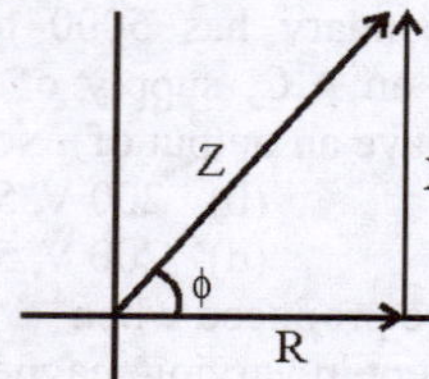

In case of DC circuit, Inductive reactance, $X_L = 0$ as DC frequency $V_{DC} = 0$ and capacitive reactance,

$$Xe = \frac{1}{C2\pi V_{DC}} = \infty$$

8. **(d)** When A.C. source is applied

$X_1 = \omega L = 2\pi fL$

$= 2\pi \times 50 \times 2 \times 10^{-3} = 0.628\Omega$

And when D.C. source is applied

$X_2 = \omega L = 2\pi fL = 0 \; (\because f = 0)$

9. **(c)** By inserting iron rod in the coil,

$\downarrow L \uparrow \;\; Z \;\; \uparrow \;\; I \;\downarrow$ so brightness

NOTES Inductive reactance, $XL = \omega L$. On inserting iron rod in the coil, L increases. As a result Impedance $Z = \sqrt{X_L^2 + R^2} = \sqrt{(\omega L)^2 + R^2}$ also increases. So, Current, $I = \frac{V}{Z}$ decreases..

10. **(d)** $V = -L\frac{di}{dt}$

Here $\frac{di}{dt}$ is + ve for $\frac{T}{2}$ time and

$\frac{di}{dt}$ is – ve for next $\frac{T}{2}$ time so

11. **(b)** The phase difference ϕ is given by

$\tan\phi = \frac{X_L}{R} = \frac{3}{3} = 1 \Rightarrow \phi = \frac{\pi}{4}$.

12. **(a)** If $\omega = 50 \times 2\pi$ then $\omega L = 20\Omega$

If $\omega' = 100 \times 2\pi$ then $\omega' L = 40\Omega$

Current flowing in the coil is

$$I = \frac{200}{Z} = \frac{200}{\sqrt{R^2 + (\omega' L)^2}} = \frac{200}{\sqrt{(30)^2 + (40)^2}}$$

$I = 4A$.

13. **(b)** Capacitive Reactance, $X_C = \frac{1}{\omega C} = \frac{1}{2\pi fC}$

$$= \frac{1}{2 \times 3.14 \times 50 \times 10 \times 10^{-6}} = \frac{1000}{3.14}$$

Peak current, $i_0 = \sqrt{2} i_{rms} = \sqrt{2}.\frac{V_{rms}}{X_C}$

$$= \sqrt{2} \times \frac{210}{1000} \times 3.14 = 0.932 \simeq 0.93A$$

14. **(d)** In AC circuit current through the capacitor leads the potential difference across it by a phase $\frac{\pi}{2}$ not π

15. **(b)** As we know, power $P = V_{rms} \cdot I_{rms} \cos\phi$

as $\cos\phi = 0$ $\quad (\because \phi = 90°)$

$\therefore$ Power consumed = 0 (in one complete cycle)

16. **(d)** $V_{rms} = \frac{200\sqrt{2}}{\sqrt{2}} = 200V$

$$I_{rms} = \frac{V_{rms}}{X_C} = \frac{200}{\frac{1}{100 \times 10^{-6}}} = 2 \times 10^{-2} = 20mA$$

17. **(a)** For series R – C circuit,

capacitive reactance,

$$Z_c = \sqrt{R^2 + \left(\frac{1}{C\omega}\right)^2}$$

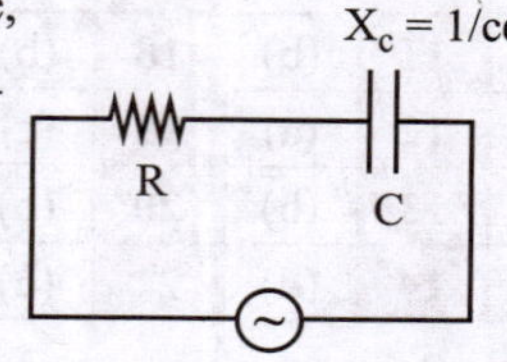

Current $i = \frac{V}{Z_c}$

$$= \frac{V}{\sqrt{R^2 + \left(\frac{1}{C\omega}\right)^2}}$$

$$V_c = iX_c = \frac{V}{\sqrt{R^2 + \left(\frac{1}{C\omega}\right)^2}} \times \frac{1}{C\omega}$$

$$V_c = \frac{V}{\sqrt{(RC\omega)^2 + 1}}$$

If we fill a di-electric material like mica instead of air then capacitance $C\uparrow \Rightarrow V_c\downarrow$

So, $V_a > V_b$

18. **(b)** $V_{rms} = \sqrt{\frac{(T/2)V_0^2 + 0}{T}} = \frac{V_0}{\sqrt{2}}$.

19. **(c)** Capacitive reactance, $X = \frac{1}{\omega C} = \frac{1}{2\pi fC}$

$\Rightarrow X \propto \frac{1}{fC}$

$\therefore \frac{X'}{X} = \frac{f}{f'} \times \frac{C}{C'} = \frac{f}{2f} \times \frac{C}{2C} = \frac{1}{4} \Rightarrow X' = \frac{X}{4}$

20. **(b)** In AC circuit,

$X_C = 25\Omega, X_L = 45\Omega$

$\therefore\ Z = \sqrt{R^2 + (X_L - X_C)^2} = \sqrt{20^2 + (45-25)^2}$

$= 20\sqrt{2}\Omega$

$\therefore I_{rms} = \frac{V_{rms}}{Z} = \frac{220}{20\sqrt{2}} = \frac{11}{\sqrt{2}} = 7.8A$

Also, $\tan\phi = \frac{X_L - X_C}{R} = \frac{45-25}{20} = \frac{20}{20} = 1$

$\therefore\ \phi = 45°$

21. **(b)** Using $Q = Q_0 e^{-Rt/2L}$

substituting necessary values,

$0.5Q_0 = Q_0 e^{-Rt/2L} \Rightarrow \frac{1}{2} = e^{-Rt/2L}$

Taking log on both sides

$\ell n\left(\frac{1}{2}\right)\ \ell n\, e^{-Rt/2L} \Rightarrow \ell n 2 = \frac{Rt}{2L}$

$t = \frac{2L\ell n2}{R} = \frac{2\times12\times10^{-3}\times0.693}{1.5} = 11.09$ ms

22. **(c)** As, D. C. voltage is supplied so impedance

$Z_1 = R = 10\ \Omega$

And $Z_2 = \sqrt{R^2 + (X_c)^2}$

$= \sqrt{(10)^2 + \left[\frac{1}{2\times\pi\times50\times\frac{10^3}{\pi}\times10^{-6}}\right]^2}$

$\Rightarrow Z_2 = \sqrt{(10)^2 + (10)^2} = 10\sqrt{2}$

$\therefore\ Z_2 > Z_1$ or $Z_1 < Z_2$

23. **(b)** Inductive reactance, $X_L = \omega L$ and

Capacitive reactance, $X_C = \frac{1}{\omega C}$

At very high frequency $X_c = 0_1\ X_L = \infty$

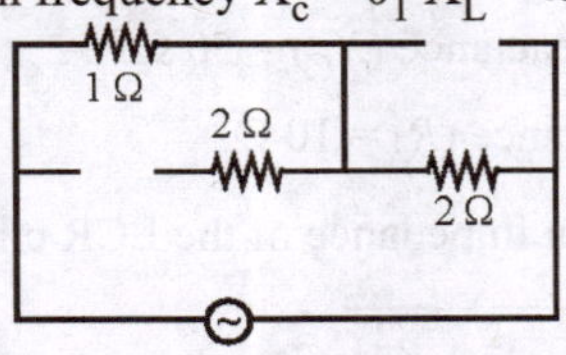

Therefore the effective impedance of the circuit,

$Z = (1 + 2)\ \Omega = 3\ \Omega$

24. **(c)** Inductance, $L = \frac{50}{\pi}$ mH

Resistance, $R = 10\ \Omega$

Capacitance, $C = \frac{10^3}{\pi}\times10^{-6}$

Inductive reactance, $X_L = \omega L = 2\pi f L$

$= 2\pi\times50\times\frac{50}{\pi}\times10^{-3} = 5\ \Omega$

Capacitive reactance, $X_C = \frac{1}{\omega C} = \frac{1}{2\pi f C}$

$= \frac{1\times\pi}{2\pi\times50\times10^3\times10^{-6}} = \frac{10^3}{100} = 10\ \Omega$

Impedance $Z = \sqrt{(X_C - X_L)^2 + R^2}$

$= \sqrt{(10-5)^2 + 10^2}$

$\Rightarrow\ Z = \sqrt{125} = 5\sqrt{5}\ \Omega$

25. **(d)** For resonance frequency $f = \frac{1}{2\pi\sqrt{LC}}$

$\Rightarrow f = \frac{1}{2\times\pi\times\sqrt{10\times10^{-3}\times1\times10^{-6}}} = \frac{10^4}{2\pi}$

$= \frac{10^4}{2\times22/7} = 1.591\times10^3$

$\therefore f = 1.591$ kHz

26. **(a)** At resonance

$X_L = X_C \Rightarrow \omega_r L = \frac{1}{\omega_r C} \Rightarrow \omega_r = \frac{1}{\sqrt{LC}}$

$\Rightarrow\ f_x = \frac{1}{2\pi\sqrt{LC}}$

$\Rightarrow\ v_0 = \frac{1}{2\pi\sqrt{10\times10\times10^{-6}}}$

$\Rightarrow\ v_0 = \frac{10^3}{10\times2\pi} = \frac{50}{\pi}$

So, if frequency of AC source is $\frac{50}{\pi}$ ie $v = \frac{50}{\pi}$ then there will only be resonance.

NOTES In series LCR circuit resonant frequency doesn't depend upon the resistance of the circuit.
Current, $i = i_0 \sin(wt \pm f)$.

27. **(a)** Impedance $z = \frac{V_{RMS}}{I_{RMS}}$

$V_{RMS} = \sqrt{V_R^2 + (V_L - V_C)^2}$

$= \sqrt{(40)^2 + (40-10)^2} = 50\ V$

$I_{RMS} = \frac{I_0}{\sqrt{2}} = \frac{10\sqrt{2}}{\sqrt{2}} = 10A \therefore Z = \frac{V_{RMS}}{I_{RMS}} = \frac{50}{10} = 5\ \Omega$

NOTES Impedance (z) is the oppsition offered by ac circuits to the flow of ac. For LCR series circuit.

Voltage $V = \sqrt{V_R^2 + (V_L - V_C)^2}$

and $Z = \sqrt{R^2 + (X_L - X_C)^2}$.

28. **(d)** For LCR series circuit resonance frequency

$\omega_0 = \frac{1}{\sqrt{LC}} = \frac{1}{\sqrt{5 \times 80 \times 10^{-6}}} = 50\ rad/s$

Half power frequency, $\omega = \omega_0 \pm \frac{R}{2L}$

i.e. $\omega_L = 50 - \frac{40}{2 \times 5} = 46\ rad/s$ and

$\omega_H = 50 + \frac{40}{2 \times 5} = 54\ rad/s$

29. **(b)** $I_1 = \frac{V}{Z}$

$I_1 = \frac{12}{\sqrt{R^2 + (X_L - X_C)^2}} = 0.2\ A$

In second case, the capacitor would provide infinite resistance but current is present in circuit, it means resistor and inductor can be present in the circuit. Since current with AC source and DC source are different, inductor must be present with resistance.

30. **(b)** As $V_L = V_C = 300\ V$, resonance will take place

$\therefore \quad V_R = 220\ V$

Current, $I = \frac{220}{100} = 2.2A$

$\therefore$ reading of $V_3 = 220\ V$

and reading of $A = 2.2\ A$

31. **(a)** $q = CV_1 \cos\omega \Rightarrow i = \frac{dq}{dt} = -\omega Cv_1 \sin\omega t$

Also, $\omega^2 = \frac{1}{LC}$ and $V = V_1 \cos\omega t$

At $t = t_1$, $V = V_2$ and $i = -\omega CV_1 \sin\omega t_1$

$\therefore \cos\omega t_1 = \frac{V_2}{V_1}$ (–ve sign gives direction)

Hence,

$i = V_1\sqrt{\frac{C}{L}}\left(1 - \frac{V_2^2}{V_1^2}\right)^{1/2}$

$= \left(\frac{C(V_1^2 - V_2^2)}{L}\right)^{1/2}$

32. **(d)** Condition for which the current is maximum in a series LCR circuit is,

$\omega = \frac{1}{\sqrt{LC}} \Rightarrow 1000 = \frac{1}{\sqrt{L(10 \times 10^{-6})}}$

$\Rightarrow \quad L = 100\ mH$

33. **(d)** From figure,

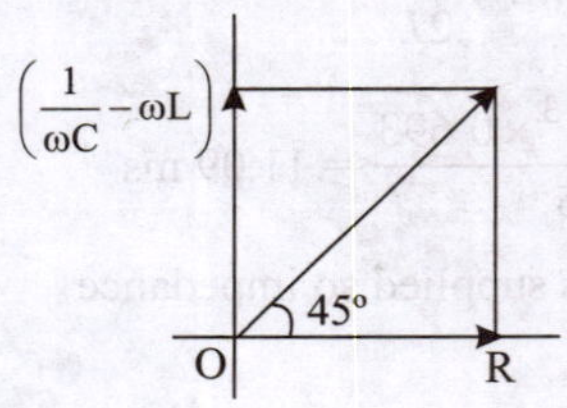

$\tan 45° = \frac{\frac{1}{\omega C} - \omega L}{R} \Rightarrow \frac{1}{\omega C} - \omega L = R$

$\Rightarrow \frac{1}{\omega C} = R + \omega L \left(\because\ \omega = \frac{2\pi}{T} = 2\mu f\right)$

$C = \frac{1}{\omega(R + \omega L)} = \frac{1}{2\pi f(R + 2\pi f\ L)}$

34. **(b)** Time constant is L/R

Given, $L = 40H$ & $R = 8\Omega$

$\therefore \tau = 40/8 = 5$ sec.

35. **(a)** Given : Supply voltage $(V_{ac}) = 200$ V

Inductive reactance $(X_L) = 50\ \Omega$

Capacitive reactance $(X_C) = 50\ \Omega$

Ohmic resistance $(R) = 10\ \Omega$.

We know that impedance of the LCR circuit (Z)

$= \sqrt{\{(X_L - X_C)^2 + R^2\}}$

$= \sqrt{\{(50-50)^2 + (10)^2\}} = 10\Omega$

36. **(d)** At resonance, $\omega L = \dfrac{1}{\omega C}$. The circuit behaves as if it contains R only. So, phase difference = 0

NOTES At resonance, impedance is minimum Zmin = R and current is maximum, given by

$$I_{max} = \frac{E}{Z_{min}} = \frac{E}{R}$$

It is interesting to note that before resonance the current leads the applied emf, at resonance it is in phase, and after resonance it lags behind the emf. LCR series circuit is also called as acceptor circuit and parallel LCR circuit is called rejector circuit.

37. **(c)** The time constant for resonance circuit,

= CR

NOTES Growth of charge in a circuit containing capacitance and resistance is given by the formula, $q = q_0(1 - e^{-t/CR})$ CR is known as time constant in this formula.

38. **(c)** Power, $P = V_{rms}/I_{rms}\cos\phi$

In purely resistive circuit voltage and current are in same phase

$\therefore \cos\phi = \cos 0° = 1$

$\therefore$ Power dissipated in maximum.

39. **(d)** The average power in the circuit where $\cos\phi$ = power factory

$$\langle P \rangle = V_{rms} \times I_{rms}\cos\phi$$

$$\phi = \pi/3 = \text{phase difference} = \frac{180}{3} = 60$$

$$V_{rms} = \frac{\frac{1}{\sqrt{2}}}{\sqrt{2}} = \frac{1}{2}\text{volt}$$

$$I_{rms} = \frac{\frac{1}{\sqrt{2}}}{\sqrt{2}} = \left(\frac{1}{2}\right)A$$

$$\cos\phi = \cos\frac{\pi}{3} = \frac{1}{2}$$

$$\langle P \rangle = \frac{1}{2} \times \frac{1}{2} \times \frac{1}{2} = \frac{1}{8}W$$

NOTES The instantaneous power is the power in circuit at any instant of time. It is equal to the product of values of alternating voltage and alternating current at that time

$Pin = EI = (E_0 \sin\omega t)(I_0 \sin\omega t)$

(in non-inductive circuit)

$Pin = (E_0 \sin\omega t)(I_0 \sin\omega t \pm \theta)$

(in inductive circuit).

40. **(b)** When L is removed,

Phase difference

$$\tan\phi = \frac{|X_C|}{R} = \tan\frac{\pi}{3} = \frac{X_C}{R} \quad ...(1)$$

When C is removed,

Phase difference

$$\tan\phi = \frac{|X_L|}{R} = \tan\frac{\pi}{3} = \frac{X_L}{R} \quad ...(2)$$

From eqs. (1) and (2), $X_L = X_C$

Since, $X_L = X_C$, the circuit is in resonance.

In this case, $Z = R$

$\therefore$ Power factor, $\cos\phi = \dfrac{R}{Z} = 1.$

41. **(a)** Power dissipated in an LCR series circuit connected to an a.c. source of emf E

$$P = E_{rms}\, i_{rms}\cos\phi = \frac{E_{rms}^2 R}{Z^2} = \frac{E_{rms}^2 R}{R^2 + \left(\omega L - \frac{1}{C\omega}\right)^2}$$

$$= \frac{\left(\frac{10}{\sqrt{2}}\right)^2 \times 50}{(50)^2 + \left(314 \times 20 \times 10^{-3} - \frac{1}{314 \times 100 \times 10^{-6}}\right)^2}$$

Solving we get, P = 0.79 W

42. **(b)**

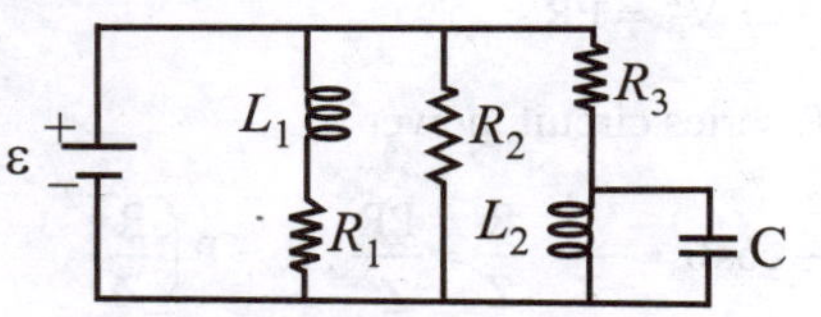

At t = 0, no current flows through R_1 and R_3

$\therefore$

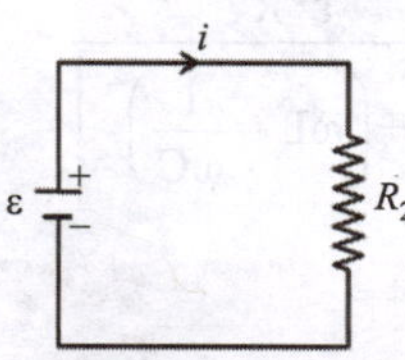

Current through battery just after the switch closed is

$$i = \frac{\varepsilon}{R_2} = \frac{18}{9} = 2A$$

43. **(a)** Given: L = 20 mH; C = 50 μF; R = 40 Ω V = 10 sin 340 t

$$\therefore \quad V_{runs} = \frac{10}{\sqrt{2}}$$

$$X_C = \frac{1}{\omega C} = \frac{1}{340 \times 50 \times 10^{-6}} = 58.8\,\Omega$$

$X_L = \omega L = 340 \times 20 \times 10^{-3} = 6.8\,\Omega$

Impedance, $Z = \sqrt{R^2 + (X_C - X_L)^2}$

$= \sqrt{40^2 + (58.8 - 6.8)^2} = \sqrt{4304}\,\Omega$

Power loss in A.C. circuit,

$P = i_{rms}^2 R = \left(\frac{V_{rms}}{Z}\right)^2 R$

$= \left(\frac{10/\sqrt{2}}{\sqrt{4304}}\right)^2 \times 40 = \frac{50 \times 40}{4304} \simeq 0.51\text{ W}$

44. (d)

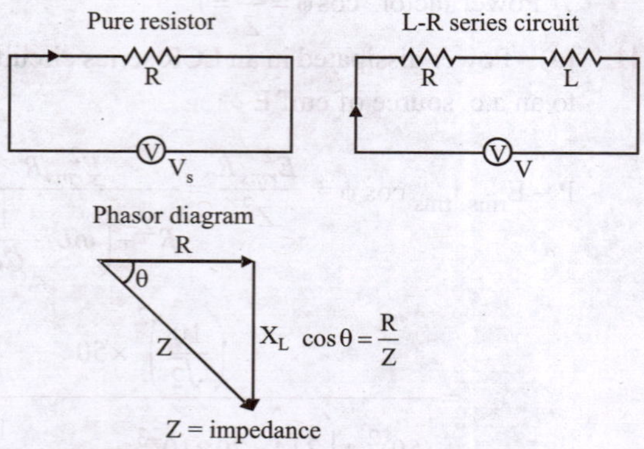

For pure resistor circuit, power

$P = \frac{V^2}{R} \Rightarrow V^2 = PR$

For L-R series circuit, power

$P = \frac{V^2}{Z}\cos\theta = \frac{V^2}{Z}.\frac{R}{Z} = \frac{PR}{Z^2}.R = P\left(\frac{R}{Z}\right)^2$

45. (d) Power dissipated in series LCR;

$$P = I^2R = \frac{\varepsilon^2}{(Z)^2}R = \frac{\varepsilon^2 R}{\left[R^2 + \left(\omega L - \frac{1}{\omega C}\right)^2\right]}$$

NOTES $Z = \sqrt{R^2 + \left(\omega L - \frac{1}{\omega C}\right)^2}$ is called the impedance of the circuit.

46. (c) The average power in the circuit over one cycle of a.c. is given by

$P_{av} = e_{rms} \times i_{rms} \times \cos\phi$

$= \frac{E_0}{\sqrt{2}} \times \frac{I_0}{\sqrt{2}} \times \cos\phi = \frac{E_0 I_0}{2}\cos\phi$

47. (a) Power dissipated $= E_{rms}.\, I_{rms}$

$= (E_{rms})(I_{rms})\cos\phi$

Hence, power dissipated depends upon phase difference.

48. (b) Power factor, $\cos\phi = \frac{R}{\sqrt{\left(\omega L - \frac{1}{\omega C}\right)^2 + R^2}}$

$= \frac{8}{\sqrt{(31-25)^2 + 8^2}} = \frac{8}{\sqrt{6^2 + 8^2}} = \frac{8}{10} = 0.8$

NOTES Power factor $= \cos\theta = \frac{R}{Z}$

For purely inductive and purely capacitive circuits, $\theta = 90°$

Power factor $= \cos\theta = \cos 90° = 0$

For non-inductive circuit, $\theta = 0°$

$\cos\theta = \cos 0° = 1$

49. (d) At resonance $L\omega = \frac{1}{C\omega}$, $\omega = \frac{1}{\sqrt{LC}}$

Current through circuit $i = \frac{E}{R}$

Power dissipated at Resonance $= i^2R$

50. (a) In secondary circuit, $P = VI$

$\Rightarrow 88 = 11000\, I$

$\Rightarrow I = \frac{88}{11 \times 10^3} = 8 \times 10^{-3}$ A

$\Rightarrow I = 8$ mA

51. (b) Transformer ratio,

$\frac{N_P}{N_S} = \frac{V_P}{V_S} = \frac{1}{2}$

$\therefore V_S : V_P = 2 : 1$

NOTES When $V_S > V_P$ step-up transformer and when $V_S < V_P$ step-down transformer.

52. (b) Power $P = V \times I$

For an ideal transformer

Input power = Output power

$V_P I_P = V_S I_S$

or, $220 \times I_P = 44 \therefore I_P = 0.2$ A

53. (a)

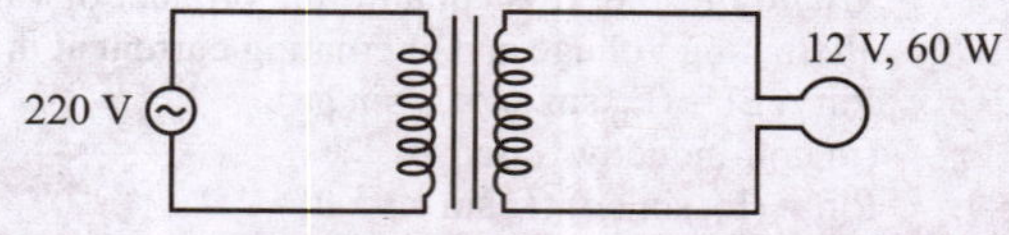

As transfer to be ideal and for ideal transformer

$P_{input} = P_{output}$

$\therefore (VI)_{in} = 60 \Rightarrow 220 \times I = 60$

$\therefore I = 0.27$ A

54. **(b)** Efficiency $\eta = \frac{V_s I_s}{V_p I_p} \Rightarrow 0.9 = \frac{V_s(6)}{3 \times 10^3}$

$\Rightarrow V_s = 450$ V

As $V_p I_p = 3000$

so $I_p = \frac{3000}{V_p} = \frac{3000}{200}$ A $= 15$A

55. **(d)** A transformer is essentially an *AC* device. *DC* source so no mutual induction between coils

$\Rightarrow E_2 = 0$ and $I_2 = 0$

56. **(d)** $\frac{V_2}{V_1} = 0.8\frac{I_1}{I_2} \Rightarrow \frac{V_2 I_2}{V_1 I_1} = 0.8$

$V_1 = 220\ V, I_2 = 2.0\ A, V_2 = 440\ V$

$I_1 = \frac{V_2\, I_2}{V_1} \times \frac{10}{8}$

$= \frac{440 \times 2 \times 10}{220 \times 8} = 5\ A$

57. **(a)** Since $\frac{V_s}{V_p} = \frac{N_s}{N_p}$

Where

N_s = No. of turns across primary coil = 50

N_p = No. of turns across secondary coil = 1500

and $V_p = \frac{d\phi}{dt} = \frac{d}{dt}(\phi_0 + 4t) = 4$

$\Rightarrow V_s = \frac{1500}{50} \times 4 = 120$ V

58. **(b)** Efficiency of the transformer

$\eta = \frac{P_{output}}{P_{input}} \times 100$

$= \frac{100}{220 \times 0.5} \times 100 = 90.9\%$

59. **(d)** When there is change of flux in the core of a transformer due to change in current around it, eddy current is produced. The direction of this current is opposite to the current which produces it, so it will reduce the main current. We laminate the core so that flux is reduced resulting in the reduced production of eddy current.

60. **(b)** $\frac{N_P}{N_s} = \frac{E_P}{E_s} = \frac{1}{25}$

$\therefore E_s = 25\ E_p$

But $E_s I_s = E_p I_p$

$25E_p \times 2 = E_p \times I_p$

$\Rightarrow I_p = 50$ A

NOTES If np is number of turns in primary coil and ns is number of turns in secondary coil, then

$\frac{E_s}{E_p} = \frac{I_p}{I_s} = \frac{n_s}{n_p} = k$

Here, I_p = current in primary
k = Transformation ratio
I_s = current is secondary
For a step up transformer, $k > 1$
For step down transformer, $k < 1$

61. **(d)** The transformer converts A.C. high voltage into A.C. low voltage, but it does not cause any change in frequency. The formula for voltage is

$\frac{E_s}{E_p} = \frac{N_s}{N_p} \Rightarrow E_s = \frac{N_s}{N_p} \times E_p$

$= \frac{5000}{500} \times 20 = 200$V

Thus, output has voltage 200 V and frequency 50 Hz.

62. **(a)** Eddy currents are produced when a metal is kept in a varying magnetic field.

8 Electromagnetic Waves

TREND ANALYSIS

TOPIC	2025		2024		2023		2022		2021		2020	
	Qns.	LOD	Qns.	LOD	Qns.	LOD	Qns.	LOD	Qns.	LOD	Qns.	LOD
Displacement Current	1	Avg.	1	Avg.	1	Easy			1	Avg.		
Electromagnetic Waves	1	Avg.	1	Avg.	1	Avg.	1	Easy	1	Avg.	1	Avg.
Electromagnetic Spectrum							1	Diff.			1	Easy

How to Code Qns for Better & Faster Revision

Type 1: Able to solve (No code Req)

Type 2: Solved with struggle or by seeing solution **(Use code NR means Needs Revision or orange stickers)**

Type 3: Tricky/ Could not solve on my own. Need help from Teacher/ Friend. **(Use code TQ means Tricky Question or red stickers)**

You can either hand write the code in boxes or else paste coloured stickers given at end of book.

Topic-8.2 Displacement Current

1. A parallel plate capacitor made of circular plates is being charged such that the surface charge density on its plates is increasing at a constant rate with time. The magnetic field arising due to displacement current is :
(a) zero at all places NCERT | Page-203, 204 | 2025
(b) constant between the plates and zero outside the plates
(c) non-zero everywhere with maximum at the imaginary cylindrical surface connecting peripheries of the plates
(d) zero between the plates and non-zero outside

2. A parallel plate capacitor is charged by connecting it to a battery through a resistor. If I is the current in the circuit, then in the gap between the plates:
(a) There is no current NCERT | Page-203, 204 | 2024
(b) Displacement current of magnitude equal to I flows in the same direction as I
(c) Displacement current of magnitude equal to I flows in a direction opposite to that of I
(d) Displacement current of magnitude greater than I flows but can be in any direction

3. An ac source is connected to a capacitor C. Due to decrease in its operating frequency NCERT | Page-203, 204 | 2023
(a) Capacitive reactance decreases
(b) Displacement current increases
(c) Displacement current decreases
(d) Capacitive reactance remains constant

4. A capacitor of capacitance 'C', is connected across an ac source of voltage V, given by $V = V_0 \sin \omega t$
The displacement current between the plates of the capacitor, would then be given by
NCERT | Page-203, 204 | 2021
(a) $I_d = V_0 \omega C \sin \omega t$ (b) $I_d = V_0 \omega C \cos \omega t$
(c) $I_d = \frac{V_0}{\omega C} \cos \omega t$ (d) $I_d = \frac{V_0}{\omega C} \sin \omega t$

Topic-8.3 Electromagnetic Waves

5. The electric field in a plane electromagnetic wave is given by $E_z = 60 \cos(5x + 1.5 \times 10^9 t)$ V/m.
NCERT | Page-206, 207 | 2025
Then expression for the corresponding magnetic field is (here subscripts denote the direction of the field):
(a) $B_y = 2 \times 10^{-7} \cos(5x + 1.5 \times 10^9 t)\ T$
(b) $B_x = 2 \times 10^{-7} \cos(5x + 1.5 \times 10^9 t)\ T$
(c) $B_z = 60 \cos(5x + 1.5 \times 10^9 t)\ T$
(d) $B_y = 60 \sin(5x + 1.5 \times 10^9 t)\ T$

6. If the ratio of relative permeability and relative permittivity of a uniform medium is 1 : 4. The ratio of the magnitudes of electric field intensity (E) to the magnetic field intensity (H) of an EM wave propagating in that medium is
(Given that $\sqrt{\frac{\mu_0}{\varepsilon_0}} = 120\pi$) : NCERT | Page-207 | Ph-II 2024
(a) $30\pi : 1$ (b) $1 : 120\pi$
(c) $60\pi : 1$ (d) $120\pi : 1$

7. The property which is not of an electromagnetic wave travelling in free space is that: NCERT | Page-206, 207 | 2024

(a) They are transverse in nature
(b) The energy density in electric field is equal to energy density magnetic field
(c) They travel with a speed equal to $\frac{1}{\sqrt{\mu_0 \varepsilon_0}}$
(d) They originate from charges moving with uniform speed

8. In a plane electromagnetic wave travelling in free space, the electric field component oscillates sinusoidally at a frequency of 2.0×10^{10} Hz and amplitude 48 V m^{-1}. Then the amplitude of oscillating magnetic field is **NCERT | Page-207 | 2023**
(Speed of light in free space = 3×10^8 m s^{-1})
(a) 1.6×10^{-9} T (b) 1.6×10^{-8} T
(c) 1.6×10^{-7} T (d) 1.6×10^{-6} T

9. The ratio of the magnitude of the magnetic field and electric field intensity of a plane electromagnetic wave in free space of permeability μ_0 and permittivity ε_0 is (Given that C - velocity of light in free space) **NCERT | Page-206, 207 | RS 2022**
(a) $\frac{\sqrt{\mu_0 \varepsilon_0}}{c}$ (b) c (c) $\frac{1}{c}$ (d) $\frac{c}{\sqrt{\mu_0 \varepsilon_0}}$

10. The magnetic field of a plane electromagnetic wave is given by **NCERT | Page-207 | RS 2022**
$\vec{B} = 3\times10^{-8}\cos(1.6\times10^3 x + 48\times10^{10} t)\hat{j}$,
then the associated electric field will be:
(a) $9\cos(1.6\times10^3 x + 48\times10^{10} t)\hat{k}$ V/m
(b) $3\times10^{-8}\cos(1.6\times10^3 x + 48\times10^{10} t)\hat{i}$ V/m
(c) $3\times10^{-8}\sin(1.6\times10^3 x + 48\times10^{10} t)\hat{i}$ V/m
(d) $9\sin(1.6\times10^3 x - 48\times10^{10} t)\hat{k}$ V/m

11. When light propagates through a material medium of relative pemittivity ϵ_r and relative permeability μ_r, the velocity of light, v is given by: (c-velocity of light in vacuum) **NCERT | Page-207 | 2022**
(a) $v = \sqrt{\frac{\mu_r}{\epsilon_r}}$ (b) $v = \sqrt{\frac{\epsilon_r}{\mu_r}}$
(c) $v = \frac{c}{\sqrt{\epsilon_r \mu_r}}$ (d) v = c

12. For a plane electromagnetic wave propagating in x-direction, which one of the following combination gives the correct possible directions for electric field (E) and magnetic field (B) respectively? **NCERT | Page-206 | 2021**
(a) $-\hat{j}+\hat{k}, -\hat{j}+\hat{k}$ (b) $\hat{j}+\hat{k}, \hat{j}+\hat{k}$
(c) $-\hat{j}+\hat{k}, -\hat{j}-\hat{k}$ (d) $\hat{j}+\hat{k}, -\hat{j}-\hat{k}$

13. The ratio of contributions made by the electric field and magnetic field components to the intensity of an electromagnetic wave is :
(c = speed of electromagnetic waves) **NCERT | Page-207 | 2020**
(a) 1 : 1 (b) 1 : c
(c) 1 : c^2 (d) c : 1

14. An EM wave is propagating in a medium with a velocity $\vec{V} = V\hat{i}$. The instantaneous oscillating electric field of this EM wave is along +y axis. Then the direction of oscillating magnetic field of the em wave will be along **NCERT | Page-206 | 2018**
(a) –z direction (b) +z direction
(c) –x direction (d) –y direction

15. In an electromagnetic wave in free space the root mean square value of the electric field is E_{rms} = 6V/m. The peak value of the magnetic field is :- **NCERT | Page-207 | 2017**
(a) 2.83×10^{-8} T (b) 0.70×10^{-8} T
(c) 4.23×10^{-8} T (d) 1.41×10^{-8} T

16. Out of the following options which one can be used to produce a propagating electromagnetic wave ? **NCERT | Page-205 | 2016**
(a) A charge moving at constant velocity
(b) A stationary charge
(c) A chargeless particle
(d) An accelerating charge

17. A radiation of energy 'E' falls normally on a perfectly reflecting surface. The momentum transferred to the surface is (C = Velocity of light) **2015**
(a) $\frac{2E}{C}$ (b) $\frac{2E}{C^2}$ (c) $\frac{E}{C^2}$ (d) $\frac{E}{C}$

18. Light with an energy flux of 25×10^4 Wm^{-2} falls on a perfectly reflecting surface at normal incidence. If the surface area is 15 cm^2, the average force exerted on the surface is : **2014**
(a) 1.25×10^{-6} N (b) 2.50×10^{-6} N
(c) 1.20×10^{-6} N (d) 3.0×10^{-6} N

19. An electromagnetic wave of frequency ν = 3.0 MHz passes from vacuum into a dielectric medium with relative permittivity ε = 4.0. Then **NCERT | Page-207 | Kar. 2013**
(a) wavelength is doubled and frequency is unchanged
(b) wavelength is doubled and frequency becomes half
(c) wavelength is halved and frequency remains unchanged
(d) wavelength and frequency both remain unchanged

20. The electric field associated with an e.m. wave in vacuum is given by $\vec{E} = \hat{i}\, 40 \cos(kz - 6 \times 10^8 t)$, where E, z and t are in volt/m, meter and seconds respectively. The value of wave vector k is : **NCERT | Page-206 | 2012**
(a) 2 m^{-1} (b) 0.5 m^{-1}
(c) 6 m^{-1} (d) 3 m^{-1}

21. The ratio of amplitude of magnetic field to the amplitude of electric field for an electromagnetic wave propagating in vacuum is equal to : **NCERT | Page-207 | 2012 M**
(a) the speed of light in vacuum
(b) reciprocal of speed of light in vacuum
(c) the ratio of magnetic permeability to the electric susceptibility of vacuum
(d) unity

22. The electric and the magnetic field associated with an E.M. wave, propagating along the +z-axis, can be represented by NCERT | Page-206 | 2011

(a) $\left[\vec{E}=E_0\hat{i}, \vec{B}=B_0\hat{j}\right]$ (b) $\left[\vec{E}=E_0\vec{k}, \vec{B}=B_0\hat{i}\right]$

(c) $\left[\vec{E}=E_0\hat{j}, \vec{B}=B_0\hat{i}\right]$ (d) $\left[\vec{E}=E_0\hat{j}, \vec{B}=B_0\hat{k}\right]$

23. Which of the following statement is false for the properties of electromagnetic waves? NCERT | Page-206 | 2010

(a) Both electric and magnetic field vectors attain the maxima and minima at the same place and same time.
(b) The energy in electromagnetic wave is divided equally between electric and magnetic vectors
(c) Both electric and magnetic field vectors are parallel to each other and perpendicular to the direction of propagation of wave
(d) These waves do not require any material medium for propagation.

24. The electric field part of an electromagnetic wave in a medium is represented by $E_x = 0$;

$$E_y = 2.5\frac{N}{C}\cos\left[\left(2\pi\times10^6\frac{rad}{m}\right)t-\left(\pi\times10^{-2}\frac{rad}{s}\right)x\right];$$

$E_z = 0$. The wave is : NCERT | Page-206 | 2009

(a) moving along x direction with frequency 10^6 Hz and wave length 100 m.
(b) moving along x direction with frequency 10^6 Hz and wave length 200 m.
(c) moving along – x direction with frequency 10^6 Hz and wave length 200 m.
(d) moving along y direction with frequency $2\pi \times 10^6$ Hz and wave length 200 m.

25. The velocity of electromagnetic radiation in a medium of permittivity ε_0 and permeability μ_0 is given by NCERT | Page-207 | 2008

(a) $\sqrt{\frac{\epsilon_0}{\mu_0}}$ (b) $\sqrt{\mu_0 \epsilon_0}$

(c) $\frac{1}{\sqrt{\mu_0 \epsilon_0}}$ (d) $\sqrt{\frac{\mu_0}{\epsilon_0}}$

26. The electric and magnetic field of an electro-magnetic wave are NCERT | Page-206 | 2007

(a) in opposite phase and perpendicular to each other
(b) in opposite phase and parallel to each other
(c) in phase and perpendicular to each other
(d) in phase and parallel to each other.

27. If $\vec{E}$ and $\vec{B}$ represent electric and magnetic field vectors of the electromagnetic waves, then the direction of propagation of the waves will be along NCERT | Page-206 | 2002

(a) $\vec{B}\times\vec{E}$ (b) $\vec{E}$
(c) $\vec{B}$ (d) $\vec{E}\times\vec{B}$

28. The electromagnetic radiations are caused by NCERT | Page-205 | 1999

(a) a stationary charge
(b) uniformly moving charges
(c) accelerated charges
(d) all of above

29. If ε_0 and μ_0 are the electric permittivity and magnetic permeability in vacuum, ε and μ are corresponding quantities in medium, then refractive index of the medium is NCERT | Page-207 | 1997

(a) $\sqrt{\frac{\varepsilon}{\varepsilon_0}}$ (b) $\sqrt{\frac{\varepsilon_0\mu}{\varepsilon\mu_0}}$ (c) $\sqrt{\frac{\varepsilon_0\mu_0}{\varepsilon\mu}}$ (d) $\sqrt{\frac{\varepsilon\mu}{\varepsilon_0\mu_0}}$

30. The oscillating electric and magnetic field vectors of electromagnetic wave are oriented along NCERT | Page-206 | 1994

(a) the same direction and in phase
(b) the same direction but have a phase difference of 90°
(c) mutually perpendicular directions and are in phase
(d) mutually perpendicular directions but has a phase difference of 90°

31. The frequency of electromagnetic wave, which is best suited to observe a particle of radius 3×10^{-4} cm is of the order of NCERT | Page-207 | 1991

(a) 10^{15} (b) 10^{14}
(c) 10^{13} (d) 10^{12}

Topic-8.4 Electromagnetic Spectrum

32. The electromagnetic radiation which has the smallest wavelength are NCERT | Page-210, 211 | Ph-II 2024

(a) X-rays (b) Gamma rays
(c) Ultraviolet rays (d) Microwaves

33. Match List-I with List-II

	List-I (Electromagnetic waves)		List-II (Wavelength)
(A)	AM radio waves	(i)	10^{-10}
(B)	Microwaves	(ii)	10^2 m
(C)	Infared radiations	(iii)	10^{-2} m
(D)	X-rays	(iv)	10^{-4} m

Choose the **correct** answer from the options given below: NCERT | Page-209 | 2022

(a) A–iii; B–ii; C–i; D–iv
(b) A–iii; B–iv; C–ii; D–i
(c) A–ii; B–iii; C–iv; D–i
(d) A–iv; B–iii; C–ii; D–i

34. The E.M. wave with shortest wavelength among the following is, **NCERT | Page-209 | Ph-II 2020**
(a) Microwaves (b) Ultraviolet rays
(c) X-rays (d) Gamma rays

35. Which colour of the light has the longest wavelength ? **NCERT | Page-209 | 2019**
(a) red (b) blue
(c) green (d) violet

36. The energy of the em waves is of the order of 15 keV. To which part of the spectrum does it belong? **2015 RS**
(a) Infra-red rays (b) Ultraviolet rays
(c) γ-rays (d) X-rays

37. The condition under which a microwave oven heats up a food item containing water molecules most efficiently is **NCERT | Page-209 | 2013**
(a) the frequency of the microwaves has no relation with natural frequency of water molecules.
(b) microwaves are heat waves, so always produce heating.
(c) infra-red waves produce heating in a microwave oven.
(d) the frequency of the microwaves must match the resonant frequency of the water molecules.

38. The decreasing order of wavelength of infrared, microwave, ultraviolet and gamma rays is **NCERT | Page-209 | 2011**
(a) microwave, infrared, ultraviolet, gamma rays
(b) gamma rays, ultraviolet, infrared, micro-waves
(c) microwaves, gamma rays, infrared, ultraviolet
(d) infrared, microwave, ultraviolet, gamma rays

39. If λ_v, λ_x and λ_m represent the wavelengths of visible light, *X*-rays and microwaves respectively, then **NCERT | Page-209 | 2005**
(a) $\lambda_m > \lambda_x > \lambda_v$
(b) $\lambda_m > \lambda_v > \lambda_x$
(c) $\lambda_v > \lambda_x > \lambda_m$
(d) $\lambda_v > \lambda_m > \lambda_x$

40. Which one of the following rays is not electromagnetic wave ? **NCERT | Page-209 | 2003**
(a) heat rays (b) *X*-rays
(c) γ-rays (d) β-rays

41. We consider the radiation emitted by the human body. Which of the following statements is true? **NCERT | Page-210 | 2003**
(a) the radiation emitted lies in the ultraviolet region and hence is not visible.
(b) the radiation emitted is in the infra-red region.
(c) the radiation is emitted only during the day.
(d) the radiation is emitted during the summers and absorbed during the winters.

42. Which of the following electromagnetic radiations has the least wavelength? **NCERT | Page-209 | 2002**
(a) gamma rays (b) infra-red
(c) ultraviolet (d) *X*-rays

43. Green-house effect is the heating up of earth's atmosphere due to **NCERT | Page-209 | 2002**
(a) green plants (b) infra-red rays
(c) *X*-rays (d) ultraviolet rays

44. The frequencies of *X*-rays, γ-rays and ultraviolet rays are respectively a, b, and c. Then **NCERT | Page-209 | 2000**
(a) $a < b, b < c$ (b) $a < b, b > c$
(c) $a > b, b > c$ (d) $a > b, b < c$

45. Which one of the following electromagnetic radiations has the smallest wavelength? **NCERT | Page-209 | 1994**
(a) ultraviolet waves (b) X-rays
(c) γ-rays (d) microwaves

46. The structure of solids is investigated by using **NCERT | Page-210 | 1992**
(a) cosmic rays
(b) X-rays
(c) γ-rays
(d) infra-red radiations

47. Pick out the longest wavelength from the following types of radiation. **NCERT | Page-209 | 1990**
(a) blue light (b) gamma rays
(c) *X*-rays (d) red light

48. Which of the following, has the longest wavelength? **NCERT | Page-209 | 1989**
(a) X-rays (b) γ-rays
(c) microwaves (d) radio waves

ANSWER KEY

1	(c)	6	(c)	11	(c)	16	(d)	21	(b)	26	(c)	31	(b)	36	(d)	41	(b)	46	(b)
2	(b)	7	(d)	12	(c)	17	(a)	22	(a)	27	(d)	32	(b)	37	(d)	42	(a)	47	(d)
3	(c)	8	(c)	13	(a)	18	(b)	23	(c)	28	(c)	33	(c)	38	(a)	43	(b)	48	(d)
4	(b)	9	(c)	14	(b)	19	(c)	24	(b)	29	(d)	34	(d)	39	(b)	44	(b)		
5	(a)	10	(a)	15	(a)	20	(a)	25	(c)	30	(c)	35	(a)	40	(d)	45	(c)		

Hints & Solutions

1. (c) When a capacitor is charging, the changing electric field between the plates produces a displacement current I_d, given by

$$I_d = \epsilon_0 \frac{d\phi_e}{dt} = \epsilon_0 \frac{d}{dt}(EA\cos 0°)$$

$$I_d = \epsilon_0 A \frac{dE}{dt}$$

The electric field (E) $= \frac{\sigma}{\epsilon_0}; I_d = \epsilon_0 A \frac{d}{dt}\left(\frac{\sigma}{\epsilon_0}\right)$

$$\Rightarrow I_d = A\left(\frac{d\sigma}{dt}\right) = \text{constant}$$

The magnetic field (B) is non-zero both inside and outside.

It peaks at $r = R$ (the periphery)

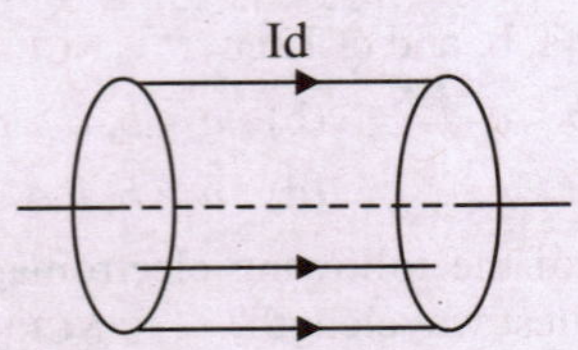

2. (b)

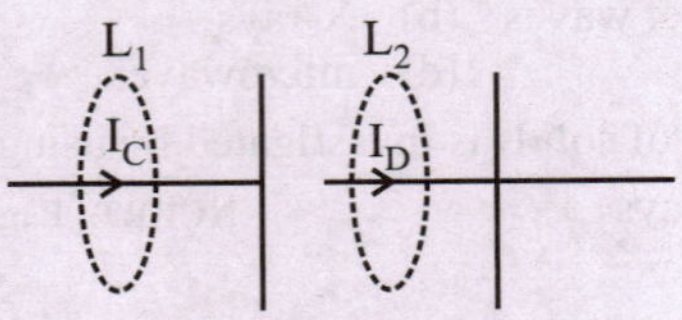

According to modified Ampere's law,

$$\oint \vec{B}.\vec{d\ell} = \mu_0 (I_C + I_D)$$

For Loop L_1, $I_C \neq 0$ and $I_D = 0$

$\Rightarrow \oint \vec{B}.\vec{dl} = \mu_0 I_C$ ... (i)

For Loop L_2, $I_C = 0$ and $I_D \neq 0$

$\Rightarrow \oint \vec{B}.\vec{dl} = \mu_0 I_D$ ... (ii)

From eqs (i) and (ii), we get

$I_C = I_D$

3. (c) Capacitive reactance $X_C = \frac{1}{\omega C} = \frac{1}{2\pi\nu C}$

Since ν decreasing X_C will increases

$\therefore$ Current, $I = \frac{E_0}{X_C}$ will decrease

As conduction current = displacement current

Therefore displacement current will decrease.

4. (b) Displacement current

$$I_d = C\frac{dV}{dt} = C\frac{d}{dt}(V_0 \sin \omega t) (\because V = V_0 \sin \omega t \text{ given})$$

$$= C (V_0\omega) \cos \omega t$$

$$\therefore I_d = V_0 \omega C \cos \omega t$$

5. (a) Given, $E_0 = 60$ V/m, $\omega = 1.5 \times 10^9$ rad/s, $k = 5\ m^{-1}$

$$c = \frac{\omega}{k} = \frac{1.5 \times 10^9}{5} = 3 \times 10^8 \text{ m/s}$$

$$B_0 = \frac{E_0}{c} = \frac{60}{3 \times 10^2} = 20 \times 10^{-8} = 2 \times 10^{-7} \text{ T}$$

$$\therefore B_y = 2 \times 10^{-7} \cos (5x + 1.5 \times 10^9 t) T$$

6. (c) For EM wave, $\frac{\mu_r}{\epsilon_r} = \frac{1}{4}$

$$\frac{E}{H} = \frac{E\mu}{B} = \nu\mu$$

$$\therefore \frac{E}{H} = \frac{1}{\sqrt{\mu\varepsilon}}\mu = \sqrt{\frac{\mu}{\varepsilon}} = \sqrt{\frac{\mu_0\mu_r}{\varepsilon_0\varepsilon_r}}$$

$$= \sqrt{\frac{\mu_0}{\varepsilon_0}}\sqrt{\frac{\mu_r}{\varepsilon_r}} = 120\pi\left(\frac{1}{2}\right) = \frac{60\pi}{1}$$

7. (d) The accelerating charge radiates EM waves and charge moving with uniform velocity produces steady state magnetic field.

8. (c) Given, Amplitude of oscillating electric field component, $E_0 = 48\ Vm^{-1}$

Using, $C = \frac{E_0}{B_0}$

Here, B_0 = Amplitude of oscillating magnetic field

$$\therefore B_0 = \frac{48}{3 \times 10^8} = 16 \times 10^{-8} = 1.6 \times 10^{-7} \text{ T}$$

9. (c) We know $|E| = C|B| \Rightarrow \frac{|B|}{|E|} = \frac{1}{C}$

10. (a) From given equation

$B_0 = 3 \times 10^{-8}$ v/m

$$\frac{E_0}{B_0} = C \Rightarrow E_0 = B_0 C = 3 \times 10^{-8} \times 3 \times 10^8$$

$= 9$ v/m

Phase of magnetic and electric field is same of travelling EM wave

$$\hat{E} = \hat{B} \times \hat{C} = \hat{j} \times \left(-\hat{i}\right) = \hat{k}$$

$\therefore \quad \vec{E} = 9 \cos (1.6 \times 10^3 x + 48 \times 10^{10} t)\ \hat{k}$ v/m

11. (c) We know that

$$V = \frac{1}{\sqrt{\mu\varepsilon}} \text{ and } C = \frac{1}{\sqrt{\mu_0\varepsilon_0}}$$

$$\text{So, } \frac{V}{C} = \frac{\sqrt{\mu_0\varepsilon_0}}{\sqrt{\mu\varepsilon}} = \frac{\sqrt{\mu_0\varepsilon_0}}{\sqrt{\mu_0\varepsilon_0\,\mu_r\varepsilon_r}} = \frac{1}{\sqrt{\mu_r\varepsilon_r}}$$

$$V = \frac{C}{\sqrt{\mu_r\varepsilon_r}}$$

12. (c) Direction of propagation along x-axis.

$$\vec{E} \times \vec{B} = \left(-\hat{j} + \hat{k}\right) \times \left(-\hat{j} - \hat{k}\right) = 2\hat{i}$$

NOTES $\vec{E} \times \vec{B}$ is always in the direction of propagation of wave.
The EM wave propagating in the positive x-direction $E = E_y = E_0 \sin(kx - \omega t)$ and $B = B_z = B_0 \sin(kx - \omega t)$

13. (a) The energy in electromagnetic wave is divided equally between the electric and magnetic field.

So, in an electromagnetic wave, half of the intensity is provided by the electric field and half by the magnetic field.

Hence, required ratio should be 1 : 1.

14. (b) As we know, $\vec{E} \times \vec{B} = \vec{V}$

$(E\hat{j}) \times (\vec{B}) = V\hat{i}$

($\because$ Electric field vector is along +y axis)

So, $\vec{B} = B\hat{k}$

i.e., direction of magnetic field vector is along +z direction.

15. (a) Given, $E_{rms} = 6$ V/m

$$\frac{E_{rms}}{B_{rms}} = c \Rightarrow B_{rms} = \frac{E_{rms}}{c} \quad \text{...(i)}$$

$$B_{rms} = \frac{B_0}{\sqrt{2}} \Rightarrow B_0 = \sqrt{2}\, B_{rms}$$

$$B_0 = \sqrt{2} \times \frac{E_{rms}}{c} \quad \text{From equation (i)}$$

$$= \frac{\sqrt{2} \times 6}{3 \times 10^8} = 2.83 \times 10^{-8} \text{ T}$$

16. (d) To generate electromagnetic waves we need accelerating charge particle.

17. (a) Momentum of light falling on reflecting surface $p = \frac{E}{C}$

As surface is perfectly reflecting so momentum reflect $p' = \frac{E}{C}$

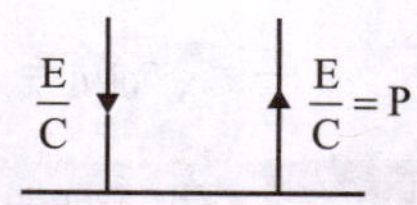

So, momentum transferred

$$= P - P' = \frac{E}{C} - \left(-\frac{E}{C}\right) = \frac{2E}{C}$$

18. (b) Average force $F_{av} = \frac{\Delta p}{\Delta t} = \frac{2IA}{c}$

($\because$ Power = F.V)

$$= \frac{2 \times 25 \times 10^4 \times 15 \times 10^{-4}}{3 \times 10^8} = 2.50 \times 10^{-6} \text{ N}$$

NOTES Pressure $= \frac{\text{Force}}{\text{Area}} = \frac{F}{A}$

Force is the rate of change of momentum $F = \frac{dp}{dt}$

But energy, $U = pc$

$\therefore$ Pressure $= \frac{1}{A}\frac{U}{c\,.\,dt} = \frac{I}{c} \left[\because \text{Intensity} = \frac{U}{Adt}\right]$

For perfectly reflecting surface, $P_r = \frac{2I}{c}$

$\therefore$ For perfectly reflecting surface, Force $= \frac{2IA}{c}$

19. (c) Given: frequency f = 2MHz, relative permittivity $\epsilon_r = 4$

From formula,

$$\text{velocity } v = \frac{c}{\sqrt{\epsilon_r}} = \frac{c}{2} \Rightarrow \lambda' = \frac{\lambda}{2}$$

[Since frequency remains unchanged]

20. (a) On comparing the given equation to

$$\vec{E} = a_0\hat{i} \cos(\omega t - kz)$$

$\omega = 6 \times 10^8$

Wave number,

$$k = \frac{2\pi}{\lambda} = \frac{\omega}{c}$$

$$k = \frac{\omega}{c} = \frac{6 \times 10^8}{3 \times 10^8} = 2\text{ m}^{-1}$$

21. (b) The average energy stored in the electric field,

$$U_E = \frac{1}{2}\varepsilon_0 E^2$$

The average energy stored in the magnetic field = U_B

$= \frac{1}{2}\frac{B^2}{\mu_0}$,

According to conservation of energy $U_E = U_B$

$$\varepsilon_0\mu_0 = \frac{B^2}{E^2}$$

$$\frac{B}{E} = \sqrt{\varepsilon_0\mu_0} = \frac{1}{c}$$

NOTES The average energy density of electric field

$$\mu_E = \frac{1}{4}\epsilon_0 E_0^2$$

$$\Rightarrow \mu_E = \frac{1}{4}\epsilon_0 (c^2 B_0^2) \quad \left[\because \frac{E_0}{B_0} = c\right]$$

$$\Rightarrow \mu_E = \frac{1}{4}E_0 B_0^2 \times \frac{1}{\mu_0 \epsilon_0} \quad \left[\because c = \frac{1}{\sqrt{\mu_0 \epsilon_0}}\right]$$

$$\Rightarrow \mu_E = \frac{1}{4}\frac{B_0^2}{\mu_0} = \mu_B$$

Thus, the average energy density of electric field equal to the average density of magnetic field.

22. **(a)** E.M. wave always propagates in a direction perpendicular to both electric and magnetic fields. So, electric and magnetic fields should be along + X and + Y–directions respectively. Therefore, option (a) is the correct option.

23. **(c)** Electromagnetic waves are the combination of mutually perpendicular electric and magnetic fields. So, option (c) is false.

24. **(b)** Comparing with the equation of wave.

$E_y = E_0 \cos(\omega t - kx)$

$\omega = 2\pi f = 2\pi \times 10^6 \quad \therefore \quad f = 10^6$ Hz

$\frac{2\pi}{\lambda} = k = \pi \times 10^{-2}\ m^{-1}, \lambda = 200$ m

25. **(c)** The velocity of electromagnetic radiation in a medium of permittivity ϵ_0 and permeability μ_0 is equal to $\frac{1}{\sqrt{\mu_0 \epsilon_0}}$.

26. **(c)** Variation in magnetic field causes electric field and vice-versa.

In electromagnetic waves, $\vec{E} \perp \vec{B}$. Both $\vec{E}$ and $\vec{B}$ are in the same phase.

In electromagnetic waves

$E = E_0 \sin(\omega t - kx)$

$B = B_0 \sin(\omega t - kx)$

The electromagnetic waves travel in the direction of $(\vec{E} \times \vec{B})$.

27. **(d)** Direction of propagation of electro-magnetic waves is perpendicular to Electric field and Magnetic field. Hence, direction is given by vector $\vec{S} = \vec{E} \times \vec{H} = \frac{\vec{E} \times \vec{B}}{\mu_0}$.

NOTES In electromagnetic waves, the rate of flow of energy crossing a unit area is described by Poynting vector. The magnitude of poynting vector

$$|\vec{S}| = \frac{1}{\mu_0} EB \sin 90° = \frac{EB}{\mu_0}$$

$$= \frac{E^2}{\mu_0 c} \quad \left(\because B = \frac{E}{c}\right)$$

28. **(c)** A stationary charge produces electric field only; an uniformly moving charge produces localised electromagnetic field; an accelerated charge produces electromagnetic radiations.

29. **(d)** We know that velocity of electromagnetic wave in vacuum $(v_0) = \frac{1}{\sqrt{\mu_0\varepsilon_0}}$ and velocity of electromagnetic wave in medium is $(v) = \frac{1}{\sqrt{\mu\varepsilon}}$.

Therefore refractive index of the medium

$$(\mu) = \frac{\text{Vel. of E.M. wave in vacuum}(v_0)}{\text{Vel. of E.M. wave in medium}(v)}$$

$$= \frac{1/\sqrt{\mu_0\varepsilon_0}}{1/\sqrt{\mu\varepsilon}} = \sqrt{\frac{\mu\varepsilon}{\mu_0\varepsilon_0}}$$

30. **(c)** The direction of oscillations of E and B fields are perpendicular to each other as well as to the direction of propagation. So, electro-magnetic waves are transverse in nature.

The electric and magnetic fields oscillate in same phase.

31. **(b)** Size of particle $= \lambda = \frac{c}{\nu}$

$$3 \times 10^{-6} = \frac{3 \times 10^8}{\nu}$$

$\nu = 10^{14}$ Hz

NOTES When frequency is higher than this, wavelength is still smaller. Resolution becomes better.

32. **(b)** γ-rays has the smallest wavelength.

33. **(c)** Radio waves has longest wavelength and χ-rays have shortest wavelength, so correct option is (c).

34. **(d)** The increasing order of wavelength of electromagnetic wave is as follows -

$$\lambda_{\gamma\text{-rays}} < \lambda_{\text{x-rays}} < \lambda_{\text{infrared}} < \lambda_{\text{microwave}} < \lambda_{\text{radiowave}}$$

Clearly, γ-rays has shortest wavelength.

35. **(a)** $\lambda_{Red} > \lambda_{Green} > \lambda_{Blue} > \lambda_{Violet.}$ Red has the longest wavelength.

36. **(d)** Energy of x-ray is (100 ev to 100 kev)

Hence energy of the order of 15 keV belongs to x-rays.

37. **(d)** Required condition : Frequency of microwaves = Resonant frequency of water molecules.

38. **(a)** The decreasing order of the wavelengths is as given below :

microwave, infrared, ultraviolet, gamma rays.

39. **(b)** We know $E = \frac{hc}{\lambda} \Rightarrow E \propto \frac{1}{\lambda}$

$\Rightarrow E_m < E_v < E_x$

$\therefore \lambda_m > \lambda_v > \lambda_x$

NOTES

Energy of electromagnetic wave, $E = \frac{hc}{\lambda}$

For *em* wave of lower energy, (ν) is small and λ is larger. For *em* wave of higher energy (ν) is large and λ is small.

40. **(d)** β ray is not electromagnetic ray.

41. **(b)** Depends on the magnitude of frequency.

42. **(a)** Gamma ray has highest frequency and lowest wavelength.

43. **(b)** Infrared rays is the cause of Green house effect. The glass transmits visible light and short infrared rays which are absorbed by plants. Then it emits long infrared rays, which are reflected back by glass.

44. **(b)** γ rays has lowest wavelength and highest frequency among them while ultraviolet ray has highest wavelength and lowest frequency.

Order of frequency : $b > a > c$

45. **(c)**

Rays	Wavelength [Range in m]
X-rays	1×10^{-11} to 3×10^{-8}
γ-rays	6×10^{-14} to 1×10^{-11}
Microwaves	10^{-3} to 0.3
Radiowaves	10 to 10^4

Wavelength of U.V. Rays ranges from 6×10^{-8} to 4×10^{-7}.

46. **(b)** X-rays are used for the investigation of structure of solids.

47. **(d)** Wavelength of red light is longest.

48. **(d)**

Rays	Wavelength [Range in m]
X-rays	1×10^{-11} to 3×10^{-8}
γ-rays	6×10^{-14} to 1×10^{-11}
Microwaves	10^{-3} to 0.3
Radio waves	10 to 10^4

9 Ray Optics and Optical Instruments

TREND ANALYSIS

TOPIC	2025		2024		2023		2022		2021		2020	
	Qns.	LOD	Qns.	LOD	Qns.	LOD	Qns.	LOD	Qns.	LOD	Qns.	LOD
Reflection of Light by Spherical Mirrors											1	Easy
Refraction							1	Avg.				
Total Internal Reflection					1	Easy	1	Avg.				
Refraction at Spherical Surfaces and by Lenses	1	Easy			2	Avg., Easy	1	Avg.	1	Avg.		
Refraction through a Prism			1	Avg.					1	Avg.	1	Easy
Optical Instruments	1	Avg.	1	Avg.			1	Diff.				

How to Code Qns for Better & Faster Revision

Type 1: Able to solve (No code Req)

Type 2: Solved with struggle or by seeing solution **(Use code NR means Needs Revision or orange stickers)**

Type 3: Tricky/ Could not solve on my own. Need help from Teacher/ Friend. **(Use code TQ means Tricky Question or red stickers)**

You can either hand write the code in boxes or else paste coloured stickers given at end of book.

Topic-9.2 Reflection of Light by Spherical Mirrors

1. Given below are two statements:

NCERT | Page-222| Ph-II 2024

Statement I: Image formation needs regular reflection and/or refraction.

Statement II: The variety in colour of objects we see around us is due to the constituent colours of the light incident on them.

In the light of the above statements, choose **the most appropriate** answer from the options given below:

(a) Statement I is correct but statement II is incorrect

(b) Statement I is incorrect but Statement II is correct

(c) Both Statement I and Statement II are correct

(d) Both Statement I and Statement II are incorrect

2. An object is placed on the principal axis of a concave mirror at a distance of 1.5 f (f is the focal length). The image will be at, NCERT | Page-225 | 2020 M

(a) $3f$ (b) $-3f$

(c) $1.5f$ (d) $-1.5f$

3. A beam of light from a source L is incident normally on a plane mirror fixed at a certain distance x from the source. The beam is reflected back as a spot on a scale placed just above the source I. When the mirror is rotated through a small angle θ, the spot of the light is found to move through a distance y on the scale. The angle θ is given by 2017

(a) $\frac{y}{x}$ (b) $\frac{x}{2y}$

(c) $\frac{x}{y}$ (d) $\frac{y}{2x}$

4. Match the corresponding entries of column-1 with column-2 (Where m is the magnification produced by the mirror):

NCERT | Page-225 | 2016

Column-1	Column-2
(P) $m = -2$	(A) Convex mirror
(Q) $m = -\frac{1}{2}$	(B) Concave mirror
(R) $m = +2$	(C) Real image
(S) $m = +\frac{1}{2}$	(D) Virtual image

(a) P → B and C, Q → B and C, R → B and D, S → A and D

(b) P → A and C, Q → A and D, R → A and B, S → C and D

(c) P → A and D, Q → B and C, R → B and D, S → B and C

(d) P → C and D, Q → B and D, R → B and C, S → A and D

5. Two plane mirrors are inclined at 70°. A ray incident on one mirror at angle θ after reflection falls on second mirror and is reflected from there parallel to first mirror. The value of θ is Kar. 2013

(a) 50° (b) 45°

(c) 30° (d) 55°

6. A rod of length 10 cm lies along the principal axis of a concave mirror of focal length 10 cm in such a way that its end closer to the pole is 20 cm away from the mirror. The length of the image is : **NCERT | Page-225 | 2012 M**
(a) 10 cm (b) 15 cm (c) 2.5 cm (d) 5 cm

7. A person is six feet tall. How tall must a vertical mirror be if he is able to see his entire length? **2000**
(a) 3 ft (b) 4.5 ft (c) 7.5 ft (d) 6 ft

8. If two mirrors are kept inclined at 60° to each other and a body is placed at the middle, then total number of images formed is **1995**
(a) six (b) five (c) four (d) three

Topic-9.3 Refraction

9. ε_0 and μ_0 are the electric permittivity and magnetic permeability of free space respectively. If the corresponding quantities of a medium are $2\varepsilon_0$ and $1.5\ \mu_0$ respectively, the refractive index of the medium will nearly be **NCERT | Page-228 | Manipur 2023**
(a) 3 (b) 2 (c) $\sqrt{2}$ (d) $\sqrt{3}$

10. A light ray falls on a glass surface of refractive index $\sqrt{3}$, at an angle 60°. The angle between the refracted and reflected rays would be: **NCERT | Page-228 | 2022**
(a) 60° (b) 90° (c) 120° (d) 30°

11. The frequency of a light wave in a material is 2×10^{14} Hz and wavelength is 5000 Å. The refractive index of material will be **NCERT | Page-228 | 2007**
(a) 1.50 (b) 3.00 (c) 1.33 (d) 1.40

12. A small coin is resting on the bottom of a beaker filled with liquid. A ray of light from the coin travels upto the surface of the liquid and moves along its surface. How fast is the light travelling in the liquid? **NCERT | Page-229 | 2007**

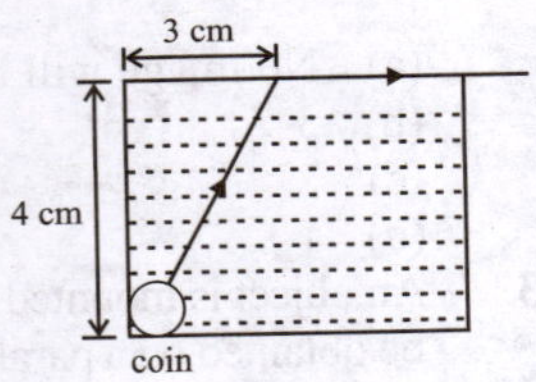

(a) 2.4×10^8 m/s (b) 3.0×10^8 m/s
(c) 1.2×10^8 m/s (d) 1.8×10^8 m/s

13. A beam of light composed of red and green rays is incident obliquely at a point on the face of rectangular glass slab. When coming out on the opposite parallel face, the red and green rays emerge from **NCERT | Page-229 | 2004**
(a) one point propagating in the same direction
(b) two points propagating in two different non-parallel directions
(c) two points propagating in two different parallel directions
(d) one point propagating in two different directions

14. An air bubble in a glass slab ($\mu = 1.5$) is 5 cm deep when viewed from one face and 2 cm deep when viewed from the opposite face. The thickness of the slab is **NCERT | Page-229 | 2000**
(a) 7.5 cm (b) 10.5 cm
(c) 7 cm (d) 10 cm

15. Wavelength of light of frequency 100 Hz **1999**
(a) 2×16^6 m (b) 3×10^6 m
(c) 4×10^6 m (d) 5×10^6 m

16. An electromagnetic radiation of frequency n, wavelength λ, travelling with velocity v in air enters in a glass slab of refractive index (μ). The frequency, wavelength and velocity of light in the glass slab will be respectively **NCERT | Page-229 | 1997**
(a) $n, \frac{\lambda}{\mu}$ and $\frac{v}{\mu}$ (b) $n, 2\lambda$ and $\frac{v}{\mu}$
(c) $\frac{n}{\mu}, \frac{\lambda}{\mu}$ and $\frac{v}{\mu}$ (d) $\frac{2\pi}{\mu}, \frac{\lambda}{\mu}$ and v

17. Light travels through a glass plate of thickness t and refractive index μ. If c is the speed of light in vacuum, the time taken by light to travel this thickness of glass is **1996**
(a) μtc (b) $\frac{tc}{\mu}$ (c) $\frac{t}{\mu c}$ (d) $\frac{\mu t}{c}$

18. One face of a rectangular glass plate 6 cm thick is silvered. An object held 8 cm in front of the first face forms an image 12 cm behind the silvered face. The refractive index of the glass is **NCERT | Page-229 | 1996**
(a) 0.4 (b) 0.8 (c) 1.2 (d) 1.6

19. Time taken by sunlight to pass through a window of thickness 4 mm whose refractive index is $\frac{3}{2}$ is **NCERT | Page-229 | 1993**
(a) 2×10^{-4} sec (b) 2×10^8 sec
(c) 2×10^{-11} sec (d) 2×10^{11} sec

20. A beam of monochromatic light is refracted from vacuum into a medium of refractive index 1.5, the wavelength of refracted light will be **NCERT | Page-229 | 1991**
(a) dependent on intensity of refracted light
(b) same
(c) smaller
(d) larger

21. Green light of wavelength 5460 Å is incident on an air-glass interface. If the refractive index of glass is 1.5, the wavelength of light in glass would be ($c = 3 \times 10^8\ ms^{-1}$) **NCERT | Page-229 | 1991**
(a) 3640 Å (b) 5460 Å
(c) 4861 Å (d) none of the above

Topic-9.4 Total Internal Reflection

22.

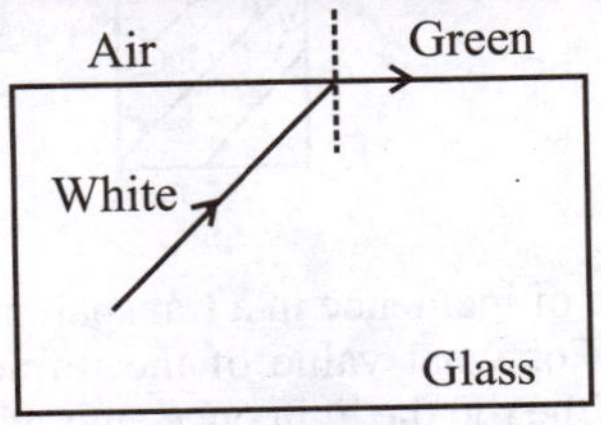

Which set of colours will come out in air for a situation shown in figure? **NCERT | Page-238 | Manipur 2023**

(a) Orange, Red and Violet
(b) Blue, Green and Yellow
(c) Yellow, Orange and Red
(d) All

23. Light travels a distance x in time t_1 in air and 10x in time t_2 in another denser medium. What is the critical angle for this medium? **NCERT | Page-230 | 2023**

(a) $\sin^{-1}\left(\frac{t_2}{t_1}\right)$ (b) $\sin^{-1}\left(\frac{10t_2}{t_1}\right)$

(c) $\sin^{-1}\left(\frac{t_1}{10t_2}\right)$ (d) $\sin^{-1}\left(\frac{10t_1}{t_2}\right)$

24. Two transparent media A and B are separated by a plane boundary. The speed of light in those media are 1.5×10^8 m/s and 2.0×10^8 m/s, respectively. The critical angle for a ray of light for these two media is: **NCERT | Page-230 | 2022**

(a) $\sin^{-1}(0.750)$ (b) $\tan^{-1}(0.500)$
(c) $\tan^{-1}(0.750)$ (d) $\sin^{-1}(0.500)$

25. In total internal reflection when the angle of incidence is equal to the critical angle for the pair of media in contact, what will be angle of refraction ? **NCERT | Page-230 | 2019**

(a) 180°
(b) 0°
(c) equal to angle of incidence
(d) 90°

26. Which of the following is not due to total internal reflection? **NCERT | Page-231 | 2011**

(a) Working of optical fibre
(b) Difference between apparent and real depth of pond
(c) Mirage on hot summer days
(d) Brilliance of diamond

27. A ray of light travelling in a transparent medium of refractive index μ, falls on a surface separating the medium from air at an angle of incidence of 45°. For which of the following value of μ the ray can undergo total internal reflection? **NCERT | Page-230 | 2010**

(a) μ = 1.33 (b) μ = 1.40
(c) μ = 1.50 (d) μ = 1.25

28. A light ray falls on a rectangular glass slab as shown. The index of refraction of the glass, if total internal reflection is to occur at the vertical face, is **NCERT | Page-230 | 2002**

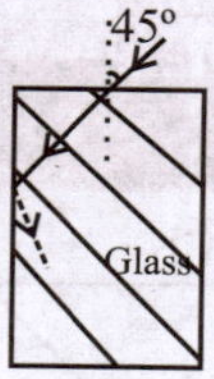

(a) $\sqrt{3}/2$

(b) $\frac{(\sqrt{3}+1)}{2}$

(c) $\frac{(\sqrt{2}+1)}{2}$

(d) $\sqrt{5}/2$

29. Light enters at an angle of incidence in a transparent rod of refractive index n. For what value of the refractive index of the material of the rod the light once entered into it will not leave it through its lateral face whatsoever be the value of angle of incidence? **NCERT | Page-228 | 1998**

(a) $n > \sqrt{2}$ (b) $n = 1$
(c) $n = 1.1$ (d) $n = 1.3$

30. A point source of light is placed 4 m below the surface of water of refractive index $\frac{5}{3}$. The minimum diameter of a disc, which should be placed over the source, on the surface of water to cut off all light coming out of water is **NCERT | Page-230 | 1994**

(a) ∞ (b) 6 m (c) 4 m (d) 3 m

Topic-9.5 Refraction at Spherical Surfaces and by Lenses

31. In a certain camera, a combination of four similar thin convex lenses are arranged axially in contact. Then the power of the combination and the total magnification in comparison to the power (p) and magnification (m) for each lens will be, respectively **NCERT | Page-238| 2025**

(a) $4p$ and $4m$ (b) p^4 and $4m$
(c) $4p$ and m^4 (d) p^4 and m^4

32. A lens is made up of 3 different transparent media as shown in figure. A point object O is placed on its axis beyond 2f. How many real images will be obtained on the other side? **NCERT | Page-235 | Manipur 2023**

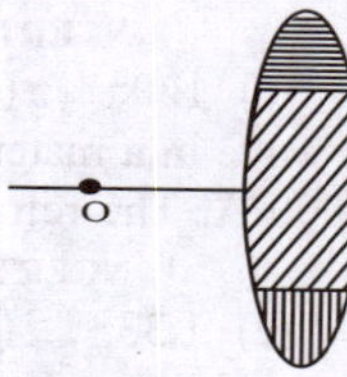

(a) No image will be formed
(b) 3
(c) 2
(d) 1

33. An object is mounted on a wall. Its image of equal size is to be obtained on a parallel wall with the help of a convex lens placed between these walls. The lens is kept at distance x in front of the second wall. The required focal length of the lens will be **NCERT | Page-235 | Manipur 2023**

(a) $\frac{x}{2}$ (b) $\frac{x}{4}$

(c) Less than $\frac{x}{4}$

(d) More than $\frac{x}{4}$ but less than $\frac{x}{2}$

34. Two thin lenses are of same focal lengths (f), but one is convex and the other one is concave. When they are placed in contact with each other, the equivalent focal length of the combination will be **NCERT | Page-238 | 2023**

(a) Zero (b) $\frac{f}{4}$

(c) $\frac{f}{2}$ (d) Infinite

35. In the figure shown here, what is the equivalent focal length of the combination of lenses (Assume that all layers are thin)? **NCERT | Page-237, 238 | 2023**

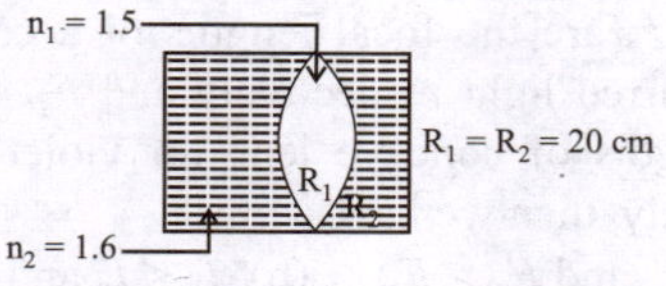

(a) 40 cm (b) –40 cm
(c) –100 cm (d) –50 cm

36. A biconvex lens has radii of curvature, 20 cm each, if the refractive index of the material of the lens is 1.5, the power of the lens is: **NCERT | Page-234 | 2022**

(a) +20D (b) + 5D
(c) infinity (d) +2D

37. A convex lens 'A' of focal length 20 cm and a concave lens 'B' of focal length 5 cm are kept along the same axis with a distance 'd' between them. If a parallel beam of light falling on 'A' leaves 'B' as a parallel beam, then the distance 'd' in cm will be **NCERT | Page-237 | 2021**

(a) 30 (b) 25 (c) 15 (d) 50

38. Two similar thin equi-convex lenses, of focal length f each, are kept coaxially in contact with each other such that the focal length of the combination is F_1. When the space between the two lenses is filled with glycerin (which has the same refractive index (= 1.5) as that of glass) then the equivalent focal length is F_2. The ratio $F_1 : F_2$ will be : **NCERT | Page-237 | 2019**

(a) 2 : 1 (b) 1 : 2 (c) 2 : 3 (d) 3 : 4

39. An equiconvex lens has power *P*. It is cut into two symmetrical halves by a plane containing the principal axis. The power of one part will be, **NCERT | Page-237 | Odisha 2019**

(a) P (b) 0 (c) $\frac{P}{2}$ (d) $\frac{P}{4}$

40. A double convex lens has focal length 25 cm. The radius of curvature of one of the surfaces is double of the other. Find the radii if the refractive index of the material of the lens is 1.5. **NCERT | Page-235 | Odisha 2019**

(a) 50 cm, 100 cm (b) 100 cm, 50 cm
(c) 25 cm, 50 cm (d) 18.75 cm, 37.5 cm

41. Two identical thin plano-convex glass lenses (refractive index 1.5) each having radius of curvature of 20 cm are placed with their convex surfaces in contact at the centre. The intervening space is filled with oil of refractive index 1.7. The focal length of the combination is **NCERT | Page-235 | 2015**

(a) –25 cm (b) –50 cm
(c) 50 cm (d) –20 cm

42. A plano convex lens fits exactly into a plano concave lens. Their plane surfaces are parallel to each other. If lenses are made of different materials of refractive indices μ_1 and μ_2 and R is the radius of curvature of the curved surface of the lenses, then the focal length of the combination is **NCERT | Page-237 | 2013**

(a) $\frac{R}{2(\mu_1 - \mu_2)}$ (b) $\frac{R}{(\mu_1 - \mu_2)}$

(c) $\frac{2R}{(\mu_2 - \mu_1)}$ (d) $\frac{R}{2(\mu_1 + \mu_2)}$

43. When a biconvex lens of glass having refractive index 1.47 is dipped in a liquid, it acts as a plane sheet of glass. This implies that the liquid must have refractive index. **NCERT | Page-235 | 2012**

(a) equal to that of glass
(b) less then one
(c) greater than that of glass
(d) less then that of glass

44. A biconvex lens has a radius of curvature of magnitude 20 cm. Which one of the following options best describe the image formed of an object of height 2 cm placed 30 cm from the lens? **NCERT | Page-235 | 2011**

(a) Virtual, upright, height = 1 cm
(b) Virtual, upright, height = 0.5 cm
(c) Real, inverted, height = 4 cm
(d) Real, inverted, height = 1cm

45. A converging beam of rays is incident on a diverging lens. Having passed through the lens the rays intersect at a point 15 cm from the lens on the opposite side. If the lens is removed the point where the rays meet will move 5 cm closer to the lens. The focal length of the lens is **NCERT | Page-235 | 2011 M**

(a) – 10 cm (b) 20 cm
(c) –30 cm (d) 5 cm

46. A lens having focal length *f* and aperture of diameter *d* forms an image of intensity *I*. Aperture of diameter $\frac{d}{2}$ in central region of lens is covered by a black paper. Focal length of lens and intensity of image now will be respectively: **2010**

(a) f and $\frac{I}{4}$ (b) $\frac{3f}{4}$ and $\frac{I}{2}$

(c) f and $\frac{3I}{4}$ (d) $\frac{f}{2}$ and $\frac{I}{2}$

47. Two thin lenses of focal lengths f_1 and f_2 are in contact and coaxial. The power of the combination is: **NCERT | Page-237 | 2008**

(a) $\sqrt{\frac{f_1}{f_2}}$ (b) $\sqrt{\frac{f_2}{f_1}}$

(c) $\frac{f_1 + f_2}{2}$ (d) $\frac{f_1 + f_2}{f_1 f_2}$

48. A boy is trying to start a fire by focusing sunlight on a piece of paper using an equiconvex lens of focal length 10 cm. The diameter of the Sun is 1.39×10^9 m and its mean distance from the earth is 1.5×10^{11} m. What is the diameter of the Sun's image on the paper? **NCERT | Page-236 | 2008**

(a) 9.2×10^{-4} m (b) 6.5×10^{-4} m
(c) 6.5×10^{-5} m (d) 12.4×10^{-4} m

49. A convex lens and a concave lens, each having same focal length of 25 cm, are put in contact to form a combination of lenses. The power in diopters of the combination is **NCERT | Page-236 | 2006**

(a) 50 (b) infinite
(c) zero (d) 25

50. An equiconvex lens is cut into two halves along (i) XOX' and (ii) YOY' as shown in the figure. Let f, f', f'' be the focal lengths of the complete lens, of each half in case (i), and of each half in case (ii), respectively. NCERT | Page-237 | 2003

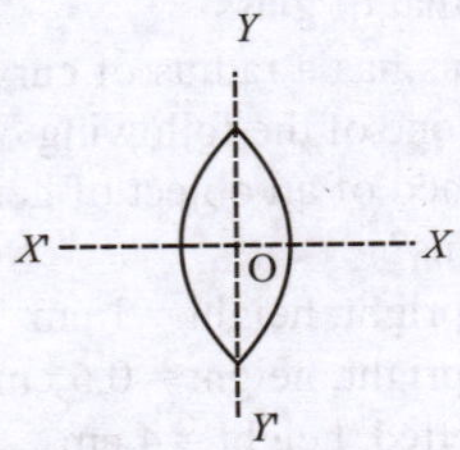

Choose the correct statement from the following
(a) $f' = 2f, f'' = 2f$ (b) $f' = f, f'' = 2f$
(c) $f' = 2f, f'' = f$ (d) $f' = f, f'' = f$

51. A convex lens is dipped in a liquid whose refractive index is equal to the refractive index of the lens. Then its focal length will NCERT | Page-235 | 2003
(a) remain unchanged
(b) become zero
(c) become infinite
(d) become small, but non-zero

52. A body is located on a wall. Its image of equal size is to be obtained on a parallel wall with the help of a convex lens. The lens is placed at a distance 'd' ahead of second wall, then the required focal length will be NCERT | Page-235 | 2002
(a) only $\frac{d}{4}$
(b) only $\frac{d}{2}$
(c) more than $\frac{d}{4}$ but less than $\frac{d}{2}$
(d) less than $\frac{d}{4}$

53. A plano-convex lens is made of material of refractive index 1.6. The radius of curvature of the curved surface is 60 cm. The focal length of the lens is NCERT | Page-235 | 1999
(a) 50 cm (b) 100 cm
(c) 200 cm (d) 400 cm

54. The focal length of converging lens is measured for violet, green and red colours. It is respectively f_v, f_g, f_r. We will get NCERT | Page-332 | 1997
(a) $f_v = f_g$ (b) $f_g > f_r$
(c) $f_v < f_r$ (d) $f_v > f_r$

55. A convex lens of focal length 80 cm and a concave lens of focal length 50 cm are combined together. What will be their resulting power? NCERT | Page-237 | 1996
(a) + 6.5 D (b) – 6.5 D
(c) + 7.5 D (d) – 0.75 D

56. If f_V and f_R are the focal lengths of a convex lens for violet and red light respectively and F_V and F_R are the focal lengths of concave lens for violet and red light respectively, then we have 1996
(a) $f_V < f_R$ and $F_V > F_R$ (b) $f_V < f_R$ and $F_V < F_R$
(c) $f_V > f_R$ and $F_V > F_R$ (d) $f_V > f_R$ and $F_V < F_R$

57. An achromatic combination of lenses is formed by joining 1995
(a) 2 convex lenses
(b) 2 concave lenses
(c) 1 convex and 1 concave lens
(d) 1 convex and 1 plane mirror

58. A lens is placed between a source of light and a wall. It forms images of area A_1 and A_2 on the wall for its two different positions. The area of the source of light is 1995
(a) $\sqrt{A_1 A_2}$ (b) $\frac{A_1 + A_2}{2}$ (c) $\frac{A_1 - A_2}{2}$ (d) $\frac{1}{A_1} + \frac{1}{A_2}$

59. Focal length of a convex lens will be maximum for 1994
(a) blue light (b) yellow light
(c) green light (d) red light

60. Focal length of a convex lens of refractive index 1.5 is 2 cm. Focal length of the lens when immersed in a liquid of refractive index of 1.25 will be NCERT | Page-235 | 1988
(a) 10 cm (b) 2.5 cm (c) 5 cm (d) 7.5 cm

Topic-9.6

Refraction through a Prism

61. A light ray enters through a right angled prism at point P with the angle of incidence 30° as shown in figure. It travels through the prism parallel to its base BC and emerges along the face AC. The refractive index of the prism is : NCERT | Page-240 | 2024

(a) $\frac{\sqrt{5}}{4}$
(b) $\frac{\sqrt{5}}{2}$
(c) $\frac{\sqrt{3}}{4}$
(d) $\frac{\sqrt{3}}{2}$

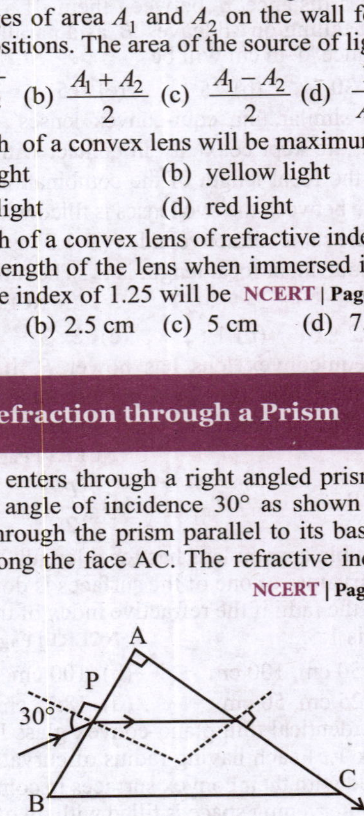

62. A horizontal ray of light is incident on the right-angled prism with prism angle 6°. If the refractive index of the material of the prism is 1.5, then the angle of emergence will be NCERT | Page-240 | Manipur 2023

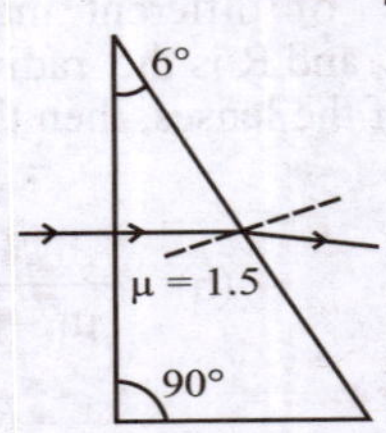

(a) 4° (b) 6°
(c) 9° (d) 10°

63. Find the value of the angle of emergence from the prism. Refractive index of the glass is $\sqrt{3}$. NCERT | Page-239 | 2021

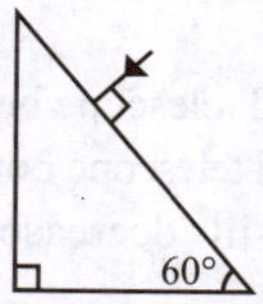

(a) 90° (b) 60° (c) 30° (d) 45°

64. A ray is incident at an angle of incidence i on one surface of a small angle prism (with angle of prism A) and emerges normally from the opposite surface. If the refractive index of the material of the prism is μ, then the angle of incidence is nearly equal to : NCERT | Page-239 | 2020

(a) $\frac{2A}{\mu}$ (b) μA (c) $\frac{\mu A}{2}$ (d) $\frac{A}{2\mu}$

65. The refractive index of the material of a prism is $\sqrt{2}$ and the angle of the prism is 30°. One of the two refracting surfaces of the prism is made a mirror inwards, by silver coating. A beam of monochromatic light entering the prism from the other face will retrace its path (after reflection from the silvered surface) if its angle of incidence on the prism is NCERT | Page-240 | 2018

(a) 60° (b) 45° (c) Zero (d) 30°

66. The angle of incidence for a ray of light at a refracting surface of a prism is 45° . The angle of prism is 60°. If the ray suffers minimum deviation through the prism, the angle of minimum deviation and refractive index of the material of the prism respectively, are : NCERT | Page-240 | 2016

(a) $45°, \frac{1}{\sqrt{2}}$ (b) $30°, \sqrt{2}$ (c) $45°, \sqrt{2}$ (d) $30°, \frac{1}{\sqrt{2}}$

67. The refracting angle of a prism is 'A', and refractive index of the material of the prism is cot(A/2). The angle of minimum deviation is : NCERT | Page-240 | 2015

(a) 180° – 2A (b) 90° – A
(c) 180° + 2A (d) 180° – 3A

68. A beam of light consisting of red, green and blue colours is incident on a right angled prism. The refractive index of the material of the prism for the above red, green and blue wavelengths are 1.39, 1.44 and 1.47, respectively. NCERT | Page-240 | 2015 RS

The prism will:

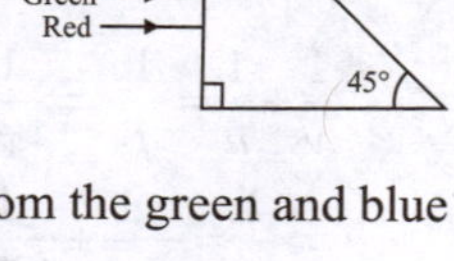

(a) separate all the three colours from one another
(b) not separate the three colours at all
(c) separate the red colour part from the green and blue colours
(d) separate the blue colour part from the red and green colours

69. The angle of a prism is 'A'. One of its refracting surfaces is silvered. Light rays falling at an angle of incidence 2A on the first surface returns back through the same path after suffering reflection at the silvered surface. The refractive index μ, of the prism is : NCERT | Page-240 | 2014

(a) 2 sin A (b) 2 cos A (c) $\frac{1}{2}\cos A$ (d) tan A

70. A ray of light is incident at an angle of incidence, *i*, on one face of prism of angle *A* (assumed to be small) and emerges normally from the opposite face. If the refractive index of the prism is μ, the angle of incidence *i*, is nearly equal to: NCERT | Page-240 | 2012

(a) μA (b) $\frac{\mu A}{2}$ (c) $\frac{A}{\mu}$ (d) $\frac{A}{2\mu}$

71. For the angle of minimum deviation of a prism to be equal to its refracting angle, the prism must be made of a material whose refractive index : NCERT | Page-240 | 2012 M

(a) lies between $\sqrt{2}$ and 1 (b) lies between 2 and $\sqrt{2}$
(c) is less than 1 (d) is greater than 2

72. A ray of light is incident on a 60° prism at the minimum deviation position. The angle of refraction at the first face (i.e., incident face) of the prism is NCERT | Page-239 | 2010 M

(a) Zero (b) 30° (c) 45° (d) 60°

73. The refractive index of the material of a prism is √2 and its refracting angle is 30º. One of the refracting surfaces of the prism is made a mirror inwards. A beam of monochromatic light enters the prism from the mirror surface if its angle of incidence of the prism is NCERT | Page-240 | 2004, 1992

(a) 30° (b) 45° (c) 60° (d) 0°

74. The refractive index of the material of the prism is $\sqrt{3}$; then the angle of minimum deviation of the prism is NCERT | Page-240 | 1999

(a) 30º (b) 45º (c) 60º (d) 75º

75. Angle of deviation (δ) by a prism (refractive index = μ and supposing the angle of prism A to be small) can be given by NCERT | Page-240 | 1994

(a) $\delta = (\mu - 1)A$ (b) $\delta = (\mu + 1)A$

(c) $\delta = \frac{\sin\frac{A+\delta}{2}}{\sin\frac{A}{2}}$ (d) $\delta = \frac{\mu - 1}{\mu + 1}A$

Topic-9.7

Optical Instruments

76. A microscope has an objective of focal length 2 cm, eyepiece of focal length 4 cm and the tube length of 40 cm. If the distance of distinct vision of eye is 25 cm, the magnification in the microscope is: NCERT | Page-242, 243, 244 | 2025

(a) 100 (b) 125 (c) 150 (d) 250

77. A small telescope has an objective of focal length 140 cm and an eye piece of focal length 5.0 cm. The magnifying power of telescope for viewing a distant object is: NCERT | Page-244, 245 | 2024

(a) 34 (b) 28 (c) 17 (d) 32

78. An astronomical refracting telescope is being used by an observer to observe planets in normal adjustment. The focal lengths of the objective and eye piece used in the construction of the telescope are 20 m and 2 cm respectively. Consider the following statements about the telescope : **RS 2022**
(A) The distance between the objective and eyepiece is 20.02 m
(B) The magnification of the telescope is (–) 1000
(C) The image of the planet is erect and diminished
(D) The aperture of eye piece is smaller than that of objective
The correct statements are :
(a) (A), (B) and (D) (b) (A), (B) and (C)
(c) (B), (C) and (D) (d) (C), (D) and (A)

79. A astronomical telescope has objective and eyepiece of focal lengths 40 cm and 4 cm respectively. To view an object 200 cm away from the objective, the lenses must be separated by a distance : **NCERT | Page-245 | 2016**
(a) 37.3 cm (b) 46.0 cm
(c) 50.0 cm (d) 54.0 cm

80. In an astronomical telescope in normal adjustment a straight black line of lenght L is drawn on inside part of objective lens. The eye-piece forms a real image of this line. The length of this image is l. The magnification of the telescope is : **NCERT | Page-245 | 2015 RS**
(a) $\frac{L}{I}-1$ (b) $\frac{L+I}{L-I}$ (c) $\frac{L}{I}$ (d) $\frac{L}{I}+1$

81. If the focal length of objective lens is increased then magnifying power of : **NCERT | Page-243-245 | 2014**
(a) microscope will increase but that of telescope decrease.
(b) microscope and telescope both will increase.
(c) microscope and telescope both will decrease.
(d) microscope will decrease but that of telescope increase.

82. The magnifying power of a telescope is 9. When it is adjusted for parallel rays the distance between the objective and eyepiece is 20 cm. The focal length of lenses are : **NCERT | Page-245 | 2012**
(a) 10 cm, 10 cm (b) 15 cm, 5 cm
(c) 18 cm, 2 cm (d) 11 cm, 9 cm

83. A microscope is focussed on a mark on a piece of paper and then a slab of glass of thickness 3 cm and refractive index 1.5 is placed over the mark. How should the microscope be moved to get the mark in focus again ? **2006**
(a) 4.5 cm downward
(b) 1 cm downward
(c) 2 cm upward
(d) 1 cm upward

84. An astronomical telescope has a length of 44 cm and tenfold magnification. The focal length of the objective lens is **NCERT | Page-245 | 1997**
(a) 4 cm (b) 40 cm
(c) 44 cm (d) 440 cm

ANSWER KEY

1	(c)	10	(b)	19	(c)	28	(a)	37	(c)	46	(c)	55	(d)	64	(b)	73	(b)	82	(c)
2	(b)	11	(b)	20	(c)	29	(a)	38	(b)	47	(d)	56	(a)	65	(b)	74	(c)	83	(d)
3	(d)	12	(d)	21	(a)	30	(b)	39	(a)	48	(a)	57	(c)	66	(b)	75	(a)	84	(b)
4	(a)	13	(c)	22	(c)	31	(c)	40	(d)	49	(c)	58	(a)	67	(a)	76	(b)		
5	(a)	14	(b)	23	(d)	32	(b)	41	(b)	50	(b)	59	(d)	68	(c)	77	(b)		
6	(d)	15	(b)	24	(a)	33	(a)	42	(b)	51	(c)	60	(c)	69	(b)	78	(a)		
7	(a)	16	(a)	25	(d)	34	(d)	43	(a)	52	(b)	61	(b)	70	(a)	79	(d)		
8	(b)	17	(d)	26	(b)	35	(c)	44	(c)	53	(b)	62	(c)	71	(b)	80	(c)		
9	(d)	18	(c)	27	(c)	36	(b)	45	(c)	54	(c)	63	(b)	72	(b)	81	(d)		

Hints & Solutions

1. **(c)** Regular reflection needs for image formation, that is why we can see our image in a mirror but not in wall. Different constituent colours in white light are responsible for the variety in colour of objects.

2. **(b)** From the new Cartesian sign convention,
Object distance, $u = -1.5f$
According to mirror formula,

$$\frac{1}{v}+\frac{1}{u}=\frac{1}{f} \Rightarrow \frac{1}{v}+\frac{1}{-1.5f}=\frac{1}{-f}$$

$$\Rightarrow \frac{1}{v}=-\frac{1}{f}+\frac{1}{1.5f}=\frac{1}{f}\left[-1+\frac{1}{1.5}\right]=\frac{-1.5+1}{1.5f}$$

$$\Rightarrow \frac{1}{v}=\frac{1}{f}\left(-\frac{1}{3}\right) \Rightarrow v=-3f$$

3. **(d)** When mirror is rotated by angle θ reflected ray will be rotated by 2θ.

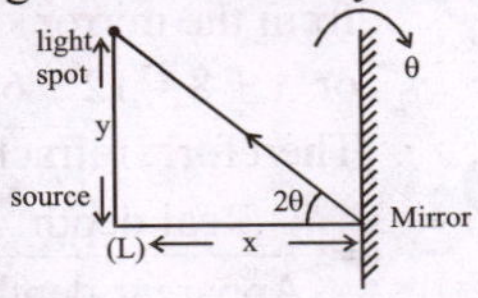

According to the condition,

$\frac{y}{x} = 2\theta \Rightarrow \theta = \frac{y}{2x}$

4. **(a)** Magnitude m = +ve ⇒ virtual image
m = –ve ⇒ real image
magnitude of magnification,
| m | > 1 ⇒ magnified image
| m | < 1 ⇒ diminished image

5. **(a)**

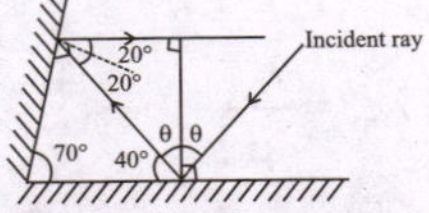

From fig. 40° + θ = 90° ∴ θ = 90° – 40° = 50°

6. **(d)** According to the condition,

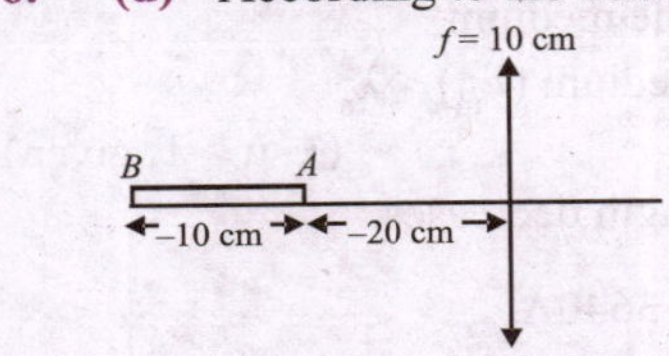

The focal length of the mirror

$-\frac{1}{f} = \frac{1}{v} + \frac{1}{u}$

For *A* end of the rod the image distance
When $u_1 = -20$ cm

$\Rightarrow \frac{-1}{10} = \frac{1}{v_1} - \frac{1}{20}; \frac{1}{v_1} = \frac{-1}{10} + \frac{1}{20} = \frac{-2+1}{20}$

$v_1 = -20$ cm
For when $u_2 = -30$ cm

$\frac{1}{f} = \frac{1}{v_2} - \frac{1}{30}$

$\frac{1}{v_2} = \frac{-1}{10} + \frac{1}{30} = \frac{-30+10}{300} = \frac{-20}{300}$

$v_2 = -15$ cm
$L = v_2 - v_1 = -15 - (-20); L = 5$ cm

7. **(a)** To see his full image in a plane mirror a person requires a mirror of at least half of his height.

8. **(b)** Angle between two mirrors (θ) = 60°. Number of images formed by the inclined mirror

$(n) = \frac{360°}{\theta} - 1 = \frac{360°}{60°} - 1 = 6 - 1 = 5$

NOTES When two plane mirrors are inclined to each other at an angle θ then number of images (n) formed $n = \frac{360}{\theta}$ when object is placed asymmetrically, and $n = \frac{360}{\theta} - 1$ when object is placed symmetrically.

9. **(d)** We know Refractive index of the medium, μ_{med}

$= \frac{\sqrt{\mu\varepsilon}}{\sqrt{\mu_0\varepsilon_0}} = \frac{\sqrt{(1.5)\mu_0(2)\varepsilon_0}}{\sqrt{\mu_0\varepsilon_0}} = \sqrt{3}$

10. **(b)** We have, $\frac{\mu_2}{\mu_1} = \frac{\sin i}{\sin r}$

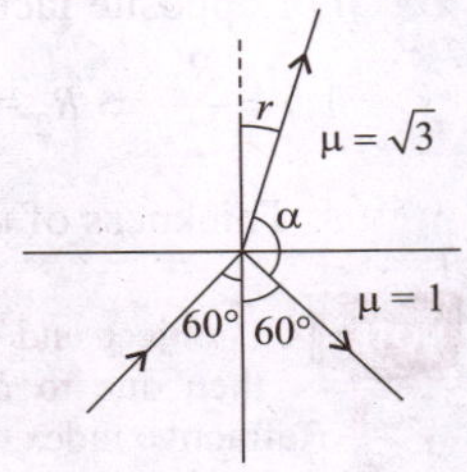

$\Rightarrow \sqrt{3} = \frac{\sin 60°}{\sin r}$

$\Rightarrow \sin r = \frac{\sqrt{3}}{2 \times \sqrt{3}}$

$\Rightarrow \sin r = \frac{1}{2} \Rightarrow r = \sin^{-1}\left(\frac{1}{2}\right) = 30°$

Now, angle between refracted ray and reflected ray = (180 – 60 – 30)° = 90°

NOTES Snell's law for two media $_1\mu_2 = \frac{\mu_1}{\mu_2} = \frac{\sin i}{\sin r}$

or, $\mu_1 \times \sin i = \mu_2 \times \sin r$ i.e., $\mu \sin\theta$ = constant

11. **(b)** By using $v = n\lambda$

Here, $n = 2 \times 10^{14}$ Hz

$\lambda = 5000 Å = 5000 \times 10^{-10}$ m

$v = 2 \times 10^{14} \times 5000 \times 10^{-10} = 10^8$ m/s

Refractive index of the material,

$\mu = c/v = \frac{3 \times 10^8}{10^8} = 3$

NOTES In case of refraction of light frequency (and hence colour) and phase do not change while wavelength and velocity will change.

12. **(d)** From pythagorus theorem,
Hypotenuse comes out to be 5 cm.

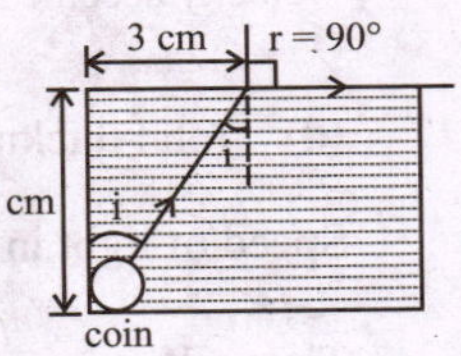

Since, $\frac{1}{\mu} = \frac{\sin i}{\sin 90°}$;

$\mu = \frac{1}{\sin i} = \frac{5}{3}$

Speed, $v = \frac{c}{\mu} = \frac{3 \times 10^8}{5/3} = 1.8 \times 10^8$ m/s

13. **(c)** Since refractive index for both the light are different, so they emerge out moving in two different parallel directions.

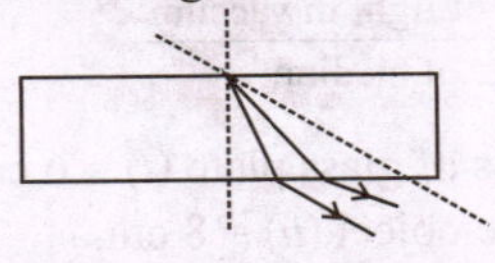

14. **(b)** $1.5 = \frac{\text{Real depth }(R_1)}{\text{Apparent depth }(5\text{ cm})}$

$\therefore R_1 = 1.5 \times 5 = 7.5$ cm

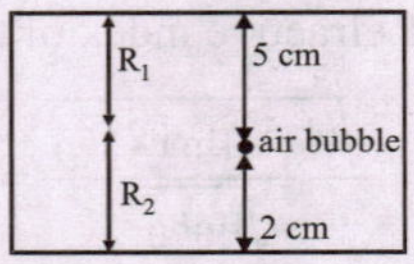

For opposite face,

$$1.5 = \frac{R_2}{2} \Rightarrow R_2 = 3.0 \text{ cm}$$

$\therefore$ Thickness of the slab $= R_1 + R_2 = 7.5 + 3 = 10.5$ cm

NOTES If object and observer are situated in different medium then due to refraction, object appears to be displaced Refractive index,

$$\mu = \frac{\text{Real depth}(h)}{\text{Apparent depth}(h')}$$

Real depth > Apparent depth

In case object is in rarer medium and object is in denser medium then, $\mu = \frac{h'}{h}$

Real depth < apparent depth

15. **(b)** Frequency (n) = 100 Hz

$v = n\lambda$

$$\lambda = \frac{3\times10^8}{100}$$

[where, velocity of light (v) = 3×10^8 m/s]

$\lambda = 3\times10^6$ m

16. **(a)** For e.m. wave entering from air to glass slab (μ), frequency remains n,

wavelength, $\lambda' = \frac{\lambda}{\mu}$ and velocity, $v' = \frac{v}{\mu}$

NOTES When electromagnetic wave enters in other medium, frequency remains unchanged while wavelength and velocity become $\frac{1}{\mu}$ times.

17. **(d)** Total thickness = t; Refrative index = μ

Speed of light in Glass plate $= \frac{c}{\mu}$

Time taken $= \frac{t}{\left(\frac{c}{\mu}\right)} = \frac{\mu t}{c}$.

where, t = thickness of glass plate

NOTES Speed of light in medium

$$v = \frac{\text{Speed of light in vaccum}}{\mu \text{ of medium}}$$

18. **(c)** Thickness of glass plate (t) = 6 cm;

Distance of the object (u) = 8 cm.

And distance of the image (v) = 12 cm.

Let x = Apparent position of the silvered surface in cm. Since the image is formed due to reflection at the silvered face and by the property of mirror image.

Distance of object from the mirror = Distance of image from the mirror

or, $x + 8 = 12 + 6 - x \Rightarrow x = 5$ cm.

Therefore, refractive index of glass

$$= \frac{\text{Real depth}}{\text{Apparent depth}} = \frac{6}{5} = 1.2$$

19. **(c)** $v_g = \frac{c}{\mu} = \frac{3\times10^8}{\frac{3}{2}} = 2\times10^8$ m/s

$$t = \frac{x}{v_g} = \frac{4\times10^{-3}}{2\times10^8} = 2\times10^{-11} \text{ s}$$

20. **(c)** From $\mu = \frac{c}{v} = \frac{n\lambda_v}{n\lambda_m}$, $\lambda_m = \frac{\lambda_v}{\mu}$

Here, c = velocity of light in medium and v = velocity of light in vacuum;

μ = refractive index of the medium.

Hence, wavelength in medium $(\lambda_m) < \lambda_a$

($\because \mu > 1$, given)

So, the required wavelength decreases.

21. **(a)** $\lambda_g = \frac{\lambda_a}{\mu} = \frac{5460}{1.5} = 3640$Å

22. **(c)** From the relation, $\sin i_C = \frac{1}{\mu}$ and $\mu \propto \frac{1}{\lambda}$

$\sin i_c \propto \lambda$

Therefore, for higher value of λ critical angle will be more, so Yellow, Orange and Red will come out to air.

The order of wavelength of different colours

V < I < B < G < Y < O < R

When a glass slab is kept over various coloured letters and seen from the top, the violet colour letters appears closer, as $\lambda_V < \lambda_R \therefore \mu_V > \mu_R$ and $M = \frac{h}{h'}$

23. **(d)** $\mu_2 \sin i_c = \mu_1$

$$\Rightarrow \sin i_c = \frac{\mu_1}{\mu_2}$$

$$\mu = \frac{c}{V}$$

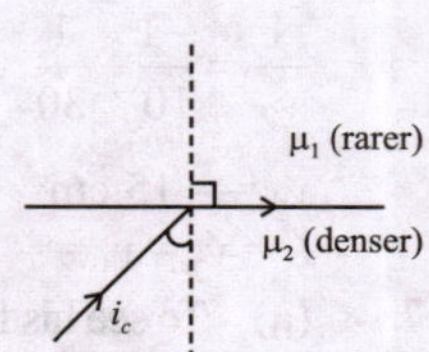

$$\therefore \sin i_c = \frac{\mu_1}{\mu_2} = \frac{V_2}{V_1}$$

$$\sin i_c = \frac{10xt_1}{t_2 x} \quad \because i_c = \sin^{-1}\left(\frac{10t_1}{t_2}\right)$$

24. **(a)** $\sin C = \frac{\mu_r}{\mu_d}$ and, $\mu \alpha \frac{1}{v}$

therefore, $\frac{\mu_r}{\mu_d} = \frac{v_d}{v_r} = \sin c = \frac{v_d}{v_r} = \frac{1.5\times10^8}{2\times10^8} = \frac{3}{4}$

so, $\theta = \sin^{-1}\left(\frac{3}{4}\right) = \sin^{-1}(0.750)$

25. **(d)** For total internal reflection when $i = i_c$, then

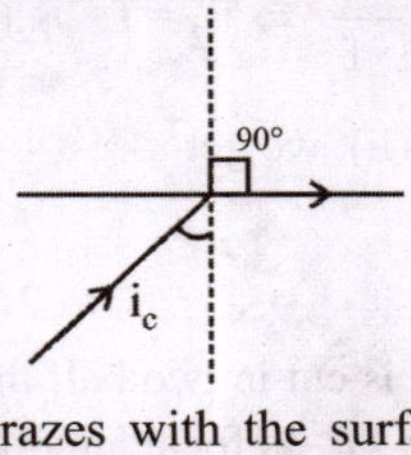

refracted ray grazes with the surface. That means the angle of refraction $r = 90°$.

26. **(b)** Difference between apparent and real depth of a pond is due to the refraction of light, not due to the total internal reflection. Other three phenomena are due to the total internal reflection.

27. **(c)** For total internal reflection,

$$\mu \geq \frac{1}{\sin C} \geq \sqrt{2} > 1.414 \Rightarrow \mu = 1.50$$

28. **(a)** For point A, ${}_a\mu_g = \frac{\sin 45°}{\sin r}$

$$\Rightarrow \sin r = \frac{1}{\sqrt{2}\,{}_a\mu_g}$$

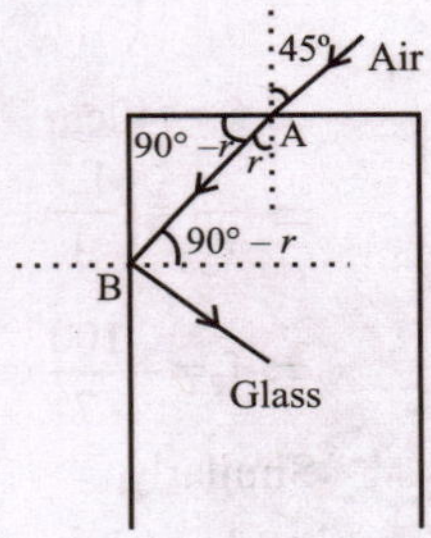

For point B, $\sin(90° - r) = {}_g\mu_a$ where, $(90° - r)$ is critical angle.

$$\therefore \cos r = {}_g\mu_a = \frac{1}{{}_a\mu_g}$$

$$\Rightarrow {}_a\mu_g = \frac{1}{\cos r}$$

$$= \frac{1}{\sqrt{1-\sin^2 r}} = \frac{1}{\sqrt{1-\frac{1}{2\,{}_a\mu_g{}^2}}}$$

$$\Rightarrow {}_a\mu_g{}^2 = \frac{1}{1-\frac{1}{2\,{}_a\mu_g^2}} = \frac{2\,{}_a\mu_g^2}{2\,{}_a\mu_g^2 - 1}$$

$$\Rightarrow 2\,{}_a\mu_g^2 - 1 = 2 \Rightarrow {}_a\mu_g = \sqrt{\frac{3}{2}}$$

29. **(a)** Let a ray of light enter at A and the refracted beam is AB. This is incident at an angle θ. For no refraction at the lateral face, $\theta > C$

or, $\sin\theta > \sin C$ But $\theta + r = 90° \Rightarrow \theta = (90° - r)$

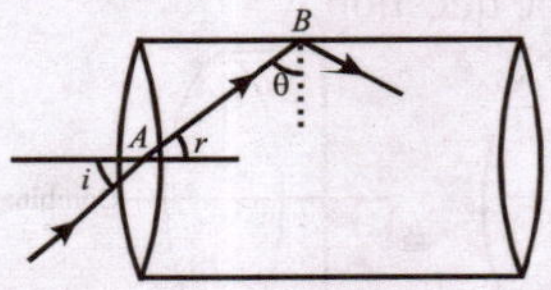

$\therefore \sin(90° - r) > \sin C$

or $\cos r > \sin C$ ..(1)

From Snell's law, $n = \frac{\sin i}{\sin r} \Rightarrow \sin r = \frac{\sin i}{n}$

$$\therefore \cos r = \sqrt{1-\sin^2 r} = \sqrt{\left(1-\frac{\sin^2 i}{n^2}\right)}$$

$\therefore$ equation (1) gives

$$\sqrt{1-\frac{\sin^2 i}{n^2}} > \sin C \Rightarrow 1-\frac{\sin^2 i}{n^2} > \sin^2 C$$

Also, $\sin C = \frac{1}{n}$

$$\therefore 1-\frac{\sin^2 i}{n^2} > \frac{1}{n^2} \text{ or } 1 > \frac{\sin^2 i}{n^2} + \frac{1}{n^2}$$

$$\text{or } \frac{1}{n^2}(\sin^2 i + 1) < 1 \text{ or } n^2 > (\sin^2 i + 1)$$

Maximum value of $\sin i = 1$

$$\therefore n^2 > 2 \Rightarrow n > \sqrt{2}$$

30. **(b)** $\sin C = \frac{1}{\mu} = \frac{1}{\frac{5}{3}} = \frac{3}{5} \Rightarrow$ So, $\tan C = \frac{3}{4}$

Now, $\tan C = \frac{r}{h}; r = h\tan C = 4 \times \frac{3}{4} = 3m$

Diameter of disc $= 2r = 6\ m$

> **NOTES** Field of vision of a swimmer or fish,
>
> Radius, $r = h \tan c = \frac{h}{\sqrt{\mu^2 - 1}}$
>
> Area, $A = \pi r^2 = \frac{\pi h^2}{(\mu^2 - 1)}$

31. **(c)** For combination of lens,

$P_{net} = P_1 + P_2 + P_3 + P_4 = P + P + P + P == 4P$

$\therefore$ Total magnification,

$m_{net} = m \times m \times m \times m = m^4$

32. **(b)** Since object distance u > 2f so all three medium will form individual real images at different points on the other side of lens

$\therefore$ Number of real images formed = 3.

33. **(a)** If object is placed at 2f then its image of equal size is to be obtained on a parallel wall with the help of a convex lens placed between the two walls.

$$\therefore 2f = x \Rightarrow f = \frac{x}{2}$$

34. **(d)** We know that

For convex lens $f_1 > 0$,

concave lens $f_2 < 0$

$$\frac{1}{f_{net}} = \frac{1}{f_1} + \frac{1}{f_2} = \frac{1}{f} - \frac{1}{f} = 0$$

$$\Rightarrow f_{net} = \infty$$

NOTES In two thin lenses of same focal length but having opposite nature are placed in contact then combination will behave like a plane glass plate.

35. **(c)** Effective focal length (f_{eff}) is

$$\frac{1}{f_{eff}} = \frac{1}{f_1} + \frac{1}{f_2} + \frac{1}{f_3}$$

From the lens maker's formula

$$\frac{1}{f} = (\mu - 1)\left(\frac{1}{R_1} - \frac{1}{R_2}\right)$$

$$\frac{1}{f_1} = (1.6 - 1)\left(\frac{1}{\infty} - \frac{1}{20}\right) = \frac{-0.6}{20} = \frac{-3}{100}$$

$$\frac{1}{f_2} = (1.5 - 1)\left(\frac{1}{20} - \frac{1}{-20}\right) = \frac{0.5}{10} = \frac{1}{20}$$

$$\frac{1}{f_3} = (1.6 - 1)\left(\frac{1}{-20} - \frac{1}{\infty}\right) = \frac{-0.6}{10} = \frac{-3}{100}$$

$$\Rightarrow \frac{1}{f_{eff}} = \frac{-3}{100} + \frac{1}{20} - \frac{3}{100} \Rightarrow \frac{1}{f_{eff}} = \frac{-1}{100}$$

$$\Rightarrow f_{eff} = -100 \text{ cm}$$

36. **(b)** We have $P = \frac{1}{f} = (\mu - 1)\left(\frac{1}{R_1} - \frac{1}{R_2}\right)$

$$= (1.5 - 1)\left(\frac{1}{20} - \frac{1}{-20}\right) \times \frac{1}{10^{-2}} = 0.5 \times \frac{2}{20} \times 100 = +5 \text{ D}$$

37. **(c)**

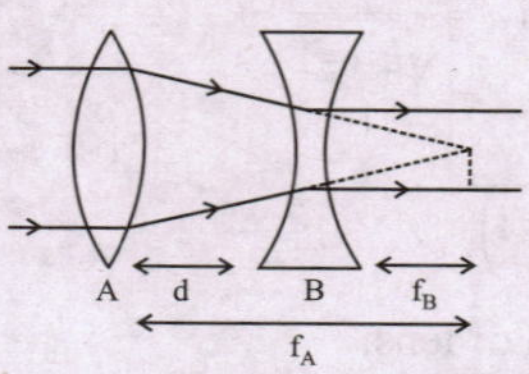

From the ray diagram light refracted from convex lens virtually meet at focus of concave lens

$\therefore$ $d = f_A - f_B = 20 - 5 = 15$ cm

38. **(b)**

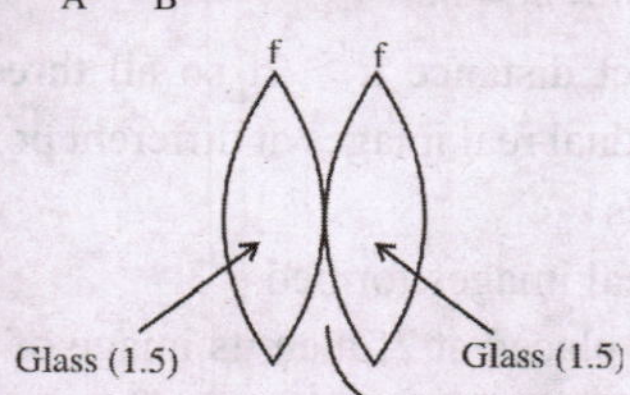

Equivalent focal length in air

$$\frac{1}{F_1} = \frac{1}{f} + \frac{1}{f} = \frac{2}{f} \Rightarrow F_1 = \frac{f}{2} \quad ...(i)$$

When glycerin is filled inside, it behaves like a concave lens of focal length (–f)

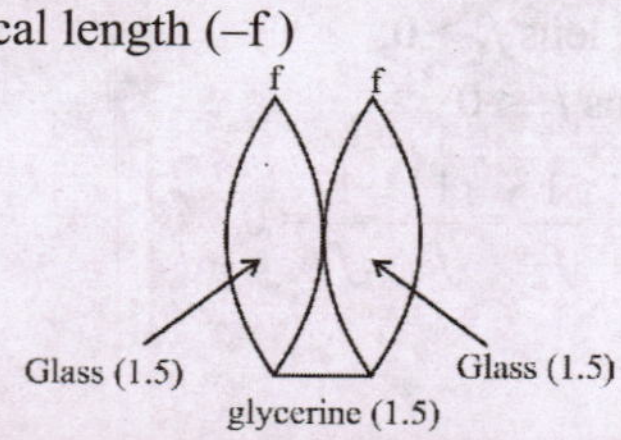

$$\therefore \frac{1}{F_2} = \frac{1}{f} + \frac{1}{f} - \frac{1}{f} \Rightarrow F_2 = f \quad ...(ii)$$

Dividing (i) by (ii), we get

$$\frac{F_1}{F_2} = \frac{1}{2}$$

39. **(a)** When lens is cut in two half then power of one part will be same. i.e., *P*, as focal length remains same.

40. **(d)** $\frac{1}{f} = (\mu - 1)\left(\frac{1}{R_1} - \frac{1}{R_2}\right)$

$$\frac{1}{25} = (1.5 - 1)\left(\frac{1}{R} + \frac{1}{2R}\right) \quad \frac{1}{25} = 0.5\left(\frac{3}{2R}\right)$$

$2R = 37.5$ cm $\qquad R = 18.75$ cm

41. **(b)** Using lens maker's formula,

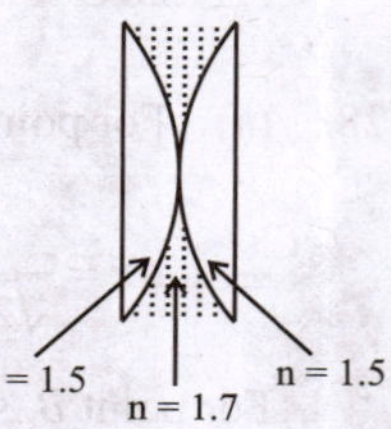

$$\frac{1}{f} = (\mu - 1)\left(\frac{1}{R_1} - \frac{1}{R_2}\right)$$

$$\Rightarrow \frac{1}{f_1} = \left(\frac{1.5}{1} - 1\right)\left(\frac{1}{\infty} - \frac{1}{-20}\right)$$

$$\Rightarrow f_1 = 40 \text{cm}$$

$$\Rightarrow \frac{1}{f_2} = \left(\frac{1.7}{1} - 1\right)\left(\frac{1}{-20} - \frac{1}{+20}\right)$$

$$\Rightarrow f_2 = -\frac{100}{7} \text{ cm}$$

Similarly,

$$\Rightarrow \frac{1}{f_3} = \left(\frac{1.5}{1} - 1\right)\left(\frac{1}{\infty} - \frac{1}{-20}\right) \Rightarrow f_3 = 40 \text{ cm}$$

Now according to the condition

$$\frac{1}{f_{eq}} = \frac{1}{f_1} + \frac{1}{f_2} + \frac{1}{f_3}$$

$$\Rightarrow \frac{1}{f_{eq}} = \frac{1}{40} + \frac{1}{-100/7} + \frac{1}{40}$$

$$\therefore \quad f_{eq} = -50 \text{ cm}$$

Therefore, the focal length of the combination is – 50 cm.

NOTES For a system of lenses net focal length F_{net}

$$= \frac{1}{f_1} + \frac{1}{f_2} + ... + \frac{1}{f_n}$$

Net power $P_{net} = P_1 + P_2 + ... + Pn$

42. **(b)** From the question,

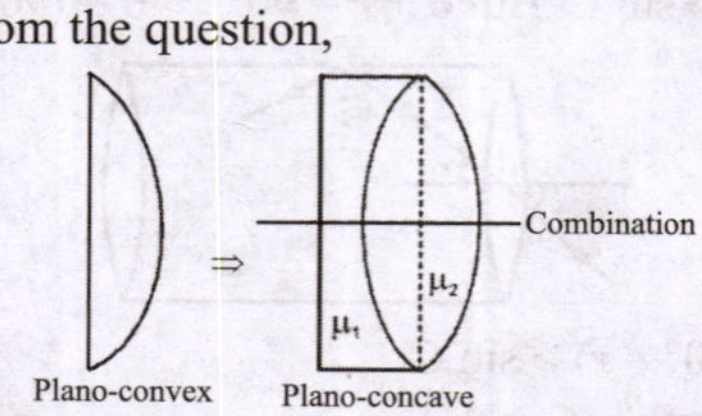

$$\frac{1}{f} = \frac{1}{f_1} + \frac{1}{f_2}$$

$$= (\mu_1 - 1)\left(\frac{1}{\infty} - \frac{1}{-R}\right) + (\mu_2 - 1)\left(\frac{1}{\infty} - \frac{1}{R}\right)$$

$$= \frac{(\mu_1 - 1)}{R} - \frac{(\mu_2 - 1)}{R} \Rightarrow \frac{1}{f} = \frac{\mu_1 - \mu_2}{R}$$

$$\Rightarrow f = \frac{R}{\mu_1 - \mu_2}$$

Hence, focal length of the combination is $\frac{R}{\mu_1 - \mu_2}$

43. (a) $\frac{1}{f} = \left(\frac{\mu_g}{\mu_m} - 1\right)\left(\frac{1}{R_1} - \frac{1}{R_2}\right)$

If $\mu_g = \mu_m$, then $\frac{1}{f} = (1-1)\left(\frac{1}{R_1} - \frac{1}{R_2}\right)$

$\Rightarrow \frac{1}{f} = 0$ $\boxed{f = \frac{1}{0} = \infty}$

This implies that the liquid must have refractive index equal to glass.

NOTES A transparent solid is invisible in a liquid of same refractive index because of no refraction.

44. (c) R = 20 cm

$h_0 = 2$; u = –30 cm

We have, $\frac{1}{f} = (\mu - 1)\left(\frac{1}{R_1} - \frac{1}{R_2}\right) = \left(\frac{3}{2} - 1\right)\left[\frac{1}{20} - \left(-\frac{1}{20}\right)\right]$

$$\Rightarrow \frac{1}{f} = \left(\frac{3}{2} - 1\right) \times \frac{2}{20} \quad \therefore f = 20 \text{ cm}$$

$$\frac{1}{f} = \frac{1}{v} - \frac{1}{u} \Rightarrow \frac{1}{20} = \frac{1}{v} + \frac{1}{30}$$

$$\frac{1}{v} = \frac{1}{20} - \frac{1}{30} = \frac{10}{600}$$

v = 60 cm

$$m = \frac{h_i}{h_0} = \frac{v}{u} \Rightarrow h_i = \frac{v}{u} \times h_0 = \frac{60}{-30} \times 2 = -4 \text{ cm}$$

So, image is inverted.

45. (c) By lens formula,

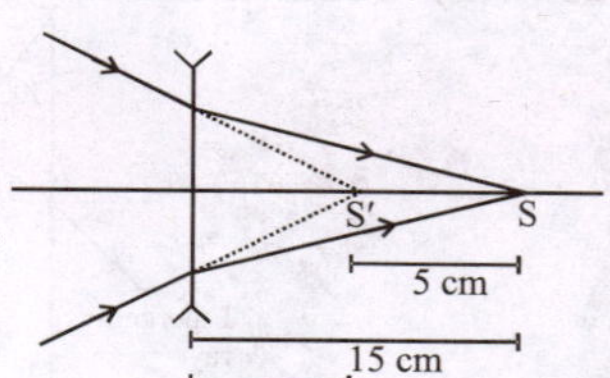

$$\frac{1}{v} - \frac{1}{u} = \frac{1}{f}$$

u = 10 cm; v = 15 cm; f = ?

Putting the values, we get

$$\frac{1}{15} - \frac{1}{10} = \frac{1}{f}$$

$$\frac{10 - 15}{150} = \frac{1}{f} \quad \therefore f = -\frac{150}{5} = -30 \text{ cm}$$

46. (c) By covering aperture, focal length does not change.

But intensity is reduced by $\frac{1}{4}$ times, as aperture diameter $\frac{d}{2}$ is covered.

$$\therefore \quad I' = I - \frac{I}{4} = \frac{3I}{4}$$

$\therefore$ New focal length = f and intensity = $\frac{3I}{4}$.

47. (d) The focal length of the combination

$$\frac{1}{f} = \frac{1}{f_1} + \frac{1}{f_2}$$

$\therefore$ Power of the combinations,

$$P = \frac{f_1 + f_2}{f_1 f_2} \quad \left(\because P = \frac{1}{f}\right)$$

48. (a) We have, $\left|\frac{v}{u}\right| = \frac{\text{Size of image}}{\text{Size of object}}$

or, Size of image = $\left|\frac{v}{u}\right| \times$ Size of object

$$= \left(\frac{10^{-1}}{1.5 \times 10^{11}}\right) \times (1.39 \times 10^9)$$

$= 0.92 \times 10^{-3}$ m $= 9.2 \times 10^{-4}$ m

$\therefore$ Diameter of the sun's image $= 9.2 \times 10^{-4}$ m.

49. (c) From the formula,

$$\frac{1}{f} = \frac{1}{f_1} + \frac{1}{f_2} = \frac{1}{25} - \frac{1}{25} = 0$$

Power of combination $= \frac{1}{f} = 0$

50. (b)

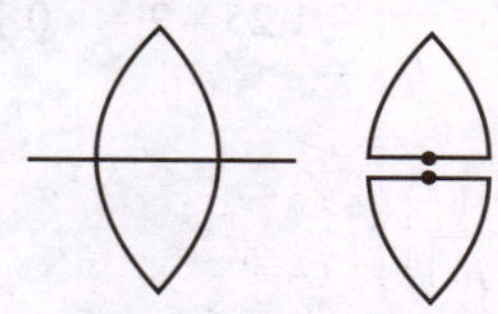

$$\frac{1}{f} = (\mu - 1)\left(\frac{1}{R_1} - \frac{1}{R_2}\right)$$

In this case, R_1 and R_2 are unchanged

So, f will remain unchanged for both pieces of the lens

$\therefore f = f'$

$$\frac{1}{f} = \frac{1}{f_1} + \frac{1}{f_2}$$

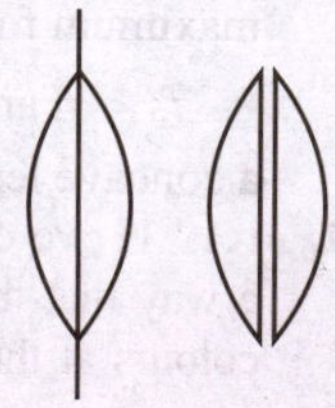

This is combination of two lenses of equal focal lengths

$$\therefore \quad \frac{1}{f} = \frac{1}{f''} + \frac{1}{f''} = \frac{2}{f''} \Rightarrow f'' = 2f.$$

NOTES A symmetric lens is cut along optical axis in two equal parts. Intensity of image formed by each part will be same as that of complete lens. Focal length is double the original for each part.

51. (c) $\frac{1}{f} = \left({}_{\ell}\mu_g - 1\right)\left(\frac{1}{R_1} - \frac{1}{R_2}\right)$

where, ${}_{\ell}\mu_g = 1$ is given.

$$\Rightarrow \frac{1}{f} = (1 - 1)\left(\frac{1}{R_1} - \frac{1}{R_2}\right) = 0 \Rightarrow f = \infty$$

52. **(b)** Using the lens formula $\frac{1}{f}=\frac{1}{v}-\frac{1}{u}$

Given $v = d$, for equal size image $|v| = |u| = d$

By sign convention $u = -d$

$\therefore \quad \frac{1}{f}=\frac{1}{d}+\frac{1}{d} \quad$ or $f=\frac{d}{2}$

53. **(b)** $R_1 = 60$ cm, $R_2 = \infty$, $\mu = 1.6$

$$\frac{1}{f}=(\mu-1)\left(\frac{1}{R_1}-\frac{1}{R_2}\right)$$

$$\frac{1}{f}=(1.6-1)\left(\frac{1}{60}\right) \Rightarrow f = 100 \text{ cm.}$$

54. **(c)** Red light has wavelength greater than violet light. Therefore focal length of lens for red is greater than for violet. ($fr > fv$)

NOTES Cauchy relation

$$\mu = A+\frac{B}{\lambda^2}+\frac{C}{\lambda^4}\cdots \quad \text{Hence } f \propto \lambda.$$

55. **(d)** We know that $\frac{1}{f}=\sum_{i=1}^{n}\frac{1}{f_i}$

$\frac{1}{f}=\frac{1}{f_1}+\frac{1}{f_2}$; $f_1 = 80$ cm, $f_2 = -50$ cm

$$\frac{1}{f}=\frac{1}{\frac{80}{100}}-\frac{1}{\frac{50}{100}} \Rightarrow P=\frac{1}{f}=1.25-2=-0.75\text{D}$$

56. **(a)** $\frac{1}{f}=(\mu-1)\left(\frac{1}{R_1}-\frac{1}{R_2}\right)$

According to Cauchy relation

$$\mu = A+\frac{B}{\lambda^2}+\frac{C}{\lambda^4}\cdots \quad \text{Hence } f \propto \lambda.$$

Hence, red light having maximum wavelength has maximum focal length.

$\therefore$ $f_v < f_r$ and also $F_v > F_r$ as focal length is negative for a concave lens.

57. **(c)** If two or more lenses are combined together in such a way that this combination produces images of different colours at the same point and of the same size, then this property is called 'achromatism'. Concave and convex type of lenses are used for this combination.

58. **(a)** Size of images = A_1 and A_2. From the displacement method, the area of the source of light = $\sqrt{A_1 A_2}$.

59. **(d)** For red light, focal length of lens is maximum because $f \propto \lambda$ and λ is maximum for red light.

60. **(c)** $$\frac{f_a}{f_l}=\frac{\left(\frac{\mu_g}{\mu_l}-1\right)}{(\mu_g-1)}=\frac{\left(\frac{1.5}{1.25}-1\right)}{1.5-1}=\frac{\frac{1}{5}}{\frac{1}{2}}=\frac{2}{5}$$

$$f_l=\frac{5}{2}f_a=\frac{5}{2}\times 2=5 \text{ cm}$$

61. **(b)** In prism, $r_1 + c = A$

$\Rightarrow \quad r_1 = 90° - c$ ... (i)

$\sin c = \frac{1}{\mu} \Rightarrow \cos c = \frac{\sqrt{\mu^2-1}}{\mu}$

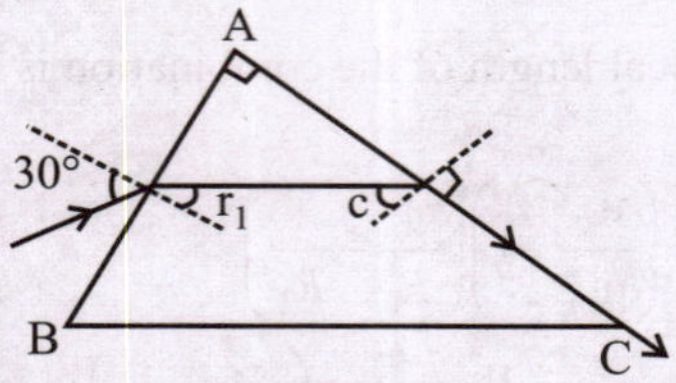

From Snell's law, on incidence surface

$1 \sin 30° = \mu \sin(r_1) \Rightarrow 1\times\frac{1}{2}=\mu\times\sin(90°-c)$

or, $\frac{1}{2}=\mu\times\frac{\sqrt{\mu^2-1}}{\mu}$ ... (ii)

On squaring eq. (ii)

$\frac{1}{4}=\mu^2-1 \Rightarrow \quad \mu^2=\frac{5}{4} \therefore \mu=\frac{\sqrt{5}}{2}$

NOTES The angle of deviation depends on the angle of incidence. At the minimum deviation δ_m, the refracted ray inside the prism becomes parallel to its base.

62. **(c)** Deviation produced by a thin prism

$$\delta=(\mu-1)A=\left(\frac{3}{2}-1\right)6°=3°$$

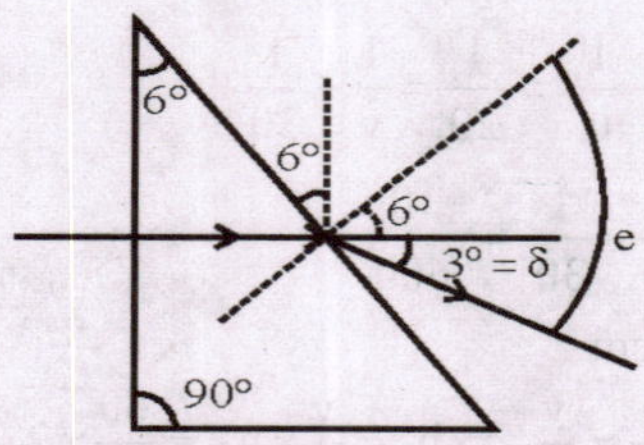

Angle of emergence, e = 6° + 3° = 9°

63. **(b)** From Snell's law at point P.

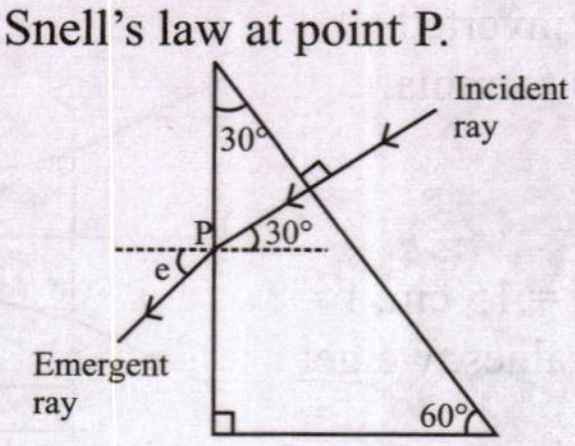

$$\frac{\sin i}{\sin r}=\frac{\mu_{air}}{\mu_{Prism}}$$

$\Rightarrow \frac{\sin 30°}{\sin e}=\frac{1}{\sqrt{3}}$ ($\angle r = \angle e$ emergent angle)

$\Rightarrow \sin e=\sqrt{3}.\frac{1}{2} \quad \therefore \angle e = 60°$

NOTES If a light ray is incident normally on first surface i.e., $\angle i = 0°$ so refraction of first surface $\angle r_1 = 0°$.
When angle of prism A > C (critical angle) then no ray emerges from second surface of the prism.

64. **(b)** Light ray emerges normally from another surface, hence, e (angle of emergence) = 0

$\therefore r_2 = 0 \quad r_1 + r_2 = A \Rightarrow r_1 = A$

$\mu_1 \cdot \sin i = \mu_2 \cdot \sin r$

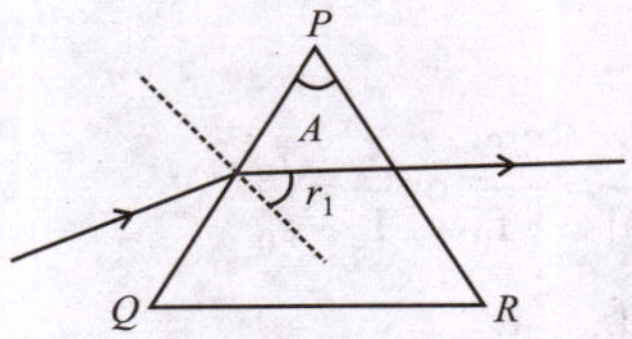

Applying Snell's law on first surface PQ

$\Rightarrow \mu_1 \cdot \sin i = \mu_2 \cdot \sin r_1 \Rightarrow \sin i = \mu \sin A$

For small angles $(\sin\theta \approx \theta) \therefore i = \mu A$

65. **(b)** For retracing the path, light ray should be normally incident on silvered face.

A = r + O ⇒ r = 30°

Applying Snell's law at point M,

$$\frac{\sin i}{\sin 30°} = \frac{\sqrt{2}}{1}$$

$$\Rightarrow \sin i = \sqrt{2} \times \frac{1}{2}$$

or, $\sin i = \frac{1}{\sqrt{2}}$ i.e., i = 45°

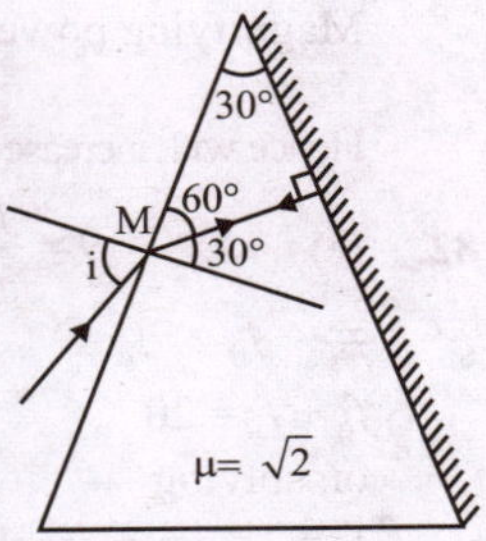

66. **(b)** Given: Angle of incidence i = 45°
angle of prism, A = 60°
Angle of minimum deviation,
$\delta_m = 2i - A = 30°$
Refractive index of material of prism.

$$\mu = \frac{\sin\left(\frac{A+\delta_m}{2}\right)}{\sin(A/2)} = \frac{\sin 45°}{\sin 30°} = \frac{1}{\sqrt{2}} \cdot \left(\frac{2}{1}\right) = \sqrt{2}$$

67. **(a)** As we know, the refractive index of the material of the prism

$$\mu = \frac{\sin\left(\frac{\delta_m + A}{2}\right)}{\sin(A/2)}$$

$$\cot A/2 = \frac{\sin\left(\frac{A+\delta_m}{2}\right)}{\sin A/2} = \frac{\cos(A/2)}{\sin(A/2)} \quad [\because \mu = \cot(A/2)]$$

$$\Rightarrow \sin\left(\frac{\delta_m + A}{2}\right) = \sin(90° + A/2)$$

$$\Rightarrow \delta_{min} = 180° - 2A$$

68. **(c)** For total internal reflection,
incident angle (i) > critical angle (i_c)

So, $\sin i > \sin i_c$

$$\sin 45° > \frac{1}{\mu}$$

$\Rightarrow \mu > \sqrt{2}$

⇒ 1.414

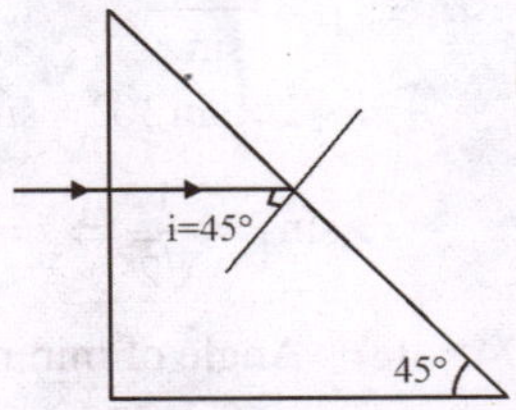

Since refractive index μ of green and voilet are greater than 1.414 so they will total internal reflected. But red colour will be refracted.

69. **(b)**

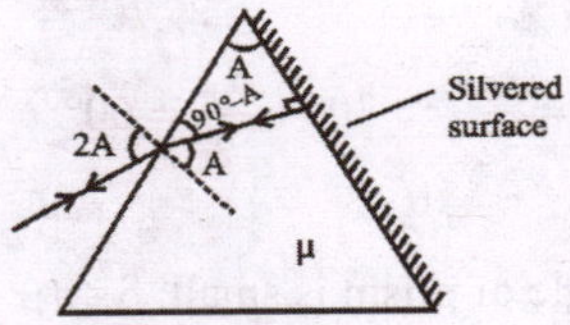

According to Snell's law $\mu = \frac{\sin i}{\sin r}$

$\Rightarrow (1) \sin 2A = (\mu) \sin A \Rightarrow \mu = 2\cos A$

70. **(a)** For normally emerge $e = 0$
Therefore $r_2 = 0$ and $r_1 = A$
Snell's Law for incident ray's
$1 \sin i = \mu \sin r_1 = \mu \sin A$
For small angle
$i = \mu A$

71. **(b)** The angle of minimum deviation is given as

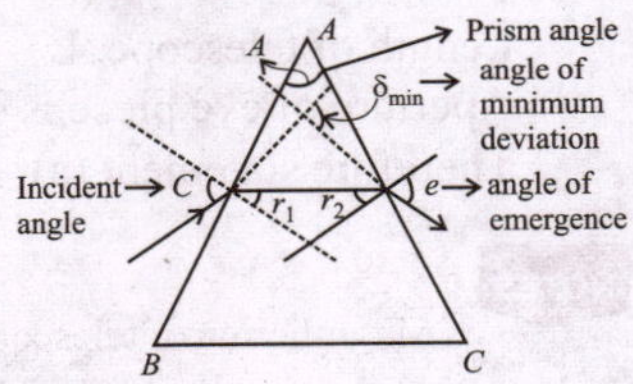

$\delta_{min} = i + e - A$

for minimum deviation

$\delta_{min} = A$ then

$2A = i + e$

in case of δ_{min}, $i = e$;

$2A = 2i \quad r_1 = r_2 = \frac{A}{2}$

$i = A = 90°$

from Snell's law

$1 \sin i = n \sin r_1$

$\sin A = n \sin\frac{A}{2}; \; 2\sin\frac{A}{2}\cos\frac{A}{2} = n\sin\frac{A}{2}$

$2\cos\frac{A}{2} = n$

when $A = 90° = i_{min}$

then $n_{min} = \sqrt{2}; i = A = 0 \; n_{max} = 2$

72. **(b)** In minimum deviation
$r_1 = r_2 = r$
$\therefore A = r_1 + r_2 = 2r$
∴ Angle of refraction

$$r = \frac{60}{2} = 30°$$

73. (b) $\angle r = 30°$ (using law of triangle)

$\Rightarrow \mu = \dfrac{\sin i}{\sin r}$

$\Rightarrow \sqrt{2} \times \sin 30° = \sin i$

$\Rightarrow \sin i = \dfrac{1}{\sqrt{2}} \Rightarrow i = 45°.$

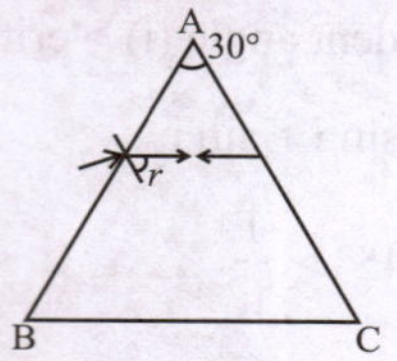

74. (c) Angle of minimum deviation

$$\mu = \frac{\sin\left(\frac{A+\delta_m}{2}\right)}{\sin\left(\frac{A}{2}\right)} \Rightarrow \sqrt{3} = \frac{\sin\left(\frac{60° + \delta_m}{2}\right)}{\sin\left(\frac{60°}{2}\right)}$$

$$\Rightarrow \sin\left(30° + \frac{\delta_m}{2}\right) = \frac{\sqrt{3}}{2} \Rightarrow 30° + \frac{\delta_m}{2} = 60°$$

$\Rightarrow \delta m = 60°.$

75. (a) When the angle of prism is small, $\delta = (\mu - 1) A$

76. (b)

77. (b) $f_0 = 140$ cm and $f_e = 5$ cm

For distant object, the magnifying power is given by

$$m = \frac{f_0}{f_e} = \frac{140}{5} = 28$$

78. (a) In normal adjustment of astronomical telescope, final image is inverted and formed at infinity.

$$\text{Magnification, } m = -\frac{f_0}{f_e} = -\frac{20}{2 \times 10^{-2}} = -1000$$

Length of telescope, $L = f_0 + f_e = 20 + 0.02 = 20.02$

Aperture of eye piece is smaller than that of objective.

Therefore statement (a), (b) and (d) are correct.

> **NOTES**
>
> Magnifiction of telescope for normal adjustment $m = \dfrac{f_0}{f_e}$
>
> For large magnification, f_o is made as large as practically possible and f_e small.

79. (d) Given: Focal length of objective, $f_0 = 40$cm

Focal length of eye – piece $f_e = 4$ cm

image distance, $v_0 = 200$ cm

Using lens formula for objective lens

$$\frac{1}{v_0} - \frac{1}{u_0} = \frac{1}{f_0} \Rightarrow \frac{1}{v_0} = \frac{1}{f_0} + \frac{1}{u_0}$$

$$\Rightarrow \frac{1}{v_0} = \frac{1}{40} + \frac{1}{-200} = \frac{+5-1}{200} \Rightarrow v_0 = 50 \text{ cm}$$

Tube length $\ell = |v_0| + f_e = 50 + 4 = 54$ cm.

> **NOTES**
>
> In a telescope, if field and eye lenses are interchanged magnification will change from $\left(\dfrac{f_o}{f_e}\right)$ to $\left(\dfrac{f_e}{f_o}\right)$ i.e., it will change from m to $\dfrac{1}{m}$ i.e., will become $\left(\dfrac{1}{m^2}\right)$ times of its initial value.

80. (c)

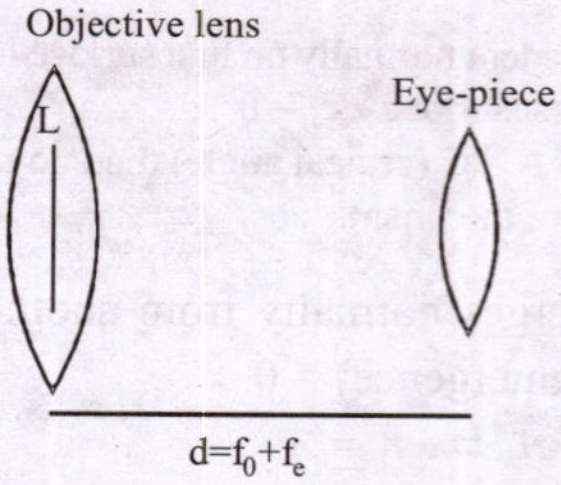

Magnification by eye piece

$$m = \frac{f}{f+u}$$

$$\frac{I}{L} = \frac{f_e}{f_e + [-(f_0 + f_e)]} = -\frac{f_e}{f_0} \text{ or } \frac{I}{L} = \frac{f_e}{f_0}$$

$$\text{Magnification, } M = \frac{f_0}{f_e} = \frac{L}{I}$$

81. (d) Magnifying power of microscope $= \dfrac{LD}{f_0 f_e} \propto \dfrac{1}{f_0}$

Hence with increase f_0 magnifyig power of microscope decreases.

Magnifying power of telescope $= \dfrac{f_0}{f_e} \propto f_0$

Hence with increase f_0 magnifying power of telescope increases.

82. (c) M.P. $= 9 = \dfrac{f_0}{f_e}$

$\Rightarrow \quad f_0 = 9f_e$ ...(i)

$f_0 + f_e = 20$ ...(ii)

on solving

$f_0 = 18$ cm = focal length of the objective

$f_e = 2$ cm = focal length of the eyepiece

83. (d) In the later case microscope will be focussed for O'. So, it is required to be lifted by distance OO'.

OO' = real depth of O – apparent depth of O.

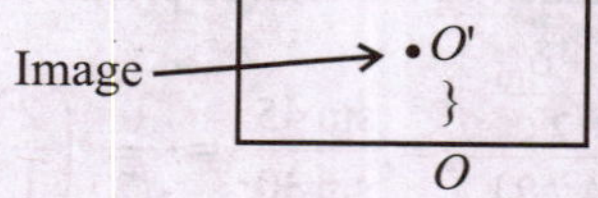

$$= 3 - \frac{3}{1.5} \quad \left[\mu = \frac{\text{real depth}}{\text{apparent depth}}\right]$$

$$= 3\left[\frac{1.5-1}{1.5}\right] = \frac{3 \times .5}{1.5} = 1 \text{ cm}$$

84. (b) Given : Length of astronomical telescope $(f_0 + f_e)$ $= 44$ cm and magnification $\left(\dfrac{f_0}{f_e}\right) = 10.$

From the given magnification, we find that $f_0 = 10f_e$. Therefore, $10f_e + f_e = 44$ or $11f_e = 44$ or $f_e = 4$. And focal length of the objective $(f_0) = 44 - f_e = 44 - 4 = 40$ cm.

> **NOTES**
>
> In case if objective and eye lens of a telescope are interchanged, it will not behave as a microscope but object appears very small.

10 Wave Optics

TREND ANALYSIS

TOPIC	2025		2024		2023		2022		2021		2020	
	Qns.	LOD	Qns.	LOD	Qns.	LOD	Qns.	LOD	Qns.	LOD	Qns.	LOD
Interference of Light Waves and Young's Experiment			1	Easy	1	Avg.	2	Easy			1	Avg.
Polarisation	2	Avg.	1	Avg.			1	Easy			1	Easy

How to Code Qns for Better & Faster Revision

Type 1: Able to solve (No code Req)

Type 2: Solved with struggle or by seeing solution **(Use code NR means Needs Revision or orange stickers)**

Type 3: Tricky/ Could not solve on my own. Need help from Teacher/ Friend. **(Use code TQ means Tricky Question or red stickers)**

You can either hand write the code in boxes or else paste coloured stickers given at end of book.

Topic-10.2 Huygens Principle

1. Which one of the following phenomena is not explained by Huygens construction of wavefront?

NCERT | Page-257 | 1988

(a) Refraction (b) Reflection
(c) Diffraction (d) Origin of spectra

Topic-10.4 Coherent and Incoherent Addition of Waves

2. The interference patter n is obtained with two coherent light source of intensity ratio n. In the interference pattern, the ration $\frac{I_{max}-I_{min}}{I_{max}+I_{min}}$ will be :-

NCERT | Page-362 | Ph-II 2016

(a) $\frac{\sqrt{n}}{(n+1)^2}$ (b) $\frac{2\sqrt{n}}{(n+1)^2}$

(c) $\frac{\sqrt{n}}{n+1}$ (d) $\frac{2\sqrt{n}}{n+1}$

3. The periodic waves of intensities I_1 and I_2 pass through a region at the same time in the same direction. The sum of the maximum and minimum intensities is:

NCERT | Page-362 | 2008

(a) $I_1 + I_2$ (b) $\left(\sqrt{I_1}+\sqrt{I_2}\right)^2$

(c) $\left(\sqrt{I_1}-\sqrt{I_2}\right)^2$ (d) $2(I_1 + I_2)$

4. Ratio of intensities of two waves are given by 4 : 1. Then the ratio of the amplitudes of the two waves is

NCERT | Page-263 | 1991

(a) 2 : 1 (b) 1 : 2 (c) 4 : 1 (d) 1 : 4

Topic-10.5 Interference of Light Waves and Young's Experiment

5. Two slits in Young's double slit experiment are 1.5 mm apart and the screen is placed at a distance of 1 m from the slits. If the wavelength of light used is 600×10^{-9} m then the fringe separation is NCERT | Page-265, 266 | 2024

(a) 4×10^{-5} m (b) 9×10^{-8} m
(c) 4×10^{-7} m (d) 4×10^{-4} m

6. Interference pattern can be observed due to superposition of the following waves: NCERT | Page-265, 266 | 2024

A. $y = a \sin \omega t$ B. $y = a \sin 2\omega t$
C. $y = a \sin (\omega t - \phi)$ D. $y = a \sin 3\omega t$

Choose the **correct** answer from the options given below.

(a) B and C (b) B and D
(c) A and C (d) A and B

7. If the monochromatic source in Young's double slit experiment is replaced by white light, then

NCERT | Page-265, 266 | 2024

(a) Interference pattern will disappear.
(b) There will be a central dark fringe surrounded by a few coloured fringes.
(c) There will be a central bright white fringe surrounded by a few coloured fringes.
(d) All bright fringes will be of equal width.

8. For Young's double slit experiment, two statements are given below:

Statement I : If screen is moved away from the plane of slits, angular separation of the fringes remains constant.

Statement II : If the monochromatic source is replaced by another monochromatic source of larger wavelength, the angular separation of fringes decreases.

In the light of the above statements, choose the correct answer from the options given below:

NCERT | Page-265, 266 | 2023

(a) Both Statement I and Statement II are true.
(b) Both Statement I and Statement II are false.
(c) Statement I is true but Statement II is false.
(d) Statement I is false but Statement II is true.

9. If the screen is moved away from the plane of the slits in a Young's double slit experiment, then the : **NCERT | Page-265, 266 | RS 2022**
(a) linear separation of the fringes decreases
(b) angular separation of the fringes increases
(c) angular separation of the fringes decreases
(d) linear separation of the fringes increases

10. In a Young's double slit experiment, a student observes 8 fringes in a certain segment of screen when a monochromatic light of 600 nm wavelength is used. If the wavelength of light is changed to 400 nm, then the number of fringes he would observe in the same region of the screen is: **NCERT | Page-265, 266 | 2022**
(a) 8 (b) 9 (c) 12 (d) 6

11. In Young's double slit experiment, if the separation between coherent sources is halved and the distance of the screen from the coherent sources is doubled, then the fringe width becomes : **NCERT | Page-265, 266 | 2020**
(a) half (b) four times
(c) one-fourth (d) double

12. In a double slit experiment, when light of wavelength 400 nm was used, the angular width of the first minima formed on a screen placed 1 m away, was found to be 0.2°. What will be the angular width of the first minima, if the entire experimental apparatus is immersed in water? ($\mu_{water} = 4/3$) **NCERT | Page-266 | 2019**
(a) 0.266° (b) 0.15° (c) 0.05° (d) 0.1°

13. In a Young's double slit experiment, if there is no initial phase difference between the light from the two slits, a point on the screen corresponding to the fifth minimum has path difference **NCERT | Page-266 | Odisha 2019**
(a) $11\frac{\lambda}{2}$ (b) $5\frac{\lambda}{2}$ (c) $10\frac{\lambda}{2}$ (d) $9\frac{\lambda}{2}t$

14. In Young's double slit experiment the separation d between the slits is 2 mm, the wavelength λ of the light used is 5896 Å and distance D between the screen and slits is100 cm. It is found that the angular width of the fringes is 0.20°. To increase the fringe angular width to 0.21° (with same λ and D) the separation between the slits needs to be changed to **NCERT | Page-266 | 2018**
(a) 1.8 mm (b) 1.9 mm
(c) 1.7 mm (d) 2.1 mm

15. Young's double slit experment is first performed in air and then in a medium other than air. It is found that 8th bright fringe in the medium lies where 5th dark fringe lies in air. The refractive index of the medium is nearly **NCERT | Page-266 | 2017**
(a) 1.59 (b) 1.69 (c) 1.78 (d) 1.25

16. The intensity at the maximum in a Young's double slit experiment is I_0. Distance between two slits is d = 5λ, where λ is the wavelength of light used in the experiment. What will be the intensity in front of one of the slits on the screen placed at a distance D = 10 d ? **NCERT | Page-266 | 2016**
(a) I_0 (b) $\frac{I_0}{4}$ (c) $\frac{3}{4}I_0$ (d) $\frac{I_0}{2}$

17. In a double slit experiment, the two slits are 1 mm apart and the screen is placed 1 m away. A monochromatic light wavelength 500 nm is used. What will be the width of each slit for obtaining ten maxima of double slit within the central maxima of single slit pattern ? **NCERT | Page-266 | 2015**
(a) 0.1 mm (b) 0.5 mm
(c) 0.02 mm (d) 0.2 mm

18. Two slits in Young's experiment have widths in the ratio 1 : 25. The ratio of intensity at the maxima and minima in the interference pattern, $\frac{I_{max}}{I_{min}}$ is: **NCERT | Page-266 | 2015 RS**
(a) $\frac{121}{49}$ (b) $\frac{49}{121}$ (c) $\frac{4}{9}$ (d) $\frac{9}{4}$

19. In the Young's double-slit experiment, the intensity of light at a point on the screen where the path difference is λ is K, (λ being the wave length of light used). The intensity at a point where the path difference is λ/4, will be: **NCERT | Page-266 | 2014**
(a) K (b) K/4 (c) K/2 (d) Zero

20. In Young's double slit experiment, the slits are 2 mm apart and are illuminated by photons of two wavelengths $\lambda_1 = 12000$Å and $\lambda_2 = 10000$Å. At what minimum distance from the common central bright fringe on the screen 2 m from the slit will a bright fringe from one interference pattern coincide with a bright fringe from the other ? **NCERT | Page-266 | 2013**
(a) 6 mm (b) 4 mm (c) 3 mm (d) 8 mm

21. In Young's double slit experiment the distance between the slits and the screen is doubled. The separation between the slits is reduced to half. As a result the fringe width **NCERT | Page-266 | Kar. 2013**
(a) is doubled
(b) is halved
(c) becomes four times
(d) remains unchanged

22. Colours appear on a thin soap film and on soap bubbles due to the phenomenon of **NCERT | Page-266 | 1999**
(a) refraction (b) dispersion
(c) interference (d) diffraction

23. Interference was observed in interference chamber where air was present, now the chamber is evacuated, and if the same light is used, a careful observer will see **NCERT | Page-266 | 1993**
(a) no interference
(b) interference with brighter bands
(c) interference with dark bands
(d) interference fringe with larger width

24. In Young's double slit experiment carried out with light of wavelength (λ) = 5000Å, the distance between the slits is 0.2 mm and the screen is at 200 cm from the slits. The central maximum is at $x = 0$. The third maximum (taking the central maximum as zeroth maximum) will be at x equal to NCERT | Page-266 | 1992

(a) 1.67 cm (b) 1.5 cm
(c) 0.5 cm (d) 5.0 cm

25. If yellow light emitted by sodium lamp in Young's double slit experiment is replaced by a monochromatic blue light of the same intensity NCERT | Page-266 | 1992

(a) fringe width will decrease
(b) finge width will increase
(c) fringe width will remain unchanged
(d) fringes will become less intense

26. In Young's experiment, two coherent sources are placed 0.90 mm apart and fringe are observed one metre away. If it produces second dark fringe at a distance of 1 mm from central fringe, the wavelength of monochromatic light used would be NCERT | Page-266 | 1991, 1992

(a) 60×10^{-4} cm (b) 10×10^{-4} cm
(c) 10×10^{-5} cm (d) 6×10^{-5} cm

27. The Young's double slit experiment is performed with blue and with green light of wavelengths 4360Å and 5460Å respectively. If x is the distance of 4th maxima from the central one, then NCERT | Page-266 | 1990

(a) x (blue) = x (green)
(b) x (blue) > x (green)
(c) x (blue) < x (green)
(d) $\frac{x(\text{blue})}{x(\text{green})} = \frac{5460}{4360}$

28. In Young's double slit experiment, the fringe width is found to be 0.4 mm. If the whole apparatus is immersed in water of refrative index $\frac{4}{3}$, without disturbing the geometrical arrangement, the new fringe width will be NCERT | Page-266 | 1990

(a) 0.30 mm (b) 0.40 mm
(c) 0.53 mm (d) 450 microns

29. Interference is possible in NCERT | Page-265 | 1989

(a) light waves only
(b) sound waves only
(c) both light and sound waves
(d) neither light nor sound waves

Topic-10.6

Diffraction

30. Angular width of the central maxima in the Fraunhofer diffraction for λ = 6000Å is θ_0. When the same slit is illuminated by another monochromatic light, the angular width decreases by 30%. The wavelength of this light is NCERT | Page-267, 268 | Odisha 2019

(a) 420 Å (b) 1800 Å (c) 4200 Å (d) 6000 Å

31. In a diffraction pattern due to a single slit of width 'a', the first minimum is observed at an angle 30° when light of wavelength 5000 Å is incident on the slit. The first secondary maximum is observed at an angle of : NCERT | Page--267 | 2016

(a) $\sin^{-1}\left(\frac{1}{4}\right)$ (b) $\sin^{-1}\left(\frac{2}{3}\right)$
(c) $\sin^{-1}\left(\frac{1}{2}\right)$ (d) $\sin^{-1}\left(\frac{3}{4}\right)$

32. For a parallel beam of monochromatic light of wavelength 'λ', diffraction is produced by a single slit whose width 'a' is of the wavelength of the light. If 'D' is the distance of the screen from the slit, the width of the central maxima will be : NCERT | Page-267, 268 | 2015

(a) $\frac{D\lambda}{a}$ (b) $\frac{Da}{\lambda}$ (c) $\frac{2Da}{\lambda}$ (d) $\frac{2D\lambda}{a}$

33. At the first minimum adjacent to the central maximum of a single-slit diffraction pattern, the phase difference between the Huygen's wavelet from the edge of the slit and the wavelet from the midpoint of the slit is : NCERT | Page-266 | 2015 RS

(a) $\frac{\pi}{2}$ radian (b) π radian
(c) $\frac{\pi}{8}$ radian (d) $\frac{\pi}{4}$ radian

34. A beam of light of λ = 600 nm from a distant source falls on a single slit 1 mm wide and the resulting diffraction pattern is observed on a screen 2 m away. The distance between first dark fringes on either side of the central bright fringe is: NCERT | Page-267, 268 | 2014

(a) 1.2 cm (b) 1.2 mm
(c) 2.4 cm (d) 2.4 mm

35. A parallel beam of fast moving electrons is incident normally on a narrow slit. A fluorescent screen is placed at a large distance from the slit. If the speed of the electrons is increased, which of the following statements is correct ? NCERT | Page-267, 268 | 2013

(a) The angular width of the central maximum of the diffraction pattern will increase.
(b) The angular width of the central maximum will decrease.
(c) The angular width of the central maximum will be unaffected.
(d) Diffraction pattern is not observed on the screen in case of electrons.

36. A parallel beam of light of wavelength λ is incident normally on a narrow slit. A diffraction pattern is formed on a screen placed perpendicular to the direction of the incident beam. At the second minimum of the diffraction pattern, the phase difference between the rays coming from the two edges of slit is NCERT | Page-267 | Kar. 2013

(a) πλ (b) 2π (c) 3π (d) 4π

37. A parallel beam of monochromatic light of wavelength 5000Å is incident normally on a single narrow slit of width 0.001 mm. The light is focussed by a convex lens on a screen placed in focal plane. The first minimum will be formed for the angle of diffraction equal to NCERT | Page-267 | 1993

(a) 0° (b) 15° (c) 30° (d) 50°

Topic-10.7

Polarisation

38. An unpolarized light beam travelling in air is incident on a medium of refractive index 1.73 at Brewster's angle. Then - **2025**

(a) reflected light is completely polarized and the angle of reflection is close to 60°
(b) reflected light is partially polarized and the angle of reflection is close to 30°
(c) both reflected and transmitted light are perfectly polarized with angles of reflection and refraction close to 60° and 30°, respectively
(d) transmitted light is completely polarized with angle of refraction close to 30°

39. The intensity of transmitted light when a polaroid sheet, placed between two crossed polaroids at 22.5° from the polarization axis of one of the polaroid, is (I_0 is the intensity of polarised light after passing through the first polaroid): **NCERT | Page-270 | 2025**

(a) $\frac{I_0}{2}$ (b) $\frac{I_0}{4}$ (c) $\frac{I_0}{8}$ (d) $\frac{I_0}{16}$

40. A beam of unpolarized light of intensity I_0 is passed through a polaroid *A*, then through another polaroid *B*, oriented at 60° and finally through another polaroid *C*, oriented at 45° relative to *B* as shown. The intensity of emergent light is: **NCERT | Page-270 | Ph-II 2024**

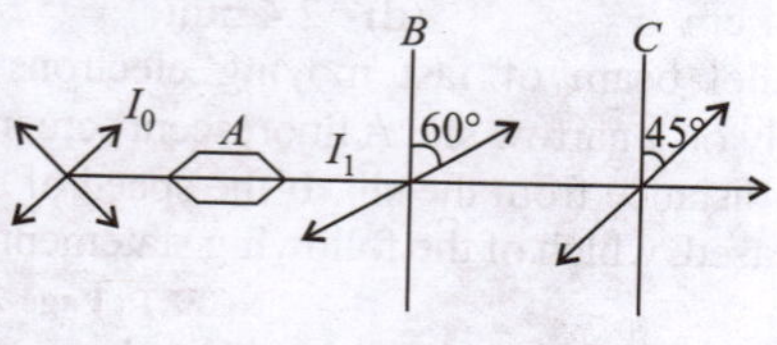

(a) $\frac{I_0}{16}$ (b) $\frac{I_0}{4}$ (c) $\frac{I_0}{2}$ (d) $\frac{I_0}{32}$

41. An unpolarised light beam strikes a glass surface at Brewster's angle. Then **2024**

(a) The reflected light will be partially polarised.
(b) The refracted light will be completely polarised.
(c) Both the reflected and refracted light will be completely polarised.
(d) The reflected light will be completely polarised but the refracted light will be partially polarised.

42. After passing through a polariser a linearly polarised light of intensity I is incident on an analyser making an angle of 30° with that of the polariser. The intensity of light emitted from the analyser will be: **NCERT | Page-270 | RS 2022**

(a) $\frac{2I}{3}$ (b) $\frac{I}{2}$ (c) $\frac{I}{3}$ (d) $\frac{3I}{4}$

43. The Brewsters angle i_b for an interface should be **2020**

(a) $30° < i_b < 45°$ (b) $45° < i_b < 90°$
(c) $i_b = 90°$ (d) $0° < i_b < 30°$

44. Unpolarised light is incident from air on a plane surface of a material of refractive index 'μ'. At a particular angle of incidence 'i', it is found that the reflected and refracted rays are perpendicular to each other. Which of the following options is correct for this situation? **2018**

(a) Reflected light is polarised with its electric vector parallel to the plane of incidence
(b) Reflected light is polarised with its electric vector perpendicular to the plane of incidence
(c) $i = \tan^{-1}\left(\frac{1}{\mu}\right)$
(d) $i = \sin^{-1}\left(\frac{1}{\mu}\right)$

45. Two Polaroids P_1 and P_2 are placed with their axis perpendicular to each other. Unpolarised light I_0 is incident on P_1. A third polaroid P_3 is kept in between P_1 and P_2 such that its axis makes an angle 45° with that of P_1. The intensity of transmitted light through P_2 is **NCERT | Page-270 | 2017**

(a) $\frac{I_0}{4}$ (b) $\frac{I_0}{8}$ (c) $\frac{I_0}{16}$ (d) $\frac{I_0}{2}$

46. Which of the following phenomenon is not common to sound and light waves ? **NCERT | Page-270 | 1988**

(a) Interference (b) Diffraction
(c) Coherence (d) Polarisation

ANSWER KEY

1	(d)	7	(c)	13	(d)	19	(c)	25	(a)	31	(d)	37	(c)	43	(b)
2	(d)	8	(c)	14	(b)	20	(a)	26	(d)	32	(d)	38	(a)	44	(b)
3	(d)	9	(d)	15	(c)	21	(c)	27	(c)	33	(b)	39	(c)	45	(b)
4	(a)	10	(c)	16	(d)	22	(c)	28	(a)	34	(d)	40	(a)	46	(d)
5	(d)	11	(b)	17	(d)	23	(d)	29	(c)	35	(b)	41	(d)		
6	(c)	12	(b)	18	(d)	24	(b)	30	(c)	36	(d)	42	(d)		

Hints & Solutions

1. **(d)** Huygen's construction of wavefront does not apply to origin of spectra which is explained by quantum theory.

2. **(d)** Let us consider, the ratio of intensities obtained from two coherent sources is $\frac{I_1}{I_2}=\frac{n}{1}$

$$\frac{I_{max}-I_{min}}{I_{max}+I_{min}}=\frac{(\sqrt{I_1}+\sqrt{I_2})^2-(\sqrt{I_1}-\sqrt{I_2})^2}{(\sqrt{I_1}+\sqrt{I_2})^2+(\sqrt{I_1}-\sqrt{I_2})^2}$$

$$=\frac{4\sqrt{I_1I_2}}{2(I_1+I_2)}=\frac{2\sqrt{\frac{I_1}{I_2}}}{\left(\frac{I_1}{I_2}+1\right)}=\frac{2\sqrt{n}}{n+1}$$

3. **(d)** The resultant intensity of two periodic waves at a point is given by

$I=I_1+I_2+2\sqrt{I_1I_2}.\cos\phi.$

Resultant intensity is maximum if

$\cos\phi=-1$

i.e. $I_{max}=I_1+I_2+2\sqrt{I_1I_2}$

Resultant intensity is minimum if

$\cos\phi=+1$

i.e, $I_{min}=I_1+I_2-2\sqrt{I_1I_2}$

Therefore, the sum of the maximum and minimum intensities is $I_{max}+I_{min}$

$=I_1+I_2+2\sqrt{I_1I_2}+I_1+I_2-2\sqrt{I_1I_2}$

$=2(I_1+I_2)$

4. **(a)** $\frac{I_1}{I_2}=\frac{a_1^2}{a_2^2}=\frac{4}{1}\quad\therefore\frac{a_1}{a_2}=\frac{2}{1}$

NOTES If w_1 and w_2 are widths of two slits from which intensities of light I_1 and I_2 emanate, then $\frac{I_1}{I_2}=\frac{a^2}{b^2}=\frac{w_1}{w_2}$; a and b are amplitudes.

5. **(d)** Fringe width (β) = Fringe separation $=\frac{\lambda D}{d}$

$$\Rightarrow\beta=\frac{600\times10^{-9}\times1}{1.5\times10^{-3}}=\frac{6\times10^{-7}}{1.5\times10^{-3}}$$

$=4\times10^{-4}$ m

6. **(c)** For interference pattern to be observed the sources must be coherent.

Hence the interference will be observed due to superposition of waves $y=a\sin(\omega t)$ and $y=a\sin(\omega t-\phi)$

7. **(c)** Central bright fringe is white as at central point on screen, path difference is zero for all wavelength. Other fringes depend on wavelength as $\beta=\frac{\lambda D}{d}$ so other fringes will be coloured.

8. **(c)** For Young's double slit experiment, angular fringe width $\alpha=\frac{\lambda}{d}$

It does not depend on the distance of screen from the slit, so statement I is correct.

Angular fringe width $\propto\lambda$

So if λ increases, angular separation of fringes increases.

Therefore, statement I is true and statement II is false.

9. **(d)** We know fringe width $\beta=\frac{\lambda D}{d}$

And if the screen is moved away from the plane of the slits i.e., as D increases B increases.

So, linear separation of fringes increases.

10. **(c)** No. of fringes(n) $=\frac{\text{length of segment (x)}}{\text{fringe width }(\beta)}$

$\Rightarrow n=\frac{x}{\beta}$

$\Rightarrow x=n\beta=$ cons.

$\Rightarrow n_1\beta_1=n_2\beta_2$

$\Rightarrow n_1\frac{\lambda_1D}{d}=n_2\frac{\lambda_2D}{d}$

$\Rightarrow n_1\lambda_1=n_2\lambda_2\Rightarrow 8\times600=400\times n_2$

$\Rightarrow n_2=12$

11. **(b)** Fringe width $\beta=\frac{\lambda D}{d}$

Here, λ = wavelength of light from coherent sources,
D = distance of screen from the coherent sources,
d = separation between coherent sources

When, $d'=\frac{d}{2}$ and $D'=2D$

New Fringe width, $\beta'=\frac{\lambda(2D)}{d/2}=\frac{4\lambda D}{d}$

$\Rightarrow\beta'=4\beta$

Fringe width becomes 4 times.

12. **(b)** For double slit experiment angular fringe width $\theta_0=\frac{\beta}{D}$

Angular fringe width (in water)

$$\theta_w=\frac{\beta}{\mu D}=\frac{\theta_0}{\mu}=\frac{0.2°}{\left(\frac{4}{3}\right)}=0.15°$$

13. (d) Path difference for destructive interference in YDSE

$\Rightarrow \quad \Delta X_n = \dfrac{(2n-1)}{2}\lambda \; n = 1, 2, 3.....$

$\Delta X_{5^{th}} = \dfrac{9\lambda}{2}$

14. (b) Angular width $= \dfrac{\lambda}{d}$

So, $0.20° = \dfrac{\lambda}{2mm} \Rightarrow \lambda = 0.20° \times 2$

Again, $0.21° = \dfrac{\lambda}{d}$

Now putting the value of λ

$d = \dfrac{0.20° \times 2mm}{0.21°}$

$\therefore \quad d = 1.9$ mm

15. (c) According to question

8th bright fringe in medium = 5th dark fringe in air

$Y_{8th\ bright} = 8\dfrac{\lambda D}{\mu d}$

$Y_{5th\ dark} = (2 \times 5 - 1)\dfrac{\lambda D}{2d} = \dfrac{9}{2}\dfrac{\lambda D}{d}$

$\Rightarrow \quad \dfrac{9}{2}\dfrac{\lambda D}{d} = 8\dfrac{\lambda D}{\mu d}$

or, refractive index $\mu = \dfrac{16}{9} = 1.78$

NOTES In YDSE for constructive or bright fringes.

$Y_n = \dfrac{nD\lambda}{d}$ for nth bright fringe and for destructive or dark fringes. $y_n = (2n-1)\dfrac{D\lambda}{2d}$ for n^{th} dark fringe.

16. (d) Let P is a point infront of one slit at which intensity is to be calculated. From figure,

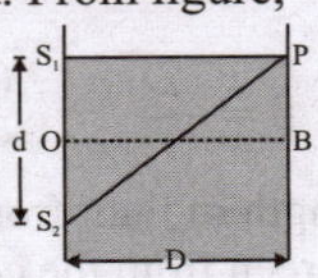

Path difference $= S_2P - S_1P$

$= \sqrt{D^2 + d^2} - D = D\left(1 + \dfrac{1}{2}\dfrac{d^2}{D^2}\right) - D$

$= D\left[1 + \dfrac{d^2}{2D^2} - 1\right] = \dfrac{d^2}{2D}$

$\Delta x = \dfrac{d^2}{2 \times 10d} = \dfrac{d}{20} = \dfrac{5\lambda}{20} = \dfrac{\lambda}{4}$

Phase difference,

$\Delta\phi = \dfrac{2\pi}{\lambda} \times \dfrac{\lambda}{4} = \dfrac{\pi}{2}$

So, resultant intensity at the desired point 'p' is

$I = I_0 \cos^2\dfrac{\phi}{2} = I_0 \cos^2\dfrac{\pi}{4} = \dfrac{I_0}{2}$

17. (d) Here, distance between two slits,

$d = 1mm = 10^{-3}$ m

distance of screen from slits, D = 1 m

wavelength of monochromatic light used,

$\lambda = 500nm = 500 \times 10^{-9}m$

width of each slit a = ?

Width of central maxima in single slit pattern $= \dfrac{2\lambda D}{a}$

Fringe width in double slit experiment $\beta = \dfrac{\lambda D}{d}$

So, required condition $\dfrac{10\lambda D}{d} = \dfrac{2\lambda D}{a}$

$\Rightarrow \quad a = \dfrac{d}{5D} = \dfrac{1}{5} \times 10^{-3}$ m $= 0.2$ mm

18. (d) The ratio of slits width $= \dfrac{}{25}$ (given)

$\therefore \quad \dfrac{I_1}{I_2} = \dfrac{25}{1}$

$I \propto A^2 \Rightarrow \dfrac{I_1}{I_2} = \dfrac{A_1^2}{A_2^2} = \dfrac{25}{1}$ or $\dfrac{A_1}{A_2} = \dfrac{5}{1}$

As we know that,

$\dfrac{A_{max}}{A_{min}} = \dfrac{A_1 + A_2}{A_1 - A_2} = \dfrac{5+1}{5-1} = \dfrac{6}{4} = \dfrac{3}{2}$

$\therefore \quad \dfrac{I_{max}}{I_{min}} = \dfrac{A_{max}^2}{A_{min}^2} = \left(\dfrac{3}{2}\right)^2 = \dfrac{9}{4}$

NOTES In YDSE, the ratio of $\dfrac{I_{max}}{I_{min}}$ is maximum when both the sources have same intensity.

19. (c) For path difference λ, phase difference = 2π rad.

For path difference $\dfrac{\lambda}{4}$, phase difference $= \dfrac{\pi}{2}$ rad.

As $K = 4I_0$ so intensity at given point where path difference is $\dfrac{\lambda}{4}$

$K' = 4I_0 \cos^2\left(\dfrac{\pi}{4}\right)\left(\cos\dfrac{\pi}{4} = \cos 45°\right) = 2I_0 = \dfrac{K}{2}$

20. (a) $\because \; Y = \dfrac{n\lambda D}{d}$

$\therefore n_1\lambda_1 = n_2\lambda_2$

$\Rightarrow n_1 \times 12000 \times 10^{-10} = n_2 \times 10000 \times 10^{-10}$

or, $n(12000 \times 10^{-10}) = (n+1)(10000 \times 10^{-10}) \Rightarrow n = 5$

$\left(\because \lambda_1 = 12000 \times 10^{-10} m;\ \lambda_2 = 10000 \times 10^{-10} m\right)$

Hence, $Y_{common} = \dfrac{n\lambda_1 D}{d}$

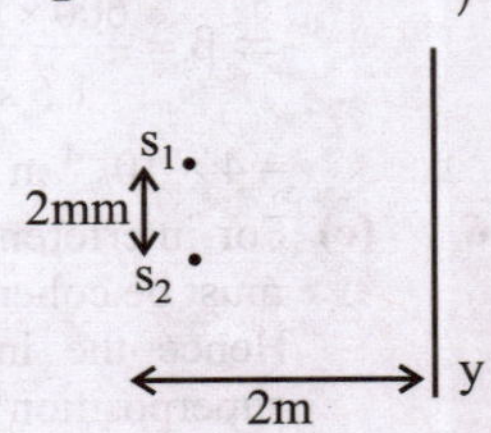

$= \dfrac{5(12000 \times 10^{-10}) \times 2}{2 \times 10^{-3}}$

$(\because d = 2$ mm and $D = 2m)$

$= 5 \times 12 \times 10^{-4}$ m
$= 60 \times 10^{-4}$ m
$= 6 \times 10^{-3}$ m $= 6$ mm

21. **(c)** Fringe width $\beta = \frac{\lambda D}{d}$;

From question $D' = 2D$ and $d' = \frac{d}{2}$

$\therefore \beta' = \frac{\lambda D'}{d'} = 4\beta$

22. **(c)** We know that the colours for which the condition of constructive interference is satisfied are observed in a given region of the film. The path difference between the light waves reaching the eye changes when the position of the eye is changed. Therefore, colours appear on a thin soap film or soap bubbles due to the phenomenon of interference.

23. **(d)** In vaccum, λ increases very slightly compared to that in air. As $\beta \propto \lambda$, therefore, width of interference fringe increases slightly.

24. **(b)** $x = (n)\lambda\frac{D}{d} = 3 \times 5000 \times 10^{-10} \times \frac{2}{0.2 \times 10^{-3}}$

$= 1.5 \times 10^{-2}$ m $= 1.5$ cm

25. **(a)** As $\beta = \frac{\lambda D}{d}$ and $\lambda_b < \lambda_y$,

Fringe width β will decrease

26. **(d)** For dark fringe $x = (2n-1)\frac{\lambda D}{2d}$

$\therefore \lambda = \frac{2xd}{(2n-1)D}$

$= \frac{2 \times 10^{-3} \times 0.9 \times 10^{-3}}{(2 \times 2 - 1) \times 1}$

$\lambda = 0.6 \times 10^{-6}$ m $= 6 \times 10^{-5}$ cm

27. **(c)** Distance of nth maxima, $x = n\lambda\frac{D}{d} \propto \lambda$

As $\lambda_b < \lambda_g \quad \therefore \quad x_{blue} < x_{green}$

NOTES Fringes with red light are thicker than those for blue light.

$\therefore \lambda_{red} > \lambda_{blue}$.

28. **(a)** $\beta' = \frac{\beta}{\mu} = \frac{0.4}{\frac{4}{3}} = 0.3$ mm

NOTES If the whole YDSE set up is taken in another medium then λ changes. So B changes.

In water $\lambda\omega = \frac{\lambda_a}{\mu_w}$

$\Rightarrow \quad \beta\omega = \frac{B_a}{\mu_w} = \frac{3}{4}B_a$.

29. **(c)** Interference is a wave phenomenon shown by both the light waves and sound waves.

30. **(c)** As $\theta = \frac{2\lambda}{a}$

$\theta_0 = \frac{2 \times 6000}{a}$ ...(i)

$\frac{\theta_1}{\theta_0} = \frac{\lambda_1}{6000}$ ...(ii)

$\Rightarrow \quad \lambda_1 = 0.7 \times 6000$ (as $\theta_1 = 0.7\,\theta_0$) $\Rightarrow 4200$ Å

31. **(d)** For the first minima,

$\theta = \frac{\eta\lambda}{a} \quad \Rightarrow \quad \sin 30° = \frac{\lambda}{a} = \frac{1}{2}$

First secondary maxima will be at

$\sin\theta = \frac{3\lambda}{2a} = \frac{3}{2}\left(\frac{1}{2}\right) \Rightarrow \theta = \sin^{-1}\left(\frac{3}{4}\right)$

32. **(d)** Linear width of central maxima Y

$= D(2\theta) = 2D\theta = \frac{2D\lambda}{a} \quad \therefore \theta = \frac{\lambda}{a}$

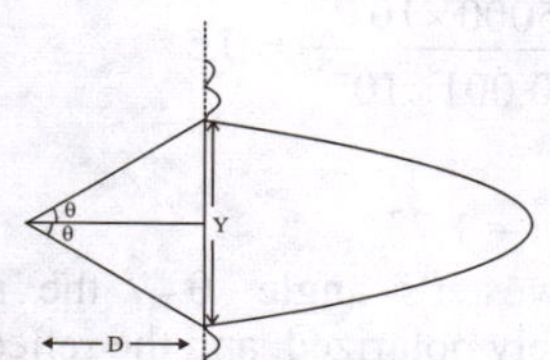

NOTES The central maxima lies between the first minima on both sides.

The angular width 'd' central maxima

$= 2\theta = \frac{2\lambda}{b}$

Linear width of central maxima $= 2D\theta = \frac{2D\lambda}{b}$

b = width of slits and D = distance between slit and screen.

33. **(b)** For first minima at P

$AP - BP = \lambda$

$AP - MP = \frac{\lambda}{2}$

So phase difference,

$\phi = \frac{2\pi}{\lambda} \times \frac{\lambda}{2} = \pi$ radian

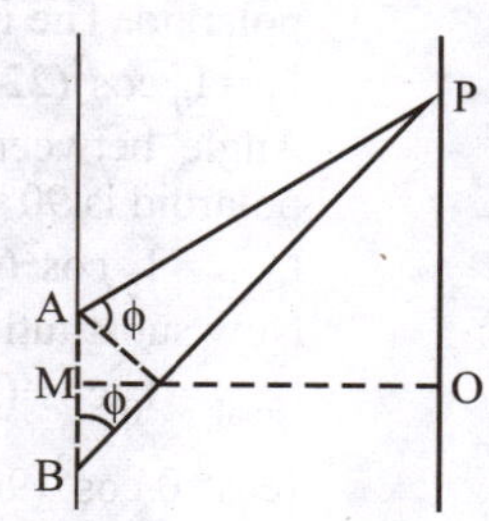

34. **(d)** Given: D = 2m; d = 1 mm = 1×10^{-3} m

$\lambda = 600$ nm $= 600 \times 10^{-6}$ m

Width of central bright fringe (= 2β)

$= \frac{2\lambda D}{d} = \frac{2 \times 600 \times 10^{-6} \times 2}{1 \times 10^{-3}}$ m $= 2.4 \times 10^{-3}$ m $= 2.4$ mm

35. (b)

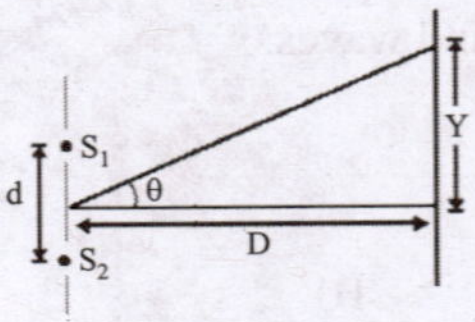

Angular width, $\theta = \frac{Y}{D} = \frac{n\lambda D}{dD}$

$\left[\because Y = \frac{D\lambda}{d}\right]$

so, $\theta = \frac{\lambda}{d}$, v $\uparrow \lambda \downarrow \theta \downarrow$

[For central maxima n = 1]

Hence, with increase in speed of electrons angular width of central maximum decreases.

36. (d) Conditions for diffraction minima are

Path diff. $\Delta x = n\lambda$ and Phase diff. $\delta\phi = 2n\pi$

Path diff. = $n\lambda = 2\lambda$

Phase diff. = $2n\pi = 4\pi$ $(\because n = 2)$

37. (c) For first minimum, $a \sin\theta = n\lambda = 1\lambda$

$$\sin\theta = \frac{\lambda}{a} = \frac{5000\times10^{-10}}{0.001\times10^{-3}} = 0.5$$

$\theta = 30°$

38. (a) Given : $n = 1.73$

At Brewester's angle θ_B, the reflected light is completely polarized, and the reflected and refracted rays are at 90° to each other.

$\tan\theta_B = n \Rightarrow \theta_B = \tan t\,(1.73)$

$\theta_B = 60°$

Using law of reflection

$i_p = r = 60°$

39. (c) Intensity after the first polaroid

$I_1 = I_0 \cos^2(0°) = I_0 \times 1 = I_0$

Intensity after the second polaroid

$I_2 = I_1 \cos^2(90°) = I_0 \times 0 = 0$

Third polaroid which is at 22.5° from the first polaraid. The intensity after the third polaroid

$I_3 = I_0 \cos^2(22.5°)$ ...(i)

Angle between the third polaraid and the second polaroid is $90° - 22.5° = 67.5°$

$I_{final} = I_3 \cos^2(67.5°)$

Now substituting the value of I_3 from eq^n ...(i)

$I_{final} = I_0 \cos^2(22.5°)\cos^2(67.5°)$

$[\cos^2\theta.\cos^2(90° - \theta) = \frac{1}{4}\cos^2(2\theta)]$

from trigonometry formula, here $\theta = 22.5°$

$$I_{final} = \frac{I_0}{4}\cos^2(2\times22.5°) = \frac{I_0}{4}\cos^2 45°$$

$$I_{final} = \frac{I_0}{8}$$

40. (a) $I_1 = \frac{I_0}{2}$

$I_B = (I_1)\cos^2 60°$

$= \frac{I_0}{2}\left(\frac{1}{2}\right)^2 = \frac{I_0}{8}$

$I_C = I_B \cos^2 45°$

$= \frac{I_0}{8} \times \left(\frac{1}{\sqrt{2}}\right)^2 = \frac{I_0}{16}$

41. (d)

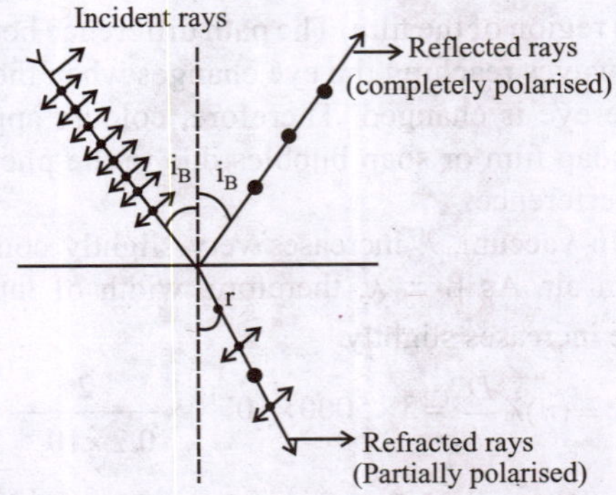

According to Brewster's law when an unpolarised light beam strikes a glass at Brewster's angle, then reflected rays are completely polarized and refracted rays are partially polarized.

NOTES When an unpolarised light beam is incident at angle of incidence equal to Brewster's angle, then the angle between reflected and refracted ray will be 90°.

42. (d) By law of Malus, Intensity

$$I' = I\cos^2\theta = I\cos^2 30° = I \times \frac{3}{4} = \frac{3I}{4}$$

43. (b) According to Brewster's law, when a beam of unpolarised light is reflected from a transparent medium of refractive index (μ_2), the reflected light is completely polarised at certain angle of incidence called the angle of polarisation (i_b).

$\tan i_b = \frac{\mu_2}{\mu_1}$

For air, $\mu_1 = 1$

$\therefore \tan i_b = \mu_2 > 1$

$\Rightarrow \tan i_b > 1 \Rightarrow 90° > i_b > 45°$

44. (b) When reflected light rays and refracted rays are perpendicular, reflected light is polarised with electric field vector perpendicular to the plane of incidence.

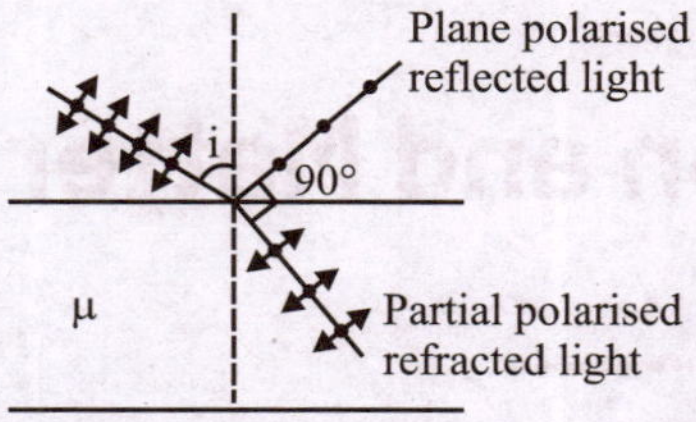

Also, tan i = μ (i = Brewester angle)

45. **(b)**

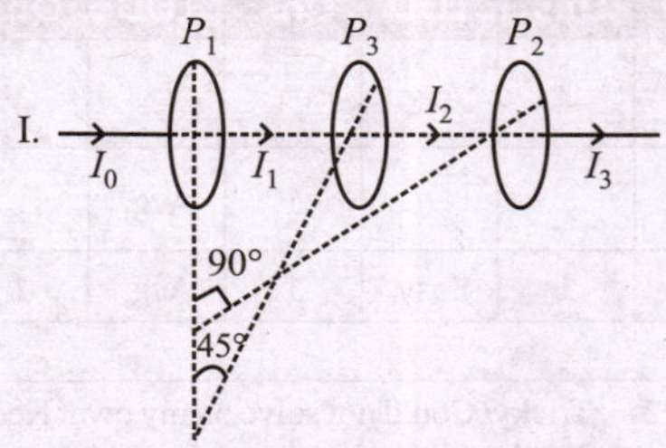

According to Malus law, $I = I_0 \cos^2 \theta$

$$I_1 = \frac{I_0}{2}$$

$$I_2 = \frac{I_0}{2}\cos^2 45°$$

$$= \frac{I_0}{2} \times \frac{1}{2} = \frac{I_0}{4}$$

$$I_3 = \frac{I_0}{4}\cos^2 45° = \frac{I_0}{8}$$

46. **(d)** Sound waves can not be polarised as they are longitudinal. Light waves can be polarised as they are transverse.

NOTES Polarisation of light is the restricting the vibration of light in a particular direction perpendicular to the direction of wave motion. Light is an electromagnetic wave and is transverse in nature.

11 Dual Nature of Radiation and Matter

TREND ANALYSIS

TOPIC	2025		2024		2023		2022		2021		2020	
	Qns.	LOD	Qns.	LOD	Qns.	LOD	Qns.	LOD	Qns.	LOD	Qns.	LOD
Einstein's Photoelectric Equation: Energy Quantum of Radiation	1	Easy			1	Avg.	3	Avg.			1	Easy
Particle Nature of Light: The Photon			1	Diffi.					1	Avg.		
Wave Nature of Matter	1	Avg.	1	Avg.			1	Easy.	1	Avg.	1	Avg.

How to Code Qns for Better & Faster Revision

Type 1: Able to solve (No code Req)

Type 2: Solved with struggle or by seeing solution **(Use code NR means Needs Revision or orange stickers)**

Type 3: Tricky/ Could not solve on my own. Need help from Teacher/ Friend. **(Use code TQ means Tricky Question or red stickers)**

You can either hand write the code in boxes or else paste coloured stickers given at end of book.

Topic-11.1 Introduction

1. In the phenomenon of electric discharge through gases at low pressure, the coloured glow in the tube appears as a result of: NCERT | Page-274 | 2008

(a) excitation of electrons in the atoms
(b) collision between the atoms of the gas
(c) collisions between the charged particles emitted from the cathode and the atoms of the gas
(d) collision between different electrons of the atoms of the gas

2. In a discharge tube ionization of enclosed gas is produced due to collisions between NCERT | Page-274 | 2006

(a) negative electrons and neutral atoms /molecules
(b) photons and neutral atoms /molecules
(c) neutral gas atoms/molecules
(d) positive ions and neutral atoms/molecules

3. Which one of the following statements is not true for cathode rays? NCERT | Page-274 | 2002

(a) Cathode rays produce heat when incident on metals
(b) Cathode rays travel in a straight line
(c) Cathode rays do not deflect in electric field
(d) Cathode rays produce fluorescence when they fall on certain meterials

4. Gases begin to conduct electricity at low pressure because 1994

(a) at low pressures gases turn of plasma
(b) colliding electrons can acquire higher kinetic energy due to increased mean free path leading to ionisation of atoms
(c) atoms break up into electrons and protons
(d) the electrons in atoms can move freely at low pressures

5. An ionization chamber with parallel conducting plates as anode and cathode has 5×10^7 electrons and the same number of singly charged positive ions per cm^3. The electrons are moving towards the anode with velocity 0.4 m/s. The current density from anode to cathode is $4\mu A/m^2$. The velocity of positive ions moving towards cathode is NCERT | Page-274 | 1992

(a) 0.4 m/s (b) 1.6 m/s
(c) zero (d) 0.1 m/s

Topic-11.2 Electron Emission

6. The nature of ions knocked out from hot surfaces is NCERT | Page-275 | 1995

(a) Protons (b) Neutrons
(c) Electrons (d) Nuclei

Topic-11.3 Photoelectric Effect

7. Photoelectric emmision occurs only when the incident light has more than a certain minimum NCERT | Page-278 | 2011

(a) power (b) wavelength
(c) intensity (d) frequency

Topic-11.4 Experimental Study of Photoelectric Effect

8. Which of the following options represent the variation of photoelectric current with property of light shown on the x-axis? NCERT | Page-278, 279 | 2025

A. 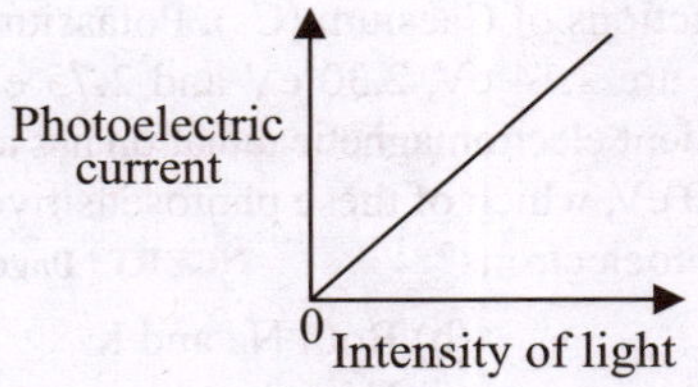

B. 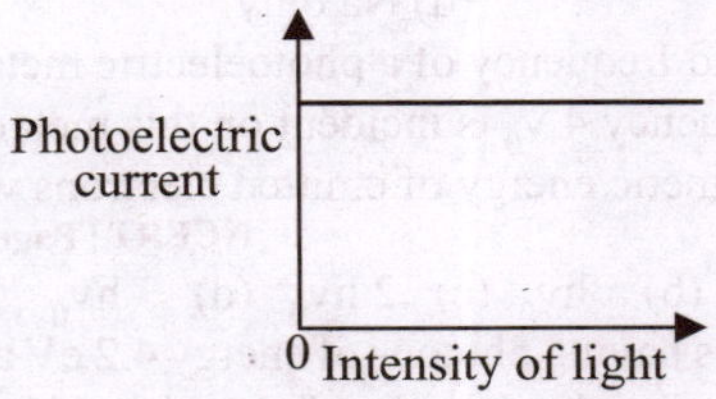

C. 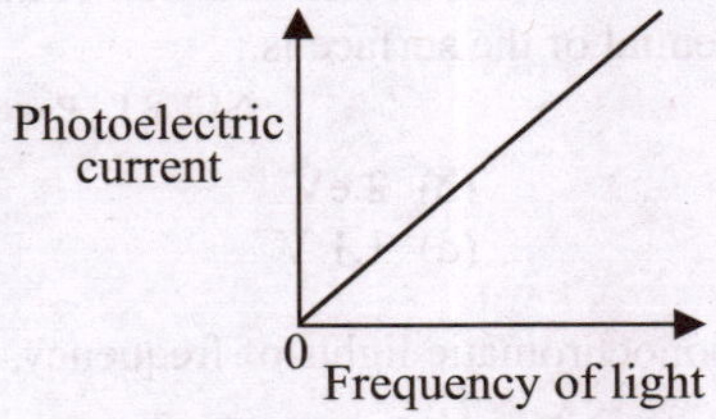

D. 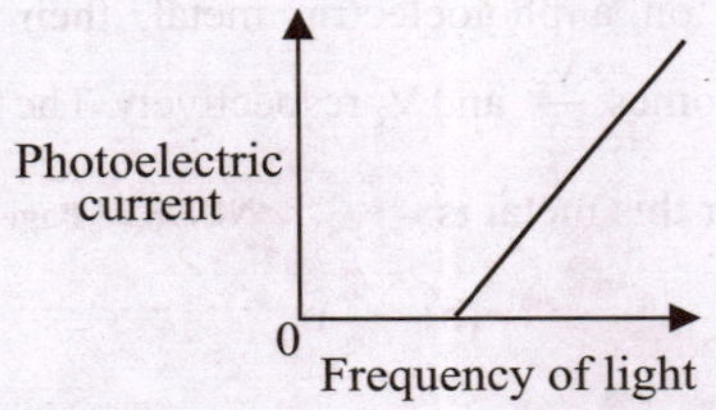

(a) A only (b) A and C
(c) A and D (d) B and C

9. The maximum kinetic energy of the emitted photoelectrons in photoelectric effects is independent of: **NCERT | Page- 279 | Manipur 2023**
(a) Frequency of incident radiation
(b) Wavelength of incident radiation
(c) Work function of material
(d) Intensity of incident radiation

10. A source of light is placed at a distance of 50 cm from a photocell and the stopping potential is found to be V_0. If the distance between the light source and photocell is made 25 cm, the new stopping potential will be **NCERT | Page-278 | Kar. 2013**
(a) $2V_0$ (b) $V_0/2$
(c) V_0 (d) $4V_0$

11. In photoelectric emission process from a metal of work function 1.8 eV, the kinetic energy of most energetic electrons is 0.5 eV. The corresponding stopping potential is **NCERT | Page-281 | 2011**
(a) 1.8 V (b) 1.2 V
(c) 0.5 V (d) 2.3 V

12. The number of photo electrons emitted for light of a frequency ν (higher than the threshold frequency ν_0) is proportional to: **NCERT | Page-278 | 2009**
(a) Threshold frequency (ν_0)
(b) Intensity of light
(c) Frequency of light (ν)
(d) $\nu - \nu_0$

13. The figure shows a plot of photocurrent versus anode potential for a photo sensitive surface for three different radiations. Which one of the following is a **correct** statement? **NCERT | Page-278 | 2009**

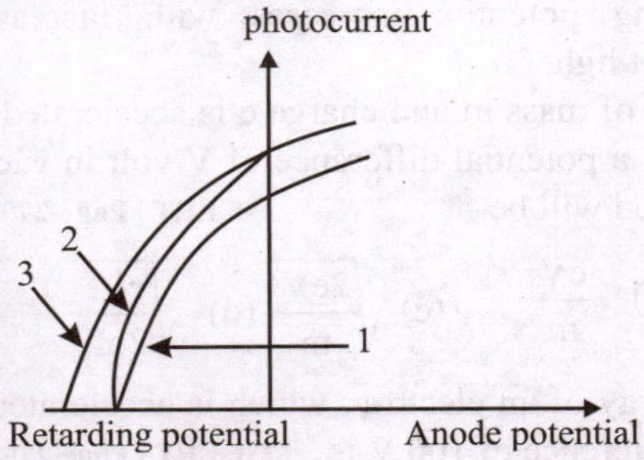

(a) Curves (1) and (2) represent incident radiations of same frequency but of different intensities.
(b) Curves (2) and (3) represent incident radiations of different frequencies and different intensities.
(c) Curves (2) and (3) represent incident radiations of same frequency having same intensity.
(d) Curves (1) and (2) represent incident radiations of different frequencies and different intensities.

14. A 5 watt source emits monochromatic light of wavelength 5000 Å. When placed 0.5 m away, it liberates photoelectrons from a photosensitive metallic surface. When the source is moved to a distance of 1.0 m, the number of photoelectrons liberated will be reduced by a factor of **NCERT | Page-279 | 2007**
(a) 8 (b) 16 (c) 2 (d) 4

15. A photo-cell employs photoelectric effect to convert **NCERT | Page-278 | 2006**
(a) change in the intensity of illumination into a change in photoelectric current
(b) change in the intensity of illumination into a change in the work function of the photocathode
(c) change in the frequency of light into a change in the electric current
(d) change in the frequency of light into a change in electric voltage

16. A photoelectric cell is illuminated by a point source of light 1m away. When the source is shifted to 2m then **NCERT | Page-278 | 2003**
(a) number of electrons emitted is a quarter of the initial number
(b) each emitted electron carries one quarter of the initial energy
(c) number of electrons emitted is half the initial number
(d) each emitted electron carries half the initial energy

17. As the intensity of incident light increases **NCERT | Page-278 | 1999**
(a) photoelectric current increases
(b) K. E. of emitted photelectrons increases
(c) photoelectric current decreases
(d) K.E. of emitted photoelectrons decreases

18. Which of the following statement is correct? NCERT | Page-278 | 1997
(a) Photocurrent increases with intensity of light
(b) Photocurrent is proportional to the applied voltage
(c) Current in photocell increases with increasing frequency
(d) Stopping potential increases with increase of incident light

19. An electron of mass m and charge e is accelerated from rest through a potential difference of V volt in vacuum. Its final speed will be NCERT | Page-279 | 1996
(a) $\frac{eV}{2m}$ (b) $\frac{eV}{m}$ (c) $\sqrt{\frac{2eV}{m}}$ (d) $\sqrt{\frac{eV}{2m}}$

20. Kinetic energy of an electron, which is accelerated in a potential difference of 100 V is NCERT | Page-279 | 1995
(a) 1.6×10^{-17} J
(b) 1.6×10^{-19} J
(c) 1.6×10^{-21} J
(d) 1.6×10^{-25} J

21. Doubly ionised helium atoms and hydrogen ions are accelerated from rest through the same potential drop. The ratio of the final velocities of the helium and the hydrogen ion is NCERT | Page-279 | 1994
(a) 1/2 (b) 2
(c) $1/\sqrt{2}$ (d) $\sqrt{2}$

22. Number of ejected photoelectron increases with increase NCERT | Page-279 | 1993
(a) in intensity of light
(b) in wavelength of light
(c) in frequency of light
(d) never

23. The cathode of a photoelectric cell is changed such that the work function changes from W_1 to W_2 ($W_2 > W_1$). If the current before and after changes are I_1 and I_2, all other conditions remaining unchanged, then (assuming $h\nu > W_2$) NCERT | Page-279 | 1992
(a) $I_1 = I_2$ (b) $I_1 < I_2$
(c) $I_1 > I_2$ (d) $I_1 < I_2 < 2I_1$

Topic-11.6
Einstein's Photoelectric Equation: Energy Quantum of Radiation

24. If ϕ is the work function of photosensitive material in eV and light of wavelength of numerical value $\lambda = \frac{hc}{e}$ metre, is incident on it with energy above its threshold value at an instant then the maximum kinetic energy of the photo-electron ejected by it at that instant (Take h - Plank's constant, c - velocity of light in free space) is (in SI units): NCERT | Page-282 | Ph-II 2024
(a) $e + 2\phi$ (b) $2e - \phi$
(c) $e - \phi$ (d) $e + \phi$

25. The work functions of Caesium (Cs), Potassium (K) and Sodium (Na) are 2.14 eV, 2.30 eV and 2.75 eV respectively. If incident electromagnetic radiation has an incident energy of 2.20 eV, which of these photosensitive surfaces may emit photoelectrons? NCERT | Page-282 | 2023
(a) Cs only (b) Both Na and K
(c) K only (d) Na only

26. The threshold frequency of a photoelectric metal is ν_0. If light of frequency $4\nu_0$ is incident on this metal, then the maximum kinetic energy of emitted electrons will be NCERT | Page-281 | 2022
(a) $4h\nu_0$ (b) $h\nu_0$ (c) $2h\nu_0$ (d) $3h\nu_0$

27. The light rays having photons of energy 4.2 eV are falling on a metal surface having a work function of 2.2 eV. The stopping potential of the surface is : NCERT | Page-281 | 2022
(a) 6.4 V (b) 2 eV
(c) 2 V (d) 1.1 V

28. When two monochromatic lights of frequency, υ and $\frac{\upsilon}{2}$ are incident on a photoelectric metal, their stopping potential becomes $\frac{V_s}{2}$ and V_s respectively. The threshold frequency for this metal is: NCERT | Page-282 | 2022
(a) 3υ (b) $\frac{2}{3}\upsilon$
(c) $\frac{3}{2}\upsilon$ (d) 2υ

29. Light of frequency 1.5 times the threshold frequency is incident on a photosensitive material. What will be the photoelectric current if the frequency is halved and intensity is doubled? NCERT | Page-281 | 2020
(a) four times (b) one-fourth
(c) zero (d) doubled

30. The work function of a photosensitive material is 4.0 eV. The longest wavelength of light that can cause photon emission from the substance is (approximately) NCERT | Page-281 | Odisha 2019
(a) 310 nm (b) 3100 nm
(c) 966 nm (d) 31 nm

31. The photoelectric threshold wavelength of silver is 3250×10^{-10}m. The velocity of the electron ejected from a silver surface by ultraviolet light of wavelength 2536×10^{-10} m is NCERT | Page-281 | 2017
(Given $h = 4.14 \times 10^{-15}$ eVs and $c = 3 \times 10^8$ ms^{-1})
(a) $\approx 0.6 \times 10^6$ ms^{-1} (b) $\approx 61 \times 10^3$ ms^{-1}
(c) $\approx 0.3 \times 10^6$ ms^{-1} (d) $\approx 6 \times 10^5$ ms^{-1}

32. Photons with energy 5 eV are incident on a cathode C in a photoelectric cell. The maximum energy of emitted photoelectrons is 2 eV. When photons of energy 6 eV are incident on C, no photoelectrons will reach the anode A, if the stopping potential of A relative to C is NCERT | Page-282 | Ph-II 2016

(a) – 1 V (b) – 3 V
(c) + 3 V (d) + 4 V

33. When a metallic surface is illuminated with radiation of wavelength λ, the stopping potential is V. If the same surface is illuminated with radiation of wavelength 2λ, the stopping potential is $\frac{V}{4}$. The threshold wavelength for the metallic surface is : **NCERT | Page-282 | 2016**

(a) 4λ (b) 5λ
(c) $\frac{5}{2}\lambda$ (d) 3λ

34. A certain metallic surface is illuminated with monochromatic light of wavelength λ. The stopping potential for photo-electric current for this light is $3V_0$. If the same surface is illuminated with light of wavelength 2λ, the stopping potential is V_0. The threshold wavelength for this surface for photo-electric effect is **NCERT | Page-282 | 2015**

(a) 4λ (b) $\frac{\lambda}{4}$
(c) $\frac{\lambda}{6}$ (d) 6λ

35. A photoelectric surface is illuminated successively by monochromatic light of wavelength λ and $\frac{\lambda}{2}$. If the maximum kinetic energy of the emitted photoelectrons in the second case is 3 times that in the first case, the work function of the surface of the material is :
(h = Planck's constant, c = speed of light) **NCERT | Page-281 | 2015 RS**

(a) $\frac{hc}{\lambda}$ (b) $\frac{2hc}{\lambda}$ (c) $\frac{hc}{3\lambda}$ (d) $\frac{hc}{2\lambda}$

36. When the energy of the incident radiation is increased by 20%, the kinetic energy of the photoelectrons emitted from a metal surface increased from 0.5 eV to 0.8 eV. The work function of the metal is: **NCERT | Page-281 | 2014**

(a) 0.65 eV (b) 1.0 eV
(c) 1.3 eV (d) 1.5 eV

37. For photoelectric emission from certain metal the cut-off frequency is ν. If radiation of frequency 2ν impinges on the metal plate, the maximum possible velocity of the emitted electron will be (m is the electron mass) **NCERT | Page-281 | 2013**

(a) $\sqrt{h\nu/m}$ (b) $\sqrt{2h\nu/m}$
(c) $2\sqrt{h\nu/m}$ (d) $\sqrt{h\nu/(2m)}$

38. Monochromatic radiation emitted when electron on hydrogen atom jumps from first excited to the ground state irradiates a photosensitive material. The stopping potential is measured to be 3.57 V. The threshold frequency of the materials is: **NCERT | Page-281 | 2012**

(a) 4×10^{15} Hz (b) 5×10^{15} Hz
(c) 1.6×10^{15} Hz (d) 2.5×10^{15} Hz

39. Two radiations of photons energies 1 eV and 2.5 eV, successively illuminate a photosensitive metallic surface of work function 0.5 eV. The ratio of the maximum speeds of the emitted electrons is : **NCERT | Page-281 | 2012 M**

(a) 1 : 4 (b) 1 : 2
(c) 1 : 1 (d) 1 : 5

40. Light of two different frequencies whose photons have energies 1 eV and 2.5 eV respectively illuminate a metallic surface whose work function is 0.5 eV successively. Ratio of maximum speeds of emitted electrons will be **NCERT | Page-281 | 2011**

(a) 1 : 4 (b) 1 : 2
(c) 1 : 1 (d) 1 : 5

41. The threshold frequency for a photosensitive metal is 3.3×10^{14} Hz. If light of frequency 8.2×10^{14} Hz is incident on this metal, the cut-off voltage for the photoelectric emission is nearly **NCERT | Page-281 | 2011 M**

(a) 2 V (b) 3 V
(c) 5 V (d) 1 V

42. The potential difference that must be applied to stop the fastest photoelectrons emitted by a nickel surface, having work function 5.01 eV, when ultraviolet light of 200 nm falls on it, must be: **NCERT | Page-281 | 2010**

(a) 2.4 V (b) – 1.2 V
(c) – 2.4 V (d) 1.2 V

43. The work function of a surface of a photosensitive material is 6.2 eV. The wavelength of incident radiation for which the stopping potential is 5 V lies in the: **NCERT | Page-282 | 2008**

(a) Ultraviolet region (b) Visible region
(c) Infrared region (d) X-ray region

44. When photons of energy hν fall on an aluminium plate (of work function E_0), photoelectrons of maximum kinetic energy K are ejected. If the frequency of the radiation is doubled, the maximum kinetic energy of the ejected photoelectrons will be **NCERT | Page-281 | 2006**

(a) 2K (b) K
(c) K + hν (d) $K + E_0$

45. A photosensitive metallic surface has work function, $h\nu_0$. If photons of energy $2h\nu_0$ fall on this surface, the electrons come out with a maximum velocity of 4×10^6 m/s. When the photon energy is increased to $5h\nu_0$, then maximum velocity of photoelectrons will be **NCERT | Page-282 | 2005**

(a) 2×10^7 m/s (b) 2×10^6 m/s
(c) 8×10^6 m/s (d) 8×10^5 m/s

46. According to Einstein's photoelectric equation, the graph between the kinetic energy of photoelectrons ejected and the frequency of incident radiation is **NCERT | Page-281 | 2004**

(a) 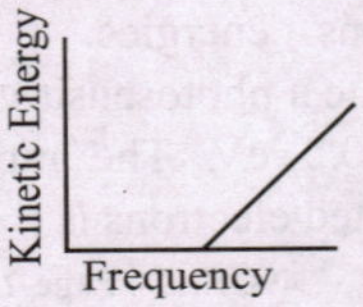

(b) 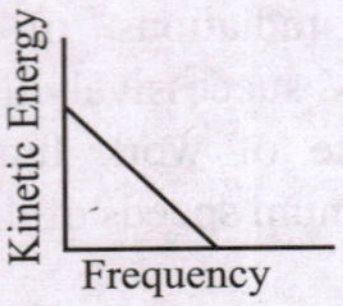

(c) 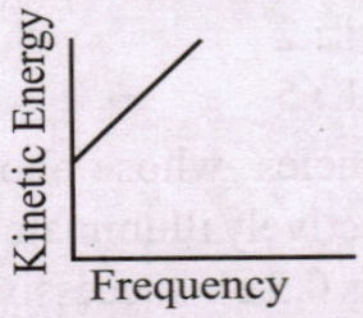

(d) 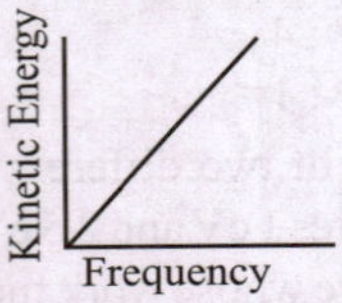

47. When ultraviolet radiation is incident on a surface, no photoelectrons are emitted. If a second beam causes photoelectrons to be ejected, it may consists of

NCERT | Page-282 | 2002

(a) infra-red waves (b) *X*-rays
(c) visible light rays (d) radio waves

48. Einstein work on the photoelectric effect provided support for the equation **NCERT | Page-281 | 2000**

(a) $E = h\nu$ (b) $E = mc^2$
(c) $E = -\frac{Rhc}{n^2}$ (d) $K.E. = \frac{1}{2}mv^2$

49. The photoelectric work function for a metal surface is 4.125 eV. The cut off wavelength for this surface is

NCERT | Page-282 | 1999

(a) 4125 Å (b) 3000 Å
(c) 6000 Å (d) 2062.5 Å

50. In a photo-emissive cell, with exciting wavelength λ, the fastest electron has speed v. If the exciting wavelength is changed to $\frac{3\lambda}{4}$, the speed of the fastest emitted electron will be **NCERT | Page-282 | 1998**

(a) $(3/4)^{1/2}$. v
(b) $(4/3)^{1/2}$. v
(c) less than $(4/3)^{1/2}$. v
(d) greater than $(4/3)^{1/2}$. v

51. Light of wavelength 5000 Å falls on a sensitive plate with photo-electric work function of 1.9 eV. The kinetic energy of the photo-electrons emitted will be

NCERT | Page-281 | 1998

(a) 0.58 eV (b) 2.48 eV
(c) 1.24 eV (d) 1.16 eV

52. If the threshold wavelength for a certain metal is 2000 Å, then the work-function of the metal is

NCERT | Page-282 | 1995

(a) 6.2 J (b) 6.2 eV
(c) 6.2 MeV (d) 6.2 keV

53. In photoelectric effect the work function of a metal is 3.5 eV. The emitted electrons can be stopped by applying a potential of –1.2 V. Then **NCERT | Page-281 | 1994**

(a) the energy of the incident photon is 4.7 eV
(b) the energy of the incident photon is 2.3 eV
(c) if higher frequency photon be used, the photoelectric current will rise
(d) when the energy of photon is 3.5 eV, the photoelectric current will be maximum

54. When light of wavelength 300 nm (nanometer) falls on a photoelectric emitter, photoelectrons are liberated. For another emitter, however, light of 600 nm wavelength is sufficient for creating photoemission. What is the ratio of the work functions of the two emitters?

NCERT | Page-282 | 1993

(a) 1 : 2 (b) 2 : 1
(c) 4 : 1 (d) 1 : 4

55. Photoelectric work function of a metal is 1eV. Light of wavelength λ = 3000 Å falls on it. The photo electrons come out with velocity **NCERT | Page-281 | 1991**

(a) 10 metres/sec (b) 10^2 metres/sec
(c) 10^4 metres/sec (d) 10^6 metres/sec

56. Ultraviolet radiations of 6.2 eV falls on an aluminium surface. K.E. of fastest electron emitted is (work function = 4.2 eV) **NCERT | Page-282 | 1989**

(a) 3.2×10^{-21} J (b) 3.2×10^{-19} J
(c) 7×10^{-25} J (d) 9×10^{-32} J

57. The threshold frequency for photoelectric effect on sodium corresponds to a wavelength of 5000 Å. Its work function is **NCERT | Page-282 | 1988**

(a) 4×10^{-19} J (b) 1 J
(c) 2×10^{-19} J (d) 3×10^{-19} J

Topic-11.7

Particle Nature of Light: The Photon

58. If c is the velocity of light in free space, the correct statements about photon among the following are:

A. The energy of a photon is E = hν.
B. The velocity of a photon is c.
C. The momentum of a photon, $p = \frac{h\nu}{c}$.
D. In a photon-electron collision, both total energy and total momentum are conserved.
E. Photon possesses positive charge.

Choose the correct answer from the options given below:

NCERT | Page-283 | 2024

(a) A and B only
(b) A, B, C and D only
(c) A, C and D only
(d) A, B, D and E only

59. The number of photons per second on an average emitted by the source of mono-chromatic light of wavelength 600 nm, when it delivers the power of 3.3×10^{-3} watt will be ($h = 6.6 \times 10^{-34}$ J s) **NCERT | Page-284 | 2021**

(a) 10^{15} (b) 10^{18} (c) 10^{17} (d) 10^{16}

60. A 200 W sodium street lamp emits yellow light of wavelength 0.6 μm. Assuming it to be 25% efficient in converting electrical energy to light, the number of photons of yellow light it emits per second is

NCERT | Page-284 | 2012

(a) 1.5×10^{20} (b) 6×10^{18}
(c) 62×10^{20} (d) 3×10^{19}

61. The momentum of a photon of energy hν will be

NCERT | Page-283 | 2011

(a) hν/c (b) c/hν
(c) hν (d) $h\nu/c^2$

62. A source S_1 is producing, 10^{15} photons per second of wavelength 5000 Å. Another source S_2 is producing 1.02 × 10^{15} photons per second of wavelength 5100Å Then, (power of S_2) (power of S_1) is equal to :

NCERT | Page-284 | 2010

(a) 1.00 (b) 1.02
(c) 1.04 (d) 0.98

63. Monochromatic light of wavelength 667 nm is produced by a helium neon laser. The power emitted is 9 mW. The number of photons arriving per sec on the average at a target irradiated by this beam is:

NCERT | Page-284 | 2009

(a) 3×10^{16} (b) 9×10^{15}
(c) 3×10^{19} (d) 9×10^{17}

64. Monochromatic light of frequency 6.0×10^{14} Hz is produced by a laser. The power emitted is 2×10^{-3} W. The number of photons emitted, on the average, by the sources per second is

NCERT | Page-284 | 2007

(a) 5×10^{16} (b) 5×10^{17}
(c) 5×10^{14} (d) 5×10^{15}

65. The momentum of a photon of energy 1 MeV in kg m/s, will be

NCERT | Page-283 | 2006

(a) 7×10^{-24} (b) 10^{-22}
(c) 5×10^{-22} (d) 0.33×10^{6}

66. The work functions for metals A, B and C are respectively 1.92 eV, 2.0 eV and 5 eV. According to Einstein's equation, the metals which will emit photoelectrons for a radiation of wavelength 4100 Å is/are

NCERT | Page-282 | 2005

(a) none (b) A only
(c) A and B only (d) all three metals

67. The 21 cm radio wave emitted by hydrogen in interstellar space is due to the interaction called the hyperfine interaction in atomic hydrogen. The energy of the emitted wave is nearly

NCERT | Page-282 | 1998

(a) 10^{-17} J (b) 1 J
(c) 7×10^{-8} J (d) 10^{-24} J

68. Momentum of a photon of wavelength λ is

NCERT | Page-283 | 1993

(a) $\frac{h}{\lambda}$ (b) zero

(c) $\frac{h\lambda}{c^2}$ (d) $\frac{h\lambda}{c}$

69. The wavelength of a 1 keV photon is 1.24×10^{-9} m. What is the frequency of 1 MeV photon?

NCERT | Page-283 | 1991

(a) 1.24×10^{15} (b) 2.4×10^{20}
(c) 1.24×10^{18} (d) $2 \times 4 \times 10^{23}$

70. Energy levels A, B, C of a certain atom correspond to increasing values of energy i.e., $E_A < E_B < E_C$. If λ_1, λ_2, λ_3 are the wavelengths of radiation corresponding to the transitions C to B, B to A and C to A respectively, which of the following relation is correct?

NCERT | Page-283 | 1990, 2005

(a) $\lambda_3 = \lambda_1 + \lambda_2$ (b) $\lambda_3 = \frac{\lambda_1\lambda_2}{\lambda_1 + \lambda_2}$

(c) $\lambda_1 + \lambda_2 + \lambda_3 =$ (d) $\lambda_3^2 = \lambda_1^2 + \lambda_2^2$

71. A radio transmitter operates at a freqency 880 kHz and a power of 10 kW. The number of photons emitted per second is

NCERT | Page-284 | 1990

(a) 1.72×10^{31} (b) 1.327×10^{25}
(c) 1.327×10^{37} (d) 1.327×10^{45}

72. The momentum of a photon of an electromagnetic radiation is 3.3×10^{-29} kgms^{-1}. What is the frequency of the associated waves ?

NCERT | Page-283 | 1990

[$h = 6.6 \times 10^{-34}$ Js; $c = 3 \times 10^{8}$ ms^{-1}]

(a) 1.5×10^{13} Hz (b) 7.5×10^{12} Hz
(c) 6.0×10^{3} Hz (d) 3.0×10^{3} Hz

73. The energy of a photon of wavelength λ is

NCERT | Page-283 | 1988

(a) $hc\lambda$ (b) $\frac{hc}{\lambda}$ (c) $\frac{\lambda}{hc}$ (d) $\frac{\lambda h}{c}$

Topic-11.8 Wave Nature of Matter

74. A photon and an electron (mass m) have the same energy E. The ratio ($\lambda_{photon}/\lambda_{electron}$) of their de-Broglie wavelengths is: (c is the speed of light)

NCERT | Page-285 | 2025

(a) $\sqrt{\frac{E}{2m}}$ (b) $c\sqrt{2mE}$

(c) $c\sqrt{\frac{2m}{E}}$ (d) $\frac{1}{c}\sqrt{\frac{E}{2m}}$

75. An electron and an alpha particle are accelerated by the same potential difference.Let λ_e and λ_α denote the de-Broglie wavelengths of the electron and the alpha particle, respectively, then: **NCERT | Page-285 | Ph-II 2024**

(a) $\lambda_e > \lambda_\alpha$ (b) $\lambda_e = 4\lambda_\alpha$
(c) $\lambda_e = \lambda_\alpha$ (d) $\lambda_e < \lambda_\alpha$

76. The graph which shows the variation of $\left(\frac{1}{\lambda^2}\right)$ and its kinetic energy, E is (where λ is de Broglie wavelength of a free particle): **NCERT | Page-285 | 2024**

(a) 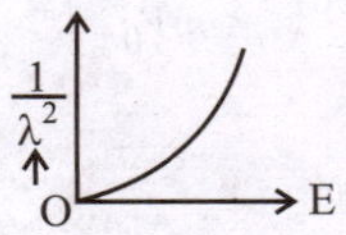

(b) 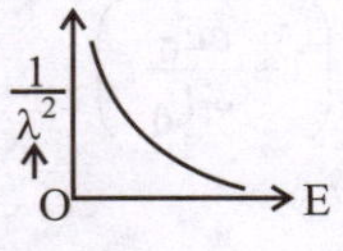

(c) 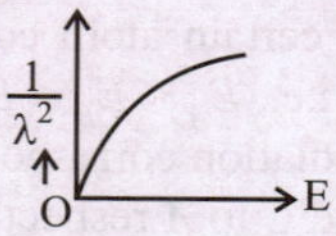

(d) 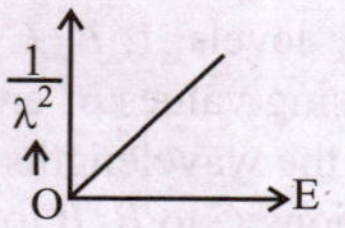

77. The de Broglie wavelength associated with an electron, accelerated by a potential difference of 81 V is given by:
NCERT | Page-285 | Manipur 2023

(a) 1.36 nm (b) 0.136 nm
(c) 13.6 nm (d) 136 nm

78. The graph which shows the variation of the de-Broglie wavelngth (λ) of a particle and its associated momentum (ρ) is:
NCERT | Page-285 | 2022

(a) 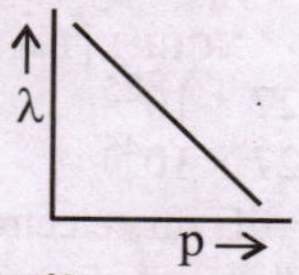

(b) 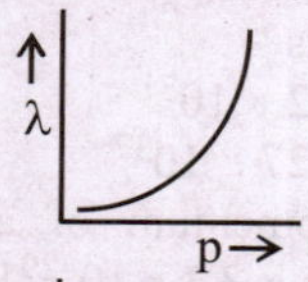

(c) 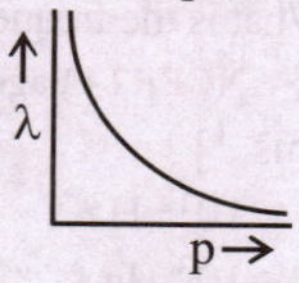

(d) 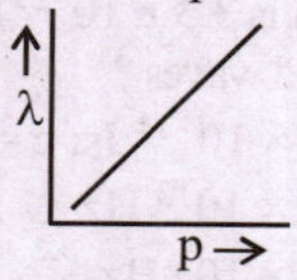

79. An electromagnetic wave of wavelength 'λ' is incident on a photosensitive surface of negligible work function. If '*m*' mass is of photoelectron emitted from the surface has de-Broglie wavelength λ_d, then
NCERT | Page-282 | 2021

(a) $\lambda = \left(\frac{2h}{mc}\right)\lambda_d^2$ (b) $\lambda = \left(\frac{2m}{hc}\right)\lambda_d^2$

(c) $\lambda_d = \left(\frac{2mc}{h}\right)\lambda^2$ (d) $\lambda = \left(\frac{2mc}{h}\right)\lambda_d^2$

80. An electron is accelerated from rest through a potential difference of V volt. If the de-Broglie wavelength of the electron is 1.227×10^{-2} nm, the potential difference is :
NCERT | Page-285 | 2020

(a) 10^2 V (b) 10^3 V
(c) 10^4 V (d) 10 V

81. An electron is accelerated through a potential difference of 10,000 V. Its de Broglie wavelength is, (nearly): ($m_e = 9 \times 10^{-31}$ kg)
NCERT | Page-285 | 2019

(a) 12.2×10^{-13} m (b) 12.2×10^{-12} m
(c) 12.2×10^{-14} m (d) 12.2 nm

82. A proton and an α–particle are accelerated from rest to the same energy. The de-Broglie wavelengths λ_p and λ_a are in the ratio
NCERT | Page-285 | Odisha 2019

(a) 4 : 1 (b) 2 : 1 (c) 1 : 1 (d) $\sqrt{2} : 1$

83. An electron of mass m with an initial velocity $\vec{V} = V_0\hat{i}$ ($V_0 > 0$) enters an electric field $\vec{E} = -E_0\hat{i}$ (E_0 = constant > 0) at t = 0. If λ_0 is its de-Broglie wavelength initially, then its de-Broglie wavelength at time t is
NCERT | Page-285 | 2018

(a) $\frac{\lambda_0}{\left(1+\frac{eE_0}{mV_0}t\right)}$ (b) $\lambda_0\left(1+\frac{eE_0}{mV_0}t\right)$

(c) λ_0 (d) $\lambda_0 t$

84. The de-Broglie wavelength of a neutron in thermal equilibrium with heavy water at a temperature T (Kelvin) and mass m, is :
NCERT | Page-285 | 2017

(a) $\frac{h}{\sqrt{3mkT}}$ (b) $\frac{2h}{\sqrt{3mkT}}$

(c) $\frac{2h}{\sqrt{mkT}}$ (d) $\frac{h}{\sqrt{mkT}}$

85. An electron of mass m and a photon have same energy E. The ratio of de-Broglie wavelengths associated with them is :
NCERT | Page-285 | 2016

(a) $\frac{1}{c}\left(\frac{E}{2m}\right)^{\frac{1}{2}}$ (b) $\left(\frac{E}{2m}\right)^{\frac{1}{2}}$

(c) $c(2mE)^{\frac{1}{2}}$ (d) $\frac{1}{c}\left(\frac{2m}{E}\right)^{\frac{1}{2}}$

86. Which of the following figures represents the variation of particle momentum and the associated de-Broglie wavelength?
NCERT | Page-285 | 2015

(a) 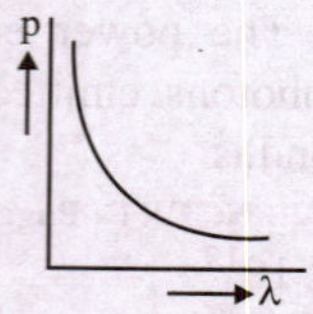

(b) 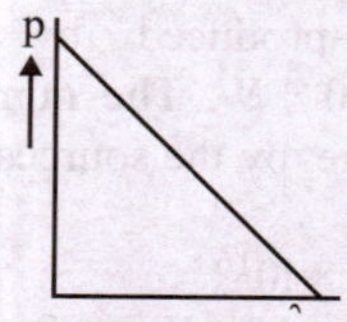

(c) 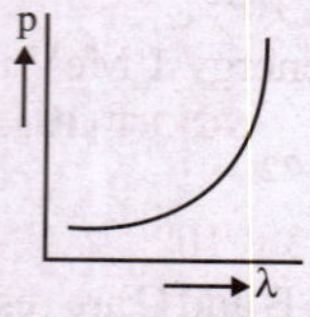

(d) 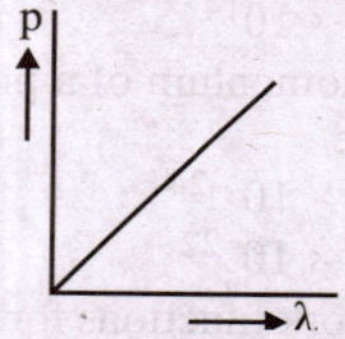

87. Light of wavelength 500 nm is incident on a metal with work function 2.28 eV. The wavelength of the emitted electron is:
NCERT | Page-281 | 2015 RS

(a) $< 2.8 \times 10^{-9}$ *m* (b) $\geq 2.8 \times 10^{-9}$ *m*
(c) $\leq 2.8 \times 10^{-12}$ *m* (d) $< 2.8 \times 10^{-10}$ *m*

88. If the kinetic energy of the particle is increased to 16 times its previous value, the percentage change in the de-Broglie wavelength of the particle is :
NCERT | Page-285 | 2014

(a) 25 (b) 75 (c) 60 (d) 50

89. The wavelength λ_e of an electron and λ_p of a photon are of same energy E are related by
NCERT | Page-285 | 2013

(a) $\lambda_p \propto \lambda_e$ (b) $\lambda_p \propto \sqrt{\lambda_e}$

(c) $\lambda_p \propto \frac{1}{\sqrt{\lambda_e}}$ (d) $\lambda_p \propto \lambda_e^2$

90. The de-Broglie wavelength of neutron in thermal equilibrium at temperature *T* is
Kar. 2013

(a) $\frac{30.8}{\sqrt{T}}$ Å (b) $\frac{3.08}{\sqrt{T}}$ Å

(c) $\frac{0.308}{\sqrt{T}}$ Å (d) $\frac{0.0308}{\sqrt{T}}$ Å

91. If the momentum of electron is changed by P, then the de Broglie wavelength associated with it changes by 0.5%. The initial momentum of electron will be : **NCERT | Page-285 | 2012 M**

(a) $200\,P$ (b) $400\,P$
(c) $\frac{P}{200}$ (d) $100\,P$

92. Electrons used in an electron microscope are accelerated by a voltage of 25 kV. If the voltage is increased to 100kV then the de–Broglie wavelength associated with the electrons would **NCERT | Page-285 | 2011**

(a) increase by 2 times
(b) decrease by 2 times
(c) decrease by 4 times
(d) increase by 4 times

93. A particle of mass 1 mg has the same wavelength as an electron moving with a velocity of 3×10^6 ms^{-1}. The velocity of the particle is: **NCERT | Page-285 | 2008**

(a) 2.7×10^{-18} ms^{-1} (b) 9×10^{-2} ms^{-1}
(c) 3×10^{-31} ms^{-1} (d) 2.7×10^{-21} ms^{-1}
(mass of electron = 9.1×10^{-31}kg)

94. Which of the following moving particles (moving with same velocity) has largest wavelength of matter waves? **NCERT | Page-285 | 2002**

(a) Electron (b) α-particle
(c) Proton (d) Neutron

95. The wavelength associated with an electron, accelerated through a potential difference of 100 V, is of the order of **NCERT | Page-285 | 1996**

(a) 1000 Å (b) 100 Å
(c) 10.5 Å (d) 1.2 Å

ANSWER KEY

1	(c)	11	(c)	21	(d)	31	(a, d)	41	(a)	51	(a)	61	(a)	71	(a)	81	(b)	91	(a)
2	(a)	12	(b)	22	(a)	32	(b)	42	(b)	52	(b)	62	(a)	72	(a)	82	(b)	92	(b)
3	(c)	13	(a)	23	(a)	33	(d)	43	(a)	53	(a)	63	(a)	73	(b)	83	(a)	93	(a)
4	(b)	14	(d)	24	(c)	34	(a)	44	(c)	54	(b)	64	(d)	74	(c)	84	(a)	94	(a)
5	(d)	15	(a)	25	(a)	35	(d)	45	(c)	55	(d)	65	(c)	75	(c)	85	(a)	95	(d)
6	(c)	16	(a)	26	(d)	36	(b)	46	(a)	56	(b)	66	(c)	76	(d)	86	(a)		
7	(d)	17	(a)	27	(c)	37	(b)	47	(b)	57	(a)	67	(d)	77	(b)	87	(b)		
8	(a)	18	(a)	28	(None)	38	(c)	48	(a)	58	(b)	68	(a)	78	(c)	88	(b)		
9	(d)	19	(c)	29	(c)	39	(b)	49	(b)	59	(d)	69	(b)	79	(d)	89	(d)		
10	(c)	20	(a)	30	(a)	40	(b)	50	(d)	60	(a)	70	(b)	80	(c)	90	(a)		

Hints & Solutions

1. **(c)** In discharge tube, collisions between the charged particles emitted from the cathode and atoms of the gas results to the coloured glow in the tube.

2. **(a)** When electrons emitted from cathode collide with gas molecules or atoms, they knock out outer electrons and produce positively charged ions. They become part of positive rays.

3. **(c)** Cathode rays are negatively charged rays. Hence, when electric field is applied to them, then they deflect in direction opposite to electric field.

4. **(b)** The ionisation requires high energy electrons.

5. **(d)** Current = $I_e + I_p$ ....(i)

I_e and I_p are current due to electrons and positively charged ions.

$I = neAV_d$

where,

$n = 5 \times 10^7/\text{cm}^3 = 5 \times 10^7 \times 10^6/\text{cm}^3$
$= 5 \times 10^{13}/\text{m}^3$

$I_e = 5 \times 10^{13} \times 1.6 \times 10^{-19} \times A \times 0.4$

$I_p = 5 \times 10^{13} \times 1.6 \times 10^{-19} \times A \times \nu$

$I = I_e + I_p$ (from equation (i))

$= 5 \times 10^{13} \times 1.6 \times 10^{-19} \times A(\nu + 0.4)$

Given, $I/A = 4 \times 10^{-6}\ A/m^2$

$4 \times 10^{-6} \times A = 5 \times 10^{-6} \times 1.6 \times A(\nu + 0.4)$

$\frac{4}{8} = \nu + 0.4 \Rightarrow 0.5 = \nu + 0.4 \Rightarrow \nu = 0.1$ m/s

6. **(c)** The nature of ions knocked out from hot surfaces are electrons.

7. **(d)** For occurence of photoelectric effect, the incident light should have frequency more than a certain minimum which is called the threshold frequency (ν_0).

> **NOTES**
> We have, $\frac{1}{2}mv^2 = h\nu - h\nu_0$
> For photoelectric effect emission $n > n_0$ where n is the frequency of the incident light.

8. **(a)** Photoelectric current (i) ∝ intensity remaining are not correct.

9. **(d)** By Einstein's photoelectric equation

$KE_{max} = hf - \phi_0$

or, $KE_{max} = \frac{hc}{\lambda} - \frac{hc}{\lambda_0}$

Clearly KE_{max} is independent of intensity of incident radiation.

10. (c) Since, stopping potential is independent of distance hence new stopping potential will remain unchanged i.e., new stopping potential = V_0.

11. (c) The stopping potential is equal to maximum kinetic energy.

12. (b) The number of photoelectrons emitted is proportional to the intensity of incident light. Saturation current $\propto$ intensity.

13. (a) Retarding potential depends on the frequency of incident radiation but is independent of intensity.

14. (d) Number of emitted electrons $N_E \propto$ Intensity $\propto \frac{1}{(\text{Distance})^2}$

Therefore, as distance is doubled, N_E decreases by (1/4) times.

15. (a) A photo-cell employs photoelectric effect to convert light energy into photoelectric current.

16. (a) Intensity $\propto$ No. of electrons emitted (N)

$$I \propto \frac{1}{r^2} \Rightarrow N \propto \frac{1}{r^2}$$

NOTES Energy crossing per unit area normally per second is called Intensity.

$$I = \frac{E}{At} = \frac{P}{A} \left(\text{Here, } P = \frac{E}{t} = \text{radiation power}\right)$$

At a distance r from a point source of power P intensity is given by $I = \frac{P}{4\pi r^2}$

$$\Rightarrow I \propto \frac{1}{r^2}$$

17. (a) K.E. of electrons emitted depends upon the frequency of incident rays rather than the intensity. While number of photo electrons emitted depends upon intensity of radiation.

18. (a) According to photoelectric effect, speed of electron (kinetic energy) emitted depends upon frequency of incident light while number of photoelectrons emitted depends upon intensity of incident light. Hence, as the intensity of light increases, the photocurrent increases.

NOTES In a photo-cell, the photocurrent has no relation with the applied voltage.
Stopping potential is the (negative) potential at which the current is just reduced to zero. It is independent of intensity of light but depends on the frequency of light similar to K.E.

19. (c) Kinetic energy of electron accelerated through a potential V = eV

$$\Rightarrow \frac{1}{2}mv^2 = eV \Rightarrow v^2 = \frac{2eV}{m} \Rightarrow v = \sqrt{\frac{2eV}{m}}$$

20. (a) Potential difference (V) = 100 V. The kinetic energy of an electron = 1 eV = 1 × (1.6 × 10^{-19})

= 1.6 × 10^{-19} J. Therefore kinetic energy in 100 volts = (1.6 × 10^{-19}) × 100 = 1.6 × 10^{-17}J.

NOTES K.E. = qV
= 1.6 × 10^{-19} × 100 J = 1.6 × 10^{-17}

21. (d) $\frac{1}{2}mv^2 = eV$, or, $v = (2eV/m)^{1/2}$

The mass of helium ion is four times that of hydrogen ion. Further, the charge on helium ion is twice that of hydrogen ion.

$$\therefore\ v_{He}/v_H = \sqrt{2}$$

22. (a) Photoelectric current is directly proportional to the intensity of incident light.

NOTES Kinetic energy of emitted photoelectrons varies with frequency of incident radiation and it does not depend upon the intensity of incident radiations. But, the number of photons emitted from the surface varies directly with intensity of incident light.

23. (a) The work function has no effect on photoelectric current so long as $h\nu > W_0$. The photoelectric current is proportional to the intensity of incident light. Since there is no change in the intensity of light, hence $I_1 = I_2$.

24. (c) From the Einstein's photoelectric equation

$$(\text{K.E.})_{max} = \frac{hc}{\lambda} - \phi_0 \text{ and } \lambda = \frac{hc}{e}$$

$$\Rightarrow (\text{K.E.})_{max} = \frac{hc}{\frac{hc}{e}} - \phi_0$$

$$\Rightarrow (\text{K.E.})_{max} = e - \phi$$

25. (a) Energy of incident radiation = 2.20 eV

Work function of Cs, ϕ_{Cs} = 2.14 eV and ϕ_K = 2.30 eV

ϕ_{Na} = 2.75 eV

Since the work function of caesium, $\phi_{Cs} < \phi_{Incident.}$ Hence photons may be emitted from caesium.

26. (d) By photoelectric equation

$h\nu = h\nu_0 + K.E_{max}$

$h \times 4\nu_0 = h\nu_0 + K.E_{max}$ $(\because \nu = 4\nu_0)$

$h \times 4\nu_0 = h\nu_0 + K.E_{max}$

$\therefore K.E_{max} = 4h\nu_0 - h\nu_0 = 3h\nu_0$

27. (c) By Einstein's photoelectric equation

$h\nu = w + eV_s$

$4.2eV = 2.2eV + e \times V_s$

$\Rightarrow 2eV = eV_s$

$\Rightarrow 2 \times 1.6 \times 10^{-19} = 1.6 \times 10^{-19} \times V_s$

$\therefore V_s = 2V$

28. (None) By Einstein's photoelectric equation,

$$hf = \phi + eV_{stopping}$$

when $f = \nu$

$$h\nu = h\nu_0 + e\frac{V_s}{2} \quad ..... (i)$$

when $f = \nu/2$

$$h\frac{\nu}{2} = h\nu_0 + eV_s \quad ..... (ii)$$

$$\Rightarrow \frac{1}{2}\left(h\nu_0 + e\frac{V_s}{2}\right) = h\nu_0 + eV_s \quad \text{[From (i)]}$$

$$\Rightarrow -\frac{h\nu_0}{2} = \frac{3}{4}eV_s$$

$$\Rightarrow \phi = h\nu_0 = +\frac{3}{2}eV_s, \text{ which is –ve.}$$

So data given is incorrect because 'ϕ' can never be negative.

29. (c) For photoelectric emission, photoelectric current, incident light frequency should be greater than threshold frequency.

Light of frequency 1.5 times the threshold frequency ν_0 incident.

$$\nu = \frac{3}{2}\nu_0$$

If frequency is halved,

$$\therefore \nu' = \frac{\nu}{2} = \frac{3}{4}\nu_0 \qquad \because \nu' < \nu_0$$

∴ No photoelectric emission will take place.

30. (a) As $E = \frac{12400}{\lambda}$

$$\lambda = \frac{12400}{4} = 3100 \text{ Å}$$

$$\therefore \quad \lambda = 310 \text{ nm}$$

31. (a, d) Both answers are correct

Given,

$\lambda_0 = 3250 \times 10^{-10}$ m

$\lambda = 2536 \times 10^{-10}$ m

$$\phi = \frac{hc}{\lambda_0} = \frac{4.14\times10^{-15}\times3\times10^8}{3250\times10^{-10}} = 3.82\text{eV}$$

$$h\nu = \frac{hc}{\lambda} = \frac{4.14\times10^{-15}\times3\times10^8}{2536\times10^{-10}} = 4.89\text{eV}$$

According to Einstein's photoelectric equation,

$$K_{max} = h\nu - \phi$$

$$KE_{max} = (4.89 - 3.82)\text{eV} = 1.077 \text{ eV}$$

$$\frac{1}{2}mv^2 = 1.077 \times 1.6 \times 10^{-19}$$

$$\Rightarrow v = \sqrt{\frac{2\times1.077\times1.6\times10^{-19}}{9.1\times10^{-31}}}$$

or, $v = 0.6 \times 10^6$ m/s or 6×10^5 m/s

32. (b) Einstein Photoelectric Equation

$$eV = \frac{1}{2}mv_{max}^2 = h\nu = \phi_0$$

According to question,

I case $\quad 2 = 5 - \phi_0 \Rightarrow = 3$ eV

II case $\quad eV = 6 - 3 = 3$ eV $\Rightarrow$ Vs = 3 V.

$\therefore V_{AC} = -3$ V

33. (d) According to Einstein's photoelectric effect,

$$eV = \frac{hc}{\lambda} - \frac{hc}{\lambda_0} \quad ...(i)$$

$$eV/4 = \frac{hc}{2\lambda} - \frac{hc}{\lambda_0} \quad ...(ii)$$

Dividing equation (i) by (ii) by

$$\Rightarrow \quad 4 = \frac{\frac{1}{\lambda} - \frac{1}{\lambda_0}}{\frac{1}{2\lambda} - \frac{1}{\lambda_0}} \text{ on solving we get,}$$

$$\lambda_0 = 3\lambda$$

> **NOTES** Threshold wavelength (λ_0) is the maximum wavelength of incident radiations required to eject the electrons from a metallic surface.
> If incident wavelength $\lambda > \lambda_0$
> ⇒ No photoelectron emission occur

34. (a) As we know,

$$eV_s = \frac{hc}{\lambda} - \Psi$$

$$3eV_o = \frac{hc}{\lambda} - \Psi \quad ...(i)$$

$$eV_o = \frac{hc}{2\lambda} - \Psi \quad ...(ii)$$

$$3eV_o = \frac{3hc}{2\lambda} - 3\Psi \quad ...(iii)$$

Multiplying eqn. (ii) by (iii) and subtracting it from eqn (i)

$$\Psi = \frac{hc}{4\lambda}$$

So, threshold wavelength,

$$\lambda_{th} = \frac{hc}{\Psi} = \frac{hc}{hc/4\lambda} = 4\lambda$$

35. (d) Photoelectric equations

$$Ek_{1max} = \frac{hc}{\lambda} - \phi \quad ...(i)$$

and $Ek_{2max} = \frac{hc}{\lambda/2} - \phi$

$$EK_{2max} = \frac{2hc}{\lambda} - \phi \quad ...(ii)$$

From question, $Ek_{2max} = 3Ek_{1max}$

Multiplying equation (i) by 3

$$3Ek_{1max} = 3\left(\frac{hc}{\lambda} - \phi\right) \quad ...(iii)$$

From equation (ii) and (iii)

$$\frac{3hc}{\lambda} - 3\phi = \frac{2hc}{\lambda} - \phi \therefore \phi \text{ (work function)} = \frac{hc}{2\lambda}$$

36. (b) According to Einstein's photoelectric equation, $h\nu = \phi_0 + K_{max}$

We have

$$h\nu = \phi_0 + 0.5 \quad ...(i)$$

and $1.2h\nu = \phi_0 + 0.8 \quad ...(ii)$

Therefore, from above two equations $\phi_0 = 1.0\,eV$.

NOTES Work function (ϕ_0) varies from metal to metal. The material is better for photoelectric emission whose work function is least. As caesium has least work function, hence it is best metal for photoelectric emission.

37. (b) From photo-electric equation,
$h\nu' = h\nu + K_{max} \quad ...(i)$

$$h.2\nu = h\nu + \frac{1}{2} mV^2_{max} \quad [\because \nu' = 2\nu]$$

$$\Rightarrow h\nu = \frac{1}{2} mV^2_{max} \Rightarrow V_{max} = \sqrt{\frac{2h\nu}{m}}$$

38. (c) $n \to 2 - 1$

$E = 10.2$ eV

$kE = E - \phi$

$Q = 10.20 - 3.57$

$h\upsilon_0 = 6.63$ eV

$$\upsilon_0 = \frac{6.63 \times 1.6 \times 10^{-19}}{6.67 \times 10^{-34}} = 1.6 \times 10^{15} \text{ Hz}$$

39. (b) According to Einsten's photoelectric effect, the K.E. of the radiated electrons

$$K.E_{max} = E - W$$

$$\frac{1}{2} mv_1^2 = (1 - 0.5)\text{ eV} = 0.5 \text{ eV}$$

$$\frac{1}{2} mv_2^2 = (2.5 - 0.5)\text{ eV} = 2 \text{ eV}$$

$$\frac{v_1}{v_2} = \sqrt{\frac{0.5}{2}} = \frac{1}{\sqrt{4}} = 1:2$$

40. (b) The maximum kinetic energy of emitted electrons is given by

$$K.E = \phi - \phi_0$$

$$K.E_1 = 1\text{ eV} - 0.5\text{ eV} = 0.5\text{ eV}$$

$$K.E_2 = 2.5\text{ eV} - 0.5\text{ eV} = 2\text{ eV}$$

$$\therefore \frac{K.E_1}{K.E_2} = \frac{0.5eV}{2eV} = \frac{1}{4} \Rightarrow \frac{v_1}{v_2} = \sqrt{\frac{1}{4}} = \frac{1}{2}$$

41. (a) $K.E. = h\nu - h\nu_{th} = eV_0$ (V_0 = cut off voltage)

$$\Rightarrow V_0 = \frac{h}{e}(8.2 \times 10^{14} - 3.3 \times 10^{14})$$

$$= \frac{6.6 \times 10^{-34} \times 4.9 \times 10^{14}}{1.6 \times 10^{-19}} \approx 2V.$$

42. (b) $K_{max} = \frac{hc}{\lambda} W = \frac{hc}{\lambda} 5.01$

$$= \frac{12375}{\lambda(\text{in Å})} - 5.01$$

$$= \frac{12375}{2000} - 5.01 = 6.1875 - 5.01 = 1.17775$$

$$\simeq 1.2 \text{ V}$$

The potential difference that must be applied to stop photoelectrons = –1.2V

43. (a) Work function, $\phi_0 = 6.2$eV.

Stopping potential $V_0 = 5$V.
As, $eV_0 = h\nu - \phi_0$

or $eV_0 + \phi_0 = \frac{hc}{\lambda}$

or $\lambda = \frac{hc}{eV_0 + \phi_0}$

$$\lambda = \frac{6.62 \times 10^{-34} Js \times 3 \times 10^8 ms^{-1}}{1.6 \times 10^{-19} \times 5J + 6.2 \times 1.6 \times 10^{-19} J}$$

$$= \frac{6.62 \times 10^{-34} \times 3 \times 10^8}{1.6 \times 10^{-19}(5 + 6.2)} m$$

$$= \frac{19.86 \times 10^{-26}}{17.92 \times 10^{-19}} = 1.10 \times 10^{-7} \text{ m.}$$

Thus the wavelength of the incident radiation lies in the ultraviolet region.

44. (c) Applying Einstein's formula for photo-electricity

$h\nu = \phi + \frac{1}{2}mv^2$; $\quad h\nu = \phi + K$

$\phi = h\nu - K$

If we use 2ν frequency then let the kinetic energy becomes K'

So, $h \cdot 2\nu = \phi + K'$

$2h\nu = h\nu - K + K'$

$K' = h\nu + K$

45. (c) We know that

$h\nu - \phi = K_{max} = \frac{1}{2}mv_{max}^2$

According to question

$\frac{5h\nu_0 - h\nu_0}{2h\nu_0 - h\nu_0} = \frac{v_2^2}{v_1^2}$

$v_2 = 2v_1 = 2 \times 4 \times 10^6 = 8 \times 10^6$ m/s.

46. (a) K.E. $= h\nu - \phi$

$\therefore$ graph of K.E. versus ν is a straight line with +ve slope h and K.E. > 0 for ν such that $h\nu > \phi$.

47. (b) Energy of photon of *X*-rays is more than energy of photon of ultraviolet rays. Because frequency of *X* rays is more than ultraviolet rays.

48. (a) Einstein work on photoelectric effect supports the equation $E = h\nu$. It is based on quantum theory of light.

49. (b) Let λ_0 be cut off wavelength.

Work function $= \frac{hc}{\lambda_0} = 4.125 \times 1.6 \times 10^{-19}$

$\lambda_0 = \frac{6.626 \times 10^{-34} \times 3 \times 10^8}{4.125 \times 1.6 \times 10^{-19}} = 3000\text{Å}$

50. (d) $\frac{1}{2}mv^2 = \frac{hc}{\lambda} - W_0$

or $\frac{hc}{\lambda} = \frac{1}{2}mv^2 + W_0$ and

$\frac{1}{2}mv_1^2 = \frac{hc}{(3\lambda/4)} - W_0 = \frac{4}{3}\left(\frac{1}{2}mv^2 + W_0\right) - W_0$

So, v_1 is greater than $v\left(\frac{4}{3}\right)^{1/2}$.

51. (a) K.E. $= h\nu - h\nu_0$

$= \frac{hc}{\lambda} - h\nu_0 = \frac{12375}{5000} - 1.9$

$= 2.48 - 1.9 = 0.58$ eV

52. (b) Threshold wavelength $(\lambda) = 2000$ Å

$= 2000 \times 10^{-10}$ m.

Work function $(W) = \frac{hc}{\lambda} = \frac{(6.6 \times 10^{34}) \times (3 \times 10^8)}{2000 \times 10^{-10}}$

$= 9.9 \times 10^{-19}\,\text{J} = \frac{9.9 \times 10^{-19}}{1.6 \times 10^{-19}} = 6.2\text{eV}$

53. (a) $h\nu = W_0 + E_k = 3.5 + 1.2 = 4.7$ eV

54. (b) $W_0 = \frac{hc}{\lambda_0}$ or $W_0 \propto \frac{1}{\lambda_0}$;

$\Rightarrow \frac{W_1}{W_2} = \frac{\lambda_2}{\lambda_1} = \frac{600}{300} = 2$

55. (d) $h\nu = W + \frac{1}{2}mv^2$

or $\frac{hc}{\lambda} = W + \frac{1}{2}mv^2$

Here $\lambda = 3000$ Å $= 3000 \times 10^{-10}$ m

and W = 1 eV = 1.6×10^{-19} joule

$\therefore \frac{(6.6 \times 10^{-34})(3 \times 10^8)}{3000 \times 10^{-10}}$

Solving we get, $v \cong 10^6$ m/s

56. (b) K.E. of fastest electron

$= E - W_0 = 6.2 - 4.2 = 2.0$ eV

$= 2 \times 1.6 \times 10^{-19} = 3.2 \times 10^{-19}$ J

57. (a) $W_0 = \frac{hc}{\lambda_0}$

$= \frac{6.63 \times 10^{-34} \times 3 \times 10^8}{5000 \times 10^{-10}} = 4 \times 10^{-19}$ J

58. (b) (A) The energy of a photon E = hν

(B) Velocity of photon is equal to velocity of light i.e. c.

(C) Since $\lambda = \frac{h}{p} \Rightarrow p = \frac{h}{\lambda} \Rightarrow p = \frac{h\nu}{c}$

(D) In photon-electron collision both total energy and total momentum are conserved.

59. (d) Power, $P = \frac{E}{t} = \frac{n}{t}\left(\frac{hc}{\lambda}\right)$

$\therefore$ No. of photons / second, $\frac{n}{t} = \frac{P}{\left(\frac{hc}{\lambda}\right)}$

$\frac{n}{t} = \frac{3.3 \times 10^{-3} \times 6 \times 10^{-7}}{6.6 \times 10^{-34} \times 3 \times 10^8} = 10^{16}$

NOTES The number of photons emitted per second from a source of monochromatic radiation of wavelength λ and power p and if E = energy of each photon then,

$$\frac{n}{t}=\frac{P}{E}=\frac{P}{h\nu}=\frac{P}{hc/\lambda}$$

60. **(a)** Give that, only 25% of 200W converter electrical energy into light of yellow colour

$$\left(\frac{hc}{\lambda}\right)\times N = 200\times\frac{25}{100}$$

Where N is the No. of photons emitted per second, h = Plank's constant, c, speed of light.

$$N=\frac{200\times 25}{100}\times\frac{\lambda}{hc}$$

$$=\frac{200\times 25\times 0.6\times 10^{-6}}{100\times 6.2\times 10^{-34}\times 3\times 10^{8}} = 1.5\times 10^{20}$$

61. **(a)** Momentum of photon, $p=\frac{h\nu}{c}$

62. **(a)** Energy emitted/sec by $S_1, P_1 = n_1\frac{hc}{\lambda_1}$

Energy emitted/sec by $S_2, P_2 = n_2\frac{hc}{\lambda_2}$

$$\therefore \frac{P_2}{P_1}=\frac{n_2}{n_1}\cdot\frac{\lambda_1}{\lambda_2}$$

$$=\frac{1.02\times 10^{15}}{10^{15}}\cdot\frac{5000}{5100}=1.0$$

63. **(a)** $\lambda = 667\times 10^{-9}$ m, $P = 9\times 10^{-3}$ W

$P=\frac{Nhc}{\lambda}$, N = No. of photons emitted/sec.

$$N=\frac{9\times 10^{-3}\times 667\times 10^{-9}}{6.6\times 10^{-34}\times 3\times 10^{8}}$$

$$=\frac{9\times 6.67\times 10^{-10}}{3\times 6.6\times 10^{-26}}=3\times 10^{16}\text{ / sec}$$

64. **(d)** Since $p = nh\nu$

$$\Rightarrow n=\frac{p}{h\nu}=\frac{2\times 10^{-3}}{6.6\times 10^{-34}\times 6\times 10^{14}}=5\times 10^{15}$$

65. **(c)** 1 MeV = $10^6\times 1.6\times 10^{-19}$ joule

Momentum of photon,

$$p=\frac{E}{c}=\frac{1.6\times 10^{-13}}{3\times 10^{8}}$$

$$p=\frac{1.6}{3}\times 10^{-21}=\frac{16}{3}\times 10^{-22}$$

$p = 5\times 10^{-22}$ kg m/sec

NOTES Rest mass of photon is zero. But its effective mass is given by

$E = mc^2 = h\nu$

$$\Rightarrow m=\frac{E}{c^2}$$

$\therefore$ Momentum of photon, $p = mc = \frac{E}{c^2}\times C$

$$\Rightarrow p=\frac{E}{c}$$

66. **(c)** $E=\frac{hc}{\lambda}$

$$\Rightarrow E=\frac{12375}{\lambda(in\ \mathring{A})}\text{eV} \qquad \Rightarrow E=\frac{12375}{4100}\text{eV}$$

$\Rightarrow E\approx 3$ eV

So, only metals having work function less than 3eV can emit photoelectrons for the incident radiation of wavelength 4100Å.

67. **(d)** $E=h\nu=\frac{hc}{\lambda}=\frac{6.6\times 10^{-34}\times 3\times 10^{8}}{0.21}$

$$=\frac{6.6}{7}\times 10^{-24}=0.94\times 10^{-24}\text{ J}$$

68. **(a)** According to de-Broglie wave equation,

$$\lambda=\frac{h}{mv}=\frac{h}{p} \qquad \therefore p=\frac{h}{\lambda}$$

69. **(b)** Here, $\frac{hc}{\lambda}=10^3$ eV and $h\nu = 10^6$ eV

Hence, $\nu=\frac{10^3 c}{\lambda}=\frac{10^3\times 3\times 10^8}{1.24\times 10^{-9}}$

$= 2.4\times 10^{20}$ Hz

70. **(b)** $(E_2-E_1)=h\nu=\frac{hc}{\lambda}$

$$\therefore \frac{hc}{\lambda_1}=(E_C-E_B),\ \frac{hc}{\lambda_2}=(E_B-E_A)$$

and $\frac{hc}{\lambda_3}=(E_C-E_A)$

Now,

$$(E_C-E_A)=(E_C-E_B)+(E_B-E_A)$$

or, $\frac{hc}{\lambda_3}=\frac{hc}{\lambda_1}+\frac{hc}{\lambda_2}$ or $\frac{1}{\lambda_3}=\frac{1}{\lambda_1}+\frac{1}{\lambda_2}$

$$\therefore \frac{1}{\lambda_3}=\frac{\lambda_1+\lambda_2}{\lambda_1\lambda_2} \text{ or } \lambda_3=\frac{\lambda_1\lambda_2}{\lambda_1+\lambda_2}$$

71. (a) No. of photons emitted per sec,

$$n = \frac{\text{Power}}{\text{Energy of photon}}$$

$$= \frac{P}{h\nu} = \frac{10000}{6.6\times10^{-34}\times880\times10^{3}} = 1.72\times10^{31}$$

72. (a) As $\lambda = \frac{h}{p}$ and $\lambda = \frac{c}{\nu}$; so

$$\nu = \frac{c}{\lambda} = \frac{cP}{h} = 3\times10^{8}\times\frac{3.3\times10^{-29}}{6.6\times10^{-34}}$$

$= 1.5 \times 10^{13}$ Hz

73. (b) Energy of a photon $E = h\nu = \frac{hc}{\lambda}$

74. (c) Photon & electron will have same energy E

$$E(\text{photon}) = \frac{hc}{\lambda} \text{ and } E(\text{electron}) = \frac{h}{\sqrt{2mE}}$$

$$\frac{\lambda_{\text{photon}}}{\lambda_{\text{electron}}} = \frac{\frac{hc}{E}}{\frac{h}{\sqrt{2mE}}} = c\sqrt{\frac{2m}{E}}$$

75. (c) de-Broglie wavelength is

$$\lambda = \frac{h}{\sqrt{2mqV}} \quad \therefore \lambda \propto \frac{1}{\sqrt{mq}}$$

$$\frac{\lambda_\alpha}{\lambda_e} = \sqrt{\frac{m_e q_e}{m_\alpha q_\alpha}}$$

$\because m_\alpha >>> m_e$

$\lambda_e = \lambda_\alpha$

76. (d) de-Broglie wavelength $\lambda = \frac{h}{P} = \frac{h}{mv}$

$= \frac{h}{\sqrt{2mE}}$ where $E = \frac{1}{2}mv^2$

Squaring both sides,

$$\lambda^2 = \frac{h^2}{4m^2E} \Rightarrow \frac{1}{\lambda^2} \propto E$$

Hence $\frac{1}{\lambda^2}$ versus E graph is a straight line passes through the origin.

77. (b) De Broglie wavelength,

$$\lambda = \frac{12.27}{\sqrt{v}}\text{Å} = \frac{12.27}{\sqrt{81}} = \frac{12.27}{9}\text{Å}$$

$\therefore \lambda = 1.36$ Å or 0.136 nm

78. (c) Wavelength, $\lambda = \frac{h}{P} \Rightarrow \lambda \propto \frac{1}{P}$

So graph will be rectangular hyperbola as shown below

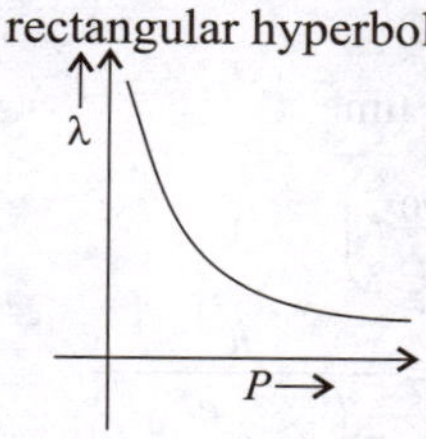

79. (d) From Einstein's photoelectric equation

$$\frac{hc}{\lambda} = \phi_0 + k$$

Where ϕ_0 = work function

k = maximum kinetic energy of photoelectrons

$$\therefore \frac{hc}{\lambda} = k = \frac{P^2}{2m} \Rightarrow P = \sqrt{\frac{2mhc}{\lambda}} \quad [\because \phi = \text{negligible}]$$

de-Broglie wavelength,

$$\lambda_d = \frac{h}{P} = \frac{h}{\sqrt{2mhc/\lambda}} \Rightarrow \sqrt{\lambda} = \lambda_d\sqrt{\frac{2mc}{h}}$$

$$\therefore \lambda = \left(\frac{2mc}{h}\right)\lambda_d^2$$

80. (c) Using, $\lambda = \frac{12.27}{\sqrt{V}}$Å

$$\Rightarrow \sqrt{V} = \frac{12.27\times10^{-10}}{1.227\times10^{-11}} = 10^2$$

[Given, $\lambda = 1.227 \times 10^{-11}$ m] $\therefore V = 10^4$ volt

81. (b) de-Broglie wavelength of electron

$$\lambda = \frac{12.27}{\sqrt{V}}\text{Å} = \frac{12.27\times10^{-10}}{\sqrt{10000}} = 12.27\times10^{-12}\text{m}$$

82. (b) As $\lambda = \frac{h}{p} = \frac{h}{\sqrt{2mK}}$

$\therefore \quad \lambda \propto \frac{1}{\sqrt{m}}$ (Kinetic energies are same)

$$\frac{\lambda_p}{\lambda_\alpha} = \sqrt{\frac{m_\alpha}{m_p}}$$

$$\therefore \quad \frac{\lambda_p}{\lambda_\alpha} = \frac{2}{1}$$

83. (a) Initial de-Brogile wavelength

$$\lambda_0 = \frac{h}{mV_0} \quad ...(i)$$

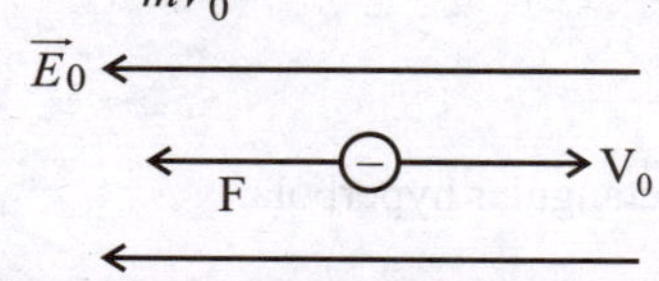

Acceleration of electron

$a = \frac{eE_0}{m}$ $(\because F = ma = eE_0)$

Velocity after time 't'

$V = \left(V_0 + \frac{eE_0}{m}t\right)$

So, $\lambda = \frac{h}{mV} = \frac{h}{m\left(V_0 + \frac{eE_0}{m}t\right)}$

$= \frac{h}{mV_0\left[1 + \frac{eE_0}{mV_0}t\right]} = \frac{\lambda_0}{\left[1 + \frac{eE_0}{mV_0}t\right]}$ ...(ii)

Dividing eqs. (ii) by (i),

de-Broglie wavelength $\lambda = \frac{\lambda_0}{\left[1 + \frac{eE_0}{mV_0}t\right]}$

84. **(a)** We know that,

de-Broglie wavelength $\lambda = \frac{\hbar}{p} = \frac{h}{\sqrt{2m(KE)}}$

K.E. of thermal neutron $= \frac{3}{2}kT$

$= \frac{h}{\sqrt{2m\left(\frac{3}{2}kT\right)}}$ $\therefore \lambda = \frac{h}{\sqrt{3mkT}}$

85. **(a)** For electron De-Broglie wavelength,

$\lambda_e = \frac{h}{\sqrt{2mE}}$

For photon E = pc

$\Rightarrow$ De-Broglie wavelength, $\lambda_{Ph} = \frac{hc}{E}$

$\therefore \frac{\lambda_e}{\lambda_{Ph}} = \frac{h}{\sqrt{2mE}} \times \frac{E}{hc} = \left(\frac{E}{2m}\right)^{1/2} \frac{1}{c}$

86. **(a)** According to De-broglie $p = \frac{h}{\lambda}$ or $P \propto \frac{1}{\lambda}$

where

P = particle momentum;

λ = de-Broglie wavelength

h = Plank's constant

$P \propto \frac{1}{\lambda}$ represents rectangular hyperbola.

87. **(b)** **Given :** Work function ϕ of metal = 2.28 eV

Wavelength of light $\lambda = 500$ nm $= 500 \times 10^{-9}$m

$KE_{max} = \frac{hc}{\lambda} - \phi$

$KE_{max} = \frac{6.6 \times 10^{-34} \times 3 \times 10^8}{5 \times 10^{-7} \times 1.6 \times 10^{-19}} - 2.28$

$KE_{max} = 2.48 - 2.28 = 0.2$ ev

$\lambda_{min} = \frac{h}{p} = \frac{h}{\sqrt{2m(KE)_{max}}}$

$= \frac{\frac{20}{3} \times 10^{-34}}{\sqrt{2 \times 9 \times 10^{-31} \times 0.2 \times 1.6 \times 10^{-19}}}$

$\lambda_{min} = \frac{25}{9} \times 10^{-9}$

$= 2.80 \times 10^{-9}$ nm $\therefore \lambda \geq 2.8 \times 10^{-9}$m

88. **(b)** As we know

$\lambda = \frac{h}{P} = \frac{h}{\sqrt{2mK}}$ $(\because P = \sqrt{2mKE})$

or $\frac{\lambda_1}{\lambda_2} = \sqrt{\frac{K_2}{K_1}} = \sqrt{\frac{16K}{K}} = \frac{4}{1}$

Therefore the percentage change in de-Broglie wavelength $= \frac{1-4}{4} \times 100 = -75\%$

89. **(d)** As $P = \frac{E}{c}$

$\lambda_p = \frac{hc}{E}$ ...(i)

$\lambda_e = \frac{h}{\sqrt{2mE}} \Rightarrow \lambda_e = \frac{h}{2mE}$ ...(ii)

From equations (i) and (ii)

$\lambda_p \propto \lambda_e^2$

NOTES

de-Broglie wavelength, $\lambda = \frac{h}{p}$

Here, h = plank's constant

p = momentum

Momentum, $p = \sqrt{2mk}$

$\therefore \lambda = \frac{h}{\sqrt{2mE}}$ (Here, E = kinetic energy)

$\therefore E = \frac{h^2}{2m\lambda^2} \Rightarrow \lambda^2 = \frac{h^2}{2mE}$

90. (a) From formula $\lambda = \dfrac{h}{\sqrt{2mKT}}$

$$= \frac{6.63\times10^{-34}}{\sqrt{2\times1.67\times10^{-27}\times1.38\times10^{-23}T}}m$$

[By placing value of h, m and k]

$$= \frac{30.8}{\sqrt{T}}\text{Å}$$

91. (a) The de-Broglie's wavelength associated with the moving electron $\lambda = \dfrac{h}{P}$

Now, according to problem

$$\frac{d\lambda}{\lambda} = -\frac{dp}{P}$$

$$\frac{0.5}{100} = \frac{P}{P'}$$

$P' = 200\ P$

92. (b) $\lambda \propto \dfrac{1}{\sqrt{V}}$

$$\Rightarrow \frac{\lambda_1}{\lambda_2} = \sqrt{\frac{V_2}{V_1}} = \sqrt{\frac{100\text{keV}}{25\text{keV}}} = 2$$

$$\Rightarrow \lambda_2 = \frac{\lambda_1}{2}$$

93. (a) Wavelength of particle

$$(\lambda_1) = \frac{h}{mv} = \frac{h}{(1\times10^{-6})\times v}$$

where v is the velocity of the particle.

Wave length of electron,

$$(\lambda_2) = \frac{h}{(9.1\times10^{-31})\times(3\times10^{6})}$$

But $\lambda_1 = \lambda_2$ (Given)

$$\therefore \frac{h}{(1\times10^{-6})\times v} = \frac{h}{(9.1\times10^{-31})\times(3\times10^{6})}$$

$$\Rightarrow v = \frac{9.1\times10^{-31}\times3\times10^{6}}{10^{-6}} = 2.73\times10^{-18}\ \text{ms}^{-1}$$

94. (a) de-Broglie wavelength $\lambda = \dfrac{h}{mv}$

For same velocity, $\lambda \propto \dfrac{1}{m}$

Out of given particles, the mass of electron is minimum, so the associated de-Broglie wavelength is maximum for electron.

95. (d) Potential difference = 100 V

K.E. acquired by electron = e (100)

$$\frac{1}{2}mv^2 = e(100) \Rightarrow v = \sqrt{\frac{2e(100)}{m}}$$

According to de Broglie's concept

$$\lambda = \frac{h}{mv} \Rightarrow \lambda = \frac{h}{m\sqrt{\frac{2e(100)}{m}}}$$

$$= \frac{h}{\sqrt{2me(100)}} = 1.2\times10^{-10} = 1.2\text{Å}$$

NOTES

de-Broglie wavelength, $\lambda_e = \dfrac{h}{p} = \dfrac{h}{\sqrt{2mE}}$

Here, h = Planck's constant

m = mass of electron

Kinetic energy, $E = eV$

Here, V = potential difference

$$\therefore \lambda_e = \frac{h}{\sqrt{2m_e eV}}$$

Substituting h = 6.63×10^{-34} Js

$e = 1.6\times10^{-19}$ C

$m_e = 9.1\times10^{-31}$ kg

we get $\lambda_e = \dfrac{12.27}{\sqrt{V}}\text{Å}$

12 Atoms

TREND ANALYSIS

TOPIC	2025		2024		2023		2022		2021		2020	
	Qns.	LOD	Qns.	LOD	Qns.	LOD	Qns.	LOD	Qns.	LOD	Qns.	LOD
Atomic Spectra			1	Avg.								
Bohr Model of the Hydrogen Atom	2	Avg.	1	Diff.	1	Easy	1	Avg.			1	Easy
The Line Spectra of the Hydrogen Atom					1	Avg.						
De Broglie's Explanation of Bohr's Second Postulate of Quantisation	1	Avg.										

How to Code Qns for Better & Faster Revision

Type 1: Able to solve (No code Req)
Type 2: Solved with struggle or by seeing solution **(Use code NR means Needs Revision or orange stickers)**
Type 3: Tricky/ Could not solve on my own. Need help from Teacher/ Friend. **(Use code TQ means Tricky Question or red stickers)**

You can either hand write the code in boxes or else paste coloured stickers given at end of book.

Topic-12.1 Introduction

1. J.J. Thomson's experiment demonstrated that

NCERT | Page-294 | 2003

(a) the e/m ratio of the cathode-ray particles changes when a different gas is placed in the discharge tube
(b) cathode rays are streams of negatively charged ions
(c) all the mass of an atom is essentially in the nucleus
(d) the e/m of electrons is much greater than the e/m of protons

2. Who indirectly determined the mass of the electron by measuring the charge of the electron?

NCERT | Page-291 | 2000

(a) Millikan (b) Rutherford
(c) Einstein (d) Thomson

Topic-12.2 Alpha-particle Scattering and Rutherford's Nuclear Model of Atom

3. When an α-particle of mass 'm' moving with velocity 'v' bombards on a heavy nucleus of charge 'Ze', its distance of closest approach from the nucleus depends on m as :

NCERT | Page-295 | 2016

(a) $\frac{1}{m}$ (b) $\frac{1}{\sqrt{m}}$ (c) $\frac{1}{m^2}$ (d) m

4. An alpha nucleus of energy $\frac{1}{2}mv^2$ bombards a heavy nuclear target of charge *Ze*. Then the distance of closest approach for the alpha nucleus will be proportional to

NCERT | Page-295 | 2010

(a) $\frac{1}{Ze}$ (b) v^2 (c) $\frac{1}{m}$ (d) $\frac{1}{v^4}$

5. In a Rutherford scattering experiment when a projectile of charge Z_1 and mass M_1 approaches a target nucleus of charge Z_2 and mass M_2, the distance of closest approach is r_0. The energy of the projectile is

NCERT | Page-295 | 2009

(a) directly proportional to $Z_1 Z_2$
(b) inversely proportional to Z_1
(c) directly proportional to mass M_1
(d) directly proportional to $M_1 \times M_2$

6. In Rutherford scattering experiment, what will be the correct angle for α-scattering for an impact parameter, $b = 0$?

NCERT | Page-294 | 1994

(a) 90° (b) 270° (c) 0° (d) 180°

Topic-12.3 Atomic Spectra

7. Given below are two statements:

Statement I: Atoms are electrically neutral as they contain equal number of positive and negative charges.

Statement II: Atoms of each element are stable and emit their characteristic spectrum.

In the light of the above statements, choose the most appropriate answer from the options given below.

NCERT | Page-296, 297 | 2024

(a) Both Statement I and Statement II are correct
(b) Both Statement I and Statement II are incorrect
(c) Statement I is correct but Statement II is incorrect
(d) Statement I is incorrect but Statement II is correct

8. The spectrum obtained from a sodium vapour lamp is an example of NCERT | Page-297 | 1995
(a) band spectrum (b) continuous spectrum
(c) emission spectrum (d) absorption spectrum

9. Which source is associated with a line emission spectrum ? NCERT | Page-297 | 1993
(a) Electric fire (b) Neon street sign
(c) Red traffic light (d) Sun

Topic-12.4
Bohr Model of the Hydrogen Atom

10. A model for quantized motion of an electron in a uniform magnetic field B states that the flux passing through the orbit of the electron is $n\left(\frac{h}{e}\right)$ where n is an integer, h is Planck's constant and e is the magnitude of electron's charge. According to the model, the magnetic moment of an electron in its lowest energy state will be (m is the mass of the electron) NCERT | Page-299 | 2025
(a) $\frac{he}{\pi m}$ (b) $\frac{he}{2\pi m}$ (c) $\frac{heB}{\pi m}$ (d) $\frac{heB}{2\pi m}$

11. A particle of mass m is moving around the origin with a constant force F pulling it towards the origin. If Bohr model is used to describe its motion, the radius r of the nth orbit and the particle's speed v in the orbit depend on n as NCERT | Page-299 | 2025
(a) $r \propto n^{\frac{1}{3}}; v \propto n^{\frac{1}{3}}$ (b) $r \propto n^{\frac{1}{3}}; v \propto n^{\frac{2}{3}}$
(c) $r \propto n^{\frac{2}{3}}; v \propto n^{\frac{1}{3}}$ (d) $r \propto n^{\frac{4}{3}}; v \propto n^{-\frac{1}{3}}$

12. Some energy levels of a molecule are shown in the figure with their wavelengths of transitions. Then: NCERT | Page-299 | Ph-II 2024

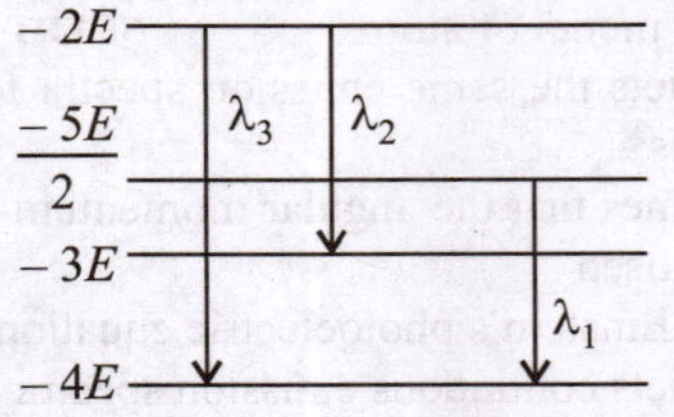

(a) $\lambda_3 > \lambda_2, \lambda_1 = 2\lambda_2$ (b) $\lambda_3 > \lambda_2, \lambda_1 = 4\lambda_2$
(c) $\lambda_1 > \lambda_2, \lambda_2 = 2\lambda_3$ (d) $\lambda_2 > \lambda_1, \lambda_2 = 2\lambda_3$

13. Match List I with List II.

List I (Spectral Lines of Hydrogen for transitions from)	List II (Wavelengths (nm))
A. $n_2 = 3$ to $n_1 = 2$	I. 410.2
B. $n_2 = 4$ to $n_1 = 2$	II. 434.1
C. $n_2 = 5$ to $n_1 = 2$	III. 656.3
D. $n_2 = 6$ to $n_1 = 2$	IV. 486.1

Choose the correct answer from the options given below: NCERT | Page-299 | 2024
(a) A-II, B-I, C-IV, D-III
(b) A-III, B-IV, C-II, D-I
(c) A-IV, B-III, C-I, D-II
(d) A-I, B-II, C-III, D-IV

14. The ground state energy of hydrogen atom is –13.6 eV. The energy needed to ionize hydrogen atom from its second excited state will be NCERT | Page-299 | Manipur 2023
(a) 1.51 eV (b) 3.4 eV
(c) 13.6 eV (d) 6.8 eV

15. The radius of inner most orbit of hydrogen atom is 5.3×10^{-11} m. What is the radius of third allowed orbit of hydrogen atom? NCERT | Page-299 | 2023
(a) 0.53 Å (b) 1.06 Å
(c) 1.59 Å (d) 4.77 Å

16. Let R_1 be the radius of the second stationary and R_2 be the radius of the fourth stationary orbit of an electron in Bohr's model. The ratio $\frac{R_1}{R_2}$ is: NCERT | Page-299 | RS 2022
(a) 1 : 1 (b) 1 : 4 (c) 1 : 2 (d) 2 : 1

17. Let T_1 and T_2 be the energy of an electron in the first and second excited states of hydrogen atom, respectively. According to the Bohr's model of an atom, the ratio $T_1 : T_2$ is: NCERT | Page-299 | 2022
(a) 4 : 1 (b) 4 : 9 (c) 9 : 4 (d) 1 : 4

18. For which one of the following, Bohr model is **not** valid ? NCERT | Page-299 | 2020
(a) Singly ionised helium atom (He^+)
(b) Deuteron atom
(c) Singly ionised neon atom (Ne^+)
(d) Hydrogen atom

19. The total energy of an electron in an atom in an orbit is –3.4 eV. Its kinetic and potential energies are, respectively: NCERT | Page-299 | 2019
(a) –3.4 eV, –3.4 eV (b) –3.4 eV, –6.8 eV
(c) 3.4 eV, –6.8 eV (d) 3.4 eV, 3.4 eV

20. The radius of the first permitted Bohr orbit, for the electron, in a hydrogen atom equals 0.51 Å and its ground state energy equals –13.6 eV. If the electron in the hydrogen atom is replaced by muon (μ^-) [charge same as electron and mass 207 m_e], the first Bohr radius and ground state energy will be, NCERT | Page-299 | Odisha 2019
(a) 2.56×10^{-13} m, –13.6 eV
(b) 0.53×10^{-13} m, –3.6 eV
(c) 25.6×10^{-13} m, –2.8 eV
(d) 2.56×10^{-13} m, –2.8 eV

21. The ratio of kinetic energy to the total energy of an electron in a Bohr orbit of the hydrogen atom, is NCERT | Page-299 | 2018
(a) 1 : 1 (b) 1 : –1 (c) 1 : –2 (d) 2 : –1

22. Two particles of masses m_1, m_2 move with initial velocities u_1 and u_2. On collision, one of the particles get excited to higher level, after absorbing energy ε. If final velocities of particles be v_1 and v_2 then we must have 2015

(a) $\frac{1}{2}m_1u_1^2 + \frac{1}{2}m_2u_2^2 = \frac{1}{2}m_1v_1^2 + \frac{1}{2}m_2v_2^2 - \varepsilon$

(b) $\frac{1}{2}m_1u_1^2 + \frac{1}{2}m_2u_2^2 - \varepsilon = \frac{1}{2}m_1v_1^2 + \frac{1}{2}m_2v_2^2$

(c) $\frac{1}{2}m_1^2u_1^2 + \frac{1}{2}m_2^2u_2^2 + \varepsilon = \frac{1}{2}m_1^2v_1^2 + \frac{1}{2}m_2^2v_2^2$

(d) $m_1^2u_1 + m_2^2u_2 - \varepsilon = m_1^2v_1 + m_2^2v_2$

23. Consider 3rd orbit of He^+ (Helium), using non-relativistic approach, the speed of electron in this orbit will be [given K = 9 × 10^9 constant, Z = 2 and h (Plank's Constant) = 6.6 × 10^{-34} J s] **NCERT | Page-299 | 2015**

(a) 1.46 × 10^6 m/s (b) 0.73 × 10^6 m/s
(c) 3.0 × 10^8 m/s (d) 2.92 × 10^6 m/s

24. An electron in hydrogen atom makes a transition $n_1 \rightarrow n_2$ where n_1 and n_2 are principal quantum numbers of the two states. Assuming Bohr's model to be valid the time period of the electron in the initial state is eight times that in the final state. The possible values of n_1 and n_2 are **Kar. 2013**

(a) $n_1 = 4$ and $n_2 = 2$ (b) $n_1 = 6$ and $n_2 = 2$
(c) $n_1 = 8$ and $n_2 = 1$ (d) $n_1 = 8$ and $n_2 = 2$

25. Electron in hydrogen atom first jumps from third excited state to second excited state and then from second excited to the first excited state. The ratio of the wavelength $\lambda_1 : \lambda_2$ emitted in the two cases is **NCERT | Page-299 | 2012**

(a) 7/5 (b) 27/20 (c) 27/5 (d) 20/7

26. An electron of a stationary hydrogen atom passes from the fifth energy level to the ground level. The velocity that the atom acquired as a result of photon emission will be : **NCERT | Page-299 | 2012**

(a) $\frac{24hR}{25m}$ (b) $\frac{25hR}{24m}$
(c) $\frac{25m}{24hR}$ (d) $\frac{24m}{25hR}$

(m is the mass of the electron, R, Rydberg constant and h Planck's constant)

27. The transition from the state $n = 3$ to $n = 1$ in a hydrogen like atom results in ultraviolet radiation. Infrared radiation will be obtained in the transition from : **NCERT | Page-299 | 2012M**

(a) $2 \rightarrow 1$ (b) $3 \rightarrow 2$
(c) $4 \rightarrow 2$ (d) $4 \rightarrow 3$

28. The wavelength of the first line of Lyman series for hydrogen atom is equal to that of the second line of Balmer series for a hydrogen like ion. The atomic number Z of hydrogen like ion is **NCERT | Page-299 | 2011**

(a) 3 (b) 4 (c) 1 (d) 2

29. An electron in the hydrogen atom jumps from excited state n to the ground state. The wavelength so emitted illuminates a photosensitive material having work function 2.75 eV. If the stopping potential of the photoelectron is 10 V, the value of n is **NCERT | Page-299 | 2011M**

(a) 3 (b) 4 (c) 5 (d) 2

30. Out of the following which one is not a possible energy for a photon to be emitted by hydrogen atom according to Bohr's atomic model? **NCERT | Page-299 | 2011M**

(a) 1.9 eV (b) 11.1 eV (c) 13.6 eV (d) 0.65 eV

31. The energy of a hydrogen atom in the ground state is – 13.6 eV. The energy of a He^+ ion in the first excited state will be **NCERT | Page-299 | 2010**

(a) –13.6 eV (b) – 27.2 eV
(c) – 54.4 eV (d) – 6.8 eV

32. The ionization energy of the electron in the hydrogen atom in its ground state is 13.6 eV. The atoms are excited to higher energy levels to emit radiations of 6 wavelengths. Maximum wavelength of emitted radiation corresponds to the transition between **2009**

(a) n = 3 to n = 1 states (b) n = 2 to n = 1 states
(c) n = 4 to n = 3 states (d) n = 3 to n = 2 states

33. The ground state energy of hydrogen atom is 13.6eV. When its electron is in the first excited state, its excitation energy is **NCERT | Page-299 | 2008**

(a) 3.4 eV (b) 6.8 eV
(c) 10.2 eV (d) 0

34. The total energy of electron in the ground state of hydrogen atom is – 13.6 eV. The kinetic energy of an electron in the first excited state is **NCERT | Page-299 | 2007**

(a) 6.8 eV (b) 13.6 eV
(c) 1.7 eV (d) 3.4 eV

35. Ionization potential of hydrogen atom is 13.6eV. Hydrogen atoms in the ground state are excited by monochromatic radiation of photon energy 12.1 eV. According to Bohr's theory, the spectral lines emitted by hydrogen will be **NCERT | Page-299 | 2006**

(a) Three (b) Four (c) One (d) Two

36. The total energy of an electron in the first excited state of hydrogen atom is about –3.4 eV. Its kinetic energy in this state is **NCERT | Page-299 | 2005**

(a) 3.4 eV (b) 6.8 eV
(c) –3.4 eV (d) –6.8 eV

37. The Bohr model of atoms **NCERT | Page-299 | 2004**

(a) predicts the same emission spectra for all types of atoms
(b) assumes that the angular momentum of electrons is quantised
(c) uses Einstein's photoelectric equation
(d) predicts continuous emission spectra for atoms

38. Energy E of a hydrogen atom with principal quantum number n is given by $E = -13.6/n^2$ eV. The energy of photon ejected when the electron jumps from $n = 3$ state to $n = 2$ state of hydrogen is approximately **NCERT | Page-299 | 2004**

(a) 1.9 eV (b) 1.5 eV
(c) 0.85 eV (d) 3.4 eV

39. In which of the following systems will the radius of the first orbit ($n = 1$) be minimum ? **NCERT | Page-299 | 2003**

(a) Hydrogen atom
(b) Doubly ionized lithium
(c) Singly ionized helium
(d) Deuterium atom

40. An electron changes its position from orbit $n = 2$ to the orbit $n = 4$ of an atom. The wavelength of the emitted radiations is (R = Rydberg's constant) **2001**

(a) $\frac{16}{R}$ (b) $\frac{16}{3R}$ (c) $\frac{16}{5R}$ (d) $\frac{16}{7R}$

41. The energy of hydrogen atom in nth orbit is E_n, then the energy in nth orbit of single ionised helium atom will be **NCERT | Page-299 | 2001**

(a) $4E_n$ (b) $E_n/4$ (c) $2E_n$ (d) $E_n/2$

42. Which of the following transitions in a hydrogen atom emits the photon of highest frequency? **NCERT | Page-299 | 2000**

(a) $n = 2$ to $n = 1$ (b) $n = 2$ to $n = 6$
(c) $n = 6$ to $n = 2$ (d) $n = 1$ to $n = 2$

43. In the Bohr model of a hydrogen atom, the centripetal force is furnished by the coulomb attraction between the proton and the electron. If a_0 is the radius of the ground state orbit, m is the mass, e is the charge on the electron and ε_0 is the vacuum permittivity, the speed of the electron is **NCERT | Page-299 | 1998**

(a) 0 (b) $\frac{e}{\sqrt{\varepsilon_0 a_0 m}}$

(c) $\frac{e}{\sqrt{4\pi\varepsilon_0 a_0 m}}$ (d) $\frac{\sqrt{4\pi\varepsilon_0 a_0 m}}{e}$

44. What is the radius of iodine atom (At. no. 53, mass no. 126) **1998**

(a) 2.5×10^{-11} m (b) 2.5×10^{-9} m
(c) 7×10^{-9} m (d) 7×10^{-6} m

45. When hydrogen atom is in its first excited level, its radius is **NCERT | Page-299 | 1997**

(a) four times its ground state radius
(b) twice
(c) same
(d) half

46. When a hydrogen atom is raised from the ground state to an excited state, **NCERT | Page-296 | 1995**

(a) P.E decreases and K.E. increases
(b) P.E. increases and K.E decreases
(c) both K.E. and P.E. decrease
(d) absorption spectrum

47. The radius of hydrogen atom in its ground state is 5.3 $\times 10^{-11}$ m. After collision with an electron it is found to have a radius of 21.2×10^{-11} m. What is the principal quantum number n of the final state of the atom **NCERT | Page-299 | 1994**

(a) $n = 4$ (b) $n = 2$
(c) $n = 16$ (d) $n = 3$

48. In terms of Bohr radius a_0, the radius of the second Bohr orbit of a hydrogen atom is given by **NCERT | Page-299 | 1992**

(a) $4\,a_0$ (b) $8\,a_0$
(c) $\sqrt{2}\,a_0$ (d) $2\,a_0$

49. The ionization energy of hydrogen atom is 13.6 eV. Following Bohr's theory, the energy corresponding to a transition between 3rd and 4th orbit is **NCERT | Page-299 | 1992**

(a) 3.40 eV (b) 1.51 eV
(c) 0.85 eV (d) 0.66 eV

50. The ground state energy of H-atom 13.6 eV. The energy needed to ionize H-atom from its second excited state. **NCERT | Page-299 | 1991**

(a) 1.51 eV (b) 3.4 eV
(c) 13.6 eV (d) 12.1 eV

51. To explain his theory, Bohr used **NCERT | Page-299 | 1989**

(a) conservation of linear momentum
(b) conservation of angular momentum
(c) conservation of quantum frequency
(d) conservation of energy

52. The ionisation energy of hydrogen atom is 13.6 eV, the ionisation energy of helium atom would be **NCERT | Page-299 | 1988**

(a) 13.6 eV (b) 27.2 eV
(c) 6.8 eV (d) 54.4 eV

Topic-12.5 The Line Spectra of the Hydrogen Atom

53. The spectral series which corresponds to the electronic transition from the levels $n_2 = 5, 6, \ldots$ to the level $n_1 = 4$ is **Ph-II 2024**

(a) Pfund series (b) Brackett series
(c) Lyman series (d) Balmer series

54. The wavelength of Lyman series of hydrogen atom appears in **Manipur 2023**

(a) Ultraviolet region
(b) Infrared region
(c) Visible region
(d) Far infrared region

55. In hydrogen spectrum, the shortest wavelength in the Balmer series is λ. The shortest wavelength in the Bracket series is **2023**

(a) 2λ (b) 4λ (c) 9λ (d) 16λ

56. The ratio of wavelengths of the last line of Balmer series and the last line of Lyman series is :- **2017**

(a) 1 (b) 4
(c) 0.5 (d) 2

57. Given the value of Rydberg constant is $10^7 m^{-1}$, the wave number of the last line of the Balmer series in hydrogen spectrum will be **2016**

(a) $0.025 \times 10^4\ m^{-1}$ (b) $0.5 \times 10^7\ m^{-1}$
(c) $0.25 \times 10^7\ m^{-1}$ (d) $2.5 \times 10^7\ m^{-1}$

58. In the spectrum of hydrogen, the ratio of the longest wavelength in the Lyman series to the longest wavelength in the Balmer series is **2015 RS, 2013**

(a) $\frac{9}{4}$ (b) $\frac{27}{5}$ (c) $\frac{5}{27}$ (d) $\frac{4}{9}$

59. Hydrogen atom in ground state is excited by a monochromatic radiation of $\lambda = 975$ Å. Number of spectral lines in the resulting spectrum emitted will be **2014**

(a) 3 (b) 2
(c) 6 (d) 10

60. When an electron jumps from the fourth orbit to the second orbit, one gets the **2000**
(a) second line of Lyman series
(b) second line of Paschen series
(c) second line of Balmer series
(d) first line of Pfund series

Topic-12.6 De Broglie's Explanation of Bohr's Second Postulate of Quantisation

61. De-Broglie wavelength of an electron orbiting in the $n = 2$ state of hydrogen atom is close to (Given Bohr radius = 0.052 nm) **NCERT | Page-302 | 2025**
(a) 0.067 nm (b) 0.67 nm
(c) 1.67 nm (d) 2.67 nm

ANSWER KEY																	
1	(b)	**8**	(c)	**15**	(d)	**22**	(b)	**29**	(b)	**36**	(a)	**43**	(c)	**50**	(a)	**57**	(c)
2	(a)	**9**	(b)	**16**	(b)	**23**	(a)	**30**	(b)	**37**	(b)	**44**	(a)	**51**	(b)	**58**	(c)
3	(a)	**10**	(b)	**17**	(c)	**24**	(a)	**31**	(a)	**38**	(a)	**45**	(a)	**52**	(d)	**59**	(c)
4	(c)	**11**	(c)	**18**	(c)	**25**	(d)	**32**	(c)	**39**	(b)	**46**	(b)	**53**	(b)	**60**	(c)
5	(a)	**12**	(d)	**19**	(c)	**26**	(a)	**33**	(c)	**40**	(b)	**47**	(b)	**54**	(a)	**61**	(b)
6	(d)	**13**	(b)	**20**	(d)	**27**	(d)	**34**	(d)	**41**	(a)	**48**	(a)	**55**	(b)		
7	(c)	**14**	(a)	**21**	(b)	**28**	(d)	**35**	(a)	**42**	(a)	**49**	(d)	**56**	(b)		

Hints & Solutions

1. **(b)** Cathode rays are streams of negatively charged ions.

2. **(a)** Millikan's experiment established that electric charge is quantised. From the values of charge (e) and specific charge $\left(\frac{e}{m}\right)$, the mass of the electron (m) could be determined.

3. **(a)** At closest distance of approach, the kinetic energy of the particle will convert completely into electrostatic potential energy.

Kinetic energy K.E. $= \frac{1}{2}mv^2$

Potential energy P.E. $= \frac{KQq}{r}$

$$\frac{1}{2}mv^2 = \frac{KQq}{r} \Rightarrow r \propto \frac{1}{m}$$

4. **(c)** Kinetic energy of alpha nucleus is equall to electrostatic potential energy of the system of the alpha particle and the heavy nucleus. That is,

$$\frac{1}{2}mv^2 = \frac{1}{4\pi\varepsilon_0}\frac{q_\alpha Ze}{r_0}$$

where r_0 is the distance of closest approach

$$r_0 = \frac{2}{4\pi\varepsilon_0}\frac{q_\alpha Ze}{mv^2}$$

$$\Rightarrow r_0 \propto Ze \propto q_\alpha \propto \frac{1}{m} \propto \frac{1}{v^2}$$

Hence, correct option is (c).

5. **(a)** The kinetic energy of the projectile is given by

$$\frac{1}{2}mv^2 = \frac{Ze(2e)}{4\pi\varepsilon_0 r_0} = \frac{Z_1Z_2}{4\pi\varepsilon_0 r_0}$$

Thus energy of the projectile is directly proportional to Z_1Z_2.

6. **(d)** Impact parameter for Rutherford scattering experiment,

$$b = \frac{Ze^2 \cot\left(\frac{\theta}{2}\right)}{4\pi \epsilon_0 k_i} = 0 \Rightarrow \cot\left(\frac{\theta}{2}\right) = 0$$

$$\Rightarrow \frac{\theta}{2} = 90° \text{ or } \theta = 180°$$

7. **(c)** Statement I is correct as atoms are electrically neutral as they contain equal number of positive and negative charges.

Statement II is incorrect as atom of most of the elements but not every atom are stable and emit characteristic spectrum. This statement is not true for every atom.

8. **(c)** Spectrum obtained from a sodium vapour lamp is emission spectrum.

NOTES A spectrum is observed, when light coming directly from a source is examined with a spectroscope.

9. **(b)** Neon street sign is a source of line emission spectrum.

10. **(b)** Magnetic moment, $M = IA = I(\pi r^2)$

$$\therefore M = \left(\frac{ev}{2\pi r}\right)(\pi r^2) \quad ...(i)$$

Given, flux = B. (πr^2) = $n\left(\frac{h}{e}\right)$

$\Rightarrow r^2 = \frac{h}{B\pi e}$ ...(ii)

($\because n = 1$)

And when charge moving in external magnetic field then $r = \frac{mv}{qB}$

$\Rightarrow \frac{v}{r} = \frac{eB}{m}$ ...(iii)

($\because q = e$)

From equation (i), (ii) and (iii),

$$M = \left(\frac{ev}{2\pi r}\right)(\pi r^2)$$

$$= \frac{e}{2\pi}\left(\frac{eB}{m}\right)\pi\left(\frac{h}{B\pi e}\right) = \frac{eh}{2\pi m}$$

11. **(c)** $\frac{mv^2}{r} = F$ (centripetal force)

$\Rightarrow v^2 = \frac{Fr}{m}$ ...(i)

Bohr's quantization condition

$mvr = \frac{nh}{2\pi} \Rightarrow v = \frac{nh}{2\pi mr}$ ...(ii)

From equation (i),

$$\frac{m}{r}\left(\frac{nh}{2\pi mr}\right)^2 = F$$

$$\Rightarrow \frac{n^2h^2}{4\pi^2 mF} = r^3 \Rightarrow r^3 \propto n^2$$

$$\Rightarrow r \propto n^{\frac{2}{3}}$$

From equation (ii)

$$v \propto \frac{n}{r} \propto \frac{n}{n^{\frac{2}{3}}} \Rightarrow v \propto n^{\frac{1}{3}}$$

12. **(d)** From Einstein's photoelectric equation,

$h\frac{c}{\lambda_1} = \frac{-5E}{2} + 4E = \frac{3}{2}E$ ..(1)

$h\frac{c}{\lambda_2} = -2E + 3E = E$ ...(2)

$h\frac{c}{\lambda_3} = -2E + 4E = 2E$ ...(3)

Comparing (2) and (3)

$\frac{1}{\lambda_3} = \frac{2}{\lambda_2} \Rightarrow \lambda_2 = 2\lambda_3$

Comparing (1) and (2)

$3\lambda_1 = 2\lambda_2 \Rightarrow \lambda_1 < \lambda_2$

13. **(b)** Energy of transition, $\Delta E = \frac{hc}{\lambda}$

$\therefore \lambda \propto \frac{1}{\Delta E}$

As, $(\Delta E)_{6-2} > (\Delta E)_{5-2} > (\Delta E)_{4-2} > (\Delta E)_{3-2}$

$\Rightarrow \lambda_{6-2} < \lambda_{5-2} < \lambda_{4-2} < \lambda_{3-2}$

A – III, B – IV, C – II, D – I

14. **(a)** Total energy of electron in hydrogen atom

$$E_\eta = \frac{-13.6}{n^2}$$

In second excited state $= \frac{-13.6}{(3)^2} = -1.51\text{eV}$

Therefore 1.51 eV energy required to ionize hydrogen atom from its second excited state.

15. **(d)** Given,

Radius of inner most orbit of hydrogen atom,

$r_1 = 5.3 \times 10^{-11}$ m

Radius of nth orbit $r_n = \frac{n^2}{Z} \Rightarrow r_n \propto n^2$

$$\therefore \frac{r_1}{r_2} = \left(\frac{n_1}{n_2}\right)^2 \Rightarrow \frac{r_1}{r_2} = \left(\frac{1}{3}\right)^2$$

$\Rightarrow r_2 = 9r_1 = 5.3 \times 10^{-11} \times 9$

$= 47.7 \times 10^{-11} = 4.77$ Å

16. **(b)** Radius, $R = 0.529 \times \frac{n^2}{z}\overset{0}{A}$

$$\therefore \frac{R_1}{R_2} = \frac{2^2}{4^2} = \frac{1}{4} = 0.25$$

17. **(c)** We have, $n_1 = 2$ and $n_2 = 3$

and, $T_n = \frac{-13.6}{n^2}$ So, $\frac{T_1}{T_2} = \left(\frac{n_2}{n_1}\right)^2 = \left(\frac{3}{2}\right)^2 = \frac{9}{4}$

18. **(c)** Bohr model is only valid for single electron species i.e., Hydrogen or hydrogen like atom – He^+, deuteron, etc. Singly ionised neon atom has more than one electron in orbit. Hence, Bohr model is not valid.

19. **(c)** According to Bohr's model of H-atom, the relation between kinetic energy, potential energy and total energy

$$K.E. = |TE| = \frac{|P.E.|}{2}$$

so, K.E. = – (T.E.) $\therefore$ K.E. = 3.4 eV

and P.E. = 2 (T.E.) $\Rightarrow$ P.E. = – 6.8 eV

20. **(d)** $r \propto \frac{1}{m}$

$R = \frac{0.51}{207} \quad \therefore\ 2.56 \times 10^{-13}$ m

$E \propto m$

$\therefore \quad (E)_\mu = -13.6 \times 207 = -2.8$ keV

21. **(b)** In a Bohr orbit of the hydrogen atom

Kinetic energy,

$k = \frac{kze^2}{2r_n}$

Total energy, $E = \frac{-kze^2}{2r_n}$

So, Kinetic energy : total energy = 1 : –1

22. **(b)** By law of conservation of energy,

$K.E_f = K.E_i$ – excitation energy (ε)

or $\frac{1}{2}mv_1^2 + \frac{1}{2}mv_2^2 = \frac{1}{2}m_1u_1^2 + \frac{1}{2}m_2u_2^2 - \varepsilon$

23. **(a)** Speed of electron in nth orbit

$v_n = \frac{2\pi\, KZe^2}{nh}$

$v = (2.19 \times 10^6 \text{ m/s}) \frac{Z}{n}$

$v = (2.19 \times 10^6) \frac{2}{3}$ (Z = 2 and n = 3)

$v = 1.46 \times 10^6$ m/s

24. **(a)** $\because T \propto n^3$

$\frac{T_1}{T_2} = \frac{8T_2}{T_2} = \left(\frac{n_1}{n_2}\right)^3$

Hence, $n_1 = 2n_2$

25. **(d)**

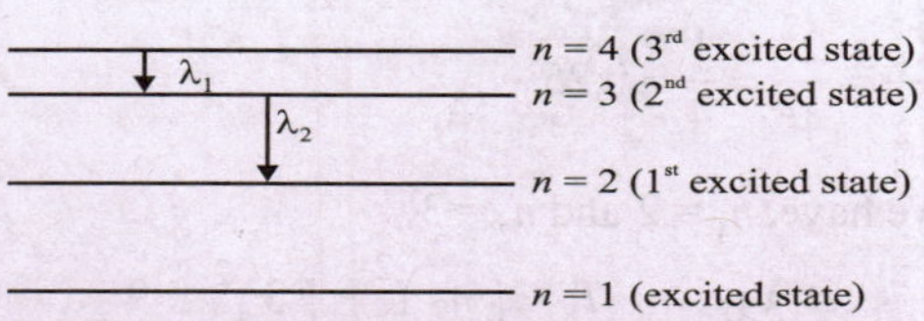

According to Rydberg formula

$\frac{1}{\lambda} = R\left[\frac{1}{n_f^2} - \frac{1}{n_i^2}\right]$

In first case, $n_f = 3$, $n_i = 4$

$\therefore \quad \frac{1}{\lambda_1} = R\left[\frac{1}{3^2} - \frac{1}{4^2}\right] = R\left[\frac{1}{9} - \frac{1}{16}\right] = \frac{7}{144}R$ ...(i)

In second case, $n_f = 2$, $n_i = 3$

$\therefore \quad \frac{1}{\lambda_2} = R\left[\frac{1}{2^2} - \frac{1}{3^2}\right] = R\left[\frac{1}{4} - \frac{1}{9}\right] = \frac{5}{36}R$ ...(ii)

Divide (ii) by (i), we get

$\frac{\lambda_1}{\lambda_2} = \frac{5}{36} \times \frac{144}{7} = \frac{20}{7}$

26. **(a)** For emission, the wave number of the radiation is given as

$\frac{1}{\lambda} = Rz^2\left(\frac{1}{n_1^2} - \frac{1}{n_2^2}\right)$

R = Rydberg constant, Z = atomic number

$= R\left(\frac{1}{1^2} - \frac{1}{5^2}\right) = R\left(1 - \frac{1}{25}\right) \Rightarrow \frac{1}{\lambda} = R\frac{24}{25}$

linear momentum

$p = \frac{h}{\lambda} = h \times R \times \frac{24}{25}$ (de-Broglie hypothesis)

$\Rightarrow mv = \frac{24hR}{25} \quad \Rightarrow v = \frac{24hR}{25m}$

27. **(d)** $\because$ The frequency of the transition $\nu \propto \frac{1}{n^2}$, when n = 1, 2, 3.

28. **(d)** For first line of Lyman series of hydrogen

$\frac{hc}{\lambda_1} = Rhc\left(\frac{1}{1^2} - \frac{1}{2^2}\right)$

For second line of Balmer series of hydrogen like ion

$\frac{hc}{\lambda_2} = Z^2Rhc\left(\frac{1}{2^2} - \frac{1}{4^2}\right)$

By question, $\lambda_1 = \lambda_2$

$\Rightarrow \left(\frac{1}{1} - \frac{1}{2}\right) = Z^2\left(\frac{1}{4} - \frac{1}{16}\right)$ or Z = 2

29. **(b)** KE_{max} = 10eV

ϕ = 2.75 eV

Total incident energy

$E = \phi + KE_{max}$ = 12.75 eV

$\therefore$ Energy is released when electron jumps from the excited state n to the ground state.

$\because E_4 - E_1 = \{-0.85 - (-13.6)$ ev$\}$

= 12.75eV

$\therefore$ value of n = 4

30. **(b)** Obviously, difference of 11.1eV is not possible.

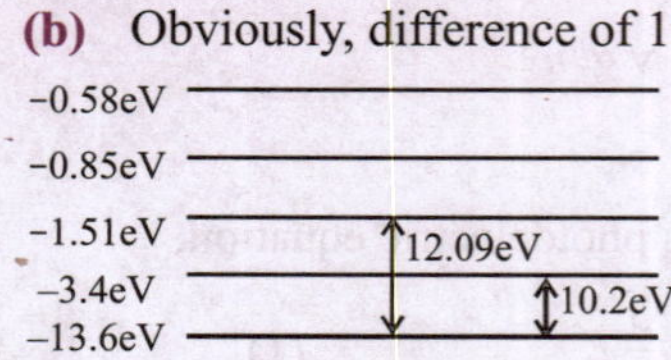

31. **(a)** Energy of a H-like atom in it's nth state is given by

$E_n = -Z^2 \times \frac{13.6}{n^2} eV$

For first excited state of He^+, n = 2, Z = 2

$\therefore \quad E_{He^+} = -\frac{4}{2^2} \times 13.6 = -13.6\,eV$

32. (c) $\frac{n(n-1)}{2}=6$

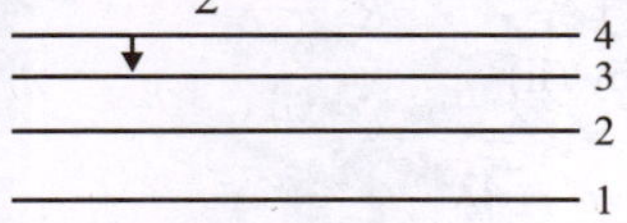

$n^2 - n - 12 = 0$

$(n-4)(n+3) = 0$ or $n = 4$

NOTES If electron falls from orbit n_2 to n_1 then the number of spectral lines emitted is given by

$$N_E = \frac{(n_2 - n_1 + 1)(n_2 - n_1)}{2}$$

If an electron falls from n^{th} orbit to ground state ($n = 1$) then number of spectral lines emitted, $N_E = \frac{n(n-1)}{2}$

33. (c) When the electron is in first excited state ($n = 2$), the excitation energy is given by

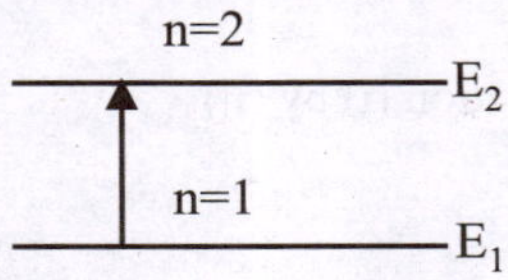

$\Delta E = E_2 - E_1$

We have, $E_n = -\frac{13.6}{n^2}\text{eV}$

$\therefore\ E_2 = -\frac{13.6}{2^2}\text{eV} = -3.4\text{eV}$

Given $E_1 = -13.6\text{eV}$

$\therefore\ \Delta E = (-3.4) - (-13.6) = 10.2$ eV.

34. (d) Energy in the first excited state

$= \frac{-13.6}{n^2} = \frac{-13.6}{2^2} = -3.4\text{eV}$

But K.E. = –(Total energy) = +3.4 eV.

35. (a) Energy of ground state 13.6 eV

Energy of first excited state

$= -\frac{13.6}{4} = -3.4\text{ eV}$

Energy of second excited state

$= -\frac{13.6}{9} = -1.5\text{ eV}$

Difference between ground state and 2nd excited state = 13.6 – 1.5 = 12.1 eV

So, electron can be excited upto 3rd orbit

No. of possible transition

$1 \to 2,\ 1 \to 3,\ 2 \to 3$

So, three lines are possible.

36. (a) $\text{K.E.} = \frac{Z^2}{n^2}(13.6\text{ eV})$

Mechanical energy $= \frac{-Z^2}{n^2}(13.6\text{ eV})$

$\therefore$ K.E. in 2nd orbital for hydrogen

= – Mechanical energy

$= \frac{(1)^2}{(2)^2}(13.6) = +3.4eV$

37. (b) In Bohr's model, angular momentum is quantised

i.e. $\ell = n\left(\frac{h}{2\pi}\right)$

38. (a) $\Delta E = E_3 - E_2 = \frac{-13.6}{3^2} + \frac{13.6}{2^2}$

$= 13.6\left\{\frac{1}{4} - \frac{1}{9}\right\}\text{eV} = 1.9\text{ eV}$

39. (b) $r \propto \frac{1}{Z}$; $Z(=3)$ is maximum for Li^{2+}.

40. (b) $\bar{\nu} = R\left(\frac{1}{n_1^2} - \frac{1}{n_2^2}\right)$, where $n_1 = 2, n_2 = 4$

$\bar{\nu} = R\left(\frac{1}{4} - \frac{1}{16}\right)$

$\frac{1}{\lambda} = R\left(\frac{12}{4 \times 16}\right) \Rightarrow \lambda = \frac{16}{3R}$

41. (a) We have $E_n = \frac{-2\pi^2 mK^2 Z^2 e^4}{n^2 h^2}$. For helium $Z = 2$.

Hence requisite answer is $4E_n$

NOTES The energy required to ionise an atom is called ionisation energy. It is energy required to make the electron Jump from the present orbit to the infinite orbit.

$$E_{\text{ionisation}} = E_\infty - E_n = 0 - \left(-13.6\frac{Z^2}{n^2}\right) = +\frac{13.6z^2\text{ eV}}{n^2}$$

42. (a) $\bar{\nu} = R\left[\frac{1}{n_1^2} - \frac{1}{n_2^2}\right]$ or $\frac{1}{\lambda} = R\left[\frac{1}{n_1^2} - \frac{1}{n_2^2}\right]$

Frequency, $\nu = Rc\left[\frac{1}{n_1^2} - \frac{1}{n_2^2}\right]$

NOTES See the greatest energy difference and also see that the transition is from higher to lower energy level. Hence, it is highest in case of $n = 2$ to $n = 1$.

43. (c) Centripetal force = Coulombian force

$\frac{mv^2}{a_0} = \frac{1}{4\pi\varepsilon_0} \cdot \frac{e \times e}{a_0^2}$

$\Rightarrow v^2 = \frac{e^2}{4\pi\varepsilon_0 . a_0 . m} \Rightarrow v = \frac{e}{\sqrt{4\pi\varepsilon_0 . a_0 . m}}$

44. (a) 53 electrons in iodine atom are distributed as 2, 8, 18, 18, 7 $\therefore n = 5$

$$r_n = (0.53\times10^{-10})\frac{n^2}{Z}$$

$$= \frac{0.53\times10^{-10}\times5^2}{53} = 2.5\times10^{-11}\text{m}$$

45. (a) $R = \frac{R_0 n^2}{Z}$

Radius in ground state $= \frac{R_0}{Z}$

Radius in first excited state $= \frac{R_0\times4}{Z}$ $(\because n=2)$

Hence, radius of first excited state is four times the radius in ground state.

46. (b) $P.E. = -\frac{KZe^2}{r}$ and $K.E. = \frac{KZe^2}{2r}$,

where, r is the radius of orbit which increases as we move from ground to an excited state. Therefore, when a hydrogen atom is raised from the ground state, it increases the value of r. As a result of this, P.E. increases (decreases in negative) and K.E. decreases.

47. (b) $r \propto n^2$

$$\therefore \frac{\text{radius of final state}}{\text{radius of initial state}} = n^2$$

$$\frac{21.2\times10^{-11}}{5.3\times10^{-11}} = n^2 \therefore n^2 = 4 \text{ or } n = 2$$

48. (a) As $r \propto n^2$, therefore, radius of 2nd Bohr's orbit $= 4\,a_0$.

49. (d) $E = E_4 - E_3$

$$= -\frac{13.6}{4^2} - \left(-\frac{13.6}{3^2}\right) = -0.85 + 1.51 = 0.66\text{ eV}$$

50. (a) Second excited state corresponds to $n = 3$

$$\therefore E = \frac{13.6}{3^2}\text{eV} = 1.51\text{eV}$$

51. (b) Bohr used conservation of angular momentum.

52. (d) $E \propto Z^2$ and Z for helium = 2

$\therefore (E)_{He} = 4\times13.6 = 54.4$ eV

53. (b) The spectral series corresponding to the electronic transition from the levels $n_2 = 5, 6,...$ to $n_1 = 4$ is Brackett series.

54. (a) Lyman series of hydrogen atom appears in Ultraviolet region. Balmer series in visible region and Paschen, Bracket and P – fund series in infrared region.

55. (b) Using $\frac{1}{\lambda} = R\left[\frac{1}{n_2^2} - \frac{1}{n_1^2}\right]$

For Balmer series, $[n_2 = 2, n_1 = \infty]$

$$\frac{1}{\lambda} = R\left[\frac{1}{4} - \frac{1}{\infty}\right] \Rightarrow \lambda = \frac{4}{R} \quad \text{... (i)}$$

And for bracket series, $(n_2 = 4, n_1 = \infty)$

$$\frac{1}{\lambda'} = R\left[\frac{1}{16} - \frac{1}{\infty}\right] \Rightarrow \lambda' = \frac{16}{R} \quad \text{... (ii)}$$

Dividing eq (i) by (ii)

$$\frac{\lambda}{\lambda'} = \frac{4/R}{16/R} \quad \therefore \lambda' = 4\lambda$$

56. (b) For last line of Balmer series :

$n_1 = 2$ and $n_2 = \infty$

$$\frac{1}{\lambda_B} = RZ^2\left[\frac{1}{n_1^2} - \frac{1}{n_2^2}\right] = R_1^2\left[\frac{1}{2^2} - \frac{1}{\infty^2}\right]$$

$$\lambda_B = \frac{4}{R} \quad \text{...(i)}$$

For last line of Lyman series : $n_1 = 1$ and $n_2 = \infty$

$$\frac{1}{\lambda_L} = RZ^2\left[\frac{1}{n_1^2} - \frac{1}{n_2^2}\right] = RZ^2\left[\frac{1}{1^2} - \frac{1}{\infty^2}\right]$$

$$\lambda_L = \frac{1}{R} \quad \text{...(ii)}$$

Dividing equation (i) by (ii)

$$\frac{\lambda_B}{\lambda_L} = \frac{\frac{4}{R}}{\frac{1}{R}}$$

Ratio of wavelengths is $\frac{\lambda_B}{\lambda_L} = 4$

57. (c) According to Bohr's theory, the wave number of the last line of the Balmer series in hydrogen spectrum, For hydrogen atom Z = 1

$$\frac{1}{\lambda} = RZ^2\left(\frac{1}{n_2^2} - \frac{1}{n_1^2}\right)$$

$$= 10^7 \times 1^2\left(\frac{1}{2^2} - \frac{1}{\infty^2}\right)$$

$\Rightarrow$ wave number $\frac{1}{\lambda} = 0.25\times10^7\text{ m}^{-1}$

NOTES Last line of the series is called series limit. For this line wavelength is minimum (λ_{min}).

For minimum wavelength, $n_2 = \infty$, $n_1 = n$

So, wavelength $\lambda_{min} = \frac{n^2}{R}$

58. (c) For Lyman series $(2 \to 1)$

$$\frac{1}{\lambda_L} = R\left[1 - \frac{1}{2^2}\right] = \frac{3R}{4}$$

For Balmer series $(3 \to 2)$

$$\frac{1}{\lambda_B} = R\left[\frac{1}{4} - \frac{1}{9}\right] = \frac{5R}{36}$$

$$\Rightarrow \frac{\lambda_L}{\lambda_B} = \frac{\frac{4}{3R}}{\frac{36}{5R}} = \frac{4}{36}\left(\frac{5}{3}\right) = \frac{5}{27}$$

NOTES The wavelength of spectral lines increases with the increases of order of the series.

$\lambda_{PFund} > \lambda_{Brackett} > \lambda_{Paschen} > \lambda_{Balmer} > \lambda_{Lymen}$

59. **(c)** For the $\lambda = 975$ Å

$$\frac{1}{\lambda} = R\left(\frac{1}{n_1^2} - \frac{1}{n_2^2}\right)$$

where R is the Rydberg constant

Solving we get $n_2 = n = 4$

($\because n_1 = 1$ ground state)

Therefore number of spectral lines

$$= \frac{n(n-1)}{2} = \frac{4(4-1)}{2} = 6$$

60. **(c)** When the electron drops from any orbit to second orbit, then wavelength of line obtained belongs to Balmer series.

61. **(b)** Given $n = 2$, $Z = 1$ for hydrogen atom

$$2\pi r = n\lambda$$

$$2\pi \times \left(0.052\frac{n^2}{Z}\right) = n\lambda \quad \therefore r = 0.052\frac{h^2}{z}$$

$$\lambda = 0.67\ nm$$

13 Nuclei

TREND ANALYSIS

TOPIC	2025		2024		2023		2022		2021		2020	
	Qns.	LOD	Qns.	LOD	Qns.	LOD	Qns.	LOD	Qns.	LOD	Qns.	LOD
Size of the Nucleus							1	Avg.				
Mass-Energy and Nuclear Binding Energy									1	Avg.	1	Avg.
Nuclear Energy			1	Diffi.							1	Avg.

How to Code Qns for Better & Faster Revision

Type 1: Able to solve (No code Req)
Type 2: Solved with struggle or by seeing solution **(Use code NR means Needs Revision or orange stickers)**
Type 3: Tricky/ Could not solve on my own. Need help from Teacher/ Friend. **(Use code TQ means Tricky Question or red stickers)**

You can either hand write the code in boxes or else paste coloured stickers given at end of book.

Topic-13.2
Atomic Masses and Composition of Nucleus

1. A nucleus represented by the symbol $^{A}_{Z}X$ has NCERT | Page-308 | 2004
(a) A protons and $(Z\text{-}A)$ neutrons
(b) Z neutrons and $(A\text{-}Z)$ protons
(c) Z protons and $(A\text{-}Z)$ neutrons
(d) Z protons and A neutrons

2. The mass number of a nucleus is NCERT | Page-308 | 2003
(a) sometimes less than and sometimes more than its atomic number
(b) always less than its atomic number
(c) always more than its atomic number
(d) sometimes equal to its atomic number

3. The volume occupied by an atom is greater than the volume of the nucleus by a factor of about NCERT | Page-307 | 2003
(a) 10^{15} (b) 10^{1} (c) 10^{5} (d) 10^{10}

4. The mass number of a nucleus is equal to the number of NCERT | Page-308 | 1995
(a) protons it contains
(b) nucleons it contains
(c) neutrons it contains
(d) electron it contains

5. The constituents of atomic nuclei are believed to be NCERT | Page-308 | 1991
(a) neutrons and protons
(b) protons only
(c) electrons and protons
(d) electrons, protons and neutrons

6. In the nucleus of $_{11}Na^{23}$, the number of protons, neutrons and electrons are NCERT | Page-308 | 1991
(a) 11, 12, 0 (b) 23, 12, 11
(c) 12, 11, 0 (d) 23, 11, 12

Topic-13.3
Size of the Nucleus

7. A nucleus of mass number 189 splits into two nuclei having mass number 125 and 64. The ratio of radius of two daughter nuclei respectively is: NCERT | Page-309 | 2022
(a) 4 : 5 (b) 5 : 4
(c) 25 : 16 (d) 1 : 1

8. If radius of the $^{27}_{12}Al$ nucleus is taken to be R_{Al}, then the radius of $^{125}_{53}Te$ nucleus is nearly: NCERT | Page-309 | 2015
(a) $\frac{5}{3}R_{Al}$ (b) $\frac{3}{5}R_{Al}$
(c) $\left(\frac{13}{53}\right)^{1/3} R_{Al}$ (d) $\left(\frac{53}{13}\right)^{1/3} R_{Al}$

9. If the nuclear radius of ^{27}Al is 3.6 Fermi, the approximate nuclear radius of ^{64}Cu in Fermi is: NCERT | Page-309 | 2012
(a) 2.4 (b) 1.2
(c) 4.8 (d) 3.6

10. Two nuclei have their mass numbers in the ratio of 1 : 3. The ratio of their nuclear densities would be NCERT | Page-309 | 2008
(a) 1 : 3 (b) 3 : 1
(c) $(3)^{1/3} : 1$ (d) 1 : 1

11. If the nucleus $^{27}_{13}Al$ has nuclear radius of about 3.6 fm, then $^{125}_{32}Te$ would have its radius approximately as NCERT | Page-309 | 2007

(a) 9.6 fm (b) 12.0 fm
(c) 4.8 fm (d) 6.0 fm

12. The radius of germanium (Ge) nuclide is measured to be twice the radius of $^{9}_{4}Be$. The number of nucleons in Ge are NCERT | Page-309 | 2006

(a) 74 (b) 75 (c) 72 (d) 73

13. The stable nucleus that has a radius half that of Fe^{56} is NCERT | Page-309 | 1997

(a) Li^7 (b) Na^{21}
(c) S^{16} (d) Ca^{40}

14. A nucleus ruptures into two nuclear parts, which have their velocity ratio equal to 2:1. What will be the ratio of their nuclear size (nuclear radius)? NCERT | Page-309 | 1996

(a) $2^{1/3}:1$ (b) $1:2^{1/3}$
(c) $3^{1/2}:1$ (d) $1:3^{1/2}$

15. The mass number of He is 4 and that for sulphur is 32. The radius of sulphur nuclei is larger than that of helium by NCERT | Page-309 | 1994

(a) $\sqrt{8}$ (b) 4 (c) 2 (d) 8

16. The nucleus which has radius one-third of the radius of Cs^{189} is NCERT | Page-309 | 1993

(a) $^{4}_{2}He$ (b) $^{7}_{3}Li$
(c) $^{141}_{56}Ba$ (d) $^{235}_{92}U$

17. The mass density of a nucleus varies with mass number A as NCERT | Page-309 | 1992

(a) A^2 (b) A
(c) constant (d) 1/A

18. The ratio of the radii of the nuclei $_{13}Al^{27}$ and $_{52}Te^{125}$ is approximately NCERT | Page-309 | 1990

(a) 6 : 10 (b) 13 : 52
(c) 40 : 177 (d) 14 : 73

Topic-13.4
Mass-Energy and Nuclear Binding Energy

19. A nucleus with mass number 240 breaks into two fragments each of mass number 120, the binding energy per nucleon of unfragmented nuclei is 7.6 MeV while that of fragments is 8.5 MeV. The total gain in the binding energy in the process is NCERT | Page-311 | 2021

(a) 216 MeV (b) 0.9 MeV
(c) 9.4 MeV (d) 804 MeV

20. The energy equivalent of 0.5 g of a substance is : NCERT | Page-310 | 2020

(a) 4.5×10^{13} J
(b) 1.5×10^{13} J
(c) 0.5×10^{13} J
(d) 4.5×10^{16} J

21. The Binding energy per nucleon of $^{7}_{3}Li$ and $^{4}_{2}He$ nuclei are 5.60 MeV and 7.06 MeV, respectively.

In the nuclear reaction $^{7}_{3}Li + ^{1}_{1}H \rightarrow ^{4}_{2}He + Q$, the value of energy Q released is : NCERT | Page-311 | 2014

(a) 19.6 MeV (b) – 2.4 MeV
(c) 8.4 MeV (d) 17.3 MeV

22. A certain mass of Hydrogen is changed to Helium by the process of fusion. The mass defect in fusion reaction is 0.02866 a.m.u. The energy liberated per a.m.u. is

(Given : 1 a.m.u = 931 MeV) NCERT | Page-311 | 2013

(a) 26.7 MeV (b) 6.675 MeV
(c) 13.35 MeV (d) 2.67 MeV

23. How does the binding energy per nucleon vary with the increase in the number of nucleons? NCERT | Page-311 | Kar. 2013

(a) Increases continuously with mass number
(b) Decreases continuously with mass number
(c) First decreases and then increases with increase in mass number
(d) First increases and then decreases with increase in mass number

24. The power obtained in a reactor using U^{235} disintegration is 1000 kW. The mass decay of U^{235} per hour is 2011

(a) 10 microgram (b) 20 microgram
(c) 40 microgram (d) 1 microgram

25. The mass of a $^{7}_{3}Li$ nucleus is 0.042 u less than the sum of the masses of all its nucleons. The binding energy per nucleon of $^{7}_{3}Li$ nucleus is nearly NCERT | Page-311 | 2010

(a) 46 MeV (b) 5.6 MeV
(c) 3.9 MeV (d) 23 MeV

26. The binding energy per nucleon in deuterium and helium nuclei are 1.1 MeV and 7.0 MeV, respectively. When two deuterium nuclei fuse to form a helium nucleus the energy released in the fusion is : NCERT | Page-311 | 2010

(a) 30.2 MeV (b) 23.6 MeV
(c) 2.2 MeV (d) 28.0 MeV

27. If M(A;Z), M_p and M_n denote the masses of the nucleus $^{A}_{Z}X$, proton and neutron respectively in units of u (1u = 931.5 MeV/c^2) and BE represents its binding energy in MeV, then NCERT | Page-311 | 2008

(a) $M(A, Z) = ZM_p + (A - Z)M_n - BE/c^2$
(b) $M(A, Z) = ZM_p + (A - Z)M_n + BE$
(c) $M(A, Z) = ZM_p + (A - Z)M_n - BE$
(d) $M(A, Z) = ZM_p + (A - Z)M_n + BE/c^2$

28. A nucleus $^{A}_{Z}X$ has mass represented by $M(A, Z)$. If M_p and M_n denote the mass of proton and neutron respectively and B.E. the binding energy in MeV, then NCERT | Page-311 | 2007

(a) B.E. = $[ZM_p + (A - Z)M_n - M(A, Z)]c^2$
(b) B.E. = $[ZM_p + ZM_n - M(A, Z)]c^2$
(c) B.E. = $M(A, Z) - ZM_p - (A - Z)M_n$
(d) B.E. = $[M(A, Z) - ZM_p - (A - Z)M_n]c^2$

29. The binding energy of deuteron is 2.2 MeV and that of ${}_2^4\text{He}$ is 28 MeV. If two deuterons are fused to form one ${}_2^4\text{He}$, then the energy released is **NCERT | Page-311 | 2006**
(a) 23.6 MeV (b) 19.2 MeV
(c) 30.2 MeV (d) 25.8 MeV

30. In the reaction, ${}_1^2\text{H} + {}_1^3\text{H} \longrightarrow {}_2^4\text{He} + {}_0^1\text{n}$, if the binding energies of ${}_1^2\text{H}$, ${}_1^3\text{H}$ and ${}_2^4\text{He}$ are respectively, *a*, *b* and *c* (in MeV), then the energy (in MeV) released in this reaction is **NCERT | Page-311 | 2005**
(a) $a + b + c$ (b) $a + b - c$
(c) $c - a - b$ (d) $c + a - b$

31. M_p denotes the mass of a proton and M_n that of a neutron. A given nucleus, of binding energy *B*, contains *Z* protons and *N* neutrons. The mass *M*(*N*, *Z*) of the nucleus is given by (*c* is the velocity of light) **NCERT | Page-311 | 2004**
(a) $M(N, Z) = NM_n + ZM_p + B/c^2$
(b) $M(N, Z) = NM_n + ZM_p - Bc^2$
(c) $M(N, Z) = NM_n + ZM_p + Bc^2$
(d) $M(N, Z) = NM_n + ZM_p - B/c^2$

32. The mass of proton is 1.0073 *u* and that of neutron is 1.0087 *u* (*u* = atomic mass unit). The binding energy of ${}_2^4\text{He}$ is **NCERT | Page-311 | 2003**
(a) 0.061 u (b) 0.0305 j
(c) 0.0305 erg (d) 28.4 MeV

33. M_n and M_p represent mass of neutron and proton respectively. If an element having atomic mass *M* has *N*-neutron and *Z*-proton, then the correct relation will be **NCERT | Page-311 | 2001**
(a) $M < [NM_n + ZM_p]$
(b) $M > [NM_n + ZM_p]$
(c) $M = [NM_n + ZM_p]$
(d) $M = N[M_n + M_p]$

34. If the binding energy per nucleon in ${}_3\text{Li}^7$ and ${}_2\text{He}^4$ nuclei are respectively 5.60 MeV and 7.06 MeV, then the energy of proton in the reaction ${}_3\text{Li}^7 + p \rightarrow 2\ {}_2\text{He}^4$ is **NCERT | Page-311 | 1994**
(a) 19.6 MeV (b) 2.4 MeV
(c) 8.4 MeV (d) 17.3 MeV

35. The energy equivalent of one atomic mass unit is **NCERT | Page-310 | 1992**
(a) 1.6×10^{-19} J
(b) 6.02×10^{23} J
(c) 931 MeV
(d) 9.31 MeV

36. The average binding energy of a nucleon inside an atomic nucleus is about **NCERT | Page-311 | 1989**
(a) 8 MeV (b) 8 eV
(c) 8 J (d) 8 ergs

Topic-13.5 Nuclear Force

37. If the nuclear force between two protons, two neutrons and between proton and neutron is denoted by F_{pp}, F_{nn} and F_{pn} respectively, then **NCERT | Page-313 | 1991**
(a) $F_{pp} \approx F_{nn} \approx F_{pn}$
(b) $F_{pp} \neq F_{nn}$ and $F_{pp} = F_{nn}$
(c) $F_{pp} = F_{nn} = F_{pn}$
(d) $F_{pp} \neq F_{nn} \neq F_{pn}$

38. Which of the following statements is true for nuclear forces? **NCERT | Page-313 | 1990**
(a) they obey the inverse square law of distance
(b) they obey the inverse third power law of distance
(c) they are short range forces
(d) they are equal in strength to electro-magnetic forces.

Topic-13.7 Nuclear Energy

39. Water is used as a coolant in a nuclear reactor because of its **Ph-II 2024**
(a) high thermal expansion coefficient
(b) high specific heat capacity
(c) low density
(d) low boiling point

40. Select the correct statements among the following :
A. Slow neutrons can cause fission in ${}_{92}^{235}U$ than fast neutrons.
B. α-rays are Helium nuclei.
C. β-rays are fast moving electrons or positrons.
D. γ-rays are electromagnetic radiations of wavelengths larger than X-rays.

Choose the **most appropriate** answer from the options given below : **NCERT | Page-315| Ph-II 2024**
(a) A, B and C only (b) A, B and D only
(c) A and B only (d) C and D only

41. ${}_{82}^{290}\text{X} \xrightarrow{\alpha} \text{Y} \xrightarrow{e^+} \text{Z} \xrightarrow{\beta^-} \text{P} \xrightarrow{e^-} \text{Q}$

In the nuclear emission stated above, the mass number and atomic number of the product Q respectively, are **NCERT | Page-314, 315 | 2024**
(a) 280, 81 (b) 286, 80
(c) 288, 82 (d) 286, 81

42. When a uranium isotope ${}_{92}^{235}\text{U}$ is bombarded with a neutron, it generates ${}_{36}^{89}\text{kr}$, three neutrons and : **NCERT | Page-315 | 2020**
(a) ${}_{40}^{91}\text{Zr}$ (b) ${}_{36}^{101}\text{Kr}$
(c) ${}_{36}^{103}\text{Kr}$ (d) ${}_{56}^{144}\text{Ba}$

43. In the given nuclear reaction, the element X is:
$^{22}_{11}\text{Na} \rightarrow \text{X} + e^{+} + \nu$
(a) $^{23}_{10}\text{Ne}$ (b) $^{22}_{10}\text{Ne}$
(c) $^{22}_{12}\text{Mg}$ (d) $^{23}_{11}\text{Na}$

44. Fusion reaction takes place at high temperature because NCERT | Page-315 | 2011
(a) nuclei break up at high temperature
(b) atoms get ionised at high temperature
(c) kinetic energy is high enough to overcome the coulomb repulsion between nuclei
(d) molecules break up at high temperature

45. In any fission process, the ratio $\dfrac{\text{mass of fission products}}{\text{mass of parent nucleus}}$ is
(a) equal to 1 NCERT | Page-315 | 2005
(b) greater than 1
(c) less than 1
(d) depends on the mass of the parent nucleus

46. Fission of nuclei is possible because the binding energy per nucleon in them NCERT | Page-315 | 2005
(a) increases with mass number at low mass numbers.
(b) decreases with mass number at low mass numbers.
(c) increases with mass number at high mass numbers.
(d) decreases with mass number at high mass numbers.

47. If in nuclear fusion process the masses of the fusing nuclei be m_1 and m_2 and the mass of the resultant nucleus be m_3, then NCERT | Page-315 | 2004
(a) $m_3 > (m_1 + m_2)$
(b) $m_3 = m_1 + m_2$
(c) $m_3 = |m_1 - m_2|$
(d) $m_3 < (m_1 + m_2)$

48. Solar energy is mainly caused due to 2003
(a) gravitational contraction
(b) burning of hydrogen in the oxygen
(c) fission of uranium present in the Sun
(d) fusion of protons during synthesis of heavier elements

49. For a nuclear fusion process, the suitable nuclei are NCERT | Page-315 | 2002
(a) any nuclei
(b) heavy nuclei
(c) light nuclei
(d) nuclei lying in the middle of the periodic table

50. It is possible to understand nuclear fission on the basis of the NCERT | Page-315 | 2000
(a) liquid drop model of the nucleus
(b) meson theory of the nuclear forces
(c) proton-proton cycle
(d) independent particle model of the nucleus

51. In nuclear reactions, we have the conservation of 2000
(a) mass only
(b) energy only
(c) momentum only
(d) mass, energy and momentum

52. The explosion of the atomic bomb takes place due to
(a) nuclear fission NCERT | Page-315 | 1999
(b) nuclear fusion
(c) scattering
(d) thermionic emission

53. Complete the equation for the following fission process :
$_{92}U^{235} + {_0}n^1 \rightarrow {_{38}}Sr^{90} + ...$ NCERT | Page-315 | 1998
(a) $_{54}X^{143} + 3\,{_0}n^1$ (b) $_{54}X^{145} + 3\,{_0}n^1$
(c) $_{57}X^{142} + 3\,{_0}n^1$ (d) $_{54}X^{142} + {_0}n^1$

54. In a fission reaction
$^{236}_{92}U \rightarrow {^{117}}X + {^{117}}Y + n + n$
the binding energy per nucleon of X and Y is 8.5 MeV whereas of ^{236}U is 7.6 MeV. The total energy liberated will be about NCERT | Page-311, 315 | 1997
(a) 2000 MeV (b) 200 MeV
(c) 2 MeV (d) 200 keV

55. Which of the following is used as a moderator in nuclear reactors? 1997
(a) Plutonium (b) Cadmium
(c) Heavy water (d) Uranium

56. Heavy water is used as a moderator in a nuclear reactor. The function of the moderator is 1994
(a) to control energy released in the reactor
(b) to absorb neutrons and stop chain reaction
(c) to cool the reactor
(d) to slow down the neutrons to thermal energies.

57. Energy released in the fission of a single $^{235}_{92}\text{U}$ nucleus is 200 MeV. The fission rate of a $^{235}_{92}\text{U}$ filled reactor operating at a power level of 5 W is NCERT | N-315 | 1993
(a) $1.56 \times 10^{-10}\,s^{-1}$ (b) $1.56 \times 10^{11}\,s^{-1}$
(c) $1.56 \times 10^{-16}\,s^{-1}$ (d) $1.56 \times 10^{-17}\,s^{-1}$

58. Solar energy is due to NCERT | Page-315 | 1992
(a) fusion reaction
(b) fission reaction
(c) combustion reaction
(d) chemical reaction

ANSWER KEY

1	(c)	7	(b)	13	(a)	19	(a)	25	(b)	31	(d)	37	(c)	43	(b)	49	(c)	55	(c)
2	(d)	8	(a)	14	(b)	20	(a)	26	(b)	32	(d)	38	(c)	44	(c)	50	(a)	56	(d)
3	(a)	9	(c)	15	(c)	21	(d)	27	(a)	33	(a)	39	(b)	45	(c)	51	(d)	57	(b)
4	(b)	10	(d)	16	(b)	22	(b)	28	(a)	34	(d)	40	(a)	46	(d)	52	(a)	58	(a)
5	(a)	11	(d)	17	(c)	23	(d)	29	(a)	35	(c)	41	(d)	47	(d)	53	(a)		
6	(a)	12	(c)	18	(a)	24	(c)	30	(c)	36	(a)	42	(d)	48	(d)	54	(b)		

Hints & Solutions

1. (c) ${}_Z^AX$ has Z protons and $(A - Z)$ neutrons

2. (d) In case of hydrogen atom,
Mass number = atomic number

3. (a) $V \propto r^3 \Rightarrow V_{atom} = \frac{r^3_{atom}}{r^3_{nuclers}} \times V_{nucleus}$

But $\frac{r_{nucleus}}{r_{atom}} = \left(\frac{1}{10^5}\right)^3$

Hence, $V_{atom} = 10^{15} \times V_{nucleus}$

4. (b) The mass number of a nucleus is equal to the number of nucleons it contains.

5. (a) Nucleus contains only neutrons and protons.

6. (a) $Z = 11$ i.e., number of protons $= 11$, $A = 23$

$\therefore$ Number of neutrons $= A - Z = 12$

Number of electron = 0 (No electron in nucleus)

Therefore 11, 12, 0.

7. (b) We know that

$R = R_0 A^{1/3}$

$$\frac{R_1}{R_2} = \left(\frac{A_1}{A_2}\right)^{1/3} = \left(\frac{125}{64}\right)^{1/3} = \frac{5}{4}$$

NOTES The volume of a nucleus varies directly with mass number A.

$V = \frac{4}{3}\pi R_0^3 A = \frac{4}{3}\pi R_0^3 A \qquad (\because R = R_0 A^{1/3})$

$\Rightarrow V \propto A$

Also, density of nucleus is constant, independent of A.

Nuclear density $(\rho) = \frac{\text{Mass of nucleus}}{\text{Volume of nucleus}}$

$$= \frac{mA}{\frac{4}{3}\pi(R_0A^{1/3})} = \frac{3m}{4\pi R_0^3} = \text{Constant}$$

Here, m = Average of mass of a nucleon $= 1.66 \times 10^{-27}$ kg

$R_0 = 1.2 \times 10^{-15} = 1.2$ fm

8. (a) As we know, $R = R_0 (A)^{1/3}$

where A = mass number

$R_{AI} = R_0 (27)^{1/3} = 3R_0$

$R_{Te} = R_0 (125)^{1/3} = 5R_0 = \frac{5}{3} R_{AI}$

9. (c) The radius of the nuclears is directly proportional to cube root of atomic number i.e. $R \propto A^{1/3}$

$R = R_0 A^{1/3}$, where R_0 is a constant of proportionality

$$\frac{R_2}{R_1} = \left(\frac{A_2}{A_1}\right)^{1/3} \Rightarrow \left(\frac{64}{27}\right)^{1/3} = \frac{4}{3}$$

where R_1 = the radius of ^{27}Al, and A_1 = Atomic mass number of Al

R_2 = the radius of ^{64}Cu and A_2 = Atomic mass number of C4

$R_2 = 3.6 \times \frac{4}{3} = 4.8\,\text{m}$

10. (d) Requird ratio of nuclear densities $= \frac{\rho_1}{\rho_2}$

$$= \frac{\left(\frac{M_1}{V_1}\right)}{\left(\frac{M_2}{V_2}\right)} = \frac{M_1}{M_2} \times \frac{V_2}{V_1} = \frac{1}{3} \times \frac{\frac{4}{3}\pi R_2^3}{\frac{4}{3}\pi R_1^3}$$

$$= \frac{1}{3} \times \left(\frac{R_2}{R_1}\right)^3 = \frac{1}{3} \times \left(\frac{R_0 M_2^{1/3}}{R_0 M_1^{1/3}}\right)^3$$

$[\therefore R = R_0 M^{1/3}]$

$$= \frac{1}{3} \times \left(\frac{M_2}{M_1}\right) = \frac{1}{3} \times \left(\frac{3}{1}\right) = 1:1$$

NOTES Nuclear density, $\rho = \frac{3m}{4\pi R_0^3}$

Here, $R_0 = 1.2 \times 10^{-15}$m m = Average of mass of a nuclean (mass of proton + mass of neutron) $= 1.66 \times 10^{-27}$ kg

This formula suggest that density of nuclear matter is same for all nuclei.

11. (d) It has been known that a nucleus of mass number A has radius

$R = R_0 A^{1/3}$,

where $R_0 = 1.2 \times 10^{-15}$m

and A = mass number

In case of ${}^{27}_{13}A\ell$, let nuclear radius be R_1

and for ${}^{125}_{32}Te$, nuclear radius be R_2

For ${}^{27}_{13}Al$, $R_1 = R_0(27)^{1/3} = 3R_0$

For ${}^{125}_{32}Te$, $R_2 = R_0(125)^{1/3} = 5R_0$

$$\frac{R_2}{R_1} = \frac{5R_0}{3R_0} = \frac{5}{3}R_1 = \frac{5}{3} \times 3.6 = 6\,\text{fm}$$

12. (c) We use the formula,

$R = R_0 A^{1/3}$

This represents relation between atomic mass and radius of the nucleus.

For berillium, $R_1 = R_0 (9)^{1/3}$

For germanium, $R_2 = R_0 A^{1/3}$

$$\frac{R_1}{R_2} = \frac{(9)^{1/3}}{(A)^{1/3}} \Rightarrow \frac{1}{2} = \frac{(9)^{1/3}}{(A)^{1/3}}$$

$$\Rightarrow \frac{1}{8} = \frac{9}{A} \Rightarrow A = 8 \times 9 = 72$$

13. (a) The nuclear radius $R = R_0 A^{1/3}$

or $R \propto A^{1/3}$ or, $R^3 \propto A \Rightarrow A \propto R^3$

$$\therefore \frac{A_2}{A_1} = \left(\frac{R_2}{R_1}\right)^3 = \left(\frac{R_1/2}{R_1}\right)^3 = \frac{1}{8}$$

$$A_2 = \frac{A_1}{8} = \frac{56}{8} = 7$$

14. (b) Applying law of conservation of momentum,

$m_1 v_1 = m_2 v_2$

$$\frac{v_1}{v_2} = \frac{m_2}{m_1}$$

As $m = \frac{4}{3}\pi r^3 \rho \Rightarrow m \propto r^3$ Hence, $\frac{m_2}{m_1} = \frac{r_2^3}{r_1^3}$

$$\therefore \frac{v_1}{v_2} = \frac{r_2^3}{r_1^3} \Rightarrow \frac{r_2}{r_1} = \left(\frac{1}{2}\right)^{\frac{1}{3}}$$

15. (c) $\frac{R_s}{R_{He}} = \left(\frac{A_s}{A_{He}}\right)^{1/3} = \left(\frac{32}{4}\right)^{1/3} = 2$

16. (b) Radius, $R = R_0 A^{1/3}$

When, A = 189,

$R = R_0 (189)^{1/3}$ ...(i)

For R = R/3

$\frac{R}{3} = R_0 (A)^{1/3}$ ...(ii)

Dividing (i) by (ii) we get,

$$\frac{R}{\frac{R}{3}} = \left(\frac{189}{A}\right)^{1/3} \Rightarrow (3)^3 = \frac{189}{A}$$

$$\Rightarrow A = \frac{189}{27} = 7$$

17. (c) The nuclear radius r varies with mass number A according to the relation

$r \propto A^{1/3}$ or $A \propto r^3$

$$\text{density} = \frac{\text{mass}}{\text{volume}}$$

Further, mass $\propto$ A

and volume $\propto r^3$

$$\therefore \frac{\text{mass}}{\text{volume}} = \text{constant}$$

18. (a) $R \propto (A)^{1/3}$

$\therefore R_{Al} \propto (27)^{1/3}$ and $R_{Te} \propto (125)^{1/3}$

$$\therefore \frac{R_{Al}}{R_{Te}} = \frac{3}{5} = \frac{6}{10}$$

19. (a) Gain in BE = (BE) of products – (BE) of reactants.

= [120 + 120] × 8.5 – [240] × 7.6

= (240) × 8.5 – 240 × 7.6

= (2040 – 1824) MeV = 216 MeV

20. (a) From the Einstein's mass-energy equivalence, the relation between the mass of a substance m and its energy E is

$E = mc^2$

Here, c = speed of light

$$\therefore E = 0.5 \times 10^{-3} \times (3 \times 10^8)^2 = 0.5 \times 10^{-3} \times 9 \times 10^{16}$$

$= 4.5 \times 10^{13}$ J.

21. (d) BE of ${}_2He^4 = 4 \times 7.06 = 28.24$ MeV

BE of ${}^7_3Li = 7 \times 5.60 = 39.20$ MeV

${}^7_3Li + {}^1_1H \rightarrow {}_2He^4 + {}_2He^4 + Q$

39.20 28.24×2(= 56.48 MeV)

Therefore, Q = 56.48 – 39.20 = 17.28 MeV.

22. (b) Mass defect Δm = 0.02866 a.m.u.

Energy = 0.02866 × 931 = 26.7 MeV

As ${}_1H^2 + {}_1H^2 \longrightarrow {}_2He^4$

Energy liberated per a.m.u = 13.35/2 MeV

= 6.675 MeV

23. (d)

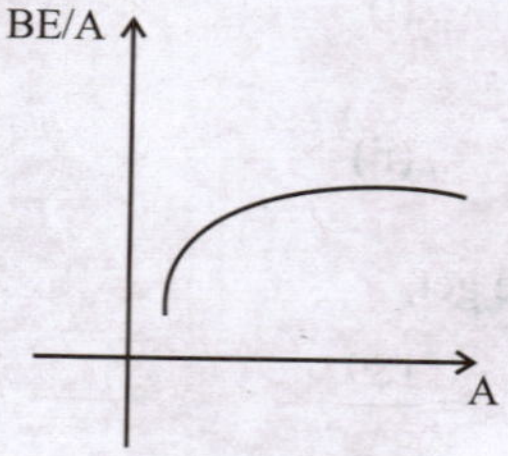

From the graph of BE/A versus mass number A it is clear that, BE/A first increases and then decreases with increase in mass number.

24. (c) $E = mc^2$; $m = \frac{E}{c^2}$

So, mass decay per second

$\frac{dm}{dt} = \frac{1}{c^2}\frac{dE}{dt} = \frac{1}{c^2}$ (Power in watt)

$= \frac{1}{(3\times10^8)^2}\times1000\times10^3$

and mass decay per hour $= \frac{dm}{dt}\times60\times60$

$= \frac{1}{(3\times10^8)^2}\times10^6\times3600 = 4\times10^{-8}$ kg

= 40 microgram

25. (b) B.E. = 0.042 × 931 ≃ 42 MeV

Number of nucleons in ^7_3Li is 7.

∴ B.E./ nucleon $= \frac{42}{7} = 6$ MeV ≃ 5.6 MeV

26. (b) Binding energy of two $_1H^2$ nuclei

$= 2(1.1 \times 2) = 4.4\ MeV$

Binding energy of one $_2He^4$ nucelus

$= 4 \times 7.0 = 28\ MeV$

∴ Energy released $= 28 - 4.4 = 23.6\ MeV$

27. (a) Mass defect $= ZM_p + (A - Z)M_n - M(A, Z)$

or, $\frac{B.E.}{c^2} = ZM_p + (A - Z)\,M_n - M(A, Z)$

∴ $M(A, Z) = ZM_p + (A - Z)M_n - \frac{B.E.}{c^2}$

28. (a) The difference in mass of a nucleus and its constituents, ΔM, is called the mass defect and is given by

$\Delta M = [ZM_p + (A-Z)M_n] - M$

and binding energy $= \Delta Mc^2$

$= [\{ZM_p + (A-Z)M_n\} - M]c^2$

29. (a) $_1D^2 \longrightarrow {}_2He^4$

Energy released = 28 – 2 × 2.2 = 23.6 MeV

NOTES Binding energy is energy released on formation of Nucleus.

30. (c) ^2_1H and ^3_1H requires a and b amount of energies for their nucleons to be separated.

$^4_2$He releases c amount of energy in its formation i.e., in assembling the nucleons as nucleus.

Hence, Energy released $= c - (a + b) = c - a - b$

31. (d) Mass defect $= \frac{B.E}{c^2}$

Mass of nucleus = Mass of proton + mass of neutron – mass defect

32. (d) $\Delta m = (2 \times 1.0074 + 2 \times 1.0087 - 4.0015)$
$= 0.0307$
$E = (\Delta m) \times 931$ MeV $= 0.0307 \times 931 = 28.5$ MeV

33. (a) Given : Mass of neutron $= M_n$
Mass of proton $= M_p$; Atomic mass of the element $= M$; Number of neutrons in the element $= N$ and number of protons in the element $= Z$. We know that the atomic mass (M) of any stable nucleus is always less than the sum of the masses of the constituent particles.
Therefore, $M < [NM_n + ZM_p]$.
X is a neutrino, when β-particle is emitted.

34. (d) Applying principle of energy conservation,
Energy of proton
= total B.E. of 2α – energy of Li^7
= 8 × 7.06 – 7 × 5.6
= 56.48 – 39.2 = 17.28 MeV

35. (c) 1 a.m.u = 931 MeV

36. (a) Average B.E./nucleon in nuclei is of the order of 8 MeV.

37. (c) Nuclear force is same between any two nucleons.

38. (c) Nuclear forces are short range attractive forces which balance the repulsive forces between the protons inside the nucleus.

39. (b) Water is used as a coolant in nuclear reactor due to its high specific heat capacity.

40. (a) (A) Slow neutron can cause fission in $^{235}_{92}$U' fast neutrons cannot cause fission because they are too quick so they scatter of atoms instead of being captured by them

(B) α-rays are Helium nuclei.

(C) β-rays are generated when neutron is converted into proton by releasing electron or proton is converted into neutron by releasing positron.

(D) Due to smaller wavelength as compared to X-rays, γ-rays have higher energies as compared to X-rays.

41. (d) $^{290}_{82}X \xrightarrow{\alpha} {}^{286}_{80}Y \xrightarrow{e^+} {}^{286}_{79}Z \xrightarrow{\beta^-}$

$^{286}_{80}P \xrightarrow{e^-} {}^{286}_{81}Q$

For the product Q, A → 286 and Z = 81.

42. **(d)** It is a nuclear fission reaction.

$${}^{235}_{92}U + {}^{1}_{0}n \longrightarrow {}^{89}_{36}Kr + {}^{144}_{56}Ba + {}^{1}_{3}n + Q$$

43. **(b)** ${}^{12}_{11}Na \to {}^{A}_{Z}X + {}^{0}_{+1}e + v$

So, $22 = A + 0 \Rightarrow A = 22$

$11 = Z + 1 \Rightarrow Z = 10 \Rightarrow$ Element is Neon

Clearly, answer is ${}^{22}_{10}Ne$

44. **(c)** When the coulomb repulsion between the nuclei is overcome then nuclear fusion reaction takes place. This is possible when temperature is too high.

45. **(c)** Binding energy per nucleon for fission products is higher relative to Binding energy per nucleon for parent nucleus, i.e., more masses are lost and are obtained as kinetic energy of fission products. So, the given ratio < 1.

46. **(d)** B.E. per nucleon is smaller for lighter as well as heavier nucleus. But fusion reaction occurs for small mass number nuclei and fission reaction occurs for larger mass number nuclei to attain reaction binding energy per nucleon.

47. **(d)** $m_3 < (m_1 + m_2)$ $(\because m_1 + m_2 = m_3 + E)$ as $E = [m_1 + m_2 - m_3]\,C^2$

NOTES Mass defect occur in both nuclear fission and fusion processes. Therefore, nuclear energy is released in both the processes as per Einstein mass energy relation $E = (\Delta m)c^2$

48. **(d)** As a result of fusion, enormous amount of heat is liberated which is the main cause of source of solar energy.

49. **(c)** For nuclear fusion process the nuclei with low mass are suitable.

50. **(a)** According to liquid drop model of nucleus, an excited nucleus breaks into lighter nuclei just like an excited drop breaks into tiny drops.

51. **(d)** In nuclear reactions, we have the conservation of mass, energy and momentum.

52. **(a)** The explosion of the atomic bomb takes place due to nuclear fission.

53. **(a)** ${}_{92}U^{235} + {}_{0}n^{1} \to {}_{38}Sr^{90} + {}_{54}Xe^{143}$

$$+ 3\,{}_{0}n^{1} + \text{energy}$$

54. **(b)** Binding energy

$= 117 \times 8.5 + 117 \times 8.5 - 236 \times 7.6$

$= 234 \times 8.5 - 236 \times 7.6$

$= 1989 - 1793.6 = 200$ MeV

Thus, in per fission of Uranium nearly 200 MeV energy is liberated

55. **(c)** Moderator used in nuclear reactor are graphite and heavy water.

NOTES Characteristics of moderator are

(i) Its atomic weight must be low

(ii) It should not absorb neutrons

(iii) It should undergo elastic collisions with neutrons and reduce their velocity.

56. **(d)** Moderator slows down the neutrons to thermal energies.

57. **(b)** Fission rate

$$= \frac{\text{total power}}{\frac{\text{energy}}{\text{fission}}} = \frac{5}{200 \times 1.6 \times 10^{-13}}$$

$= 1.56 \times 10^{11}\ s^{-1}$

58. **(a)** Fusion reaction.

14

Semiconductor Electronics : Materials, Devices and Simple Circuits

--------------TREND ANALYSIS --------------

TOPIC	2025		2024		2023		2022		2021		2020	
	Qns.	LOD	Qns.	LOD	Qns.	LOD	Qns.	LOD	Qns.	LOD	Qns.	LOD
Extrinsic Semiconductor									1	Avg		
p-n Junction											1	Easy
Semiconductor Diodes			1	Diffi.			1	Avg				
Application of Junction Diode as a Rectifier	1	Avg			1	Avg.	1	Avg				
Special Purpose p-n Junction Diodes			1	Avg	1	Avg			1	Avg		
Digital Electronics and Logic Gates	1	Avg	2	Easy	1	Avg	1	Avg	1	Avg	1	Avg

How to Code Qns for Better & Faster Revision

Type 1: Able to solve (No code Req)
Type 2: Solved with struggle or by seeing solution **(Use code NR means Needs Revision or orange stickers)**
Type 3: Tricky/ Could not solve on my own. Need help from Teacher/ Friend. **(Use code TQ means Tricky Question or red stickers)**

You can either hand write the code in boxes or else paste coloured stickers given at end of book.

Topic-14.3

Intrinsic Semiconductor

1. A piece of copper and other of germanium are cooled from the room temperature to 80 K, then 1993
   (a) resistance of each will increase
   (b) resistance of copper will decrease
   (c) resistance of copper will increase while that of germanium will decrease
   (d) resistance of copper will decrease while that of germanium will increase

2. At absolute zero, Si acts as NCERT | Page-327 | 1988
   (a) non-metal (b) metal
   (c) insulator (d) none of these

3. The intrinsic semiconductor becomes an insulator at NCERT | Page-327 | 2001
   (a) 0ºC (b) 0 K (c) 300 K (d) –100ºC

Topic-14.4

Extrinsic Semiconductor

4. A p-type extrinsic semiconductor is obtained when germanium is doped with NCERT | Page-331 | Manipur 2023
   (a) Arsenic (b) Boron
   (c) Antimony (d) Phosphorous

5. The electron concentration in an *n*-type semiconductor is the same as hole concentration in a *p*-type semiconductor. An external field (electric) is applied across each of them. Compare the currents in them. NCERT | Page-330, 331 | 2021
   (a) No current will flow in *p*-type, current will only flow in *n*-type
   (b) Current in *n*-type = current in *p*-type
   (c) Current in *p*-type > current in *n*-type
   (d) Current in *n*-type > current in *p*-type

6. For a p-type semiconductor, which of the following statements is **true** ? NCERT | Page-331 | 2019
   (a) Electrons are the majority carriers and trivalent atoms are the dopants.
   (b) Holes are the majority carriers and trivalent atoms are the dopants.
   (c) Holes are the majority carriers and pentavalent atoms are the dopants.
   (d) Electrons are the majority carriers and pentavalent atoms are the dopants.

7. In a n-type semiconductor, which of the following statement is true? NCERT | Page330 | 2013
   (a) Electrons are minority carriers and pentavalent atoms are dopants.
   (b) Holes are minority carriers and pentavalent atoms are dopants.

(c) Holes are majority carriers and trivalent atoms are dopants.
(d) Electrons are majority carriers and trivalent atoms are dopants.

8. If a small amount of antimony is added to germanium crystal **NCERT | Page-330 | 2011**
(a) it becomes a p–type semiconductor
(b) the antimony becomes an acceptor atom
(c) there will be more free electrons than holes in the semiconductor
(d) its resistance is increased

9. Pure Si at 500K has equal number of electron (n_e) and hole (n_h) concentrations of 1.5×10^{16} m^{-3}. Doping by indium increases n_h to 4.5×10^{22} m^{-3}. The doped semiconductor is of **NCERT | Page-332 | 2011 M**
(a) n–type with electron concentration $n_e = 5 \times 10^{22}$ m^{-3}
(b) p–type with electron concentration $n_e = 2.5 \times 10^{10}$ m^{-3}
(c) n–type with electron concentration $n_e = 2.5 \times 10^{23}$ m^{-3}
(d) p–type having electron concentration $n_e = 5 \times 10^{9}$ m^{-3}

10. Which one of the following statement is FALSE ? **NCERT | Page-330, 331 | 2010**
(a) Pure Si doped with trivalent impurities gives a p-type semiconductor
(b) Majority carriers in a n-type semiconductor are holes
(c) Minority carriers in a p-type semiconductor are electrons
(d) The resistance of intrinsic semiconductor decreases with increase of temperature

11. Which of the following when added acts as an impurity into silicon produced n-type semi-conductor? **NCERT | Page-330 | 1999**
(a) P (b) Al (c) B (d) Mg

12. To obtain a *p*-type germanium semiconductor, it must be doped with **NCERT | Page-331 | 1997**
(a) arsenic (b) antimony
(c) indium (d) phosphorus

13. When arsenic is added as an impurity to silicon, the resulting material is **NCERT | Page-330 | 1996**
(a) *n*-type semiconductor
(b) *p*-type semiconductor
(c) *n*-type conductor
(d) insulator

14. Which of the following, when added as an impurity, into the silicon, produces *n*-type semi-conductor? **NCERT | Page-330 | 1995**
(a) Phosphorous (b) Aluminium
(c) Magnesium (d) Both b and c

15. When *n*-type semiconductor is heated **NCERT | Page-330 | 1989**
(a) number of electrons increases while that of holes decreases
(b) number of holes increases while that of electrons decreases
(c) number of electrons and holes remain same
(d) number of electrons and holes increases equally.

Topic-14.5

p-n Junction

16. The increase in the width of the depletion region in a p-n junction diode is due to : **NCERT | Page-335 | 2020**
(a) reverse bias only
(b) both forward bias and reverse bias
(c) increase in forward current
(d) forward bias only

17. In a p-n junction diode, change in temperature due to heating **NCERT | Page-336 | 2018**
(a) Affects only reverse resistance
(b) Affects only forward resistance
(c) Affects the overall V-I characteristics of p-n junction
(d) Does not affect resistance of p-n junction

18. The barrier potential of a p-n junction depends on: **NCERT | Page-333 | 2014**
(A) type of semi conductor material
(B) amount of doping
(C) temperature
Which one of the following is correct ?
(a) (A) and (B) only (b) (B) only
(c) (B) and (C) only (d) (A), (B) and (C)

19. In an unbiased p-n junction, holes diffuse from the p-region to n-region because of **NCERT | Page-333 | Kar. 2013**
(a) the potential difference across the p-n junction
(b) the attraction of free electrons of n-region
(c) the higher hole concentration in p-region than that in n-region
(d) the higher concentration of electrons in the n-region than that in the p-region

20. Barrier potential of a *p-n* junction diode does not depend on **NCERT | Page-334 | 2003**
(a) doping density (b) diode design
(c) temperature (d) forward bias

21. In a *p-n* junction **NCERT | Page-333 | 2002**
(a) The potential of the *p* and *n*-sides becomes higher alternately
(b) The *p*-side is at higher electrical potential than the n side
(c) The *n*-side is at higher electrical potential than the *p*-side
(d) Both the *p* and *n*-sides are at the same potential

22. A depletion layer consists of **NCERT | Page-333 | 1999**
(a) electrons (b) protons
(c) mobile ions (d) immobile ions

23. In a junction diode, the holes are due to **NCERT | Page-334 | 1999**
(a) protons (b) extra electrons
(c) neutrons (d) missing electrons

24. The cause of the potential barrier in a *p-n* diode is

NCERT | Page-334 | 1998

(a) depletion of positive charges near the junction
(b) concentration of positive charges near the junction
(c) depletion of negative charges near the junction
(d) concentration of positive and negative charges near the junction

25. The depletion layer in the *p-n* junction region is caused by

NCERT | Page-333 | 1991

(a) drift of holes
(b) diffusion of charge carriers
(c) migration of impurity ions
(d) drift of electrons

Topic-14.6

Semiconductor Diode

26. Choose the correct circuit which can achieve the bridge balance.

NCERT | Page-335, 336 | 2024

(a)

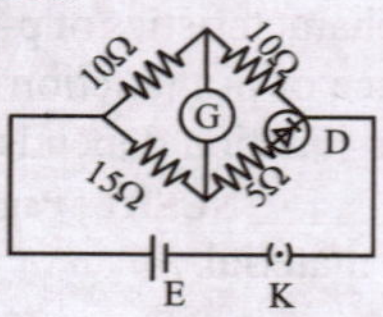

(b)

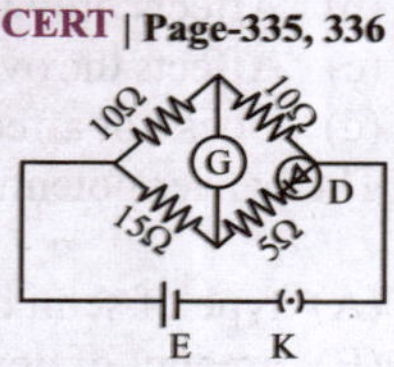

(c)

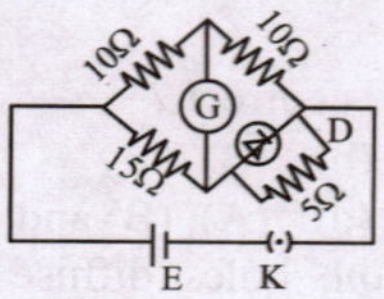

(d)

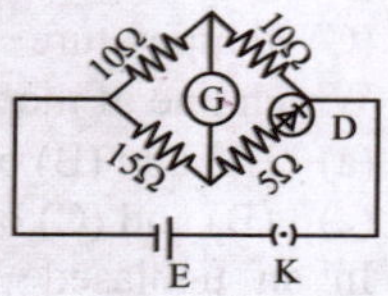

27.

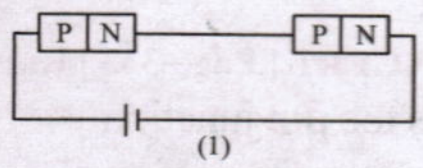

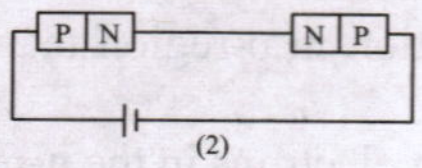

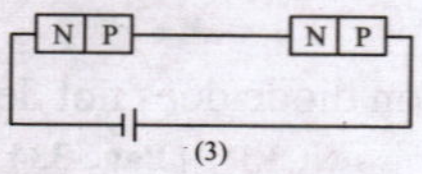

In the given circuits (1), (2) and (3), the potential drop across the two p-n junctions are equal in:

NCERT | Page-334 | 2022

(a) Circuit (2) only
(b) Circuit (3) only
(c) Both circuits (1) and (3)
(d) Circuit (1) only

28. Which one of the following represents forward bias diode ?

NCERT | Page-334 | 2017, 2006, 2000

(a)

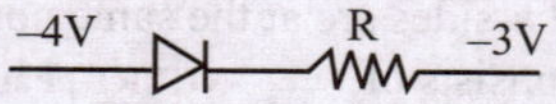

(b)

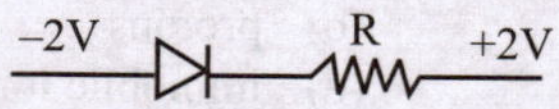

(c)

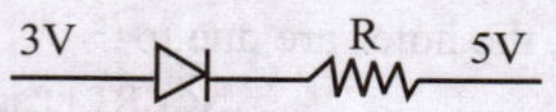

(d)

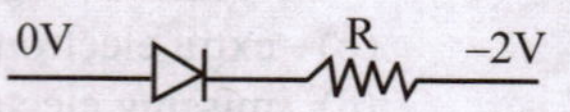

29. Consider the junction diode as ideal. The value of current flowing through AB is :

NCERT | Page-334 | 2016

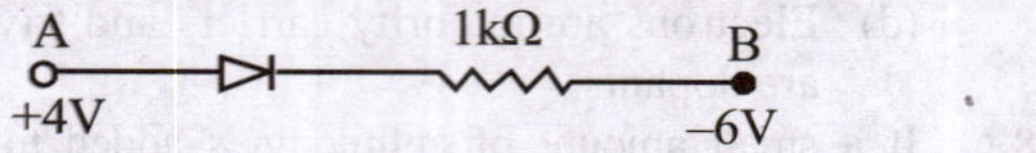

(a) 0 A (b) 10^{-2} A (c) 10^{-1} A (d) 10^{-3} A

30. In the given figure, a diode D is connected to an external resistance R = 100 Ω and an e.m.f. of 3.5 V. If the barrier potential developed across the diode is 0.5 V, the current in the circuit will be:

NCERT | Page-335 | 2015 RS

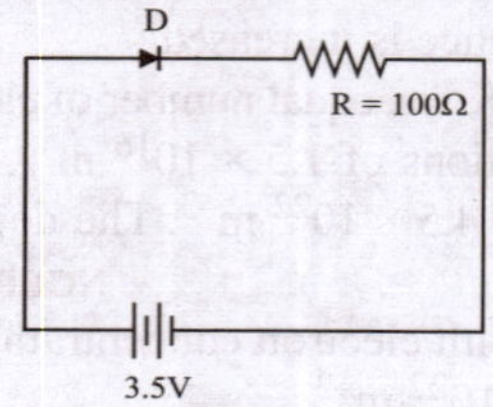

(a) 40 mA (b) 20 mA (c) 35 mA (d) 30 mA

31. Two ideal diodes are connected to a battery as shown in the circuit. The current supplied by the battery is :

NCERT | Page-334 | 2012

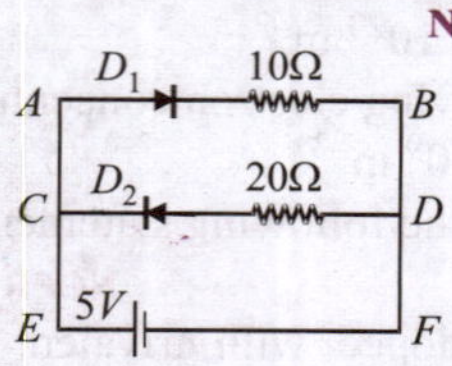

(a) 0.75 A (b) zero (c) 0.25 A (d) 0.5 A

32. In forward biasing of the p–n junction

NCERT | Page-335 | 2011

(a) the positive terminal of the battery is connected to p–side and the depletion region becomes thick
(b) the positive terminal of the battery is connected to n–side and the depletion region becomes thin
(c) the positive terminal of the battery is connected to n–side and the depletion region becomes thick
(d) the positive terminal of the battery is connected to p–side and the depletion region becomes thin

33. In the following figure, the diodes which are forward biased, are

NCERT | Page-334 | 2011 M, 2004

(A)

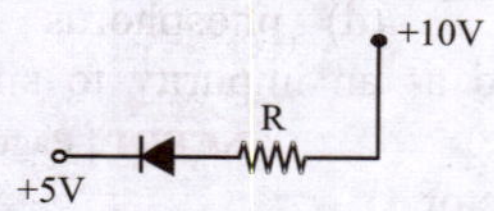

(B)

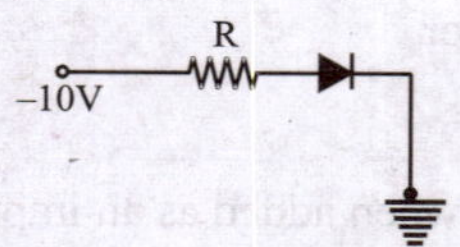

(C)

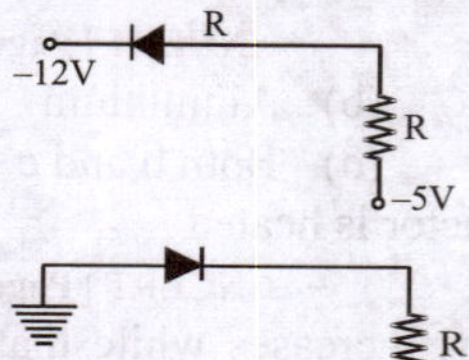

(D)

(a) (C) only (b) (C) and (A)
(c) (B) and (D) (d) (A), (B) and (D)

34. Application of a forward bias to a p–n junction
NCERT | Page-333 | 2005
(a) widens the depletion zone
(b) increases the potential difference across the depletion zone
(c) increases the number of donors on the n side
(d) increases the electric field in the depletion zone

35. Reverse bias applied to a junction diode
NCERT | Page-336 | 2003
(a) increases the minority carrier current
(b) lowers the potential barrier
(c) raises the potential barrier
(d) increases the majority carrier current

36. A d.c. battery of V volt is connected to a series combination of a resistor R and an ideal diode D as shown in the figure below. The potential difference across R will be
NCERT | Page-334 | 2002

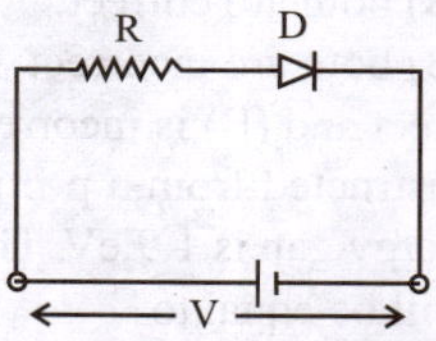

(a) 2V when diode is forward biased
(b) Zero when diode is forward biased
(c) V when diode is reverse biased
(d) V when diode is forward biased

37. In forward bias, the width of potential barrier in a p-n junction diode **NCERT | Page-335 | 1999**
(a) increases (b) decreases
(c) remains constant (d) first '1' then '2'

38. A semi-conducting device is connected in a series circuit with a battery and a resistance. A current is found to pass through the circuit. If the polarity of the battery is reversed, the current drops to almost zero. The device may be **NCERT | Page-336 | 1998**
(a) a p-n junction
(b) an intrinsic semi-conductor
(c) a p-type semi-conductor
(d) an n-type semi-conductor

39. When a p-n junction diode is reverse biased the flow of current across the junction is mainly due to
NCERT | Page-336 | 1994
(a) diffusion of charges
(b) drift of charges
(c) depends on the nature of material
(d) both drift and diffusion of charges

40. p-n junction is said to be forward biased, when
NCERT | Page-335 | 1988
(a) the positive pole of the battery is joined to the p-semiconductor and negative pole to the n-semiconductor
(b) the positive pole of the battery is joined to the n-semiconductor and p-semiconductor joined to negative pole of the battery
(c) the positive pole of the battery is connected to n-semiconductor and p-semiconductor is connected to the positive pole of the battery
(d) a mechanical force is applied in the forward direction

Topic-14.7

Application of Junction Diode as a Rectifier

41. A full wave rectifier circuit with diodes (D_1) and (D_2) is shown in the figure. If input supply voltage $V_{in} = 220 \sin (100 \pi t)$ volt, then at $t = 15$ msec
NCERT | Page-338, 339 | 2025

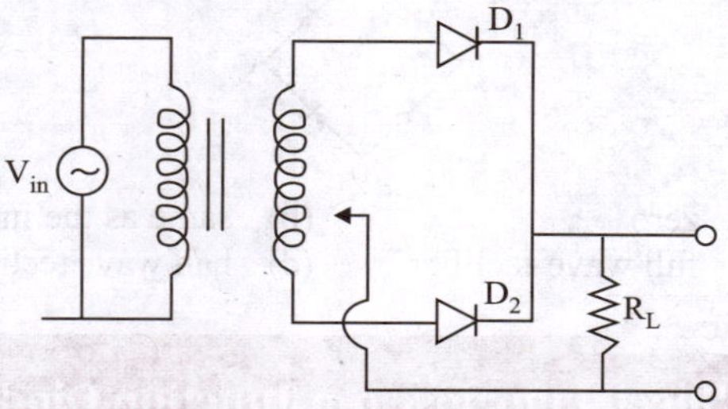

(a) D_1 is forward biased, D_2 is reverse biased
(b) D_1 is reverse biased, D_2 is forward biased
(c) D_1 and D_2 both are forward biased
(d) D_1 and D_2 both are reverse biased

42. A full wave rectifier circuit consists of two p-n junction diodes, a centre–tapped transfor er, capacitor and a load resistance. Which of these components remove the ac ripple from the rectified output? **NCERT | Page-338, 339 | 2023**
(a) A centre–tapped transformer
(b) p–n junction diodes
(c) Capacitor
(d) Load resistance

43. In half wave rectification, if the input frequency is 60 Hz, then the output frequency is would be:
NCERT | Page-338 | 2022
(a) 30 Hz (b) 60 Hz (c) 120 Hz (d) Zero

44. If in a p-n junction, a square input signal of 10 V is applied as shown, then the output across R_L will be
NCERT | Page-334 | 2015

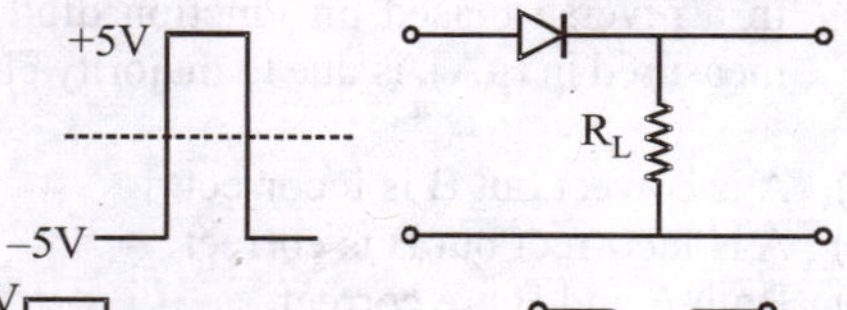

(a) 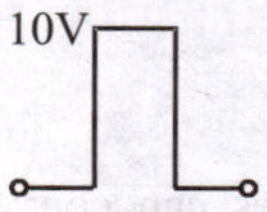

(b) 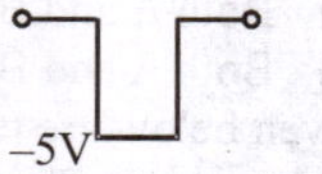

(c) 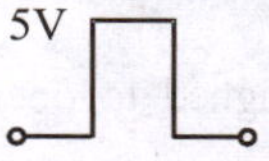

(d) 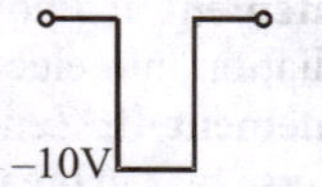

45. The peak voltage in the output of a half-wave diode rectifier fed with a sinusoidal signal without filter is 10V. The d.c. component of the output voltage is **NCERT | Page-338 | 2004**
(a) $20/\pi$ V (b) $10/\sqrt{2}$ V (c) $10/\pi$ V (d) 10V

46. If a full wave rectifier circuit is operating from 50Hz mains, the fundamental frequency in the ripple will be **NCERT | Page-339 | 2003**
(a) 100 Hz (b) 25 Hz (c) 50 Hz (d) 70.7 Hz

47. An alternating current can be converted into direct current by a **NCERT | Page-338 | 2001**
(a) transformer (b) dynamo
(c) motor (d) rectifier

48. In the diagram, the input is across the terminals A and C and the output is across B and D. Then the output is **NCERT | Page-330 | 1994**

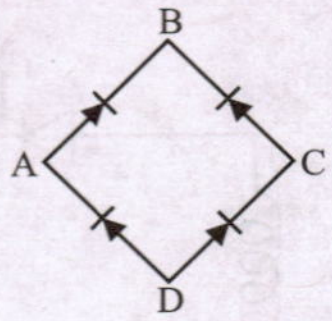

(a) zero (b) same as the input
(c) full wave rectifier (d) half wave rectifier

Topic-*
Special Purpose p-n Junction Diodes

49.

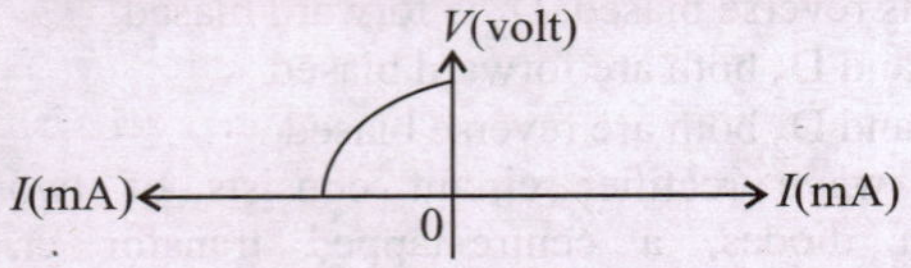

Ph-II 2024

The I-V characteristics shown above are exhibited by a
(a) Light emitting diode (b) Zener diode
(c) Photodiode (d) Solar cell

50. Consider the following statements A and B and identify the correct answer:

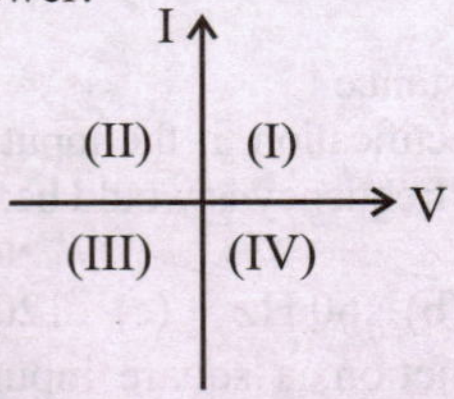

A. For a solar-cell, the I-V characteristics lies in the IV quadrant of the given graph.
B. In a reverse biased pn junction diode, the current measured in (μA), is due to majority charge carriers. **2024**

(a) A is correct but B is incorrect
(b) A is incorrect but B is correct
(c) Both A and B are correct
(d) Both A and B are incorrect

51. Given below are two statements: **2023**
Statement I: Photovoltaic devices can convert optical radiation into electricity.
Statement II: Zener diode is designed to operate under reverse bias in breakdown region.
In the light of the above statements, choose the **most appropriate** answer from the options given below.
(a) Both Statement I and Statement II are correct
(b) Both Statement I and Statement II are incorrect
(c) Statement I is correct but Statement II is incorrect
(d) Statement I is incorrect but Statement II is correct

52. The incorrect statement about the property of a Zener diode is: **RS 2022**
(a) p and n regions of zener diode are heavily doped
(b) Zener voltage remains constant at breakdown
(c) It is designed to operate under reverse bias
(d) Depletion region formed is very wide

53. Consider the following statements (A) and (B) and identify the correct answer. **2021**
(A) A zener diode is connected in reverse bias, when used as a voltage regulator.
(B) The potential barrier of p-n junction lies between 0.1 *V* to 0.3 *V*.
(a) (A) is incorrect but (B) is correct.
(b) (A) and (B) both are correct.
(c) (A) and (B) both are incorrect
(d) (A) is correct and (B) is incorrect.

54. An LED is constructed from a p-n junction diode using GaAsP. The energy gap is 1.9 eV. The wavelength of the light emitted will be equal to **Odisha 2019**
(a) 654×10^{-11} m (b) 10.4×10^{-26} m
(c) 654 nm (d) 654 Å

55. The given graph represents V - I characteristic for a semiconductor device.

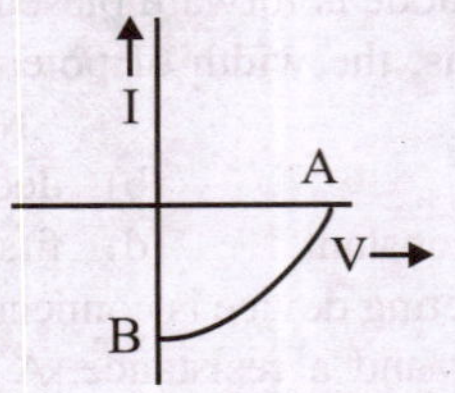

Which of the following statement is **correct** ? **2014**
(a) It is V - I characteristic for solar cell where, point A represents open circuit voltage and point B short circuit current.
(b) It is a for a solar cell and point A and B represent open circuit voltage and current, respectively.
(c) It is for a photodiode and points A and B represent open circuit voltage and current, respectively.
(d) It is for a LED and points A and B represent open circuit voltage and short circuit current, respectively.

56. A zener diode, having breakdown voltage equal to 15V, is used in a voltage regulator circuit shown in figure. The current through the diode is **2011M**

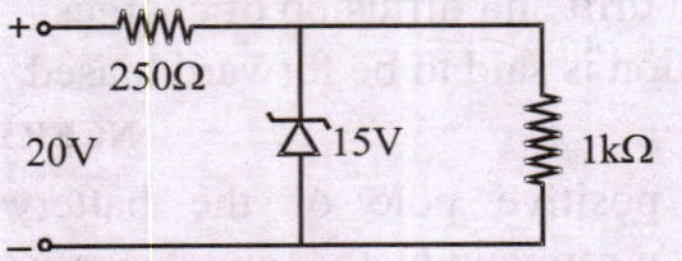

(a) 10 mA (b) 15 mA (c) 20 mA (d) 5 mA

57. Zener diode is used for **2005**
(a) amplification
(b) rectification
(c) stabilisation
(d) producing oscillations in an oscillator

58. In a *p-n* junction photocell, the value of the photo-electromotive force produced by monochromatic light is proportional to **2004**
(a) the voltage applied at the *p-n* junction
(b) the barrier voltage at the *p-n* junction
(c) the intensity of the light falling on the cell
(d) the frequency of the light falling on the cell

Topic-*
Digital Electronics and Logic Gates

59. The output (Y) of the given logic implementation is similar to the output of an/a ________ gate. **2025**

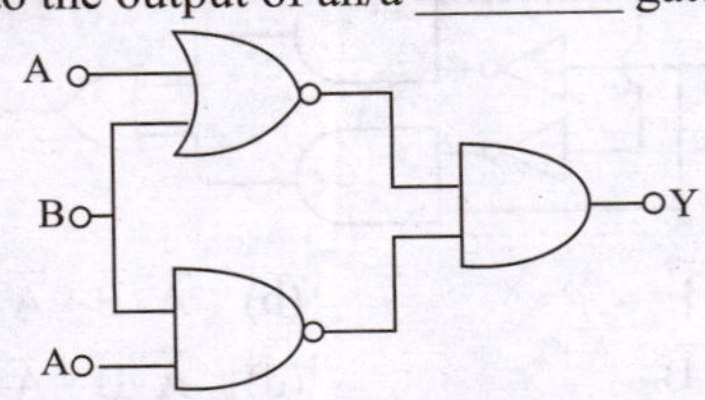

(a) AND (b) NAND (c) OR (d) NOR

60. When the output of an OR gate is applied as input to a NOT gate, then the combination acts as a **2024**
(a) NAND gate (b) NOR gate
(c) AND gate (d) OR gate

61. The output *Y* for the inputs *A* and *B* of the given logic circuit is : **2024**

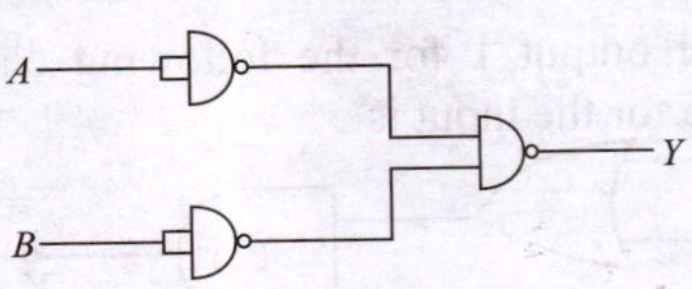

(a) $A \cdot B$ (b) $\overline{A} \cdot \overline{B}$ (c) $A + B$ (d) $\overline{A} + \overline{B}$

62. A logic circuit provides the output Y as per the following truth table : **2024**

A	B	Y
0	0	1
0	1	0
1	0	1
1	1	0

The expression for the output Y is :
(a) $A.B+\overline{A}$ (b) $A.\overline{B}+\overline{A}$
(c) $\overline{B}$ (d) B

63. The output (Y) of the given logic gate is similar to the output of an/a **2024**

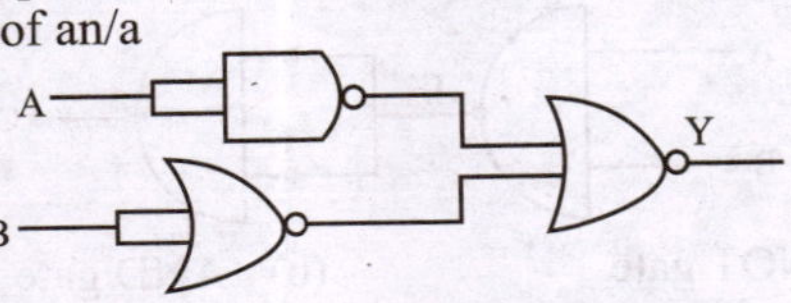

(a) NAND gate (b) NOR gate
(c) OR gate (d) AND gate

64. The given circuit is equivalent to **Manipur 2023**

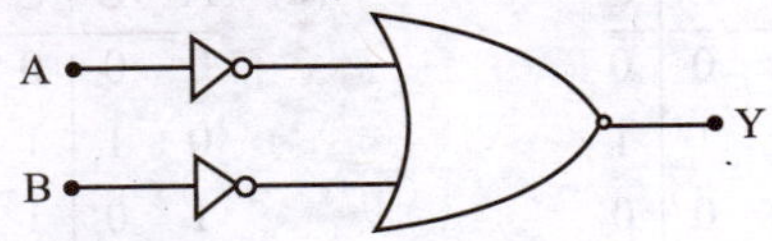

(a) 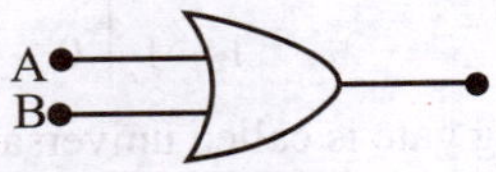

(b) 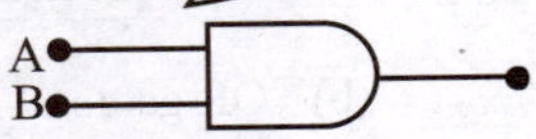

(c) 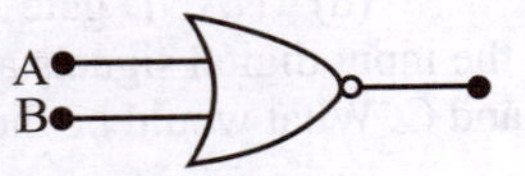

(d) 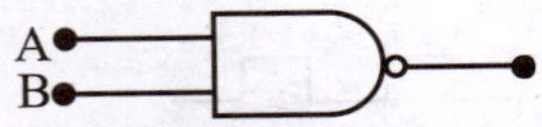

65. For the following logic circuit, the truth table is **2023**

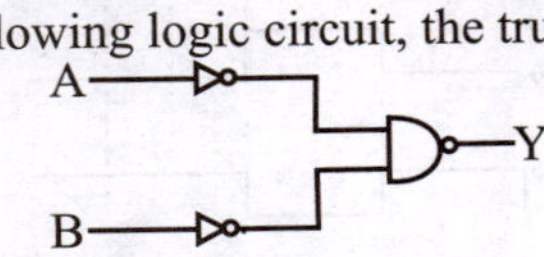

(a)

A	B	Y
0	0	1
0	1	1
1	0	1
1	1	0

(b)

A	B	Y
0	0	0
0	1	1
1	0	1
1	1	1

(c)

A	B	Y
0	0	1
0	1	0
1	0	1
1	1	0

(d)

A	B	Y
0	0	0
0	1	0
1	0	0
1	1	1

66.

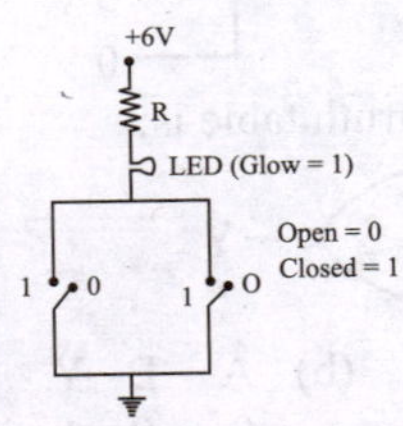

Identify the equivalent logic gate represented by the given circuit : **RS 2022**
(a) NAND (b) OR (c) NOR (d) AND

67.

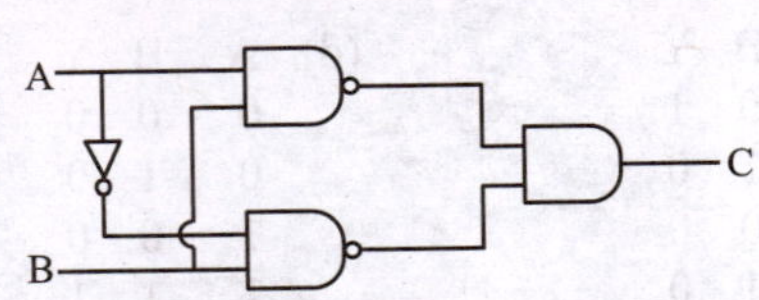

The truth table for the given logic circuit is: **2022**

(a)

A	B	C
0	0	1
0	1	0
1	0	0
1	1	1

(b)

A	B	C
0	0	1
0	1	0
1	0	1
1	1	0

(c)

A	B	C
0	0	0
0	1	1
1	0	0
1	1	1

(d)

A	B	C
0	0	0
0	1	1
1	0	1
1	1	0

68. Which of the following gate is called universal gate? **Ph-II 2022**

(a) NOT gate (b) OR gate
(c) AND gate (d) NAND gate

69. For the given circuit, the input digital signals are applied at the terminals A, B and C. What would be the output at the terminal y? **2021**

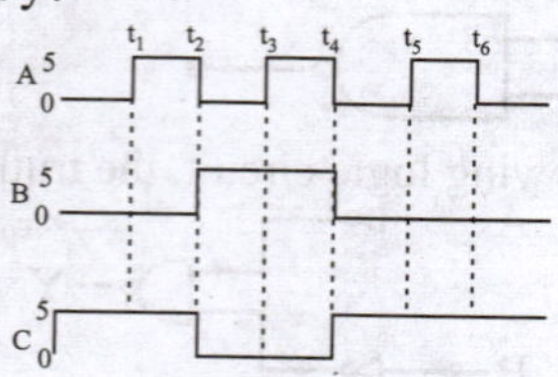

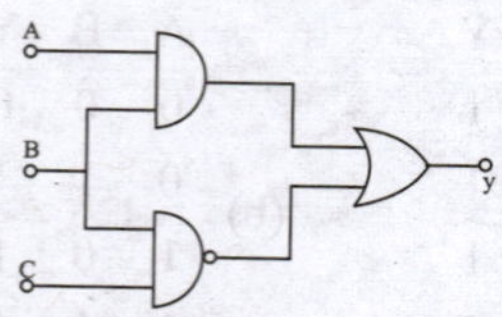

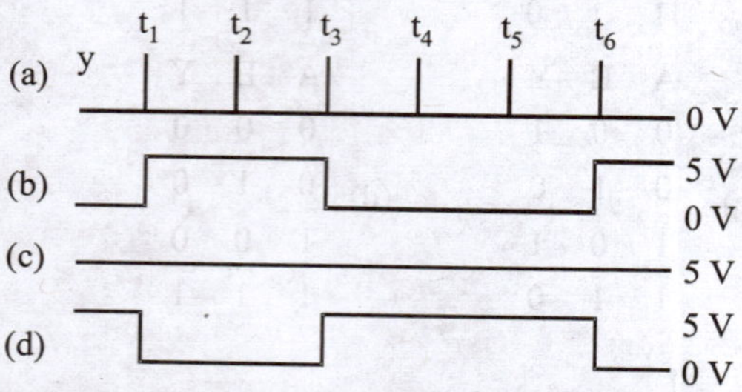

70. For the logic circuit shown, the truth table is: **2020**

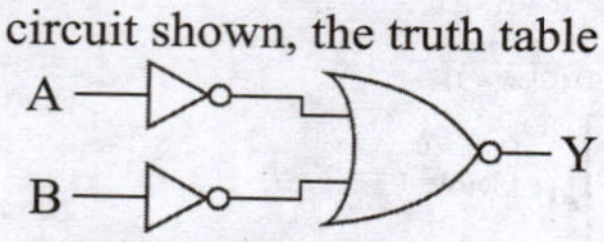

(a)

A	B	Y
0	0	0
0	1	1
1	0	1
1	1	1

(b)

A	B	Y
0	0	1
0	1	1
1	0	1
1	1	0

(c)

A	B	Y
0	0	1
0	1	0
1	0	0
1	1	0

(d)

A	B	Y
0	0	0
0	1	0
1	0	0
1	1	1

71.

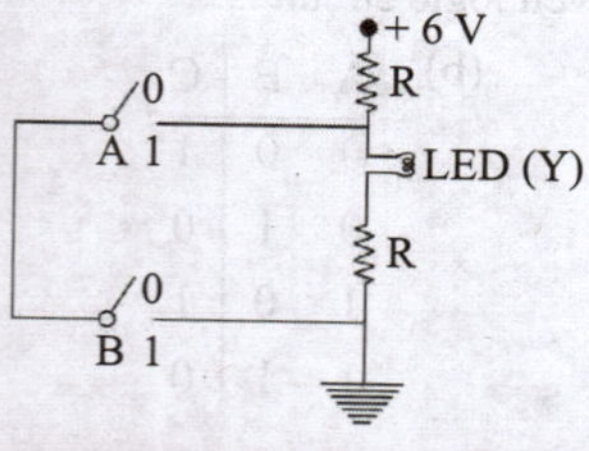

The **correct** Boolean operation represented by the circuit diagram drawn is : **2019**

(a) AND (b) OR (c) NAND (d) NOR

72. The circuit diagram shown here corresponds to the logic gate, **Odisha 2019**

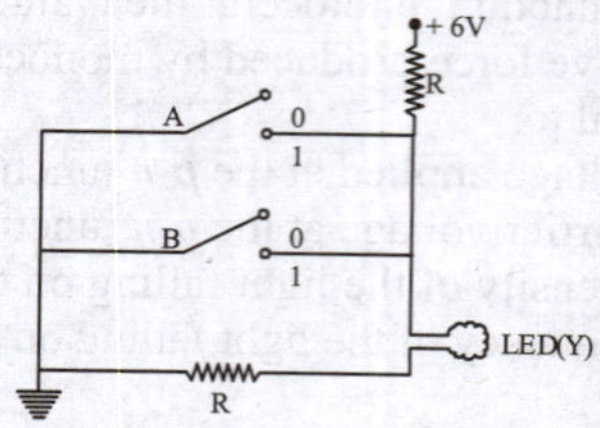

(a) NAND (b) NOR
(c) AND (d) OR

73. In the combination of the following gates the output Y can be written in terms of inputs A and B as **2018**

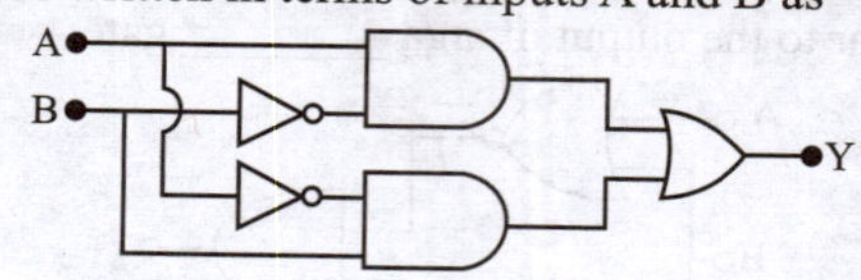

(a) $\overline{A \cdot B}$ (b) $A \cdot \overline{B} + \overline{A} \cdot B$
(c) $\overline{A + B}$ (d) $\overline{A \cdot B} + A \cdot B$

74. The given electrical network is equivalent to : **2017, 2006, 2000**

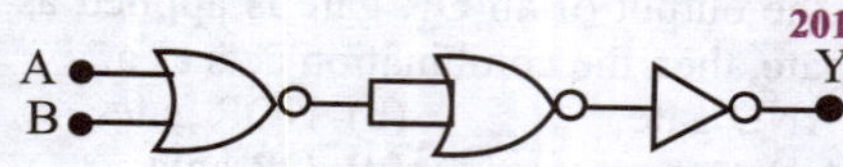

(a) OR gate (b) NOR gate
(c) NOT gate (d) AND gate

75. To get output 1 for the following circuit, the correct choice for the input is **2016, 2012, 2010**

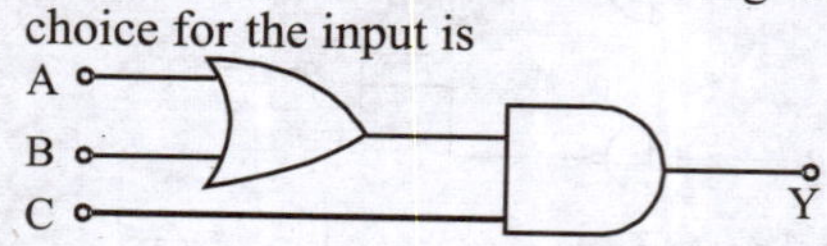

(a) A = 0, B = 1, C = 0 (b) A = 1, B = 0, C = 0
(c) A = 1, B = 1, C = 0 (d) A = 1, B = 0, C = 1

76. Which logic gate is represented by the following combination of logic gate ? **2015**

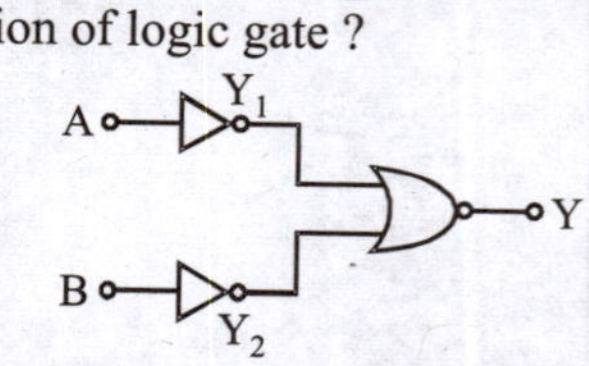

(a) NAND (b) AND (c) NOR (d) OR

77. The output from a NAND gate is divided into two in parallel and fed to another NAND gate. The resulting gate is a **Kar. 2013**

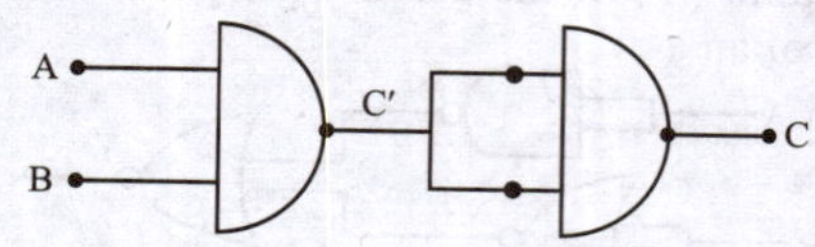

(a) NOT gate (b) AND gate
(c) NOR gate (d) OR gate

78. The output(X) of the logic circuit shown in figure will be **2013**

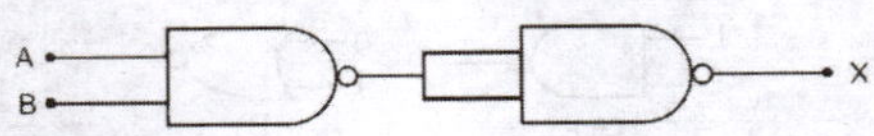

(a) $X = \overline{A.B}$ (b) $X = A.B$

(c) $X = \overline{A+B}$ (d) $X = \overline{\overline{A}} + \overline{\overline{B}}$

79. The figure shows a logic circuit with two inputs A and B and the output C. The voltage wave forms across A, B and C are as given. The logic circuit gate is : **2012**

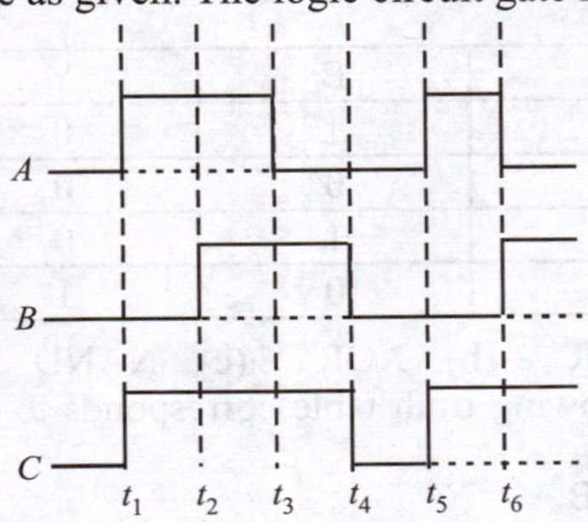

(a) OR gate (b) NOR gate

(c) AND gate (d) NAND gate

80. Symbolic representation of four logic gate are shown as **2011, 2009**

(i) 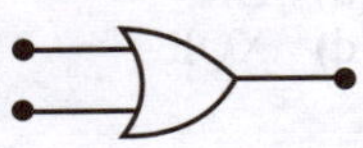 (ii) 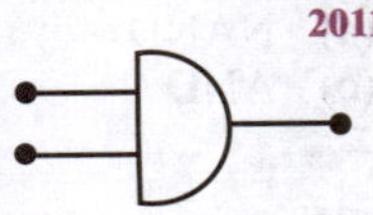

(iii) 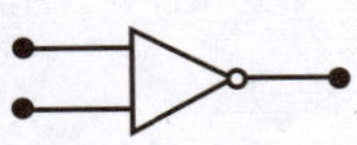 (iv) 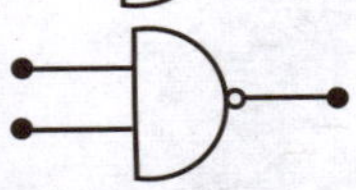

Pick out which ones are for AND, NAND and NOT gates, respectively

(a) (ii), (iii) and (iv) (b) (iii), (ii) and (i)

(c) (iii), (iii) and (iv) (d) (ii), (iv) and (iii)

81. The device that can act as a complete electronic circuit is **2010**

(a) junction diode (b) integrated circuit

(c) junction transistor (d) zener diode

82. The following Figure shows a logic gate circuit with two inputs A and B and the output Y. The voltage waveforms of A, B and Y are given :

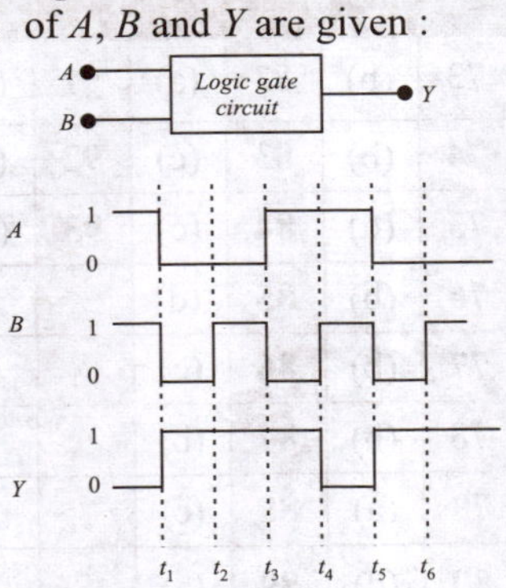

The logic gate is : **2010**

(a) NAND gate (b) NOR gate

(c) OR gate (d) AND gate

83. The circuit **2008**

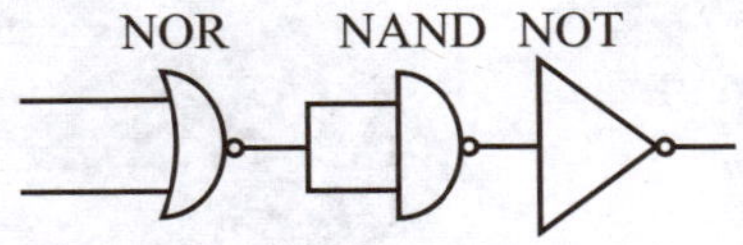

is equivalent to

(a) AND gate (b) NAND gate

(c) NOR gate (d) OR gate

84. In the following circuit, the output Y for all possible inputs A and B is expressed by the truth table **2007**

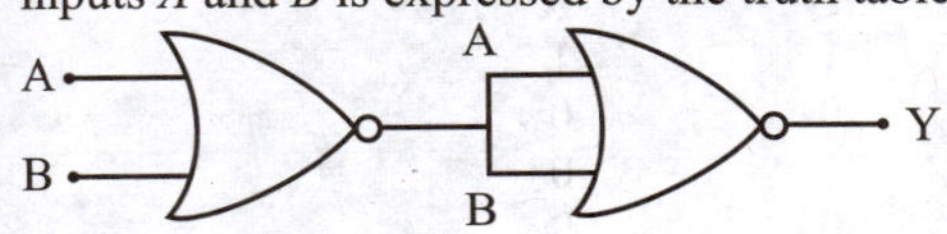

(a)

A	B	Y
0	1	1
0	1	1
1	0	1
1	1	0

(b)

A	B	Y
0	0	1
0	1	0
1	0	0
1	1	0

(c)

A	B	Y
0	0	0
0	1	1
1	0	1
1	1	1

(d)

A	B	Y
0	0	0
0	1	0
1	0	0
1	1	1

85. The following figure shows a logic gate circuit with two inputs A and B and the output C. The voltage waveforms of A, B and C are as shown below

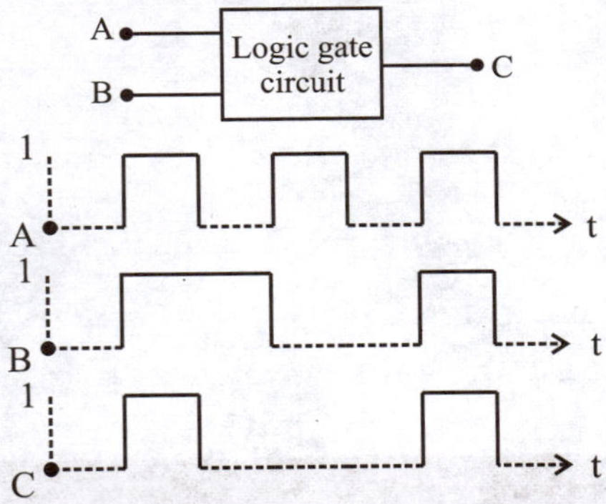

The logic circuit gate is **2006**

(a) NAND gate (b) NOR gate

(c) OR gate (d) AND gate

86. The output of OR gate is 1 **2004**

(a) if either input is zero

(b) if both inputs are zero

(c) if either or both inputs are 1

(d) only if both inputs are 1

87. Following diagram performs the logic function of **2003**

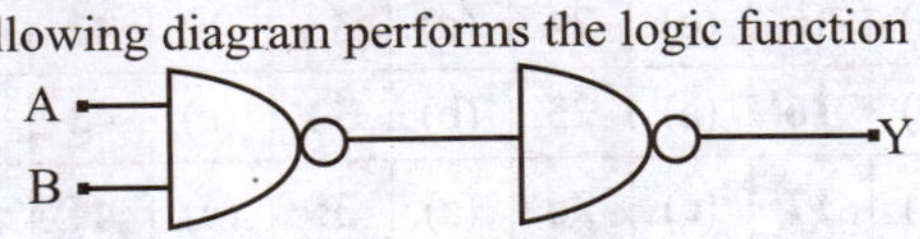

(a) XOR gate (b) AND gate

(c) NAND gate (d) OR gate

88. Which gate is represented by the following truth table ? 2001, 1998, 1994

A	B	Y
0	0	1
1	0	1
0	1	1
1	1	0

(a) XOR (b) NOT
(c) NAND (d) AND

89. A gate has the following truth table

P	Q	R
1	1	1
1	0	0
0	1	0
0	0	0

The gate is 2000
(a) AND (b) NOR
(c) OR (d) NAND

90. The following circuit represents 1999

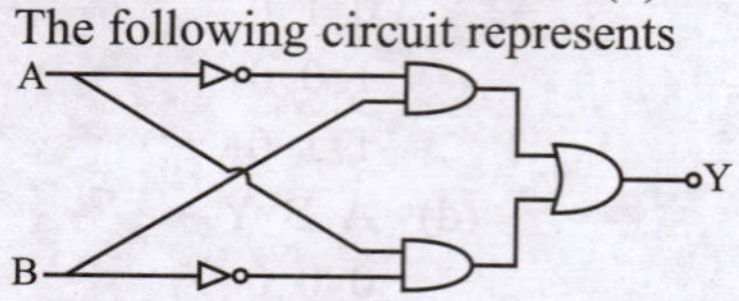

(a) OR gate (b) XOR gate
(c) AND gate (d) NAND gate

91. Which of the following gates will have an output of 1? 1998

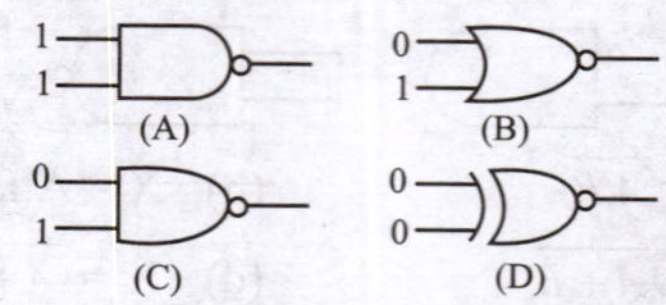

(a) D (b) A (c) B (d) C

92. The following truth table belongs to which of the following four gates? 1997

A	B	Y
1	1	0
1	0	0
0	1	0
0	0	1

(a) NOR (b) XOR (c) NAND (d) OR

93. The following truth table corresponds to the logic gate 1991

A	B	X
0	0	0
0	1	1
1	0	1
1	1	1

(a) NAND (b) OR
(c) AND (d) XOR

ANSWER KEY

1	(d)	10	(b)	19	(c)	28	(d)	37	(b)	46	(a)	55	(a)	64	(b)	73	(b)	82	(a)	91	(d)
2	(c)	11	(a)	20	(b)	29	(b)	38	(a)	47	(d)	56	(d)	65	(b)	74	(b)	83	(c)	92	(a)
3	(a)	12	(c)	21	(b)	30	(d)	39	(b)	48	(c)	57	(c)	66	(b)	75	(d)	84	(c)	93	(b)
4	(b)	13	(a)	22	(d)	31	(d)	40	(a)	49	(d)	58	(c)	67	(b)	76	(b)	85	(d)		
5	(d)	14	(a)	23	(d)	32	(d)	41	(b)	50	(a)	59	(d)	68	(d)	77	(b)	86	(c)		
6	(b)	15	(d)	24	(d)	33	(b)	42	(c)	51	(a)	60	(b)	69	(c)	78	(b)	87	(b)		
7	(b)	16	(a)	25	(b)	34	(c)	43	(b)	52	(d)	61	(c)	70	(d)	79	(a)	88	(c)		
8	(c)	17	(c)	26	(a)	35	(c)	44	(c)	53	(d)	62	(c)	71	(c)	80	(d)	89	(a)		
9	(d)	18	(d)	27	(c)	36	(d)	45	(c)	54	(c)	63	(d)	72	(b)	81	(b)	90	(b)		

Hints & Solutions

1. **(d)** Copper is a conductor, so, its resistance decreases on decreasing temperature as thermal agitation decreases whereas germanium is semiconductor, therefore, on decreasing temperature resistance increases.
2. **(c)** Semiconductors are insulators at room temperature.

NOTES At absolute zero temperature (OK) conduction band of semiconductor is completely empty i.e., $\sigma = 0$. So semiconductor behaves as an insulator.

3. **(a)** At 0K, motion of free electrons stop. Hence conductivity becomes zero. Therefore, at 0K intrinsic semiconductor becomes insulator.
4. **(b)** To make a p–type extrinsic semiconductor, an intrinsic semiconductor is doped with trivalent intrinsic impurity like Gallium, Indium, Boron, Aluminium, etc.
5. **(d)** Current $\frac{I_n}{Ip} = \frac{\mu_e}{\mu_h} \because \mu_e > \mu_h \therefore I_n > I_p$

Therefore current in n-type > Current in p-type

NOTES The mobility of electron is 2-3 times the mobility of holes.

6. **(b)** In p-type semiconductor, trivalent impurities are added to intrinsic semiconductor, which creates holes which are majority charge carriers.
7. **(b)** In a n-type semiconductor holes are minority carriers and pentavalent atoms are dopants.
8. **(c)** When small amount of antimony (pentavalent) is added to germanium crystal then crystal becomes n-type semi conductor. Therefore, there will be more free electrons than holes in the semiconductor.
9. **(d)** $n_i^2 = n_e n_h$

$(1.5 \times 10^{16})^2 = n_e (4.5 \times 10^{22})$

$\Rightarrow \quad n_e = 0.5 \times 10^{10}$

or $\quad n_e = 5 \times 10^9$

Given $\quad n_h = 4.5 \times 10^{22} \Rightarrow n_h >> n_e$

$\therefore$ Semiconductor is p-type and

$n_e = 5 \times 10^9\ m^{-3}$.

10. **(b)** Majority carriers in an n-type semiconductor are electrons.

NOTES The current due to minority carriers in the junction diode is independent of the applied voltage. It only depends upon the temperature of the diode.

11. **(a)** *n*-type of silicon semiconductor is formed when impurity is mixed with pentavalent atom. Out of given choices only phosphorus is pentavalent.
12. **(c)** p-type germanium semiconductor is formed when it is doped with a trivalent impurity atom.

NOTES Doping is the process of addition of impurity. When pure or intrinsie semiconductor is mixed with small amounts of certain specific impurities with valency different from that of the parent material the number of mobile electrons/holes transtically changes.

13. **(a)** Arsenic contains 5 electrons in its outermost shell. When Arsenic is mixed with silicon there is one electron extra in silicon crystal. Hence, such type of semi conductor is *n*-type semi conductor.
14. **(a)** Phosphorous (P) is pentavalent and silicon is tetravalent. Therefore, when silicon is doped with pentavalent impurity, it forms a n-type semiconductor.
15. **(d)** Due to heating, when a free electron is produced then simultaneously a hole is also produced.
16. **(a)** Due to reverse biasing, the width of the depletion region increases in a *p-n* junction diode.
17. **(c)** On heating, number of electron-hole pairs increases, so overall resistance of diode will change.

Hence forward biasing and reversed biasing both are changed.

18. **(d)** The barrier potential of a p-n junction depends on amount of doping, type of semiconductor material and temperature.
19. **(c)** In p-region of p-n junction holes concentration > electrons concentration and in n-region of p-n junction.

electrons concentration > holes concentration.

20. **(b)** Barrier potential does not depend on diode design. Barrier potential depends upon temperature, doping density, and forward biasing.

NOTES Potential barrier is the potential difference created across the P-N junction due to the diffusion of electrons and holes.

21. **(b)** For conduction, *p-n* junction must be forward biased. For conduction *p*-side should be connected to higher potential and *n*-side to lower potential.
22. **(d)** Depletion layer is formed by diffusion of holes and electrons from *p*-type semiconductor to n-type semiconductor and vice-versa. Hence, donor and acceptor atom get positive and negative charge leading to formation of *p-n* junction. The donor and acceptor are immobile.
23. **(d)** Holes are produced due to missing of electrons.
24. **(d)** During the formation of a junction diode, holes from *p*-region diffuse into *n*-region and electrons from n-region diffuse into *p*-region. In both cases, when an electron meets a hole, they cancel the effect of each other and as a result, a thin layer at the junction becomes free from any of charge carriers. This is called depletion layer. There is a potential gradient in the depletion layer,

negative on the *p*-side and positive on the *n*-side. The potential difference thus developed across the junction is called potential barrier.

25. (b) The depletion layer in the *p-n* junction region is caused by diffusion of charge carriers.

26. (a) For bridge balance,

$$\frac{R_1}{R_2}=\frac{R_3}{R_4}\Rightarrow\frac{10}{15}=\frac{10}{R_4}$$

$\therefore R_4 = 15\Omega$

As, R_4 is a combination of 5Ω resistance and diode resistance.

∴ The diode should be in forward biased and connected in series with 5Ω resistance.

27. (c) In option (b), left P-N junction diode is forward biased whereas right one is Reverse biased so obviously both will have different potential difference.

NOTES A P-N junction acts like a voltage controlled Switch. When forward biased it acts like ON switch and Off switch when reverse biased.

28. (d) 

In forward bias, $V_1 > V_2$ i.e., in figure (d) p-side is at higher potential w.r.t. n-side of p-n junction diode.

29. (b) Since Diode is in forward bias, so the value of current flowing through AB

$$i=\frac{\Delta V}{R}=\frac{4-(-6)}{1\times10^3}=\frac{10}{10^3}=10^{-2}\text{A}$$

30. (d) Current $I=\frac{V}{R}=\frac{(3.5-0.5)}{100}$A

[∵ Barrier potential $V_B = 0.5$V]

$$=\frac{3}{100}=30\text{mA}$$

NOTES On an average the potential barrier in P-N junction is v 0.5 V and the width of depletion region 10^{-6} m. Barrier electric field, $E=\frac{V}{d}=\frac{0.5}{10-6}=5\times10^5$ V/m

31. (d) Here D_1 is in forward bias and D_2 is in reverse bias so, D_1 will conduct and D_2 will not conduct. Thus, no current will flow through *DC*.

$$I=\frac{V}{R}=\frac{5}{10}=\frac{1}{2}\text{Amp.}$$

32. (d) In forward biasing of the p-n junction, the positive terminal of the battery is connected to p-side and the negative terminal of the battery is connected to n-side. The depletion region becomes thin.

NOTES An ideal junction diode when forward biased offers zero resistance. Voltage drop across such a junction diode is zero. In reverse bias diode offers infinite resistance and voltage drop across it is equal to voltage applied.

33. (b) Only in (A) and (C) diodes are forward biased as p-type should be at higher potential and n-type at lower potential.

34. (c)

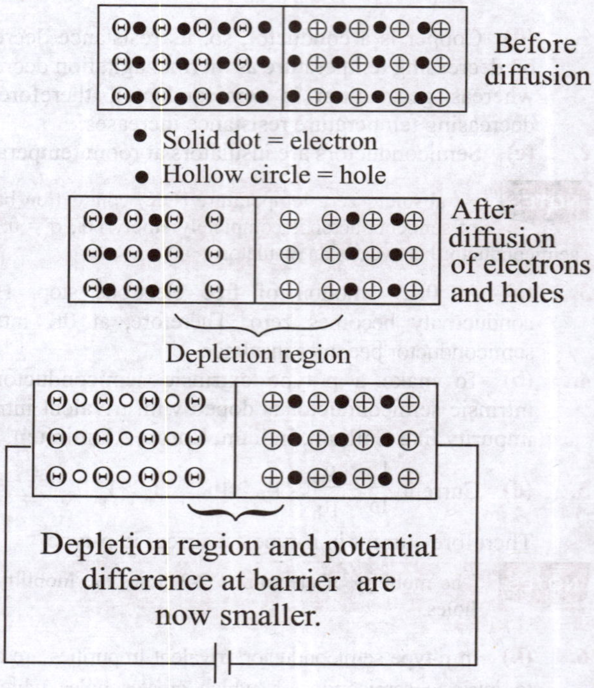

Number of donors is more because electrons from –ve terminal of the cell pushes (enters) the n side and decreases the number of uncompensated pentavalent ion due to which potential barrier is reduced. The neutralised pentavalent atom are again in position to donate electrons.

35. (c) In reverse biasing, the conduction across the *p-n* junction does not take place due to majority carriers but takes place due to minority carriers if the voltage of external battery is large. The size of the depletion region increases thereby increasing the potential barrier.

36. (d) In forward biasing, the diode conducts. For ideal junction diode, the forward resistance is zero; therefore, entire applied voltage occurs across external resistance R i.e., there occurs no potential drop, so potential across R is V in forward biased.

37. (b) We know that in forward bias of *p-n* junction diode, when positive terminal is connected to *p*-type diode, the repulsion of holes takes place which decreases the width of potential barrier by striking the combination of holes and electrons.

38. (a) In reverse bias, the current through a *p-n* junction is almost zero.

39. (b) When *p-n* junction is reverse biased, the flow of current is due to drifting of minority charge carriers across the junction.

40. (a) For forward biasing of *p-n* junction, the positive terminal of external battery is to be connected to p-semiconductor and negative terminal of battery to the *n*-semiconductor.

41. (b) Given : $t = 15$ msec

putting it in given equation

$V_{In} = 220 \sin (100\pi \times 15 \times 10^{-3})$

$= 220 \sin\left(\frac{3\pi}{2}\right) = -220\text{V}$

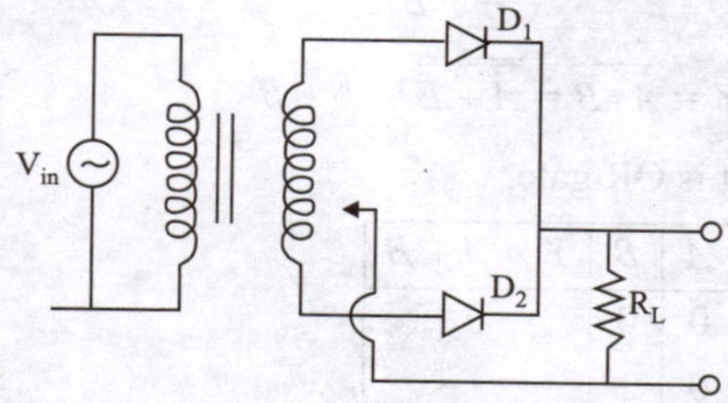

Hence, diode D_1 will be in reverse bias and D_2 will be in forward bias.

42. (c) Capacitor is used to remove the ac ripple from the rectified output.

43. (b)

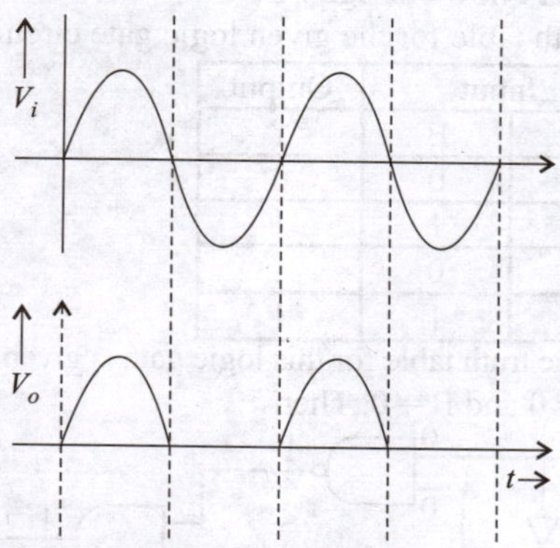

We can see from graph, of input and output of Half Wave Ractifier

$T_{input} = T_{output}$

$\Rightarrow f_{input} = f_{output} \quad \left[\because T = \frac{1}{f}\right]$

44. (c) Here P-N junction diode rectifies half of the ac wave i.e., acts as half wave rectifier. During + ve half cycle Diode → forward biased output across will be

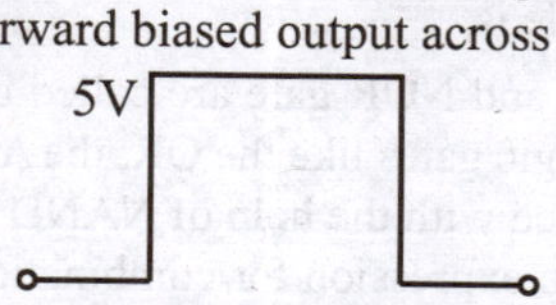

During –ve half cycle Diode → reverse biased output will not obtained.

45. (c) $V = \frac{V_0}{\pi} = \frac{10}{\pi} V$

46. (a) In case of full wave rectifier,

Fundamental frequency = 2 × mains frequency

= 2 × 50 = 100Hz.

47. (d) An alternating current can be converted into direct current by a rectifier.

48. (c) The given circuit is a circuit of full wave rectifier.

49. (d) The I-V characteristics of solar cell is given as

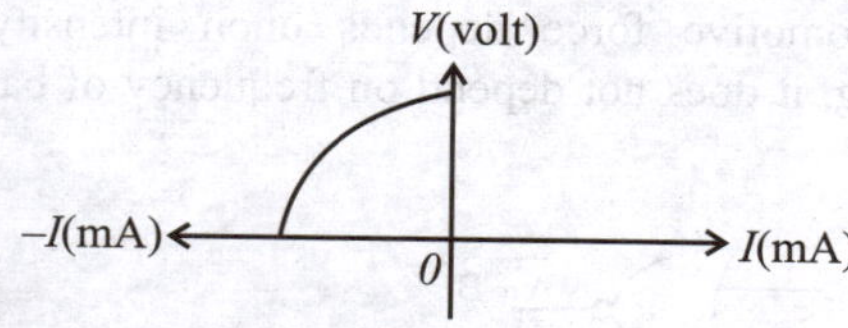

50. (a) In reverse biased *pn* junction diode, the current measured in (μA), is due to minority charge carrier. Solar cell I - V characteristics is as shown in figure.

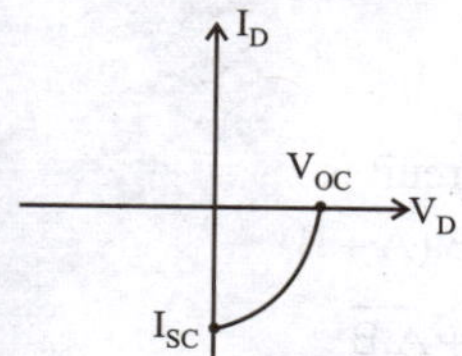

51. (a) Statement I : Photovoltaic devices convert optical radiation into electricity.

Statement II : Zener diode is designed to operate under reverse bias in breakdown region.

52. (d) Zener diode is fabricated by heavily doping both P- and N- sides of the junction. Due to this, depletion region formed is very thin.

NOTES In the forward biasing condition, the zener diode works as ordinary diode.

53. (d) • A zener diode is connected in reverse biased when used as voltage regulator as in reverse biased after breakdown, voltage becomes constant.

• Potential barrier of silicon diode is nearly 0.7V.

54. (c) $\lambda = \frac{12400}{E_g} \text{Å} = \frac{12400}{1.9} = 6526\text{Å} \simeq 654$ nm

55. (a) The given graph represents V-I characteristics of solar cell.

56. (d) Voltage across zener diode is constant.

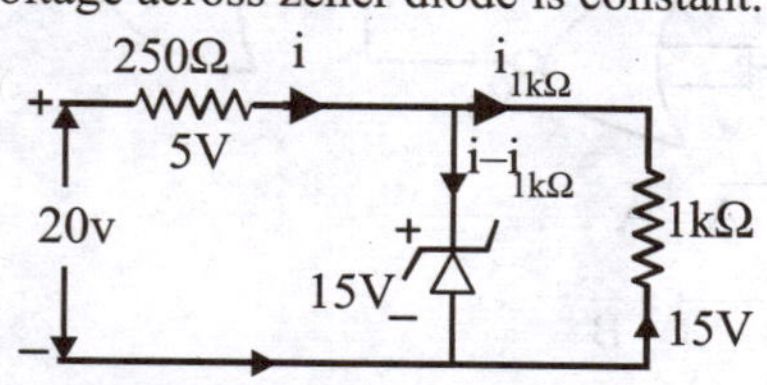

Current in 1kΩ resistor,

$(i)_{1k\Omega} = \frac{15\text{volt}}{1k\Omega} = 15$ mA

Current in 250Ω resistor,

$(i)_{250\Omega} = \frac{(20-15)\text{V}}{250\Omega} = \frac{5\text{V}}{250\Omega}$

$= \frac{20}{1000}\text{A} = 20$ mA

$\therefore \ (i)_{zener\,diode} = (20-15) = 5\text{mA}.$

57. (c) At a certain reverse bias voltage, zener diode allows current to flow through it and hence maintains the voltage supplied to any load. It is used for stabilisation.

58. **(c)** Electromotive force depends upon intensity of light falling, it does not depend on frequency of barrier voltage.

59. **(d)**

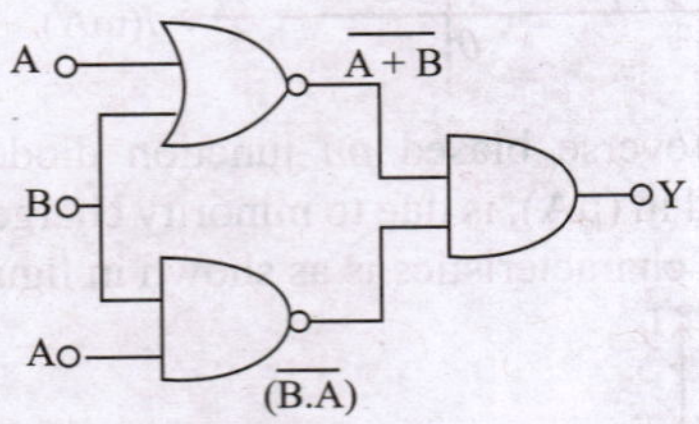

The output of the logic circuit is

$Y = (\overline{A+B}).(\overline{A.B}) = \overline{A}.\overline{B}.(\overline{A+B})$

$= \overline{A}.\overline{B}.\overline{A} + \overline{A}.\overline{B}.\overline{B} = \overline{A}.\overline{B} + \overline{A}.\overline{B}$

$= \overline{A}.\overline{B} = (\overline{A+B})$

So, the given circuit behaves as NOR gate

60. **(b)** When the output of an OR gate is applied to input of a NOT gate, then the combination acts as NOR gate.

61. **(c)**

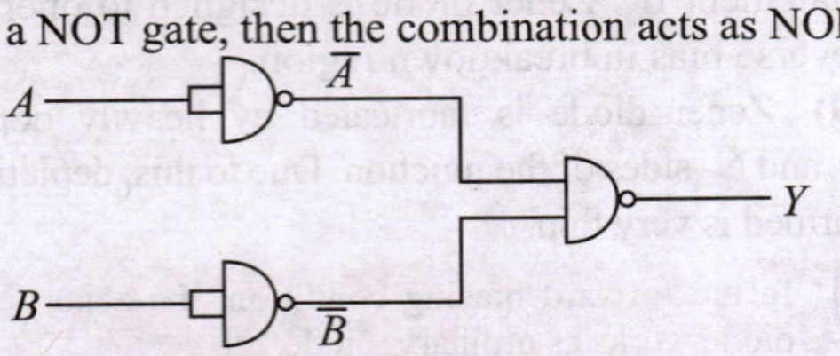

$Y = \overline{\overline{A}.\overline{B}} = \overline{\overline{A}} + \overline{\overline{B}} = A + B$

62. **(c)** As per given truth table, output is independent of input A

$\therefore$ Output $Y = \overline{B}$

63. **(d)**

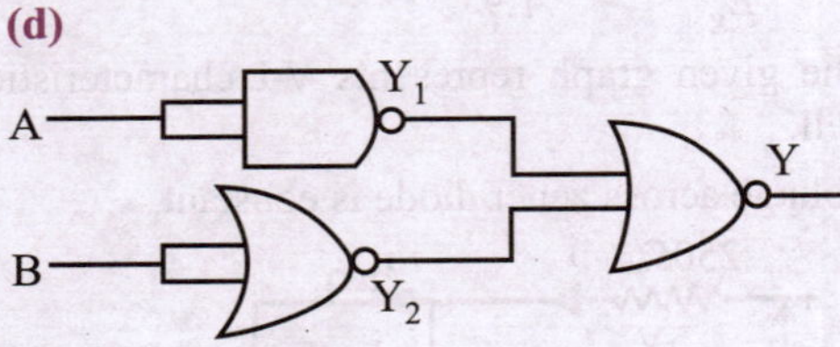

$Y_1 = \overline{A \cdot A} = \overline{A}$

$Y_2 = \overline{B + B} = \overline{B}$

$\therefore$ Output, $Y = \overline{Y_1 + Y_2} = \overline{\overline{A} + \overline{B}} = \overline{\overline{A}} \cdot \overline{\overline{B}}$
$= A \cdot B$

$\therefore$ The given logic gate is similar to AND gate.

64. **(b)** Truth table of given logic circuit is as follows

A	B	$\overline{A}$	$\overline{B}$	$X = \overline{A} + \overline{B}$	$Y = \overline{X}$
0	0	1	1	1	0
1	1	0	0	0	1
0	1	1	0	1	0
1	0	0	1	1	0

Find out put is same as that of and gate out put so given circuit is equivalent to and gate.

65. **(b)**

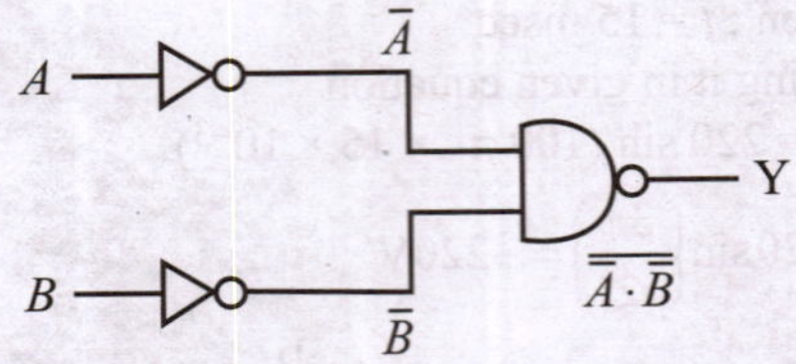

$Y = \overline{\overline{A} \cdot \overline{B}} = \overline{\overline{A + B}} = A + B$

It is OR gate.

A	B	$Y = A + B$
0	0	0
0	1	1
1	0	1
1	1	1

66. **(b)** When both inputs given as "O" means open circuit, no current flows throgh LED.

The truth table for the given logic gate circuit.

Input		Output
A	B	Y
0	0	0
0	1	1
1	0	1
1	1	1

67. **(b)** The truth table for this logic gate is given as

Let A = 0 and B = 0. Then

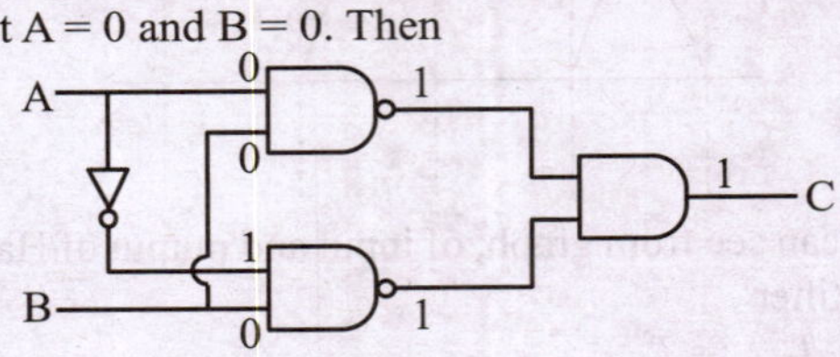

Similarly,

$A = 0, B = 1 \Rightarrow C = 0$

$A = 1, B = 0 \Rightarrow C = 1$

$A = 1, B = 1 \Rightarrow C = 0$

So, truth table is

A	B	C
0	0	1
0	1	0
1	0	1
1	1	0

68. **(d)** NAND and NOR gate are called universal because all type of logic gates like the OR, the AND and the NOT can be realised with the help of NAND and NOR gate.

69. **(c)** Boolean expression for combination of logic gates

Output, $y = A.B + \overline{B.C}$

Time duration	Input Signals			Output Signal (y)		
	A	B	C	AB	$\overline{B.C}$	$y = A.B + \overline{B.C}$
$0 - t_1$	0	0	1	0	1	1
$t_1 - t_2$	1	0	1	0	1	1
$t_2 - t_3$	0	1	0	0	1	1
$t_3 - t_4$	1	1	0	1	1	1

$t_4 - t_5$	0	0	1	0	1	1
$t_5 - t_6$	1	0	1	0	1	1
$t_6 - t_7$	0	0	1	0	1	1

Output y is high (1) so output at the terminal y, $v_0 = 5V$

70. **(d)**

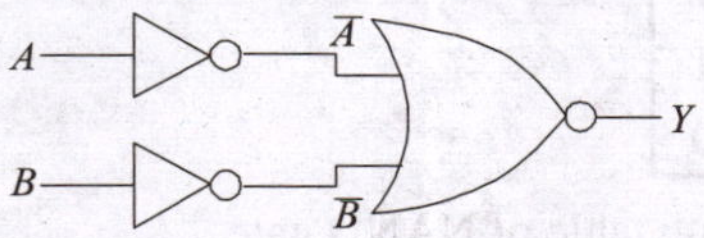

$Y = \overline{\overline{A}+\overline{B}} = \overline{\overline{A \cdot B}} = A \cdot B \Rightarrow$ AND Gate

Truth Table is :

A	B	Y
0	0	0
0	1	0
1	0	0
1	1	1

71. **(c)** From the given logic circuit, LED will glow, when voltage across LED is high.

Truth Table

A	B	Y
0	0	1
0	1	1
1	0	1
1	1	0

Boolean expression $Y = \overline{A.B}$

i.e.,circuit represents NAND gate.

72. **(b)** The bulb will light only when A = 0 and B = 0. Hence it is NOR gate.

73. **(b)**

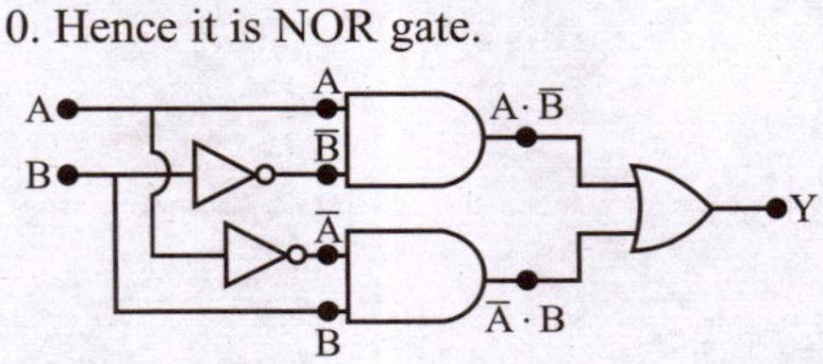

$Y = (A \cdot \overline{B} + \overline{A} \cdot B)$

74. **(b)**

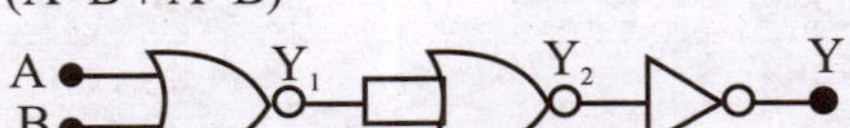

$Y_1 = \overline{A+B}$

$Y_2 = \overline{Y_1 + Y_1} = \overline{Y_1} = A + B$

$Y = \overline{Y_2} = \overline{A+B}$ i.e. NOR gate

75. **(d)** The Boolen expression for the given combination is output $Y = (A + B) \cdot C$

Truth table

A	B	C	$Y = (A+B) \cdot C$
0	0	0	0
1	0	0	0
0	1	0	0
0	0	1	0
1	1	0	0
0	1	1	1
1	0	1	1
1	1	1	1

Hence, A = 1, B = 0, C = 1

76. **(b)** First two gates are NOT gates and the last gate is NOR gate.

Thus, $Y_1 = \overline{A}$, $Y_2 = \overline{B}$ and $Y = \overline{Y_1 + Y_2}$

The truth table corresponding to this is as follows:

A	B	$Y_1 = \overline{A}$	$Y_2 = \overline{B}$	$Y_1 + Y_2$	$Y = \overline{Y_1 + Y_2}$	A . B
0	0	1	1	1	0	0
0	1	1	0	1	0	0
1	0	0	1	1	0	0
1	1	0	0	0	1	1

Thus the combination of gate represents AND gate.

77. **(b)** $C' = \overline{A \cdot B} \Rightarrow C = \overline{\overline{A \cdot B}} = A \cdot B$

Hence the resultant gate is AND gate.

78. **(b)**

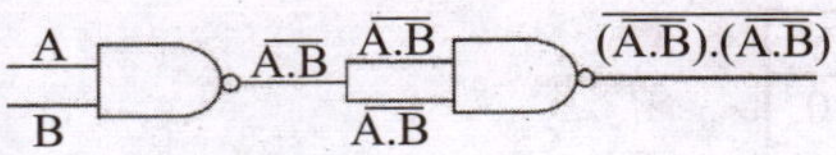

$= \overline{\overline{A.B}} = A.B$

i.e., output X = A.B

Truth Table

A	B	X
0	0	0
1	0	0
0	1	0
1	1	1

79. **(a)**

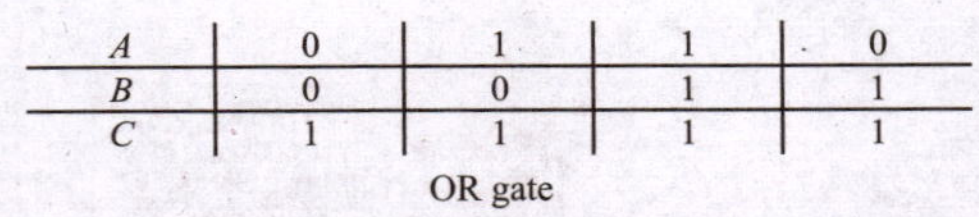

80. **(d)** Figure (i) represents OR gate. Figure (ii), (iv) and (iii) represents AND, NAND and NOT gates respectively.

81. **(b)** Integrated circuit can act as a complete electronic circuit.

82. **(a)** From the given waveforms, the truth table is as follows.

A	B	Y
1	1	0
0	0	1
0	1	1
1	0	1

The above truth table is for NAND gate.

Therefore, the logic gate is NAND gate.

83. **(c)** Let A and B be inputs and Y the output.

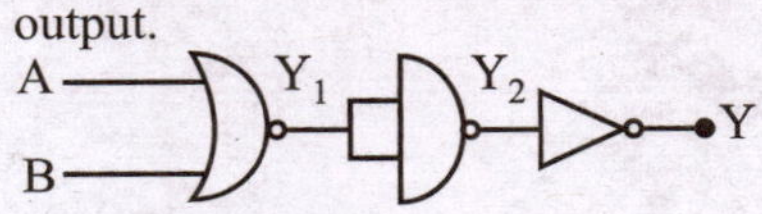

Then $Y_1 = \left(\overline{A+B}\right) = \overline{A}.\overline{B}$

(By De-Morgan's theorem)

$Y_2 = \overline{Y_1} = \left(\overline{\overline{A}.\overline{B}}\right) = \overline{\overline{A}} + \overline{\overline{B}} = A + B$

$\therefore Y = \overline{Y_2} = \left(\overline{A+B}\right)$

Hence, the given circuit is equivalent to a NOR gate.

NOTES NOR and NAND gates is a universal gates as they can be used to perform the basic logic functions. AND, OR and NOT gate.

84. **(c)** 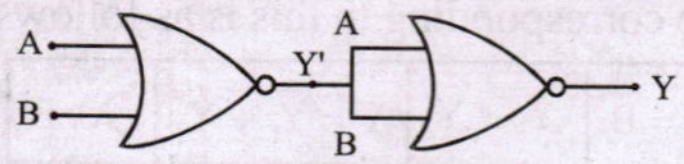

$Y' = \overline{A+B}$; $Y = \overline{\overline{A+B}} = A+B$

Therefore truth table :

A	B	Y
0	0	0
0	1	1
1	0	1
1	1	1

85. **(d)** On the basis of given graph, following table is possible.

A	B	Y
0	0	0
1	1	1
0	1	0
1	0	0

It is the truth table of AND gate.

86. **(c)** Output will be one if A or B or both are one. $Y = A + B$

87. **(b)**

88. **(c)** $Y = \overline{A.B}$

A	B	Y
0	0	1
0	1	1
1	0	1
1	1	0

Which is truth table of NAND gate.

89. **(a)** P, Q and R are related as $R = P.Q$ which is relation of AND gate.

90. **(b)** Output of upper AND gate = $A \cdot \overline{B}$

Output of lower AND gate = $\overline{A}.B$

$\therefore$ Output of OR gate, $Y = A.\overline{B} + B.\overline{A}$

This is Boolean expression for XOR gate.

91. **(d)**

92. **(a)** The given truth table is of (OR gate + NOT gate) $\equiv$ NOR gate

93. **(b)** This truth table is of identity, $X = A + B$, hence OR gate.

15 Experimental Skills

TREND ANALYSIS

TOPIC	2025		2024		2023		2022		2021		2020	
	Qns.	LOD	Qns.	LOD	Qns.	LOD	Qns.	LOD	Qns.	LOD	Qns.	LOD
Vernier Calipers - its use to Measure the Internal and External Diameter and Depth of a Vessel	1	Diff.	1	Avg.								
To Determine Thickness/Diameter of Thin Sheet/Wire with the help of Screw Gauge									1	Avg.	1	Easy.

How to Code Qns for Better & Faster Revision

Type 1: Able to solve (No code Req)
Type 2: Solved with struggle or by seeing solution **(Use code NR means Needs Revision or orange stickers)**
Type 3: Tricky/ Could not solve on my own. Need help from Teacher/ Friend. **(Use code TQ means Tricky Question or red stickers)**

You can either hand write the code in boxes or else paste coloured stickers given at end of book.

Topic
Experimental Skills

1. Consider the diameter of a spherical object being measured with the help of a Vernier callipers. Suppose its 10 Vernier Scale Divisions (V.S.D.). are equal to its 9 Main Scale Divisions (M.S.D.) The least division in the M.S. is 0.1 cm and the zero of V.S. is at $x = 0.1$ cm when the jaws of Vernier callipers are closed. **2025**
If the main scale reading for the diameter is M = 5 cm and the number of coinciding vernier division is 8, the measured diameter after zero error correction, is
(a) 5.18 cm (b) 5.08 cm
(c) 4.98 cm (d) 5.00 cm

2. The pitch of an error free screw gauge is 1 mm and there are 100 divisions on the circular scale. While measuring the diameter of a thick wire, the pitch scale reads 1 mm and 63rd division on the circular scale coincides with the reference line. The diameter of the wire is: **Ph-II 2024**
(a) 1.63 cm (b) 0.163 cm
(c) 0.163 m (d) 1.63 m

3. In a vernier callipers, (N + 1) divisions of vernier scale coincide with N divisions of main scale. If 1 MSD represents 0.1 mm, the vernier constant (in cm) is: **2024**
(a) $\frac{1}{10N}$ (b) $\frac{1}{100(N+1)}$
(c) 100 N (d) 10 (N + 1)

4. A screw gauge gives the following readings when used to measure the diameter of a wire
Main scale reading : 0 mm
Circular scale reading : 52 divisions
Given that 1 mm on main scale corresponds to 100 divisions on the circular scale. The diameter of the wire from the above data is **2021**
(a) 0.052 cm (b) 0.52 cm
(c) 0.026 cm (d) 0.26 cm

5. A screw gauge has least count of 0.01 mm and there are 50 divisions in its circular scale. The pitch of the screw gauge is: **2020**
(a) 0.25 mm (b) 0.5 mm
(c) 1.0 mm (d) 0.01 mm

6. The main scale of a vernier callipers has n divisions/cm. n divisions of the vernier scale coincide with $(n-1)$ divisions of main scale. The least count of the vernier callipers is, **Odisha, 2019**
(a) $\frac{1}{n(n+1)}$ cm (b) $\frac{1}{(n+1)(n-1)}$ cm
(c) $\frac{1}{n}$ cm (d) $\frac{1}{n^2}$ cm

7. A student measured the diameter of a small steel ball using a screw gauge of least count 0.001 cm. The main scale reading is 5 mm and zero of circular scale division coincides with 25 divisions above the reference level. If screw gauge has a zero error of –0.004 cm, the correct diameter of the ball is **2018**
(a) 0.521 cm (b) 0.525 cm
(c) 0.529 cm (d) 0.053 cm

8. In a vernier calliper N divisions of vernier scale coincides with (N – 1) divisions of main scale (in which length of one division is 1 mm). The least count of the instrument should be **1994**
(a) N
(b) N – 1
(c) 1/10 N
(d) 1/N – 1

ANSWER KEY																			
1	(c)	2	(b)	3	(b)	4	(a)	5	(b)	6	(d)	7	(c)	8	(c)				

Hints & Solutions

1. **(c)** Least count of main scale
10 V.S.D. = 9 M.S.D.
Given 1 M.S.D. = 0.1 cm
Positive zero error = 0.1 cm (Given)
1 V.S.D. = 0.9 M.S.D.
1 V.S.D. = 0.9 × 0.1 cm
∴ V.S.D. = 0.09 cm
L.C. = 1 M.S.D. – 1 V.S.D.
= (0.1 – 0.09) cm = 0.01 cm
Total reading = 5.08 cm
Apply zero error correction = (5.08 – 0.1) cm
= 4.98 cm

2. **(b)** Least count of screw gauge
$$= \frac{\text{Pitch}}{\text{No. of divisions on circular scale}}$$
$$\Rightarrow \text{Least count} = \frac{1}{100} = 0.01 \text{ mm}$$
Final reading = MSR + CSR × L.C.
= 1 mm + (63) (0.01) mm
= 1.63 mm = 0.163 cm

3. **(b)** Vernier Constant V. C. = MSD –VSD ...(i)
where MSD = Main scale division and VSD = Vernier scale division
Given : (N + 1) VSD = N MSD
$$\text{VSD} = \left(\frac{\text{N}}{\text{N}+1}\right)\text{MSD} \quad \text{...(ii)}$$
From eq. (i) and (ii)
$$\text{V.C.} = (\text{MSD}) - \frac{\text{N}}{\text{N}+1}(\text{MSD})$$
$$= \text{MSD}\left(1 - \frac{\text{N}}{\text{N}+1}\right) = \frac{\text{MSD}}{\text{N}+1}$$
$$= \frac{0.01}{\text{N}+1} = \frac{1}{100(\text{N}+1)}$$

4. **(a)** $\text{Least count LC} = \frac{\text{Pitch}}{\text{CSD}} = \frac{1}{100} = 0.01\text{mm}$
= 0.001 cm
∴ Diameter of the wire = MSR + (CSR × LC)
= 0 + (52 × 0.001 cm)
= 0.052 cm

5. **(b)** Least count of screw gauge = 0.01 mm
Least count
$$= \frac{\text{Pitch}}{\text{No. of divisions on circular scale}}$$
$$\Rightarrow 0.01 \text{ mm} = \frac{\text{Pitch}}{50} \Rightarrow \text{Pitch} = 0.5 \text{ mm}$$

6. **(d)** n VSD = $(n – 1)$ MSD
$$1 \text{ VSD} = \frac{(n-1)}{n}\text{ MSD}$$
$$\text{L.C.} = 1 \text{ MSD} - 1 \text{ VSD} = 1 \text{ MSD} - \frac{(n-1)}{n}\text{MSD}$$
$$\frac{1}{n}\text{MSD} = \frac{1}{n} \times \frac{1}{n} = \frac{1}{n^2}$$

7. **(c)** Diameter of the ball
= MSR + CSR × (least count) – zero error
= 0.5 cm + 25 × 0.001 – (–0.004)
= 0.5 + 0.025 + 0.004 = 0.529 cm

8. **(c)** Least count = 1 MSD – 1 VSD
$$= 1 \text{ MSD} - \left(\frac{\text{N}-1}{\text{N}}\right)\text{MSD}$$
$$= \frac{1}{\text{N}}\text{MSD} = \frac{1}{\text{N}} \times \frac{1}{10}\text{cm} = \frac{1}{10\text{N}}$$

NR	NR	NR	NR	NR	NR	NR	NR	NR	NR	NR	NR	NR	NR	NR	NR	NR	NR	NR	NR		NR	NR	NR	NR	NR	NR
NR	NR	NR	NR	NR	NR	NR	NR	NR	NR	NR	NR	NR	NR	NR	NR	NR	NR	NR	NR	NR	NR	NR	NR	NR	NR	NR
NR	NR	NR	NR	NR	NR	NR	NR	NR	NR	NR	NR	NR	NR	NR	NR	NR	NR	NR	NR	NR	NR	NR	NR	NR	NR	NR
NR	NR	NR	NR	NR	NR	NR	NR	NR	NR	NR	NR	NR	NR	NR	NR	NR	NR	NR	NR	NR	NR	NR	NR	NR	NR	NR
NR	NR	NR	NR	NR	NR	NR	NR	NR	NR	NR	NR	NR	NR	NR	NR	NR	NR	NR	NR	NR	NR	NR	NR	NR	NR	NR
NR	NR	NR	NR	NR	NR	NR	NR	NR	NR	NR	NR	NR	NR	NR	NR	NR	NR	NR	NR	NR	NR	NR	NR	NR	NR	NR
NR	NR	NR	NR	NR	NR	NR	NR	NR	NR	NR	NR	NR	NR	NR	NR	NR	NR	NR	NR	NR	NR	NR	NR	NR	NR	NR
NR	NR	NR	NR	NR	NR	NR	NR	NR	NR	NR	NR	NR	NR	NR	NR	NR	NR	NR	NR	NR	NR	NR	NR	NR	NR	NR
NR	NR	NR	NR	NR	NR	NR	NR	NR	NR	NR	NR	NR	NR	NR	NR	NR	NR	NR	NR	NR	NR	NR	NR	NR	NR	NR
NR	NR	NR	NR	NR	NR	NR	NR	NR	NR	NR	NR	NR	NR	NR	NR	NR	NR	NR	NR	NR	NR	NR	NR	NR	NR	NR
NR	NR	NR	NR	NR	NR	NR	NR	NR	NR	NR	NR	NR	NR	NR	NR	NR	NR	NR	NR	NR	NR	NR	NR	NR	NR	NR
NR	NR	NR	NR	NR	NR	NR	NR	NR	NR	NR	NR	NR	NR	NR	NR	NR	NR	NR	NR	NR	NR	NR	NR	NR	NR	NR
NR	NR	NR	NR	NR	NR	NR	NR	NR	NR	NR	NR	NR	NR	NR	NR	NR	NR	NR	NR	NR	NR	NR	NR	NR	NR	NR
NR	NR	NR	NR	NR	NR	NR	NR	NR	NR	NR	NR	NR	NR	NR	NR	NR	NR	NR	NR	NR	NR	NR	NR	NR	NR	NR
NR	NR	NR	NR	NR	NR	NR	NR	NR	NR	NR	NR	NR	NR	NR	NR	NR	NR	NR	NR	NR	NR	NR	NR	NR	NR	NR
NR	NR	NR	NR	NR	NR	NR	NR	NR	NR	NR	NR	NR	NR	NR	NR	NR	NR	NR	NR	NR	NR	NR	NR	NR	NR	NR
NR	NR	NR	NR	NR	NR	NR	NR	NR	NR	NR	NR	NR	NR	NR	NR	NR	NR	NR	NR	NR	NR	NR	NR	NR	NR	NR
NR	NR	NR	NR	NR	NR	NR	NR	NR	NR	NR	NR	NR	NR	NR	NR	NR	NR	NR	NR	NR	NR	NR	NR	NR	NR	NR
NR	NR	NR	NR	NR	NR	NR	NR	NR	NR	NR	NR	NR	NR	NR	NR	NR	NR	NR	NR	NR	NR	NR	NR	NR	NR	NR
NR	NR	NR	NR	NR	NR	NR	NR	NR	NR	NR	NR	NR	NR	NR	NR	NR	NR	NR	NR	NR	NR	NR	NR	NR	NR	NR
NR	NR	NR	NR	NR	NR	NR	NR	NR	NR	NR	NR	NR	NR	NR	NR	NR	NR	NR	NR	NR	NR	NR	NR	NR	NR	NR
TQ	TQ	TQ	TQ	TQ	TQ	TQ	TQ	TQ	TQ	TQ	TQ	TQ	TQ	TQ	TQ	TQ	TQ	TQ	TQ	TQ	TQ	TQ	TQ	TQ	TQ	TQ
TQ	TQ	TQ	TQ	TQ	TQ	TQ	TQ	TQ	TQ	TQ	TQ	TQ	TQ	TQ	TQ	TQ	TQ	TQ	TQ	TQ	TQ	TQ	TQ	TQ	TQ	TQ
TQ	TQ	TQ	TQ	TQ	TQ	TQ	TQ	TQ	TQ	TQ	TQ	TQ	TQ	TQ	TQ	TQ	TQ	TQ	TQ	TQ	TQ	TQ	TQ	TQ	TQ	TQ
TQ	TQ	TQ	TQ	TQ	TQ	TQ	TQ	TQ	TQ	TQ	TQ	TQ	TQ	TQ	TQ	TQ	TQ	TQ	TQ	TQ	TQ	TQ	TQ	TQ	TQ	TQ
TQ	TQ	TQ	TQ	TQ	TQ	TQ	TQ	TQ	TQ	TQ	TQ	TQ	TQ	TQ	TQ	TQ	TQ	TQ	TQ	TQ	TQ	TQ	TQ	TQ	TQ	TQ
TQ	TQ	TQ		TQ	TQ	TQ	TQ	TQ	TQ	TQ	TQ	TQ	TQ	TQ	TQ	TQ	TQ	TQ	TQ	TQ	TQ	TQ	TQ	TQ	TQ	TQ
TQ	TQ	TQ	TQ	TQ	TQ	TQ	TQ	TQ	TQ	TQ	TQ	TQ	TQ	TQ	TQ	TQ	TQ	TQ	TQ	TQ	TQ	TQ	TQ	TQ	TQ	TQ